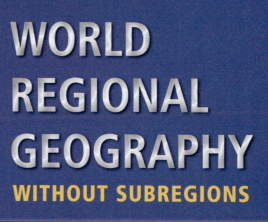

WORLD REGIONAL GEOGRAPHY
WITHOUT SUBREGIONS

Global Patterns, Local Lives   Fifth Edition

LYDIA MIHELIČ PULSIPHER
Professor of Geography
University of Tennessee

ALEX PULSIPHER
Geographer and
Independent Scholar

with the assistance of
CONRAD "MAC" GOODWIN
Anthropologist/Archaeologist and
Independent Scholar

W. H. Freeman and Company
New York

To Hedayat, Betsy, Leila, and Ayesha Amin Arsala, for their constant work for democracy, peace, prosperity, and understanding in Afghanistan and their unending willingness to help us write more accurately about that country

EXECUTIVE EDITOR: **Steven Rigolosi**

DEVELOPMENTAL EDITOR: **Barbara Muller**

DIRECTOR OF MARKETING: **John Britch**

PROJECT EDITOR: **Leigh Renhard**

COVER AND TEXT DESIGNER: **Vicki Tomaselli**

SENIOR ILLUSTRATION COORDINATOR: **Bill Page**

ASSISTANT EDITOR: **Kerri Russini**

MAPS: **University of Tennessee, Cartographic Services Laboratory: Will Fontanez, Director; Martha Bostwick, Lead Cartographer**

PHOTO EDITOR: **Ted Szczepanski**

PHOTO RESEARCHER: **Alex Pulsipher**

PRODUCTION MANAGER: **Julia De Rosa**

COMPOSITION: **Sheridan Sellers, W. H. Freeman and Company**

PRINTING AND BINDING: **RR Donnelley**

Library of Congress Control Number: 2010941183

ISBN-13: 978-1-4292-3244-9
ISBN-10: 1-4292-3244-7

Printed in the United States of America
Second printing

W. H. Freeman and Company
41 Madison Avenue
New York, NY 10010
Houndmills, Basingstoke RG21 6XS, England
www.whfreeman.com/geography

Lydia Mihelič Pulsipher is a cultural-historical geographer who studies the landscapes of ordinary people through the lens of archaeology, geography, and ethnography. She has contributed to several geography-related exhibits at the Smithsonian Museum of Natural History in Washington, D.C., including "Seeds of Change," which featured the research she and Conrad Goodwin have done on the eastern Caribbean. Lydia Pulsipher has ongoing research projects in the eastern Caribbean (historical archaeology) and in central Europe, where she is interested in various aspects of the post-Communist transition. Her graduate students have studied human ecology in the Caribbean and border and national identity issues in several central European countries. She has taught cultural, gender, European, North American, and Mesoamerican geography at the University of Tennessee at Knoxville since 1980; through her research, she has given many students their first experience in fieldwork abroad. Previously she taught at Hunter College and Dartmouth College. She received her B.A. from Macalester College, her M.A. from Tulane University, and her Ph.D. from Southern Illinois University.

Alex Pulsipher is an independent scholar and geographer in Knoxville, Tennessee, where he is studying the diffusion of green technologies in the United States. In the early 1990s, while a student at Wesleyan University in Connecticut, Alex spent time in South Asia working for a sustainable development research center. He then completed his bachelor's degree at Wesleyan University, where he wrote his undergraduate thesis on the history of Hindu nationalism. Beginning in 1995, Alex contributed to the research and writing of the first edition of this textbook. In 1999 and 2000, he traveled to South America, Southeast Asia, and South Asia, where he collected information for the second edition of the text and for the Web site. In 2000 and 2001, he wrote and designed maps for the second edition. He participated in the writing of the fourth edition and in restructuring the content of, and creating photo essays and maps for, the fifth edition. Alex has an M.A. in geography from Clark University.

Conrad McCall Goodwin In the writing of *World Regional Geography*, Lydia and Alex Pulsipher were assisted in many ways by Lydia's husband, **Conrad "Mac" Goodwin,** an anthropologist and historical archaeologist. Mac holds a B.A. in anthropology from the University of California at Santa Barbara, an M.A. in historical archaeology from the College of William and Mary, and a Ph.D. in archaeology from Boston University. He specializes in sites created during the European colonial era in North America, the Caribbean, and the Pacific. He has particular expertise in the archaeology of agricultural systems, gardens, domestic landscapes, and urban spaces. When not working on the textbook, Mac is a master organic gardener and slow-food chef.

BRIEF CONTENTS

CONTENTS

CONTENTS

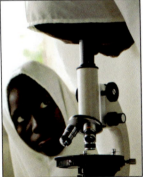

CHAPTER 10

Southeast Asia 430

CHAPTER 11

Oceania: Australia, New Zealand, and the Pacific 472

Over the past four editions of this text, we have sought to portray the rich diversity of human life across the world and to humanize geographic issues by representing the daily lives of women, men, and children in various regions of the globe. Our goal has always been to create a book that makes global patterns of trade and consumption meaningful for students by showing how these patterns affect the regions of the world and ordinary people at the local level. In other words, we have tried to craft a text that explains and illustrates global patterns while helping readers engage with the way these patterns affect individuals. The fifth edition continues in this tradition and complements it with improvements to make the text as current, instructive, and visually appealing as possible.

NEW TO THE FIFTH EDITION

Thematic Concepts

Teaching world regional geography is never easy. Many instructors have found that focusing their courses on a few key ideas makes their teaching more effective and helps students retain information. With that goal in mind, we have identified nine thematic concepts that provide a few basic hooks on which students can hang their growing knowledge of the world and each of its regions. These thematic concepts are, in alphabetical order:

- **Climate change:** What are the indications that climate change is under way? How are places, people, and ecosystems in a particular region vulnerable to the shifts that climate change may

bring? How are people and governments in the region responding to the threats posed by global warming? Which human activities contribute significant amounts of greenhouse gases?

- **Democratization:** What is the history and current status of democracy in a region? Which factors are working for or against the political freedoms that make democracy possible? What are the difficulties associated with establishing democracy?

- **Development:** How do shifts in economic, social, and other dimensions of development affect human well-being? What paths have been charted by the so-called "developed" world and how are they relevant, or irrelevant, to the rest of the world? What new homegrown solutions are emerging from the so-called less-developed countries?

- **Food:** How do food production systems impact environments and societies in a region? How has the use of new agricultural technologies impacted farmers? How have changes in food production created pressure on regions to urbanize?

- **Gender:** How do the lives and livelihoods of men and women differ and how do gender roles influence societies in a region? To what extent do men and women differ in their contributions to family and community well-being? What is the nature of the disparities in income, education, and rights that persist between the genders?

- **Globalization:** How has a particular region been impacted by globalization, historically and currently? How are lives changing as flows of people, ideas, products, and resources become more global?

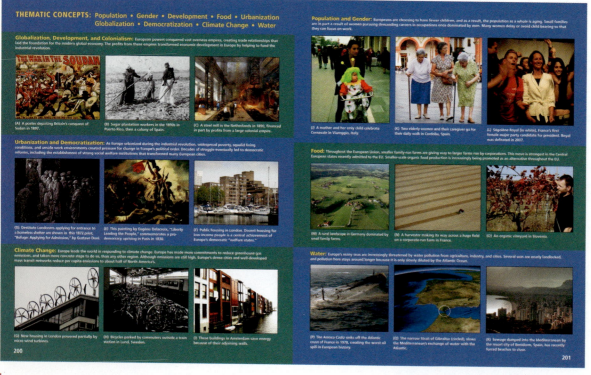

THEMATIC CONCEPTS: Population • Gender • Development • Food • Urbanization
Globalization • Democratization • Climate Change • Water

Globalization, Development, and Colonialism: European powers conquered vast overseas empires, creating trade relationships that laid the foundation for the modern global economy. The profits from these empires transformed economic development in Europe by helping to fund the industrial revolution.

(A) A poster depicting Britain's conquest of Sudan in 1897.

(B) Sugar plantation workers in the 1890s in Puerto Rico, then a colony of Spain.

(C) A steel mill in the Netherlands in 1890, financed in part by profits from a large colonial empire.

Urbanization and Democratization: As Europe urbanized during the industrial revolution, widespread poverty, squalid living conditions, and unsafe work environments created pressure for change in Europe's political order. Decades of struggle eventually led to democratic reforms, including the establishment of strong social welfare institutions that transformed many European cities.

(D) Destitute Londoners applying for entrance to a homeless shelter are shown in this 1872 print, "Refuge: Applying for Admission," by Gustave Doré.

(E) This painting by Eugène Delacroix, "Liberty Leading the People," commemorates a pro-democracy uprising in Paris in 1830.

(F) Public housing in London. Decent housing for low-income people is a central achievement of Europe's democratic "welfare states."

Climate Change: Europe leads the world in response to climate change. Europe has made more commitments to reduce greenhouse gas emissions, and taken more concrete steps to do so, than any other region. Although emissions are still high, Europe's dense cities and well-developed mass transit networks reduce per capita emissions to about half of North America's.

(G) New housing in London powered partially by micro wind turbines.

(H) Bicycles parked by commuters outside a train station in Lund, Sweden.

(I) These buildings in Amsterdam save energy because of their adjoining walls.

Population and Gender: Europeans are choosing to have fewer children, and as a result, the population as a whole is aging. Small families are in part a result of women pursuing demanding careers in occupations once dominated by men. Many women delay or avoid child bearing so that they can focus on work.

(J) A mother and her only child celebrate Carnevale in Viareggio, Italy.

(K) Two elderly women and their caregiver go for their daily walk in Cordoba, Spain.

(L) Ségolène Royal (in white), France's first female major party candidate for president. Royal was defeated in 2007.

Food: Throughout the European Union, smaller family-run farms are giving way to larger farms run by corporations. This move is strongest in the Central European states recently admitted to the EU. Smaller-scale organic food production is increasingly being promoted as an alternative throughout the EU.

(M) A rural landscape in Germany dominated by small family farms.

(N) A harvester making its way across a huge field on a corporate-run farm in France.

(O) An organic vineyard in Slovenia.

Water: Europe's many seas are increasingly threatened by water pollution from agriculture, industry, and cities. Several seas are nearly landlocked, and pollution here stays around longer because it is only slowly diluted by the Atlantic Ocean.

(P) The Amoco Cadiz sinks off the Atlantic coast of France in 1978, creating the worst oil spill in European history.

(Q) The narrow Strait of Gibraltar (circled), slows the Mediterranean's exchange of water with the Atlantic.

(R) Sewage dumped into the Mediterranean by the resort city of Benidorm, Spain, has recently forced beaches to close.

200

201

- **Population:** What are the major forces driving population growth or decline in a region? How have changes in gender roles influenced population growth? How are changes in life expectancy, family size, and the age of the population influencing population change?
- **Urbanization:** Which forces are driving urbanization in a particular region? How have cities responded to growth? How is the region affected by the changes that accompany urbanization—for example, changes in employment, education, and access to health care?
- **Water:** How do issues of water scarcity, water pollution, and water management affect people and environments in a particular region? Water issues are often discussed in the context of global warming or food production.

Thematic Concepts are not necessarily presented in the same order in each regional chapter. Rather, the content of each chapter focuses on the themes of particular importance to that region.

Photo Essays

Photos are a rich source of geographic information, and the number of photos in the fifth edition has been greatly expanded. Early in each regional chapter, a **Thematic Concepts Photo Essay** explains how each of the nine thematic concepts applies to that particular region and also how two or more concepts relate to one another (see page xv). Subsequent **Photo Essays** illustrate particular concepts. For example, a photo essay about urbanization might include a map of urban patterns in the region and photos that illustrate various aspects of current urban life in that part of the world. Because circumstances vary so greatly between regions, these concepts are treated according to the dominant patterns in the region. Many of these Photo Essays can be found in animated form on GeographyPortal (see page xix for more information).

Restructured Chapters

Each chapter includes a variety of features to support the teaching and learning of world regional geography.

Things to Remember: At the close of every main section, a few concise statements review the important points of the section. These statements emphasize some key themes while encouraging students to think through the ways in which the material illustrates these points.

> **THINGS TO REMEMBER**
>
> 1. Democracy has transformed the politics of the region, yet problems remain. Elections are sometimes poorly or unfairly run, with results frequently contested. Elected governments are at times challenged by citizen protests or with a coup d'état, because policies are unpopular with large segments of the population, powerful elites, or the military.
>
> 2. The drug trade has emerged as a major obstacle to democracy, with huge amounts of illicit cash being used to pay off elected officials, civil servants, police forces, and the military.
>
> 3. The Internet plays an important role in the politics of the region, which is more technologically advanced than the world's other developing regions.

On-Page Glossary of Key Terms: Terms important to the chapter content are boldfaced on first usage and defined in a glossary in the page margins. These terms are listed at the end of the chapter with the page numbers on which they are defined. They are also listed alphabetically in the glossary at the end of the book.

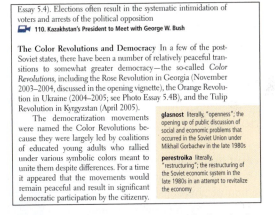

Critical Thinking Questions: At the end of each chapter, a series of questions encourages broader analysis of the chapter content. These questions could be used for assignments, group projects, or class discussion.

Consistent Base Maps: This edition focuses on improving further what has often been cited as a principal strength of this text: high-quality, relevant, and consistent maps. To help students make conceptual connections and to compare regions, every chapter contains the following maps:

- Regional map at the beginning of each chapter
- Political map of the region
- Climate map of the region
- Human impact on the biosphere
- Vulnerability to climate change
- Democratization and conflict

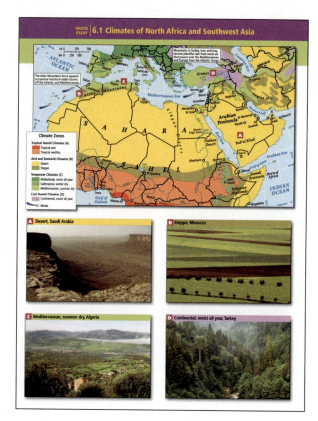

- Population density
- Human well-being
- Urbanization

New to the fifth edition are the maps on vulnerability to climate change, democratization and conflict, and human well-being. Also new are the photos that accompany many of these maps.

New Photos

An ongoing aim of this text has been to awaken the student to the circumstances of people around the world, and photos are a powerful way to accomplish this objective. The fifth edition promotes careful attention to photos by including in Chapter 1 a short lesson on photo interpretation. Students are encouraged to use these skills as they look at every photo in the text, and instructors are encouraged to use the photos as lecture themes and to generate analytical class discussions.

The fifth edition has more than twice as many photos as the previous edition, and each new photo was chosen to complement a thematic concept or situation described in the text. All photos are numbered and referenced in the text, making it easier for students to integrate the text with the visuals as they read. Moreover, the photos—like all the book's graphics, including the maps—have been given significantly more space and prominence in the page layout. The result is a more visually engaging, dynamic, and instructive text. All photo credits are included at the end of the book according to photo numbers.

Videos   📹   at www.whfreeman.com/geographyvideos

Over 300 videos clips (an average of 27 per chapter) are available with the fifth edition. All videos are 2 to 6 minutes long and cover key issues discussed in the text. They can help instructors gain further expertise or can be used to generate class discussion.

Each video is keyed to the text with the icon 📹 at the point in the discussion where it is most relevant. These videos, along with a multiple-choice quiz for each video, can be accessed at www.whfreeman.com/geographyvideos. The quiz for each video has questions that can be automatically graded and entered into an instructor grade book.

To access the videos, students need a password that can be bundled free with this textbook. For more information, please contact your W. H. Freeman sales representative.

Up-to-Date Content

Because the world is constantly changing, it is essential that a world regional geography text be as current as possible. To that end, the fifth edition discusses the BP oil spill in the Gulf of Mexico; the earthquakes in Chile and Haiti; the varying effects of the global recession on regions, countries, and individuals; civil unrest in once-stable Thailand; the new influence of Arabic media outlets, such as Al Jazeera, which now affects thinking around the world; and the consequences of global climate change on Pacific Island nations. Some of the major content areas of the book that have been updated include:

- Domestic and global implications of the U.S. political, economic, and military stances.
- The role of terrorism in the realignment of power, globally and locally.

- Immigration and the ways it is changing countries economically and culturally.

- The global economic recession and its effect on migrants, labor outsourcing, and job security in importing and exporting countries.

- Continuing conflicts and growing turmoil in Southwest Asia and North Africa (the Middle East).

- Recent economic crises in the European Union that may bring about significant reorganization with consequences for the original EU members, new and potential member states, and the global community.

- Changing gender roles, particularly in developing countries.

- The increasing role and influence of Islam around the world.

- Climate change and its environmental, political, and economic implications.

THE ENDURING VISION: GLOBAL AND LOCAL PERSPECTIVES

The Global View

In addition to the new features and enhancements to the text, we retain the hallmark features that have made past editions of this text successful for instructors and students. For the fifth edition, we continue to emphasize global trends and the interregional linkages that are changing lives throughout the world, including those trends related to changing gender roles. The following connections are explored in every chapter, as appropriate:

- The **multifaceted economic links** among world regions. These include (1) the effects of colonialism; (2) trade; (3) the role in the world economy of transnational corporations, such as Walmart, BP, and Cisco; (4) the influence of regional trade organizations, such as ASEAN and NAFTA; and (5) the changing roles of the World Bank and the International Monetary Fund as the negative consequences of structural adjustment programs become better understood.

- **Migration.** Migrants are changing economic and social relationships in virtually every part of the globe. The societies they leave are changed radically by their absence, just as the host societies are changed by their presence. The text explores the local and global effects of foreign workers in places such as Japan, Europe, Africa, the Americas, and Southwest Asia, and the increasing number of refugees resulting from conflicts around the world. Also discussed are long-standing migrant groups, such as the Overseas Chinese and the Indian diasporas.

- The **mass communications and marketing techniques** that are promoting **world popular culture** across regions. The text integrates coverage of popular culture and its effects in discussions of topics such as tourism in the Caribbean and Southeast Asia, and the blending of Western and traditional culture in places such as South Africa, Malawi, Japan, and China.

- **Gender issues,** which are covered in every chapter with the aim of addressing more completely the lives of ordinary people. Gender is intimately connected to other patterns, such as internal and global migration, and these connections and other region-wide gender patterns are illustrated in a variety of maps and photos and in vignettes that illustrate gender roles as played out in the lives of individuals. The lives of children, especially their roles in families, are also covered, often in concert with the treatment of gender issues.

The Local Level

Our approach pays special attention to the local scale—a town, a village, a household, an individual. Our hope is, first, that stories of individual people and families will make geography interesting and real to students and, second, that seeing the effects of abstract processes and trends on ordinary lives will dramatize the effects of these developments for students. Reviewers have mentioned that students particularly appreciate the personal vignettes, which are often stories of real people (with names changed). For each region, we examine the following local responses:

- **Cultural change:** We look closely at changes in the family, gender roles, and social organization in response to urbanization, modernization, and the global economy.

- **Impacts on well-being:** Ideas of what constitutes well-being differ from culture to culture, yet broadly speaking, people everywhere try to provide a healthy life for themselves in a community of their choosing. Their success in doing so is affected by local conditions, global forces, and their own ingenuity.

- **Issues of identity:** Paradoxically, as the world becomes more tightly knit through global communications and media, ethnic and regional identities often become stronger. The text examines how modern developments such as the Internet are used to reinforce particular cultural identities.

- **Local attitudes toward globalization:** People often have ambivalent reactions to global forces: They are repelled by the seeming power of these forces, fearing effects on their own lives and livelihoods and on local traditional cultural values, but they are also attracted by the economic opportunities that may emerge from greater global integration. The text looks at how the people of a region react to cultural or economic globalization.

ONE VISION, TWO VERSIONS: WITH OR WITHOUT SUBREGIONAL COVERAGE

To better serve the different needs of diverse faculty and curricula, the fifth edition is available in two versions.

World Regional Geography with Subregions
(1-4292-3241-2)

The fifth edition continues to employ a consistent structure for each chapter. Each chapter beyond the first is divided into three

parts: I: The Geographic Setting; II: Current Geographic Issues; III: Subregions.

The subregion sections provide a descriptive characterization of particular countries and places within the region that expands on coverage in the main part of the chapter. For example, in the sub-Saharan Africa chapter, the West, Central, East, and Southern Africa subregions are considered, providing additional insights into differences in well-being and in social and economic issues across the African continent.

World Regional Geography without Subregions
(1-4292-3244-7)

The briefer version provides essentially the same main text coverage as the version described above, omitting only the subregional sections. The result is a text that is much shorter, a useful alternative for teachers constrained by time. This version contains all the types of pedagogy found in the main version.

FOR THE INSTRUCTOR

The authors have taught world regional geography many times and understand the need for quick, accessible aids to instruction. Many of the new features were designed to streamline the job of organizing the content of each class session, with the goal of increasing student involvement through interactive discussions. Ease of instruction and active student involvement were the principal motivations behind the book's key features—the thematic concepts, photo essays, and content maps that facilitate region-to-region comparisons, and the wide selection of videos.

Computerized Test Bank (With Subregions: 1-4292-4610-3; Without Subregions: 1-4292-7399-2)

The fifth edition **Computerized Test Bank,** expanded from the original test bank created by Jason Dittmer, University College London, and Andy Walter, West Georgia University. Powered by Diploma, the Computerized Test Bank is designed to match the pedagogical intent of the text and offers more than 5000 test questions (multiple-choice, short-answer, matching, true/false, and essay) in a format that makes it easy to edit, add, and reorder questions. Word files for the computerized test bank can be downloaded from the Instructor Resources on the book's companion Web site or accessed via the Instructor's Resource CD (described below).

Map Builder: Custom Mapping Program

Instructors use maps in the classroom in many different ways, and we have tried to make the book's maps easy to use for instructors and students by creating a flexible, friendly, online mapping tool. W. H. Freeman and Maps.com have built a revolutionary, first-of-its-kind teaching tool for instructors. It is a program that allows instructors to create, display, and print *custom maps*. The program permits in-class display of thousands of different custom maps and allows instructors to zoom in on any portion of a map they want to emphasize. With only a few mouse clicks, instructors can choose the data from selected maps that they would like to show on a single map. For example, instructors can compare climate and agriculture in North America by displaying together the continental climate zone and dairying. These maps can be printed in full color or displayed as slides in the classroom. In addition, the book's companion Web site offers map-building exercises for each chapter. For a demonstration of the Map Builder program and its accompanying exercises, please ask your W. H. Freeman sales representative for a guided tour, or visit www.whfreeman.com/pulsipher5e.

Instructor's Resource CD-ROM (1-4292-4611-1)

To help instructors create their own Web sites and orchestrate dynamic lectures, the CD contains:

- All **text images** in PowerPoint and JPEG formats, with enlarged labels for better projection quality.

- **PowerPoint lecture outlines,** by Rebecca Johns, University of South Florida. The main themes of each chapter are outlined and enhanced with images from the book, providing a pedagogically sound foundation on which to build personalized lecture presentations.

- **Instructor's Resource Manual,** by Jennifer Rogalsky, State University of New York, Geneseo, and Helen Ruth Aspaas, Virginia Commonwealth University. The resource manual contains suggested lecture outlines, points to ponder for class discussion, and ideas for exercises and class projects. It is offered as chapter-by-chapter Word files to facilitate editing and printing.

- **Test Bank,** expanded from the original test bank created by Jason Dittmer, University College London, and Andy Walter, West Georgia University. The Test Bank is designed to match the pedagogical intent of the text and offers more than 5000 test questions (multiple-choice, short-answer, matching, true/false, and essay) in a Word format that makes it easy to edit, add, and reorder questions. A **computerized test bank** (powered by Diploma) with the same content is also available.

- **i>Clicker Questions,** by Elizabeth Leppman, Walden University. Written in PowerPoint for easy integration into lecture presentations, i>Clicker questions allow instructors to jump-start discussions, illuminate important points, and promote better conceptual understanding during lectures.

Instructor's Web Site (password protected)

www.whfreeman.com/pulsipher5e

In addition to all of the resources available on the *Instructor's CD-ROM*, this password-protected Web site also includes:

- **Syllabus-posting** functionality.
- An integrated **grade book** that records students' performance on online quizzes and video quizzes.
- The W. H. Freeman **World Regional Geography DVD**, with 35 projection-quality videos and an accompanying instructor's video manual.
- A link to the **Geography Faculty Lounge,** an online community of geographers offering shared resources.

Course Management

All instructor and student resources are also available via **Black-Board, WebCT, Angel, Moodle, Sakai, and Desire2Learn** to enhance your course. W. H. Freeman offers a course cartridge that populates your site with content tied directly to the book.

W. H. Freeman World Regional Geography 2-DVD Set

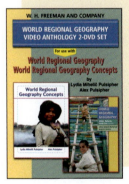

This DVD, available free to adopters of the fifth edition, builds on the book's purpose of putting a face on geography by giving students and instructors access to the fascinating personal stories of people from all over the world. The DVD contains 35 projection-quality video clips from 3 to 7 minutes in length. An **instructor's video manual** is also included on the DVD.

FOR THE STUDENT

World Regional Geography Online/ Book Companion Site

www.whfreeman.com/pulsipher5e

This free companion Web site offers a wealth of resources to support the textbook, including:

- **Chapter quizzes:** These multiple-choice quizzes help students assess their mastery of each chapter.

- **Thinking Critically About Geography activities:** These activities, fully updated for the fifth edition, allow students to explore a set of current issues, such as deforestation, human rights, or free trade, and see how geography helps clarify our understanding of them. Linked Web sites are matched with a series of questions or brief activities that enable students to think about the ways they are connected to the places and people they read about in the text.

- **Map Builder software and Map Builder exercises:** The Map Builder program allows students to create layered thematic maps on their own, while Map Builder Exercises offer a specific activity for each chapter in the fifth edition.

- **Map Learning exercises:** Students use this interactive feature to identify and locate countries, cities, and the major geographic features of each region.

- **Blank outline maps:** Printable maps of the world and of each region, for use in note-taking or exam review or both, as well as for preparing assigned exercises.

- **Flashcards:** Matching exercises teach vocabulary and definitions.

- **Audio pronunciation guide:** Spoken guide of place names, regional terms, and names of historical figures.

- **World recipes and cuisines:** From *International Home Cooking*, the United Nations International School Cookbook.

GeographyPortal

www.yourgeographyportal.com

This comprehensive resource is designed to offer a complete solution for today's classroom. GeographyPortal offers all of the instructor and student resources listed above, as well as premium resources available only on the Portal:

- An **eBook** of *World Regional Geography*, fifth edition, complete and customizable. Students can quickly search the text and personalize it just as they would the printed version, using highlighting, bookmarking, and note-taking features.

- A **Guide to using Google Earth** for the novice, plus step-by-step **Google Earth** exercises for each chapter.

- Selected articles from *Focus on Geography* magazine (one for each chapter in the textbook) and accompanying **quizzes** for each article.

- **Animated photo essays** with sound, easily downloadable as .mp3 files.

- **Physical geography videos** for instructors who like to cover physical geography topics in more detail.

- An **online observation journal.**

- Online **news feeds** for highly respected magazines such as *Scientific American* and *The Economist*.

- An online version of the **Mapping Workbook and Study Guide** (see below).

For more information or to schedule a demonstration, please contact your W. H. Freeman sales representative.

Mapping Workbook and Study Guide (1-4292-5349-5)

Jennifer Rogalsky, State University of New York, Geneseo, and **Helen Ruth Aspaas,** Virginia Commonwealth University

The *Mapping Workbook and Study Guide* retains the exercises that have made it a useful pedagogical tool over four editions. Mapping exercises help students understand and explain geographic patterns by employing the skills geographers routinely use. The Important Places list asks students to enter country names, provinces, cities, and physical features on the blank maps that appear at the end of each chapter.

Rand McNally's Atlas of World Geography, 176 pages

This atlas contains:

- Fifty-two physical, political, and thematic maps of the world and continents; 49 regional, physical, political, and thematic maps; and dozens of metropolitan-area inset maps.

- Geographic facts and comparisons covering topics such as population, climate, and weather.

- A section on common geographic questions, a glossary of terms, and a comprehensive 25-page index.

ACKNOWLEDGMENTS

Fifth Edition

Gillian Acheson
Southern Illinois University, Edwardsville

Greg Atkinson
Tarleton State University

Robert Begg
Indiana University of Pennsylvania

Richard Benfield
Central Connecticut State University

Fred Brumbaugh
University of Houston, Downtown

Deborah Corcoran
Missouri State University

Kevin Curtin
George Mason University

Lincoln DeBunce
Blue Mountain Community College

Scott Dobler
Western Kentucky University

Catherine Doenges
University of Connecticut at Stamford

Jean Eichhorst
University of Nebraska, Kearney

Brian Farmer
Amarillo College

Eveily Freeman
Ohio State University

Hari Garbharran
Middle Tennessee State University

Abe Goldman
University of Florida

Angela Gray
University of Wisconsin, Oshkosh

Ellen Hansen
Emporia State University

Nick Hill
Greenville Technical College

Johanna Hume
Alvin Community College

Edward Jackiewicz
California State University, Northridge

Rebecca Johns
University of South Florida, St. Petersburg

Suzanna Klaf
Ohio State University

Jeannine Koshear
Fresno City College

Brennan Kraxberger
Christopher Newport University

Heidi Lannon
Santa Fe College

Angelia Mance
Florida Community College, Jacksonville

Meredith Marsh
Lindenwood University

Linda Murphy
Blinn Community College

Monica Nyamwange
William Paterson University

Adam Pine
University of Minnesota, Duluth

Amanda Rees
Columbus State University

Benjamin Richason
St. Cloud State University

Amy Rock
Kent State University

Betty Shimshak
Towson University

Michael Siola
Chicago State University

Steve Smith
Missouri Southern State University

Jennifer Speights-Binet
Samford University

Emily Sturgess Cleek
Drury University

Gregory Taff
University of Memphis

Catherine Veninga
College of Charleston

Mark Welford
Georgia Southern University

Donald Williams
Western New England College

Peggy Robinson Wright
Arkansas State University, Jonesboro

Fourth Edition

Robert Acker
University of California, Berkeley

Joy Adams
Humboldt State University

John All
Western Kentucky University

Jeff Allender
University of Central Arkansas

David L. Anderson
Louisiana State University, Shreveport

Donna Arkowski
Pikes Peak Community College

Jeff Arnold
Southwestern Illinois College

Richard W. Benfield
Central Connecticut University

Sarah A. Blue
Northern Illinois University

Patricia Boudinot
George Mason University

Michael R. Busby
Murray State College

Norman Carter
California State University, Long Beach

Gabe Cherem
Eastern Michigan University

Brian L. Crawford
West Liberty State College

Phil Crossley
Western State College of Colorado

Gary Cummisk
Dickinson State University

Kevin M. Curtin
University of Texas at Dallas

Kenneth Dagel
Missouri Western State University

Jason Dittmer
University College London

Rupert Dobbin
University of West Georgia

James Doerner
University of Northern Colorado

Ralph Feese
Elmhurst College

Richard Grant
University of Miami

Ellen R. Hansen
Emporia State University

Holly Hapke
Eastern Carolina University

Mark L. Healy
Harper College

David Harms Holt
Miami University

Douglas A. Hurt
University of Central Oklahoma

Edward L. Jackiewicz
California State University, Northridge

Marti L. Klein
Saddleback College

Debra D. Kreitzer
Western Kentucky University

Jeff Lash
University of Houston, Clear Lake

Unna Lassiter
California State University, Long Beach

Max Lu
Kansas State University

Donald Lyons
University of North Texas

Shari L. MacLachlan
Palm Beach Community College

Chris Mayda
Eastern Michigan University

Armando V. Mendoza
Cypress College

Katherine Nashleanas
University of Nebraska, Lincoln

Joseph A. Naumann
University of Missouri at St. Louis

Jerry Nelson
Casper College

Michael G. Noll
Valdosta State University

Virginia Ochoa-Winemiller
Auburn University

Karl Offen
University of Oklahoma

Eileen O'Halloran
Foothill College

Ken Orvis
University of Tennessee

Manju Parikh
College of Saint Benedict and Saint John's University

Mark W. Patterson
Kennesaw State University

Paul E. Phillips
Fort Hays State University

Rosann T. Poltrone
Arapahoe Community College

Waverly Ray
MiraCosta College

Jennifer Rogalsky
State University of New York, Geneseo

Gil Schmidt
University of Northern Colorado

Yda Schreuder
University of Delaware

Tim Schultz
Green River Community College

Sinclair A. Sheers
George Mason University

D. James Siebert
North Harris Montgomery Community College, Kingwood

Dean Sinclair
Northwestern State University

Bonnie R. Sines
University of Northern Iowa

Vanessa Slinger-Friedman
Kennesaw State University

Andrew Sluyter
Louisiana State University

Kris Runberg Smith
Lindenwood University

Herschel Stern
MiraCosta College

William R. Strong
University of North Alabama

Ray Sumner
Long Beach City College

Rozemarijn Tarhule-Lips
University of Oklahoma

Alice L. Tym
University of Tennessee, Chattanooga

James A. Tyner
Kent State University

Robert Ulack
University of Kentucky

Jialing Wang
Slippery Rock University of Pennsylvania

Linda Q. Wang
University of South Carolina, Aiken

Keith Yearman
College of DuPage

Laura A. Zeeman
Red Rocks Community College

Third Edition

Kathryn Alftine
California State University, Monterey Bay

Donna Arkowski
Pikes Peak Community College

Tim Bailey
Pittsburg State University

Brad Baltensperger
Michigan Technological University

Michele Barnaby
Pittsburg State University

Daniel Bedford
Weber State University

Richard Benfield
Central Connecticut State University

Sarah Brooks
University of Illinois at Chicago

Jeffrey Bury
University of Colorado, Boulder

Michael Busby
Murray State University

Norman Carter
California State University, Long Beach

Gary Cummisk
Dickinson State University

Cyrus Dawsey
Auburn University

Elizabeth Dunn
University of Colorado, Boulder

Margaret Foraker
Salisbury University

Robert Goodrich
University of Idaho

Steve Graves
California State University, Northridge

Ellen Hansen
Emporia State University

Sophia Harmes
Towson University

Mary Hayden
Pikes Peak Community College

R. D. K. Herman
Towson University

Samantha Kadar
California State University, Northridge

James Keese
California Polytechnic State University

Phil Klein
University of Northern Colorado

Debra D. Kreitzer
Western Kentucky University

Soren Larsen
Georgia Southern University

Unna Lassiter
California State University, Long Beach

David Lee
Florida Atlantic University

Anthony Paul Mannion
Kansas State University

Leah Manos
Northwest Missouri State University

Susan Martin
Michigan Technological University

Luke Marzen
Auburn University

Chris Mayda
Eastern Michigan University

Michael Modica
San Jacinto College

Heather Nicol
State University of West Georgia

Ken Orvis
University of Tennessee

Thomas Paradis
Northern Arizona University

Amanda Rees
University of Wyoming

Arlene Rengert
West Chester University of Pennsylvania

Benjamin Richason
St. Cloud State University

Deborah Salazar
Texas Tech University

Steven Schnell
Kutztown University

Kathleen Schroeder
Appalachian State University

Roger Selya
University of Cincinnati

Dean Sinclair
Northwestern State University

Garrett Smith
Kennesaw State University

Jeffrey Smith
Kansas State University

Dean Stone
Scott Community College

Selima Sultana
Auburn University

Ray Sumner
Long Beach City College

Christopher Sutton
Western Illinois University

Harry Trendell
Kennesaw State University

Karen Trifonoff
Bloomsburg University

David Truly
Central Connecticut State University

Kelly Victor
Eastern Michigan University

Mark Welford
Georgia Southern University

Wendy Wolford
University of North Carolina at Chapel Hill

Laura Zeeman
Red Rocks Community College

Second Edition

Helen Ruth Aspaas
Virginia Commonwealth University

Cynthia F. Atkins
Hopkinsville Community College

Tim Bailey
Pittsburg State University

Robert Maxwell Beavers
University of Northern Colorado

James E. Bell
University of Colorado, Boulder

Richard W. Benfield
Central Connecticut State University

John T. Bowen, Jr.
University of Wisconsin, Oshkosh

Stanley Brunn
University of Kentucky

Donald W. Buckwalter
Indiana University of Pennsylvania

Gary Cummisk
Dickinson State University

Roman Cybriwsky
Temple University

Cary W. de Wit
University of Alaska, Fairbanks

Ramesh Dhussa
Drake University

David M. Diggs
University of Northern Colorado

Jane H. Ehemann
Shippensburg University

Kim Elmore
University of North Carolina at Chapel Hill

Thomas Fogarty
University of Northern Iowa

James F. Fryman
University of Northern Iowa

Heidi Glaesel
Elon College

Ellen R. Hansen
Emporia State University

John E. Harmon
Central Connecticut State University

Michael Harrison
University of Southern Mississippi

Douglas Heffington
Middle Tennessee State University

Robert Hoffpauir
California State University, Northridge

Catherine Hooey
Pittsburg State University

Doc Horsley
Southern Illinois University, Carbondale

David J. Keeling
Western Kentucky University

James Keese
California Polytechnic State University

Debra D. Kreitzer
Western Kentucky University

Jim LeBeau
Southern Illinois University, Carbondale

Howell C. Lloyd
Miami University of Ohio

Judith L. Meyer
Southwest Missouri State University

Judith C. Mimbs
University of Tennessee, Chattanooga

Monica Nyamwange
William Paterson University

Thomas Paradis
Northern Arizona University

Firooza Pavri
Emporia State University

Timothy C. Pitts
Edinboro University of Pennsylvania

William Preston
California Polytechnic State University

Gordon M. Riedesel
Syracuse University

Joella Robinson
Houston Community College

Steven M. Schnell
Northwest Missouri State University

Kathleen Schroeder
Appalachian State University

Dean Sinclair
Northwestern State University

Robert A. Sirk
Austin Peay State University

William D. Solecki
Montclair State University

Wei Song
University of Wisconsin, Parkside

William Reese Strong
University of North Alabama

Selima Sultana
Auburn University

Suzanne Traub-Metlay
Front Range Community College

David J. Truly
Central Connecticut State University

Alice L. Tym
University of Tennessee, Chattanooga

First Edition

Helen Ruth Aspaas
Virginia Commonwealth University

Brad Bays
Oklahoma State University

Stanley Brunn
University of Kentucky

Altha Cravey
University of North Carolina at Chapel Hill

David Daniels
Central Missouri State University

Dydia DeLyser
Louisiana State University

James Doerner
University of Northern Colorado

Bryan Dorsey
Weber State University

Lorraine Dowler
Penn State University

Hari Garbharran
Middle Tennessee State University

Baher Ghosheh
Edinboro University of Pennsylvania

Janet Halpin
Chicago State University

Peter Halvorson
University of Connecticut

Michael Handley
Emporia State University

Robert Hoffpauir
California State University, Northridge

Glenn G. Hyman
International Center for Tropical Agriculture

David Keeling
Western Kentucky University

Thomas Klak
Miami University of Ohio

Darrell Kruger
Northeast Louisiana University

David Lanegran
Macalester College

David Lee
Florida Atlantic University

Calvin Masilela
West Virginia University

Janice Monk
University of Arizona

Heidi Nast
DePaul University

Katherine Nashleanas
University of Nebraska, Lincoln

Tim Oakes
University of Colorado, Boulder

Darren Purcell
Florida State University

Susan Roberts
University of Kentucky

Dennis Satterlee
Northeast Louisiana University

Kathleen Schroeder
Appalachian State University

Dona Stewart
Georgia State University

Ingolf Vogeler
University of Wisconsin, Eau Claire

Susan Walcott
Georgia State University

These books have been a family project many years in the making. Lydia Pulsipher came to the discipline of geography at the age of 5, when her immigrant father, Joe Mihelič, hung a world map over the breakfast table in their home in Coal City, Illinois, where he was pastor of the New Hope Presbyterian Church. They soon moved to the Mississippi Valley of eastern Iowa, where Lydia's father, then a professor in the Presbyterian theological seminary in Dubuque, continued his geography lessons on the passing landscapes whenever Lydia accompanied him on Sunday trips to small country churches. Lydia's sons, Anthony and Alex, got their first doses of geography in the bedtime stories she told them. For plots and settings, she drew on Caribbean colonial documents she was then reading for her dissertation. They first traveled abroad and learned about the hard labor of field geography when as 12- and 8-year-olds they were expected to help with the archaeological and ethnographic research conducted by Lydia and her colleagues on the eastern Caribbean island of Montserrat. It was Lydia's brother, John Mihelič, who first suggested that Lydia, Alex, and Mac write a book like this one, after he too came to appreciate geography. John has been a loyal cheerleader during the process, as have our extended family and friends in Knoxville, Montserrat, San Francisco, Slovenia, and beyond.

Graduate students and faculty colleagues in the Geography Department at the University of Tennessee have been generous in their support, serving as helpful impromptu sounding boards for ideas. Ken Orvis, especially, has advised us on the physical geography sections of all editions. Yingkui (Philippe) Li provided information on glaciers and climate change; Russell Kirby wrote one of the vignettes based on his research in Vietnam; Toby Applegate, Alex Pulsipher (in his capacity as an instructor), Michelle Brym, and Sara Beth Keough have helped us understand how to better assist instructors; and Ron Kalafsky, Tom Bell, Margaret Gripshover, and Micheline Van Riemsdijk have chatted with the authors many times on specific and broad issues related to this textbook.

Maps for this edition were conceived by Mac Goodwin and Alex Pulsipher and produced by Will Fontanez and the University of Tennessee cartography shop staff: Matt Kookogey and Tracey Pollock; and by Martha Bostwick and Mike Powers, lead cartographer, at Maps.com. Alex Pulsipher chose the photos and produced the Thematic Concept Photo Essays and the Photo Essays.

Liz Widdicombe and Sara Tenney at W. H. Freeman were the first to persuade us that together we could develop a new direction for *World Regional Geography*, one that included the latest thinking in geography written in an accessible style. In accomplishing this goal, we are especially indebted to our first developmental editor, Susan Moran, and to the W. H. Freeman staff for all they have done to ensure that this book is well written, beautifully designed, and well presented to the public.

We would also like to gratefully acknowledge the efforts of the following people at W. H. Freeman: Steven Rigolosi, executive editor for the fifth edition; Barbara Muller, developmental editor; Leigh Renhard, project editor; John Britch, director of marketing; Anna Paganelli, copyeditor; Diana Blume, design manager; Bill Page, senior illustration coordinator; Julia De Rosa, production manager; Sheridan Sellers, composition and layout; and Kerri Russini, assistant editor.

Given our ambitious new photo program, we are especially grateful for Vicki Tomaselli's brilliant work and responsiveness as designer for the fifth edition, as well as Ted Szczepanski's guidance and direction as our photo editor for the fifth edition. We are also grateful to the supplements authors, who have created what we think are unusually useful, up-to-date, and labor-saving materials for instructors who use our book. Finally, we owe a debt of gratitude to eagle-eyed Ola Johansson (University of Pittsburgh), who read the entire book in page proof and helped us improve the final work in innumerable ways.

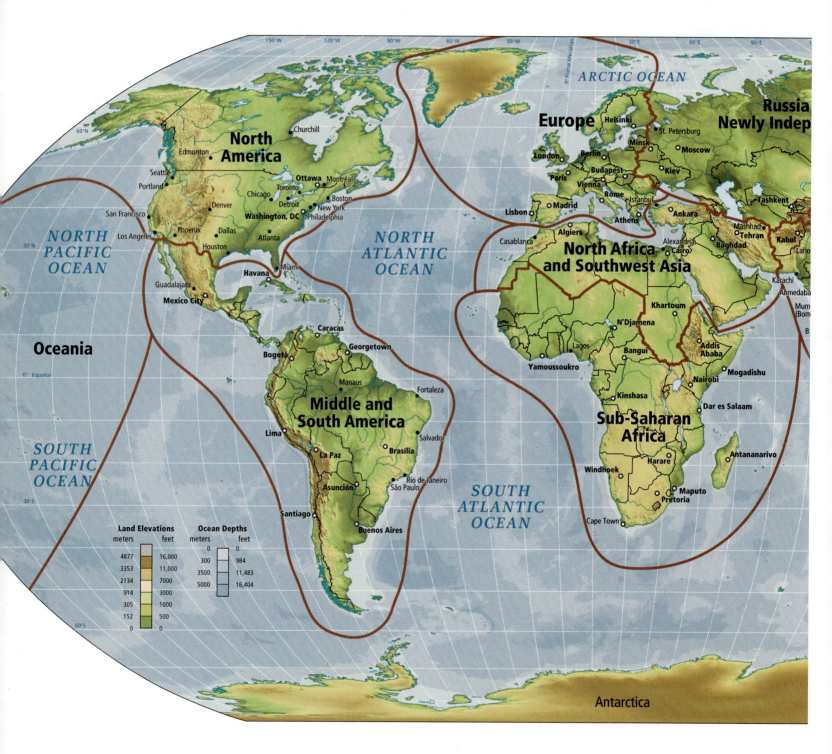

THEMATIC CONCEPTS
Population • Gender • Development
Food • Urbanization • Globalization
Democratization • Climate Change • Water

Geography:
An Exploration
of Connections

FIGURE 1.1 Regions of the world.

Where Is It? Why Is It There? Why Does It Matter?

Where are you? You may be in a house or a library or sitting under a tree on a fine fall afternoon. You are probably in a community (perhaps a college or university), and you are in a country (perhaps the United States) and a region of the world (perhaps North America, Southeast Asia, or the Pacific). Why are you where you are? Some answers are immediate, such as "I have an assignment to read." Other explanations are more complex, such as your belief in the value of an education, your career plans, and your or someone's willingness to sacrifice to pay your tuition. Even past social movements that opened up higher education to more than a fortunate few may help explain why you are where you are.

The questions *where* and *why* are central to geography. Think about a time you had to find the site of a party on a Saturday night, the location of the best grocery store, or the fastest and safest route home. You were interested in location, spatial relationships, and connections between the environment and people. Those are among the interests of geographers.

Geographers seek to understand why different places have different sights, sounds, smells, and arrangements of features. They study what has contributed to the look and feel of a place, to the standard of living and customs of its people, and to the way people in one place relate to people in other places. Furthermore, geographers often think on several scales from the local to the global. For example, when choosing the best location for a new grocery store, a geographer might consider the socioeconomic circumstances of the neighborhood, traffic patterns for the broader area in which the neighborhood is located, as well as the store's location relative to the main population concentrations for the whole city and to competitors. She could also consider national or even international transportation routes, possibly to determine cost-efficient connections to suppliers.

To make it easier to understand a geographer's many interests, try this exercise. Draw a map of your most familiar childhood landscape. Relax, and recall the objects and experiences that were most important to you in that place. If the place was your neighborhood, you might start by drawing and labeling your home. Then fill in other places you encountered regularly, such as your backyard, your best friend's home, or your school. For example, Figure 1.2

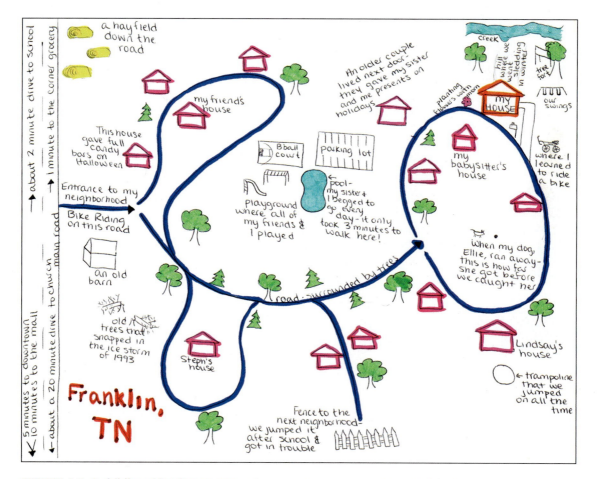

FIGURE 1.2 A childhood landscape map. Julia Stump drew this map of her childhood landscape in Franklin, Tennessee, as an exercise in Lydia Pulsipher's world geography class.
[Courtesy of Julia Stump.]

shows the childhood landscape of Julia Stump in Franklin, Tennessee.

Consider how your map reveals the ways in which your life was structured by space. What is the **scale** of your map? That is, how much space did you decide to illustrate on the map? The amount of space your map covers may represent the degree of freedom you had as a child, or how aware you were of the world around you. Were there places you were not supposed to go? Does your map reveal, perhaps subtly, such emotions as fear, pleasure, or longing? Does it indicate your sex, your ethnicity, or the makeup of your family? Did you use symbols to show certain features? In making your map and analyzing it, you have engaged in several aspects of geography:

- Landscape observation
- Description of the earth's surface and consideration of the natural environment
- Spatial analysis (the study of how people, objects, or ideas are related to one another across space)
- The use of different scales of analysis (your map probably shows the spatial features of your childhood at a detailed local scale)
- Cartography (the making of maps)

As you progress through this book and this course, you will acquire geographic information and skills. Perhaps you are planning to travel to other lands or thinking about investing in East Asian timber stocks. Maybe you are searching for a good place to market an idea or trying to understand current events in your town within the context of world events. Knowing how to practice geography will make your task easier and more engaging.

What Is Geography?

Geography is the study of our planet's surface and the processes that shape it. Yet this definition does not begin to convey the fascinating interactions of human and environmental forces that have given the earth its diverse landscapes and ways of life.

Geography, as an academic discipline, is unique in that it links the physical sciences—such as geology, physics, chemistry, biology, and botany—with the social sciences—such as anthropology, sociology, history, economics, and political science. Physical geographers have generally focused on how the earth's physical processes work independently of humans, but increasingly many are interested in how physical processes may affect humans, and how these processes are affected by humans in return. Human geography is the study of the various aspects of human life that create the distinctive landscapes and regions of the world. Physical and human geography are often tightly linked. For example, geographers might aim to understand:

- How and why people came to occupy a particular place
- How people use the physical aspects of that place (climate, landforms, and resources) and then modify them to suit their particular needs

- How people may create environmental problems
- How people interact with other places, far and near.

Geographers usually specialize in one or more fields of study, or subdisciplines. Some of these particular types of geography are mentioned over the course of the book. Despite their individual specialties, geographers often cooperate in studying the interactions between people and places. For example, in the face of increasing global warming, climate geographers, cultural geographers, and economic geographers work together to understand the spatial distribution of carbon dioxide emissions, the cultural practices that might be changed to limit such emissions, and the economic effects of these potential changes.

> **scale (of a map)** the proportion that relates the dimensions of the map to the dimensions of the area it represents; also, variable-sized units of geographical analysis
>
> **lines of longitude (meridians)** the distance in degrees east and west of Greenwich, England; lines of longitude run from pole to pole (the line of longitude at Greenwich is 0° and is known as the prime meridian)

Many geographers specialize in a particular region of the world, or even in one small part of a region. Regional geography is the analysis of the geographic characteristics of a particular place, the size and scale of which can vary radically. The study of a region can reveal connections among physical features and ways of life, as well as connections to other places. These links are key to understanding the present and the past and are essential in planning for the future. This book follows a "world regional" approach, focusing on general knowledge about specific regions of the world. We will see just what geographers mean by *region* a little later in this chapter.

Geographers' Visual Tools

Among geographers' most important tools are maps, which they use to record, analyze, and explain spatial relationships, as you did on your childhood landscape map. Geographers who specialize in depicting geographic information on maps are called cartographers.

Understanding Maps

Figure 1.3 explains the various features of maps. Different *scales of imagery* are demonstrated using photographs, a satellite image, and several maps. Read the captions carefully to understand the scale being depicted in each image. Throughout this book, you will encounter different kinds of maps at different scales. Some will show physical features such as landforms or climate patterns at the regional or global scale. Others will show aspects of human activities at these same regional or global scales—for example, the routes taken by drug traders. Yet other maps will show settlement or cultural features at the scale of countries or regions, or cities or even local neighborhoods.

Longitude and Latitude

Most maps contain lines of latitude and longitude, which enable a person to establish a position on the map relative to other points on the globe. **Lines of longitude** (also called meridians) run from

FIGURE 1.3 Understanding Maps: Legend

Being able to read a map legend is crucial to understanding the maps in this book. The colors in the legend convey information about different areas on the map. In the population density map below the lowest density (0–3 persons per square mile) is colored pale yellow. A part of North America that has this density is shown in the map inset to the right of the legend. On the far right is a picture of this area. This is also done for two other densities (27–260) and above 2000 people per square mile.

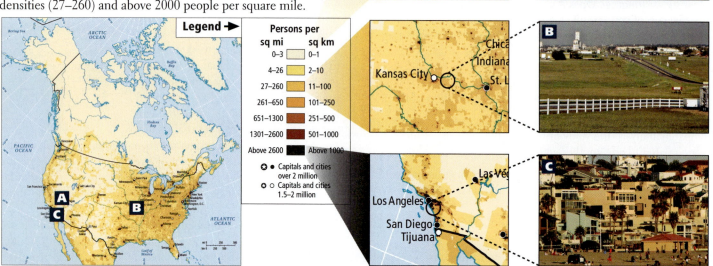

Understanding Maps: Choropleth versus Isorhythmic

There are two main types of maps presented in this book. In choropleth maps, the shapes visible on the map correspond to predefined areas (for example countries, counties, census tracts) that are shaded according to the measurement of a statistical variable—for example, population density. On map (D) you can see examples of predefined areas, the blocky yellowish shapes around Las Vegas, which in this case are census tracts that the U.S. government uses to gather data on the population. The map maker assigned each census tract a color based on the density of people in that tract. Another way to represent data spatially is to make an isorhythmic map, in which the shapes visible on the map correspond to areas defined by data patterns. Instead of fitting the data into predefined areas, the data define the areas. If map (D) had been done as an isorythmic map it would look something like map (E), an isorythmic population density map. The maker of this map created shapes that correspond to each level of population density represented in the legend above. In this particular case, it involved using other data, such as satellite images that show where people really are living, and a good deal of estimation. In this book we create maps using the method that best reflects the data we are trying to display. Since population density data are collected according to census tracts or other predefined boundaries, we use choropleth maps in maps of population density. However, when the data used to make a map are not collected according to predefined boundaries (for example, the elevation data used to create the maps found at the beginning of each chapter), we use isorhythmic maps.

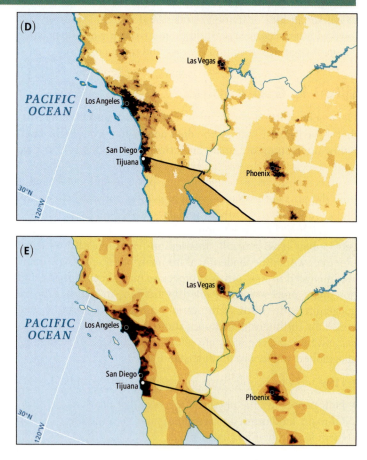

Understanding Maps: Scale

Maps often display information at different spatial scales, which means that lengths, areas, distances, and sizes can appear dramatically different on otherwise similar maps. This book often combines maps at several different scales with photographs taken by people on the surface of the earth and photos taken by satellites or astronauts in space. All of these visual tools convey information at a spatial scale. Here are some of the scales you might encounter in this book. The scale is visible on each map.

(**F**) Here is most of the region we define as Middle and South America, with just the tip of South America missing, at a scale of 1:200,000,000, meaning that 1 inch on the map equals 200,000,000 inches in reality. This is roughly the scale of the half-page world maps, such as Figure 1.11, that appear throughout this book.

(**G**) Here is a close-up from the previous map, focusing on the Caribbean sea, at a scale of 1:45,000,000. The larger islands are clearly shown, some smaller islands are too small to identify clearly. This is roughly the scale of the full page maps for Chapter 3, such as Photo Essay 3.5 on page 164 (population).

(**H**) The eastern Caribbean at a scale of 1:15,000,000. The shape of the islands and the locations of some capital cities are clearly visible.

(**I**) Zooming in more, we arrive at a map of Guadeloupe and Dominica, in the eastern Caribbean, at a scale of 1:3,000,000. This scale makes it possible to show towns and a few roads and rivers.

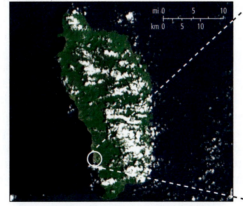

(**J**) A satellite image of Dominica at a scale of 1:950,000. The town of Roseau (circled) can just barely be seen. The dashed lines indicate that the next picture (**K**) was taken within this circle. The white puffs are clouds.

(**K**) A photo overlooking part of the town of Roseau, capital of Dominica. It's hard to give a precise scale for this picture because the houses at the bottom are about 2000 feet closer to the photographer than the shoreline. The cruise boat is at a scale of roughly 1:3200.

Understanding Maps: Conventions in This Textbook

Rivers are darker blue. Lakes and oceans are lighter blue.

Major roads are in red.

Railroads are in black.

Cities are given an icon and print size indicating how large they are.

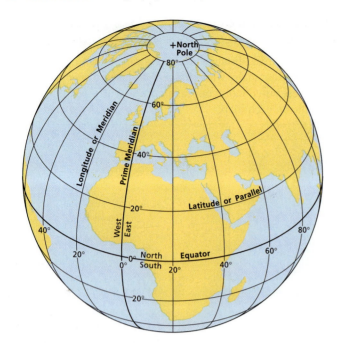

FIGURE 1.4 Summary of longitude and latitude. Lines of longitude (meridians) extend from pole to pole. The distance between them on the globe decreases steadily toward the poles, where they all meet. Lines of latitude (parallels) are equally spaced north and south of the equator and intersect the longitude lines at right angles. The only line of latitude that spans the complete circumference of the earth is the equator; all other lines of latitude describe ever-smaller circles heading away from the equator. [Adapted from *The New Comparative World Atlas* (Maplewood, N.J.: Hammond, 1997), p. 6.]

pole to pole; **lines of latitude** (also called parallels) circle the earth parallel to the equator (**Figure 1.4**).

Both latitude and longitude lines describe circles, so there are 360 degrees (designated with the symbol °) in each circle of latitude and 180 degrees in each pole to pole semicircle of longitude. Each degree spans 60 minutes (designated with the symbol ′), and each minute has 60 seconds (designated with the symbol ″). Keep in mind that these are measures of relative space on a circle, not measures of time. They do not even represent real distance because the circles of latitude get successively smaller to the north and south of the equator, until they become a virtual dot at the poles.

The globe is also divided into hemispheres. The prime meridian, 0° longitude, was arbitrarily located during the European colonial era on a line that runs from the North Pole through Greenwich, England, to the South Pole. The half of the globe's surface west of the prime meridian is called the Western Hemisphere; the half to the east is called the Eastern Hemisphere. The longitude lines both east and west of the prime meridian are labeled from 1° to 180° by their direction and distance in degrees from the prime meridian. For example, 20 degrees east longitude would

lines of latitude (parallels) the distance in degrees north or south of the equator; lines of latitude run parallel to the equator

map projections the various ways of showing the spherical earth on a flat surface

be written as 20° E. The longitude line at 180° runs through the Pacific Ocean and is used as the international date line; the calendar day officially begins when midnight falls at this line.

The equator divides the globe into the Northern and Southern Hemispheres. Latitude is measured from 0° at the equator to 90° at the North Pole and South Pole, respectively.

Lines of longitude and latitude form a grid that can be used to designate the location of a place. In Figure 1.3I, you can see that the island of Marie-Galante lies just south of the parallel (line of latitude) at 16° N and just about 16′ west of the 61st west meridian. Hence, the position of the northernmost point on Marie-Galante's north coast is approximately 16° N by 61° 16′ W. Google Earth gives a more precise location of 16° 0′ 26″ N by 61° 16′ 33″ W.

Map Projections

Printed maps must solve the problem of showing the spherical earth on a flat piece of paper. Imagine drawing a map of the earth on an orange, peeling the orange, and then trying to flatten out the orange-peel map and transfer it exactly to a flat piece of paper. The various ways of showing the spherical surface of the earth on flat paper are called **map projections**. All projections create some distortion. For maps of small parts of the earth's surface, the distortion is minimal. Developing a projection for the whole surface of the earth that minimizes distortion is much more challenging.

The *Mercator projection* (**Figure 1.5A**) is popular, but geographers rarely use it because of its gross distortion near the poles. To make his flat map, the Flemish cartographer Gerhardus Mercator (1512–1594) stretched out the poles, depicting them as lines equal in length to the equator! As a result, Greenland, for example, appears about as large as Africa, even though it is only about one-fourteenth Africa's size. Nevertheless, Mercator's projection is still useful for navigation because it portrays the shapes of landmasses more or less accurately, and because a straight line between two points on this map gives the compass direction between them.

Goode's interrupted projection (**Figure 1.5B**) flattens the earth rather like an orange peel, thus preserving some of the size and shape of the landmasses. In this projection, the oceans are split. The *Robinson projection* (**Figure 1.5C**) shows the longitude lines curving toward the poles to give an impression of the earth's curvature, and it has the advantage of showing an uninterrupted view of land and ocean; however, as a result, the shapes of landmasses are slightly distorted. In this book we often use the Robinson projection for world maps.

Maps are not unbiased. Most currently popular world map projections reflect the European origins of modern cartography. Europe or North America is usually placed near the center of the map, where distortion is minimal; other population centers, such as East Asia, are placed at the highly distorted periphery. For a less-biased study of the modern world, we need world maps that center on different parts of the globe. For example, much of the world's economic activity is now taking place in and around Japan, Korea, China, Taiwan, and Southeast Asia. Discussions of the world economy require maps that focus on these regions and include other parts of the world in the periphery. Another source of bias in maps is the convention that north is always at the top

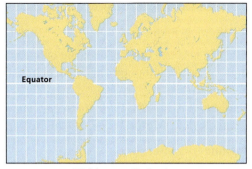

(A) Mercator Projection

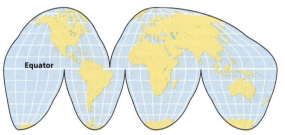

(B) Goode's Interrupted Homolosine Projection

(C) Robinson Projection

FIGURE 1.5 Three common map projections. [Mercator and Robinson projections adapted from *The New Comparative World Atlas* (Maplewood, N.J.: Hammond, 1997), pp. 6–7. Goode's interrupted homolosine projection adapted from *Goode's World Atlas,* 22nd ed. (Rand McNally, 2010), p. 13.

of the map. Some cartographers think that this can lead to a subconscious assumption that the Northern Hemisphere is somehow superior to the Southern Hemisphere.

Geographic Information Science (GIS)

The acronym GIS is now widespread and usually refers to **Geographic Information Science,** meaning the body of science that underwrites multiple spatial analysis technologies and keeps them at the cutting edge. GIS is multidisciplinary, using techniques from cartography (mapmaking), geodesy (measuring the earth's surface), and photogrammetry (the science of making reliable measurements, especially by using aerial photography). But additional sciences, such as cognitive psychology and spatial statistics (geomatics or geoinformatics) are increasingly used to give greater depth and breadth to three-dimensional spatial analysis.

GIS, then, can be used in medicine to analyze the human body, in engineering to analyze mechanical devices, in architecture to analyze buildings, in archaeology to analyze sites above and below ground, and in geography to analyze the earth's surface and the space above and below the earth's surface.

GIS is a burgeoning field in geography, with wide practical applications in government and business and in efforts to assess and improve human and environmental conditions. GIS can also apply to the computerized analytical systems that are the tools of this newest of spatial sciences.

The widespread use of GIS, particularly by governments and corporations, has changed the way information is collected, stored, and analyzed. These changes raise serious social questions. Who owns such information? Who should have access to the information? What rights do people have over the storage, analysis, and distribution of information on their location and movements? Many governments, for example, store information about land ownership using GIS. Should this information reside in the public domain? Should individuals have the right to have their location-based information suppressed from public view? Should a government or corporation have the right to sell information about a person's geographic footprint to anyone without special permission? Progress on these societal questions has not kept pace with the technological advances in GIS.

Photographs are another visual tool used by geographers to help them understand or explain a geographic issue or depict the character of a place. Interpreting a photo to extract its geographical information can sometimes be like detective work. In Figure 1.6 are some points to keep in mind as you look at the pictures throughout this book; try them out first with the photo in Figure 1.6.

The Region as a Concept

A **region** is a unit of the earth's surface with distinct physical and human features. We could speak of a desert region, a region that produces rice, or a region experiencing ethnic violence. In this book, it is rare for any two regions to be defined by the same set of indicators. For example, the region of the southern United States might be defined by its distinctive vegetation, architecture, ways of speaking, foods, and historical experience. Meanwhile Siberia, in eastern Russia, could be defined primarily by its climate, vegetation, remoteness, and sparse settlement.

Another issue in defining regions is that their boundaries are rarely crisp. The more closely we look at the border zones, the less distinct the divisions appear. Consider the case of the boundary between the United States and Mexico (see Figure 1.7 on page 9). The clearly delineated political border does not mark a separation between cultures or economies. In a wide band extending over both sides of the border, there is a blend of Native American, Spanish colonial, Mexican, and Anglo-American cultural features. Languages, place-names, food customs, music, and family organization are only a few examples of this blend. And the economy of the

Geographic Information Science the body of science that underwrites multiple spatial analysis technologies and keeps them at the cutting edge

region a unit of the earth's surface that contains distinct patterns of physical features and/or of human development

(a) **Landforms:** Notice the lay of the land and the landform features. Are there any indications about the ways the landforms and humans have influenced each other? Is environmental stress visible?

(b) **Vegetation:** Notice whether the vegetation indicates a wet or dry, or warm or cold environment. Can you recognize specific species? Does the vegetation appear to be natural or is it influenced by human use?

(c) **Material culture:** Are there buildings, tools, clothing, foods, or vehicles that give clues about the cultural background, wealth, values, or aesthetics of the people who live where the picture was taken?

(d) What do the people in the photo suggest about the situation pictured?

(e) Can you see evidence of the global economy (for example, goods that probably were not produced locally)?

(f) **Location:** Can you tell where the picture was taken or narrow down the possible locations that might be depicted?

You can use this system to analyze any of the photos in this book or elsewhere. Practice by analyzing the photos in this book before you read their captions. Here is an example of how you could do this with the photo below (Figure 1.6).

(a) **Landforms:**

1. The flat horizon suggests a plain or river delta. Environmental stress is visible in several places.

2. This fire could be burning out of control. It seems no one is tending it.

3. This looks like the charred remains of a tree trunk.

4. This black oily-looking liquid doesn't look natural. Could it be crude oil? That would explain the burning.

5. What are these lumps on the ground?

6. This could be a tree trunk coming out of one of the lumps. Could the lumps be what is left of trees that have burned?

What would have caused all this landscape transformation? Maybe an oil spill? Maybe the oil caught fire?

(b) **Vegetation:**

7. This looks like a palm tree. There are quite a few palm trees and other trees.

Must be in the tropics and be fairly wet and warm.

(c) **Material culture:** See (d).

There is nothing here but a single person. The whole area might be abandoned.

(d) **People:**

8. The clothing on this person doesn't look like he made it. It looks mass produced.

This suggests that he has access to goods produced some distance away, maybe in a nearby city. Or he could buy things in a market where imported goods are sold.

9. His worn flip-flops and callused feet suggest he walks a lot.

He might not have a car.

(e) **Global economy:** See d.

(f) **Location:** Somewhere tropical where there could have been an oil spill. Hint: Use this book! See Figure 6.21 to see where the member countries of OPEC are. This suggests that the photo could be of Venezuela or Nigeria. Suggestion: Read Chapter 7!

FIGURE 1.6 Oil development and the environment. On December 3, 2003, part of an oil pipeline burst in Rukpokwu state in Nigeria. No action was taken by the owner of the pipeline, Shell Oil, or the Nigerian government for a week. Fires burned for more than 6 weeks, and no cleanup was attempted afterward. About 740 acres of once-fertile farmland and fish ponds were destroyed. This picture was taken in 2004.

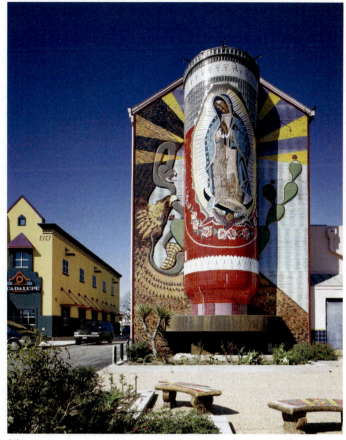

FIGURE 1.7 Cultural patterns along the border of the United States and Mexico. (**A**) Along the border, there is a wide band of cultural blending in which language, food, religion, and architectures show influences of both U.S. and Mexican culture. (**B**) A mural on Guadalupe Street in San Antonio, Texas, that portrays the Virgin of Guadalupe, the patron saint of Mexico.

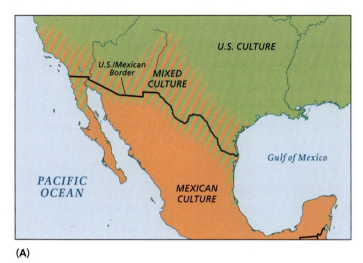

(A)

(B)

border zone depends on interactions across a broad swath of territory. In fact, as a result of trade agreements (NAFTA; see Chapters 2 and 3), the Mexican national economy is becoming more closely connected to the economies of the United States and Canada than to those of its close neighbors in Middle America (the countries that make up Central America and the Caribbean).

Why, then, does this book place Mexico in Middle America? Although northern Mexico has much in common with parts of the southwestern United States, overall Mexico still has more in common with Middle America than with North America. The use of Spanish as its official language ties Mexico to Middle and South America and separates it from the United States and Canada, where English dominates (except in Québec, where French is the dominant language). These language patterns are symbolic of the larger cultural and historical differences between the two regions, which will be discussed in Chapters 2 and 3.

If regions are so difficult to define and describe, why do geographers use them? To discuss the whole world at once would be impossible, so geographers seek a reasonable way to divide the world into manageable parts. There is nothing sacred about the criteria or the boundaries we use. They are just practical aids to learning. In defining the world regions for this book, we have considered such factors as physical features, political boundaries, cultural characteristics, history, how the places now define themselves, and what the future may hold.

This book organizes the material into four *scales of analysis:* the global scale, the world regional scale, the subregional scale, and the local scale (**Figure 1.8**). Notice that whenever geographers use the word *scale*, whether on maps or in text, they are referring to the relative size of an area. At the **global scale**, explored in this chapter, the entire world is treated as a single area—a unity that is more and more relevant as our planet operates as a global system. We use the term **world region** for the largest divisions of the globe, such as East Asia and North America (Figure 1.1). We have defined ten world regions, each of which is covered in a separate chapter:

- North America
- Middle and South America
- Europe
- Russia and the post-Soviet States
- North Africa and Southwest Asia
- Sub-Saharan Africa
- South Asia
- East Asia
- Southeast Asia
- Oceania

global scale the level of geography that encompasses the entire world as a single unified area

world region a part of the globe delineated according to criteria selected to facilitate the study of patterns particular to the area

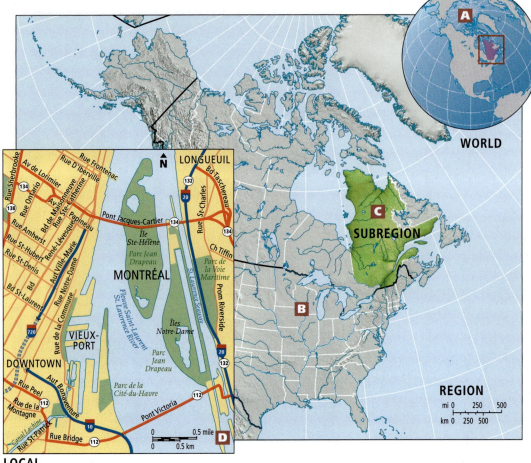

Each regional chapter considers the interaction of human and physical geography in relation to cultural, social, economic, population, environmental, and political topics. Because the world regions are all so large, they are divided into **subregions,** which can be further subdivided to reach the **local scale.**

Most of us are familiar with geography at the local scale—where we live and work, whether in a city, town, or rural area. Local geography shapes our lifestyles and the culture we create together. Throughout this book, life at the local scale is illustrated with vignettes about people in a wide range of places. Often their lives reflect trends that can be tracked across several world regions or may even occur at the global scale.

The regions and local places discussed in this book vary dramatically in size and complexity. A region can be relatively small, such as Europe, or very large, as in the case of East Asia. At the local scale, a place can be a backyard in Polynesia, a neighborhood in Rome, or a town in Kenya.

In summary, regions have the following traits:

- A region is a unit of the earth's surface that contains distinct environmental or cultural patterns.

subregions smaller divisions of the world regions delineated to facilitate the study of patterns particular to the areas

local scale the level of geography that describes the space where an individual lives or works; a city, town, or rural area

- Regional demarcations are used for various purposes, so regional definitions are fluid.

- No two regions are necessarily defined by the same set of indicators.

- Regions can vary greatly in size (scale).

- The boundaries of regions are usually indistinct and hard to agree upon.

Thematic Concepts and Their Role in This Book

Within the world regional framework, this book is also organized around nine thematic concepts of special significance in the modern world—concepts that operate in every world region and interact in numerous ways:

- Population
- Gender
- Development
- Food
- Urbanization
- Globalization
- Democratization
- Water
- Climate change

The following sections explain each of these nine thematic concepts, how they tie in with the main concerns of geographers, and how they are related to each other. This information has a special relevance in each of the regional chapters, where the thematic concepts provide a useful framework for your rapidly accumulating knowledge of the world. The nine concepts are not presented in a particular order in each chapter, but rather are discussed as they naturally come up in the coverage of issues. For example, in some regions water is chiefly a scarcity issue; in others, managing water resources equitably under conditions of climate change is the main concern.

 **THINGS TO REMEMBER**

1. Geography is the study of the earth's surface and the processes that shape it. There is a continual interaction between physical and human processes. Geographers seek to understand why a

particular place is the way it is, with a particular sight, sound, smell, and arrangement of features. Maps and photographs are two key tools geographers use to display and interpret information.

2. A region is a unit of the earth's surface that has a combination of distinct physical and/or human features, but the complex of features can vary from region to region, and regional boundaries are rarely clear and crisp.

3. This book focuses on nine thematic concepts—population, gender, development, food, urbanization, globalization, democratization, water, and climate change—which provide a consistent set of issues pursued throughout the book, issues that often interact with or affect each other. For example, changing gender patterns can affect rates of population growth.

POPULATION

To study population is to study the growth and decline of numbers of people on earth, their distribution across the earth's surface, age and sex distributions, migration patterns and what makes people move (see Thematic Concepts Parts A, B, C on page 12). Over the last several hundred years, the global human population has boomed, but growth rates are now slowing in most societies and in a few have even moved into negative growth. Much of this pattern has to do with changes in economic development and gender roles that have reduced incentives for large families.

Global Patterns of Population Growth

It took between 1 million and 2 million years (at least 40,000 generations), for humans to evolve and to reach a global population of 2 billion, which happened around 1945. Then, remarkably, in just 64 years—by the year 2009—the world's population more than tripled to 6.7 billion (Figure 1.9). What happened to make the population grow so quickly in such a short time?

The explanation lies in changing relationships between humans and the environment. For most of human history, fluctuating food availability, natural hazards, and disease kept human death rates high, especially for infants. Periodically, there were even crashes in human population—for example, the Black Death, a pandemic throughout Europe and Asia in the 1300s.

An astonishing upsurge in human population began about 1500, at a time when the technological, industrial, and scientific revolutions were beginning in some parts of the world. Human

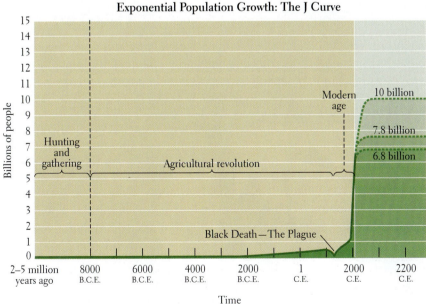

FIGURE 1.9 Exponential growth of the human population. The curve's J shape is a result of successive doublings of the population: it starts out nearly flat, but as doubling time shortens, the curve bends ever more sharply upward. Note that B.C.E. (before the common era) is equivalent to B.C. (before Christ); C.E. (common era) is equivalent to A.D. (anno domini).

Population: Rapid global population growth has occurred over the last several hundred years, but growth rates are now slowing and some societies are even shrinking. Changes in economic development patterns, government policy, access to health care, and gender roles have reduced incentives for large families. These and other factors also shape the distribution and movement of human populations.

(A) As growth slows, many populations are rapidly aging. These retired men are in Portugal.

(B) Families with only one child are now common in Japan and elsewhere in East Asia.

(C) The most rapidly growing populations are poor and rural. These women are in Congo (Kinshasa).

Gender: Many places are moving toward greater equality between the genders. As more women pursue educational and employment opportunities outside the home, birth rates are declining. Meanwhile, economic development and politics are becoming transformed by the increasing participation of women.

(D) Female high school student in Tanzania.

(E) Israeli military police.

(F) A woman votes in Congo (Kinshasa).

Development: Parts of the world (often labeled "the developing world") are shifting from lower-value and labor-intensive raw materials–based economies to higher-value and higher-skill–based manufacturing and service economies. This shift depends in part on the availability of social services, such as education and health care, that enable people to contribute to economic growth.

(G) A young boy cultivating by hand in Uganda.

(H) Students in a science class in Dakar, Senegal.

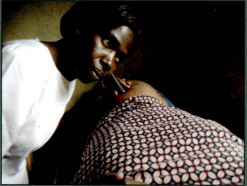

(I) A nurse delivers prenatal care to a pregnant woman in South Africa.

Food: So far, food production systems are keeping pace with global population growth, in part by shifting away from labor-intensive, small-scale, subsistence agriculture toward mechanized, chemically intensive, large-scale, commercial agriculture. This process increases productivity, but at the cost of environmental degradation that threatens further growth in food production. Moreover, many farmers are unable to afford the chemicals and machinery required for commercial agriculture and have to give up farming as a result.

(J) A crop duster sprays pesticides in Texas.

(K) A tractor is used for harvesting papayas in Brazil.

(L) Mechanized farming, as practiced above in Oregon, often exposes soils to wind and rain, resulting in erosion over time.

Urbanization: Changes in food production are pushing people out of rural areas, while the development of manufacturing and service economies is pulling them into cities. Living standards increase for some rural migrants, as access to jobs, health care, and education often improves. However, many are forced into vast slums with poor housing and inadequate access to water or social services.

(M) Many migrants to Dhaka, Bangladesh, work as bicycle rickshaw drivers.

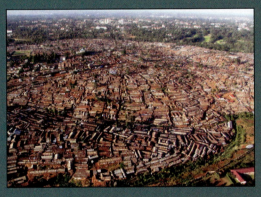

(N) Kibera slum in Nairobi, Kenya, is home to more than 1 million people.

(O) The pressures of life in urban slums break up many families. These orphans are in Kibera slum in Nairobi, Kenya.

Globalization: Local self-sufficiency is giving way to global interdependence as goods, money, and people move across vast distances faster and on a larger scale than ever before. Influences from afar are transforming even seemingly isolated societies.

(P) In Guangdong, China, a woman manufactures laptop computer cases destined for the United States.

(Q) An Indian construction worker next to an English-language advertisement for the building he works on in Dubai, United Arab Emirates.

(R) A Masai herder in Tanzania displays his new cell phone.

13

Democratization: Authoritarianism, based on the authority of the state or community leaders, is giving way to more democratic systems in which each individual is given a greater voice in how governments are run. This shift is strongly linked to the growth of political freedoms, such as the right to protest and take action against injustice, especially through the media and the legal system.

(S) Women line up to vote in Yemen.

(T) Riot police disperse a peaceful protest for better access to water in Uganda.

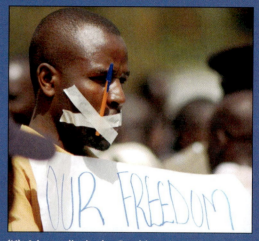

(U) A journalist in the Gambia protests censorship of the media.

Water: Fresh water is becoming scarce as human impacts on the environment increase. Pressure to reduce water use and pollution are rising as conflicts intensify over access to water for drinking and irrigation, and over the resources of aquatic ecosystems.

(V) For many urban dwellers, such as this boy in Kenya, a communal spigot is the only source of water.

(W) Dams can provide electricity and water for irrigation, but they flood potentially vast areas, and downstream river flows are forever altered.

(X) Irrigated rice terraces, such as these in China, feed billions of people worldwide.

Climate Change: Human activities that emit large amounts of carbon dioxide, methane, and other greenhouse gases are trapping heat in the atmosphere. The industrialized and rapidly industrializing countries, who are responsible for most of these emissions, are attempting to reduce their output of greenhouse gases. Meanwhile, the poorest countries of the world are highly vulnerable to the changes in climate brought about by global warming.

(Y) This coal-fired power plant in Russia spews tons of carbon dioxide into the atmosphere each day.

(Z) Global warming is bringing higher temperatures and greater acidity to the oceans, threatening coral reefs and the fishing industries they support.

(AA) Global warming is likely to increase flooding, especially in low-lying coastal areas such as Bangladesh (shown above).

B The highest birth rates in the world are in Afghanistan and sub-Saharan Africa. Low levels of economic development and gender roles that restrict women's access to education and employment give Afghanistan the fourth highest birth rate in the world. This extended Afghan family has just arrived in Kabul and is waiting to register as returned refugees from Pakistan.

C The world's largest area of extremely high population density is found in South Asia in the lower Ganges and Brahmaputra river basins of India and Bangladesh. Rural settlements can be quite dense, as in this fishing village in Bangladesh. However, these high-density areas are now growing much more slowly than in the past. The governments of India and Bangladesh have had considerable success in reducing the population growth rates by making birth control more available.

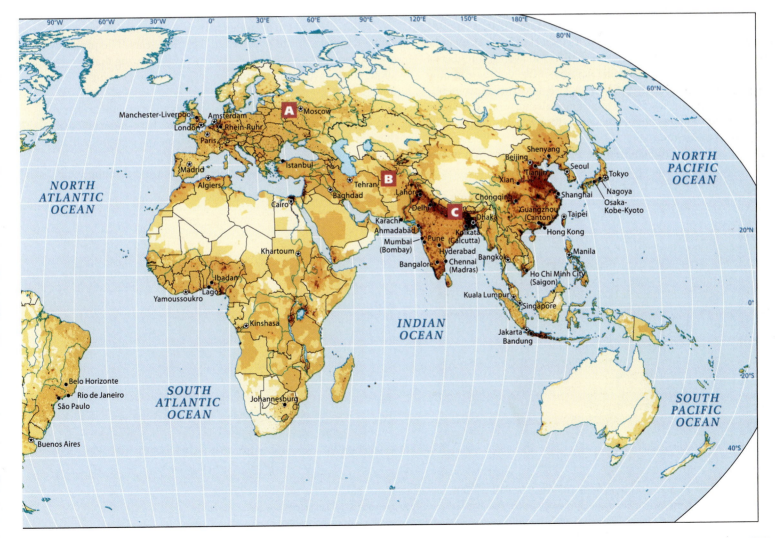

relationship in a given population between the number of people being born, the **birth rate**, and the number dying, the **death rate**, without regard to the effects of migration.

The rate of natural increase is expressed as a percentage per year. For example, in 2008, Austria (in Europe) had a population of 8.4 million people. The annual birth rate was 9 per 1000 people, and the death rate was 9 per 1000 people. Therefore, the annual rate of natural increase was 0 per 1000 (9 − 9 = 0), or 0 percent. In 2008, 15 percent of Austrians were under age 15, and 15 percent were over 65 years of age. By 2009, those over age 65 had increased to 17 percent of the population. Austria is aging rapidly.

For comparison, consider Jordan (in Southwest Asia). In 2008, Jordan had 5.8 million people. The birth rate was 28 per 1000, and the death rate was 4 per 1000. Thus the annual rate of natural increase was 24 per 1000 (28 − 4 = 24), or 2.4 percent per year. At this rate, Jordan will double its population in just 29 years. As you might expect, the population of Jordan has many young people: 37 percent are under 15 years of age, and only 3 percent are over 65.

Total fertility rate (TFR) is another term used to indicate trends in population. The TFR is the average number of children a woman in a specific country is likely to have during her reproductive years (15–49). The TFR for Austria's women is 1.4 and for Jordan's is 3.6. As education rates for women increase and as they postpone childbearing into their late 20s, total fertility rates tend to decline.

Another powerful contributor to population growth is **migration.** In Europe, for example, the rate of natural increase is quite low, but the region's economic power attracts immigrants from throughout the world. In 2005, international migration accounted for 85 percent of the European Union's population growth. All across the world people are on the move, seeking to improve their circumstances; often they are fleeing war or natural disasters. Jordan has experienced unusually high rates of immigration, mostly by people seeking refuge from conflicts in Palestine and Iraq. It is hard to get accurate figures, but of the 5.8 million people in the country, perhaps as many as 70 percent are from outside the country. As a result, native Jordanians are a minority in their own land.

Age and Sex Structures

The age distribution, or age structure, of a population is the proportion of the total population in each age-group. The sex structure is the proportion of males and females in each age-group. Age and sex structures reflect past and present social conditions and can help predict future population trends.

The **population pyramid** is a graph that depicts age and sex structures. Consider the population pyramids for Austria and Jordan (**Figure 1.10**), which reveal the age and sex differences between these two countries. As we have noted, 37 percent in Jordan are under age 15. Notice toward the bottom of the Jordan pyramid that the largest groups are in the age categories 0 through 14. Notice also that the pyramid tapers in at the very bottom, indicating that fewer births occurred in the last 4 years. The pyramid also tapers off sharply as it rises above age 39, showing that before about 40 years ago, the population in Jordan was growing much more slowly.

In contrast, Austria's pyramid has an irregular vertical shape that tapers toward the bottom. The narrow base indicates that there are now fewer people in the younger age categories than in young adulthood or middle age, and that those over 70 greatly outnumber the youngest (ages 0 to 4). This age distribution tells

---

birth rate the number of births per 1000 people in a given population, per unit of time, usually per year

death rate the ratio of total deaths to total population in a given population, usually expressed in numbers per 1000 or in percentages

total fertility rate (TFR) the average number of children that women in a country are likely to have at the present rate of natural increase

migration movement of people from a place or country to another, often for safety or economic reasons

population pyramid a graph that depicts the age and gender structures of a country

---

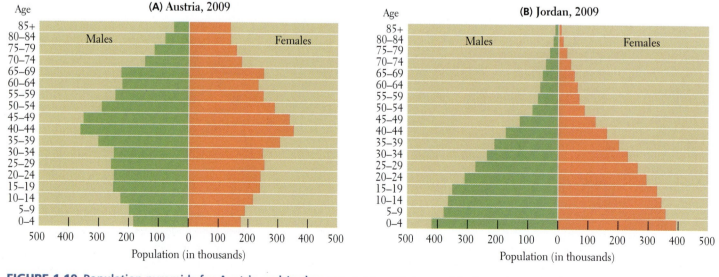

FIGURE 1.10 Population pyramids for Austria and Jordan. [Adapted from U.S. Bureau of the Census, International Data Base (IDB), at http://www.census.gov/ipc/www/idb/informationGateway.php.; 2009 pyramid for Jordan http://www.census.gov/ipc/www/idb/country/joportal.html; 2009 pyramid for Austria http://www.census.gov/ipc/www/idb/country/auportal.html.]

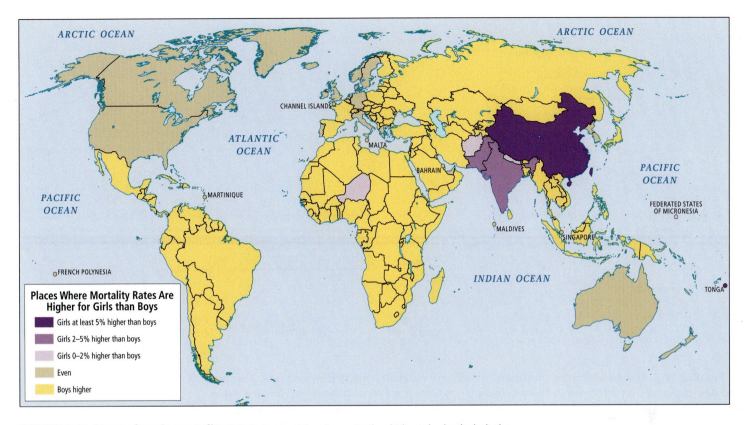

ARCTIC OCEAN

ARCTIC OCEAN

ATLANTIC
OCEAN

CHANNEL ISLANDS

MALTA

BAHRAIN

PACIFIC
OCEAN

FEDERATED STATES
OF MICRONESIA

PACIFIC
OCEAN

MARTINIQUE

MALDIVES

SINGAPORE

INDIAN OCEAN

FRENCH POLYNESIA

TONGA

**Places Where Mortality Rates Are
Higher for Girls than Boys**

- Girls at least 5% higher than boys
- Girls 2–5% higher than boys
- Girls 0–2% higher than boys
- Even
- Boys higher

FIGURE 1.11 Young female mortality rates. In countries shown in the darkest shades (color), the mortality rate for girls is abnormally high. The darker the color, the greater the risk to girls. [Adapted from Joni Seager, *The State of Women in the World Atlas* (New York: Penguin, 1997), p.35; and updated from the Department of Economic and Social Affairs of the United Nations Secretariat, Population Division, *World Population Prospects: The 2008 Revision,* at http://esa.un.org/unpp/p2k0data.asp.]

us that many Austrians now live to an old age and that in the last several decades Austrian couples have been choosing to have only one child, or none. If this trend continues, Austrians, like many Europeans, Japanese, and even Chinese, will need to support and care for large numbers of elderly people, and those responsible will be an ever-declining group of working-age people.

Population pyramids also reveal sex differences within populations. Look closely at the right (female) and left (male) halves of the pyramids in Figure 1.10. In several age categories the sexes are not evenly balanced on both sides of the line. In the Austrian pyramid, there are more women than men near the top (especially age 80 and older). In Jordan, the sex imbalances occur at younger ages. For instance, in the age category of 25 to 29, there are about 270,000 women and about 300,000 men.

Demographic research on the reasons behind statistical sex differences is relatively new, so explanations for imbalances are only now being proposed. In Austria, of course, the predominance of elderly women reflects both the deaths among male soldiers in World War II and the trend in countries with long life expectancies for women to live about 5 years longer than men (a trend still poorly understood). But the sex discrepancies in younger populations have different explanations. The normal ratio worldwide is for about 95 females to be born for every 100 males. Because baby boys are somewhat weaker than girls, the ratio normally evens out naturally within the first 5 years. However, in several places over the last

100 years, the ratio of females to males at birth has declined further. In parts of South and East Asia, the ratio is as low as 80 females to 100 males (see Thematic Concepts Parts HH, II, JJ on page 15). The widespread cultural preference for boys over girls (discussed in later chapters) appears to increase as couples choose to have fewer children. Some female fetuses are aborted. Yet other factors influence the survival of baby girls once born. In societies afflicted by poverty, all females, including baby girls, are sometimes fed less well than males and receive less health care; hence, females are more likely to die in early childhood (Figure 1.11).

Population Growth Rates and Wealth

Despite a wide range of variation, regions with slow population growth rates usually tend to be affluent, and regions with fast growth rates tend to have widespread poverty. The reasons for this difference are complicated; again, Austria and Jordan are useful examples.

Austria has an annual **gross domestic product (GDP) per capita** of $33,700. This figure represents the total production of goods and services in a country in a given year divided by the population. (All GDP per capita figures cited in this book are adjusted for purchasing power parity (PPP) so they represent comparable purchasing

gross domestic product (GDP) per capita the market value of all goods and services produced by workers and capital within a particular country's borders and within a given year, divided by the number of people in the country

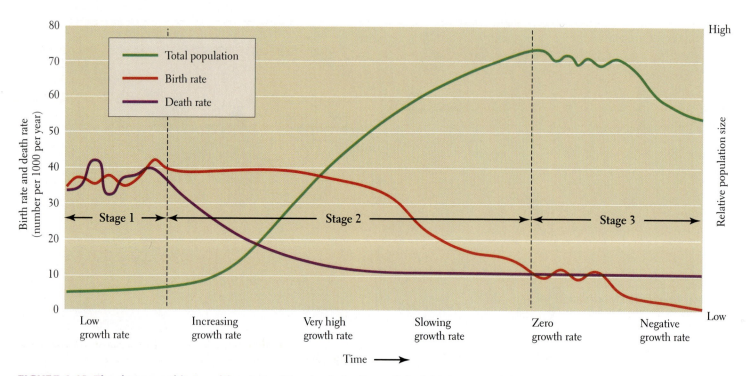

FIGURE 1.12 The demographic transition. In traditional societies (Stage 1), both birth rates and death rates are usually high (left vertical axis), and population numbers (right vertical axis) remain low and stable. With advances in food production, education, and health care (Stage 2), death rates usually drop rapidly, but strong cultural values regarding reproduction remain, often for generations, so birth rates drop much more slowly, with the result that for decades or longer, the population continues to grow significantly. When changed social and economic circumstances enable most children to survive to adulthood and it is no longer necessary to produce a cadre of family labor (Stage 3), population growth rates slow and may eventually drift into negative growth. At this point, demographers say that the society has gone through the demographic transition. [Adapted from G. Tyler Miller, Jr., *Living in the Environment,* 8th ed. (Belmont, Calif.: Wadsworth, 1994), p. 218.]

demographic transition
the change from high birth and death rates to low birth and death rates that usually accompanies a cluster of other changes, such as change from a subsistence to a cash economy, increased education rates, and urbanization

subsistence economy
circumstances in which a family produces most of its own food, clothing, and shelter

cash economy an economic system in which the necessities of life are purchased with monetary currency

power.) Austria has a very low infant mortality rate of 3.7 per 1000. Its highly educated population is 100 percent literate and is employed largely in advanced industry and services. Large amounts of time, effort, and money are required to educate a child to compete in this economy. Moreover, it is highly likely that a child will survive to adulthood and hence be available to care for aging parents. No surprise, then, that many Austrian couples choose to have only one or two children.

By contrast, Jordan has a GDP per capita of $5,530 and a high infant mortality rate of 24 per 1000. Although more than 80 percent of Jordanians live in cities, much of the everyday work is still done by hand, so each new child is a potential contributor to the family income at a young age. There is also a much greater risk of children not surviving into adulthood, so having more children ensures that someone will be there to provide care for aging parents.

Circumstances in Jordan have changed rapidly over the last 25 years. In 1985, Jordan's per capita GDP was $993, infant mortality was 77 per 1000, and women had an average of 8 children. Now Jordanian women have an average of only 3.7 children. Geographers would say that Jordan is going through a **demographic transition**, meaning that a period of high birth and death rates is giving way to a period of much lower birth and death rates. This transition occurs gradually as people make the shift from subsistence to cash economies and from rural to urban ways of life. In the beginning phases, when death rates decline and birth rates have not yet followed suit, population numbers can increase rapidly as attitudes toward optimal family size adjust (see the green line in Figure 1.12). Eventually, when low, or no, growth is achieved, population may also tend to decline, but from a much higher level than was the case at the beginning of the transition.

In a **subsistence economy**, a family, usually in a rural setting, produces most of its own food, clothing, and shelter, so there is little need for cash; many children to share the work are an asset; and it is not expensive to prepare a child for a competent adulthood. Today, subsistence economies are disappearing as people seek cash with which to pay school fees and buy food and goods, such as television sets and bicycles. In a **cash economy**, which tends to be urban but may be rural, skilled workers, well-trained specialists, and even farm laborers are paid in money. Each child needs years of education to qualify for a good cash-paying job and does not contribute to the family budget while in school. Having many children, therefore, is a drain on the family's resources. Perhaps most important, cash economies are more likely to have better health care, increasing the likelihood that each child will survive to adulthood.

1. Rapid global population growth has occurred over the last several hundred years, and although growth will continue for many years, rates are now slowing in many places and in a few places the population is even shrinking. As the growth rate slows, many populations are rapidly aging.

GENDER

Note the difference between the terms *gender* and *sex*. **Gender** indicates how a particular social group defines the differences between the sexes. **Sex** is the biological category of male or female but does not indicate how males or females may behave or identify themselves. Gender definitions and accepted behavior for the sexes can vary greatly from one social group to another. Here we consider gender.

For women, the historical and modern global gender picture is puzzlingly negative. In nearly every culture, in every region of the world, and for a great deal of recorded history,

women have had (and still have) an inferior status. Exceptions are rare, although the intensity of this second-class designation varies considerably. Around the world, people of both sexes still routinely accept the idea that males are more productive and intelligent than females. In nearly all cultures, families prefer boys over girls because, as adults, boys will have greater earning capacity (Table 1.1), will have more power in society, and will perpetuate the family name (because of patrilineal naming customs). This preference for boys has some unexpected

> **gender** the sexual category of a person
>
> **sex** the biological category of male or female

TABLE 1.1	Comparisons of male and female income in countries where average education levels are higher for females than for males		
Country (HDI rank)	Female income (PPP[a], 2005 U.S.$)	Male income (PPP, 2005 U.S.$)	Female income as percent of male income
Austria (15)	18,397	40,000	46
Barbados (31)	12,868	20,309	63
Canada (4)	25,448	40,000	64
Japan (8)[b]	17,802	40,000	45
Jordan (86)	2566	8270	31
Kuwait (33)	12,623	36,403	35
Poland (37)	10,414	17,493	60
Russia (67)	8476	13,581	62
Saudi Arabia (61)[c]	4031	25,678	16
Sweden (6)	29,044	36,059	81
United Kingdom (16)	26,242	40,000	66
United States (12)	25,005	40,000	63

[a] PPP, purchasing power parity, is the amount that the local currency equivalent of U.S.$1 will purchase in a given country.
[b] The education level for males in 2005 was 2% higher than for females.
[c] The education level was reported as equal in 2005.

Source: United Nations Human Development Report, 2007–2008 (New York: United Nations Development Programme), Table 28, "Gender-related development index," pp. 326–329, and Table 29, "Gender inequality in education," pp. 330–333.

side effects. When families are limited to one child, fewer girls may be born and eventually there will be a shortage of marriageable women, leaving many men without the hope of forming a family (see Thematic Concepts Parts HH, II, JJ on page 15). On average, females have less access to education, medical care, and even food. They start work at a younger age and work longer hours than males. The puzzling question of how and why women became subordinate to men has not yet been well explored because, oddly enough, few thought the question significant until recently.

Gender Roles

Geographers have begun to pay more attention to **gender roles**—the socially assigned roles for males and females—in different culture groups. In virtually all parts of the world, and for at least tens of thousands of years, the biological fact

gender roles the socially assigned roles of males and females

of maleness and femaleness has been translated into specific roles for each sex. The activities assigned to men and to women can vary greatly from culture to culture and from era to era. Nevertheless, there are some striking consistencies around the globe and over time.

Men are expected to fulfill public roles and women are expected to fulfill private roles. Certainly there are exceptions in every culture, and in developed countries customs are changing, but generally, men work outside the home in positions such as executives, animal herders, hunters, farmers, warriors, or government leaders. Women keep house, bear and rear children, care for the elderly, grow and preserve food and prepare the meals, among many other tasks. In nearly all cultures, women are defined as dependent on men—their fathers, husbands, brothers, or adult sons—even when the women may produce most of the family sustenance. The current popularity of and critical acclaim for the U.S. television show *Mad Men* demonstrates the persistence of this view.

Gender Issues

Because their activities are focused on the home, women tend to marry early. One quarter of the girls in developing countries are mothers before they are 18. This is crucial in that pregnancy is the leading cause of death among girls 15 to 19 worldwide, primarily because immature female bodies are not ready for the stress of pregnancy and birth. Globally, babies born to women under 18 have a 60 percent greater chance of dying in infancy than do those born to women over 18. Typically women also have less access to education than men (globally, 70 percent of youth who leave school early are girls) and are less likely to have access to information and paid employment, and hence less access to wealth and political power. When they do work outside the home (as is the case increasingly in every world region), women tend to fill lower-paid positions, such as laborers, service workers, or lower-level professionals. And even when they work outside the home, most women retain their household duties, so they work a *double day*.

Gender and sex categories can be confusing in matters such as how physical differences between males and females may affect their social roles. For example, women's physical capabilities are somewhat limited during pregnancy and nursing—and, for some, during menstruation—but from the age of about 45, women are no longer subject to these limits. Most contribute in some significant way to the well-being of their adult children and grandchildren, clearly a social role. A growing number of evolutionary biologists suggest that the evolutionary advantage of menopause in midlife is that it gives women the time and energy and freedom to help succeeding generations thrive. This notion—sometimes labeled the *grandmother hypothesis*—seems to have worldwide validity (**Figure 1.13**). Although grandfathers can play similar nurturing roles, women tend to live 3 to 5 years longer than men, which also can increase their usefulness to progeny.

Generally speaking, men have larger muscles, can lift heavier weights, and can run faster than women (but not necessarily for longer periods). In some physical exercises, the average woman has more endurance and is capable of more

FIGURE 1.13 Grandmothers fill nurturing and economic roles for their grandchildren. In every culture and community worldwide, grandmothers contribute to the care and education of their grandchildren. This commonplace phenomenon has played an essential role in human evolution. Here a Peruvian grandmother tends to her grandchild in the Andes.

precise movements than the average man. However, though average physical sex differences exist, in most cultures they carry greater social significance than the biological facts would warrant. Cultural ideas about masculinity and femininity, proper gender roles, and sexual orientation vary widely among groups. Yet within groups these ideas, handed down from generation to generation, have enormous effects on the everyday lives of men, women, and children. Perhaps more than for any other culturally defined human characteristic, significant agreement exists that gender is important; but just how gender roles are defined varies greatly across places and over time.

In the military and in the world of sports, gender equity is a rising issue. The Israeli military pioneered gender integration, requiring service by both males and females (see **Thematic Concepts Part E** on page 12). In many countries, women volunteers in the military are now commonplace. In sports, true equity has been elusive. Take for example the Olympic Games: between 2000 and 2008, female athletes increased their participation in the summer Olympic Games from 30 to 42 percent of the competitors. In some countries (China, for example), more than half the Olympic competitors were female (**Figure 1.14**). And yet thus far, women and men athletes generally do not compete with each other, so the question of how either sex would do in nongendered sports endeavors remains open.

Traditional notions of gender are now being challenged everywhere. In many countries, including conservative Muslim countries, females are acquiring education at higher rates than males (see **Thematic Concepts Part D** on page 12). Eventually, this fact should make women competitive with men for jobs and roles in public life as policy makers and government officials, not just voters (see **Thematic Concepts Parts F, S** on pages 12 and 14). Unless discrimination persists, women should also begin to earn pay equal to that of men (see Table 1.1). Research data suggest that there is a ripple effect that benefits the whole group when developing countries pay attention to girls:

- Girls in developing countries who get 7 or more years of education marry 4 years later than average and have 2.2 fewer children.

- An extra year of secondary schooling on average boosts a girl's lifetime income by 15 to 25 percent.

- The children of educated mothers are healthier and more likely to finish secondary school.

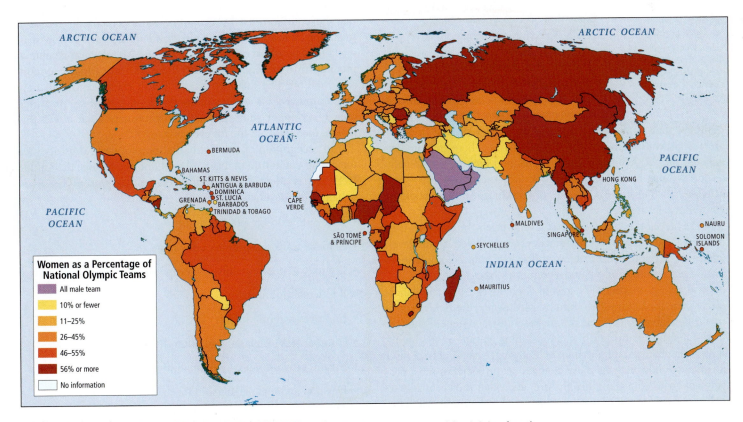

FIGURE 1.14 Women's participation in the 2004 Olympics. Traditional notions of femininity, female roles, and female strength, speed, and endurance are being strongly challenged by Olympic athletes. In the first modern Olympic Games, held in 1896, women were barred from participating. In 2004, women made up 44 percent of the Olympic competitors. In 2008 in Beijing, there were 4746 women participants, or just over 42 percent of the athletes. [Adapted from Joni Seager, *The State of Women in the World Atlas* (New York: Penguin, 2003), pp. 50–51; updated from http://multimedia.olympic.org/pdf/en_report_1000.pdf.]

- When women and girls earn income, 90 percent of their earnings is invested in the family, as compared to just 40 percent of males' earnings.

Considering only women's perspectives on gender, however, misses half the story. Men are also affected by strict gender expectations, often negatively. For most of human history, young men have borne a disproportionate share of burdensome physical tasks and dangerous undertakings. Until recently, only young men left home to migrate to distant, low-paying jobs. Overwhelmingly, it has been young men who die in wars or suffer physical and psychological injuries from combat.

This book will return repeatedly to the question of gender disparities in an effort to investigate this most perplexing cross-cultural phenomenon. Our examination will also reveal that many societies are addressing gender inequalities as they go through the development process. In every country on earth, what it means to be a man or a woman is being renegotiated, as traditional ways are modified by modernization.

THINGS TO REMEMBER

1. Gender—the sexual category of a person—is both a biological and a cultural phenomenon. Gender indicates how a particular social group defines the differences between the sexes. Sex is the biological category of male or female; it does not indicate how males or females may behave or identify themselves.

2. There are significant social gender disparities throughout the world, yet many places are moving toward greater equality between the sexes.

DEVELOPMENT

The economy is the forum in which people make their living, and resources are what they use to do so. Extraction (mining and agriculture), industrial production, and services (including knowledge production) are three types of economic activities, or *sectors of the economy*. Generally speaking, as people in a society shift from extractive activities, such as farming and mining, to industrial and service activities, their material standards of living rise—a process typically labeled **development**. *Extractive resources* are resources that must be mined from the earth's surface (mineral ores) or grown from its soil (timber and plants). There are also *non-material human resources*, such as skills and brainpower. Usually resources must be transformed to produce useful products (such as refrigerators or bread) or bodies of knowledge (such as books or computer software).

development usually describes economic changes like greater productivity of agriculture and industry that lead to better standards of living or simply to increased mass consumption

The development process has several facets. In many parts of the world, especially in poorer societies (often labeled "underdeveloped" or "developing"), there are shifts away from labor-intensive, low-wage, agriculturally based economies (see Thematic Concepts Part G on page 12) toward higher-wage but still labor-intensive manufacturing economies. Meanwhile, the richest countries (often labeled "developed") are lessening their dependence on labor-intensive manufacturing and shifting toward more highly skilled mechanized production or knowledge-based service and technology industries. As these changes occur, societies must provide adequate education, health care, and other social services to help their people contribute to economic development (see Thematic Concepts Parts H, I on page 12).

Millennium Development Goals

Until recently, the term *development* was used to describe only economic changes that lead to higher standards of living defined primarily by rates of material consumption. These changes often accompany the greater productivity in agriculture and industry that comes from such technological advances as mechanization and computerization. Yet, as we shall see, people's actual well-being may be compromised by economic advancement if it requires migrating far from home, living in crowded, unsafe shantytowns, or working in hostile or unhealthy conditions. Merely raising national productivity may benefit primarily those who are already economically well off. The majority may be left in circumstances that are little improved, or even worsened. Development as measured only by economic gains may entail environmental side effects that reduce the quality of life for everyone.

In recognition of the reality that development must include social and environmental as well as economic factors, the United Nations has adopted eight *Millennium Development Goals* on behalf of the developing countries:

- Eradicate extreme poverty and hunger
- Achieve universal primary education
- Promote gender equality and empower women
- Reduce child mortality
- Improve maternal health
- Combat HIV, malaria, and other diseases
- Ensure environmental sustainability
- Develop a global partnership for development

Measures of Economic Development

The most popular economic measure of development is **gross domestic product (GDP) per capita**. Recall that GDP is the total value of all goods and services produced in a country in a given year. When GDP is divided by the number of people in the country, the result is GDP per capita. This figure is often used as a crude indicator of how well people are living in a given country.

Using GDP per capita as a measure of overall well-being has several disadvantages, however. First is the matter of wealth distribution. Because GDP per capita is an average, it can hide the fact that a country has a few fabulously rich people and a mass of abjectly poor people. For example, a GDP per capita of U.S.$20,000 would be meaningless if a few lived on millions per year and most lived on less than $5000 per year.

Second, the purchasing power of currency varies widely around the globe. A GDP of U.S.$15,000 per capita in Barbados might represent a middle-class standard of living, whereas the same amount in New York City could not buy even basic food and shelter. Because of these purchasing power variations, in this book GDP per capita figures have been adjusted for **purchasing power parity (PPP)**. PPP is the amount that the local currency equivalent of U.S.$1 will purchase in a given country. For example, according to the *Economist*, on February 4, 2009, a Big Mac at McDonald's in the United States cost U.S.$3.54. In Norway it cost the equivalent of U.S.$7.79, and in China it cost U.S.$1.83. Of course, for the consumer in China, where annual per capita income averages about $5400, this would be a rather expensive meal.

A third disadvantage of using GDP per capita is that it measures only what goes on in the **formal economy**—all the activities that are officially recorded as part of a country's production. Many goods and services are produced outside formal markets, in the **informal economy**. Here, work is often traded for *in-kind* payments (food, or housing, for example) or for cash payment that is not reported to the government as taxable income. It is estimated that one-third or more of the world's work takes place in the informal economy. Examples of workers in this category include anyone who contributes to her/his own or someone else's well-being through unpaid services such as housework, gardening, herding, animal care, or elder and child care. *Remittances* sent home by migrants, if they are sent through banks or similar financial institutions, become part of the formal economy of the receiving society because records are kept and taxes levied. If they are transmitted in "off the books" (perhaps illegal) ways, such as via mail or cash, they are part of the informal economy.

There is a gender aspect to informal economies. Researchers studying all types of societies and cultures have shown that, on average, women perform about 60 percent of all the work done—much of it unpaid and in the informal economy. Yet only the work women are paid for in the formal economy appears in the statistics, so GDP per capita figures ignore much of the work women do. Statistics also neglect the contributions of millions of men and children who work in the informal economy as subsistence farmers, traders, service people, or seasonal laborers.

Geographic Patterns of Human Well-Being

Some development experts, such as the Nobel Prize–winning economist Amartya Sen, advocate a broader definition of development that includes measures of **human well-being.** This term generally means a healthy and socially rewarding standard of living in an environment that is safe. The following section explores the three

measures of human well-being that are used in this book: GDP per capita (calculated as PPP); the United Nations Human Development Index (HDI); and a comparison of **female earned income as a percent of male earned income (F/MEI).**

Global GDP per capita PPP is mapped in Figure 1.15A. Comparisons between regions and countries are possible, but as discussed above, GDP per capita figures ignore all aspects of development other than economic ones. For example, there is no way to tell from GDP per capita figures how fast a country is consuming its natural resources, or how well it is educating its young or maintaining its environment. Therefore, along with the traditional GDP per capita figure, geographers increasingly use several other measures of development. This book uses three maps: **GDP per capita PPP** (Figure 1.15A); the **United Nations Human Development Index (HDI)** (Figure 1.15B), which calculates a country's level of well-being by considering income adjusted to PPP, data on life expectancy at birth, and data on educational attainment; and to assess how well a country is doing in ensuring gender equality, we map female earned income as a percent of male earned income (Figure 1.15C). Together, these measures reveal some of the subtleties and nuances of well-being and make comparisons between countries somewhat more valid. Because these more sensitive indices (HDI and F/MEI) are also more complex than the purely economic GDP per capita, they are all still being refined by the United Nations. All the Human Well-Being maps in the regional chapters include a thumbnail global map so that you can conveniently see how a region compares with other parts of the world.

A geographer looking at these maps might make the following observations:

- The GDP per capita figures (Figure 1.15A) show a wide range of difference across the globe with very obvious concentrations of high and low GDP per capita. The most populous parts of the world—China and India—rank rather low, but sub-Saharan Africa ranks the lowest and this is true for nearly all countries in that region.

- The HDI rank map (Figure 1.15B) shows a similar pattern, but look closely. Middle and South America, Southeast Asia, and Italy and Spain rank a bit higher on HDI than they do on GDP, and several countries in Africa as well as Iran rank lower on HDI than on GDP, thus illustrating the disconnect between GDP per capita and human well-being.

purchasing power parity (PPP) the amount that the local currency equivalent of U.S.$1 will purchase in a given country

formal economy all aspects of the economy that take place in official channels

informal economy all aspects of the economy that take place outside official channels

human well-being various measures of the extent to which people are able to obtain a healthy life in a community of their choosing

female earned income as a percent of male earned income (F/MEI) a measure of pay equity that shows average female earned income as a percent of average male earned income

GDP per capita PPP the market value of all goods and services produced by modern workers and capital within a particular country's borders and within a given year, divided by the number of people in the country and adjusted for purchasing power parity

United Nations Human Development Index (HDI) the ranking of countries based on three indicators of well-being: life expectancy at birth, educational attainment, and income adjusted to purchasing power parity

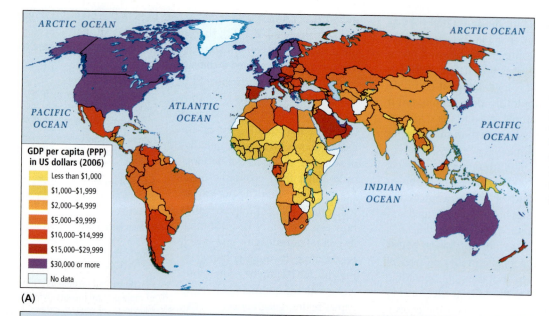

(A)

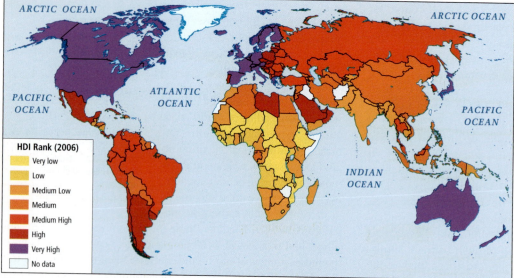

(B)

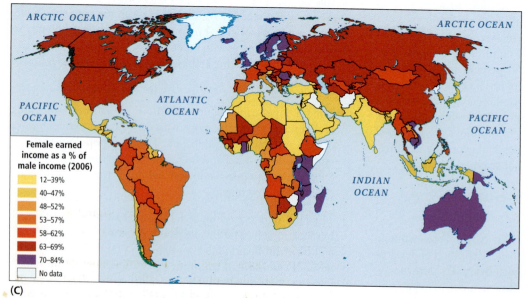

(C)

FIGURE 1.15 Global maps of human well-being. (A) Gross domestic product (GDP) per capita, adjusted for purchasing power parity (PPP); (B) human development index (HDI); (C) female earned income as percent of male earned income (F/MEI), adjusted for purchasing power parity (PPP). [Maps adapted from data: "Human development indices," (Maps A and B): Table 2, pages 28–32; (Map C): Table 5, pages 41–44 at http://hdr.undp.org/en/media/HDI_2008_EN_Tables.pdf.]

- The map of female earned income as a percent of male earned income (Figure 1.15C) shows, first of all, that nowhere on earth are average female earnings equal to average male earnings. Some poor parts of the world (East Africa, Vietnam, Romania, and Ghana in West Africa) have closer gender equity in pay than some of the richest, but of course pay scales for both genders are low. Only parts of Europe are in the category closest to equity—70 to 84 percent. Pay inequality is most extreme in North Africa and Southwest Asia, South Asia, and parts of Southeast Asia.

Sustainable Development and Political Ecology

The United Nations defines **sustainable development** as the effort to improve present living standards in ways that will not jeopardize those of future generations. Destroying resources (as in deforestation) or poisoning them (as in pollution of water and air), may deprive future generations of needed resources. Sustainability has only recently gained recognition as an important goal, well after the developed parts of the world had already achieved high standards of living based on high consumption. But sustainability is particularly important for the vast majority of the earth's people who do not yet enjoy an acceptable level of well-being. Without sustainable development strategies, their efforts to improve living standards will increasingly encounter degraded or scarce resources.

Geographers who study the interactions among development, politics, human well-being, and the environment are called **political ecologists**. They are known for asking the "Development for whom?" question, meaning, "Who is actually benefiting from so-called development projects?" Political ecologists examine how the power relationships in a society affect the ways in which development proceeds. For instance, in a Southeast Asian country, the clearing of forests to grow oil palm trees might at first seem to benefit the country. It would earn profits for the growers and raise tax revenue for the government through the sale of palm oil. However, these gains must be balanced against the loss of forest resources and soil fertility that result when a single plant species replaces a multispecies, multistory tropical forest. Moreover, whole ways of life are lost when forest dwellers are forced to migrate to crowded cities, where their woodland skills are useless.

This same country could measure its development by improvements in average human well-being and environmental quality, and in the potential for sustaining those improvements into the future. By these standards, oil palm plantation development might appear less attractive, since only a few benefit, while the majority of citizens, the environment, and future generations lose.

Human Impact on the Biosphere

Concerns over the sustainability of development grew out of increasing awareness that human life on earth was having a negative effect on environments. It is now known that from the beginning of human life, in seeking to improve their own living conditions by producing better and more abundant food, shelter, clothing, and other material needs, people have overused resources, sometimes even eliminating plant and animal species altogether, and poisoned water and soil and the air.

Photo Essay 1.2 (see pages 28–29) shows a global map of the relative intensity of human impact on the **biosphere**, the global ecological system that integrates all living things and their relationships. The map includes insets that show particular trouble spots in South America, Europe, and Southeast Asia. Of the nine thematic concepts stressed in this book, this figure focuses on the interactions of water, development, and food.

> **sustainable development** improvement of standards of living in ways that will not jeopardize those of future generations
>
> **political ecologist** a geographer who studies power allocations in the interactions among development, human well-being, and the environment
>
> **biosphere** the global ecological system that integrates all living things and their relationships

THINGS TO REMEMBER

1. Development refers to the rise in material standards of living that usually accompanies the shift from extractive activities, such as farming and mining, to industrial and service activities.

2. For development to happen, social services, such as education and health care, are necessary to enable people to contribute to economic growth.

FOOD

Food production, essential to human life, has undergone many changes over time: from hunting and gathering; to labor intensive, small-scale, subsistence agriculture (see discussion of the development of agriculture on page 56); to modern mechanized, chemically intensive, large-scale, commercial agriculture; and most recently, to organic farming. Modern processes of food production, distribution, and consumption have greatly increased the supply and, to some extent, the security of food systems, but they have evolved at the expense of environments; this may threaten further changes in food production.

Modern Food Production and Vulnerability

For most of human history, people produced nearly all they needed for food, clothing, and shelter. They subsisted. However, over the past five centuries of increasing global interaction and trade, people have become ever more removed from their sources of food. Today, dense settlement and occupational specialization mean that food is increasingly mass-produced. Far fewer people work in agriculture, and food is usually purchased with money.

Now most humans work for cash to buy food and other necessities. A side effect of this dependence on money is that the **food security** of individuals and families can be threatened by economic and natural disruptions. Lost jobs can mean lost access to food. Natural disasters can cause food prices to rise out of reach. And as countries get more involved with the global economy, their food markets are more vulnerable to changes in distant places. For example, too much production in rich countries can be threatening to poor countries, if oversupply causes prices to drop and forces farmers to sell at below their costs of production, perhaps sinking into debt and economic ruin.

Modernized agriculture usually relies on machines, large tracts of land, irrigation, chemical fertilizers, pesticides, and genetically engineered seeds. The change to this kind of agriculture is called the **green revolution.** When this type of agricultural system is successfully implemented, the results, at least in the beginning, are soaring production levels and very high profits. But too often the new systems displace traditional farmers (small holders, wage laborers, or renters of land), who are then forced by poverty to migrate to cities in search of work. The produce of the new modern farms usually is exported or goes to feed distant burgeoning cities.

Farming that meets human needs without harming the environment or depleting water and soil resources is called **sustainable agriculture.** This term and this process are related to **carrying capacity,** which refers to the maximum number of people a given place can support sustainably. So just how sustainable are the world's present food production systems? The answer is unclear. Technology has increased food production remarkably, especially over the last several decades. In the 25 years between 1965 and 1990, total food production rose between 70 and 135 percent, depending on the region. But population also rose quickly during this period, so the gains per capita were much less. Nutrition levels rose in parts of East and Southeast Asia and Latin America, but they fell in South Asia, Southwest Asia and North Africa and in sub-Saharan Africa (see Figure 1.16 on page 30).

By 2007, growth in global agricultural production was slowing, but overall the global system is still capable of producing more than enough food for all. However, according to the United Nations Food and Agriculture Organization, already one-fifth of humanity subsists on a diet too low in total calories and vital nutrients to sustain adequate health and normal physical and mental development. This massive hunger problem is partly due to poor distribution systems and political corruption, but ultimately food goes to those who have the money to pay for it.

A crisis in food security began to develop in 2007 when there was a spike in world corn prices brought about by speculators in alternative energy. Thinking that corn would be an ideal raw material to make ethanol, they invested heavily in corn. As a result, global corn prices rose beyond the reach of those

food security the ability of a state to consistently supply a sufficient amount of basic food to the entire population

green revolution increases in food production brought about through the use of new seeds, fertilizers, mechanized equipment, irrigation, pesticides, and herbicides

sustainable agriculture farming that meets human needs without poisoning the environment or using up water and soil resources

carrying capacity the maximum number of people that a given territory can support sustainably with food, water, and other essential resources

Humans have had enormous impacts on the biosphere. The map and insets show varying levels of human impact on the biosphere as of 2002. The impacts depicted here are derived from a synthesis of hundreds of studies. High impact areas are associated with intense urbanization. Medium-to-high impact areas are associated with roads, railways, agriculture, or other intensive land uses. Low-to-medium impact areas are experiencing biodiversity loss and other disturbances related to human activity. For more details, go to www.whfreeman/pulsipher.com. [Adapted from United Nations Environment Program, 2002, 2003, 2004, 2005, 2006 (New York: United Nations Development Program), at http://maps.grida.no/go/graphic/human_impact_year_1700_approximately and http://maps.grida.no/go/graphic/human_impact_year_2002.]

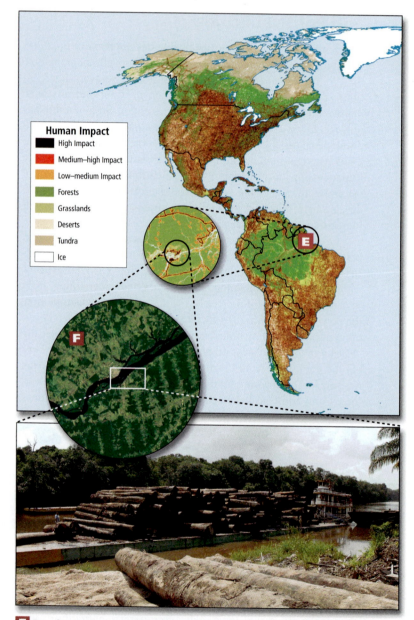

Human Impact
- ■ High Impact
- ■ Medium–high Impact
- ■ Low–medium Impact
- ■ Forests
- ■ Grasslands
- ■ Deserts
- ■ Tundra
- □ Ice

E **Development and Deforestation:** In the Brazilian Amazon, deforestation often occurs in regularized spatial patterns, such as the "fishbone" pattern (see satellite image inset **F**). This pattern results from regulations guiding the location of roads used for settlement and logging. Whole logs are brought by road to rivers where they are put on barges and taken to a port for export.

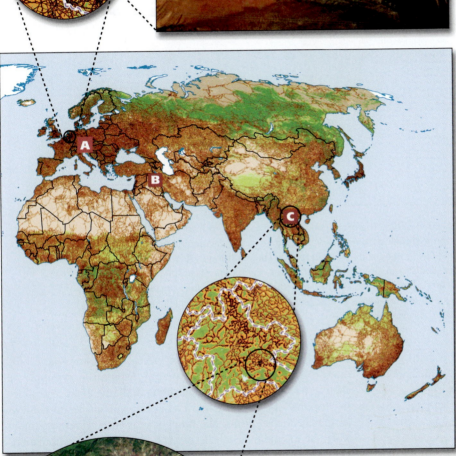

A **Development and Mining:** Perhaps no other human activity has as striking an impact on the landscape as mining. This open pit coal mine is located in one of the most industrialized and densely inhabited parts of Germany.

B **War and political conflict** can have a devastating effect on the environment. Iraqi fire fighters work to contain a fire at an oil facility set off by rocket fire.

C **Food and Deforestation:** Farmers practicing shifting cultivation plant "hill rice" in Laos. Shifting cultivation is an ancient technique that can be sustained indefinitely given sufficient land and fallow periods long enough for forest to regrow—20 years or more. Today, more and more forest is being turned over to short-fallow—3 to 6 years cultivation (see inset **D**) resulting in loss of habitat and biodiversity.

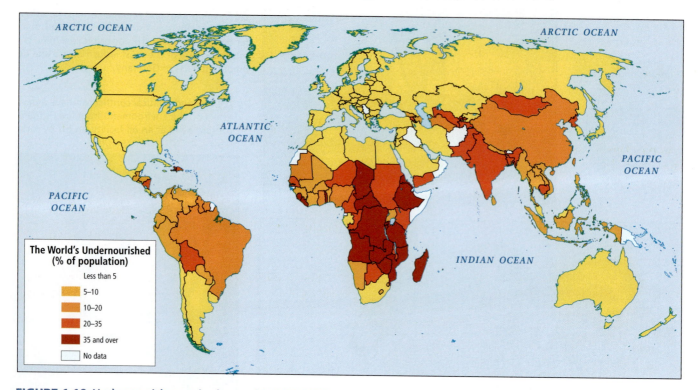

FIGURE 1.16 Undernourishment in the world, 2003–2005. The proportion of people suffering from undernourishment—the lack of adequate nutrition to meet their daily needs—has declined in the developing world over the past several years. However, hundreds of millions of people are still affected by chronic hunger. As you can see from the map, people in much of sub-Saharan Africa, South Asia, Mongolia, North Korea, Cambodia, Nicaragua, Bolivia, and Haiti and the Dominican Republic in the Caribbean suffer the most. [Adapted from the Food and Agriculture Organization of the United Nations, FAO Statistics Division, Rome 2009: Map 14, year 2003–2005, at http://www.fao.org/economic/ess/publications-studies/statistical-yearbook/fao-statistical-yearbook-2007-2008/g-human-welfare/en/.]

who depended on corn as a dietary staple. Then the sharp price rise in oil in 2008 and the recession of 2008–2009 added about 17 percent to the cost of basic foods globally. When oil prices rise, all foods produced and transported with machines get more expensive. The global recession, started by rising oil prices but exacerbated by the global credit bubble, contributed to food shortages for the poor mostly because lost jobs meant that remittances from migrant workers stopped; families no longer had money with which to buy food. These episodes called into question food security and the sustainability of current food production systems. In the developing countries, family economies were so ruined that they stopped sending children to school, sold important assets, and went without food, to the detriment of their long-term health. United Nations statistics, such as those related to the Millennium Development Goals (see page 24), show real reversals of progress in human well-being in 2007–2008.

Scientists from many disciplines estimate that within the next 50 years the world will reach the limit of its carrying capacity because of growing environmental problems such as water scarcity and global climate change. Yet to adequately feed the population projected for 2050, global food output must increase by 70 percent.

genetic modification (GM) in agriculture, the practice of splicing together the genes from widely divergent species to achieve particular desirable characteristics

erosion the process by which fragmented rock and soil are moved over a distance, primarily by wind and water

Will technological advances make present agriculture more productive and unused land more useful? A promising example of such an advance is **genetic modification (GM)**, the practice of splicing together the genes from widely divergent species to achieve particular characteristics like drought resistance. However, many worry about the side effects of such agricultural manipulation. For example, Europeans, fearing that the worldwide move to mass-produced and scientifically manipulated food could lead to unforeseen genetic consequences or catastrophic failures, have sought (unsuccessfully) to keep all such products out of Europe. Others are concerned that genetic modification of food is being led by large food corporations that then seek to patent and protect the GM crops they develop by restricting the use of traditional seeds by surrounding farmers who cannot afford or do not wish to use the GM seeds. Still, as water becomes an ever more central concern in food production, the genetic manipulation of crop plants to make them more resistant to the stresses of drought may override objections to GM crops. And there is some unverified evidence that the largest GM corporations may make new improved seeds available at low cost to farmers in the poorest parts of the world.

Maintenance of soil fertility is also an issue in food security. Many of the most agriculturally productive parts of North America, Europe, and Asia have already suffered moderate to serious losses of soil through **erosion**. Globally, soil erosion, chemical poisoning, and other problems related to food production affect about 7 million square miles (2 billion hectares), putting at risk the livelihoods of a billion people (see **Thematic Concepts Parts J, K, L** on page 13). The main causes of soil degradation are overgrazing, deforestation, and overuse of irrigation and agricultural chemicals.

Food Production and Gender

Food production is also affected by gender. Historically, most of the world's family food producers and processors have been women. Nonetheless, on family farms everywhere, the male of the farm family is usually acknowledged as the "farmer," and his wife is considered mostly a helper in the enterprise. A closer look reveals that in many if not most cases, the females of these families produce the family food in gardens and animal pens and process it for storage and consumption, while the males tend to produce and sell the cash crops. Yet agricultural modernization is often guided by European or American officials who make the assumption that the local males are the farmers of greatest significance. As a result, men may be the only ones taught the new systems, and too often women are left out of the training. Unless more is done to help them learn new agricultural methods, women, especially young women, may join the flow to the cities where they hope to earn the cash needed by their extended families in the countryside to buy food and other necessities.

THINGS TO REMEMBER

1. Modern processes of food production, distribution, and consumption have greatly increased the supply and, usually, the security of food systems, at least in the short term. But they have evolved at the expense of environments, and this fact may threaten further changes in food production.

2. Many farmers are unable to afford the chemicals and machinery required for commercial agriculture or new genetically modified seeds and have to give up farming as a result.

3. Sustainable agriculture is farming that meets human needs without harming the environment or depleting water and soil resources.

URBANIZATION

In 1700, fewer than 7 million people, or just 10 percent of the world's total population today, lived in cities, and only five cities had populations of several hundred thousand people. By 2008, the world had been transformed by **urbanization**. Now about half of the world's population lives in cities; there are more than 400 cities of more than 1 million and about 25 cities of more than 10 million people (see **Thematic Concepts Parts M, N, O** on page 13).

Why Are Cities Growing?

For some time, changes in food production have been pushing people out of rural areas, while the development of manufacturing and service economies and the possibility of earning cash incomes is pulling them into cities. This process is called the **push/pull phenomenon of urbanization**. Numerous cities, especially in poorer parts of the world, have been unprepared for the massive inflow of rural migrants, many of whom now live in slums with poor housing and inadequate access to food, water, education, and social services. Often a substantial portion of the migrants' cash income goes to support their still-rural families. Rural development projects such as mechanized farming may also push people to cities and trigger rural decline, when these projects fail to provide jobs for workers formerly employed in traditional labor-intensive agriculture. Migrants are also pulled to cities because urban areas present the lure of higher wages and a more interesting life in a new place.

Patterns of Urban Growth

In the past, most migrants in cities were young males, but increasingly they are young females whose more traditional farm and household duties have disappeared with modernization. Cities offer women more than better paying jobs. They also provide access to education, better health care, and more personal freedom. Nonetheless, young women are particularly vulnerable in harsh urban environments. Raised in sheltered conditions and initially possessing little education and few skills, they can be unwittingly pressed into the sex trade or involuntary servitude.

The most rapidly growing cities are in developing countries in Middle and South America, Africa, and in many parts of Asia. Because of the large numbers and poverty of the newcomers, these cities have not been able to provide the necessary housing and services for healthy living. The result is a particular settlement pattern that is indicative both of the rapid growth and of the disparity of wealth among urban dwellers. Typically a city sprawls out from a small affluent core, often the oldest part, where there are upscale businesses, fine old buildings, banks, shopping centers, and the upper-class residences of several thousand people. Surrounding these elite landscapes are millions of poor people living in vast stretches of **slums** (also called *barrios, favelas, hutments, shantytowns, ghettos,* and *tent villages*) that are built out of any materials the residents can commandeer: cardboard, corrugated metal, masonry, scraps of wood and plastic. There are no building codes; no organized provision of utilities (electricity may be pirated from the municipal grid); no plumbing, sewers, or clean water; far too few schools; and only informal nonscheduled transportation services. People may be sleeping

urbanization the movement of people from rural areas to cities

push/pull phenomenon of urbanization conditions, such as political instability, that encourage (push) people to leave rural areas, and urban factors, such as job opportunities, that encourage (pull) people to move to the urban area

slums densely populated areas characterized by crowding, run-down housing, and poverty

on the street just a few blocks from soaring modern skyscrapers (Photo Essay 1.3A, B; see also Thematic Concepts Parts, M, N, O on page 13). In these circumstances, millions struggle through everyday life to provide for themselves and their families. Violent gangs of young men may assert control in some such settlements, yet remarkably, in many settlements there are examples of self-initiated community development efforts. In some instances, even recently arrived migrants have successfully lobbied local governments for social support services, such as job training, day-care centers, and medical care.

Those who are financially able to come to urban areas for education, once they complete their studies, tend to find employment in modern industries and business services. They constitute the new middle class and leave their imprint on urban landscapes via the high-rise apartments they occupy and the shops and entertainment facilities they frequent (see Photo Essay 1.3A, D). Cities such as Mumbai in India, São Paulo in Brazil, Cape Town in South Africa, and Shanghai in China are now home to this more educated group of new urban residents, many of whom may have started life on farms and in villages.

THINGS TO REMEMBER

1. Today about half of the world's population live in cities; there are more than 400 cities with more than 1 million and about 25 cities of more than 10 million people.

2. Changes in food production are pushing people out of rural areas, while the development of manufacturing and service economies is pulling them into cities. For some, living standards improve, but many are forced into slums with inadequate food, water, and social services.

GLOBALIZATION

Throughout the world, local self-sufficiency is giving way to global interdependence and international trade. When the recession began in the developed world in 2008, it was soon clear that global linkages were so strong that economic disruptions in the United States and Europe resulted in powerful ripple effects reaching around the world. Foreclosures in the U.S. housing market meant that European banks failed and Chinese factory workers lost their jobs. These connections between distant regions are known as **interregional linkages**. The term **globalization** encompasses the worldwide changes brought about by many types of these interregional linkages and flows (see Figure 1.17 on page 34) that reach well beyond economics. Attitudes and values are modified as a result of global connections; ethnic identity may be reinforced or erased; personal or collegial relationships established between people who will never actually meet. Globalization is the most complex and far-reaching of the thematic concepts described in this book (see Thematic Concepts Parts P, Q, R, Y, Z, AA, BB, CC, DD, EE, FF, GG on pages 13–15).

interregional linkages economic, political, or social connections between regions, whether contiguous or widely separated

globalization the growth of interregional and worldwide linkages and the changes these linkages are bringing about

Urbanization and Urban Areas. The color of the country indicates the percentage of the population living in urban areas. The circles represent the populations of the world's largest urban areas in 2006 (blue circle) and projected in 2020 (black circle).

A Cities have always been centers of innovation, entertainment, and culture, in large part because they attract both money and talented people. These young residents of Shanghai are engaging in a new fad, parachuting off one of the new skyscrapers that now dominate the city's skyline.

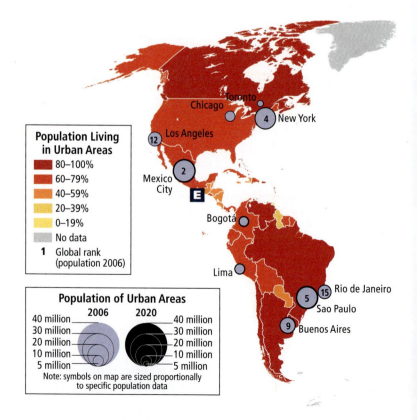

Population Living in Urban Areas
- 80–100%
- 60–79%
- 40–59%
- 20–39%
- 0–19%
- No data
- **1** Global rank (population 2006)

Population of Urban Areas

	2006	2020	
40 million			40 million
30 million			30 million
20 million			20 million
10 million			10 million
5 million			5 million

Note: symbols on map are sized proportionally to specific population data

B Recent migrants sleep on a sidewalk in Dhaka. Many of the world's fastest-growing cities are attracting more people than they can support with their existing housing and infrastructure.

C Employment is the greatest "pull" factor drawing people into cities. Most recent migrants work at physically demanding, low-paying, and often hazardous jobs. These women are handing buckets full of cement up a scaffolding on a construction site in Delhi.

D Tokyo Disneyland, Cinderella's Castle. Like most of the world's great cities, Tokyo has long been a major conduit for globalization. Tokyo has been the world's largest city since 1970 and is projected to retain that distinction for the foreseeable future.

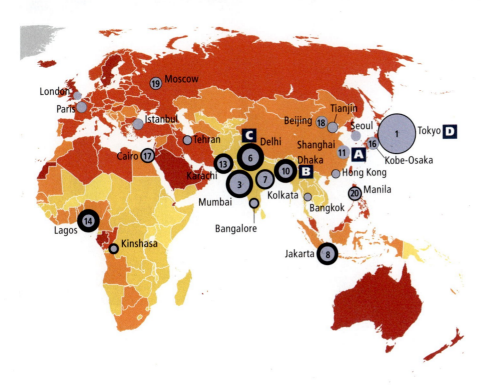

E Some cities struggle with major environmental problems. Mexico City, currently the world's second largest city, suffers from periodic flooding due in part to its location on an old now sinking lake bed and its antiquated drainage and sewage infrastructure.

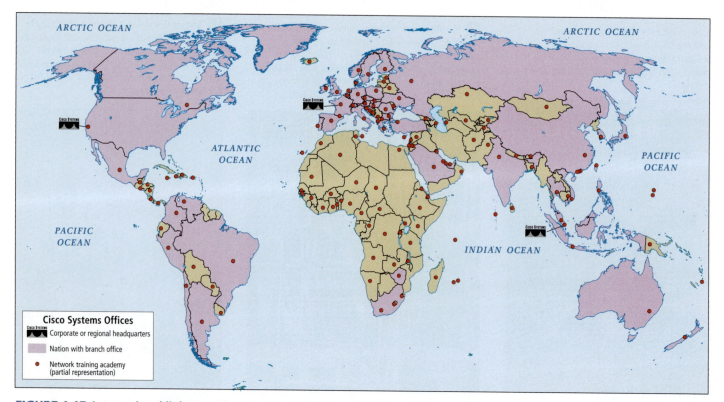

FIGURE 1.17 Interregional linkages: Cisco Systems' global network. Cisco provides hardware and services for Internet networking. This map shows several levels of Cisco's activities: corporate and regional headquarters, countries with branch offices, and the locations of network training academies. Notice the uneven distribution of Cisco Systems offices.

What Is the Global Economy?

The **global economy** includes the parts of any country's economy that are involved in global flows of resources—mined minerals, agricultural commodities, manufactured products, money, and people and their ideas. Most of us participate in the global economy every day. For example, books like this can be made from trees cut down in Southeast Asia or Siberia and shipped to a paper mill in Oregon. Many books are now printed in Asia because labor costs are lower there. Such long-distance movement of resources and products has grown tremendously in the past 500 years and remains possible because fuel costs for water transport are low. Globalization is not new. It existed at least 2500 years ago, when silk and other goods were traded along Central Asian land routes that connected Greece and then Rome in the Mediterranean with China.

European colonization was an early expansion of globalization. Starting in about 1500, European countries began extracting resources from distant parts of the world they had conquered. The colonizers organized systems to process those resources into higher-value goods to be traded wherever there was a market. Sugarcane, for example, was grown on Caribbean and Brazilian plantations with slave labor from Africa (**Figure 1.18**) and made locally into crude sugar, molasses, and rum. These products were then shipped to Europe and North America where it was further refined and sold

> **global economy** the worldwide system in which goods, services, and labor are exchanged
>
> **multinational corporation** a business organization that operates extraction, production, and/or distribution facilities in multiple countries

FIGURE 1.18 European use of colonial resources. Among the first global economic institutions were Caribbean plantations like Old North Sound on Antigua, shown here in an old painting. In the eighteenth century, thousands of sugar plantations in the British West Indies, subsidized by the labor of slaves, provided huge sums of money for England and helped fund the Industrial Revolution.

at considerable profit. The global economy grew as each region not only produced goods for export, rather than just for local consumption, but at the same time became increasingly dependent on imported food, clothing, machinery, energy, and knowledge.

The new wealth derived from the colonies, and the ready access to global resources, led to Europe's *Industrial Revolution, a series of innovations and ideas that changed the way goods were produced.* No longer was one woman producing the cotton or wool for cloth, spinning thread, weaving the thread into cloth, and sewing a garment. Instead, these separate tasks were spread out among many workers, often in distant places, with some people specializing in producing the fiber and others in spinning, weaving, or sewing. These innovations were followed by labor-saving improvements such as mechanized reaping, spinning, weaving, and sewing.

This larger-scale mechanized production accelerated globalization as it created a demand for raw materials and a need for markets in which to sell finished goods. European colonies in the Americas, Africa, and Asia provided both. For example, in the British Caribbean colonies, hundreds of thousands of African slaves wore garments made of cloth woven in England from cotton grown in British India. The sugar they produced on British-owned plantations with iron equipment from British foundries was transported to European markets in ships made in the British Isles of trees and resources from various parts of the world.

Until the early twentieth century, much of the activity of the global economy took place within the colonial empires of a few European nations. By the 1960s, global economic and political changes had brought an end to these empires, and now almost all colonial territories are independent countries. Nevertheless, the global economy persists in the form of banks and **multinational corporations** such as Shell, Walmart, Bechtel, and Cisco (see Figure 1.17) that operate across international borders. These corporations extract resources from many places, make products in factories located where they can take advantage of cheap labor and transportation facilities, and market their products wherever they can make the most profit. Their global influence, wealth, and importance to local economies enable the multinationals to influence the economic and political affairs of the countries in which they operate.

Workers in the Global Economy

Personal Vignette Olivia lives near Soufrière on St. Lucia, an island in the Caribbean (**Figure 1.19A**). Soufrière was once a quiet fishing village, but now it hosts cruise-ship passengers several times a week. Olivia is 60. She, her daughter Anna, and her three grandchildren live in a wooden house surrounded by a leafy green garden dotted with fruit trees. Anna has a tiny shop at the side of the house, from which she sells various small everyday items and preserves that she and her mother make from the garden fruits.

On days when the cruise ships dock, Olivia strolls down to the market shed on the beach with a basket of papayas, bags of roasted peanuts, and rolls of cocoa paste made from cacao beans picked in a neighbor's yard. She calls out to the passengers as they near the shore, offering her spices and snacks for sale. In a good week she makes U.S.$50. Her daughter makes about U.S.$100 per week in the

FIGURE 1.19 Workers in the global economy. (A) Soufrière, St. Lucia, as seen from a cruise ship. **(B)** Setiya working at the construction job in Malacca, Malaysia, before being expelled from the country, along with 500,000 other foreign workers, during the recession that began in 2008. **(C)** The trailer at the back of Tanya's lot where her daughter lives.

shop and is constantly looking for other ways to earn a few dollars. She often makes necklaces for tourists or takes in laundry.

Olivia and Anna support their family of five on about U.S.$170 a week (U.S.$8840 per year). From this income they pay rent on the house, the electric bill, and school fees for the granddaughter who will go to high school in the capital next year and perhaps college if she succeeds. They also buy clothes for the children and whatever food (primarily flour and sugar) they cannot grow themselves. Their livelihood puts them at or above the standard of living of most of their neighbors.

In Malacca, Malaysia, 30-year-old Setiya, an illegal immigrant from Tegal, Indonesia, is laying concrete blocks for a new tourist hotel (**Figure 1.19B**). Like a million other Indonesians attracted by Malaysia's booming economy, he snuck into Malaysia. He arrived at night by boat from Sumatra across the Strait of Malacca. The trip was expensive and he willingly risked arrest because in Malaysia average wages are four times higher than at home.

This is Setiya's second trip to Malaysia. On his first trip he had signed up for an overseas employment program sponsored by the Malaysian Ministry of Manpower to work legally on a Malaysian oil palm plantation. Upon arrival, however, Setiya found that he would have to work for 3 months just to pay off his boat fare. Not one to give in easily, he quietly caught a bus to another city and found a construction job earning U.S.$10 a day (about U.S.$2600 a year). Although working illegally, he was able to send enough money to his family in Indonesia to pay for their food, school fees, and a new corrugated metal roof for their house.

In 2008, Malaysia announced it was expelling 500,000 foreigners like Setiya. The country was suffering growing unemployment due to the global recession, and its leaders wanted to save more jobs for locals by expelling foreign workers. But the crisis affected Indonesia even more severely, making life there so difficult for workers and their families that young fathers (like Setiya) once again risked trips on leaky boats to illegally enter Malaysia and Singapore.

Tanya is a 50-year-old grandmother in North Carolina with a son in high school and a married daughter. Tanya works at a fast-food restaurant, making less than U.S.$6 an hour. She had been earning U.S.$8 an hour sewing shirts at a textile plant until it closed and moved to Indonesia. Her husband is a delivery truck driver for a snack-food company.

Between them, Tanya and her husband make $27,000 a year, but from this income they must cover their mortgage and car loan and meet regular monthly expenses for food and utilities. In addition, they help their daughter, Rayna, who quit school after 11th grade and married a man who is now out of work. They and their baby live at the back of the lot in an old mobile home that Tanya and her family once occupied (**Figure 1.19C**).

With Tanya's now lower wage ($4000 less a year), there will not be enough money to pay the college tuition for her son. He had hoped to be an engineer and would have been the first in the family to go to college. For now, he is working at the local gas station.

These people, living worlds apart, are all part of the global economy. Workers around the world are paid startlingly different rates for jobs that require about the same skill level. Varying costs of living and varying local standards of wealth make a difference in how people live and regard their own situation. Though Tanya's family has the highest income by far, compared with their neighbors they live in poverty, and their hopes for the future are dim. Olivia's family, on the other hand, is not well off, but they do not think of themselves as poor because they have what they need, others around them live in similar circumstances, and their children seem to have a future. They can subsist on local resources, and the tourist trade promises continued cash income. But their subsistence depends on circumstances beyond their control; in an instant, the cruise-line companies can choose another port of call. Setiya, by far the poorest, seems trapped by his status as an illegal worker, which robs him of many of his rights. Still, the higher pay that he can earn in Malaysia offers him a possible way out of poverty. [*Adapted from Lydia Pulsipher's field notes, 1992–2000 (Olivia and Tanya), and Alex Pulsipher's field notes, 1999–2008 (Setiya)*]. ∎

The Debate over Globalization and Free Trade

The term **free trade** refers to the unrestricted international exchange of goods, services, and capital. Free trade has not yet been achieved. Currently, all governments impose some restrictions on trade to protect their own national economies from foreign competition, although such restrictions are far fewer than in the 1980s. Restrictions take two main forms: *tariffs* and *import quotas*. Tariffs are taxes imposed on imported goods that increase the cost of those goods to the consumer, thus giving price advantage to locally made competing goods. Import quotas set limits on the amount of a given good that may be imported over a set period of time, thus curtailing supply.

These and other forms of trade protection are subjects of contention. Proponents of free trade argue that the removal of all tariffs and quotas encourages efficiency, lowers prices, and gives consumers more choices. Companies can sell to larger markets and take advantage of mass-production systems that lower costs further. As a result, they can grow faster, thereby providing people with jobs and opportunities to raise their standard of living. These pro–free trade arguments have been quite successful, and in recent decades restrictions on trade imposed by individual countries were greatly reduced. Several *regional trade blocs* have been formed; these are associations of neighboring countries that agree to lower trade barriers among one another. The main ones are the North American Free Trade Agreement (NAFTA), the European Union (EU), the Southern Common Market (Mercosur) in South America, and the Association of Southeast Asian Nations (ASEAN).

A main global institution supporting free trade is the **World Trade Organization (WTO)**, whose stated mission is to lower trade barriers and establish ground rules for international trade. Related institutions, the *World Bank* (officially named the International Bank for Reconstruction and Development) and the *International Monetary Fund* (IMF), both make loans to countries that need money to pay for economic development projects. Before approving a loan, the World Bank or the IMF may require a borrowing country to reduce and eventually remove tariffs and import quotas. These requirements are part of larger *structural adjustment policies*

(SAPs) that the IMF imposes on countries seeking loans, such as the requirement to close government enterprises and to reduce government services, mostly to the detriment of the poor. SAPs have become highly influential and controversial in virtually every region of the world. Chapter 3 and Chapter 7 provide a detailed explanation of SAPs and their effects in specific regions.

Those opposed to free trade and the SAPs that promote it argue that a less-regulated global economy can lead to rapid cycles of growth and decline that only increase global wealth disparity and can wreak havoc on smaller national economies (**Figure 1.20**). Labor unions point out that as corporations relocate factories and services to poorer countries where wages are lower, jobs are lost in richer countries. In the poorer countries, multinational corporations often work with governments to prevent workers from organizing labor unions that could bargain for **living wages**, wages that support a minimum healthy life. Environmentalists argue that in newly industrializing countries, which often lack effective environmental protection laws, multinational corporations tend to use highly polluting and unsafe production methods to lower costs. Many fear that a "race to the bottom" in wages, working conditions, government services, and environmental quality is underway as countries compete for profits and potential investors.

In response to the now widely recognized failures of SAPs and the overemphasis on the power of markets to guide development,

the IMF and the World Bank replaced SAPs with "Poverty Reduction Strategy Papers," or PRSPs. Each needy country works with World Bank and IMF personnel to design a broad-based plan for both economic growth and poverty reduction. PRSPs still push market-based solutions, aim toward reducing the role of government in the economy, and are highly bureaucratic, but they do focus on poverty reduction rather than just "development" per se. They also promote broader participation in civil society and include the possibility that all or some of a country's debt be "forgiven" (paid off by the IMF and the World Bank), thus alleviating one of the worst side effects—bankrupting debt—that stopped progress in the poorest countries.

Fair trade, proposed as an alternative to free trade, seeks to provide a fair price to producers and to uphold environmental and safety standards in the workplace. Some investor profits are sacrificed in order to provide markets for producers from developing countries. For example, "fair trade" coffee and chocolate are now marketed to North America and Europe. Prices are somewhat higher for consumers, but the extreme profits of middlemen are eliminated and growers of coffee and cocoa beans receive living wages and improved working conditions.

In evaluating free trade and globalization, consider how many of the things you own or consume were produced in the global economy—computer, clothes, furniture, appliances, car, and foods. These products are cheaper for you to buy, and your standard of living is higher as a result of lower production costs and competition among many global producers (see Thematic Concepts Part P on page 13). However, you or someone you know may have lost a job because a company moved to another location where labor and resources are cheaper. You may be concerned that the products you buy so cheaply were made under harsh conditions by underpaid workers (even children), or that resources were used unsustainably or are not safe to use. High levels of pollution may have occurred in the manufacturing and transport processes. Given all these factors, consider the advantages and drawbacks of both free trade and fair trade.

> **free trade** the movement of goods and capital without government restrictions
>
> **World Trade Organization (WTO)** a global institution made up of member countries whose stated mission is the lowering of trade barriers and the establishment of ground rules for international trade
>
> **living wages** minimum wages high enough to support a healthy life
>
> **fair trade** trade that values equity throughout the international trade system; now proposed as an alternative to free trade

FIGURE 1.20 Anti-WTO demonstration by Korean farmers and students (2005). The demonstrators were protesting WTO regulations that reduced farmers' freedom and income.

THINGS TO REMEMBER

1. Globalization encompasses many types of worldwide and interregional flows and linkages, especially the ways in which goods, capital, labor, and resources are exchanged among distant and very different places.

2. Local self-sufficiency is giving way to global interdependence and international trade worldwide as information, goods, and people move across vast distances faster and on a larger scale than ever before.

3. Globalization is the most complex of the thematic concepts in this book. Understanding globalization is critical to grasping geography today.

DEMOCRATIZATION

For geographers studying globalization, an area of increasing interest is **democratization**, the transition toward political systems guided by competitive elections. Also of interest is how democracy is advancing at smaller scales, such as local and state governments, and how social movements, international organizations, and a free media play a role in the reallocation of power to the voting and even nonvoting public (see Thematic Concepts Parts S, T, U on page 14). Some evidence suggests that globalization may give ordinary people a greater voice in how societies are run. The idea is that with the expansion of worldwide channels of communication, more democratic systems of government will gradually take hold and replace **authoritarianism**. An authoritarian form of government subordinates individual freedom to the power of the state or to elitist regional and local leaders (see Thematic Concepts Part T on page 14). By contrast, in more democratic systems of government, individuals not only have access to information and participate in elections but also enjoy more economic and political freedoms, such as the right to be an entrepreneur, the right to protest government policies, to marshal public support through the media for particular programs, and to take action against injustice, especially through legal systems.

democratization the transition toward political systems guided by competitive elections

authoritarianism a political system based on the power of the state or of elitist regional and local leaders

The Expansion of Democracy

The twentieth century saw a steady expansion of democracy throughout the world, with more and more countries holding elections of their leaders—at least at the national level. While democracy has become an ideal to which most countries aspire, there is little agreement on just what constitutes democratic institutions. If elections are the most important component of democracy, should all adults be part of the voting public? Can the principles of a specific religion be part of democratic constitutions? Should majority rule be the final criterion, or might majorities manage to oppress minorities? Perhaps government by consensus would be fairer. Is it possible to have democratic government at the national level and yet quite authoritarian or even corrupt rule at the local level? What about forms of democratic participation other than voting, such as the ability to speak openly to leaders, to protest, to lobby for or against particular laws? Are these also essential components of democracy? The map in Photo Essay 1.4 is an attempt by the *Economist* magazine to depict the global pattern of democratization and to link that pattern with the occurrence of political conflict. Most conflicts happen in places that are authoritarian, not democratic.

In recent decades the right to vote has expanded in Middle and South America, sub-Saharan Africa, Central Europe, and the post-Soviet states. What factors encouraged the public in all these places to seek a more active role in their own governance? Here are some of the most widely agreed-on factors necessary for democratization to flourish:

- Broad prosperity: As countries become wealthier, and as prosperity—access to more than the bare essentials of life—is

---

The map and accompanying photo essay show how countries with lower levels of democratization also suffer the most from violent conflict. Countries are colored according to their score on a "democracy index" created by the *Economist* magazine, which uses a combination of statistical indicators to capture elements crucial to the process of democratization. These elements include the ability of a country to hold peaceful elections that are accepted as fair and legitimate; the ability of people to participate in elections and other democratic processes; the strength of civil liberties (such as a free media and the right to hold political gatherings and peaceful protests); and the ability of governments to enact the will of their citizens free of corruption. Also displayed on the map are major conflicts initiated or in progress since 1990 that have resulted in at least 10,000 casualties. Aspects of the connections between democratization and armed conflict are explored in the photo captions.

Democratization and Conflict
Democratization Index

- Full democracy
- Flawed democracy
- Hybrid regime
- Authoritarian regime
- No data

Armed Conflicts and Genocides with High Death Tolls Since 1990

- ! Ongoing conflict
- ✳ 10,000–100,000 deaths
- ✴ 100,000–1,000,000 deaths
- ✴ over 1,000,000 deaths

A A low level of democratization has frustrated attempts to end the ongoing violence in Colombia between rebel groups, private militias (called paramilitaries), and the government. In recent years, the administration of Colombian president Álvaro Uribe (shown below) has been implicated in the use of paramilitaries to assassinate leaders of the political opposition and labor union organizers. Repressive tactics like these diminish the potential to resolve disputes through peaceful democratic processes and help maintain support for violent confrontation.

B Violence in Sudan's Darfur region relates directly to the country's very low level of democratization. Disputes over control of resources, whether they be land, water, or Sudan's oil resources, often turn violent because there are no functioning democratic institutions capable of resolving conflicts. Shown on the left are members of the Sudan Liberation Movement Army, which opposes government-backed attacks on civilians in Darfur.

C Bringing democracy to Iraq was a main justification for the U.S. invasion and subsequent occupation of the country. Iraq's newly empowered parliament responded to widespread opposition to the continuing U.S. military presence by setting a deadline of 2011 for the withdrawal of U.S. forces, which the U.S. has agreed to honor. These men in Baghdad are protesting the presence of U.S. troops in Iraq.

D New Zealand is among the world's most democratized nations. Shown above are members of the Service and Food Workers Union, making use of their well-protected civil liberties by staging a rally for better wages and working conditions for its largely immigrant and female members. The protection of civil liberties is a crucial aspect of democratization.

E Democratic processes often become compromised during times of war. Shown here is a woman and her child running from the burning compound of presidential candidate Jean-Pierre Bemba's bodyguards in Kinshasa, Congo. The compound was attacked by demonstrators who wanted Congo's recent elections, the first in 40 years, postponed because of the country's ongoing civil war and accusations that the vote counting was being rigged. The civil war in Congo has resulted in 5.4 million deaths so far, mostly due to disease and malnutrition among people displaced by the violence.

shared by a broad segment of the population (usually the "middle class"), there is generally a shift toward giving more power to citizens through free elections of leaders. Still widely debated is whether general prosperity must occur before truly stable democracy can be established. And will prosperity necessarily lead to democracy?

- **Education:** Better-educated people tend to want a stronger voice in how they are governed. Although democracy has spread to countries with relatively undereducated populations, leaders in such places sometimes become more authoritarian once elected.

- **Civil society:** Institutions that encourage a sense of unity and informed common purpose among the general population are widely seen as supportive of democracy. Such institutions can include academia, unions, political parties, community service organizations such as Rotary and Lions Clubs, the media, nongovernmental organizations (discussed below), human rights organizations, and in some cases, religious organizations.

Certainly the right to vote in national elections is a basic component of democratization, but in many ways it is only the beginning of true democratic participation. Perhaps more essential are components such as the right to influence the laws that are proposed and passed at the national, state, or local level and the right to choose the candidates for national or local office. Without input in these areas, citizens may be left with no attractive options in the voting booth. Such participatory rights are more likely to be exercised by citizens who enjoy basic prosperity; likewise, an educated population is more likely to exercise these rights with wisdom and understanding. The institutions of **civil society** are increasingly recognized as essential to democratization. These institutions inform and involve the citizenry and divert attention from narrow individualistic concerns (Photo Essay 1.4D).

Democratization and Geopolitics

As the map in Photo Essay 1.4 shows, democratization at the global level has not yet been achieved (in fact, it is a goal that is not shared by all). A possible explanation is that democratization is often at odds with **geopolitics**, the strategies that countries use to ensure that

their own interests are served in relations with other countries. Geopolitics was perhaps most obvious during the *Cold War era*, the period from 1946 to the early 1990s when the United States and its allies in Western Europe faced off against the Union of Soviet Socialist Republics (USSR) and its allies in Eastern Europe and Central Asia. Ideologically, the United States promoted a version of free market **capitalism**—an economic system based on the private ownership of the means of production and distribution of goods, driven by the profit motive and characterized by a competitive marketplace. By contrast, the USSR and its allies favored what was called **communism** but was actually a state-controlled economy, a socialized system of public services and a centralized government in which citizens participated indirectly through the Communist Party.

The Cold War became a race to attract the loyalties of unallied countries and to arm them. Sometimes the result was that unsavory dictators were embraced as allies by one side or the other. Eventually, the Cold War influenced the internal and external policies of virtually every country on earth, often oversimplifying complex local issues into a contest of democracy versus communism.

In the post–Cold War period of the 1990s, geopolitics shifted. The Soviet Union dissolved, creating many independent states, nearly all of which began to implement some democratic and free market reforms. Globally, countries jockeyed for position in what looked as if it might become a new era of trade and amicable prosperity, rather than war. But throughout the 1990s, while developed countries enjoyed unprecedented prosperity, conflicts over political power and access to resources errupted in Africa, Southeast and Southwest Asia, and southeastern Europe (see Photo Essay 1.4 B, E). Too often these disputes erupted into bloodshed and systematic attempts to remove (**ethnically cleanse**) or kill (**genocide**) all members of a particular ethnic or religious group.

The terrorist attacks on the United States on September 11, 2001, ushered in a new geopolitical era that is still evolving. Because of the size and the geopolitical power of the United States, the attacks and the U.S. reactions to them affected virtually all international relationships, public and private. The ensuing adjustments are directly or indirectly affecting the daily lives of billions of people around the world.

International Cooperation

So far there has been no serious effort to engage the principles of democratization at the global scale. However, the free movement of people, goods, and capital across national borders is a trend that favors international cooperation over national self-interest. It also creates a need for some way to enforce laws governing business, trade, and human rights at the international level.

The prime example today of international cooperation is the **United Nations (UN)**, an assembly of 192 member states. The member states sponsor programs and agencies, focusing on scientific research, humanitarian aid, economic development, general health and well-being, and peacekeeping assistance in "hot spots" around the world. Thus far, countries have been unwilling to relinquish *sovereignty*, the right of a country to conduct its internal affairs as it sees fit without interference from outside.

civil society the social groups and traditions that function independently of the state and its institutions

geopolitics the use of strategies by countries to ensure that their best interests are served

capitalism an economic system based on the private ownership of the means of production and distribution of goods, driven by the profit motive and characterized by a competitive marketplace

communism an ideology, based largely on the writings of the German revolutionary Karl Marx, that calls on workers to unite to overthrow capitalism and establish an egalitarian society in which workers share what they produce

ethnically cleanse the deliberate removal of an ethnic group from a particular area by forced migration

genocide the deliberate destruction (killing) of an ethnic, racial, or political group

United Nations (UN) an assembly of 192 member states that sponsors programs and agencies that focus on scientific research, humanitarian aid, planning for development, fostering general health, and peacekeeping assistance

nongovernmental organization (NGO) an association outside the formal institutions of government in which individuals, often from widely differing backgrounds and locations, share views and activism on political, social, economic, or environmental issues

Consequently, the United Nations has limited legal authority and often can enforce its rulings only through persuasion. Even in its peacekeeping mission, there are no true UN forces. Rather, troops from member states wear UN designations on their uniforms and take orders from temporary UN commanders. The World Bank, the International Monetary Fund, and the World Trade Organization, discussed earlier (see page 36), are important international organizations that affect economies and trade practices throughout the world, but only through negotiation, not enforcement.

Nongovernmental organizations (NGOs) are an increasingly important embodiment of globalization. In such associations, individuals, often from widely differing backgrounds and locations, agree on political, economic, social, or environmental goals. For example, some NGOs work to protect the environment (for instance, the World Wildlife Fund). Others, such as Doctors Without Borders, provide medical care to those who need it most. The Red Cross and Red Crescent provide emergency relief after disasters (Figure 1.21). The educational efforts of an NGO such as Rotary International can raise awareness among the global public of important issues, such as childhood vaccinations.

NGOs can be an important component of civil society, yet there is some concern that the power of huge international NGOs might undermine democratic processes, especially in small countries. Some critics feel that NGO officials are a powerful do-gooder elite that does not interact sufficiently well with local people. A frequent target of such criticism is OXFAM International, a group of 13 NGOs that is the world leader in emergency famine relief. OXFAM was a major provider of relief after the Indian Ocean tsunami (2004) and during the conflicts in Lebanon and Israel in 2006 and Gaza in 2009. It has now expanded to cover long-term efforts to reduce poverty and injustice, which OXFAM sees as the root causes of famine. This more politically active role has brought OXFAM into conflict with local officials and with WTO policies on trade. At the local level, the best NGOs solicit input from a wide range of individuals—a feature

FIGURE 1.21 Help from a nongovernmental organization (NGO). Members of the Kenyan Red Cross move an injured person to safety during post-election violence in Nairobi, Kenya, in January of 2008.

that political scientists consider essential to building the capacity for participatory democracy.

THINGS TO REMEMBER

1. **Democratization** is the transition toward political systems that are guided by competitive elections in which individuals have a great voice in how their governments are run.

2. Democratic systems are gradually, but not inexorably, replacing authoritarian regimes worldwide.

3. The globalization of communication systems and information exchange is involving ordinary people in public debates, which may lead to increasing democratization.

WATER

Water is emerging as the major resource issue of the twenty-first century. Increasing water scarcity is a significant cause for concern and has been closely linked to the activities of humans: population growth, skyrocketing per capita demand for clean water as modernization proceeds, water pollution, the inequitable allocation of available water, and increasing drought as a result of climate change (see Thematic Concepts Parts V, W, X on page 14). Leaps in per capita demand are connected to increased per capita consumption of all types of products that require water in the production process: food, manufactured goods, and even services (for example, cleaning and entertainment). *Water pollution*

is a side effect of rising standards of living and of population growth. Inequitable allocation can result when water becomes a commodity rather than a free good and is priced too high for some to afford, or when political disputes separate people from their customary sources of water. Drought and flooding, previously only occasionally blamed on human agency, are now thought to be one of the spreading side effects of global climate change (see Thematic Concepts Parts EE, FF, GG on page 15).

When clean fresh water is scarce, it takes on a political role and can become a pawn in international power struggles. Because it is such a basic necessity, water disputes are proliferating in those situations where rivers cross borders and upstream users either use more than their perceived fair share, block water flows with dams, or return used water to streams in a polluted condition.

Calculating Water Use Per Capita

According to the second UN World Water Development Report, released in 2006, depending on body size, activity level, and ambient temperature, each human requires an average of 20 to 50 liters (50 liters is 0.05 cubic meters [m^3]) of clean water per day, or 18.25 m^3 per year for basic domestic needs: drinking, cooking, and bathing/cleaning. (There are 1000 liters, or 263 gallons, in a cubic meter.) Only a small percentage of people make do with this amount. For example, at the very low per capita GDP of under $1000 per year, domestic water use is perhaps as low as about 10 m^3 per capita per year. At higher income levels of, say, $33,000 per year, domestic water use is more like 200 m^3 per capita per year, or 20 times that of the poor. But actual consumption of water is much higher than these figures indicate because water is used to process nearly all the goods and services consumed by each person on earth.

virtual water the volume of water used to produce all that a person consumes in a year

water footprint the water used to meet a person's basic needs for a year, added to the person's annual virtual water

Every apple eaten or cup of coffee drunk requires many liters of water for production and distribution. The volume of water used to produce all that a person consumes in a year is called **virtual water,** and it must be added to the water used to meet annual basic needs in order to arrive at a person's total annual **water footprint.** The more one consumes, the larger one's virtual water footprint. Table 1.2 shows the amounts of water used to produce some commonly consumed products.

Like domestic consumption, personal water footprints vary widely according to standards of living and rates of consumption (Figure 1.22). Moreover, the amount of virtual water used to produce one ton of a specific product varies widely country to country due to climate conditions but also to agricultural and industrial technology. The virtual water content of a product is a measure of the volume of water lost to use through the production of the product. For example, on average, to produce 1 ton of corn in the United States requires 489 m^3 of virtual water, whereas in India the same amount of corn requires 1935 m^3 of virtual water; in Mexico 1744 m^3; and in the Netherlands, just 408 m^3. In the case of corn, water is lost to evapotranspiration in the field, to the evaporation of standing irrigation water, and to evaporation as water flows to and from the field. A further component of virtual water is that the water that becomes polluted in the production process is also lost to use.

Because of globalization, food and other goods are now produced all over the world, and as a result a person's water footprint may extend to very distant locations. Individual consumers are usually quite unaware of the size and geographic reach of their virtual water footprint. Countries in arid areas may actually decide that because of extreme water scarcity, the costs of producing food and industrial products are too high, so using virtual water from elsewhere is preferable. Their imported products are then said to have a *virtual water component.*

For help calculating your individual water footprint, go to the water footprint Web site: http://www.waterfootprint.org/?page=files/WaterFootprintCalculator.

Who Owns Water?
Who Gets Access to It?

We know that life cannot exist without water, but who owns the world's water? And how do people get access to water? Though most readers may assume that water is a human right and that we pay for water only to cover the cost of getting it to our homes in a potable condition, in fact water has become the third most valuable commodity in the world, after oil and electricity. Water in wells and running in streams and rivers is increasingly claimed and managed for profit by multinational corporations. This privatization of water is happening primarily in developing countries where for-profit water companies can strike deals with governments and elite landowners. The poor must suddenly pay substantial fees for clean water or use polluted undesirable water sources. Water has been privatized and commodified (assigned a monetary value) under the rationale that private enterprise will supply a better product for the price consumers pay, but regulations have been lax and enforcement often nonexistent.

TABLE 1.2	The global average virtual water content of everyday products*

Product[a]	Virtual water content (in liters)
1 potato	25
1 cup tea	35
1 slice of bread	40
1 apple	70
1 glass of beer	75
1 glass of wine	120
1 egg	135
1 cup of coffee	140
1 glass of orange juice	170
1 lb of chicken meat	2000
1 hamburger	2400
1 lb of cheese	2500
1 pair of bovine leather shoes	8000

* Virtual water is the volume of water used to produce a product.

[a] To see the virtual water content of additional products, go to http://www.waterfootprint.org/?page=files/productgallery

Source: Arjen Y. Hoekstra and Ashok K. Chapagain, *Globalization of Water—Sharing the Planet's Freshwater Resources,* Blackwell: Malden, MA, 2008, p. 15, Table 2.2.

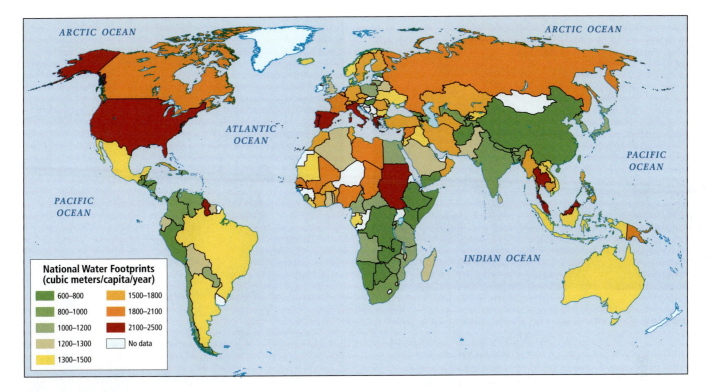

FIGURE 1.22 National water footprints, 2004. Average national water footprint per capita (m³ per capita per year). The color green indicates that the nation's water footprint is equal to or smaller than the global average. Countries in red have a water footprint beyond the global average.
[*UN World Water Development Report 2: Water—A Shared Responsibility.* Published jointly in 2006 by the UN Educational, Scientific and Cultural Organization (UNESCO), Paris; and Berghahn Books, New York, pp. 391–392.]

There are also informal markets for water. In many developing countries where water was once free for the fetching from creek or spring, small-scale private businesses now distribute water at a price that amounts to as much as 11 percent of the income of the poorest consumers; and this water is not necessarily free of germs or pollutants.

About one-sixth of the world's population do not have access to clean drinking water, and dirty water kills more people annually than armed conflict—over 5 million people (especially children in less developed countries) die each year from dirty water. In Europe and the United States, people in search of quality water buy it in a bottle. Over $100 billion is spent annually on bottled water—a primary indication that water is now a product, not a right; yet there are few standards that guarantee the quality of that bottled water. Also, water has to come from somewhere. In reality, bottled water is extracted from natural environments that are degraded in the process. Plant and animal species are deprived of natural habitats by large-scale water extraction; there is also the matter of the environmental impact of all those millions of plastic water bottles.

Water Quality

Modern methods of water use can lower its quality. For example, water used in industry or water sent through hydropower turbines on dammed rivers can be heated to unnatural temperatures. If the warmed water is then released into rivers or lakes, it can kill sensitive plant and aquatic species downstream. Water used in mining or in the treatment of coal for burning in power plants is often infused with mineral pollutants that are highly toxic, especially when ingested or dried and blown about by the wind. Tourism, one of the world's largest industries, uses fresh water in large quantities, polluting much of it in the process. In the Mediterranean and the Caribbean, coastal waters are polluted by sewage and runoff from a variety of tourism facilities.

As people move from rural agricultural work to industry or service sector jobs in cities, they begin to use more resources per capita, and they draw them from a wider and wider area. People who once fetched water by hand from nearby village wells may now draw water that has been piped hundreds of miles into apartments or shantytown spigots.

For most urban shantytown dwellers, coping with limited access to water and sanitation is often the most persistent challenge. Typically, whether in Asian, African, or Middle and South American urban areas, people acquire their water with a pail from a communal spigot (see Thematic Concepts Part V on page 14; Figure 1.23A). Usually this water should be boiled before use, even for bathing, and truly safe drinking water must be purchased in sealed containers at prohibitively high cost. These quality and access problems help explain why people are often chronically ill and why so many children die before the age of 5 from waterborne diseases.

(A)

FIGURE 1.23 Access to water. (A) A young man carries water to his home in the hills outside Kabul, Afghanistan. The city is home to 3.4 million people, but it has no central sewage system, and only 18 percent of its people have access to city water piped into their homes. **(B)** A boy drinks water from a temporary pool that formed outside the slum where he lives in Kenya. Rains often bring flooding to urban slums in Africa, which rarely have any planned drainage system.

(B)

Urban development also affects the management of waste-water. Housing in slum areas is often self-built of scavenged materials, and sanitation systems are absent. Kitchen waste and human waste get deposited in paths, streets, and nearby ravines. Even the sewage from modern high-rise apartments and hotels is often pumped untreated into a nearby river, swamp, or ocean. This method of waste disposal causes serious health hazards and widespread ecological damage. Building adequate wastewater collection and water purification systems in cities already housing several million inhabitants is often prohibitively costly.

Urban flooding can be a dangerous result of inadequate wastewater management. Buildings and roads made of concrete, asphalt, and steel often cover formerly wooded land with impervious surfaces. Rainwater quickly runs off the hard surfaces and collects in low places instead of being absorbed into the ground. In urban slums, flooding can spread polluted water over wide areas, carrying it into homes and places where children play (**Figure 1.23B**). Epidemics of diseases such as cholera can spread rapidly.

Access to water also has a gendered aspect. In developing countries, women in their capacity as homemakers, caretakers, and gardeners are usually the family members responsible for procuring water, often carrying it great distances. As much as one-quarter of a woman's time may be spent on water collection, time that cannot be spent on income-earning efforts. When water must be paid for, it is often women who must come up with the money. Yet because of their lack of political power, women have little say in molding water policy—where spigots should be located, how water should be allotted or priced, how sanitation should be provided and wastewater handled. Because women

usually remain economically subservient to the males in their families, they are left without formal rights to water, this most essential resource. Just recognizing the role of women in family and community water use could improve water access for all and uplift the status of women in their own communities.

THINGS TO REMEMBER

1. Water is emerging as the major resource issue of the twenty-first century. Increasing water scarcity is linked to population growth, skyrocketing per capita demand for clean water, water pollution, the inequitable allocation of available water, and increasing drought as a result of climate change.

2. Public pressure to reduce use and pollution of water is rising as conflicts intensify over access to water for drinking, industrial uses, irrigation for food production, leisure activities, and the maintenance of aquatic ecosystems.

3. Several international companies now own rights to water and water systems throughout the world, usually at the expense of individuals and local communities that traditionally had rights to that water, often as a free resource.

GLOBAL CLIMATE CHANGE

Planet Earth is continually undergoing climate change, a slow shifting of climate patterns due to general cooling or warming of the atmosphere. The present trend of global warming—which refers to the observed warming of the earth's surface and climates in recent decades—is extraordinary because it appears to be due primarily to human agency and may be happening more quickly than climate changes in the past. The markedly increased amounts of carbon dioxide, methane, nitrous oxide, and other gases—the so-called greenhouse gases—that are trapping heat in the atmosphere are a symptom of growing human impacts on earth environments (see Thematic Concepts Parts Y, Z, AA on page 14). Most scientists now agree that there is an urgent need to reduce these impacts. A key problem in achieving this reduction is that those most responsible for global warming (the industrialized and rapidly industrializing countries) are the least vulnerable to the changes it causes in the physical environment. Meanwhile, the most vulnerable to these changes (the poorest countries of the world) are the least responsible for the growth in greenhouse gases, and hence the least able to effect any reduction in emissions (see Figure 1.25 on page 47).

Humans and the Environment

The same forces that created globalization have had enormous impacts on physical environments. Mass consumption of resources has altered the environment most profoundly, but even less wasteful human ways of life have environmental effects. Photo Essay 1.2 on pages 28–29 shows some of the impacts of humans on the planet's land surface. The intensity and nature of the impact varies greatly, but human impact can be found virtually everywhere.

Mounting awareness of environmental impacts has prompted numerous proposals to limit the damage, from relying on technological advances to reducing resource consumption. Halting and reversing environmental damage may be the greatest challenge our species has yet faced, in part because our societies have become so transformed by our intensive use of the earth's resources that going back is enormously difficult. The issue at the forefront now is global warming.

Resource use has become so skewed that the relatively rich minority of the world's population (about 20 percent) consumes more than 80 percent of the available world resources. The poorest 80 percent of the population are left with less than 20 percent of the resources.

Human consumption of natural resources is increasingly being examined through the concept of the ecological footprint. This is a method of estimating the amount of biologically productive land and sea area needed to sustain a human population at its current standard of living. It is particularly useful for drawing comparisons. For example, the worldwide average biologically productive area per person—in other words, one individual's ecological footprint—is about 4.5 acres. However, in the United States, ecological footprints average about 24 acres, and in China about 4 acres. You can calculate your own footprint

at http://www.earthday.net/Footprint/index.asp. A similar concept more related to global warming is the *carbon footprint,* which measures the greenhouse gas emissions that a person's activities produce; see http://www.nature.org/initiatives/climatechange/calculator/to calculate your or your family's carbon footprint.

Consequences of Global Climate Change

Greenhouse gases—carbon dioxide, methane, nitrous oxide and other gases—exist naturally in the atmosphere. It is their heat-trapping ability that makes the earth warm enough for life to exist. Increase their levels, as humans are doing now, and the earth becomes warmer still.

Over the last several hundred years, humans have greatly intensified the release of greenhouse gases. Electricity generation, vehicles, industrial processes, and the heating of homes and businesses all burn large amounts of CO_2-producing fossil fuels such as coal, natural gas, and oil. Even the large-scale raising of grazing animals contributes methane through the animals' flatulence. Unusually large quantities of greenhouse gases from these sources are accumulating in the earth's atmosphere, and their presence has already led to significant warming of the planet's climate.

Widespread deforestation worsens the situation. Living forests take in CO_2 from the atmosphere, release the oxygen, and store the carbon in their biomass. As more trees are cut down and their wood used for fuel, more carbon enters the atmosphere, less is taken out, and less is stored. As much as 30 percent of the buildup of CO_2 in the atmosphere results from the loss of trees and other forest organisms. The remaining 70 percent comes from the use of fossil fuels.

Nitrous oxide is perhaps even more damaging to the atmosphere and harder to control than CO_2. It occurs naturally in the soil and plants and is a necessary component of agriculture, but it is now in oversupply as a result of long-term use of nitrogen fertilizers. Nitrous oxide is eroding the earth's natural sun shield, the ozone layer, but it is not easy to reduce human production of nitrous oxide because nitrogen fertilizers, whether organic or chemical, are crucial to food production.

Climatologists and other scientists are documenting long-term global warming and cooling trends by examining evidence in tree rings, fossilized pollen and marine creatures, and glacial ice. These data indicate that the twentieth century was the warmest century in 600 years and the decade of the 1990s was the hottest since the late nineteenth century. Evidence is mounting that these are not normal fluctuations. It is estimated that, at present

> **climate change** a slow shifting of climate patterns due to the general cooling or warming of the atmosphere
>
> **global warming** the predicted warming of the earth's climate as atmospheric levels of greenhouse gases increase
>
> **greenhouse gases** gases, such as carbon dioxide and methane, released into the atmosphere by human and natural actions
>
> **ecological footprint** the biologically productive area per person needed to sustain a particular standard of living

rates of emissions, by 2100 average global temperatures could rise between 2.5°F and 10°F (about 2°C to 5°C).

Although it is not clear just what the consequences of such a rise in temperature will be, it is clear that the effects will not be uniform across the globe. One prediction is that the glaciers and polar ice caps will melt, causing a corresponding rise in sea level. In fact, this phenomenon is already observable (Figure 1.24). Satellite imagery analyzed by scientists at the National Aeronautics and Space Administration (NASA) shows that between 1979 and 2005—just 26 years—the polar ice caps shrank by about 23 percent. The melting released thousands of trillions of gallons of meltwater into the oceans. If this trend continues, at least 60 million people in coastal areas and on low-lying islands could be displaced by rising sea levels. High mountain glaciers also are the chief source of water for many of the world's large rivers. Reduced flow in these rivers will affect millions of people on all continents.

Scientists also forecast a shift of warmer climate zones northward in the Northern Hemisphere and southward in the Southern Hemisphere. This pattern is already observable: robins have been sighted in Alaska, and the range of the mosquito that carries the West Nile virus is spreading to the north and south. Such climate shifts might lead to the displacement of huge numbers of people, because the zones where specific crops can grow would change dramatically. Animal and plant species that cannot adapt rapidly to the change will disappear. Another effect of global warming could be a shift in ocean currents. The result would be more chaotic and severe weather, such as hurricanes, and possible changes in climate for places such as Western Europe should the North Atlantic drift current shift. Photo Essay 1.5 on pages 48–49 shows a map of vulnerability to climate change and photos of the types of problems that are already observable. Notice that the pattern of greatest vulnerability covers the world regions that are the poorest (compare the maps in Photo Essay 1.5 with those in Figure 1.15).

The largest producers of total greenhouse gas emissions (Figure 1.25) are the industrialized countries. The United States produces the most (22.9 percent); China is second (13.2 percent); Russia is third (7.1 percent); Japan is fourth (4.4 percent) and India fifth (4.0 percent). In emissions per capita, the United Arab Republic leads with 56.7 tons of CO_2 per capita, followed by Jamaica, Bahrain, and Paraguay. The United States is eighth at 23 tons per capita. China is far down the list at 3.4 tons per capita.

For the period 1859–1995, developed countries produced roughly 80 percent of the greenhouse gases from industrial sources, and developing countries produced 20 percent. But by 2007, the developing countries were catching up, accounting for nearly 30 percent of total emissions. As developing nations industrialize over the next century and continue to cut down their forests, they will release more and more greenhouse gases every year. If present patterns hold, greenhouse gas contributions by the developing countries will exceed those of the developed world by 2040.

In 1992, an agreement known as the **Kyoto Protocol** was drafted. The protocol called for scheduled reductions in greenhouse gas emissions by the highly industrialized countries of North America, Europe, East Asia, and Oceania. The agreement also encouraged, though it did not require, developing countries to

Kyoto Protocol an amendment to a United Nations treaty on global warming, the Protocol is an international agreement, adopted in 1997 and in force in 2005, that sets binding targets for industrialized countries for the reduction of emissions of greenhouse gases

(A)

(B)

FIGURE 1.24 Effects of global warming: The Muir Glacier (Alaska) in 1941 (top) and 2004 (bottom). Both photos were taken from the same vantage point. Geologist Bruce Molnia, with the U.S. Geological Survey, reports that the glacier retreated 7 miles (12 km) and thinned more than 875 yards (800 m).

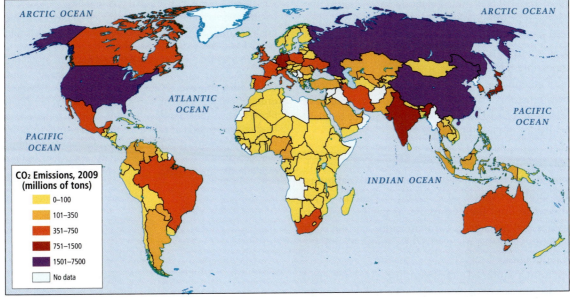

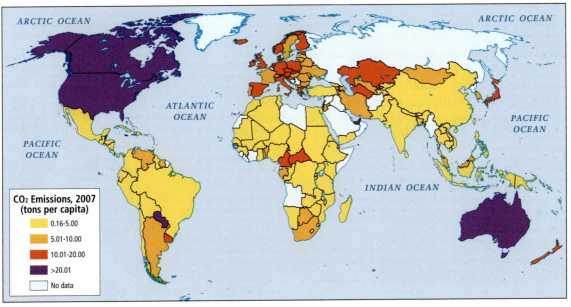

FIGURE 1.25 Greenhouse gas emissions around the world, 2006. (A) Total emissions by millions of tons, 2006. The United States leads the entire world (22.9 percent of the world total), with China second (13.2 percent), Russia third (7.1 percent), Japan fourth (4.4 percent), and India fifth (4.0 percent). These top five countries contribute about 52 percent of the world's greenhouse gas emissions. (B) Tons of emissions per capita. The United Arab Republic is the top per capita emitter at 56.7 tons of CO₂ per capita, followed by Jamaica, Bahrain, and Paraguay. The U.S. is eighth at 23 tons per capita. China is far down on the list at 3.4 tons per capita. [Adapted from the UN Department of Economic and Social Affairs, Statistics Division, 2009, at http://unstats.un.org/unsd/environment/air_greenhouse_emissions.htm.]

curtail their emissions. One hundred eighty-three countries had signed the agreement by 2009. The only developed country that had not signed was the United States, the world's biggest producer of greenhouse gases.

In December 2009 in Copenhagen, 181 countries attempted to sharply limit CO₂ emissions to 350 parts per million molecules (ppmm) of atmosphere by 2020. (In 2009, we were already at 390 ppmm, whereas 200 years earlier, at the beginning of the

industrial revolution in 1809, we were at 275 ppmm.) Negotiations were made difficult when wealthy nations did not offer to curb emissions sufficiently and were reluctant to help less developed countries with the costs of transitioning to lower emissions. The United States offered to cut its 1990 emissions by 4 percent by 2020; but climate scientists said a 40 percent reduction was necessary if the world was to get on track to realize the 350 ppmm goal. Statisticians who analyzed the promises to curb emissions by all of the 181 assembled nations found that rather than reach a reduction to 350 ppmm, the figure by 2100 would be a catastrophic 770 ppmm.

In the end, reaching agreement on specific emissions reductions proved impossible. However, the roughly 30 countries that emit 90 percent of greenhouse gases agreed that they would begin to curb their own emissions, help developing countries achieve clean-energy economies, and give these countries money to adapt to the diverse hardships of climate change.

The so-called *cap and trade* strategies, now practiced in Europe, did not become part of the Copenhagen agreement. Under these strategies, the amount of greenhouse gases that industries can lawfully emit is "capped" at a certain level and companies either find ways to limit their emissions or pay for emission permits. Companies that manage to curb emissions can sell such permits. Most scientists say this market-based scheme would not achieve desired reductions.

Evidence for human-induced global warming is now accepted virtually worldwide. Serious disagreements hinge only on how to respond. In the United States, the scientific community generally sees the currently agreed-upon reductions as far too low. They call for drastically stepped-up energy conservation and for more research into alternatives such as solar, wind, and geothermal energy. However, as research into so-called *green energy* proceeds, there is increasing evidence that all forms of alternative energy have negative impacts of their own. Scientists have invented the term *energy sprawl* to highlight the amount of land area required to produce 1 *terawatt* (1 trillion watt-hours) of energy with different energy sources. One terawatt from coal requires 9.7 square kilometers; from wind power, 72 square kilometers; and from corn-based ethanol, 350 square kilometers. Recent research shows that switchgrass (*Panicum virgatum*), not corn, may be the most efficient biofuel plant, but if we went to any of these alternative sources for a major portion of energy use, clearly the impact on earth's landscapes would be formidable.

Such calculations inevitably lead us back to conservation as the best strategy for avoiding unwelcome impacts. This book will repeatedly visit the factors shaping the vulnerability of regions to global warming.

THINGS TO REMEMBER

1. Planet Earth is continually undergoing climate change, a slow shifting of climate patterns resulting from general cooling or warming of the atmosphere. The present trend of global warming appears to be due primarily to human agency and may be happening more quickly than climate changes in the past.

2. The ecological footprint is the amount of biologically productive land and sea area needed to sustain a human population at its current standard of living. Increasingly, such footprints stretch across the globe.

3. Human activities that emit large amounts of carbon dioxide, methane, and other "greenhouse gases" are trapping heat in the atmosphere, causing what is now widely recognized as global warming. Industrialized countries and those that are rapidly industrializing are responsible for most of these emissions, and it is the poorest countries of the world that are most vulnerable to the changes in climate brought about by global warming.

PHOTO ESSAY | 1.5 Vulnerability to Climate Change GEOGRAPHY PORTAL ANIMATION

This map shows overall human vulnerability to climate change based on a combination of human and environmental factors. Areas shown in darkest red are vulnerable to floods, hurricanes, droughts, sea level rise or other impacts related to climate change. Vulnerability results when a population is exposed to an impact that it is sensitive to and has little resilience towards. For example, many populations are exposed to drought, but generally speaking, the poorest populations are the most sensitive. However sensitivity to drought can be compensated for if adequate relief and recovery systems are in place, such as emergency water and food distribution systems. These systems lend an area a level of resilience that can reduce its overall vulnerability to climate change. Hence, a place's vulnerability to climate change can be thought of as a result of its sensitivity, exposure, and resilience in the face of multiple climate impacts.

F Climate change and hurricanes. Climatologists predict that hurricanes will increase in intensity as the planet warms. Indeed recent decades have seen an increase in powerful storms. Poverty (high sensitivity) and inadequate recovery systems (low resilience) make much of Central America particularly vulnerable to many hurricane-related impacts. Shown here is flood damage along the Choluteca River, Honduras, caused by Hurricane Mitch in 1999. Over 9000 people died in the storm, making Mitch the second most deadly hurricane in history.

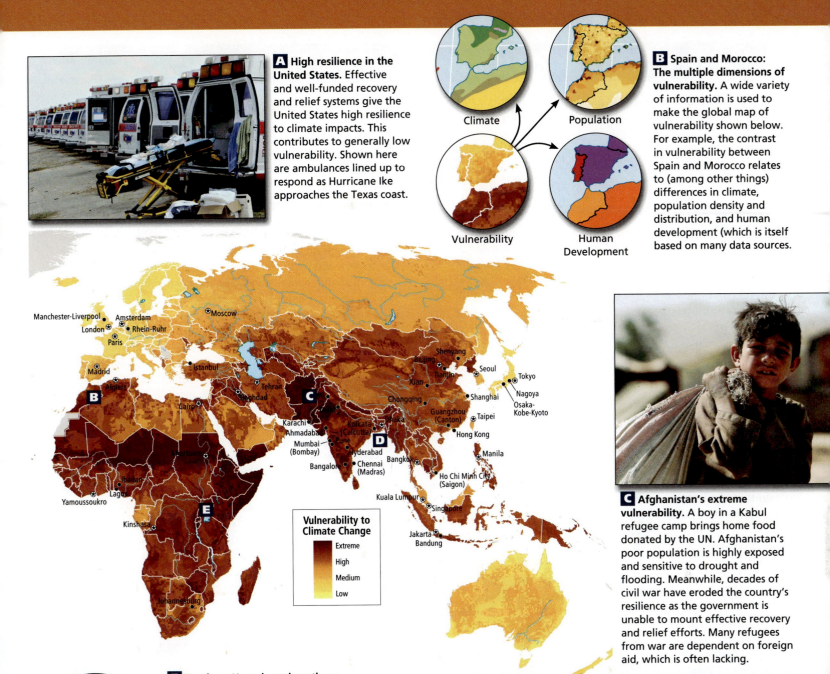

A **High resilience in the United States.** Effective and well-funded recovery and relief systems give the United States high resilience to climate impacts. This contributes to generally low vulnerability. Shown here are ambulances lined up to respond as Hurricane Ike approaches the Texas coast.

Climate

Population

Vulnerability

Human Development

B **Spain and Morocco: The multiple dimensions of vulnerability.** A wide variety of information is used to make the global map of vulnerability shown below. For example, the contrast in vulnerability between Spain and Morocco relates to (among other things) differences in climate, population density and distribution, and human development (which is itself based on many data sources.

Vulnerability to Climate Change

Extreme

High

Medium

Low

C **Afghanistan's extreme vulnerability.** A boy in a Kabul refugee camp brings home food donated by the UN. Afghanistan's poor population is highly exposed and sensitive to drought and flooding. Meanwhile, decades of civil war have eroded the country's resilience as the government is unable to mount effective recovery and relief efforts. Many refugees from war are dependent on foreign aid, which is often lacking.

E **Northern Uganda and southern Sudan.** The situation here is very similar to Afghanistan; however, sensitivity to drought is somewhat lower as access to water is better and poverty is less widespread. Left: Rebels in southern Sudan's civil war have complicated relief efforts. Below: Refugees in northern Uganda pick up bits of donated grain that have been dropped.

D **Bangladesh: Moderate resilience.** Rural Bangladeshis line up for food and water after a hurricane. Advances in government-led disaster recovery have increased Bangladesh's resilience. This has reduced the vulnerability of its poor population, which is highly exposed and sensitive to sea level rise, flooding, hurricanes (cyclones), drought, and other disturbances that climate change could create or intensify.

PHYSICAL GEOGRAPHY PERSPECTIVES

We examine now the two divisions of the discipline of geography—**physical geography** and **human geography**. Physical geographers are concerned with the processes that shape the earth's landforms and its climate. Human geographers are interested in the spatial patterns created by the economic, social, and cultural practices of people. Both groups of geographers are interested in how human practices interact with physical patterns, and their work frequently is complementary. For example, in studying the sources of conflict in Sudan, a physical geographer may look at the landforms and climate and basic resources available and at how these features are allocated geographically. A human geographer may examine the social history of the place and sketch in the role of religion, the ways in which gender is defined, and how power is now allocated. Both will acknowledge that the two perspectives are needed to understand the context of the conflict.

Of particular interest to physical geographers are two components of the physical environment: landforms and climate.

Landforms: The Sculpting of the Earth

The processes that create the world's varied **landforms**—mountain ranges, continents, and the deep ocean floor—are some of the most powerful and slow-moving forces on earth. Originating deep beneath the earth's surface, these internal processes can move entire continents, often taking hundreds of millions of years to do their work. Many of the earth's features, however, such as a beautiful waterfall or a rolling plain, are formed by external processes. These more rapid and delicate processes take place on the surface of the earth. The processes that constantly shape and reshape the earth's surface are studied by geomorphologists.

Plate Tectonics

Two key ideas related to internal processes in physical geography are the *Pangaea hypothesis* and **plate tectonics**. The Pangaea hypothesis was first suggested by geophysicist Alfred Wegener in 1912. It proposes that all the continents were once joined in a single vast continent called Pangaea (meaning "all lands"), which fragmented over time into the continents we know today (**Figure 1.26**). As one piece of evidence for his theory, Wegener pointed to the neat fit between the west coast of Africa and the east coast of South America.

For decades, most scientists rejected Wegener's hypothesis. We now know, however, that the earth's continents have been assembled into supercontinents at least three different times, only to break apart again. All of this activity is made possible by plate tectonics, a process of continental motion discovered in the 1960s, long after Wegener's time.

According to plate tectonics, the earth's surface is composed of large plates that float on top of an underlying layer of molten rock. The plates are of two types. Oceanic plates are dense and relatively thin, and they form the floor beneath the oceans. Continental plates are thicker and less dense. Much of their surface rises above the oceans, forming continents. These massive plates drift slowly, driven by the circulation of the underlying molten rock flowing from hot regions deep inside the earth to cooler surface regions and back. The creeping movement of tectonic plates fragmented and separated Pangaea and created the continents we know today (see Figure 1.26E).

Plate movements influence the shapes of major landforms, such as continental shorelines and mountain ranges. Huge mountains have piled up on the leading edges of the continents as the plates carrying them collided with other plates, folding and warping in the process. Hence, the theory of plate tectonics accounts for the long, linear mountain ranges extending from Alaska to Chile in the Western Hemisphere and from Southeast Asia to the European Alps in the Eastern Hemisphere. The highest mountain range in the world, the Himalayas of South Asia, was created when what is now India, at the northern end of the Indian-Australian Plate, ground into Eurasia. The only continent that lacks these long, linear mountain ranges is Africa. Often called the "plateau continent," Africa is believed to have been at the center of Pangaea and to have moved relatively little since the breakup.

Humans encounter tectonic forces most directly as earthquakes and volcanoes. Plates slipping past each other create the catastrophic shaking of the landscape we know as an earthquake. When plates collide and one slips under the other, this is known as *subduction*. Volcanoes arise at zones of subduction or sometimes in the middle of a plate, where gases and molten rock (called magma) can rise to the earth's surface through fissures and holes in the plate. Volcanoes and earthquakes are particularly common around the edges of the Pacific Ocean, an area known as the **Ring of Fire** (see **Figure 1.27** on page 52).

Landscape Processes

The landforms created by plate tectonics have been further shaped by external processes, which are more familiar to us because we can observe them daily. One such process is **weathering**. Rock, exposed to the onslaught of sun, wind, rain, snow, ice, and the effects of life-forms, fractures and decomposes into tiny pieces. These particles then become subject to another external process, erosion. During erosion, wind and water carry away rock particles and any associated decayed organic matter and deposit them in new locations. The deposition of eroded material can raise and flatten the land around a river, where periodic flooding spreads huge quantities of silt. As small valleys between hills are filled in by silt, a **floodplain** is created. Where rivers meet the sea, floodplains often fan out roughly in the shape of a triangle, creating a **delta**. External processes tend to smooth out the dramatic mountains and valleys created by internal processes.

physical geography the study of the earth's physical processes: how they work, how they affect humans, and how they are affected by humans

human geography the study of various aspects of human life that create the distinctive landscapes and regions of the world

landforms physical features of the earth's surface, such as mountain ranges, river valleys, basins, and cliffs

plate tectonics the scientific theory that the earth's surface is composed of large plates that float on top of an underlying layer of molten rock; the movement and interaction of the plates create many of the large features of the earth's surface, particularly mountains

Ring of Fire the tectonic plate junctures around the edges of the Pacific Ocean; characterized by volcanoes and earthquakes

weathering the physical or chemical decomposition of rocks by sun, rain, snow, ice, and the effects of life-forms

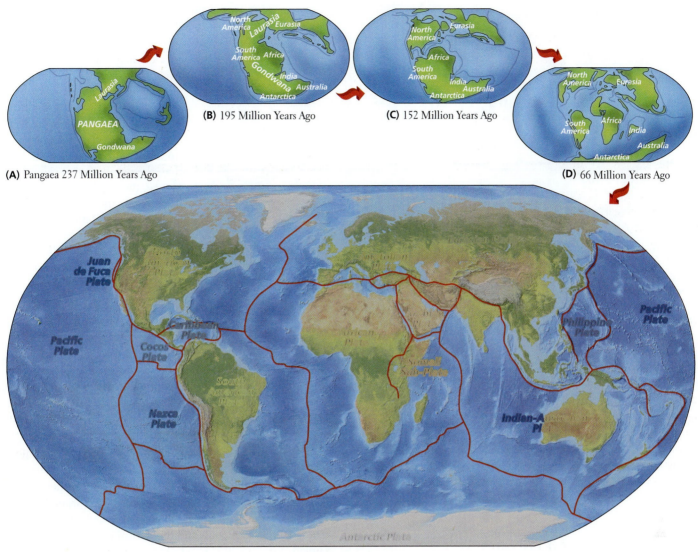

(A) Pangaea 237 Million Years Ago

(B) 195 Million Years Ago

(C) 152 Million Years Ago

(D) 66 Million Years Ago

(E) Modern World

FIGURE 1.26 The breakup of Pangaea. The modern world map **(E)** depicts the current boundaries of the major tectonic plates. Pangaea is only the latest of several global configurations that have coalesced and then fragmented over the last billion years. [Adapted from Frank Press, Raymond Siever, John Grotzinger, and Thomas H. Jordan, *Understanding Earth,* 4th ed. (New York: W. H. Freeman, 2004), pp. 42–43.]

Human activity often contributes to external landscape processes. By altering the vegetative cover, agriculture and forestry expose the earth's surface to sunlight, wind, and rain. These agents in turn increase weathering and erosion. Flooding becomes more common because the removal of vegetation limits the ability of the earth's surface to absorb rainwater. As erosion increases, rivers may fill with silt, and deltas may extend into the oceans.

Climate

The processes associated with climate are generally more rapid than those that shape landforms. Weather, the short-term and spatially limited expression of climate, can change in a matter of minutes. **Climate** is the long-term balance of temperature and precipitation that keeps weather patterns fairly consistent from year to year. By this definition, the last major global climate change took place about 15,000 years ago, when the glaciers of the last ice age began to melt. As we have seen, human activity is producing a new global climate change in our own time.

Energy from the sun gives the earth a temperature range hospitable to life. The earth's atmosphere, oceans, and land surfaces absorb huge amounts of solar energy. The atmosphere traps much of that energy at the earth's surface, insulating the earth from the deep cold of space. Solar energy is also the engine of climate. The most intense, direct sunlight falls in a broad band stretching about 30 degrees north and south of the equator (the tropics). The highest average temperatures on earth occur within this band. Moving away from the equator, sunlight, striking the earth's surface at an angle, is less intense, and average temperatures drop.

floodplain the flat land around a river where sediment is deposited during flooding

delta the triangular-shaped plain of sediment that forms where a river meets the sea

climate the long-term balance of temperature and precipitation that characteristically prevails in a particular region

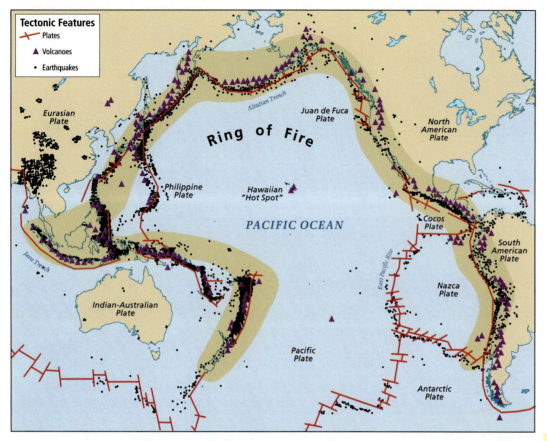

Eurasian Plate

Aleutian Trench

Juan de Fuca Plate

North American Plate

Ring of Fire

Philippine Plate

Hawaiian "Hot Spot"

PACIFIC OCEAN

Cocos Plate

South American Plate

Java Trench

Nazca Plate

Indian-Australian Plate

East Pacific Rise

Pacific Plate

Antarctic Plate

FIGURE 1.27 The Ring of Fire. Volcanic formations encircling the Pacific Basin form the Ring of Fire, a zone of frequent earthquakes and volcanic eruptions. [Adapted from http://vulcan.wr.usgs.gov/Glossary/PlateTectonics/Maps/map_plate_tectonics_world.html; and Frank Press, Raymond Siever, John Grotzinger, and Thomas H. Jordan, *Understanding Earth*, 4th ed. (New York: W. H. Freeman, 2004), p. 27.]

Temperature and Air Pressure

The wind and weather patterns we experience daily are largely a product of complex patterns of air temperature and *air pressure*. To understand air pressure, think of air as existing in a particular unit of space—for example, a column of air above a square foot of the earth's surface. Air pressure is the amount of force (due to the pull of gravity) exerted by that column on that square foot of surface. Air pressure and temperature are related: the gas molecules in warm air are relatively far apart and are associated with low air pressure. In cool air, the gas molecules are relatively close together (dense) and are associated with high air pressure.

As a unit of cool air is warmed by the sun, the molecules move farther apart. The air becomes less dense and exerts less pressure. Air tends to move from areas of higher pressure to areas of lower pressure, creating wind. If you have been to the beach on a hot day, you may have noticed a cool breeze blowing in off the water. Land heats up (and cools down) faster than water, so on a hot day, the air over the land warms, rises, and becomes less dense than the air over the water.

orographic rainfall rainfall produced when a moving moist air mass encounters a mountain range, rises, cools, and releases condensed moisture that falls as rain

monsoon a wind pattern in which in summer months, warm, wet air coming from the ocean brings copious rainfall, and in winter, cool, dry air moves from the continental interior toward the ocean

This causes the cooler, denser air to flow inland. At night the breeze often reverses direction, blowing from the now cooling land onto the now relatively warmer water.

These air movements have a continuous and important influence on global weather patterns. Over the course of a year, continents heat up and cool off much more rapidly than the oceans that surround them. Hence, the wind tends to blow from the ocean to the land during summer and from the land to the ocean during winter. It is almost as if the continents were breathing once a year, inhaling in summer and exhaling in winter.

Precipitation

Perhaps the most tangible way we experience changes in air temperature and density is through rain or snow. Precipitation occurs primarily because warm air holds more moisture than cool air. When this moist air rises to a higher altitude, its temperature drops, which reduces its ability to hold moisture. The moisture condenses into drops to form clouds and may eventually fall as rain or snow.

Several conditions that encourage moisture-laden air to rise influence the pattern of precipitation observed around the globe. When moisture-bearing air is forced to rise as it passes over mountain ranges, the air cools and the moisture condenses to produce rainfall (Figure 1.28). This process, known as **orographic rainfall**, is most common in coastal areas where wind blows moist air from above the ocean onto the land and up the side of a coastal mountain range. Most of the moisture falls as rain as the cooling air rises along the coastal side of the range. On the inland side, the descending air warms and ceases to drop its moisture. The drier side of a mountain range is said to be in the *rain shadow*. Rain shadows may extend for hundreds of miles across the interiors of continents, as they do on the Mexican Plateau, east of California's Pacific coastal ranges, or north of the Himalayas of Eurasia.

Near the equator, moisture-laden tropical air is heated by the strong sunlight and rises to the point where it releases its moisture as rain. This produces the "rain belt" in equatorial areas in Africa, Southeast Asia, and South America. Neighboring nonequatorial areas also receive some of this moisture when seasonally shifting winds blow the rain belt north and south of the equator. The huge downpours of the Asian summer monsoon are an example.

In the **monsoon** season, the Eurasian continental landmass heats up during the summer, causing the overlying air to expand, become less dense, and rise. The somewhat cooler, yet moist, air of the Indian Ocean is drawn inland. The effect is so powerful that the equatorial rain belt is sucked onto the land (see Figure 8.5

[winter and summer monsoons] on page 435). The result is tremendous, sometimes catastrophic, rains throughout virtually all of South and Southeast Asia and much of coastal and interior East Asia. Similar forces pull the equatorial rain belt south during the Southern Hemisphere's summer.

Much of the moisture that falls on North America and Eurasia is *frontal precipitation* caused by the interaction of large air masses of different temperatures and densities. These masses develop when air stays over a particular area long enough to take on the temperature of the land or sea beneath it. Often when we listen to a weather forecast, we hear about warm fronts or cold fronts. A front is the zone where warm and cold air masses come into contact, and it is always named after the air mass whose leading edge is moving into an area. At a front, the warm air tends to rise over the cold air, carrying warm clouds to a higher altitude. Rain or snow may follow. Much of the rain that falls along the outer edges of a hurricane is the result of frontal precipitation.

Climate Regions

Geographers have several systems for classifying the world's climates that are based on the patterns of temperature and precipitation just described. This book uses a modification of the widely known Köppen classification system, which divides the world into several types of climate regions, labeled A, B, C, D, and E on the climate map in Photo Essay 1.6. As you look at the regions on this map, examine the photos, and read the accompanying climate descriptions, the importance of climate to vegetation becomes evident. Each regional chapter includes a climate map; when reading these maps, refer to the verbal descriptions in Photo Essay 1.6, as necessary. Keep in mind that the sharp boundaries shown on climate maps are in reality much more gradual transitions.

THINGS TO REMEMBER

1. Physical geography focuses on the processes that shape the earth's landforms and its climate, and on how human practices interact with physical patterns.

2. The processes that create the world's varied landforms are some of the most powerful and slow-moving forces on earth. The *Pangaea hypothesis* and *plate tectonics* are two key ideas related to the internal processes.

3. Climate is the long-term balance of temperature and precipitation that keeps weather patterns fairly consistent from year to year. Weather is the short-term and spatially limited expression of climate that can change in minutes.

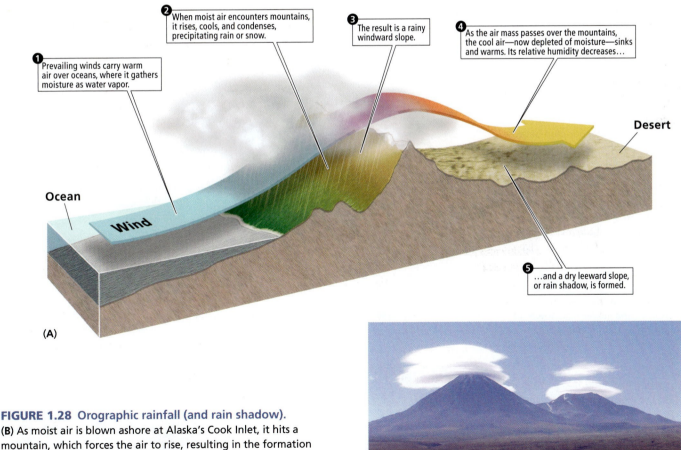

❶ Prevailing winds carry warm air over oceans, where it gathers moisture as water vapor.

❷ When moist air encounters mountains, it rises, cools, and condenses, precipitating rain or snow.

❸ The result is a rainy windward slope.

❹ As the air mass passes over the mountains, the cool air—now depleted of moisture—sinks and warms. Its relative humidity decreases…

❺ …and a dry leeward slope, or rain shadow, is formed.

Ocean

Wind

Desert

(A)

FIGURE 1.28 Orographic rainfall (and rain shadow).
(B) As moist air is blown ashore at Alaska's Cook Inlet, it hits a mountain, which forces the air to rise, resulting in the formation of a cloud. [**(A)** Adapted from Frank Press, Raymond Siever, John Grotzinger, and Thomas H. Jordan, *Understanding Earth,* 4th ed. (New York: W. H. Freeman, 2004), p. 281; **(B)** Jake-4d.]

(B)

(A) Tropical Humid Climates. In *tropical wet climates,* rain falls predictably every afternoon and usually just before dawn. The *tropical wet/dry climate,* also called a *tropical savanna,* has a wider range of temperatures and a wider range of rainfall fluctuation than the tropical wet climate.

(B) Arid and Semiarid Climates. *Deserts* generally receive very little rainfall (2 inches or less per year). Most of that rainfall comes in downpours that are extremely rare and unpredictable. *Steppes* have climates similar to those of deserts, but are more moderate. They usually receive about 10 inches more rain per year than deserts and are covered with grass or scrub.

(C) Temperate Climates. Areas with temperate climates are moist all year and have short, mild winters and long, hot summers. *Subtropical climates* differ from midlatitude climates in that subtropical winters are dry. *Mediterranean climates* have moderate temperatures but are dry in summer and wet in winter.

(D) Cool Humid Climates. Stretching across the broad interiors of Eurasia and North America are *continental climates,* with either dry winters (northeastern Eurasia) or moist all year (North America and north-central Eurasia). Summers in cool humid climates are short but can have very warm days.

(E) Coldest Climates. *Arctic* and *high-altitude climates* are by far the coldest and are also among the driest. Although moisture is present, there is little evaporation because of the low temperatures. The Arctic climate is often called *tundra,* after the low-lying vegetation that covers the ground. The high-altitude version of this climate, which may occur far from the Arctic, is more widespread and subject to greater daily fluctuations in temperature. High-altitude microclimates, such as those in the Andes and the Himalayas, can vary tremendously depending on factors such as available moisture, orientation to the sun, and vegetation cover. As one ascends in altitude, the climate changes loosely mimic those found as one moves from lower to higher latitudes. These changes are known as temperature-altitude zones (see Figure 3.7 on page 145).

Note to reader: Red letter designations on photos (A–J) correspond to the red letters on the map, not with black climate classification letters on the figure legend (A–E).

A Tropical wet, Hawaii

B Tropical wet/dry, Yucatán, Mexico

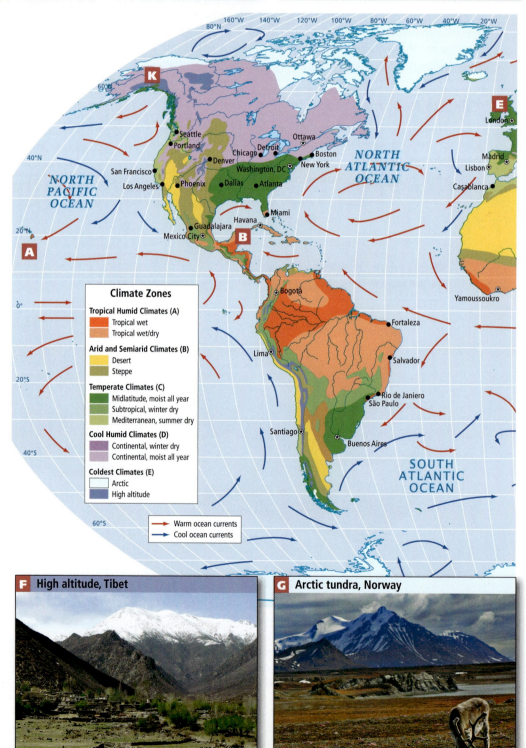

Climate Zones

Tropical Humid Climates (A)
- Tropical wet
- Tropical wet/dry

Arid and Semiarid Climates (B)
- Desert
- Steppe

Temperate Climates (C)
- Midlatitude, moist all year
- Subtropical, winter dry
- Mediterranean, summer dry

Cool Humid Climates (D)
- Continental, winter dry
- Continental, moist all year

Coldest Climates (E)
- Arctic
- High altitude

→ Warm ocean currents
→ Cool ocean currents

F High altitude, Tibet

G Arctic tundra, Norway

C Desert, Namibia

D Steppe, Mongolia

E Midlatitude, moist all year, United Kingdom

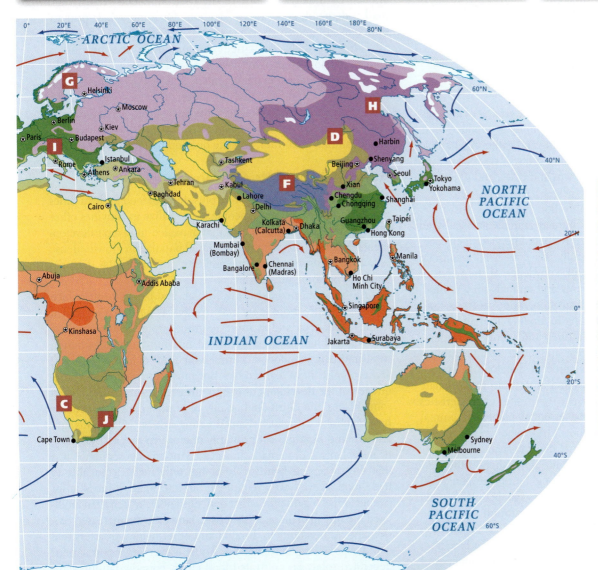

K Continental, moist all year, Alaska

H Continental, winter dry, Russia

I Mediterranean, summer dry, Italy

J Subtropical, winter dry, South Africa

HUMAN GEOGRAPHY PERSPECTIVES

Human geographers are interested in the economic, social, and cultural practices of a people and in the spatial patterns these factors create. Often human geographers need to look to the past in order to understand the present. One way to grasp the importance of modern human impacts on the physical environment, and on climate specifically, is to examine the long-term relationship between humans and the plants and animals they eat. The tending of plants and animals for sustenance—agriculture—provides a compelling illustration of how human interactions with the physical environment can transform not only nature, but also human society and ways of life.

Agriculture: Early Human Impacts on the Physical Environment

Agriculture includes animal husbandry, or the raising of animals, as well as the cultivation of plants. The ability to produce food, as opposed to the dependence on hunting and gathering, has had long-term effects on human population growth, rates of natural resource use, the development of towns and cities, and ultimately the development of civilization.

Where and when did plant cultivation and animal husbandry first develop? Very early humans hunted animals and gathered plants and plant products (seeds, fruits, roots, fibers) for their food, shelter, and clothing. To successfully use these wild resources, humans developed an extensive folk knowledge of the needs of the plants and animals they favored. The transition from hunting and gathering to tending animals in pens and pastures and plants in gardens, orchards, and fields was probably a gradual process that took thousands of years and arose from this intimate familiarity with the characteristics of these essential plant and animal resources. Those who knew the most were more likely to succeed.

Genetic studies support the view that at varying times between 8000 and 20,000 years ago, people in many different places around the globe independently learned to develop especially useful plants and animals through selective breeding, a process known as **domestication.** One of the consequences of domestication was that once plants and animals were selectively bred by humans, they soon became dependent on the humans tending them and could no longer survive on their own. This time of change in the economic base of human society is sometimes known as the **Neolithic Revolution,** a period characterized by the expansion of agriculture and the making of polished stone tools. The map in **Figure 1.29** shows several well-known centers of domestication. A continuing process of agricultural innovation proceeded in many places outside these centers.

agriculture the practice of producing food through animal husbandry, or the raising of animals, and the cultivation of plants

domestication the process of developing plants and animals through selective breeding to live with and be of use to humans

Neolithic Revolution a period 20,000 to 8000 years ago characterized by the transition from hunting and gathering to agriculture, accompanied by the making of polished stone tools; often called the first agricultural revolution

Why did agriculture and animal husbandry develop in the first place? Certainly the desire for more secure food resources played a role, but the opportunity to trade may have been just as important. Many of the known locations of agricultural innovation lie near early trade centers. There, people would have had access to new information and new plants and animals brought by traders, and would have needed products to trade. Perhaps, then, agriculture was at first a profitable hobby for hunters and gatherers that eventually, because of the desire for food security and market demand, grew into a "day job" for some—their primary source of sustenance.

This link between agriculture and trade also provides a glimpse of how cities may have emerged at trading crossroads. People were attracted to these centers, and some developed occupations that served the needs of others who gathered to trade. Increasing density encouraged more specialized services, trade, and yet more settlers. Eventually, discord pointed to the need for organization and control, and the development of government.

Agriculture and Its Consequences

Agriculture made amassing of surplus stores of food for lean times possible and allowed some people to specialize in activities other than food procurement. It also may have led to several developments now regarded as problems: rapid population growth, dense settlement, extreme social inequalities, environmental degradation, and occasionally, famine.

As groups turned to raising animals and plants for their own use or for trade, more labor was needed. Populations expanded to meet this need and as more resources were used to produce food, natural habitats were destroyed, and hunting and gathering were gradually abandoned. Through the study of human remains, archaeologists have learned of a previously unrecognized consequence of the development of agriculture. At some times and in some places, the nutritional quality of human diets may actually have declined as people stopped eating diverse wild plants and animals and began to eat primarily one or two species of domesticated plants and, perhaps, less meat. Another consequence was that the storage of food surpluses not only made it possible to trade food, as we have mentioned, but also made it possible for people to live together in larger concentrations, which then facilitated the spread of disease. Moreover, land clearing increased erosion and vulnerability to drought and other natural disasters that could wipe out an entire harvest. Thus, as ever-larger populations depended solely on cultivated food crops, episodic famine actually became more common.

Cultural Geography

An important component of human geography is the study of culture. Culture is an important distinguishing characteristic of human societies. It comprises everything people use to live on earth that is not directly part of biological inheritance. **Culture** is represented by the ideas, materials, methods, and social

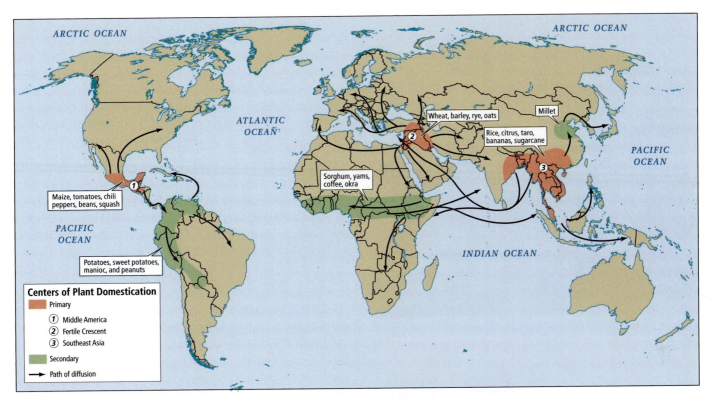

FIGURE 1.29 The origins of agriculture. Scientists have identified six main areas of the world where agriculture emerged. For lengthy periods, people in these different places tended plants and animals and selected for the genetic characteristics they valued. This knowledge eventually spread around the world. Domesticated plants and animals were then further adapted to new locations. This selection and adaptation process continues in the present. [Adapted from Terry G. Jordan-Bychkov, Mona Domosh, Roderick P. Neumann, and Patricia L. Price, *The Human Mosaic*, 10th ed. (New York: W. H. Freeman, 2006), pp. 274–275.]

arrangements that people have invented and passed on to subsequent generations, such as the innovation of agriculture. Culture includes language, music, gender roles, belief systems, and moral codes (for example, those prescribed in Confucianism, Islam, and Christianity).

Ethnicity and Culture: Slippery Concepts

A group of people who share a location, a set of beliefs, a way of life, a technology, and usually a common ancestry and sense of common history form an **ethnic group**. The term *culture group* is often used interchangeably with ethnic group. The concepts of culture and ethnicity are imprecise, however, especially as they are popularly used. For instance, as part of the modern globalization process, migrating people often move well beyond their customary cultural or ethnic boundaries to cities or even distant countries. In these new places they take on new ways of life or even new beliefs, yet they still may identify with their culture of origin.

For example, long before the U.S. war in Iraq, the Kurds in Southwest Asia were asserting their right to create their own country in the territory where they traditionally lived as nomadic herders. (This area is now claimed by Syria, Iraq, Iran, and Turkey; see Figure 6.36 [subregion map] on page 362). Many Kurds who actively support the cause of the herders are now urban dwellers living and working in modern settings in Turkey, Iraq, or even London. Although these people think of themselves as ethnic

Kurds and are so regarded in the larger society, they do not follow the traditional Kurdish way of life (**Figure 1.30**). Hence, we could argue that these urban Kurds have a new identity within the Kurdish culture or ethnic group.

Another problem with the concept of culture is that it is often applied to a very large group that shares only the most general of characteristics. For example, one often hears the terms American culture, African-American culture, or Asian culture. In each case, the group referred to is far too large to share more than a few broad characteristics.

It might fairly be said, for example, that U.S. culture is characterized by beliefs that promote individual rights, autonomy, and individual responsibility. But when we look at specifics, just what constitutes the rights and responsibilities of the individual are quite debatable. In fact, American culture encompasses many subcultures that share some of the core set of beliefs, but disagree over parts of the core and over a host of other matters. The same is true, in varying degrees, for all other regions of the world.

Cultural Markers

Members of a particular culture group may share features, such as language and common values, that help define the group. These shared features are called *cultural markers*.

culture all the ideas, materials, and institutions that people have invented to use that are not directly part of our biological inheritance

ethnic group a group of people who share a set of beliefs, a way of life, a technology, and usually a geographic location

(A)

(B)

FIGURE 1.30 What does it mean to be Kurdish? **(A)** In rural areas away from the war zone in Iraq, being Kurdish often means leading a fairly peaceful, agriculturally based life. Here a young Kurdish woman in Semalka, northern Iraq, brings home fuel for cooking. **(B)** In urban areas, Kurdish identity tends to be more politicized. Here Kurds in Istanbul, Turkey, celebrate their new year holiday of Newroz. According to Kurdish legend, the day marks a deliverance of the Kurds from a tyrant. Now it is a day when Kurds show support for their main political party, the DTP or "Democratic Society Party," which works for greater recognition of the Kurdish language and Kurdish autonomy in southeastern Turkey.

Values

Occasionally you will hear someone say, "After all is said and done, people are all alike," or "People ultimately all want the same thing." It is a heartwarming sentiment, but an oversimplification. True, we all want food, shelter, health, love, and acceptance, but culturally, people are not all alike, and that is one of the qualities that makes the study of geography interesting. We would be wise not to expect

or even to want other people to be like us. It is often more fruitful to look for the reasons behind differences among people than to search hungrily for similarities. Cultural diversity has helped humans to be successful and adaptable animals. The various cultures serve as a bank of possible strategies for responding to the social and physical challenges faced by the human species. The reasons for differences in behavior from one culture to the next are usually complex, but they are often related to differences in values.

Consider this example that contrasts the values and *norms* (accepted patterns of behavior based on values) held by modern urban individualistic culture with those held by rural community-oriented culture. One recent rainy afternoon, a beautiful 40-something Asian woman walked alone down a fashionable street in Honolulu, Hawaii. She wore high-heeled sandals, a flared skirt that showed off her long legs, and a cropped blouse that allowed a glimpse of her slim waistline. She carried a laptop case and a large fashionable handbag. Her long, shiny black hair was tied back. Everyone noticed and admired her because she exemplified an ideal Honolulu businesswoman: beautiful, self-assured, and rich enough to keep herself well-dressed.

In the village of this woman's grandmother—whether it be in Japan, Korea, Taiwan, or rural Hawaii—her clothes would breach a widespread traditional value that no individual should stand out from the group. Furthermore, the dress that exposed her body to open assessment and admiration by strangers of both sexes would signal that she lacked modesty. The fact that she walked alone down a public street—unaccompanied by her father, husband, or female relatives—might even indicate that she was not a respectable woman. Thus a particular behavior may be admired when judged by one set of values and norms but be considered questionable or even disreputable when judged by another.

If culture groups have different sets of values and standards, does that mean that there are no overarching human values or standards? This question increasingly worries geographers, who try to be sensitive both to the particularities of place and to larger issues of human rights. Those who lean too far toward appreciating difference could be led to the tacit acceptance of inhumane behavior, such as the oppression of minorities, domestic violence against women, or even torture and genocide. Acceptance of difference does not preclude judgments about the value of certain extreme customs or points of view. Nonetheless, although it is important to take a stand against cruelty of all sorts, deciding when and where to take that stand is rarely easy.

Religion and Belief Systems

The religions of the world are formal and informal institutions that embody value systems. Most have roots deep in history, and many include a spiritual belief in a higher power (such as God or Allah) as the underpinning for their value systems. Today religions often focus on reinterpreting age-old values for the modern world. Some formal religious institutions—such as Islam, Buddhism, and Christianity—proselytize; that is, they try to extend their influence by seeking converts. Others, such as Judaism and Hinduism, accept converts only reluctantly. Informal religions, often called belief systems, have no formal central doctrine and no firm policy on who may or may not be a practitioner.

Religious beliefs are often reflected in the landscape. For example, settlement patterns often demonstrate the central role of religion in community life: village buildings may be grouped around a mosque or synagogue, or an urban neighborhood may be organized around a Catholic church. In some places, religious rivalry is a major feature of the landscape. Certain spaces may be clearly delineated for the use of one group or another, as in Northern Ireland's Protestant and Catholic neighborhoods.

Religion has also been used to wield power. For example, during the era of European colonization, religion was a way to impose a change of attitude on conquered people. And the influence lingers. Photo Essay 1.7 on page 60 shows the distribution of the major religious traditions on earth today; it demonstrates some of the religious consequences of colonization. Note, for instance, the distribution of Roman Catholicism in parts of the Americas, Africa, and Southeast Asia, all places colonized by European Catholic countries.

Religion can also spread through trade contacts. In the seventh and eighth centuries, Islamic people used a combination of trade and political power (and less often, actual conquest) to extend their influence across North Africa, throughout Central Asia, and eventually into South and Southeast Asia.

The history and distribution of belief patterns throughout the world is complex. The distribution of major religions has changed many times over the course of history. Moreover, any world map is too small in scale to convey detailed religious spatial patterns, such as where two or more religious traditions intersect at the local level. And as the world's cultural traditions become increasingly mixed and urban life grows, *secularism*, a way of life informed by values that do not derive from any one religious tradition, is spreading.

Language

Language is one of the most important criteria used in delineating cultural regions. The modern global pattern of languages (Figure 1.31) reflects the complexities of human interaction and isolation over several hundred thousand years. But the map does not begin to depict the actual details of language distribution. Between 2500 and 3500 languages are spoken on earth today, some by only a few dozen people in isolated places. Many

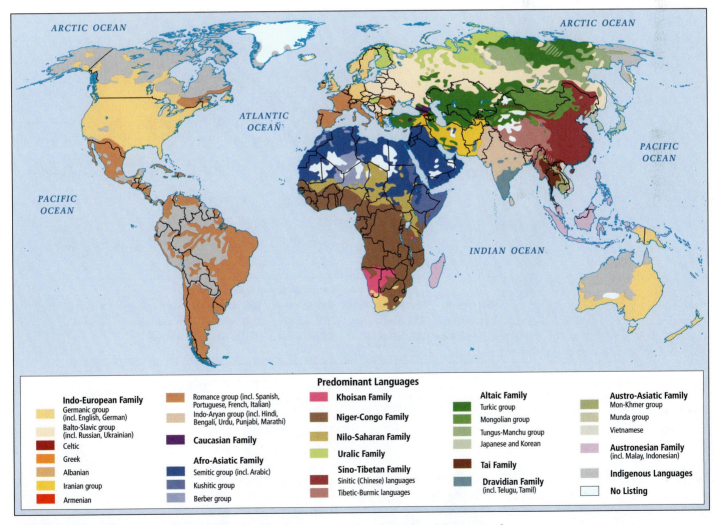

FIGURE 1.31 The world's major language families. Distinct languages (Spanish and Portuguese, for example) are part of a larger language group (Romance languages), which in turn is part of a language family (Indo-European). [Adapted from *Oxford Atlas of the World* (New York: Oxford University Press, 1996), p. 27.]

The small symbols on the map indicate a localized concentration of a particular religion within an area where another religion is predominant.

A Islam, Egypt

B Hinduism, India

C Indigenous religion, New Guinea

Predominant Religions and Belief Systems

- Buddhism
- Hinduism
- Confucianism
- Indigenous religions
- Roman Catholicism
- Orthodox and other Eastern churches
- Protestantism
- Sunni Islam
- Shi'ite Islam
- No listing
- Mixed Christian
- Mormon
- ▲ Roman Catholicism
- ● Protestantism
- ✶ Judaism
- ■ Shintoism
- ◆ Sikhism

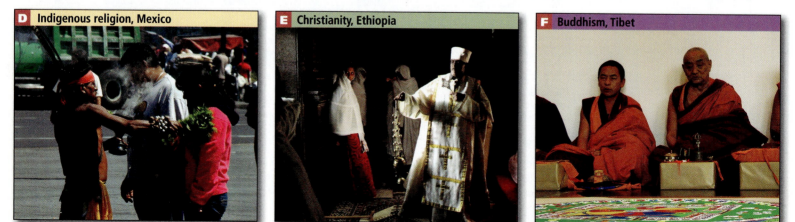

D Indigenous religion, Mexico

E Christianity, Ethiopia

F Buddhism, Tibet

languages have several *dialects*—regional variations in grammar, pronunciation, and vocabulary.

The geographic pattern of languages has continually shifted over time as people have interacted through trade and migration. The pattern changed most dramatically around 1500, when the age of European exploration and colonization began. Beginning at about that time, the languages of European colonists began to replace the languages of the colonized people. For this reason, large patches of English, Spanish, Portuguese, and French occur in the Americas, Africa, Asia, and Oceania. In North America, European languages largely replaced Native American languages. In Middle and South America, Africa, and Asia, by contrast, European and native tongues coexisted. Many people became bilingual or trilingual. Today, with increasing trade and instantaneous global communication, a few languages have become dominant. Arabic is an important language of international trade, as are English, Spanish, Chinese, and Hindi. At the same time, other languages are becoming extinct because children no longer learn them.

English is now the dominant world language, largely because the British colonial empire introduced English as a second language to many places around the globe. U.S. economic influence is another factor that led to the dominance of English, and it is currently the second language of an estimated 1.5 billion people. The need for a common world language in the computerized information age is reinforcing English as the primary *lingua franca*, or language used by people of different languages.

Material Culture and Technology

Material culture comprises all the things that people make and use: clothing, houses and office buildings, axes, guns, computers, earthmoving equipment (from hoes to work animals to bulldozers), books, musical instruments, domesticated plants and animals, agricultural and food-processing equipment—the list is virtually endless. A group's *material culture* reflects its technology, which is the integrated system of knowledge, skills, tools, and methods upon which a culture group bases its way of life.

Housing is a good example of how material culture reveals a particular culture group's way of life. With its distinctive architecture, electrical and plumbing systems, and landscaping, the typical North American suburban ranch house silently reveals a great deal about the culture's values. It reflects the nuclear family structure (mother, father, children) that remains an ideal in North American society, although it now constitutes less than 25 percent of the region's families. The ranch house also reflects values about privacy (multiple bedrooms) and gender roles (Mom's special spaces may be the kitchen and laundry room; Dad's, the TV room, the garage, and the toolshed). This house embodies a certain level of affluence and leisure, equipped as it is with labor-saving devices and conveniences and set apart by a green lawn that requires constant maintenance. It also symbolizes ideas about private property, polite neighborliness, social mobility, and transportation (Figure 1.32).

Race

Like ideas about gender roles, ideas about race affect human relationships everywhere on earth. However, according to the science of biology, all people now alive on earth are members of one species, *Homo sapiens sapiens*. Biologically, race is a meaningless concept; the characteristics we popularly identify as **race** markers—skin color, hair texture, face, and body shapes—have no significance as biological categories. For any supposed *racial trait*, such as skin color, there are wide variations within human groups. In addition, many invisible biological characteristics, such as blood type and DNA patterns, cut across skin color distributions and other so-called *racial attributes* and are shared across what are commonly viewed as different races. In fact, over the last several thousand years there has been such massive gene flow among human populations that no modern group presents a discrete set of biological characteristics. Although we may look quite different, from the biological point of view we are all closely related.

> **material culture** all the things, living or not, that humans use
>
> **race** a social or political construct that is based on apparent characteristics such as skin color, hair texture, and face and body shape, but that is of no biological significance

(A) (B)

FIGURE 1.32 Ranch houses. House types can show remarkable uniformity even across great distances when cultural patterns remain consistent. **(A)** A ranch house with a two-car garage opening onto the street in California. **(B)** A similar house in Florida.

multiculturalism the state of relating to, reflecting on, or being adapted to diverse cultures

It is likely that some of the easily visible features of particular human groups evolved to help them adapt to environmental conditions. For example, biologists have shown that darker skin evolved in regions close to the equator, where sunlight is most intense (Figure 1.33). All humans need the nutrient vitamin D, and sunlight striking the skin helps the body absorb vitamin D. Too much of the vitamin, however, can result in improper kidney functioning. Dark skin absorbs less vitamin D than light skin and thus would be a protective adaptation in equatorial zones. In higher latitudes, where the sun's rays are more dispersed, light skin facilitates the sufficient absorption of vitamin D. Similar correlations have been observed between skin color, sunlight, and another essential vitamin, folate, which if deficient can result in birth defects. For an excellent lecture on skin color by geneticist Nina Jablonski, see: http://www.ted.com/talks/lang/eng/nina_jablonski_breaks_the_illusion_of_skin_color.html.

Many physical characteristics, such as big ears, deep-set eyes, or high cheekbones, do not serve any apparent adaptive purpose. They are probably the result of random chance and ancient inbreeding within isolated groups. Similarly, there is no evidence that any "race" has particularly strong ability in any activity, such as high math ability or athletic ability, for example. Such characteristics may seem to be concentrated in particular groups, but in fact they are present in individuals in all human populations. When these characteristics are particularly strong in a certain group, this is the result of enhancement by complex cultural practices.

While race is biologically meaningless, it has acquired enormous social and political significance. Over the last several thousand years, humans from different parts of the world have increasingly encountered each other in situations of unequal power. Some researchers have suggested that European colonizers adopted *racism*—the negative assessment of unfamiliar, often darker-skinned, people—to justify taking land and resources away from those supposedly inferior beings. But racism is hardly peculiar to Europe. Humans have long committed atrocities against their own kind, often in the name of race, ethnicity, or even gender. Race and its implications in North America will be covered in Chapter 2, and the topic will be discussed in several other world regions as well.

Recognizing all the ills that have emerged from racism and similar prejudices, we need not infer that human history has been marked primarily by conflict and exploitation. Actually, humans have probably been so successful because of a strong inclination toward *altruism*, the willingness to sacrifice one's own well-being for the sake of others. On a small scale, altruism can be found in the sacrifices individuals make to help their family, neighbors, and community. On a larger scale, it includes charitable giving to help anonymous people in need. It is probably our capacity for altruism that causes us such deep distress over the relatively infrequent occurrences of inhumane behavior.

Globalization and Cultural Change

Some indications suggest that the diversity of culture is fading as trends and fads circle the globe via the instant communication now available. American fast food, popular music, and clothing styles can now be found from Mongolia to Mozambique. At the same time, a wide variety of ethnic music, textiles, cuisines, and dress from distant places now graces the lives of consumers in the United States and across the world. As globalization proceeds, people migrate and ideas spread. Inevitably, some measure of cultural homogeneity will occur, resulting in more overall similarity between culture groups.

Are we all drifting toward a common material culture, and perhaps even similar ways of thinking? Possibly, but there are also countervailing trends. It is now possible for people to reinforce their feeling of cultural identity with a particular group through the same channels that are encouraging homogenization. Consider the example of the people of Aceh in Southeast Asia.

Aceh is a province within the country of Indonesia, but it has a distinctive cultural identity. The people of Aceh desire greater autonomy—the right to control their own affairs and especially to retain control of their own resources. However, the central Indonesian government views all resources as national, not local, and in 2003 the government sent military troops to enforce its interests, with fatal consequences to the Acehese.

To build awareness of their plight and a sense of identity among Acehese migrants worldwide, a few Acehese established several Web sites and Internet chat groups. Then, after a giant tsunami devastated Aceh in December 2004, the cultural solidarity already built

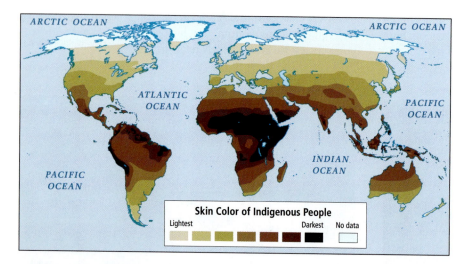

FIGURE 1.33 Skin color map for indigenous people as predicted from multiple environmental factors. Skin plays a twin role with respect to the sun: protection from excessive UV radiation and absorption of enough sunlight to trigger the production of vitamin D. [Map courtesy of UNEP/GRID-Arendal at http://maps.grida.no/go/graphic/skin-colour-map-indigenous-people, Emmanuelle Bournay, cartographer. Data source: G. Chaplin, *Geographic Distribution of Environmental Factors Influencing Human Skin Coloration*, American Journal of Physical Anthropology 125:292–302, 2004; map updated in 2007.]

through the Internet helped raise funds efficiently and distribute assistance to victims. Hence the Internet has helped maintain and strengthen the ties that bind the Acehese together, especially during times of adversity.

The ability to communicate easily over the Internet and to travel quickly can reaffirm cultural identity, but these abilities also enhance the conditions necessary for **multiculturalism,** the state of relating to, reflecting on, or being adapted to several cultures. For example, the young Acehese Webmaster actually lives in the United States, where he earns a living painting scenes from his tropical homeland and also of the mansions and portraits of U.S. corporate executives.

THINGS TO REMEMBER

1. Human geography centers on the spatial patterns created by the economic, social, and cultural practices of people and their interactions with physical patterns. Human geographers often look to the past in order to understand the present.

2. Components of human geography include agriculture, past and present; ethnicity and race; religion and belief systems; value systems; language; material culture and technology; population patterns; economic and political systems; globalization and change in all these categories.

Critical Thinking Questions

1. Some people argue that it is acceptable for people in the United States to consume at high levels because their consumerism keeps the world economy going. What are the weaknesses in this idea?

2. What are the causes of the huge increases in migration, legal and illegal, that have taken place over the last 25 years?

3. What would happen in the global marketplace if all people had a living wage?

4. As people live longer and decide not to raise large families, how can they beneficially spend the last 20 to 30 years of their lives?

5. Given the threats posed by global warming, what are the most important steps to take now?

6. As you read about differing ways of life, values, and perspectives on the world, reflect on the appropriateness of force as a way of resolving conflicts.

7. What are some possible careers that would address one or more of the issues raised in particular chapters?

8. Reflect on the reasons why some people have much and others have little.

9. What would be some of the disadvantages and advantages of abolishing gender roles in any given culture?

10. Consider the ways that access to the Internet enhances prospects for world peace and the ways that it contributes to discord.

Chapter Key Terms

agriculture 56
authoritarianism 38
biosphere 27
birth rate 16
capitalism 40
carrying capacity 28
cash economy 20
civil society 40
climate change 45
climate 51
communism 40
culture 57
death rate 18
delta 51
democratization 38
demographic transition 20
development 24
domestication 56
ecological footprint 45
erosion 31
ethnically cleanse 40

ethnic group 57
fair trade 37
female earned income as a percent of male earned income (F/MEI) 25
floodplain 51
food security 28
formal economy 25
free trade 36
gender 21
gender roles 22
genetic modification (GM) 30
genocide 40
Geographic Information Science 7
geopolitics 40
global economy 34
global scale 9
global warming 45
globalization 32
green revolution 28

greenhouse gases 45
gross domestic product (GDP) 19
gross domestic product (GDP) per capita PPP 25
human geography 50
human well-being 25
informal economy 25
interregional linkages 32
Kyoto Protocol 46
landforms 50
lines of latitude 6
lines of longitude 3
living wages 37
local scale 10
map projections 6
material culture 61
migration 18
monsoon 52
multiculturalism 62

multinational corporations 63
Neolithic Revolution 56
nongovernmental organization (NGO) 40
orographic rainfall 52
physical geography 50
plate tectonics 50
political ecologist 27
population pyramid 18
purchasing power parity (PPP) 25
push/pull phenomenon of urbanization 31
race 61
rate of natural increase (RNI) 16
region 7
Ring of Fire 50
scale (of a map) 3
sex 21

slums 31
subregions 10
subsistence economy 20
sustainable agriculture 28
sustainable development 27
total fertility rate (TFR) 18
United Nations (UN) 40
United Nations Human Development Index (HDI) 25
urbanization 31
virtual water 42
water footprint 42
weathering 50
world region 9
World Trade Organization (WTO) 36

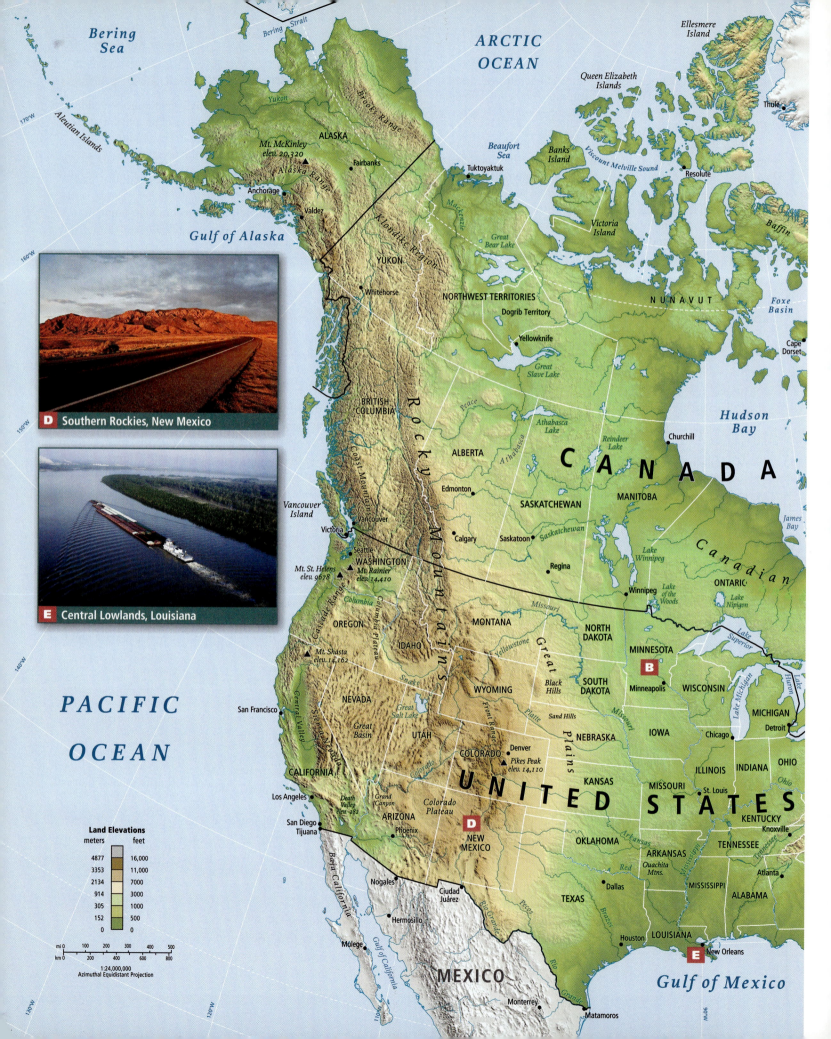

Bering Sea

ARCTIC OCEAN

Bering Strait

Ellesmere Island

Thule

Aleutian Islands

170°W

Yukon

Brooks Range

ALASKA

Mt. McKinley
elev. 20,320

Fairbanks

Alaska Range

Klondike Region

Anchorage

Valdez

Gulf of Alaska

160°W

150°W

Tuktoyaktuk

Beaufort Sea

Banks Island

Viscount Melville Sound

Victoria Island

Resolute

Mackenzie

Great Bear Lake

YUKON

Whitehorse

NORTHWEST TERRITORIES

Dogrib Territory

Yellowknife

Great Slave Lake

Peace

Athabasca

Athabasca Lake

BRITISH COLUMBIA

ALBERTA

Edmonton

Calgary

SASKATCHEWAN

Saskatoon

Saskatchewan

Regina

MANITOBA

Reindeer Lake

Lake Winnipeg

Winnipeg

Lake of the Woods

Queen Elizabeth Islands

Baffin

NUNAVUT

Foxe Basin

Cape Dorset

C A N A D A

Hudson Bay

James Bay

Canadian

ONTARIO

Lake Nipigon

Lake Superior

Churchill

Vancouver Island

Victoria

Vancouver

Seattle

Mt. St. Helens
elev. 9678

Mt Rainier
elev. 14,410

WASHINGTON

OREGON

Coast Ranges

Cascade Range

Columbia

Columbia Plateau

Mt. Shasta
elev. 14,162

Snake

IDAHO

MONTANA

Yellowstone

WYOMING

Missouri

NORTH DAKOTA

SOUTH DAKOTA

Black Hills

MINNESOTA

Minneapolis

B

WISCONSIN

Missouri

Chicago

Detroit

Lake Michigan

Lake Huron

MICHIGAN

PACIFIC OCEAN

140°W

130°W

San Francisco

Sierra Nevada

Central Valley

NEVADA

Great Basin

Great Salt Lake

UTAH

CALIFORNIA

Los Angeles

Death Valley elev. -282

San Diego

Tijuana

Grand Canyon

ARIZONA

Phoenix

Colorado

Colorado Plateau

NEW MEXICO

D

Nogales

COLORADO

Denver

Pikes Peak
elev. 14,110

Front Range

Platte

Great Plains

NEBRASKA

Sand Hills

KANSAS

IOWA

MISSOURI

St. Louis

ILLINOIS

INDIANA

OHIO

Ohio

KENTUCKY

Knoxville

TENNESSEE

U N I T E D   S T A T E S

OKLAHOMA

Arkansas

ARKANSAS

Ouachita Mtns.

Red

MISSISSIPPI

ALABAMA

Atlanta

Baja California

Ciudad Juárez

Rio Grande

Pecos

TEXAS

Dallas

Brazos

Houston

LOUISIANA

E

New Orleans

Mississippi

Nogales

Hermosillo

Gulf of California

Mulege

MEXICO

Monterrey

Matamoros

Rio Grande

Gulf of Mexico

Land Elevations

meters	feet
4877	16,000
3353	11,000
2134	7000
914	3000
305	1000
152	500
0	0

mi 0   100   200   300   400   500

km 0   200   400   600   800

1:24,000,000
Azimuthal Equidistant Projection

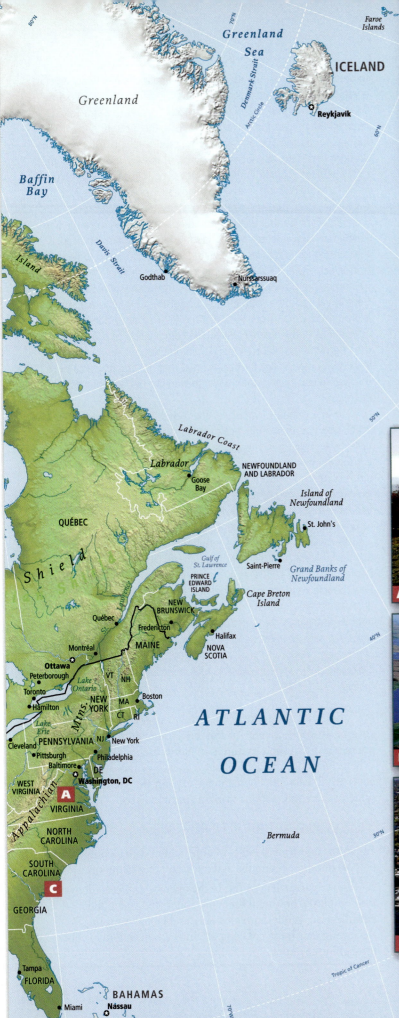

THEMATIC CONCEPTS

Population • Gender • Development
Food • Urbanization • Globalization
Democratization • Climate Change • Water

North America

A Appalachian Mountains, Virginia

B Central Lowlands, Minnesota

C Coastal Lowlands, Charleston, South Carolina

FIGURE 2.1 Regional map of North America.

North America (Figure 2.1) is one of the largest and wealthiest regions of the world; it is also a region that encompasses many environments and a complex array of local cultures and economic activities that are related to each other across wide distances.

Global Patterns, Local Lives
Javier Aguilar, a 39-year-old father of three, has worked as an agricultural laborer in California's Central Valley for 20 years. He and hundreds of thousands like him tend the fields of crops, such as tomatoes, artichokes, onions, garlic, lettuce, broccoli, avocados, almonds, olives, and grapes (Figure 2.2) that feed the nation, especially during the winter months. Aguilar used to make $8.00 an hour, but now he is unemployed and standing in a church-sponsored food line. "If I don't work, [we] don't live. And all the work is gone here," he says, grimly.

In Mendota, also in California's Central Valley and once the "Cantaloupe Center of the World," young Hispanic men wait on street corners to catch a van to the fields. None come. In the winter of 2009, the unemployment rate in Mendota was 35 percent and rising. Mayor Robert Silva says his community is dying on the vine, and he sees the trouble spreading. Many small businesses are closing. Silva worries about the drug use, alcohol abuse, family violence, and malnutrition that can accompany severe unemployment in any community.

This level of unemployment in the heart of the nation's biggest producer of fruit and vegetables is the product of six factors: a naturally arid environment, much of which is farmable only with irrigation; a 3-year, wide-ranging drought extending east into the Great Plains; mounting competition for the scarce water normally brought into the region in canals for crop irrigation; and an increasing national and global recession. A fifth contributing factor is the global energy crisis that increases the costs of production and transport. The sixth factor is the "eat local" movement that is sweeping the nation—the aim is to eat organically grown food from one's own locality and thereby improve local economies and reduce carbon emissions from transcontinental shipping.

The drought is in its third year. Rain falls occasionally but officials say there is a growing total deficit of moisture, probably related to natural dry cycles, but also to global climate change, which scientists find is worsening drought conditions globally. Even before the economic recession and the "eat local" movement struck and reduced the market for California produce, the drought had forced farmers to remove 100,000 acres from production. In 2009, with the drought persisting, as much as 1 million of the 4.7 million acres once cultivated and irrigated in the Central Valley will lie fallow. Eventually, as many as 80,000 jobs and as much as $2.2 billion may be lost in California agriculture and related industries.

Water for Central Valley agriculture is brought in from the San Joaquin and Sacramento rivers through a series of dams and canals that stretch for 500 miles. This water, which is also used to produce hydroelectricity in 11 power plants, must serve much of northern California, the growing cities in the San Francisco Bay Area, and the Central Valley. In the southern reaches of the Central Valley, water is brought in canals from the Colorado River to serve domestic, agricultural, and industrial needs. The combination of drought and urban growth has placed added strain on this complex system, and in 2009, Central Valley farmers learned that, for the first time since the system was built in the 1930s, they might get no water for irrigation.

David Leonhardt, writing for the *New York Times* in March 2009, said that the confluence of troubles in California's Central Valley was likely to hurt most the low-wage Hispanic male agricultural workers because they have so little to fall back on in terms of savings, education, or skills. Hispanic women are more likely to remain employed because they are in domestic and caregiving jobs and have somewhat higher levels of education. However, according to Leonhardt, over the long term, unskilled men like Aguilar may actually benefit from this recession, if the Obama Administration's plan to facilitate adult education is fulfilled. During the Great Depression of the 1930s, as a result of President Roosevelt's New Deal, people went back to school and the nation's high school graduation rates jumped from 20 percent to 60 percent. *[Adapted from: Jesse McKinley, "Drought Adds to Hardships in California," New York Times, February 21, 2009; David Leonhardt, "Job Losses Show Breadth of Recession," New York Times, March 3, 2009; and Catherine Rampell, "As Layoffs Surge, Women May Pass Men in Job Force," New York Times, February 5, 2009.]* ■

The nine Thematic Concepts on pages 68–69, which are covered in each chapter of this book, are illustrated here with photo essays that show examples in North America of each theme. Where relevant, interactions between two or more of these concepts are illustrated. The captions explain the pictures and how some of these interactions work in specific places.

THINGS TO REMEMBER

1. North America is a large, wealthy region with a complex array of local cultures, economic activities, and environments.

2. The story of Javier Aguilar in the opening vignette illustrates the complexity of several thematic concepts—water, food, globalization, gender, urbanization, and global warming. Reviewing how these concepts are linked here will help you throughout the book.

Figure 2.2 Grape harvest in California. Soter Mineral Springs Vineyard harvest in 2008. [Craig Camp.]

I THE GEOGRAPHIC SETTING

Terms in This Chapter

The world region discussed in this chapter consists of Canada and the United States (see Figure 2.1). The term *North America* is used to refer to both countries. *Native American* is the term used for all aboriginal Americans, whether now in Canada or the United States (all those whose ancestors were in the Americas when Europeans and other groups from the so-called Old World arrived). Other terms relate to the growing cultural diversity in this region. The text uses the term **Hispanic** to refer to all Spanish-speaking people from Middle and South America, although their ancestors may have been African, European, Asian, or Native American. When writing about the Southwest, where Latino is the preferred term, it is used instead of Hispanic. In Canada, the **Québecois**, the French Canadians living in Québec, are an ethnic group that is distinct from the rest of Canada. They are the largest of an increasingly complex mix of minorities in that country, most of whom are still content simply to be called Canadians.

Physical Patterns

The continent of North America is a huge expanse of mountain peaks, ridges and valleys, expansive plains, long winding rivers, myriad lakes, and extraordinarily long coastlines. Here the focus is on a few of the most significant landforms.

Landforms

A wide mass of mountains and basins, known as the Rocky Mountain zone, dominates western North America (see Figure 2.1D). It stretches down from the Bering Strait in the far north, through Alaska, and into Mexico. This zone formed about 200 million years ago when, as part of the breakup of the supercontinent Pangaea (see Figure 1.26 on page 51), the Pacific Plate pushed against the North American Plate, thrusting up mountains. These plates still rub against each other, causing earthquakes along the Pacific coast of North America.

The much older, and hence more eroded, Appalachian Mountains stretch along the eastern edge of North America from New Brunswick and Maine to Georgia (see Figure 2.1A). This range resulted from very ancient collisions between the North American Plate and the African Plate.

Between these two mountain ranges lies a huge central lowland of undulating plains that stretches from the Arctic to the Gulf of Mexico (see Figure 2.1B and E). This landform was created by the deposition of deep layers of material eroded from the mountains and carried to this central North American region by wind and rain and by the rivers flowing east and west into what is now the Mississippi drainage basin.

During periodic ice ages over the last 2 million years, glaciers have covered the northern portion of North America. In the most recent ice age (between 25,000 and 10,000 years ago), the glaciers, sometimes as much as 2 miles (about 3 kilometers) thick, moved south from the Arctic, picking up soil and scouring

depressions in the land surface. When the glaciers later melted, these depressions filled with water, forming the Great Lakes. Thousands of smaller lakes, ponds, and wetlands that stretch from Minnesota and Manitoba to the Atlantic were formed in the same way. Melting glaciers also dumped huge quantities of soil throughout the central United States. This soil, often many meters deep, provides the basis for large-scale agriculture, but remains susceptible to wind and water erosion.

East of the Appalachians, a coastal lowland stretches from New Brunswick to Florida (see Figure 2.1C). This lowland then sweeps west to the southern reaches of the central lowland along the Gulf of Mexico. In Louisiana and Mississippi, much of this lowland is filled in by the Mississippi river delta, a low, flat, swampy transition zone between land and sea. The delta was formed by massive loads of silt deposited during floods over the past 150-plus million years by the Mississippi, North America's largest river system. The delta deposit originally began at what is now the junction of the Mississippi and Ohio rivers at Cairo, Illinois; slowly, as ever more sediment was deposited, the delta advanced 1000 miles (1600 kilometers) into the Gulf of Mexico.

Over the centuries, human activities such as deforestation, deep plowing, and heavy grazing have led to erosion and added to the silt load of the rivers (Figure 2.3 on page 70). At the same time, the construction of levees along riverbanks has drastically reduced flooding. Because of this flood control, much of the silt that used to be spread widely across the lowlands during floods is being carried to the southern part of the Mississippi delta—a low, flat zone characterized by swamps, lagoons, and sandbars. The silt is destroying wetlands and the extra weight is causing the delta to sink into the Gulf of Mexico as the silt also extends farther out into deep waters.

Climate

The landforms over this continental expanse influence the movement and interaction of air masses and contribute to its enormous climate variety (Photo Essay 2.1 on page 71). Along the southern West Coast of North America, the climate is generally mild (Mediterranean)—dry and warm in summer, cool and moist in winter. North of San Francisco, the coast receives moderate to heavy rainfall. East of the Pacific coastal mountains, climates are much drier because as the moist air sinks into the warmer interior lowlands, it tends to hold its moisture (see Figure 1.28 on page 53). This interior region is increasingly arid as it stretches east across the Great Basin and Rocky Mountains. Many dams and reservoirs for irrigation projects have been built to make agriculture and urbanization possible. Because of the low level of rainfall, however, efforts to extract water for agriculture and urban settlements are exceeding the capacity of ancient underground water basins (**aquifers**) to replenish themselves.

Hispanic term used to refer to all Spanish-speaking people from Middle and South America, although their ancestors may have been black, white, Asian, or Native American

Québecois French Canadians living in Québec; an ethnic group distinct from the rest of Canada, they all are citizens of Canada

aquifers natural underground reservoirs

Food and Water: North American farms are highly mechanized, chemically intensive and extremely productive. But they also cause soil erosion and, in the drier parts of the region, their dependence on irrigation makes them heavy users of scarce water. Bodies of water are also being polluted by fertilizers and pesticides that wash off North American farms when it rains.

(A) A combine harvests wheat in Minnesota.

(B) Severe soil erosion on a farm in Iowa.

(C) An irrigated lettuce field in Arizona. Fertilizer makes the water appear white in places.

Global Warming and Urbanization: North America produces the most greenhouse gases of any world region, due in part to its spread-out pattern of urbanization. Widely dispersed urban development forces people to drive long distances, and the dominant housing type, single-family detached dwellings, uses large amounts of energy for heating and cooling due to poor insulation.

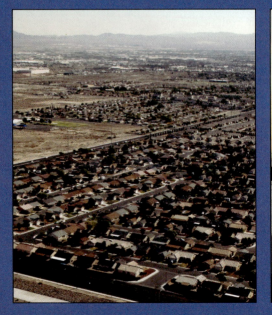

(D) The suburbs of Las Vegas in desert Nevada: the city was North America's fasting growing urban area until the recession of 2008.

(E) Icicles form on the edge of a poorly insulated roof in Pennsylvania.

(F) A traffic jam in Southern California.

Population and Gender: The North American population is aging both because people are living longer and because women are having fewer children. Most women now work outside the home, and frequently delay childbearing until later in life, resulting in declining birth rates. Hence, an increasingly elderly population is now dependent on an ever-shrinking base of working-age people.

(G) A female Seattle firefighter in training.

(H) A single-child family in California.

(I) An elderly man in a Gainesville, Florida, hospital.

Democratization: North America has a long history of democratic government, and saw two major expansions of democracy during the nineteenth and twentieth centuries. Women gained the right to vote in 1884 in Canada and 1920 in the U.S., although women remain underrepresented in the political systems of both countries. In 1965, the Supreme Court ruled that the practices that had effectively denied African-Americans the right to vote in much of the southern U.S. were discriminatory, and barred them. The U.S. has long advocated for democratization throughout the world, although other U.S. interests have often taken priority in its foreign policy.

(J) Women protesting for the right to vote in Washington, D.C., in 1917.

(K) The 1963 March on Washington, a key moment in the African-American civil rights movement.

(L) An Iraqi soldier votes in his country's national elections in 2005.

Globalization and Development: North America is a strong advocate of globalization, especially through the free trade agreement known as NAFTA. Globalization has transformed economic development in North America, with many labor-intensive industries moving abroad to take advantage of cheaper labor. While North America's economic focus has shifted to highly skilled and service-sector jobs, many remaining labor-intensive industries, such as construction and certain types of agriculture, draw immigrants, both legal and illegal, who are willing to work hard for lower wages than American workers.

(M) An abandoned textile facility in New Jersey, emblematic of jobs lost to foreign competition.

(N) A radiologist reviews data at a hospital in the United States. Radiology is typical of North America's highly skilled jobs.

(O) Cuban immigrants work at a construction site in the United States.

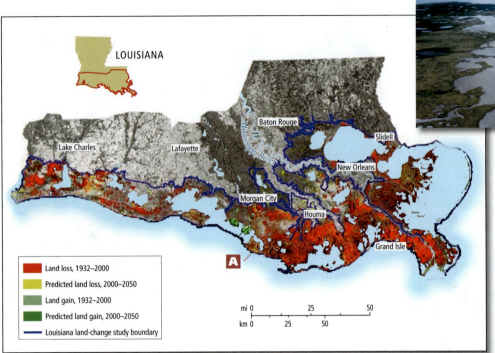

LOUISIANA

Lake Charles · Lafayette · Baton Rouge · Slidell · New Orleans · Morgan City · Houma · Grand Isle · Mississippi River

■ Land loss, 1932–2000
■ Predicted land loss, 2000–2050
■ Land gain, 1932–2000
■ Predicted land gain, 2000–2050
— Louisiana land-change study boundary

mi 0 · 25 · 50
km 0 · 25 · 50

Figure 2.3 Wetland loss in Louisiana. The Louisiana coastline and the lower Mississippi River basin are vital to the nation's interests. They are the end point for the vast Mississippi drainage basin. They provide coastal wildlife habitats, recreational opportunities, transportation lanes that connect the vast interior of the country with the ocean, and access to offshore oil and gas (see pages 72–73). Most important, the wetlands provide a buffer against damage from hurricanes. However, Louisiana has lost one-quarter of its total wetlands over the last century, largely due to human impacts on natural systems. The remaining 3.67 million acres constitute 14 percent of the total wetland area in the lower 48 states. **(A)** A section of vanishing wetland in Terrebonne Parish, Louisiana. [Map: USGS/National Wetlands Research Center.]

On the eastern side of the Rocky Mountains, the main source of moisture is the Gulf of Mexico. When the continent is warming in the spring and summer, the air masses above it rise, sucking in warm, moist, buoyant air from the Gulf. This air interacts with cooler, drier, heavier air masses moving into the central lowland from the north Photo Essay 2.1) and west, often creating violent thunderstorms and tornadoes. Generally, central North America is wettest in the eastern (Photo Essay 2.1B) and southern parts and driest in the north and west (Photo Essay 2.1C). Along the Atlantic coast, moisture is supplied by warm, wet air above the Gulf Stream, a warm ocean current that flows up the eastern seaboard from the tropics.

The large size of the North American continent creates wide temperature variations. Because land heats up and cools off more rapidly than water, temperatures in the continental interior are hotter in the summer and colder in the winter than in coastal areas where temperatures are moderated by the oceans.

THINGS TO REMEMBER

1. North America has two main mountain ranges, the Rockies and the Appalachian; expansive plains between the two ranges; long, winding rivers—the Mississippi with its feeder rivers is one of the most extensive systems in the world; myriad lakes, the largest of which are the Great Lakes; and extraordinarily long coastlines along the Atlantic and Pacific Oceans.

2. The size and variety of landforms influence the movement and interaction of air masses and create enormous climate variation and wide seasonal temperature changes. Because land

heats up and cools off more rapidly than water, temperatures in the continental interior are higher in the summer and colder in the winter than in coastal areas.

Environmental Issues

Before European colonization of North America, which began soon after 1500, the environmental impact of humans was relatively low. Though North America was by no means a pristine paradise when Europeans arrived, subsequently environmental change took place at an ever-increasing rate. To make way for European-style farms, industries, mines, and cities, millions of acres of forests and grasslands that had served as habitats for native plants and animals were cleared. This was particularly true in the area that became the United States. Though widespread forest clearing for agriculture is now rare, other human activities are having significant effects on habitats across the region.

Logging

Logging continues in many of North America's forested areas, including the northern Pacific coast where it provides most of the

Wet Pacific air blows into coastal mountains, bringing rain and moderate temperatures. Continental effect makes winters colder and summers hotter.

Climate Zones

Tropical humid climates (A)
- Tropical wet/dry

Arid and semiarid climates (B)
- Steppe
- Desert

Temperate climates (C)
- Midlatitude, moist all year
- Subtropical, winter dry
- Mediterranean, summer dry

Cool humid climates (D)
- Continental, moist all year

Coldest climates (E)
- Arctic
- High altitude

- Winds
- Ocean currents

Wet Pacific air blows into coastal mountains, bringing rain and moderate temperatures.

Continental effect makes winters colder and summers hotter.

Gulf Stream current brings warm tropical water, warming the air over the Atlantic coast.

Moist Gulf of Mexico air brings rain and moderate temperatures.

mi 0   250   500
km 0   250   500

A Arctic, Alaska

B Temperate, midlatitude, Maryland

C Desert, Utah

While parts of North America remain relatively less impacted by humans, much of this region has seen low to medium impacts, and the parts where most people live are highly impacted.

Photo (**A**) shows an old growth cedar in the Olympic National Park. Clearcutting, in which all trees on a plot of land are cut down (**B**), has become an increasingly controversial method of logging. Since 1971 over 30 percent of the forests on Washington State's Olympic peninsula (**C** and **D**) have been clearcut. Even more controversial is the logging of old-growth forests that have never been cut. This practice has been declining in recent years, but still continues in some places.

E In remote areas, such as Alaska, mining and oil industries have the largest impacts. Photo (**E**) shows the Alaskan pipeline, which runs for 800 miles.

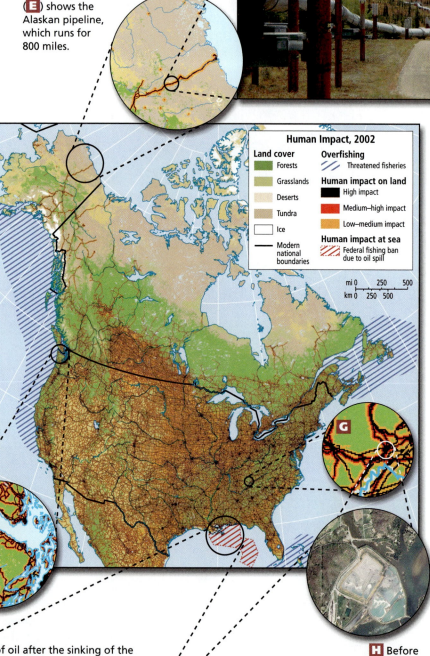

Human Impact, 2002

Land cover	Overfishing
Forests	Threatened fisheries
Grasslands	**Human impact on land**
Deserts	High impact
Tundra	Medium–high impact
Ice	Low–medium impact
Modern national boundaries	**Human impact at sea**
	Federal fishing ban due to oil spill

mi 0   250   500
km 0 250 500

In December of 2008 (**G–I**) a billon and a half gallons of coal ash from a powerplant spilled over 300 acres of river and shoreline near Kingston, Tennessee. A dike (**H**)that had been holding in the ash since the 1960s burst after a heavy rain, releasing heavy metals and radioctivity (a natural property of much coal) into the river. Thousands of fish were killed and drinking water supplies degraded. Coal supplies about half of the electric power used in the US.

F A pelican is cleaned of oil after the sinking of the Deepwater Horizon oil rig, in April of 2010 off the coast of Louisiana, resulted in the largest oil spill in U.S. history.

H Before

I After

construction lumber and an increasing amount of the paper used in Canada and the United States. Lumber and wood products are also important exports to Asia. The lumber industry is responsible directly and indirectly for hundreds of thousands of jobs across North America.

As the forests are depleted, environmentalists in North America harshly criticize the logging industry. Much attention is focused on **clear-cutting**, the dominant logging method used in the Pacific Northwest, the Northeast, and Southeast. In this method, all trees on a given plot of land are cut down, regardless of age, health, or species (Photo Essay 2.2A, B, C, D). Clear-cutting destroys wild animal and plant habitats, thereby reducing species diversity. Moreover, it is an inefficient use of particular species of trees, adds CO_2 to the atmosphere, and leaves the soil susceptible to erosion.

In the Pacific Northwest the battle lines have been drawn between those who make a living from logging or related activities and those who make a living from occupations that rely on the beauty of Pacific Northwest forested landscapes—both urban and rural people who are often advocates of strict environmental protection.

Oil Drilling

In many remote interior areas and coastal regions of North America, oil and gas drilling are by far the largest and most environmentally damaging industries. Oil-drilling operations have had a dramatic effect on humans and on land and water environments. Along the northern coast of Alaska, the Trans-Alaska Pipeline runs southward for 800 miles to the port of Valdez. Often running above ground to avoid shifting as the earth freezes and thaws, the pipeline constitutes a major ecological disruption (Photo Essay 2.2E). It interferes with caribou migrations and poses a constant threat of oil spills. All Alaskan citizens receive a yearly rebate of several thousand dollars from oil revenues, which tends to quiet protests over threats to the environment by oil extraction. The Inuit, the indigenous people in the Arctic lands of North America (and the Sami, Komi, and Sakha in the Arctic lands of Eurasia) are attempting to influence the extraction of newly discovered oil and gas fields in their homelands so that surface habitats are not destroyed in the process of extraction.

In April of 2010 the world's second largest oil spill to date began just off the Gulf Coast of North America near Louisiana and Mississippi (Photo Essay 2.2 Map). British Petroleum (BP), a large multinational corporation, was responsible for the spill, and because of increasingly lax regulation over the previous decade of the oil extraction industry in the United States, BP was unprepared to stop the spill and had at best outdated technology for capturing the spilled oil and for cleaning up the hundreds of square miles of contaminated ocean. By June 10, estimates of the daily rate of spill had risen to between 50,000 barrels (2,100,000 gallons) and 100,000 barrels (4,200,000 gallons) per day. Every day the predicted environmental impact became more dire as realization grew that all biotic life in the vicinity of the spill (including the entire column of water from the surface to the ocean floor), encompassing many aspects of the human food chain, would be deeply affected for decades. Of immediate concern were the birds and sea life that were dying from direct exposure to thick, black, toxic oil (Photo Essay 2.2F), and the thousands of coastal jobs in fishing, tourism, and related industries that were lost, perhaps permanently.

Coal Mining and Coal Use

In many remote interior areas of North America, coal mining is also a large and environmentally damaging industry. Huge piles of mining waste called "tailings" can pollute waterways and threaten communities that depend on well water. But the U.S. dependence on coal to generate electricity creates environmental concerns beyond the harm to landscapes and habitats done by strip mining. When coal is burned, the smoke and ash must be "scrubbed" to remove harmful chemicals that create air pollution, but finding a safe place to store the removed chemicals is difficult. Left to dry out, the material will become airborne and contribute to air pollution, so it is stored wet. In December 2008, a large earthen-walled pond of wet coal ash burst after a heavy rainstorm and poisonous sludge spilled over 300 acres of beautiful lakeshore and forest land in rural Tennessee (Photo Essay 2.2G, H, I). The sludge, containing heavy metals and harmful chemicals, ruined the ecology of the immediate area and threatened hundreds of square miles with water and air pollution. Cleanup will cost well over $1 billion.

Other Threats to Habitat

An important aspect of urbanization is **urban sprawl** (discussed more fully on page 98; see also Thematic Concepts Part D on page 68). In many areas, and for several decades, middle- and upper-income urbanites have sought lower-density suburban neighborhoods. Farms that were once highly productive are giving way to expansive low-density urban and suburban residential developments where pavement, golf courses, office complexes, and shopping centers cover the landscape. In the process, natural habitats are being degraded even more intensely than they were by farming. The loss of farmland and natural habitat in the urban fringe affects recreational land and our ability to produce affordable food for urban populations.

As North American plants and animals have been forced into ever smaller territories, many have died out entirely and been replaced by nonnative species (the domestic cat, for example). Estimates vary, but at least 4000 nonnative species have been introduced into North America inadvertently or purposely. An example is the Asian snakehead fish, which is rapidly invading the Potomac River, where it eats baby bass and fiercely competes with native fish for food. 🎞 **29. SNAKEHEAD REPORT**

Because of the complex interdependence of the biological world, the introduction, depletion, or extinction of species will undoubtedly have long-term negative effects for life on earth. Though many of the specific effects are not yet well understood, every day brings new evidence of the symbiotic and dependent relationships among species. Climate change will undoubtedly lead to further extinctions, as remaining wild habitats are transformed and many

> **clear-cutting** the cutting down of all trees on a given plot of land, regardless of age, health, or species
>
> **urban sprawl** the encroachment of suburbs on agricultural land surrounding cities

North America's wealth and its well-developed emergency response systems give it high resilience that reduces its overall vulnerability. However, certain regions are highly exposed to temperature increases, drought, hurricanes, and sea level rise.

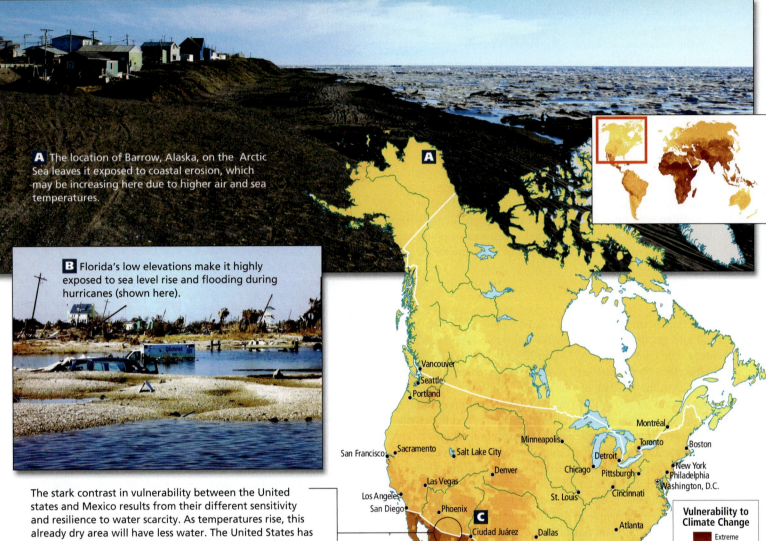

A The location of Barrow, Alaska, on the Arctic Sea leaves it exposed to coastal erosion, which may be increasing here due to higher air and sea temperatures.

B Florida's low elevations make it highly exposed to sea level rise and flooding during hurricanes (shown here).

Vulnerability to Climate Change
Extreme
High
Medium
Low

The stark contrast in vulnerability between the United states and Mexico results from their different sensitivity and resilience to water scarcity. As temperatures rise, this already dry area will have less water. The United States has much better water infrastructure, reducing its sensitivity to drought. Meanwhile more developed emergency response systems in the United States increase resilience to water shortage. With such drastic differences, the U.S.-Mexico border could become an even more contentious zone as the climate changes.

C Irrigation. Higher temperatures raise the rate at which plants loose water, increasing the need for irrigation. Farms that get their water from shrinking aquifers are less able to increase irrigation, and hence are highly sensitive to the higher temperatures and drought that climate change is bringing.

D Hurricanes gain strength with warmer temperatures. Many low-lying coastal cities are highly exposed and sensitive, though their resilience varies. Shown here are evacuees from Hurricane Katrina, an event which showed the weaknesses in local, state, and federal emergency response systems.

plant and animal populations are left without a place in which to thrive. 🖾 **30. GLACIER NATIONAL PARK**

Climate Change and Air Pollution

With only 5 percent of the world's population, North America produces 26 percent of the greenhouse gases released globally by human activity. This large share can be traced to North America's high consumption of fossil fuels, which in turn is related to its oil-dependent industrial and agricultural processes, the heating and cooling of its homes and offices, and its dependency on automobiles. As discussed in Chapter 1, greenhouse gases contribute to global warming and hence bring profound changes to the planet's climates (see pages 45–48).

Canada's government was one of the first in the world to commit to reducing the consumption of fossil fuels. Until recently, the United States resisted such moves, fearing damage to its economy. Both countries are now exploring alternative sources of energy, such as solar, wind, nuclear, and geothermal power. 🖾 **31. ENERGY REPORT** 🖾 **32. GREEN BUILDING**

Neither Canada nor the United States has so far been able to reduce its levels of greenhouse gas emissions, or even the rate at which these emissions are growing. Meanwhile, the move to alternative fuels has been slow, though a number of promising solar, wind, and other alternative energy projects are underway in both countries. Indications are that new forms of energy will define life in the twenty-first century.

Both countries are vulnerable to the effects of climate change, such as more frequent and violent hurricanes spawned in warming oceans, more numerous and destructive tornadoes in the Great Plains, sea level rise due to the melting of glaciers, and further drying of arid zones, which then require irrigation if cultivation is to continue (**Photo Essay 2.3**). Indigenous people in Alaska and northern Canada are also already affected by the warming of the oceans. Ice floes and coastal zones are melting. The polar bear population is endangered, and the rising sea levels are forcing at least 26 coastal villages to relocate inland, at an estimated cost of $130 million per village.

In addition to climate change, most greenhouse gases contribute to various forms of air pollution, such as smog and thermal inversion. *Smog* is a combination of industrial emissions and car exhaust that frequently hovers as a yellow-brown haze over many North American cities, causing a variety of health problems. In Los Angeles, the intensity of the smog is due in large part to the city's warm land temperatures and its West Coast seaside location (**Figure 2.4D** on page 76). This often results in a *thermal inversion*, which occurs when a mass of warm air settles over cooler air that blows in from the ocean and prevents the cool air from rising and dissipating the pollution. The inversion is held in place, often for days, by the mountains that surround the city.

The burning of fossil fuels also releases sulfur dioxide and nitrogen oxides into the air. **Acid rain** (and acid snow) is created when these gases dissolve in falling precipitation, making the precipitation acidic. Acid rain can kill trees (**Figure 2.4C**) and, when concentrated in lakes, streams, and snow cover, can destroy fish and wildlife. It also speeds up the corrosion of buildings, bridges, and other structures.

The United States, with its large population and extensive industry, is responsible for the vast majority of acid rain in North America. Due to continental weather and wind patterns, however, the area most affected by acid rain encompasses a wide area on both sides of the eastern U.S.–Canada border (see **Figure 2.4 Map**). The eastern half of the continent, which includes the entire eastern seaboard from the Gulf Coast to Newfoundland, is significantly affected by acid rain.

acid rain falling precipitation that has formed through the interaction of rainwater or moisture in the air with sulfur dioxide and nitrogen oxides emitted during the burning of fossil fuels, making it acidic

Water Resource Depletion, Pollution, and Marketization

People who live in the humid eastern part of North America find it difficult to believe that water is becoming scarce even there. Consider the case of Ipswich, Massachusetts, where the watershed is drying up as a result of overuse. There, innovators are saving precious water through conservation strategies in their homes and businesses (see the *Rainwater Cash* video). As populations grow and per capita water usage increases, it has become necessary to look farther and farther afield for sufficient water resources. New York City, for example, obtains most of its water from the distant Catskill Mountains in upstate New York. In the Southeast, Atlanta, Georgia, with over 5 million residents, absorbs water so insatiably that downstream users in Alabama and Florida have sued Atlanta for depriving them of their rights to water. Disputes over water rights exist around the world and are difficult to resolve, as we will see in our discussion of other world regions. 🖾 **34. RAINWATER CAR WASH**

Depletion In North America, water becomes increasingly precious the farther west one goes. On the Great Plains, rainfall is highly variable year to year, and in any given year it may be too sparse to support healthy crops and animals. To make farming more secure and predictable, taxpayers across the continent have subsidized the building of pumps and stock tanks for farm animals and aqueducts and reservoirs for crop irrigation.

Irrigation is increasingly based on "fossil water" that has been stored over the millennia in natural underground reservoirs called *aquifers*. The *Ogallala aquifer* (**Figure 2.5** on page 77) underlying the Great Plains is the largest in North America. In parts of the Ogallala, water is being pumped out at rates that exceed natural replenishment by 10 to 40 times.

Irrigation of the fruit and vegetable crops in California accounts for some of the water that goes into the *virtual water footprint* of U.S. consumers, as discussed in Chapter 1 (on page 42 and Table 1.2 on page 42). Billions of dollars of federal and state funds have paid for massive engineering projects that bring water from Washington, Oregon, Colorado, and northern California to farms in central and southern California. This water also supplies the cities of southern California, which are built on land that was once desert. Water is pumped from hundreds of miles away and over entire mountain ranges. California uses more energy to move water than some states use for all purposes. Moreover, irrigation in Southern California deprives Mexico of

A Polluted runoff from farms, cities, and industrial facilities has left 40 percent of rivers in the United States too polluted for fishing or swimming. Agricultural runoff includes fertilizers, pesticides, and soil washing off fields and into bodies of water.

B These substances add nutrients to the water, resulting in blooms of algae, which reduce the available oxygen in water as they decay. This process, also known as eutrophication, can kill fish and severely degrade water quality in rivers and coastline.

C Air pollution can kill trees, especially at high altitudes where both acid rain and acidic fog are present. Essential nutrients are stripped away from leaves and from soil, making trees more vulnerable to cold winter weather, pests, and other stresses. This dead Fraser fir tree is in Mount Rogers State Park in Virginia.

D Air pollution can result in a toxic brew called "smog." Nitrogen dioxide and other air pollutants are transformed by sunlight into a mix of chemicals that are particularly harmful to children, the elderly, and people with heart and lung ailments. The city of Los Angeles is infamous for its brown cloud of smog, which is visible in the photo at the horizon. Pollution from millions of automobiles, numerous industries, and the port complex at Long Beach is made worse by "thermal inversions." Heavy cold air sitting above warm air traps smog over the city.

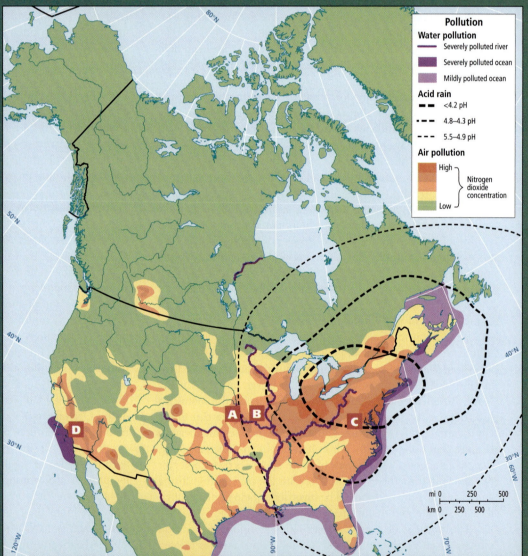

Pollution

Water pollution
— Severely polluted river
▬ Severely polluted ocean
▬ Mildly polluted ocean

Acid rain
▬ ▬ <4.2 pH
▬▬▬ 4.8–4.3 pH
- - - 5.5–4.9 pH

Air pollution
High
Low — Nitrogen dioxide concentration

FIGURE 2.4 Air and water pollution in North America. This map shows two aspects of air pollution, as well as polluted rivers and coastal areas. Red and yellow indicate concentrations of nitrogen dioxide (NO_2), a toxic gas that comes primarily from the combustion of fossil fuels by motor vehicles and power plants. This gas interacts with rain to produce nitric acid, a major component of acid rain, as well as toxic organic nitrates that contribute to urban "smog." The map also shows polluted coastlines (including all coastline from Texas to New Brunswick) as well as severely polluted rivers, which include much of the Mississippi River and its tributaries.

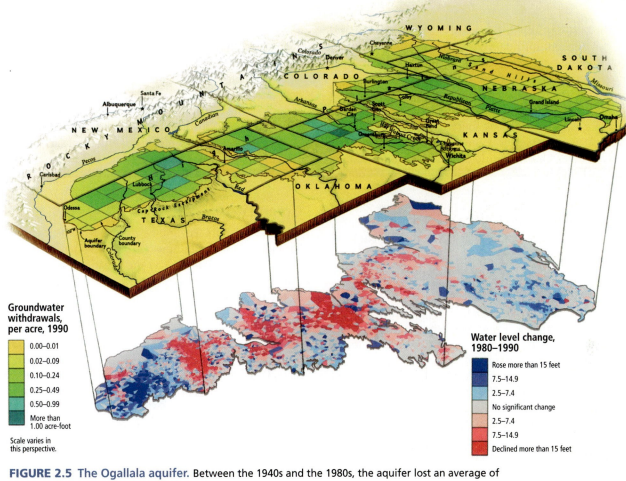

Groundwater withdrawals, per acre, 1990

- 0.00–0.01
- 0.02–0.09
- 0.10–0.24
- 0.25–0.49
- 0.50–0.99
- More than 1.00 acre-foot

Scale varies in this perspective.

Water level change, 1980–1990

- Rose more than 15 feet
- 7.5–14.9
- 2.5–7.4
- No significant change
- 2.5–7.4
- 7.5–14.9
- Declined more than 15 feet

FIGURE 2.5 The Ogallala aquifer. Between the 1940s and the 1980s, the aquifer lost an average of 10 feet (3 meters) of water overall, and more than 100 feet (30 meters) of water in some parts of Texas. During the 1980s, abundant rain and snow meant less decline in the aquifer. However, in this area the climate fluctuates from moderately moist to very dry. When a drought began in mid-1992 (continuing until late 1996), and large agribusiness firms pumped water for irrigation to supplement precipitation, water levels in the aquifer declined an average of 1.35 feet per year during the mid-1990s. [Graphic adapted from *National Geographic* (March 1993): 84–85, with supplemental information from High Plains Underground Water Conservation District 1, Lubbock, Texas, at http://www.hpwd.com; and Erin O'Brian, Biological and Agricultural Engineering, National Science Foundation Research Experience for Undergraduates, Kansas State University, 2001.]

this much-needed resource. The mouth of the Colorado River (which is in Mexico) used to be navigable; now it is dry and sandy and only a mere trickle of water gets to Mexico.

Increasingly, citizens in western North America are recognizing that the use of scarce water for irrigated agriculture or for keeping lawns and golf courses green in desert environments is unsustainable and uneconomical. Conflicts over transporting water from wet regions to dry ones, or from sparsely inhabited locations to urban areas, are ongoing and have halted some new water projects. However, government subsidies have kept water artificially cheap and, in the past, new water supplies have always been found and harnessed. Hence, there is little incentive to change.

Pollution In the United States, 40 percent of rivers are too polluted for fishing and swimming, and over 90 percent of *riparian areas* (the interface between land and flowing surface water) have been lost or degraded. Pollution in the rivers of North America comes mainly as storm-water runoff from urban and suburban developments, agricultural fields, and industrial sites. A recently

discovered type of water pollution is that of trace elements of drugs, such as male and female hormones, that are excreted by humans and not removed during water purification processes. These elements then make their way into rivers and lakes where they enter the food system in drinking water or through fish.

35. DRUGS AND WATER SUPPLY

In the 1970s, scientists studying coastal areas began noticing *dead zones* where water is so polluted that it supports almost no life. Dead zones occur near the mouths of major river systems that have been polluted by fertilizers washed from farms and lawns when it rains (see **Figure 2.4A, B**). A large dead zone is in the Gulf of Mexico near the mouth of the Mississippi, but similar zones have been found in all U.S. coastal areas. Even Canada, where much lower population density means that rivers are generally cleaner, has dead zones on its western coast.

Marketization The answer to the question, "Who owns our water?" is usually answered with "We do," but to what extent is water really public property? Americans are used to paying for water, but the cost has usually been just high enough to cover extraction, purification, and delivery in pipes. However, water scarcity and pollution are beginning to change the way water is viewed. For example, in the last few years, in response to aggressive advertising, the public has been buying bottled water even when tap water is perfectly safe. Beverage companies supplying this demand for bottled water seek sources of fresh water at the lowest price. A number of North American communities with abundant fresh water have agreed to sell water for this use only to find that such massive water withdrawals from local aquifers cause subsidence, loss of aquatic habitats, and the depletion and pollution of natural wells and springs. Eventually prices for their own water rise. Realizing belatedly that they can no longer control their access to water, expensive litigation against the water bottling companies is often the only recourse.

Depleted Fisheries

Abundant fish was the major attraction for the first wave of Europeans to North America. There is evidence that even before Columbus, fishermen from Europe's Atlantic coast came to the Grand Banks, offshore of Newfoundland and Maine, to take huge catches of cod and other fish. The fishing lasted for at least 500 years. Now the fish stocks of the Grand Banks are seriously depleted by modern fishing vessels (some from outside North America), which use equipment that destroys the marine ecosystem. Fisheries in the Pacific Northwest and the Gulf of Mexico are also threatened by pollution and overfishing. Eating fish from inland waterways is also problematic, as rivers and streams have become polluted with chemicals from industry and agriculture.

📹 **254. SCIENTISTS STUDYING DECLINE OF TARPON FISH**

North American fishing industries are also challenged by competition from fish farming, a rapidly growing industry in many parts of North America. There is a global connection in this practice: farmed fish are fed with millions of tons per year of wild ocean fish from around the world. The harvesting of these wild fish poses a threat to oceanic resources in countless places around the globe where the poor are dependent on fish for basic nutrition and modest incomes.

Hazardous Waste

Hazardous wastes are those wastes that pose a risk to public health or the environment. They are produced by a wide range of industries in North America, such as nuclear power generation, weapons manufacturing, mineral mining and drilling, and waste incineration. While industry produces 80 percent of all liquid hazardous waste, small businesses and private homes also contribute through the use of cleaning and paint products, computers, printers, weed killers, and gasoline. Disposal, particularly subject to the not-in-my-backyard (NIMBY) phenomenon, is often done secretly.

Each year the United States, with a population of roughly 307 million, generates five times the amount of hazardous waste generated by the entire European Union (with a population of about 491 million). Canada (population 33.7 million) generates much less hazardous waste per capita than the United States,

although its citizens generate several times the global average of about 2.2 pounds (1 kilogram) per person, per day.

The disposal of hazardous waste within the United States has a geographic pattern. Sociologist Robert D. Bullard, who studies environmental justice issues, has shown that a disproportionate amount of hazardous waste is disposed of in the South and in locations inhabited by poor minority people. Bullard writes, "Nationally, 60 percent of African-Americans and 50 percent of Hispanics live in communities with at least one uncontrolled toxic-waste site."

Green Living

As awareness of the environmental issues (some discussed above) has increased in North America, citizens often ask what they can do in daily life to ameliorate the situation. Experts warn that a simple solution, such as *greener living*, will not be sufficient to fix the problems. Rather, we must recognize that environmental degradation is in large part human induced. The present intensification of global climate change and all the negative effects of deforestation, dam construction, extractive activities from agriculture to mining, and the burning of fossil fuels—first coal and later oil and gas—are all manipulations of the environment by humans beginning thousands of years ago. Even in ancient Greece, Plato warned that deforestation for building and agriculture was hurting water quality. Therefore, beyond individual greener living, which is certainly helpful, collective action at the global scale will be needed to protect and restore life on earth.

THINGS TO REMEMBER

1. North America produces 26 percent of the world's greenhouse gases, even though it has only 5 percent of the world's population.

2. Water pollution is a major problem, especially in the United States, where 40 percent of the rivers are too polluted for fishing or swimming.

3. The United States, population over 307 million, generates five times the amount of hazardous waste generated by the entire European Union, population nearly 500 million.

4. Major environmental issues for North America include logging, especially clear-cutting; mining and petroleum drilling; and hazardous waste disposal.

Human Patterns over Time

The human history of North America is a series of arrivals and dispersals of people across the vast continent. In prehistoric times, humans came from Eurasia via Alaska. Beginning in the 1600s, waves of European immigrants, enslaved Africans, and their descendants, spread over the continent, primarily from east to west. Today, immigrants are coming mainly from all of Asia and from Middle and South America. They are arriving mainly in the Southwest and West. In addition, internal migration is still a defining characteristic of life for most North Americans, who are among the world's most mobile people.

The Peopling of North America

Recent evidence suggests that humans first came to North America from northeastern Asia by at least 25,000 years ago and perhaps earlier, most arriving during an ice age. At that time, the global climate was cooler, polar ice caps were thicker, and sea levels were lower. The Bering land bridge, a huge low landmass more than 1000 miles (1600 kilometers) wide, connected Siberia to Alaska. Bands of hunters crossed by foot or small boats into Alaska and traveled down the west coast of North America.

By 15,000 years ago, humans had reached nearly to the tip of South America and had spread deep into that continent. By 10,000 years ago, global temperatures began to rise, and as the ice caps melted, the Bering land bridge sank beneath the sea. In North America a midcontinent corridor opened through the glaciers, allowing more people to pass to the south. Eventually, humans occupied virtually every climate region throughout the Americas.

Over thousands of years, the people settling in the Americas hunted wild animals and plants, they domesticated plants, created paths and roads, cleared forests, burned grasslands, built permanent shelters, and sometimes created elaborate social systems. The drift away from hunting and gathering to food strategies based on domesticated plants is thought to be closely linked to a more settled lifestyle and population growth in North America. Around 200 C.E., some domesticated plants, particularly corn (maize), were introduced from Mexico into what is now the U.S. southwestern desert. However, long before then, in about 4000 B.C.E., people in central North America were planting and harvesting at least seven seed plants, including sunflowers, chenopods, knotweed, maygrass, and an edible gourd. Probably because these plants never gained favor with Europeans, the nutritional value and substantial contribution these plants made to Native American diets in North America have gone unrecognized.

The surpluses these domesticated crops provided allowed some community members to engage in activities other than food production, making possible large, city-like regional settlements. For example, by 1000 years ago, the urban settlement of Cahokia (in what is now central Illinois) covered 5 square miles (12 square kilometers) and was home to an estimated 20,000 people (Figure 2.6). Here people could specialize in crafts, trade, or other activities beyond the production of basic necessities.

The Arrival of Europeans North America was completely transformed by the sweeping occupation of the continent by Europeans. In the sixteenth century, Italian, Portuguese, and English explorers came ashore along the eastern seaboard of North America, and the Spanish sent the De Soto expeditions (1539–1542) from Florida into the heartland of North America on a course that wound through much of the southeast. In the early seventeenth century, the British established colonies along the Atlantic coast in what is now Virginia (1607) and Massachusetts (1620). Over the next two centuries, colonists from northern Europe built villages, towns, port cities, and plantations along the eastern coast. By the mid-1800s, they had occupied most Native American lands throughout the Appalachian Mountains and into the central part of the continent.

Disease, Technology, and the Native Americans The rapid expansion of European settlement was facilitated by the vulnerability of Native American populations to European diseases. Having long been isolated from the rest of the world, Native Americans had no immunity to diseases such as measles and smallpox. Transmitted by Europeans and Africans who had built up immunity to them, these diseases killed around 90 percent of Native Americans within the first 100 years of contact. It is now thought that diseases spread by De Soto's expedition so decimated populations in the North American interior that fields and villages were abandoned, the forest grew back and later explorers erroneously assumed the land had never been occupied.

Technologically advanced European weapons, trained dogs, and horses also took a large toll. Often the Native Americans had only bows and arrows. Some Native Americans in the Southwest acquired horses from the Spanish and learned to use them in warfare against the Europeans, but their other technologies could not compete. Numbers reveal the devastating effect of European

FIGURE 2.6 Precontact life in Cahokia (an artist's interpretation).
Cahokia Mounds in present-day southern Illinois was a settlement for about 700 years (700–1400 C.E.). The settlement consisted of at least 120 mounds. The largest of these mounds were platforms for important buildings and also served the purpose of rising above the plane so that a person could see the surrounding landscape in panorama. At its peak (1050–1150), the settlement had perhaps 20,000 people and covered 5 square miles (approximately 13 square kilometers). Evidence of human sacrifice and a robust wooden stockade hints at warfare. Artifacts of many types were made of materials from a wide trading area that ran from the Appalachian Mountains west to the Great Plains. A number of circular wooden post structures are thought to be astronomical calculators, one 410 feet in diameter. The agricultural economy was based on corn, squash, and seed-bearing plants such as sunflower, maygrass, and other plants only recently recognized as important in Native American diets. After 1400, the settlement was abandoned for unknown reasons, and the settlers dispersed gradually in many directions.

settlement on Native American populations. Roughly 18 million Native Americans lived in North America in 1492. By 1542, after only a few Spanish expeditions, only half that number survived. By 1907, slightly more than 400,000, or just 2 percent, remained.

The European Transformation

European settlement erased many of the landscapes familiar to Native Americans and imposed new ones that fit the varied physical and cultural needs of the new occupants. As a result, different subregions developed as settlement proceeded, and they still exist today.

> **infrastructure** road, rail, and communication networks and other facilities necessary for economic activity

The Southern Settlements European settlement of eastern North America began with the Spanish in Florida in the mid-1500s and the establishment of the British colony of Jamestown in Virginia in 1607. By the late 1600s, the colonies of Virginia, the Carolinas, and Georgia were cultivating cash crops such as tobacco and rice on large plantations.

To secure a large, stable labor force, Europeans brought enslaved African workers into North America beginning in 1619. Within 50 years, enslaved Africans were the dominant labor force on some of the larger southern plantations. By the start of the Civil War in 1861, slaves made up about one-third of the population in the southern states and were often a majority in the plantation regions. North America's largest concentrations of African-Americans are still in the southeastern states (**Figure 2.7**).

The plantation system concentrated wealth in the hands of a small class of landowners, who made up just 12 percent of southerners in 1860. Planter elites kept taxes low and invested their money in Europe or the more prosperous northern colonies, instead of in **infrastructure** at home. As a result, the road, rail, and communication networks, and other facilities necessary for economic growth were not built.

More than half of southerners were poor white farmers. Both they and the slaves lived simply, so their meager consumption did not provide a demand for goods. Hence, there were few market towns and almost no industries. Plantations tended to be self-sufficient and generated little *multiplier effect*. Enterprises like shops, garment making, small restaurants and bars, small manufacturing, and transport and repair services that normally "spin off" from, or serve, main industries simply did not develop in the South. Competition between the weak southern economy and the stronger, more diversified northern economies was a main cause of the Civil War (1861–1865), perhaps equal to the abolition movement to free enslaved Africans and their descendants. After the war, the victorious North returned to promoting its own industrial development, the plantation economy declined, and the South sank deep into poverty. This subregion remained economically and socially underdeveloped well into the 1970s (Figure 2.8).

The Northern Settlements Throughout the seventeenth and eighteenth centuries, relatively poor subsistence farming communities dominated agriculture in the colonies of New England and southeastern Canada. There were no plantations and few slaves, and not many cash crops were exported. Farmers lived in interdependent communities that prized education, ingenuity, self-sufficiency, and thrift.

At first, incomes were augmented with exports of timber and animal pelts. Some communities depended heavily on the rich fishing grounds of the Grand Banks off Newfoundland and Maine. By the late 1600s, New England was implementing ideas and technology from Europe that led to the first industries. By the 1700s, diverse industries were supplying markets in North America and also exporting to British plantations in the Caribbean. These industries included metalworks, and pottery, glass, and textile factories in Massachusetts, Connecticut, and Rhode Island.

By the early 1800s, women made up a substantial part of the labor force. In 1822, "factory girls" were working in the textile mills of Lowell, Massachusetts, and living as single women in company-owned housing. Southern New England, especially the region around Boston, became the center of manufacturing in North America. It drew largely on young male and female immigrant labor from French Canada and Europe.

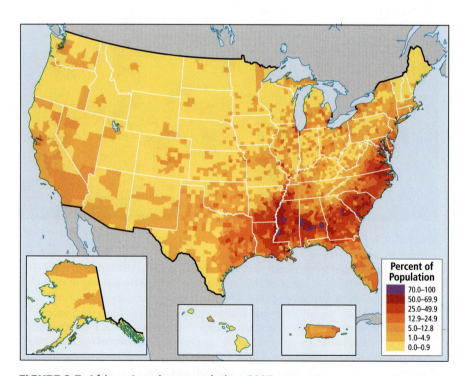

FIGURE 2.7 African-American population, 2007. Many African-Americans live in cities across the United States, but the majority continue to live in the southeastern parts of the country. More African-Americans are returning to the Southeast than are leaving, many retiring on pensions earned elsewhere. [U.S. Census Bureau, "Percent of the Total Population Who Are Black or African-American Alone: 2007," from the 2005–2007 *American Community Survey*, M0202]

Percent of Population
70.0–100
50.0–69.9
25.0–49.9
12.9–24.9
5.0–12.8
1.0–4.9
0.0–0.9

The Mid-Atlantic Economic Core The colonies of New York, New Jersey, Pennsylvania, and Maryland eventually surpassed in population and in wealth New England and southeastern Canada. This mid-Atlantic region benefited from more fertile soils, a slightly warmer climate, multiple deepwater harbors, and better access to the resources of the interior. By the end of the Revolutionary War in 1783, the mid-Atlantic region was on its way to becoming the **economic core**, or the dominant economic region, of North America.

Both agriculture and manufacturing in the region grew and diversified in the early nineteenth century, drawing immigrants from much of northwestern Europe. As farmers prospered, they bought mechanized equipment, appliances, and consumer goods, many of these items made in nearby cities. Port cities such as New York, Philadelphia, and Baltimore prospered as the intermediary for trade between Europe and the vast American continental interior.

By the mid-nineteenth century, the economy of the core was increasingly based on the steel industry, which diffused westward to Pittsburgh and the Great Lakes industrial cities of Cleveland, Detroit, and Chicago. The steel industry further stimulated the mining of deposits of coal and iron ore throughout the region and beyond. Steel became the basis for mechanization, and the region was soon producing heavy farm and railroad equipment, including steam engines.

By the early twentieth century, the economic core stretched from the Atlantic to St. Louis, Chicago, and Milwaukee, and from Ottawa to Washington, D.C. Soot-laden urban landscapes became badges of progress (Figure 2.9) as industry dominated North America economically and politically well into the mid-twentieth century. Most other areas produced food and raw materials for the core's markets and depended on the core's factories for manufactured goods.

FIGURE 2.9 The economic core in the early twentieth century. A group of young boys overlooking a landscape of steel mills and other industries in Homestead, Pennsylvania, in 1907.

Expansion West of the Mississippi

The east-to-west trend of settlement continued as land in the densely settled eastern parts of the continent became too expensive for new immigrants. By the 1840s, immigrant farmers from central and northern Europe, as well as some older eastern seaboard settlers, were pushing their way west across the Mississippi River (Figure 2.10).

economic core the dominant economic region within a larger region

The Great Plains Much of the land west of the Mississippi River was dry grassland or prairie, and farmers, accustomed to the well-watered fields and forests of Europe and eastern North America, had to adapt to the lack of trees and water. The soil usually proved very fertile in wet years, and the area became known as the region's breadbasket. Farms here succeeded well into the early twentieth century, but the naturally semi-arid character of this land eventually created an ecological disaster for Great Plains farmers. In the 1930s, after 10 especially dry years, a series of devastating dust storms blew away most of the topsoil; animals died, and entire crops were lost. This hardship was made worse by the widespread economic depression of the 1930s. Many Great Plains farm families packed up what they could and left during what became known as the Dust Bowl era, heading west to California and other states on the Pacific Coast. In the 1940s, changing weather patterns and new farming and irrigation technology helped the Dust Bowl region revive for a time.

The Mountain West and Pacific Coast As the Great Plains filled, other settlers were alerted to the possibilities farther west. By the 1840s, settlers were coming to the valleys of the Rocky Mountains, to the Great Basin, and to the well-watered and fertile coastal zones of what was then known as the Oregon Territory and California. News of the discovery of gold in California in 1849 created the *Gold Rush*, drawing thousands with the prospect of getting rich quick. The vast majority of gold seekers were unsuccessful, however, and

FIGURE 2.8 Lebeau Plantation, an antebellum plantation home in St. Bernard Parish, south of New Orleans. Lebeau plantation in St. Bernard Parish was built in 1854 and was once the largest house south of New Orleans. It has served as a hotel, a brick factory, a casino, and a private residence. It was being restored when Hurricane Katrina hit, and now sits abandoned once again.

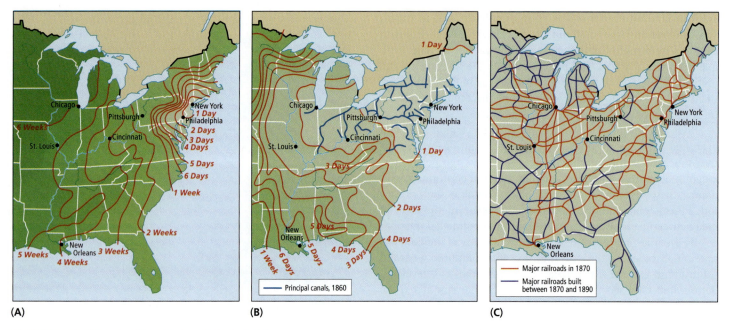

(A)   **(B)**   **(C)**

FIGURE 2.10 Nineteenth-century transportation. (A) Travel times from New York City, 1800. It took a day to travel by wagon from New York City to Philadelphia and a week to go to Pittsburgh. **(B)** Travel times from New York City, 1857. The travel time from New York to Philadelphia was now only 2 or 3 hours and to Pittsburgh less than a day because people could go part of the way via canals (dark blue). Via the canals, the Great Lakes, and rivers, they could easily reach principal cities along the Mississippi River, and travel was less expensive and onerous. **(C)** Railroad expansion by 1890. With the building of railroads, which began in the decade before the Civil War, the mobility of people and goods increased dramatically. By 1890, railroads crossed the continent, though the network was most dense in the eastern half. [Adapted from James A. Henretta, W. Elliot Brownlee, David Brody, and Susan Ware, *America's History,* 2nd ed. (New York: Worth, 1993), pp. 400–401; and James L. Roark, Michael P. Johnson, Patricia Cline Cohen, Sarah Stage, Alan Lawson, and Suan M. Hartmann, *The American Promise: A History of the United States,* 3rd ed. (Boston: Bedford/St. Martin's, 2005), p. 601.]

by 1852, they had to look for employment elsewhere. Farther north, logging eventually became a major industry.

The extension of railroads across the continent in the nineteenth century facilitated the transportation of manufactured goods to the West and raw materials to the East. Today, the coastal areas of this region, often called the Pacific Northwest, have booming, diverse, high-tech economies and growing populations. Perhaps in response to their history, residents of the Pacific Northwest are on the forefront of so many efforts to reduce human impacts on the environment that the region has been nicknamed "Ecotopia."

The Southwest People from the Spanish colony of Mexico first colonized the Southwest at the end of the sixteenth century. Their settlements were sparse, and as immigrants from the United States expanded into the region, Mexico found it increasingly difficult to maintain control. By 1850, nearly the entire Southwest was under U.S. control.

By the twentieth century, a vibrant agricultural economy was developing in central and southern California, made possible by massive government-sponsored irrigation schemes. There the mild Mediterranean climate made it possible to grow vegetables almost year-round. With the advent of refrigerated railroad cars, fresh California vegetables could be sent to the major population centers of the East during all seasons of the year, drastically changing American food habits. Across the country, local growers found it increasingly difficult to compete in the marketplace, and Americans

became dependent on food from distant locations. Meanwhile, Southern California's economy rapidly diversified to include oil, entertainment, and a variety of engineering- and technology-based industries.

European Settlement and Native Americans

As settlement relentlessly expanded west, Native Americans living in the eastern part of the continent, who had survived early encounters with Europeans, were occupying land that European settlers wished to use. During the 1800s, almost all the surviving Native Americans were forcibly relocated west to relatively small reservations with few resources. The largest relocation, in the 1830s, involved the Choctaw, Seminole, Creek, Chickasaw, and Cherokee of the southeastern states. These people had already adopted many European methods of farming, building, education, government, and religion. Nevertheless, they were rounded up by the U.S. Army and marched to Oklahoma, along a route that became known as the Trail of Tears because of the more than 4000 who died along the way (**Figure 2.11**).

As European settlers occupied the Great Plains and prairies, many of the reservations were further shrunk or relocated onto ever less desirable land. Today, reservations cover just over 2 percent of the land area of the United States.

In Canada the picture is somewhat different. Reservations now cover 20 percent of Canada, mostly due to the creation of the

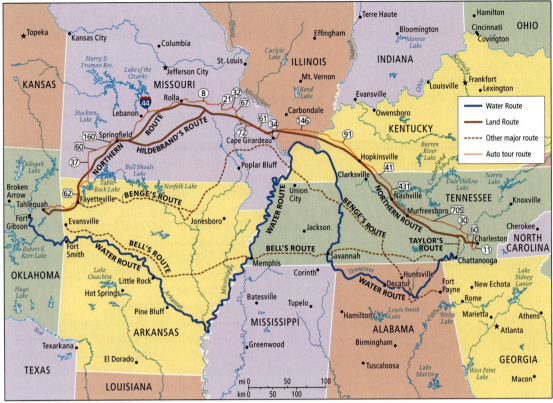

FIGURE 2.11 Trail of Tears. The Trail of Tears map shows the major routes along which the Cherokee were moved west in 1838–1839. Small groups of Cherokee would forage for food as they proceeded, so the map is only a general representation of the routes. [Adapted from the National Parks Service Trail of Tears National Historic Trail brochure, at http://ngeorgia.com/history/trailoftearsmap2.html.]

Nunavut territory in 1999 and the Dogrib territory in 2003, both in the far north of Canada (see Figure 2.1 on pages 64–65). The Nunavut and Dogrib stand out as having won the right to legal control of these lands. In contrast to the United States, it had been unusual for native groups in Canada to have legal control of their territories.

Many, if not most, reservations in North America have insufficient resources to support their populations at the standard of living enjoyed by other citizens. After centuries of mistreatment, many Native Americans still live in poverty, and as in all communities under severe stress, rates of addiction and violence are high, especially in the United States.

Nonetheless, in the last several decades, the availability of secondary and higher education has led to internally generated development. Several tribes now enjoy increasing affluence on their reservations as the result of various enterprises including manufacturing industries; the extraction of fossil fuel, uranium, and other mineral deposits underneath reservations; or the opening of gambling casinos to the wider public. For an increasing number of tribes, these activities have produced substantial income and meant a return to self-sufficiency. But, like all Americans, native groups have found that with development comes not only greater affluence but also the problems of corruption and, especially in mining areas, environmental pollution.

A measure of the resurgence of Native American societies is the fact that populations overall have expanded from a low of 400,000 in 1907 to almost 4 million in 2006, to some extent because more now claim native ancestry. The majority live in the United States.

The Changing Regional Composition of North America

The subregions of European-led settlement still remain in North America, but they are now less distinctive. The economic core region is less dominant in industry, which has spread to other parts of the continent. Some regions that were once dependent on agriculture, logging, or mineral extraction now have high-tech industries as well. The West Coast in particular has boomed with a high-tech economy and a rapidly growing population. The West Coast also benefits from trade with Asia, which now surpasses trade with Europe in volume and value.

THINGS TO REMEMBER

1. In 1492, roughly 18 million Native Americans lived in North America. By 1542, after only a few Spanish expeditions, there were half that many. By 1907, slightly more than 400,000, or just 2 percent, remained. By 2006, there were about 4 million.

2. European settlement imposed new landscapes that fit the varied physical and cultural needs of the new occupants. As a result, different subregions developed as settlement proceeded, and they still exist today.

3. Perhaps the most significant and enduring European social impact on North America was the institution of the plantation system and the concomitant enslavement of millions of Africans.

II CURRENT GEOGRAPHIC ISSUES

The huge regional economy and still-plentiful resources privilege North America, but the region faces complex challenges posed by globalization, increasingly diverse populations, and rising energy and environmental concerns. These long-standing issues must be understood now in the context of a globalized world in which economic downturns impact life everywhere, and the conflicts and disaffections arising from wide disparities of wealth and opportunity now affect all world citizens. Canada and the United States, both world leaders of opinion and of economic development, have special responsibilities during hard times.

Political Issues

North Americans are—for the most part—passionate political animals. They feel free to express opinions and are deeply interested in how power and money are allocated in their communities. Political issues in this region are very much linked to national identity—the ways that citizens want to think about their country, the image they would like it to have. Here we will look at just a few defining issues of the present day and especially at the ways in which the two countries are similar and different in their political culture. Of course, political and economic issues are intimately intertwined, so in the economic issues section that follows this, there are a number of topics that have political angles. The terrorist attacks on the United States in 2001 and their aftermath have influenced world events drastically, so the attacks and the wars that followed are discussed first.

The Wars in Afghanistan and Iraq

Immediately after the attacks in the United States on September 11, 2001 (9/11), the international community extended warm sympathy to the country and generally supported the strategies of President George W. Bush. In the fall of 2001, his administration, with the advice and consent of Congress, launched what was then called the War on Terror. The first target was Afghanistan, thought to be host to Osama bin Laden and the elusive Al Qaeda network, which had openly claimed credit for masterminding the 9/11 attacks. The aim was to capture bin Laden—an aim still unaccomplished (as of this writing). Although U.S. forces in Afghanistan were joined by NATO forces (including Canadian troops), the war has proved difficult to resolve because of heavy resistance from tribal leaders within the country, and those who support them, militant Muslim insurgents in adjacent Pakistan. When, in the spring of 2003, President Bush brought the War on Terror to Iraq, U.S. troop support was diverted and the momentum in Afghanistan was lost.

Early in its tenure, the administration of President Barack Obama began to develop another way of looking at terrorism. Taking note of its very serious threat to the United States, President Obama noted that terrorism is one of the downsides of the powerful changes brought by globalization, which can breed stark inequalities and lack of hope. In circumstances where democracy does not exist, people have no chance to mold their own futures and no peaceful political outlets for their frustrations. In 2009, the Obama administration announced a phased withdrawal from Iraq and

reluctantly decided to continue the war in Afghanistan, citing U.S. security: if the United States and NATO were to walk away from the troubles in Afghanistan, the most radical elements would flourish, recruit new terrorists, and pose an increased threat to the U.S. homeland. Despite an increasingly skeptical American public, U.S. policy was to support the government of Hamid Karzai, shakily reelected in 2009; step up military action with increased troops; provide strong support to Afghan society and to Afghan economic development; protect women and children; and foster democratization (see Photo Essay 2.4B on page 86).

As of March 2010, a total of 1693 NATO coalition forces had been killed in Afghanistan, including 146 Canadians and 1025 from the United States. By March 2010, 5568–8360 Afghan civilians had been killed as a direct result of the U.S.-led military action.

📹 **39. 9-11 CONSEQUENCES REPORT**

📹 **40. SECURITY AND PERSONAL LIBERTY REPORT**

The United States under the Bush Administration had justified its post–9/11 invasion of Iraq as one of its and previous administrations' many efforts to spread democracy worldwide (Photo Essay 2.4 on pages 86–87). The removal of Iraq's dictatorial president, Saddam Hussein, was based on the since-disproven claims that he was involved in the 9/11 events and had weapons of mass destruction (chemical and nuclear) ready to use. When the original reasons for declaring war were shown to be baseless, and democratization was so clearly not happening, his popularity sank at home and abroad, as did the worldwide popularity and reputation of the country.

Although several Iraqi elections were held, resistance to the U.S. occupation grew, and tension among Iraq's different political, ethnic, and religious groups brought the country to the brink of full-scale civil war. A reputable survey of Iraqi war-related civilian deaths, based on the NGO Iraq Body Count, substantiated that at least 104,427 Iraqis had died by March 2010, with suicide bombers still striking nearly daily. Altogether, as many as 1.3 million Iraqis, many of them children, have died as an indirect result of the economic and civil disruptions that followed the U.S. invasion. More than 4379 U.S. armed service members had been killed and 31,669 seriously wounded by the middle of February 2010, and $900 billion had been spent or approved for spending.

Further damaging to the U.S. reputation and the goal of spreading democracy was the U.S. decision not to treat prisoners captured in the War on Terror according to the *Geneva Conventions* (international treaties that protect the rights of prisoners of war). Repeated discoveries that the U.S. military and civilian contractors had tortured prisoners sparked outrage both at home and abroad, with many pointing out that overall the U.S. responses to 9/11 may have made the country more vulnerable to terror, not less, because of the loss of international support that resulted.

The manner in which the war in Iraq was conceived and carried out led to widespread diplomatic isolation of the United States in the world community. The United Nations and nearly all major U.S. allies advised against attacking Iraq. Canada refused to join the so-called *coalition of the willing* (made up of the United Kingdom plus a small group of countries dependent on the

United States) for the following reasons: a UN security resolution on military action was lacking; diplomatic efforts to disarm Iraq appeared to be succeeding, and forcing a regime change would set a dangerous precedent. Instead, Canada became a major donor of humanitarian and reconstruction assistance to Iraq. The failure of the United States to substantiate any of its prewar justifications for the invasion resulted in high levels of opposition within the United States as well. By the presidential election of 2008, over 60 percent of U.S. respondents to polls conducted by CNN and NBC reported that the Iraq War was not worth fighting. Barack Obama won the 2008 election largely on the basis of his promise to change policies regarding the two wars, health care, energy, and the regulation of financial institutions. ▶ 41. U.S. IMAGE REPORT

Dependence on Oil Questioned

There was always a suspicion—at home and abroad—that the war in Iraq was related to U.S. plans to insure access to that country's considerable oil and gas resources to ensure that the United States had sufficient secure supplies. The 9/11 attacks in 2001 had alerted the U.S. public to the fact that roughly 30 percent of the oil used in the United States came from countries along the Persian Gulf, where the terrorists had originated, plus Algeria. By 2008, this figure was down slightly to 29.6 percent (Figure 2.12).

Since 2001, the United States has bought oil and gas from a wider range of countries and the percentages imported vary from month to month but have decreased slightly in recent years. While it is doubtful that all, or even most, oil-producing countries would participate in disrupting flows of oil to the United States, primarily because they are too dependent on the income, this rate of dependency makes Americans uncomfortable. Whether or not oil lay at the heart of the initial Iraq war strategy, the course of the war made it plain that Iraqi oil and gas would never be a secure source of fuel for the United States. Finding ways to be less dependent on petroleum products became a renewed concern when oil and gas prices rose sharply in 2008, with gasoline reaching $5 a gallon in several parts of both Canada and the United States.

Analysts now agree that although oil prices are likely to fluctuate, the general trend will be upward, partly because overall peak production will be reached and producers will keep the supply tight, but primarily because demand from the fast-growing Asian economies of India and China will push prices higher. Canada, with large reserves of oil, much of it sold to the United States, is only marginally better off because its own demands for energy are high.

Concerns over rising oil and gas prices in North America have coincided with rising public awareness of how the burning of gas and oil contributes to climate change. When family

A An oil derrick in Alberta, Canada. The United States's largest source of foreign oil is Canada.

B A gas station in Nogales, Mexico, run by Pemex, Mexico's state-owned oil company. Mexico is the second-largest source of oil imports for the United States.

FIGURE 2.12 Sources of average daily crude oil imported to the United States in September, 2009. The United States is dependent on crude oil from many locations around the world. This map shows the percentage of average daily crude oil (millions of barrels) imported to the United States from specific nations in September 2009. In 2008, Canada supplied 22 percent of U.S. petroleum imports. The country sources and percentages have remained roughly the same for the past 5 years. [Data source: U.S. Energy Information Administration, "Crude Oil and Total Petroleum Imports Top 15 Countries," at http://www.eia.doe.gov/pub/oil_gas/petroleum/data_publications/company_level_imports/current/import.html.]

incomes were threatened by a major economic downturn in late 2008 right after the gasoline price increase, the American public suddenly became focused on curing the country's "addiction" to petroleum products. The sales of SUVs and other "gas guzzlers" declined sharply, as did the sale of most consumer products and services. This sharp decrease in U.S. demand had a ripple effect across all oil-producing countries, whose incomes then dropped sharply. It was becoming clear that dependence on oil by either producers or consumers was unwise.

Relationships Between Canada and the United States

Citizens of Canada and the United States share many characteristics and concerns. Indeed, in the minds of many people—especially those in the United States—the two countries are one. Yet that is hardly the case. Three key factors characterize the interaction between Canada and the United States: *asymmetries*, *similarities*, and *interdependencies*.

Asymmetries

Asymmetry means "lack of balance." Although the United States and Canada occupy about the same amount of space (Figure 2.13), much of Canada's territory is cold and sparsely inhabited. The U.S. population is about ten times the Canadian population. And Canada's economy is one of the largest and most productive in the world, producing U.S.$1.3 trillion purchasing power parity (PPP) in goods and services in 2008. But it is dwarfed by the U.S. economy, which is more than ten times larger, $14.8 trillion PPP in 2008, though in the final quarter it shrank by 6.2 percent.

FIGURE 2.13 Political map of North America. Note that the U.S. state of Hawaii is not shown here because it is not physically part of North America and has much in common with Oceania; it is discussed in Chapter 11.

PHOTO ESSAY | # 2.4 Democratization and Conflict in North America

It is sometimes argued that the United States uses its power abroad primarily to promote democracy. There is some truth to this. For example, the largest recipient of U.S. foreign aid since 2002 has been Iraq, where the United States strongly promoted democratization during the Iraq War. However, officials who helped plan the war point out that U.S. strategic and economic interests played a much larger role than democratization as a motive of U.S. involvement. The same is true elsewhere. The strategic and economic interests of the United States generally determine where U.S. foreign aid goes and where the most military bases are. For example, if promoting democracy was the primary goal of the United States abroad, sub-Saharan Africa would be the major region of focus for the United States. While the United States has been increasing its presence in this region in recent years, the map below shows that sub-Saharan Africa receives relatively little U.S. aid and has few U.S. military bases.

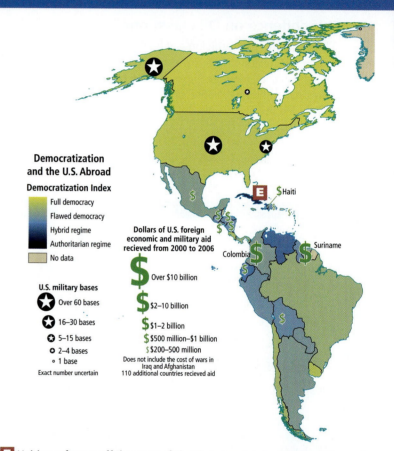

Democratization and the U.S. Abroad

Democratization Index

- Full democracy
- Flawed democracy
- Hybrid regime
- Authoritarian regime
- No data

U.S. military bases
- Over 60 bases
- 16–30 bases
- 5–15 bases
- 2–4 bases
- 1 base
- Exact number uncertain

Dollars of U.S. foreign economic and military aid recieved from 2000 to 2006
- Over $10 billion
- $2–10 billion
- $1–2 billion
- $500 million–$1 billion
- $200–500 million

Does not include the cost of wars in Iraq and Afghanistan
110 additional countries recieved aid

E Haitian refugees off the coast of Florida being detained by the U.S. Coast Guard. Haiti is far from being a functional democracy and is plagued by economic and political instability, but compared to Israel, Egypt, or Afghanistan, it receives relatively little U.S. foreign or military aid.

A U.S. forces destined for Iraq wait to board aircraft at Ramstein Air Base, one of 260 U.S. bases in Germany. The large U.S. military presence in Europe is a legacy of World War II and the Cold War era that followed. Their purpose was to discourage potential aggression from the (now defunct) Soviet Union against U.S. allies in Western Europe.

B Afghan women line up to vote in the country's first legitimate elections in decades. Democratization is a major goal of U.S. foreign aid in Afghanistan, but in the contest of the larger strategic interests of the "War on Terror." Afghanistan became a focus for the United States abroad only after the attacks of September 11th, which were were masterminded by the Al Qaida terrorist network that was based, in part, in Afghanistan.

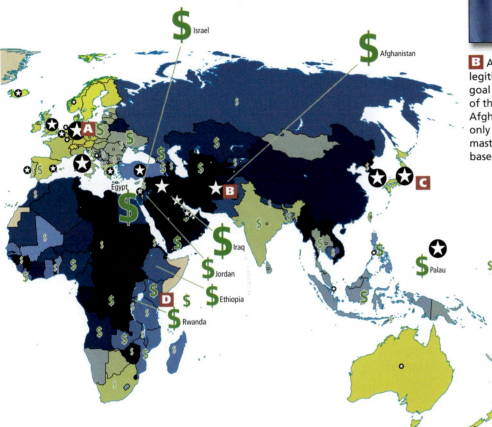

C Japanese officials surrender aboard a U.S. battleship at the end of World War II. After the war both Japan and its occupied territories in South Korea were reorganized as democracies by the United States, and numerous military bases were established. However the bases are first and foremost a projection of U.S. power designed to counter any future agression by China, North Korea, the former Soviet Union, or by Japan and South Korea themselves.

D A Kenyan throws teargas back at riot police during several days of violence that followed the presidential elections in 2008. Democracy is fragile throughout much of sub-Saharan Africa, and elections are often plagued by violence. While the United States is giving increasing support to democracy in sub-Saharan Africa, U.S. foreign aid to sub-Saharan Africa is relatively small and historically has done little to promote democracy in the region. Military bases are relatively few.

There is also asymmetry in international affairs. The United States is an economic, military, and political superpower preoccupied with maintaining a world leadership role. Canada is only an afterthought in U.S. foreign policy, in part because the country is so secure an ally. But for Canada, managing its relationship with the United States is the top foreign policy priority. As former Canadian Prime Minister Pierre Trudeau once told the U.S. Congress, "Living next to you is in some ways like sleeping with an elephant: no matter how friendly and even-tempered the beast, one is affected by every twitch and grunt." In 2008, for example, the robust debt-free Canadian economy slowed abruptly, due in large part to the slump in U.S. consumer spending, which hurt Canadian exports such as energy and cars.

Similarities

Notwithstanding the asymmetries, the United States and Canada have much in common. Both are former British colonies that also experienced settlement and exploration by the French. From their common British colonial experience, they developed comparable democratic political traditions. Both are federations (of states or provinces), and both are representative democracies. Their legal systems are also similar.

Not the least of the features they share is a 4200-mile (6720-kilometer) border, which until 2009 contained the longest sections of unfortified border in the world. For years, the Canadian border had just 1000 U.S. border guards, while the Mexican border, which is half as long, had nearly 10,000 agents. In 2009, the Obama administration decided to equalize surveillance on the two borders for national security reasons. Where some rural residents used to pass unobserved in and out of Canada and the United States many times in the course of a day simply by going about their usual activities, now there are drone aircraft with night-vision cameras and cloud-piercing radar scanning the landscape for smugglers, illegal immigrants, and terrorists. Canadians and Americans, who grew up thinking of the border as nonexistent, lament the intrusive technology, the barricades, and the delays. Tourism and commerce are also negatively affected by these efforts to keep both countries safer. **42. U.S. BORDER SECURITY REPORT**

Well beyond the border, Canada and the United States share many other landscape similarities. Their cities and suburbs look much the same. The billboards that line their highways and freeways advertise the same brand names. Shopping malls and satellite business districts have followed suburbia into the countryside, encouraging similar patterns of mass consumption and urban sprawl. The two countries also share similar patterns of ethnic diversity that developed in nearly identical stages of immigration from abroad.

Interdependencies

Canada and the United States are perhaps most intimately connected by their long-standing economic relationship. The two countries engage in mutual tourism, direct investment, migration, and most of all, trade. By 2005, that trade relationship had evolved into a two-way flow of U.S.$1 trillion annually (**Figure 2.14**). Each country is the other's largest trading partner. Canada sells 80 percent of its exports to the United States and buys 54 percent of its

imports from the United States. The United States, in turn, sells 21.4 percent of its exports to Canada and buys 15.7 percent of its imports from Canada.

Notice, however, that asymmetry exists even in the realm of interdependencies: Canada's smaller economy is much more dependent on the United States than the reverse. Nonetheless, if Canada were to disappear tomorrow, as many as 1 million American jobs would be threatened.

Democratic Systems of Government: Shared Ideals, Different Trajectories

Canada and the United States have similar democratic systems of government, but there are differences in the way power is divided between the federal government and provincial or state governments. There are also differences in the way the division of power has changed since each country achieved independence.

Both countries have a federal government, in which a union of states (provinces, in Canada) recognizes the sovereignty of a central authority while many governing powers are retained by state/provincial or local governments. In both Canada and the United States, the federal government has an elected executive branch, elected legislatures, and an appointed judiciary. In Canada, the executive branch is more closely bound to follow the will of the legislature. At the same time, the Canadian federal government has more and stronger powers (at least constitutionally) than does the U.S. federal government.

Over the years, both the Canadian and U.S. federal governments have moved away from the original intentions of their constitutions. Canada's originally strong federal government has become somewhat weaker. This change is largely in response to demands by provinces, such as the French-speaking province of Québec, for greater autonomy over local affairs.

Meanwhile, the initially more limited federal government in the United States has expanded its powers. The U.S. federal government's original source of power was its mandate to regulate trade between states. Over time, this mandate has been interpreted ever more broadly. Now the U.S. federal government powerfully affects life even at the local level. This power is exercised primarily through its ability to dispense federal tax monies in programs such as grants for school systems, federally assisted housing, military bases, and the rebuilding of interstate highways. Money for these programs is withheld if state and local governments do not conform to federal standards. This practice has made some poorer states dependent on the federal government. However, it has also encouraged some state and local governments to enact more enlightened laws than they might have done otherwise. For example, in the 1960s, the federal government promoted civil rights for African-American citizens by requiring states to end racial segregation in schools in order to receive federal support for their school systems.

THINGS TO REMEMBER

1. As of the end of the first decade of this century, U.S. military involvement in Iraq seemed to be lessening as the elected Iraqi government took on greater responsibilities, but both Canada and the United States continued their humanitarian

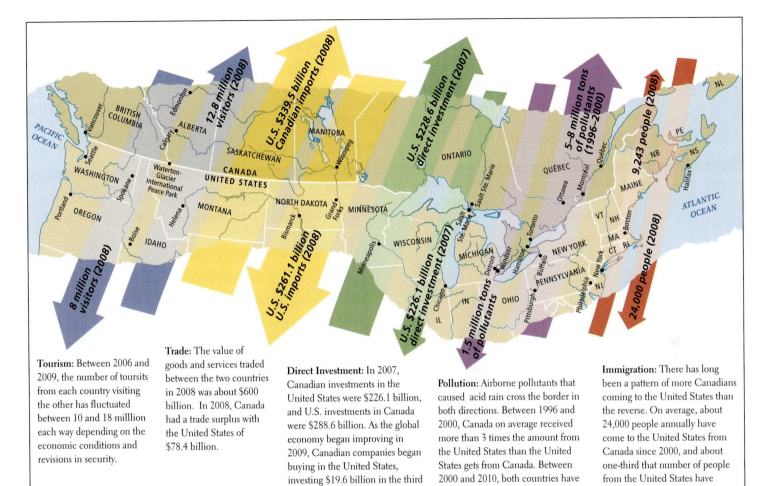

Tourism: Between 2006 and 2009, the number of toursits from each country visiting the other has fluctuated between 10 and 18 milllion each way depending on the economic conditions and revisions in security.

Trade: The value of goods and services traded between the two countries in 2008 was about $600 billion. In 2008, Canada had a trade surplus with the United States of $78.4 billion.

Direct Investment: In 2007, Canadian investments in the United States were $226.1 billion, and U.S. investments in Canada were $288.6 billion. As the global economy began improving in 2009, Canadian companies began buying in the United States, investing $19.6 billion in the third quarter, while the United States invested only $7.1 billion in Canada in the same time period.

Pollution: Airborne pollutants that caused acid rain cross the border in both directions. Between 1996 and 2000, Canada on average received more than 3 times the amount from the United States than the United States gets from Canada. Between 2000 and 2010, both countries have made efforts (promises) to reduce the pollutants; Canada apparently has been more successful in these efforts, but the numbers are not yet available.

Immigration: There has long been a pattern of more Canadians coming to the United States than the reverse. On average, about 24,000 people annually have come to the United States from Canada since 2000, and about one-third that number of people from the United States have moved to Canada.

FIGURE 2.14 Transfers of tourists, goods, investment, pollution, and immigrants between the United States and Canada. Canada and the United States have the world's largest trading relationship. The flows of goods, money, and people across the long Canada–U.S. border are essential to both countries. However, because of its relatively small population and economy, Canada is more reliant on the United States than the reverse. All amounts shown are in U.S. dollars. [Adapted from *National Geographic* (February 1990): 106–107, and augmented with data from the Office of Travel and Tourism Industries, at http://tinet.ita.doc.gov/outreachpages/inbound.country_in_north_america.canada.html; http://internationaltrade.suite101.com/article.cfm/canadas_top_exports_imports; http://import-export.suite101.com/article.cfm/canadian_imports_exports_2008; and http://www.bea.gov/newsreleases/international/fdi/2009/pdf/fdi08.pdf; http://www.thaindian.com/newsportal/business/canada-gets-record-fdi-since-tech-bubble_10045881.html; http://www.thaindian.com/newsportal/business/canadian-companies-on-buying-spree-in-us_100281211.html; http://canadaonline.about.com/gi/o.htm?zi=1/XJ/Ya&zTi=1&sdn=canadaonline&cdn=newsissues&tm=20&f=10&tt=14&bt=1&bts=1&zu=http%3A//www.ec.gc.ca/acidrain/acidfact.html; http://en.wikipedia.org/wiki/Immigration_to_the_United_States#Origin; http://www.answerbag.com/q_view/958850.]

and reconstruction aid. In Afghanistan, however, the conflict grew more intense, and both the United States and Canada retained military forces on the ground and continued to provide humanitarian and other assistance.

2. The United States, as credibly argued by many, uses its military and economic power to promote and protect both democracy and free trade abroad, particularly in Southwest Asia, sub-Saharan Africa, Latin America, and elsewhere.

3. Canada and the United States, though very different in size of population, have similar histories, landscapes, and systems of government.

4. The rise in oil prices in 2007–2008 raised concerns about foreign dependence on petroleum products. Additionally, the rising concern over climate change focused attention on the uses of fossil fuels in transport vehicles and in other aspects of American life.

Economic Issues

The economic systems of Canada and the United States, like their political systems, have much in common. Both countries evolved from societies based mainly on family farms. Then came an era of industrialization, followed by a move to primarily service-based economies (which, as of 2008, represent 70 percent of GDP in Canada and nearly 80 percent of GDP in the United States). Both countries have important technology sectors and economic influence that reach worldwide. While Canada and the United States have similar democratic governments and face similar problems, they often take very different approaches to issues such as unemployment, health care, and international relations.

North America's Changing Food Production Systems

North America benefits from an abundant supply of food, and it is an important producer of food for foreign as well as domestic consumers (Figure 2.15). At one time, exports of agricultural products were the backbone of the North American economy. However, because of growth in other sectors, agriculture and related industries now account for less than 4.8 percent of the region's GDP, and because both countries are involved in the global economy, food is increasingly imported. 📹 **43. FOOD GLOBALIZATION**

The shift to mechanized agriculture in North America has brought about sweeping changes in employment and farm

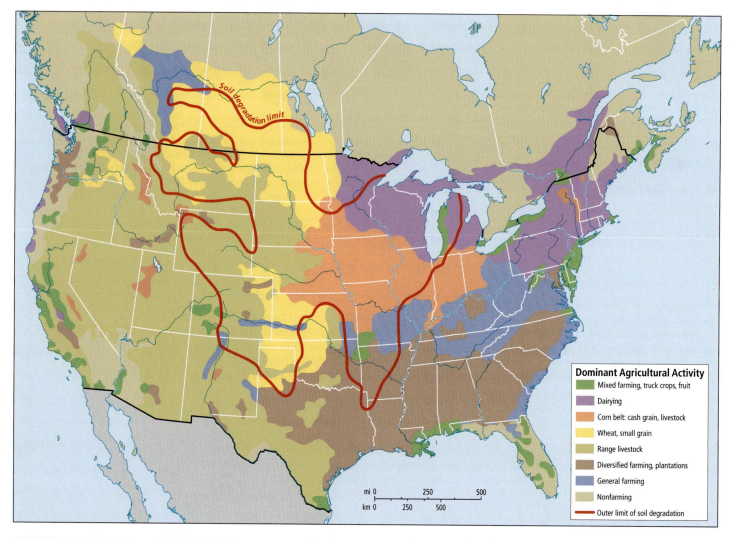

Dominant Agricultural Activity
- Mixed farming, truck crops, fruit
- Dairying
- Corn belt: cash grain, livestock
- Wheat, small grain
- Range livestock
- Diversified farming, plantations
- General farming
- Nonfarming
- Outer limit of soil degradation

FIGURE 2.15 Agriculture in North America. Throughout much of North America, some type of agriculture is possible. The major exceptions are the northern reaches of Canada and Alaska and the dry mountain and basin region (the Continental Interior) lying between the Great Plains and the Pacific coastal zone. However, in some marginal areas, such as southern California, southern Arizona, and the Utah Valley, irrigation is needed for cultivation. The soil degradation limit (red line) circumscribes the area most affected by soil erosion, chemical pollution, and loss of fertility. These problems also occur in smaller patches across the country. (Hawaii is not included here because it is covered in Oceania, Chapter 11.) [Adapted from Arthur Getis and Judith Getis, eds., *The United States and Canada: The Land and the People* (Dubuque, Iowa: William C. Brown, 1995), p. 165.]

ownership. In 1790, agriculture employed 90 percent of the American workforce; in 1890 it employed 50 percent. Until 1910, thousands of highly productive family-owned farms, spread over much of the United States and southern Canada, provided for most domestic consumption and the majority of all exports. Today, agriculture employs less than 1 percent of the U.S. workforce and less than 2 percent of the Canadian.

Farms have become highly mechanized operations that need few workers but require huge investments in land, machinery, fertilizers, and pesticides to be profitable. Large **agribusiness** corporations, with better access to loans and cash, have an advantage over individual farmers, who now produce only 27 percent of the region's agricultural output.

Corporate agriculture provides a wide variety of food at low prices for North Americans. However, in many rural areas, corporate farms have depressed local economies and created social problems. Communities in places such as rural Alberta and Saskatchewan in Canada and Iowa, Nebraska, and Kansas in the United States were once made up of farming families with similar middle-class incomes, similar social standing, and a commitment to the region. Today, farm communities are increasingly composed of a few wealthy farmer-managers amid a majority of poor, often migrant Hispanic or Asian laborers working on corporate farms and in food-processing plants for wages that are too low to provide a decent standard of living.

Under the assumption that the family farm has inherent value, the Canadian and U.S. federal governments and some states and provinces are increasing efforts to protect family farms and the rural communities of which they are a part. The states of Iowa, Kansas, Minnesota, Missouri, Nebraska, North and South Dakota, Oklahoma, and Wisconsin all have laws that restrict corporate involvement in agriculture. In parts of New England, the South, and the West, there is a revival of small family produce farms that supply organic vegetables and fruits to restaurants and to a citizenry willing to pay premium prices for food grown without chemical fertilizers and pesticides.

Food Production and Sustainability North American corporate agriculture relies heavily on the use of highly modified seeds and chemical fertilizers, pesticides, and herbicides to increase crop yields. These farming methods can contaminate food, pollute nearby rivers, and even affect distant coastal areas (see page 76). In addition, many North American farming areas have lost as much as one-third of their topsoil due to farming methods that create soil erosion.

Part of the problem may be the shift to big business agriculture. Compared with owner-operated farms in which the farmers have a personal stake in the farm's long-term sustainability (**Figure 2.16**), corporate farms are more oriented toward short-term profits and hence are less likely to take environmental precautions. However, consumers are also to blame. Despite the new movement to buy food grown organically, most consumers still prefer food that is cheaper to food grown by more sustainable methods.

Perhaps an even more controversial aspect of corporate agriculture is **genetically modified organisms (GMOs)**, in which the DNA of crop plants is modified with genes spliced from other, often

FIGURE 2.16 A small family farm in Allamakee County in northeast Iowa practicing a number of techniques that make agriculture more sustainable. To curb soil erosion, the farmers contour rows of crops so that they are perpendicular to the slope of the land, resulting in curve-shaped fields. Trees are also left in the valley bottoms to prevent the formation of gullies when it rains. To maintain soil fertility and further minimize soil erosion, crops are alternated between corn (brown in the photo) and alfalfa (green in the photo).

quite different species. GMOs produce varieties that are bigger, or more productive, or are resistant to pests and diseases. Some animals, too, are being genetically modified. However, the long-term effects on the human body of genetic manipulation of food crops and animals are unknown. The potential for allergic reactions, for example, is just being explored. Also, strong corporate control of GMO seeds could render farmers beholden to the corporations that produce them, unable any longer (due to fine-print legal restrictions) to save their own seeds for replanting. Also, the impact of GMO organisms on the genetics of the natural world is only just being explored by researchers. Some countries to which North America has traditionally exported food view GMO products as dangerous. In a number of cases, Europe has refused to import U.S. agricultural products that are genetically modified.

Concerns are also growing about the sustainability and safety of food imported into North America. Currently, 80 percent of seafood, 45 percent of fresh fruit, and 17 percent of fresh vegetables consumed in the United States are imported. In addition to a number of food safety scares related to food imported from China and elsewhere, much of this food is grown in dry climates with irrigation and hence has an unusually high *virtual water* component (see Table 1.2 on page 42). Importing this food from around the

agribusiness the business of farming conducted by large-scale operations that produce, package, and distribute agricultural products

genetically modified organisms (GMOs) animals and crop plants in which the DNA is modified

globe also contributes to climate change, as large amounts of fossil fuels are burned for transport.

Changing Transportation Networks and the North American Economy

Many of North America's current transformations are related to changes in technology and trade. Manufacturing is fading in economic importance, just as agriculture did earlier, and a vibrant new service economy is emerging. Based on technology and information exchange, new economic centers are springing up across the continent. Essential to their productivity is an extensive network of road and air transportation that enables high-speed delivery of people and goods.

The development of the inexpensive mass-produced automobile in the 1920s radically changed North American transportation. Soon trucks were delivering cargo more quickly and conveniently than the railroads could, and the number of miles of railroad track decreased between 1930 and 1980. The Interstate Highway System, a 45,000-mile (72,000-kilometer) network of high-speed, multilane roads, was begun in the 1950s and completed in 1990. Because this network was connected to the vast system of local roads, it made the truck delivery of manufactured products quicker, cheaper when gas was cheap, and more flexible than rail. Thus the highway system made possible the dispersal of industry and related services into suburban and rural locales across the country, where labor, land, and living costs were lower.

> **North American Free Trade Agreement (NAFTA)** a free trade agreement made in 1994 that added Mexico to the 1989 economic arrangement between the United States and Canada
>
> **service sector** economic activity that involves the sale of services

After World War II, air transportation also served economic growth in North America. Its primary niche is business travel because face-to-face contact remains essential to American business culture despite the growth of telecommunications and the Internet. Because many industries are widely dispersed in numerous medium-size cities, air service is organized as a *hub-and-spoke network*. Hubs are strategically located airports, such as those in Atlanta, Memphis, Chicago, and Dallas. These airports are used as collection and transfer points for passengers and cargo continuing to smaller locales. Most airports are also located near major highways, which provide an essential link for high-speed travel and cargo shipping.

In North America, flying as a way of life for personal as well as business travel may be up for some revisions. In 2007, airlines in North America carried nearly 769 million passengers, more than double the continental population. But by 2008, the economic downturn and concern over airfares and the effect of fuel emissions on climate change reduced air travel by as many as 41 million trips, affecting revenues not only for airlines, but for ground transportation, hotels, restaurants, and government tax coffers. Air cargo was down by 7.6 percent.

The New Service and Technology Economy

North America's job market has become oriented more toward knowledge-intensive jobs requiring education, knowledge of international issues, and specialized professional training in technology and management. Meanwhile, low-skill, mass-production industrial jobs are increasingly being moved abroad. **▶ 46. U.S. GEOGRAPHY REPORT**

Decline in Manufacturing Employment By the 1960s, the geography of manufacturing was changing. In the Old Economic Core, higher pay and benefits and better working conditions won by labor unions led to increased production costs. Many companies began moving their factories to the southeastern United States, where wages were lower due to the absence of labor unions, and where the warmer climate and less strict environmental regulations meant lower energy costs. **▶ 44. U.S. LABOR TRANSITION REPORT**

In 1994, the **North American Free Trade Agreement (NAFTA)** was passed. In response, many manufacturing industries, such as clothing, electronic assembly, and auto parts manufacturing, facilitated by electronic communication and transport infrastructure, began moving farther south to Mexico or overseas, where labor was vastly cheaper. Further, employers could save because laws mandating environmental protection and safe and healthy workplaces were absent or less strictly enforced.

An equally significant factor in the decline of manufacturing employment is automation. The steel industry provides an illustration. In 1980, huge steel plants, most of them in the economic core, employed more than 500,000 workers. At that time, it took about 10 person-hours and cost about $1000 to produce 1 ton of steel. Spurred by more efficient foreign competitors in the 1980s and 1990s, the North American steel industry applied new technology to lower production costs, improve efficiency, and increase production. By 2006, steel was being produced at the rate of 0.44 person-hour per ton and at a cost of about $165 per ton. As a result, the steel industry in the United States has reorganized with much steel produced in small, highly efficient mini-mills. In total, the steel industry now employs fewer than half the workers it did in 1980. Throughout North America, far fewer people are now producing more of a given product at a far lower cost than was the case 20 years ago. Therefore, although the share of the GDP produced by manufacturing has declined over the last three decades, the actual amount of industrial production expanded until the recession of 2008, and it is projected to rise again when the recession ends.

Growth of the Service Sector The economic base of North America is now a broad and varied **service sector.** Here people are engaged in the sale of services such as transportation, utilities, wholesale and retail trade, health, leisure, maintenance, finance, government, information, and education.

As of 2008, in both Canada and the United States, about three-fourths of the GDP and a majority of the jobs were in the service sector. In Canada, 76 percent of workers, and in the United States 77 percent, now work in services. High-paying jobs exist in all the service categories, but low-paying jobs are more common. The largest private employer in the United States is Walmart (1.4 million), where the average wage is $12 an hour or $24,000 a year, full time. Walmart was known for not providing health care or retirement benefits until pressured to do so by public demonstrations.

Service jobs are often connected in some way to international trade. They involve the processing, transport, and trading of agricultural and manufactured products and information that are either imported to or exported from North America. Hence, international events can shrink or expand the numbers of these jobs.

An important subcategory of the service sector involves the creation, processing, and communication of information—what is often labeled the "knowledge economy." The knowledge economy includes workers who manage information, such as those employed in finance, journalism, higher education, research and development, and many aspects of health care.

Industries that rely on the use of computers and the Internet to process and transport information—banks, software companies, and medical technology companies, for instance—are increasingly called **information technology (IT)** industries. They are freer to locate where they wish than were the manufacturing industries of the Old Economic Core, which depended on locally available steel and energy, especially coal. Because IT industries depend on highly skilled thinkers and technicians, they often locate near major universities and research institutions.

The Internet is emerging as an economic force more rapidly in North America than in any other world region. It was here that the Internet was first widely available. Though North America has only 5 percent of the world's population, it accounts for 22 percent of the world's Internet users. As of 2008, roughly 73 percent of the population of the United States used the Internet, as did 85 percent of the Canadian population, compared to 50 percent in the European Union and 15 percent for the world as a whole. The total economic impact of the Internet in North America is hard to assess, but retail Internet sales increase every year, and during the recession of 2008–2009 in the United States, while overall purchases were down, online purchases increased about 8 percent over those in 2007. Online sales of digital music alone rose 27 percent in 2008. Despite overall increases, in 2009 the recession, identity theft, and Internet fraud were deterrents to online purchases by the general public. In Canada during the recession, online searches increased dramatically but actual purchases still tended to be made in stores. In 2007, Internet sales in Canada were less than 5 percent of total retail sales.

Internet-based social networking (Facebook, MySpace, Twitter, YouTube, and others) has now moved well beyond mere personal communication to play a rapidly increasing role in marketing, not only for large retail firms, but especially for small businesses started during the recession. Social networking also played a very large role in recruiting volunteers and cash contributions during the 2008 national elections.

However, a **digital divide** has also developed as important portions of the population are not yet able to afford computers and Internet connections. For the poor, the elderly, rural dwellers, and many women and minorities, the public library or county courthouse may be their only access to computers and the Internet.

Globalization and the Economy

North America is wealthy, technologically advanced, and hugely influential in the global economy. The economy of the United States is a massive force in the world, close to that exerted by the whole of the European Union. Canada and the United States have some of the highest per capita incomes in the world. North America's advantageous position in the global economy is reflected in the pro-globalization policies of free trade promoted by major corporations and the governments of Canada and the United States.

This was not always the case and it may no longer be the case, if present trends continue. Before its rise to prosperity and global dominance, trade barriers were important aids to North American development. For example, when it achieved independence from Britain in 1776, the new U.S. government imposed tariffs and quotas on imports and gave subsidies to domestic producers. This protected fledgling domestic industries and commercial agriculture, allowing its economic core region to flourish.

Now, as wealthy and globally competitive exporters, both Canada and the United States see tariffs and quotas in other countries as obstacles to North America's economic expansion abroad and are usually powerful advocates for the reduction of trade barriers worldwide. Critics of free trade policies point out a number of fallacies in the present North American position on free trade. First, many poorer countries still need tariffs and quotas, just as North America once did. Furthermore, both countries, contrary to their own free trade precepts, still give significant subsidies to their farmers. These subsidies make it possible for North American farmers to sell their crops on the world market at such low prices that farmers elsewhere are hurt, or even driven out of business, as has been the case among Mexican small farmers who now must compete with U.S. corporate farms that have relocated to Mexico under NAFTA agreements. Beyond agriculture, the critics add, in North America, the benefits of free trade go mostly to large manufacturers and businesses and their managers, while many workers end up losing jobs to cheaper labor overseas or see their income stagnate.

Occasionally during economic recessions, when job losses in American industry are especially severe, the United States still installs tariffs to protect a particular industry. For example, in 2009 tariffs against Chinese-produced tires were introduced to save some American tire companies and alleviate the ever-growing trade imbalance with China—the United States was then spending $4.46 in China for every $1.00 the Chinese spent in the United States. This move elicited strong complaints from China because they too had unemployment problems to deal with. Meanwhile, a number of other U.S. industries put in their own claims for protective tariffs, threatening to start a chain reaction of protectionism.

Free trade is promoted regionally through NAFTA and globally through the United Nations, the World Bank, the International Monetary Fund (IMF), and the World Trade Organization (WTO) (see the discussion in Chapter 1 on pages 36–37).

NAFTA Trade between the United States and Canada has long been relatively unrestricted. The process of trade barrier reduction

information technology (IT) the part of the service sector that relies on the use of computers and the Internet to process and transport information; includes banks, software companies, medical technology companies, and publishing houses

digital divide the discrepancy in access to information technology between small, rural, and poor areas and large, wealthy cities that contain major government research laboratories and universities

formally began with the Canada–U.S. Free Trade Agreement of 1989. The creation of NAFTA in 1994 brought in Mexico as well. The major long-term goal of NAFTA is to increase the amount of trade among Canada, the United States, and Mexico. Today, it is the world's largest trading bloc in terms of the GDP of its member states. Since 1990, exports among the three countries have increased by more than 300 percent in value. NAFTA exports to the world economy by value have increased by about 300 percent for the United States and Canada and 600 percent for Mexico.

The effects of NAFTA are hard to assess because it is difficult to tell whether many changes are due to the actual agreement or to other changes in regional and global economies. However, a few things are clear. NAFTA has increased trade, and many U.S., Canadian, and Mexican companies are making higher profits because they have larger markets. For example, Walmart, the world's largest retailer, expanded aggressively into Mexico after NAFTA was passed. Mexico now has more Walmarts (1262) than any country except the United States (Figure 2.17). However, NAFTA seems to have worsened the long-standing tendency for the United States to spend more money on imports than it earns from exports. This imbalance is called a **trade deficit**. Between 1993 and 2004, the value of U.S. exports to Canada and Mexico increased 77 percent, while the value of imports increased 137 percent.

trade deficit the extent to which the money earned by exports is exceeded by the money spent on imports

NAFTA has also resulted in a net loss of around 1 million jobs in the United States. Increased imports from Canada and Mexico have displaced about 2 million U.S. jobs, while increased exports to these countries have created only about 1 million jobs. Some new NAFTA-related jobs do pay up to 18 percent more than the average North American wage. However, those jobs are usually in different locations from the ones that were lost, and the people who take them tend to be younger and more highly skilled than those who lost jobs. Former factory workers often end up with short-term contract jobs or low-skill jobs that pay the minimum wage and carry no benefits.

NAFTA and Immigration from Mexico In Mexico, NAFTA appears to have increased exports and levels of foreign investment. However, these gains have been concentrated in only a few firms along the U.S. border and have not increased growth in the Mexican economy. Moreover, stiff competition from U.S. agribusiness has resulted in about 1.3 million job losses in the Mexican agricultural sector. These job losses, in turn, fuel legal and illegal immigration from Mexico to the United States, which increased from 350,000 people per year in 1992 to 1.2 million people (arriving either legally or illegally) in 2006. From 2007 through 2009, immigration from Mexico slowed by an estimated 42 percent, due first to a crackdown on illegal entry and then to the global economic recession.

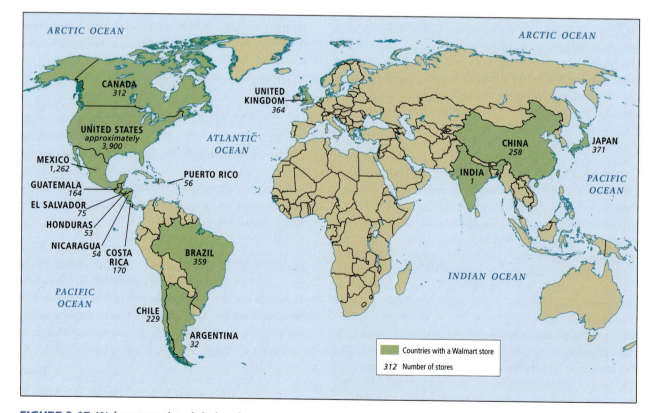

FIGURE 2.17 Walmart on the global scale. At the end of July 2009, Walmart had nearly 4400 store operations in the United States and over 3760 in 15 other countries. Walmart draws it products from over 70 countries, deals with 61,000 U.S. businesses, and is instrumental in 3 million U.S. jobs. Walmart itself employs over 1.4 million workers in the United States and 694,000 foreign workers. [Adapted from http://walmartstores.com/FactsNews/NewsRoom/9350.aspx.]

As the drawbacks and benefits of NAFTA are being assessed, talk of extending it to the entire Western Hemisphere has stalled. Such an agreement, which would be called the Free Trade Area of the Americas (FTAA), would have its own drawbacks and benefits. A number of countries, such as Brazil, Bolivia, Ecuador, and Venezuela, are wary of being overwhelmed by the U.S. economy. Even in the absence of such an agreement, trade between North America and Middle and South America is growing faster than trade with Asia and Europe. This emerging trade is discussed in Chapter 3.

The Asian Link to Globalization NAFTA is only one way in which the North American economy is becoming globalized. The lowering of trade barriers has encouraged the growth of trade between North America and Asia. One huge category of trade with Asia is the seemingly endless variety of goods imported from China—everything from underwear to the chemicals used to make prescription drugs. China's lower wages make its goods cheaper than similar products imported from Mexico, despite Mexico's proximity and membership in NAFTA. Indeed, many factories that first relocated to Mexico from the southern United States have now moved to China to take advantage of its enormous supply of cheap labor. While recent concerns about lead and other toxins in Chinese imports have made many North Americans wary of these goods, trade with China promises to remain quite robust for some time.

Asian investment in North America is also growing. For example, Japanese and Korean automotive companies located plants in North America to be near their most important pool of car buyers—commuting North Americans. They often located their plants in the rural mid-South of the United States or in southern Canada close to arteries of the Interstate Highway System. Here the Asian companies found a ready, inexpensive labor force. These workers could access high-quality housing in rural settings within a commute of 20 miles (32 kilometers) or so from secure automotive jobs that paid reasonably well and included health and retirement benefit packages. The global economic downturn beginning in 2007 eventually cut U.S. auto sales for all companies by 30 percent, threatening the continued viability of American automotive companies like Ford and General Motors and causing major layoffs and restructuring. The downturn also affected Asian automotive companies in North America. Toyota North America, for example, employed 38,000 North Americans in 2007, reduced its labor force in 2008 to about 30,000 through attrition, and in 2009 offered buyouts to another 18,000 workers. In an effort to avoid layoffs, hours were reduced and executive pay was cut by 25 percent or more.

Although business shrank in 2009, Japanese and Korean carmakers were able to succeed in North America even as U.S. auto manufacturers such as General Motors were struggling. This was primarily because more advanced Japanese automated production systems, requiring fewer but better-educated workers, produce higher-quality cars, which sell better both in North America and globally. In 2009, the Korean company Kia opened a new automated plant in West Point, Georgia. Toyota has plants in Woodstock and Cambridge, Ontario, with a total of 5000 employees, and plants in Georgetown, Kentucky (7500 employees), and Princeton, Indiana (1000 employees), where the Camry, its most popular model, is turned out at the rate of over 500,000 cars per year. Toyota stopped production of eight models in January 2010 for several weeks to investigate reports of faulty acceleration mechanisms.

Some predict that foreign carmakers will eventually take over the entire North American market, while others predict that the old American car companies will be restructured to turn out more competitive, smaller, better-built, and more fuel-efficient cars.

IT Jobs Face New Competition from Developing Countries By the early 2000s, globalization was resulting in the *offshore outsourcing* of information technology (IT) jobs. A range of jobs, from software programming to telephone-based customer-support services, were shifted to lower-cost areas outside North America. By the middle of 2003, an estimated 500,000 IT jobs had been outsourced, and forecasts are that 3.3 million more will follow by 2020. By 2009, IT jobs were still being created at a faster rate than they were ending due to the recession, but the overall trend pointed to fewer and fewer of the world's total IT jobs in North America. New IT centers are developing in India, China, Southeast Asia, the Baltic states in North Europe, Central Europe, and Russia. In these areas, large pools of highly trained, English-speaking young people work for wages that are less than 20 percent of the wages of their American counterparts. Some argue that rather than depleting jobs, outsourcing will actually help job creation in North America by saving corporations money, which will then be reinvested in new ventures. The viability of this argument remains to be seen.

◢ **45. U.S. COMPETITIVENESS**

Repercussions of the 2008 Global Economic Downturn

The severe worldwide economic downturn that began in 2007 and gained speed in the latter part of 2008 came on the heels of a long global upward trajectory of economic expansion and increasing prosperity for some, but also increasing disparity of wealth and mounting consumer debt for many. The trajectory abruptly turned into a steep downward spiral in the last year of the presidency of George W. Bush.

During his second term in office, President Bush's unpopularity was exacerbated by increasing evidence of mismanagement on several domestic fronts, especially unfettered federal spending that drastically increased budget deficits. The dwindling of demand for American-built cars (Saab, Chrysler, and certain models by General Motors and Ford) made the auto industry shrink sharply in 2008, causing several auto-supply companies to fail, with ricocheting job losses. In the same year, what had been a booming housing industry collapsed because underregulated banks had allowed millions of buyers to purchase homes with mortgages that were well beyond their means. This collapse had worldwide ripple effects because American banks had bundled many such mortgages together and quickly resold them to other (often foreign) banks, which then began to fail when hundreds of thousands of home buyers could no longer afford their mortgage payments. The situation worsened as the value of mortgaged homes sank rapidly, often well below what

social safety net the services provided by the government—such as welfare, unemployment benefits, and health care—that prevent people from falling into extreme poverty

even solvent owners owed on the loan. Foreclosures put thousands of families out of their homes and threatened the future of millions of other families. The stock market fell by nearly half, wiping out the savings and pensions of millions of Americans. By September of 2009, the United States had lost 6.9 million jobs during the recession, causing a sharp drop in U.S. consumption, which further affected world markets. Canada did not experience bank failures, but because so much of Canada's economy is linked to exports and imports from the United States (see Figure 2.14 on page 89), Canada did suffer a high rate of job losses. As firms tried to economize to stay afloat, mechanization was often a solution, meaning that the job losses would be permanent (Figure 2.18).

The global ramifications of the economic crisis that began with the banking crisis continued to ripple throughout 2010. U.S. job losses continued, and by March of 2010, the *real unemployment*—the rate that includes all workers willing to work (as opposed to all who are actively looking for work)—exceeded 14 percent.

Soon it was apparent that many industries—notably the automotive, home construction, and home furnishings industries—had expanded on the false hope that consumers for their products would inevitably materialize. When consumers suddenly stopped buying out of fear or because of loss of income, these industries

and the smaller businesses that served them also closed. The effects reached well beyond North America, into the Middle East, Europe, East Asia, Middle and South America, and Africa, where industrial and construction sites were shut down and millions of workers let go. Migrant laborers were especially affected; in every world region, they were sent home to devastated families who had once survived and even prospered on the migrants' remittances.

The Social Safety Net: Canadian and U.S. Approaches

The Canadian and U.S. governments have responded differently to the displacement of workers by economic change. For many decades, Canada has spent more per capita than the United States on social programs that lighten the financial burdens of working people in times of economic crisis. These policies have made the financial lives of working people more secure. However, until the sharp rises in health-care costs in the United States happened, Canada's higher taxes and greater regulation also made Canada slightly less attractive than the United States to new businesses.

In contrast to Canada, the United States provides little job protection or long-term government unemployment assistance. Many poor rural and urban areas in the United States that have lost jobs have experienced increases in ill health, violent crime, drug abuse, and family disintegration. All of these problems are associated with declining incomes and an inadequate **social safety net**, the services provided by the government—such as welfare, unemployment benefits, and health care—that prevent people from falling into extreme poverty. Nonetheless, for years the prevailing political position has been that the U.S. system offers lower taxes for businesses, which in turn attract new investment and new jobs, the benefits of which will *trickle down* to those most in need. The fit of this position to the reality of what actually happens has long been challenged, but in a recession the flaws of the trickle-down position become glaringly apparent. By mid-2009, those who had lost their jobs and their homes were forming tent camps on the fringes of a number of cities in Florida, California, and Washington state; and over 1 million U.S. schoolchildren were homeless by the end of 2009. In Canada, where the recession arrived somewhat later, popular opinion polls continue to support a strong social safety net as something that sets Canada apart from the United States, but recent conservative governments, perhaps seeking to make Canada more attractive to business, have reduced unemployment benefits, which are now below the average for developed countries. However, the Canadian health-care system remains strong with considerable public support.

Two Health-Care Systems In the mid-2000s, some U.S. firms began eyeing Canada as a desirable place to relocate, precisely because the country's social safety net included a government-sponsored health-care system. At present, among those U.S. firms that provide benefits, about 53 percent of the cost of every new employee is health insurance—a cost that places U.S. firms at a disadvantage in the global economy.

The contrasts between the two health-care systems are striking. The Canadian system is heavily subsidized and covers 100 percent of the population. In the United States, health care is largely

FIGURE 2.18 High-Tech Mechanization. Manufacturing in North America is an increasingly technical and mechanized process, requiring a small number of educated and well-trained workers. Often a worker's job is to maintain machinery that executes the routine tasks that large numbers of manufacturing workers used to perform. These people are maintaining equipment at a food-processing plant in Canada.

TABLE 2.1	Health-related indexes for Canada and the United States						
Country	Health care cost as a percentage of GDP	Percentage of population with no insurance	Deaths per 1000	Infant mortality per 1000 live births	Maternal mortality per 100,000 live births	Live expectancy at birth (years)	Annual health expenditures per capita (PPP* U.S.$)
Canada	11.8	0	7.4	5	7	79.8	$3173
United States	15.4	15.3	8.2	6.6	11	77.4	$6096

* PPP: Purchasing power parity.

Sources: United Nations Human Development Report 2007–2008 (New York: United Nations Development Programme),
Table 6 and Table 10; "Crude Death Rate (per 1,000 population)," *United Nations World Population Prospects: The 2008 Revision,*
at http://data.un.org/Data.aspx?q=world+population&d=PopDiv&f=variableID%3A65%3BcrID%3A900; and *Income, Poverty, and
Health Insurance Coverage in the United States: 2007,* U.S. Census Bureau, p. 19, at http://www.census.gov/prod/2008pubs/p60-235.pdf.

private and relatively expensive. The government subsidizes care for the elderly, disabled, children, veterans, and the poor, though for these groups the quality of care is often much lower than in Canada. About 37 million, or 15.3 percent of the U.S. population, are left with no health-care coverage. At the same time, the United States spends 15.4 percent of its GDP on health care, while Canada spends only 11.8 percent and gets better outcomes, outranking the United States on most indicators of overall health (Table 2.1). The taxes to cover Canada's system are substantial, but everyone shares these taxes. Because employers need not pay for private care plans, their taxes for government-provided health care are offset by the lower overall costs of creating new jobs.

📹 **48. SICKO REPORT**

Gender in National Politics and Economics

North America has some powerful political contradictions with regard to gender. Women cast the deciding votes in the U.S. presidential election of 1996, voting overwhelmingly for Bill Clinton. In the 2006 interim elections, 55 percent of women voted for the Democratic Party candidates, registering strong antiwar sentiment and making it possible for the Democratic Party to gain many seats in Congress. In 2008, Barack Obama won 56 percent of the female vote. Deciding factors in the election included such women's issues as family leave, health care, equal pay, day care, reproductive rights, and ending the war in Iraq. 📹 **49. WOMEN VOTING REPORT**

Women as Lawmakers Despite their powerful role as voters, of the 538 people in the U.S. Congress as of 2008, only 91, or 17 percent, were women. Gender equity is somewhat closer at the state level, where 24.2 percent of legislators were women as of the 2008 elections. In the Canadian Houses of Parliament, 21 percent of the members are women, and at the provincial level, 20 percent of elected officials are women. Both the United States and Canada are well behind several other countries in their percentages of women in national legislatures; in Argentina, Costa Rica, Denmark, Finland, Germany, Iceland, Mozambique, the Netherlands, Norway, South Africa, and Sweden at least 30 percent of the lawmakers are women. In 2008, Hillary Clinton was the first woman to have a serious chance at becoming the U.S. president. In Canada, women also have not yet led the country. Canada's governor-general (a largely ceremonial post appointed by the sovereign of the United Kingdom) is a woman, Michaëlle Jean (Figure 2.19).

Women in Government and Business On the economic side, in the United States, men still hold 70 percent of the top executive positions in government and business. On average, U.S. female workers earn only about 64 cents for every dollar that male workers earn (65 cents in Canada). It is estimated that if women and men earned equal wages, the poverty rate would be reduced by half. In the highly educated professions, women earn about 84 cents for every dollar a male earns. In both countries, however, women tend to work more hours than men, so their hourly earnings are actually even lower in proportion to men's.

For the first time in history, women now represent more than half of the North American labor force, with most working for

FIGURE 2.19 Michaëlle Jean, Canada's Governor General. Canada is both a parliamentary democracy and a constitutional monarchy that recognizes the sovereign of the United Kingdom as the head of state, though not as the head of government. Canada's governor general is appointed by the Queen to "carry out Her Majesty's duties." Here, Governor General Michaëlle Jean (left) is pictured with Luiz Inácio "Lula" da Silva, president of Brazil.

metropolitan areas cities of 50,000 or more and their surrounding suburbs and towns

suburbs populated areas along the peripheries of cities

megalopolis an area formed when several cities expand so that their edges meet and coalesce

male managers. In both countries, women entrepreneurs are increasingly active, starting close to half of all new businesses. Women-owned businesses tend to be small, however, and less financially secure than those in which men have dominant control. This is true in part because it is harder for women business owners to obtain loans and to obtain large contracts, facts that are addressed by special government credit programs and contract opportunities for women business owners.

In secondary and higher education, North American women have equaled or exceeded the level of men in most categories. In 2008 in the United States, 33 percent of women age 25 and over held an undergraduate degree, compared to just 26 percent of men. This imbalance is likely to increase because U.S. women age 25 to 29 were receiving 7 percent more undergraduate degrees than men. The 2010 census may show an even greater increase in women's educational achievements over those of men.

THINGS TO REMEMBER

1. North American farms have become highly mechanized operations that need few workers. To be profitable, however, they require huge investments in land, machinery, fertilizers, and pesticides.

2. In the twentieth century, the mass production of inexpensive automobiles and trucks, the Interstate Highway System, and a dramatic increase in air transportation fundamentally changed how people and goods move across the continent.

3. North America's economic base is now a service sector that is broad and varied, and includes transportation, utilities, wholesale and retail trade, health, leisure, maintenance, finance, government, information, and education. As a result, the job market has become more oriented toward knowledge-intensive jobs that require more education and specialized professional training in technology and management.

4. North America is a major promoter of globalization and free trade throughout the world. NAFTA is a major expression of this policy, yet NAFTA has led to job losses in the United States and also job losses in Mexico because North American agricultural products now undersell those of traditional Mexican farmers. The large increases in legal and illegal immigration into the United States can be attributed partly to this economic disruption in Mexico.

5. The social safety nets in Canada and the United States differ significantly in the areas of job security and retirement benefits, and especially in the health-care systems.

6. Both Canada and the United States have a lower proportion of women legislators than does Europe. Women cast the deciding votes in the U.S. presidential elections of 1996 and 2008, yet only 17 percent of Congressional representatives are women. Women earn significantly less than men in the same jobs and hold only 30 percent of executive positions, usually only at the midlevel.

Sociocultural Issues

North American attitudes about urbanization, immigration, race, ethnicity, and religion are changing rapidly. Shifting gender roles are also redefining the North American family, and an aging population is raising new concerns about the future.

Urbanization and Sprawl

A dramatic change in urban spatial patterns has transformed the way most North Americans live. Since World War II, North America's urban populations have increased by about 150 percent, but the amount of land they occupy has increased almost 300 percent (Photo Essay 2.5). This is primarily because of suburbanization, a companion process to urbanization.

Today, close to 80 percent of North Americans live in **metropolitan areas,** cities of 50,000 or more plus their surrounding suburbs and towns. Most of these people live in car-dependent suburbs built since World War II that bear little resemblance to the central cities of the past.

In the nineteenth and early twentieth centuries, cities in Canada and the United States consisted of dense inner cores and less dense urban peripheries that graded quickly into farmland. Starting in the early 1900s, central cities began losing population and investment while urban peripheries, the **suburbs,** began growing. Workers were drawn by the opportunity to raise their families in single-family homes with large lots in secure and pleasant surroundings. Some continued to work in the city, traveling to and from, usually on streetcars. After World War II, suburban growth accelerated dramatically as cars became affordable to more workers.

As North American suburbs grew and spread out, nearby cities eventually coalesced into a single urban mass. The term **megalopolis** was originally coined to describe the 500-mile (800-kilometer) band of urbanization stretching from Boston through New York City, Philadelphia, and Baltimore, to the south of Washington, D.C. Other megalopolis formations in North America include the San Francisco Bay Area, Los Angeles and its environs, the region around Chicago, and the stretch of urban development from Eugene, Oregon, to Vancouver, British Columbia.

This pattern of urban sprawl requires residents to drive automobiles to complete most daily activities such as grocery shopping or commuting to work. Increasing air pollution and emission of greenhouse gases that contribute to climate change are two major side effects of the dependence on vehicles that comes with urban sprawl. Another important environmental consequence is habitat loss brought about by the invasion of farmland, forest, grassland, and desert by suburban development.

Farmland and Urban Sprawl Urban sprawl drives farmers from land that is located close to urban areas, because farmland on the urban fringe is very attractive to real estate developers. The land is cheap compared to urban land, and it is easy to build roads and houses on it since farms are generally flat and already cleared. As farmland is turned into suburban housing, property taxes go up, and soon surrounding farmers can no longer afford to keep their land, so they sell it to housing developers. In North America each

North America is highly urbanized, with 79 percent of the population living in cities. Some of the wealthiest cities in the world are located here, though only a few have achieved high levels of "livability." Many cities, especially those in the United States, are characterized by sprawling development patterns that require high levels of dependence on the automobile.

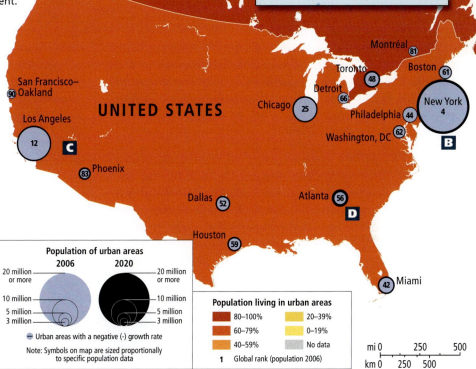

CANADA

UNITED STATES

Montréal 81
Toronto 48
Boston 61
Detroit 66
Chicago 25
New York 4
Philadelphia 44
Washington, DC 62

San Francisco–Oakland 90
Los Angeles 12
Phoenix 83
Dallas 52
Atlanta 56
Houston 59
Miami 42

A Vancouver, Canada, is consistently ranked the most "livable" city in North America, and among the top four in the world. It is a leader in controlling sprawl. Over the past 10 years the use of public transportation has risen by 50 percent while the use of cars has fallen by 30 percent.

B New York City is the second wealthiest city in the world, the financial capital of North America, and its largest city. It is known for its extensive mass transit system (shown here), its high population density (for North America), and its world-class music and arts scene. In terms of livability New York ranks toward the middle for the region.

Population of urban areas

2006	2020
20 million or more	20 million or more
10 million	10 million
5 million	5 million
3 million	3 million

Urban areas with a negative (-) growth rate

Note: Symbols on map are sized proportionally to specific population data

Population living in urban areas

80–100%	20–39%
60–79%	0–19%
40–59%	No data

1 Global rank (population 2006)

mi 0   250   500
km 0   250   500

C Las Vegas, Nevada, was the fastest growing large city in North America between 2000 and 2007, with 31.8 percent growth. However, the recession that began in 2008 slowed growth considerably. In most rankings of livability Las Vegas comes in toward the lower end for North America.

D Morning traffic in Atlanta, Georgia, which added almost a million people between 2000 and 2007, more than any other city in North America. It is among the least dense and most sprawling cities in the region, with almost universal dependence on cars for transport. In most rankings of livability, Atlanta comes in toward the lower middle for North America.

year, 2 million acres of agricultural and forest lands make way for urban sprawl.

Advocates of farmland preservation argue that beyond food and fiber, farms also provide economic diversity, soul-soothing scenery, and even habitat for some wildlife. For example, the town of Pittsford, New York, a suburb of Rochester, decided that farms were a positive influence on the community. Mark Greene's 400-acre, 200-year-old farm lay at the edge of town. As the population grew, the chances of the farm remaining in business for another generation looked dim. As land prices rose, so did property taxes, and the Greene family could not meet their tax payments. Residents in the new suburban homes sprouting up on what had been neighboring farms pushed local officials to halt normal farm practices, such as noisy nighttime harvesting or planting, spreading smelly manure, or importing bees to pollinate fruit trees. Pittsford, however, decided to stand by the farmers by issuing $410 million in bonds so that it could pay Greene and six other farmers for promises that they would not sell their 1200 acres to developers, but would continue to farm them.

Pittsford's solution could become a trend throughout North America. Urban sprawl is now an issue virtually everywhere, and open flat and rolling spaces are most vulnerable. The term *smart growth* has been coined for a range of policies aimed at stopping sprawl by making existing urban areas more livable. The Smart Growth movement pushes for convenient, attractive, and affordable multicultural neighborhoods, with more walking and less traffic (lower CO_2 emissions), shopping that is accessible on foot, usable open spaces, and mass transit. Ironically, tax dollars spent on commuter highways actually promote sprawl and increase CO_2 emissions. The same money could be better spent on urban public transportation. An urban sprawl index that ranks cities on the severity of

brownfields old industrial sites whose degraded conditions pose obstacles to redevelopment

the problem can be found at http://www.smartgrowthamerica.org/sprawlindex/sprawlexecsum.html.

Sprawl can also affect deserts. A striking example is Phoenix, Arizona, one of the fastest-growing areas in the United States, yet one that presents numerous drawbacks, especially lack of water. The map in Figure 2.20 shows how Phoenix has sprawled into surrounding farmland and desert. Phoenix now has 3.5 million people and is so short of water that scientists are forecasting doom. More than a third of Phoenix's water is diverted from the Colorado River, several hundred miles to the west. Numerous conservation strategies are in play, and all citizens are expected to participate. Per capita water usage is down 20 percent over what it was 20 years ago.

Inner-City Abandonment and Reoccupation The same forces that drive urban sprawl leave behind huge tracts of abandoned inner-city land. Old industrial sites that once held factories or rail yards are called **brownfields.** Because they are often contaminated with chemicals and covered with obsolete structures, they can be very expensive to redevelop for other uses.

Also left behind in the inner cities are the least-skilled and least-educated citizens. In the late 1980s, 70 of the 100 largest U.S. cities had middle-class majorities, most of whom were white. By 2000, many of these people had moved to the suburbs and almost half of the largest U.S. cities had nonwhite majorities that were, on average, poor. The 2010 census is likely to show a continuation of this trend. The majority population in these cities is a mixture of African-Americans, Asians, Hispanics, and other groups who identify themselves as nonwhite. Some are relatively affluent and are leading the regeneration of old city centers. Others are in great need of the very services—health care, schools, and social support (including churches, synagogues, and mosques)—that have moved to the suburbs.

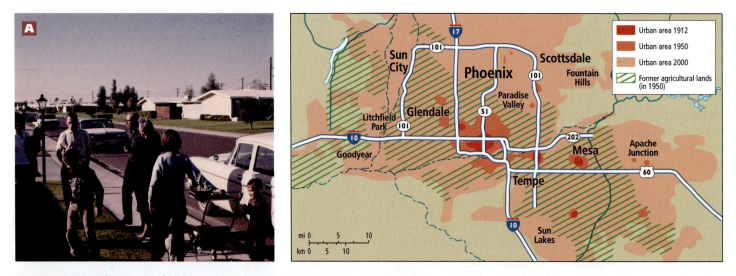

FIGURE 2.20 Urban sprawl in Phoenix, Arizona. Phoenix grew rapidly between 1950 and 2000. **(A)** The photo was taken in 1964 in Sun City, then a new car-oriented suburb. To see an informative animation of the rate and extent of urban sprawl in Phoenix, go to http://www.amnh.org/sciencebulletins/bio/v/sprawl.20050218/.

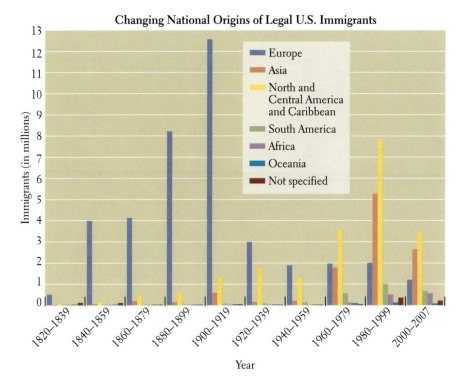

Changing National Origins of Legal U.S. Immigrants

FIGURE 2.21 Changing national origins of legal U.S. immigrants. Beginning in 1920, the number of immigrants from Europe began declining significantly from earlier years, a trend that continues. From the 1960s on, the number of immigrants from the Americas and Asia began increasing significantly, a trend that continues, but declined somewhat after 2000. [Data source: Department of Homeland Security, "Persons Obtaining Legal Permanent Resident Status by Region and Selected Country of Last Residence: Fiscal Years 1820 to 2007," *Yearbook of Immigration Statistics: 2007,* Immigrants, Table 2, at http://www.dhs.gov/ximgtn/statistics/publications/LPR07.shtm.]

In the 1980s, a movement labeled *New Urbanism* emerged, focused on creating new places to live by fostering European-like dense settlements of refined architecture and urban design with housing of mixed types and prices, in close proximity to work, shopping, restaurants, and entertainment. Walking or mass transit is preferred over the automobile, and cultural diversity is considered an amenity. In the 2000s, the New Urbanism trend is blending with other trends: the revival of old inner cities, green living, lower energy use, and sustainable food production and distribution. Across the continent there are efforts to bring organic farmers into old urban market squares to supply healthy locally grown meat and produce and other rural craft amenities to city dwellers.

Gentrification of old, urban residential districts is increasingly common as more affluent people invest substantial sums of money in renovating old houses and apartments, often displacing poor inner-city residents in the process. The effect of gentrification on the displaced poor appears to be somewhat less harsh in Canada than in the United States, primarily because of Canada's stronger social safety net that better ensures housing and social services. Some U.S. cities (for example, Knoxville, Tennessee; Portland, Oregon; and Charlotte, North Carolina) have initiated New Urbanism projects to rehouse urban poor people in pleasant, newly built, walkable neighborhoods with conveniently located services.

Immigration and Diversity

Immigration has played a central role in populating both the United States and Canada, and most of the region's people have roots in some other part of the globe, the majority in Europe. However, new waves of migration from Middle and South America and parts of Asia promise to make this a region where eventually the population will be a mix of European and non-European descent (Figure 2.21). Most major North American cities are already characterized by wide ethnic diversity. In Canada, with its relatively small population, the recent surge in immigration has led to near majorities of foreign-born residents in a number of leading cities.

50. IMMIGRATION AND POPULATION REPORT

In the United States, the spatial pattern of immigration is also changing. For decades, immigrants settled mainly in coastal or border states such as New York, Florida, Texas, and California. However, since about 1990, immigrants are increasingly settling in interior states such as Illinois, Colorado, and Utah (Figure 2.22).

Immigration and cultural diversity are topics of increasing public debate in the United States. By 2008, polls showed that a majority of U.S. residents felt that future immigration should be controlled. At the same time, a clear majority also felt that immigration is a strength of the country, and 76 percent supported the idea that illegal immigrants should have a chance to become citizens. A somewhat smaller but vocal group insisted that illegal immigrants should be deported, regardless of the circumstances. Consider below some of the main issues in the immigration debate in the United States.

gentrification the renovation of old urban districts by middle-class investment, a process that often displaces poorer residents

Do new immigrants cost U.S. taxpayers too much money? Repeated studies have shown that over the long run, immigrants contribute more to the U.S. economy than they cost. Legal immigrants have passed an exhaustive screening process that assures they will not pose any sort of threat to the country and that they will be self-supporting. Most such immigrants start to work and pay taxes within a week or two of their arrival in the country. Immigrants who draw on taxpayer-funded services such as welfare tend to be legal refugees fleeing a major crisis in their homeland and are dependent only in the first few years after they arrive. More than one third of immigrant families are firmly within the middle class, with incomes of $45,000 or higher. Even illegal immigrants play important roles as payers of payroll taxes, sales taxes, and indirect property taxes through rent. Perhaps most noteworthy is the role of immigrants in support of the elderly. As the

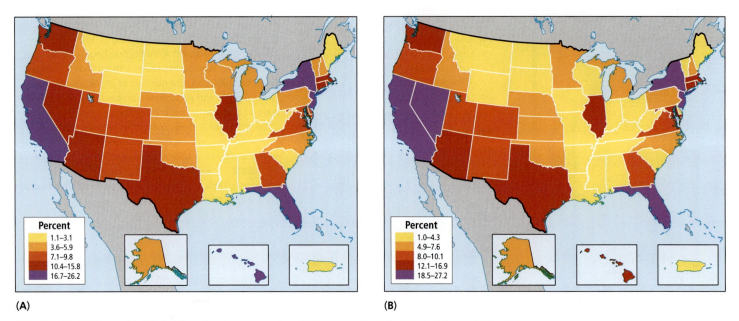

(A)   **(B)**

FIGURE 2.22 Percent of total foreign-born people within each state in 2000 (A) and 2007 (B).

[U.S. Census Bureau, "Percent of People Who Are Foreign Born: 2007," from the 2005–2007 *American Community Survey*, M0501]

U.S. population ages and the base of native-born young workers shrinks, Social Security contributions by young new immigrant workers will provide essential support for the elderly. On a more personal level, immigrant women are commonly the low-paid caregivers for the elderly in the home and in institutions.

A 2004 study by the National Institutes of Health (NIH) reports that on average, immigrants are healthier and live longer than native U.S. residents. Hence, they represent less drain on the health-care and social service systems. The NIH attributes this difference to a stronger work ethic, a healthier lifestyle that includes more daily physical activity, and the more nutritious eating patterns of new residents compared with U.S. society at large. Unfortunately, these healthier practices tend to diminish the longer immigrants are in the country and the healthier status does not carry over to their children. A study in 2009 found that immigrant children are especially vulnerable to television advertisements for food and that 98 percent of the ads they view are for products high in fat, sodium, and sugar; as a result, like American children generally, more than 36 percent of immigrant children are overweight or obese.

Do immigrants take jobs away from U.S. citizens? The least-educated, least-skilled American workers are the most likely to find themselves competing with immigrants for jobs. In a local area, a large pool of immigrant labor can drive down wages in fields like roofing, landscaping, and general construction. Immigrants with little education now fill many of the very lowest paid service and agricultural jobs.

It is often argued that U.S. citizens have rejected these jobs because of their low pay, and hence immigrants are needed to fill them. Others argue that these jobs might pay more, and hence be more attractive to U.S. citizens, if there were not a large pool of immigrants ready to do the work for less pay. Research has failed to clarify the issue. Some studies show that immigrants have driven down wages by 7.4 percent for U.S. natives without a high school diploma. A 2006 study by the National Bureau of

Economic Research found that not only do immigrants not crowd out employment for natives, they actually increase total productivity because they promote competition.

Professionals in the United States occasionally compete with highly trained immigrants for jobs, but such competition usually occurs in occupations where there is a scarcity of native-born people who are trained to fill these positions. The computer engineering industry, for example, regularly recruits abroad in such places as India, where there is a surplus of highly trained workers. In this case it is unclear whether these skilled immigrants are driving down wages. Until the recession caused jobs to be eliminated, it was quite clear that there were not enough sufficiently trained Americans to fill the available positions.

📺 **253. SKILLED FOREIGN WORKERS IN U.S. MAY HAVE TO LEAVE**

Are too many immigrants being admitted to the United States? Several circumstances contribute to increasing calls for curtailing immigration into the United States. First, there is often a lack of awareness about just how rigorous the screening process is for legal immigrants. Second, after a mid-twentieth-century lull, the immigration rate has accelerated again (see Figure 2.21), particularly beginning in the 1990s. Immigrants and their children accounted for 78 percent of the U.S. population growth in the 1990s. Legal and undocumented immigration into North America slowed in 2008 and 2009, due partly to the recession and partly to stricter enforcement of immigration laws. At 2006 rates, by the year 2050 the U.S. population will have exceeded 420 million, but according to the latest figures that may not happen. The 2010 census will provide new insights. 📺 **47. AMERICAN CENSUS REPORT**

A third reason for the call to limit immigration is that undocumented (illegal) immigration has reached unprecedented levels over the past 30 years. In February 2009, Homeland Security reported that total immigrants numbered 31.3 million in the United States, and of this total, 11.6 million were undocumented. Notably, the latter number represents a decline of about 200,000 from the previous year. Undocumented immigrants tend to lack skills, and they are not screened for criminal backgrounds as are all legal immigrants. The Migration Policy Institute (http://www.migrationinformation.org/USfocus/), a nonpartisan, nonprofit think tank that studies migration worldwide, points out, however, that many undocumented migrants were once legal, but failed to renew their papers. Only tiny percentages have committed offences; and for well over half of those arrested, the sole charge is that they are illegal immigrants, not that they have committed other crimes. Statistically, undocumented migrants are actually less likely to participate in criminal behavior than the general population. Research shows that the decision to migrate illegally into North America is largely a reluctant one undertaken because of a severe lack of economic opportunity at home. The trip is often agonizing, as the following vignette illustrates.

51. IMMIGRATION LABOR SHORTAGE REPORT

Personal Vignette Sheriff's deputy Michael Walsh works along the Arizona–Mexico border. Walsh describes a recent encounter with two bereaved young Mexican men holding the body of their relative: "Mostly you just find skeletons in the desert. This time there was a lot more emotion, family emotion…. Obviously they were close, they were crying. You have a name, you know that he had a brother, a cousin, a family in Mexico…."

Matias Garcia, age 29, died after walking 32 miles through the desert. He was a Zapotec Indian who lived near Oaxaca (in southern Mexico) with his wife and three children, as well as his parents, younger brothers, and several cousins. Matias's cash crop of chili peppers was ruined by a spring frost just as they were ripening, leaving him in debt. He reluctantly decided to risk a trip to the United States to work in some vineyards where he had worked on and off since he was a teenager. From there he could send money home to his family to keep the house in repair and to send his children to school. But it takes money just to cross the border, and it took months for Matias, his younger brother, and a cousin to save the necessary amount.

Since the NAFTA agreement of 1994, the border has been more carefully patrolled, so the men decided to attempt to cross the less-patrolled but more perilous Arizona desert on a route known as the "Devil's Highway." May is one of the hottest and driest months in this part of the Americas. The men tried to avoid the worst of the heat by walking at night, but they failed to reach the highway on the Arizona side by dawn. They ran out of water, the sun became especially hot, and Matias began having seizures. His brother and cousin carried him, desperately looking for the highway, but he died shortly before they found it and could flag down someone to call for help. That's where Deputy Walsh found the two men grieving over Matias's body. *(Find out more about Matias Garcia, why he migrated, how he died, and what his death has meant back in his village. Watch the Frontline/World video "Mexico: A Death in the Desert" on the web at http://www.pbs.org/frontlineworld/stories/mexico/.)* ∎

In 2006, federal and state laws were proposed to criminalize any assistance to undocumented immigrants. Opponents of the legislation organized large street demonstrations in cities across the United States. Some states, such as Illinois, Washington, Idaho, and New Mexico, all with large labor needs, support undocumented immigrants with programs that help them and their children. By contrast, Virginia, Kentucky, South Carolina, and Arizona have passed stern laws regarding illegal immigrants. Along the U.S.–Mexico border, the work of some 10,000 border guards was judged insufficient by conservative private militia groups, such as the Minutemen, who placed themselves along the border as vigilante enforcers, until they themselves were arrested. Similar sentiment has led to the creation of the U.S.–Mexico border barrier. Because North America is full of second- and third-generation immigrants, because there is a continuing need for immigrants' services as workers, and because the summary expulsion of immigrants would breech human rights, immigration promises to remain a major political issue across the continent for many years to come.

Gender Patterns Among Migrants to the United States Immigrants to the United States and Canada tend to be equally divided between men and women. In 2008 there were nearly 19 million female immigrants in the United States, making up 12 percent of total females. Nearly 4 out of 5 were between the ages of 18 and 64 and they tended to become citizens more often than male immigrants. An unspecified majority of immigrant women had at least a high school diploma but their incomes tended to be lower than immigrant men and lower than comparably educated native-born women. About a quarter of immigrant women were Mexican; a third of all immigrant men were Mexican.

Race and Ethnicity in North America

Despite strong scientific evidence to the contrary, people across the world still perceive skin color and other visible anatomical features to be significant markers of intelligence and ability. As discussed previously (see Chapter 1 on pages 61–62), the science of biology tells us there is no validity to such assumptions. The same is true for **ethnicity**, which is the cultural counterpart to race, in that people may ascribe overwhelming (and unwarranted) significance to cultural characteristics, such as religion or family structure or gender customs. Hence, race and ethnicity are very important sociocultural factors not because they *have* to be, but because people *make* them so.

> **ethnicity** the quality of belonging to a particular culture group

Numerous surveys show that large majorities of Americans of all backgrounds favor equal opportunities for minority groups. Nonetheless, in both the United States and Canada, many middle-class African-Americans, Native Americans, and Hispanics (and to a lesser extent, Asian-Americans) report experiencing both overt and covert discrimination that affects them economically as well as socially and psychologically. And indeed, even a cursory examination of statistics on access to health care, education, and financial services shows that on average, Americans have quite uneven experiences based on their racial and ethnic characteristics.

Despite the increasing diversity of North America, in this region the term *race* has usually been used in relation to deeply embedded discrimination against African-Americans. Prejudice has clearly hampered the ability of African-Americans to reach social and economic equality with Americans of other ethnic backgrounds. In the United States, and somewhat less so in Canada, despite the removal of legal barriers to equality, African-Americans as a group still experience lower life expectancies, higher infant mortality rates, lower levels of academic achievement, higher poverty rates, and greater unemployment than other groups.

U.S. Population by Race and Ethnicity, 1950, 2006–2008, and 2050 (projected)

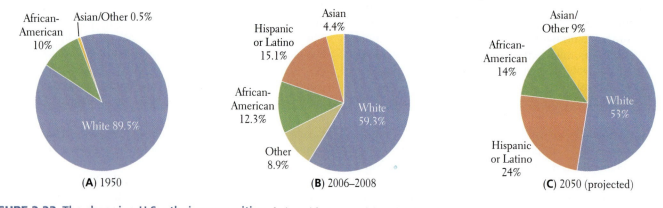

(A) 1950 **(B)** 2006–2008 **(C)** 2050 (projected)

FIGURE 2.23 The changing U.S. ethnic composition. [Adapted from Jorge del Pinal and Audrey Singer, "Generations of Diversity: Latinos in the United States," *Population Bulletin* 52 (October 1997): 14; and U.S. Census Bureau, "Race by Sex, for the United States, Urban and Rural, 1950, and for the United States, 1850 to 1940," Census of Population: 1950, Volume 2, Part 1, United States Summary, Table 36, 1953, at http://www2.census.gov/prod2/decennial/documents/21983999v2p1ch3.pdf.]

In 2001, Hispanics overtook African-Americans as the largest minority group in the United States. Because of a higher birth rate and a high immigration rate, the Hispanic population increased by 58 percent in the 1990s. In 2000, Asian-Americans made up only 3.6 percent of the U.S. population, although their population had increased by 48 percent since 1990. **Figure 2.23** shows the changes in the ethnic composition of the U.S. population in 1950 and 2004 and projected in 2050.

In the United States, there are increasing discrepancies in income among Asian-Americans, Euro-Americans, Native Americans, Hispanics, and African-Americans (**Figure 2.24**). Yet, over the past few decades, many non–Euro-Americans have joined the middle class, achieving success in the highest ranks of government and business. In particular, African-Americans have achieved advanced education in large numbers, and more than one-third now live in the suburbs. Nonetheless, overall, African-Americans, Hispanics, and Native Americans as groups remain the country's poorest people.

Is anything other than prejudice holding back some African-Americans, Hispanics, and Native Americans? Some argue yes. They point to the experience of first- and second-generation people of African descent, such as President Barack Obama (who is half African and half Euro-American) or African-Caribbean immigrants, such as Maya Angelou, Patrick Ewing, Colin Powell, or Canada's Governor General, Michaëlle Jean, who have been notably successful even though their experience of discrimination and a family history of slavery is parallel to that in North America. A similar observation can be made about North Americans of Chinese and South Asian background. Although these groups often started out poor and suffered severe discrimination, they are now among the most prosperous in North America (see Figure 2.24).

Observations such as these have led to the argument that some African-Americans, Hispanics, Native Americans, and persistently disadvantaged Anglo-Americans suffer from a *culture of poverty*. The argument is that poverty and low social status breed the perception among their victims that there is no hope for them

and hence no point in trying to succeed. This perception is supported by the fact that, relative to wealthier North Americans of all races, the poorest, no matter what their background, have fewer opportunities to get a decent education, a well-paying job, or a nice home to live in.

A major aspect of the culture of poverty is the single-parent family. In 2005, while only 23 percent of Euro-American children

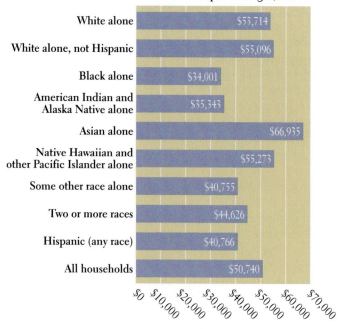

FIGURE 2.24 Median household income in the past 12 months by race and ethnicity, 2007. Amounts were reported in 2007 dollars, adjusted for inflation. [U.S. Census Bureau, Income, Earnings, and Poverty Data from the 2007 *American Community Survey*, 2008, at http://www.census.gov/prod/2008pubs/acs-09.pdf.]

and 17 percent of Asian-American children lived in single-parent families, 65 percent of African-American, 49 percent of American Indian, and 36 percent of Hispanic children did. In these situations, children usually stay with their mothers, and fathers often are not active in their support and upbringing. The enormous responsibilities of both child rearing and breadwinning are often left in the hands of undereducated young mothers who, themselves in need, are unable to help their children advance.

However, another explanation of persistent poverty among some of North America's minorities is that it is part of a larger problem of economic and spatial segregation based on class. In both the United States and Canada, the increasingly prosperous middle class, of whatever race or ethnicity, has moved to the suburbs. Hence, the very poor rarely have the chance to associate with models of success, and the successful no longer know anyone who is poor. Evidence for this class-based explanation is that when privileged Americans of all racial backgrounds share middle- and upper-class neighborhoods, workplaces, places of worship, and marriages, attention paid to skin color decreases markedly.

Meanwhile, the poor of all races see the material evidence of the success of others all around them but have little access to the life choices that made that success possible.

Religion

Because so many early immigrants to North America were Christian in their home countries, Christianity is currently the predominant religious affiliation claimed in North America. Seventy-six percent of Americans identified themselves as Christians in 2008. Nonetheless, virtually every medium-sized city has at least one synagogue, mosque, and Buddhist temple. In some localities, adherents of Judaism, Islam, or Buddhism are numerous enough to constitute a prominent cultural influence (Figure 2.25).

There are many versions of Christianity in North America, and their geographic distributions are closely linked to the settlement patterns of the immigrants who brought them (see Figure 2.25). Roman Catholicism dominates in regions where Hispanic, French, Irish, and Italian people have settled—in southern

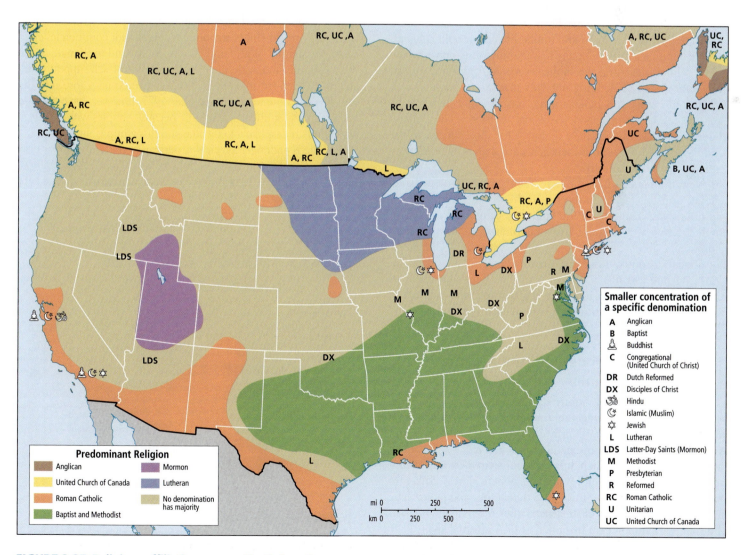

FIGURE 2.25 Religious affiliations across North America. [Adapted from Jerome Fellmann, Arthur Getis, and Judith Getis, *Human Geography* (Dubuque, Iowa: Brown & Benchmark, 1997), p. 164.]

Louisiana, the Southwest, and the far Northeast in the United States, and in Québec and other parts of Canada. Lutheranism is dominant where Scandinavian people have settled, primarily in Minnesota and the eastern Dakotas. Mormons dominate in Utah.

Baptists, particularly Southern Baptists and other evangelical Christians, are prominent in the religious landscapes of the South, which is known as the Bible Belt. Christianity—especially the Baptist version—is such an important part of community life in the South that newcomers to the region are frequently asked what church they attend. (This question is rarely asked in most other parts of the continent.)

According to data published by the U.S. Census Bureau, people in the United States who claim no religion ("nones") made up 15 percent of the population in 2008. The group includes atheists, agnostics, and the majority who simply claim no religion by choice.

The proper relationship of religion and politics has long been a controversial issue in the United States. This is true in large part because the framers of its Constitution, in an effort to ensure religious freedom, supported the idea that church and state should remain separate. In the past three decades, however, many conservative Christians have successfully pushed for a closer integration of religion and public life. Their political goals include banning abortion for any reason, promoting prayer in the public schools, teaching the biblical version of creation instead of evolution, and contending not only that sexual orientation is a matter of choice not biochemistry, but also that marriage by gays and lesbians damages heterosexual marriage. The policies of conservative Christians have met with the most success in the southern United States, but their goals are shared by a minority scattered across the country.

New immigrants have brought their own faiths and belief systems, and they are contributing to the debate about religion and public life. Some leave the faith they came with and adopt another. Some 15 percent of Hispanic immigrants have left their traditional Catholic faith and are now evangelical Christians. The long-term outcome of these contentions in American religious and political life is not yet apparent. Reputable national surveys have consistently indicated that a substantial majority of Americans favor the continued separation of church and state and personal choice in belief and behavior.

nuclear family a family consisting of a married father and mother and their children

Gender and the American Family

The family has repeatedly been identified as the institution most in need of support in today's fast-changing and ever more impersonal North America. A century ago, most North Americans lived in extended families of several generations. Families pooled their incomes and shared chores. Aunts, uncles, cousins, siblings, and grandparents were almost as likely to provide daily care for a child as were the mother and father. The **nuclear family**, consisting of a married father and mother and their children, is a rather recent invention of the industrial age.

Beginning after World War I, and especially after World War II, many young people left their large kin groups on the farm and migrated to distant cities, where they established new nuclear families.

Soon suburbia, with its many similar single-family homes, seemed to provide the perfect domestic space for the emerging nuclear family.

This small, compact family suited industry and business, too, because it had no firm ties to other relatives and hence was portable. Many North Americans born since 1950 have moved as many as ten times before reaching adulthood. The grandparents, aunts, and uncles who were left behind missed helping raise the younger generation, and they had no one to look after them in old age. Nursing homes for the elderly proliferated.

In the 1970s, the whole system began to come apart. It was a hardship to move so often. Suburban sprawl meant onerous commutes to jobs for men and long, lonely days at home for women. Women began to want their own careers, and rising consumption patterns made their incomes increasingly useful to family economies. By the 1980s, 70 percent of the females born between 1947 and 1964 were in the workforce, compared with 30 percent of their mothers' generation.

Once employed, however, women could not easily move to a new location with an upwardly mobile husband. Nor could working women manage all of the family's housework and child care as well as a job. Some married men began to handle part of the household management and child care, but the demand for commercial child care grew sharply. With family no longer around to strengthen the marital bond, and with the new possibility of self-support available to women in unhappy marriages, divorce rates rose into the mid-1970s. Those who married after 1975, however, have a slowly declining rate of divorce, primarily because couples with a college education are less likely to divorce and this group is increasing.

There is no longer a typical American household; there is only an increasing diversity of household forms (**Figure 2.26**). In 1960, the nuclear family—households consisting of married couples with

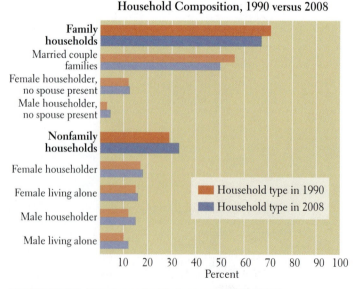

FIGURE 2.26 U.S. households by type, 1990–2008. [Data source: U.S. Census Bureau, "Households and Persons per Household by Type of Household: 1990 to 2008," Population: Households, Families, Group Quarters, Table 61, 2010, at http://www.census.gov/compendia/statab/cats/population/households_families_group_quarters.html.]

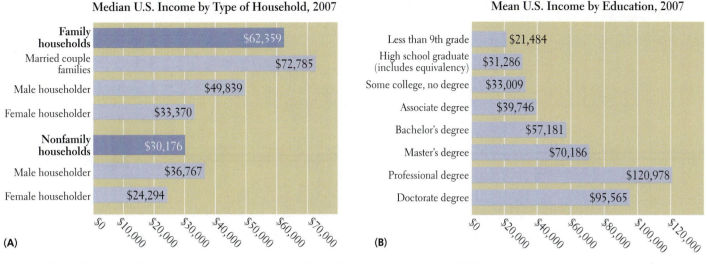

FIGURE 2.27 Median U.S. income by type of household and by level of education, 2007.

[Data sources: (A) U.S. Census Bureau, "Money Income of Households—Distribution by Income Level and Selected Characteristics: 2007," Income, Expenditures, Poverty, & Wealth: Household Income, Table 676, 2010, at http://www.census.gov/compendia/statab/cats/income_expenditures_poverty_wealth/household_income.html; (B) U.S. Census Bureau, "Mean Earnings by Highest Degree Earned: 2007," Education: Educational Attainment, Table 227, 2010, at http://www.census.gov/compendia/statab/cats/education/educational_attainment.html.]

children—were 74.3 percent of U.S. households. This number declined to 56 percent by 1990 and to 50 percent by 2008. Family households headed by a single person (female or male) rose from 10.7 percent in 1960 to 17 percent in 2008. Nonfamily households (unrelated by blood or marriage) rose from 15 percent of the total in 1960 to 33 percent in 2008. One result of America's aging population is that more Americans than ever are living alone—about one-quarter of the population at any given time; the majority of these single-person households are over the age of 45 and were once part of a nuclear family that dissolved due to divorce or death.

Some of these new family forms do not provide well for the welfare of children. In 2006, more than 29 percent of U.S. children lived in single-parent households. Although most single parents are committed to rearing their children well, the responsibilities can be overwhelming. Single-parent families tend to be hampered by economic hardship and lack of education (**Figure 2.27**). The vast majority are headed by young women whose incomes, on average, are more than one-third lower than those of single male heads of household (Figure 2.27). Low income and low levels of education are closely linked, as the graph in Figure 2.27B demonstrates. A result of these patterns of single-parent households where education levels and income are low is that children are disproportionately poor. In 2007 in the United States, 18 percent of children lived in poverty, whereas only 11.4 percent of adults did. In Canada, 14.7 percent of children were poor. In Sweden, by comparison, 2.4 percent of children lived in poverty; in Ireland, 12.4 percent; and in Poland, 12.7 percent.

THINGS TO REMEMBER

1. North America has some of the world's wealthiest cities, most of which are in the eastern United States. Nearly 80 percent of North Americans now live in metropolitan areas.

2. Since World War II, North America's urban populations have increased by about 150 percent, but the amount of land they occupy has increased almost 300 percent, a phenomenon known as urban sprawl.

3. A steady increase in migration from Middle and South America and parts of Asia promises to make North America a region where most people are of non-European descent.

4. Statistics on access to health care, education, and financial services by race and ethnicity show that on average, Americans have quite uneven experiences based on their racial and ethnic characteristics.

5. There is no longer a typical American household; there is only an increasing diversity of forms.

6. A large percentage of children in North America live in poverty, compared to other developed countries.

Population Patterns

The population map of North America (**Photo Essay 2.6**) shows the uneven distribution of the more than 340 million people who live on the continent. Canadians account for just under one-tenth (34 million) of North America's population. They live primarily in southeastern Canada, close to the border with the United States. The population of the United States is almost 307 million, with the greatest concentration of people shifting away from the Old Economic Core into other regions of the country that are now growing much faster (see **Figure 2.28** on page 109). Just how this trend will play out, however, is uncertain. In 2004, the U.S. Census Bureau projected that between 2000 and 2030 the Northeast and Middle West would grow by under 10 percent while the South and West would grow by more than 40 percent.

Population density patterns in North America reflect not only recent social and economic conditions but also reveal circumstances of the past and hint at coming trends.

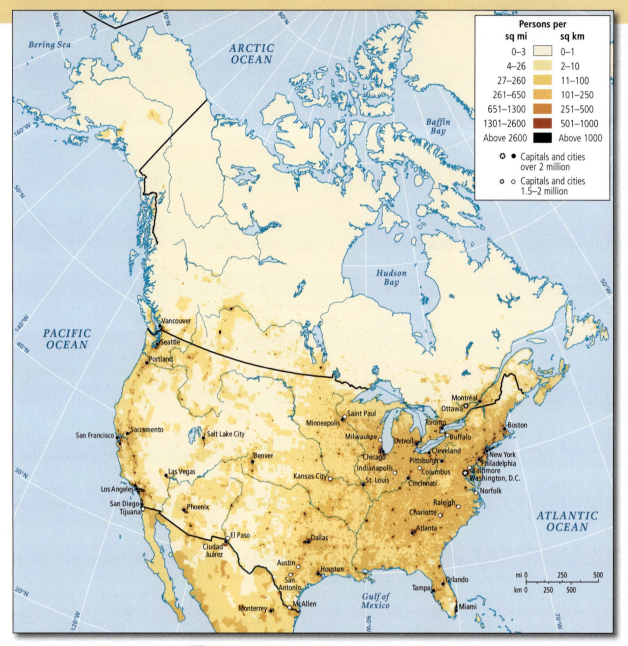

Persons per

sq mi		sq km
0–3		0–1
4–26		2–10
27–260		11–100
261–650		101–250
651–1300		251–500
1301–2600		501–1000
Above 2600		Above 1000

✪ ● Capitals and cities over 2 million
✪ ○ Capitals and cities 1.5–2 million

A North America's population is rapidly aging, and by 2050, one in five people will be over the age of 65.

B Populations are shrinking in many small farming towns in the Middle West and other parts of the interior of North America.

C Roughly 5000 immigrants arrive in North America each day. Increasingly immigrants are found in cities and small towns across the country.

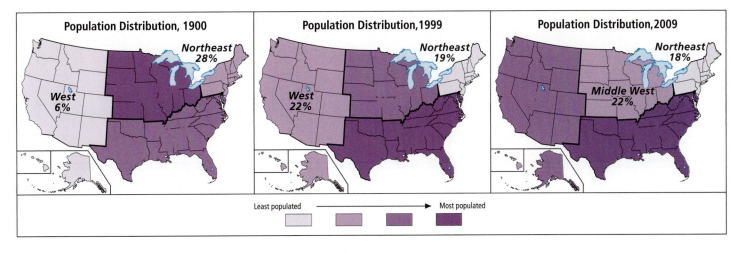

FIGURE 2.28 Percentage of population by region in the United States, 1900, 1999, and 2009. Notice how much the percentage has increased in the West and decreased in the Northeast and Middle West over the past 100 years. The trends continued during the first decade of the new century. [Adapted from "America's Diversity and Growth," *Population Bulletin* (June 2000): 12. See http://www.statcan.ca/english/research/11F0019MIE/11F0019MIE2005254.pdf, and http://www.census.gov/popest/states/NST-ann-est.html.]

This general trend may persist, but the economic recession, water shortages, and the energy crisis are limiting expansion, especially in the West, but also in the South.

In many farm towns and rural areas in the *Middle West* (the large central farming region of North America), populations are shrinking. As family farms are consolidated under corporate ownership, labor needs are decreasing and young people are choosing better-paying careers in cities. (Notice the changes in the area labeled *Middle West* in Figure 2.28.) Middle Western cities are growing only modestly yet becoming more ethnically diverse, with rising populations of Hispanics and Asians in places such as Indianapolis, St. Louis, and Chicago. **52. LATINO POLITICAL POWER**

In the mountainous interior, settlement is generally light (see Photo Essay 2.6). The principal reasons for this low population density are rugged topography, lack of rain, and in northern or high altitude zones, a growing season that is too short to sustain agriculture. Some population clusters exist in irrigated agricultural areas, such as in the Utah Valley, and near rich mineral deposits and resort areas. The gambling economy and frenetic construction activity generated by real estate speculation account for several knots of dense population at the southern end of the region. Until the recession beginning in 2007, Las Vegas, Nevada, was the fastest-growing city in the United States. But by mid-2009, approximately 67,000 homes in Las Vegas were in foreclosure, tourist arrivals were sharply down, and hotel construction projects were halted (see Photo Essay 2.5C on page 99).

Along the Pacific coast, a band of growing population centers stretches north from San Diego to Vancouver and includes Los Angeles, San Francisco, Portland, and Seattle. These are all port cities engaged in trade around the **Pacific Rim** (all the countries that border the Pacific Ocean). Over the past several decades, these North American cities have become centers of technological innovation.

The rate of natural increase in North America (0.6 percent per year) is low, less than half the rate of the rest of the Americas (1.4 percent). Still, North Americans are adding to their numbers fast enough through births and immigration that they could reach 440 million by 2050. Many of the important social issues now being debated in North America are linked to changing population patterns—issues such as legal and illegal immigration, which language should be used in public schools and in government offices, the geography of voting patterns, urbanization, cultural diversity, mobility, and the social effects of an aging population.

Pacific Rim a term referring to all the countries that border the Pacific Ocean

Mobility and Aging in North America

North Americans move more often than most other people. Every year, almost one-fifth of the U.S. population and two-fifths of Canada's relocate. Some are changing jobs; others are attending school or retiring to a warmer climate, a smaller city or town; others are merely moving across town or to the suburbs or to the countryside. Still other people are arriving from outside the region as immigrants (see Figure 2.21 on page 101).

Urbanization remains a powerful force behind this mobility. Dynamic economies and the search for lower production and living costs are drawing employers, employees, and retirees to urban areas. In Canada, they are going to cities in the southeast and on the west coast. In the United States, people are moving to the South, Southwest, and Pacific Northwest. Cities in these areas have sprouted satellite or "edge" cities around their peripheries, often based on businesses dealing with technology and international trade. Suburbanization and urban sprawl are important side effects of this urbanization that have wider implications, such as the use of farmland for residential development and increased

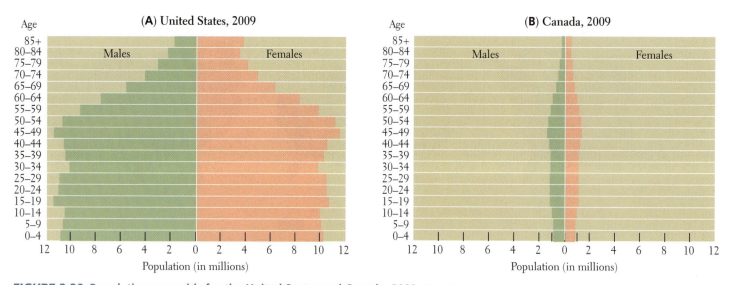

FIGURE 2.29 Population pyramids for the United States and Canada, 2009. The "baby boomers," born between 1947 and 1964, constitute the largest age group in North America, as indicated by the wider middle portion of these population pyramids. [Adapted from U.S. Census Bureau, Population Division, "Population Pyramids of the United States" and "Population Pyramids of Canada," International Data Base, 2009, at http://www.census.gov/ipc/www/idb/pyramids.html.]

emissions of greenhouse gases as people commute longer distances. Urban sprawl is discussed on pages 98–100.

During the twentieth century, the number of older North Americans grew rapidly. One in 25 individuals was over the age of 65 in 1900; by 2008, the number was 1 in 8. By 2050, when most of the current readers of this book will be over 65, 1 in 5 North Americans will be elderly. The number of elderly people will shoot up especially fast between the years 2010 and 2030. This spike will result from the marked jump in birth rate that took place after World War II, from 1947 to 1964. The so-called *baby boomers* born in those years constitute the largest age-group in North America, as can be seen in Figure 2.29. As this group reaches age 65 and retires, the outflow of money from Social Security (the pool into which all workers pay to provide support for the elderly) and

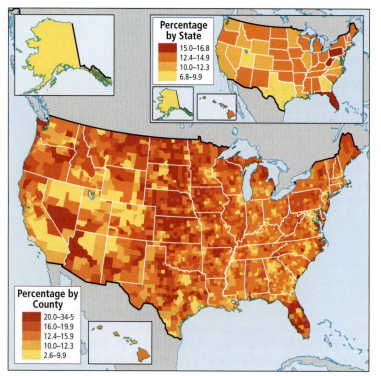

(A)

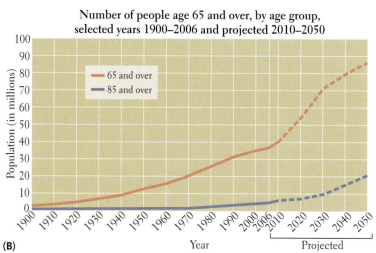

(B)

FIGURE 2.30 Distribution of elderly people in the United States. (A) The map shows the distribution of those over 65 by county and state in 2006. **(B)** The graph shows that the elderly population is projected to increase significantly over the next 40 years and that the elderly will be living longer. [Adapted from Federal Interagency Forum on Aging-Related Statistics, "Number of Older Americans," Population Indicator 1, at http://www.agingstats.gov/agingstatsdotnet/Main_Site/Data/2008_Documents/Population.aspx.]

private pensions will be high, and medical costs are expected to leap upward.

Because the boomers had fewer children than their parents, there will be fewer native-born people of working age to pay taxes and make pension fund contributions. For this reason, immigrants and their offspring are important to the system because their tax payments will help support the aging boomers. Moreover, immigrant caregivers are filling in for the fewer native-born people in caring for and providing companionship to elderly kin. Most families will be unable to afford assisted living and residential care for their elderly kin—in 2009, such care cost between $30,000 and $70,000 per year for one person. Once the boomers begin to retire in large numbers, living arrangements will likely change to reflect people's efforts to find humane and economical means to care for two or more elderly family members (**Figures 2.30** and **2.31**).

Aging populations in developed countries present us with an as yet unresolved dilemma. On the one hand, it is widely agreed that population growth should be reduced to lessen the environmental impact of human life on earth, especially that of the societies that consume the most. On the other hand, slower population growth means that there will be fewer working-age people to keep the economy going and to provide the financial and physical help the increasing number of elderly people will require.

Geographic Patterns of Human Well-Being The maps of human well-being for North America tell a quick and somewhat misleading story (**Figure 2.32**). First of all, we learn from Map A that both countries in this region have very high average per capita GDP (PPP) figures and that in the global context (see global inset) these two countries are among the most wealthy. But this is a classic example of why average GDP figures are not a sufficient measure of well-being, because we now know that there is much disparity of wealth in North America that is not revealed by country-level GDP figures. This map does not show how the wealth is distributed geographically or between social classes. It gives no hint that 18 percent of children are classed as poor or that in these countries the richest 10 percent have more than 14 times the wealth of the poorest 10 percent.

Virtually the same problem is presented by Map B, which shows the ranks of these two countries on the United Nations Human Development Index (UNHDI). Because the data are presented for the entire countries, not at the province or state level, the fact that many parts of both

(A)

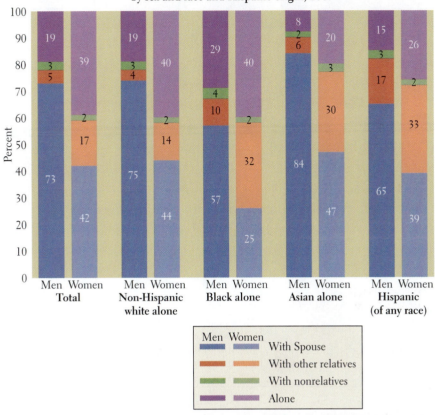

Living arrangements of the population age 65 and over, by sex and race and Hispanic origin, 2007

Legend:
Men	Women	
		With Spouse
		With other relatives
		With nonrelatives
		Alone

(B)

FIGURE 2.31 Aging patterns. As North America's population ages, more people are becoming interested in alternative living arrangements that avoid the social isolation experienced by so many elderly people, especially women **(A)**. Cohousing communities are becoming popular, especially among seniors, because in addition to private homes they include extensive common facilities such as common kitchens, group dining halls, and other community space that encourages interaction. Shown here **(B)** is the eighth anniversary party of Mosaic Commons, a cohousing community in Berlin, Massachusetts. [Adapted from Federal Interagency Forum on Aging-Related Statistics, "Living Arrangements," Population Indicator 5, at http://www.agingstats.gov/agingstatsdotnet/ Main_Site/Data/2008_Documents/Population.aspx.]

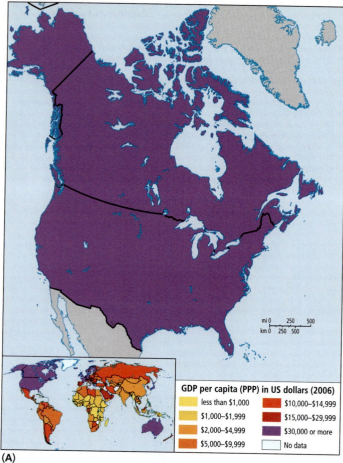

(A)

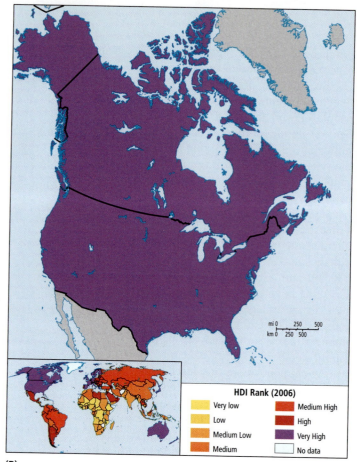

(B)

GDP per capita (PPP) in US dollars (2006)
- less than $1,000
- $1,000–$1,999
- $2,000–$4,999
- $5,000–$9,999
- $10,000–$14,999
- $15,000–$29,999
- $30,000 or more
- No data

HDI Rank (2006)
- Very low
- Low
- Medium Low
- Medium
- Medium High
- High
- Very High
- No data

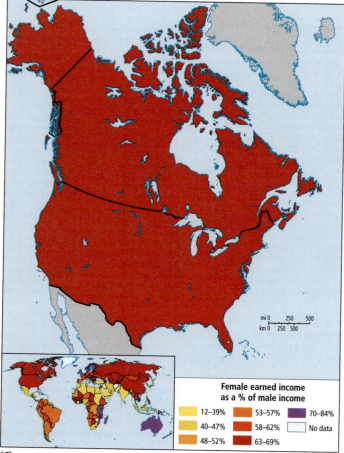

Female earned income as a % of male income
- 12–39%
- 40–47%
- 48–52%
- 53–57%
- 58–62%
- 63–69%
- 70–84%
- No data

(C)

FIGURE 2.32 Maps of human well-being. (A) Gross domestic product. **(B)** The UN Human Development Index. **(C)** Female earned income as a percent of male earned income. [Maps adapted from data: "Human development indices," (Maps A and B): Table 2, pages 28–32; (Map C): Table 5, pages 41–44 at http://hdr.undp.org/en/media/HDI_2008_EN_Tables.pdf.]

countries provide very well for their citizens in terms of income, health care, and education boosts the average figures. Glossed over is the reality that serious pockets of poverty, ill health, and illiteracy exist in virtually every state and province (but less so in Canada because of its more robust safety net).

Map C, which shows female earned income as a percentage of male earned income (F/MEI), reveals that North America lags behind other places on earth (see the global inset). Of those with comparable per capita GDP figures, Australia and North Europe (Scandinavia and the U.K.) are doing better at equalizing the pay of equally qualified female and male workers. But, of course, a look at the legend of Map C and the global inset reveals that nowhere on earth has gendered pay discrimination been eliminated.

THINGS TO REMEMBER

1. North America's population is rapidly aging, and by 2050, one in five people will be over the age of 65. There will be fewer young people to work, pay taxes, and take care of the elderly.

2. North Americans are highly mobile. Every year, almost one-fifth of the U.S. population and two-fifths of Canada's move to a different location, for a variety of reasons.

Reflections on North America

It is easy to rhapsodize about North America: the sheer size of the continent, its great wealth in natural and human resources, its superior productive capacities, and its powerful political position in the world are attributes possessed by no other world region. North America enjoys this prosperity, privilege, and power as a result of fortunate circumstances, the hard work of its inhabitants, the diversity and creativity of its largely immigrant populations, astute planning and law writing on the part of early founders, and access to worldwide resources and labor at favorable prices. Perhaps the most important factor in North America's success has been its democratic institutions, in particular the Canadian and U.S. Constitutions, which allow for individual freedom and flexibility as times and circumstances change.

Yet life has not been good to everyone in North America, nor has the influence of North America on other parts of the world always been benign. Canada and the United States were created from lands forcibly taken from indigenous peoples. Many people of non-European background continue to suffer from racial prejudice and a "culture of poverty." Settlers of all origins and their descendants had, and continue to have, a significant negative impact on the continent's environments. The standard of living now expected by all North Americans promises to increase the strain on these environments.

There is no guarantee that North America will continue its leadership role into the future. In fact, in the post–9/11 world, challenges to that leadership occur in every corner of the globe. The recent U.S. tendency to use military force for its own ends has been a major cause for concern around the world. The global economic downturn in 2008 and the subsequent collapse of banks in the United States, Europe, and Asia have been laid to inadequate regulation in the United States. North American models for development—based on democracy and on assumptions of rich and inexhaustible resources—are being challenged as inappropriate for much of the rest of the world. Shortly after his election in late 2008, President Barack Obama began to address all of these aforementioned issues. It is too early to assess his success. As subsequent chapters will show, societies elsewhere are beginning to prosper without following North American economic policies or democratic institutions.

Critical Thinking Questions

1. Discuss the ways in which North American culture is adapted to the high rate of spatial (geographic) mobility engaged in by Americans.

2. North America's population is changing in many ways. Explain why the aging of the American population is of interest and discuss how you personally are likely to be affected.

3. North American family types are changing. Explain the general patterns and discuss whether or not the nuclear family is or should be the sought-after norm.

4. The influence of globalization is now felt in small, even isolated, places in North America. Pick an example from the text or from your own experience and explain at least four ways in which this place is now connected to the global economy or global political patterns.

5. North Americans profit from having undocumented workers produce goods and services. Explain why this is the case and discuss if and how the situation should be changed. What is the argument for doing nothing or very little?

6. People like to say that the attacks on 9/11 changed everything for North Americans (especially for those living in the United States). What do they mean by this? How were U.S. attitudes toward the rest of the world modified? Do you think these modifications will be useful in the long run?

7. When you compare the economies of the United States and Canada and the ways they are related, what are the most important factors to mention?

8. When U.S. farmers or producers get subsidies from the federal government, what is intended? Why do producers and farmers in poor countries say that this practice hurts them?

9. Examine the connections between urban sprawl, rising consumption of gasoline, and rising air pollution in North American urban areas, and then propose a solution that you could abide by yourself.

10. The United States and Canada have quite different approaches to treatment of citizens who experience difficulties in life. Explain those different approaches and the apparent philosophies behind them.

Chapter Key Terms

acid rain 75
agribusiness 91
aquifers 67
brownfields 100
clear-cutting 73
digital divide 93
economic core 81
ethnicity 103
genetically modified organisms (GMOs) 91

gentrification 101
Hispanic 67
information technology (IT) 93
infrastructure 80
megalopolis 98
metropolitan areas 98
North American Free Trade Agreement (NAFTA) 92
nuclear family 106

Pacific Rim 109
Québecois 67
service sector 92
social safety net 96
suburbs 98
trade deficit 94
urban sprawl 73

A Sierra Madre, Mexico

B Andes, Chile

C Tierra del Fuego, Argentina

D Volcano, Montserrat

Western Sahara

Canary Islands

MAURITANIA
Nouakchott

CAPE VERDE
☆ **Praia**

SENEGAL
Dakar

THE GAMBIA **Banjul**

GUINEA-BISSAU

★ Cayenne

ATLANTIC OCEAN

• Belém

Fortaleza •

Z I L

• Recife

Land Elevations

meters	feet
4877	16,000
3353	11,000
2134	7000
914	3000
305	1000
152	500
0	0

mi 0 ... 200 ... 400 ... 600
km 0 ... 200 400 ... 600 800 1000

1:37,000,000
Azimuthal Equidistant Projection

Salvador •

⊕ Brasília

Brazilian Highlands

Belo Horizonte •

G

Rio de Janeiro •
São Paulo •

Curitiba •

THEMATIC CONCEPTS
Population • Gender • Development
Food • Urbanization • Globalization
Democratization • Climate Change • Water

Middle and South America

G Rio de Janeiro, Brazil

H Yucatan Lowlands, Mexico (with Mayan temple)

I Pampas, Argentina

FIGURE 3.1 Regional map of Middle and South America.

F Amazonian Basin, Brazil (from the air)

E Amazonian Basin, Brazil (from space)

115

Global Patterns, Local Lives   The boat trip down the Aguarico River in Ecuador took me into a world of magnificent trees, river canoes, and houses built high up on stilts to avoid floods. I was there to visit the Secoya, a group of 350 indigenous people locked in negotiations with the U.S. oil company Occidental Petroleum over its plans to drill for oil on Secoya lands. Oil revenues supply 40 percent of the Ecuadorian government's budget and are essential to paying off its national debt. The government had threatened to use military force to compel the Secoya to allow drilling.

The Secoya wanted to protect themselves from pollution and cultural disruption. As Colon Piaguaje, chief of the Secoya, put it to me, "A slow death will occur. Water will be poorer. Trees will be cut. We will lose our culture and our language, alcoholism will increase, as will marriages to outsiders, and eventually we will disperse to other areas. Given all the impending changes, Chief Piaguaje asked Occidental to use the highest environmental standards in the industry. He also asked the company to establish a fund to pay for the educational and health needs of the Secoya people.

Like the Secoya, indigenous peoples around the world are facing environmental and cultural disruption arising from economic development efforts (**Figure 3.2**). Chief Piaguaje based his predictions for the future on what has happened in other parts of the Ecuadorian Amazon that have already experienced several decades of oil development.

The U.S. company Texaco was the first major oil developer to establish operations in Ecuador. From 1964 to 1992, its pipelines and waste ponds leaked almost 17 million gallons of oil into the Amazon Basin, enough to fill about 1900 fully loaded oil tanker trucks, or 35 Olympic-size swimming pools. Although Texaco sold its operations to the government and left Ecuador in 1992, its oil wastes continue to leak into the environment from hundreds of open pits (**Figure 3.3**).

In 1993, some 30,000 people sued Texaco in New York State, where the company (now owned by and called Chevron) is headquartered, for damages from the pollution. Those suing were both indigenous people and settlers who had established farms along Texaco's service roads. Several epidemiological studies concluded that oil contamination has contributed to higher rates of childhood leukemia, cancer, and spontaneous abortions among people who live near pollution created by Texaco.

Seven years after my visit to the Secoya, many of Chief Piaguaje's worries have been borne out. Occidental did establish a fund to help the Secoya deal with the disruption of oil development.

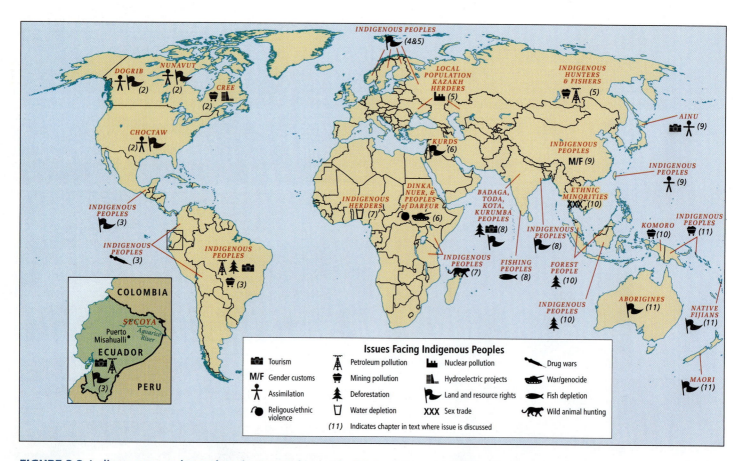

FIGURE 3.2 Indigenous peoples and environmental issues in this text. Issues relating to indigenous peoples are mentioned in many places in this book. In all cases, the issues are in some way related to interactions with the outside world and usually to uses of their resources by the outside world.

FIGURE 3.3 Pollution from oil development in Ecuador. A worker samples one of the several hundred open waste pits that Texaco left behind in the Ecuadorian Amazon. Wildlife and livestock trying to drink from these pits are often poisoned or drowned. After heavy rains, the pits overflow, polluting nearby streams and wells.

However, part of this fund was distributed to individual households, with varying results. Some people invested their money in ecotourism and other commercial enterprises and have prospered. Others simply spent their money and are now working for those who invested. Such employer–employee relationships are new to the Secoya, changing what was once an egalitarian culture.

Oil development has had many negative effects on the environment. Air and water pollution has increased rates of illness. The wildlife that the Secoya used to depend on, such as tapirs, have disappeared almost entirely due to overhunting by new settlers from the highlands who are working in the oil industry.

In 2002, the Ecuadorian suit against Chevron was dismissed by the U.S. Court of Appeals. It was refiled in Ecuador in 2003, and by 2009 it looked as if Chevron might lose the case and have to pay damages assessed now at $27 billion. [Alex Pulsipher's field notes; Amazon Watch, 2006; Oxfam America, 2005; Juan Forero, "Rain Forest Residents, Texaco Face Off In Ecuador," National Public Radio, April 30, 2009, at http://www.npr.org/templates/story/story.php?storyId=103233560.] ■

The rich resources of Middle and South America have attracted outsiders since the first voyage of Christopher Columbus in 1492. Europe's encounter with this region marked a major expansion of the global economy. However, for several hundred years Middle and South America occupied a disadvantaged position in global trade, supplying cheap raw materials that aided the Industrial Revolution in Europe but impoverished the region. In recent years, some countries of this region—Ecuador, Mexico, Bolivia, Brazil, Venezuela—have taken control of their own resources and with investment from local profits and from Europe, East Asia, and North America, are now supporting new and more profitable manufacturing and service industries. Meanwhile trade blocs within the region are making countries better able to prosper from trade with each other.

Middle and South America differ from North America in several ways. Physically, this world region is larger and has more diverse environments. Culturally, there are larger **indigenous** populations and more highly stratified social systems based on class, race, and gender. Politically, the region includes more than three dozen independent countries with models of self-government ranging from socialist Cuba to capitalist Chile. Economically, the gap between rich and poor is much wider than in North America. Yet there are many commonalities within the region, most arising from shared experience as colonies of Spain or Portugal.

The nine Thematic Concepts on pages 118–119, covered in each chapter of this book are illustrated here with photo essays that show examples of how each concept is experienced in the region of Middle and South America. In several cases, interactions between two or more of these concepts are illustrated. The captions explain the pictures and how the interactions work in specific places.

indigenous native to a particular place or region

THINGS TO REMEMBER

1. When compared with North America, the region of Middle and South America has the following characteristics: it is larger and more diverse physically; it has a far larger indigenous population; and it consists of more than three dozen countries with differing political institutions.

2. The region of Middle and South America has far greater income and wealth disparity than North America.

Climate Change and Deforestation: The rapid loss of forests in this region is a major driver of climate change. Forests absorb huge amounts of carbon dioxide in their bodies, which is released when forests are cleared through burning. Many alternatives to deforestation are being promoted throughout the region.

(A) Forest being cleared for agriculture in Brazil's Amazon Basin.

(B) An ecotourism lodge in Brazil is one example of an alternative to deforestation.

(C) A man harvests acai berries in Brazil. Sustainable agriculture, especially tree cropping, is another alternative to deforestation.

Water: In this region of abundant water resources, poorly planned urban development and the pollution of many water bodies have resulted in a water crisis. Few people have access to sanitation, and the dumping of waste in rivers is widespread.

(D) Iguazu Falls on the border between Brazil and Argentina. This part of the region has more freshwater resources than any other.

(E) An open sewer in a suburb of Brasilia, capital of Brazil.

(F) An old tire in a degraded mangrove outside of Rio de Janeiro, Brazil.

Population: A population explosion occurred during the early twentieth century as improved health care lowered death rates. However, birth rates started to decline by the mid-1980s as women began to delay childbearing in order to pursue more work outside the home, and as populations became more urbanized, which reduced the need for large families to operate farms. Contraception also became more available and widely used.

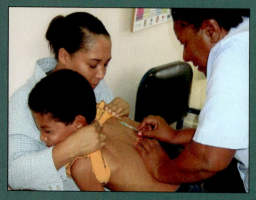

(G) Vaccination in the Dominican Republic.

(H) A woman commutes to work in Mexico City.

(I) A family-planning clinic in Guatemala.

Food and Urbanization: A shift toward large-scale agriculture has forced many small farmers, who cannot afford the investment in machinery or chemicals, off their land and into the cities. Many governments now see large-scale agriculture as the only way to feed the region's large urban populations. However, popular movements have developed in opposition to the trend.

(J) A sugarcane harvester in Cuba. This machine does the work of 600 laborers.

(K) Mexico City, the region's largest city at over 20 million people, and a major destination for migrants.

(L) Members of Brazil's "landless movement" march in Brasília for access to land.

Globalization, Development, and Democratization: The economic integration of this region with the global economy has left many poor people without decent-paying jobs and with only minimal access to health care and education. A democratic backlash against these patterns of economic development has brought many national leaders to power who question globalization.

(M) Artwork on the U.S.–Mexico border barrier, focusing on the forces of globalization that have made so many Mexicans migrate to the U.S.

(N) Women working in an export-oriented factory, or *maquiladora*, in northern Mexico.

(O) Venezuelan president Hugo Chávez came to power on a wave of anti-globalization sentiment.

Gender: Traditional gender roles are being challenged as more women work in jobs once occupied only by men. Meanwhile, urbanization is putting new pressures on the extended family and also opening up new opportunities for women.

(P) A mother and daughter negotiate their roles in a marketplace in Chiapas, Mexico.

(Q) A conference commemorating 25 years of women's service in the Brazilian air force.

(R) Michelle Bachelet, president of Chile (2006–2010).

I THE GEOGRAPHIC SETTING

Terms in This Chapter

In this book, **Middle America** refers to Mexico, Central America (the narrow ribbon of land that extends south of Mexico to South America), and the islands of the Caribbean (Figure 3.4).

South America refers to the vast continent south of Central America. The term *Latin America* is not used in this book because it describes the region only in terms of the Roman (Latin-speaking) origins of the former colonial powers of Spain and Portugal. It ignores the region's large indigenous, African, Asian, and Northern European populations, as well as the many mixed cultures, often called mestizo cultures, that have emerged here. In this chapter, we use the term *indigenous groups* or *peoples* rather than *Native Americans* to refer to the native inhabitants of the region.

> **Middle America** in this book, a region that includes Mexico, Central America, and the islands of the Caribbean
>
> **South America** the continent south of Central America

Physical Patterns

Middle and South America extend south from the midlatitudes of the Northern Hemisphere across the equator through the Southern Hemisphere, nearly to Antarctica (Figure 3.1 on pages 114–115). This vast north-south expanse combines with variation in altitude to create the wide range of climates in the region.

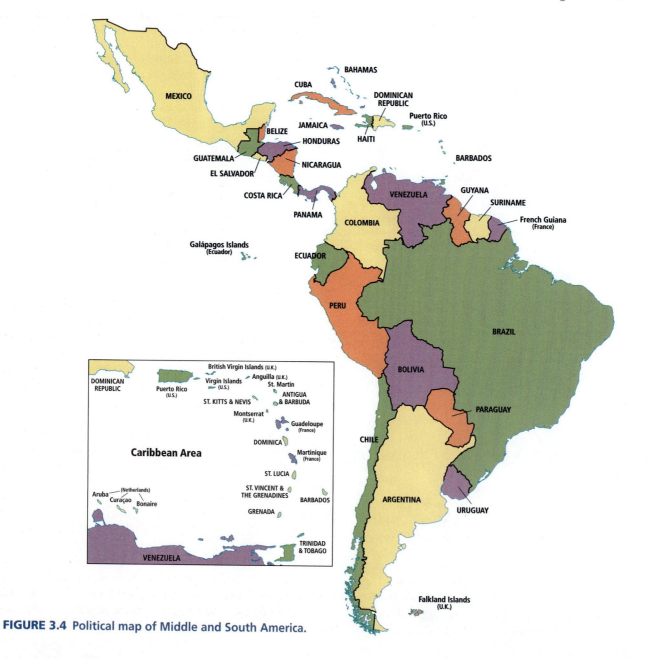

FIGURE 3.4 Political map of Middle and South America.

Tectonic forces have shaped the primary landforms of this huge territory to form an overall pattern of highlands to the west and lowlands to the east.

Landforms

There are a wide variety of landforms in Middle and South America, and this variety accounts for the many different climatic zones in the region. But for ease in learning, landforms are here divided into just two categories: highlands and lowlands.

Highlands A nearly continuous chain of mountains stretches along the western edge of the American continents for more than 10,000 miles (16,000 kilometers) from Alaska to Tierra del Fuego at the southern tip of South America. The middle part of this long mountain chain is known as the Sierra Madre in Mexico (see Figure 3.1A on page 114), by various names in Central America, and as the Andes in South America (see Figure 3.1B, C on page 114). It was formed by a lengthy **subduction zone** that runs thousands of miles along the western coast of the continents (see Figure 1.21 on page 41). Here, two oceanic plates—the Cocos Plate and the Nazca Plate—plunge beneath three continental plates—the North American Plate, the Caribbean Plate, and the South American Plate.

In a process that continues today, the leading edge of the overriding plates crumples to create mountain chains. In addition, molten rock from beneath the earth's crust ascends to the surface through fissures in the overriding plate to form volcanoes. Such volcanoes are the backbone of the continental divide that runs through Middle America and the Andes of South America (see Figure 3.1 on pages 114–115). Although these volcanic and earthquake-prone highlands have been a major barrier to transportation, communication, and settlement, people now live close to quiescent volcanoes and in earthquake-prone zones, which can pose deadly hazards (for example, the 2010 earthquake in Chile).

The chain of high and low mountainous islands in the eastern Caribbean is also volcanic in origin, created as the Atlantic Plate thrusts under the eastern edge of the Caribbean Plate. It is not unusual for volcanoes to erupt in this active tectonic zone. On the island of Montserrat, for example, people have been living with an active, and sometimes deadly, volcano for more than a decade (see Figure 3.1D on page 115). Eruptions have taken the form of violent blasts of superheated rock, ash, and gas (known as pyroclastic flows) that move down the volcano's slopes with great speed and force. The unusually strong earthquake in Haiti in January 2010 was also the result of plate tectonics.

Lowlands Vast lowlands extend over most of the land to the east of the western mountains. In Mexico, east of the Sierra Madre, a coastal plain borders the Gulf of Mexico (see Figure 3.1H on page 115). Farther south, in Central America, wide aprons of sloping land descend to the Caribbean coast. In South America, a huge wedge of lowlands, widest in the north, stretches from the Andes east to the Atlantic Ocean. These South American lowlands are interrupted in the northeast and the southeast by two modest highland zones: the Guiana Highlands and the Brazilian Highlands (see Figure 3.1G on page 115). Elsewhere in the lowlands, grasslands cover huge, flat expanses, including the llanos of Venezuela, Colombia, and Brazil and the pampas of Argentina (see Figure 3.1I on page 115).

The largest feature of the South American lowlands is the Amazon Basin, drained by the Amazon River and its tributaries (see Figure 3.1E, F on page 115). This basin lies within Brazil and the neighboring countries to its west. Here, the earth's largest remaining expanse of tropical rain forest gives the Amazon Basin global significance as a reservoir of **biodiversity**. Hundreds of thousands of plant and animal species live here.

The basin's water resources are also astounding. Twenty percent of the earth's flowing surface waters exist here, running in rivers so deep that ocean liners can steam 2300 miles (3700 kilometers) upriver from the Atlantic Ocean all the way to Iquitos, jokingly referred to as Peru's "Atlantic seaport." The vast Amazon River system starts as streams high in the Andes. These streams eventually unite as rivers that flow eastward toward the Atlantic. Once they reach the flat land of the Amazon Plain, their velocity slows abruptly, and fine soil particles, or **silt**, sink to the riverbed. When the rivers flood, silt and organic material transported by the floodwaters renew the soil of the surrounding areas, nourishing millions of acres of tropical forest. Not all of the Amazon Basin is rain forest, however. Variations in weather and soil types, as well as human activity, have created grasslands and seasonally dry deciduous tropical forests in some areas.

Climate

From the jungles of the Caribbean and the Amazon to the high, glacier-capped peaks of the Andes to the parched moonscape of the Atacama Desert, the climate variety of Middle and South America is enormous (see Photo Essay 3.1 on page 123). Climates are essentially the result of interactions between temperature and moisture. In this region the wide range of temperatures reflects both the great distance the landmass spans on either side of the equator and the tremendous variations in altitude across the region's landmass (the highest point in the Americas is Aconcagua in Argentina, at 22,841 feet [6962 meters]). Patterns of precipitation are affected both by the local shape of the land and by global patterns of wind and ocean currents that bring moisture in varying amounts.

Temperature-Altitude Zones Four main **temperature-altitude zones** are commonly recognized in the region (see Figure 3.5 on page 122). As altitude increases, the temperature of the air decreases by about 1°F per 300 feet (1°C per 165 meters) of elevation. Thus temperatures are highest in the lowlands, which are known in Spanish as the *tierra caliente*, or "hot land." The *tierra caliente* extends up to about 3000 feet (1000 meters), and in some parts of the region these lowlands cover wide expanses. Where moisture is adequate, tropical rain forests thrive, as does a wide range of tropical crops, such as bananas, sugarcane, cacao, and pineapples. Many coastal areas of the *tierra caliente*, such as northeastern Brazil, have become zones of plantation agriculture that support populations of considerable size.

> **subduction zone** a zone where one tectonic plate slides under another
>
> **biodiversity** the variety of life forms to be found in a given area
>
> **silt** fine soil particles
>
> **temperature-altitude zones** regions of the same latitude that vary in climate according to altitude

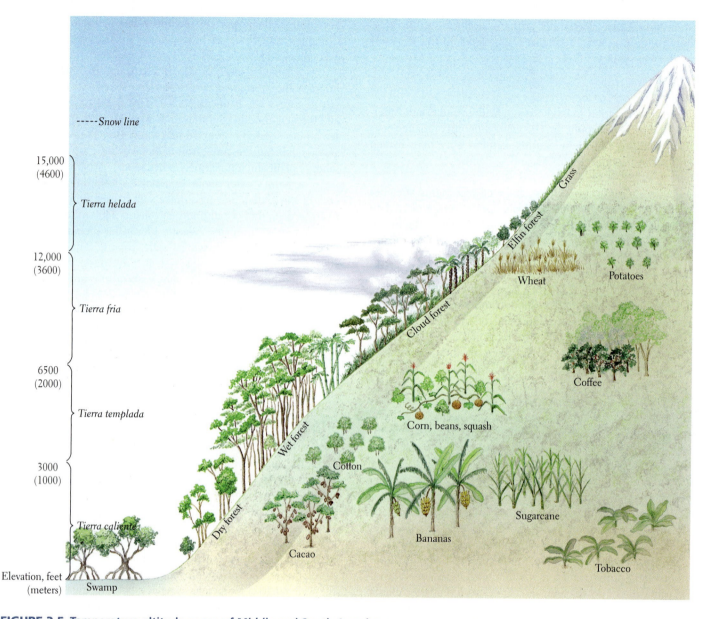

FIGURE 3.5 Temperature-altitude zones of Middle and South America. Temperatures tend to decrease as altitude increases. The natural vegetation on mountain slopes also changes, as shown along the slope of the mountain here. The same is true for crops, some of which are suited to lower, warmer elevations and some to higher, cooler ones. [Illustration by Tomo Narashima, based on fieldwork and a drawing by Lydia Pulsipher.]

Between 3000 and 6500 feet (1000 to 2000 meters) is the cooler *tierra templada* ("temperate land"). The year-round, spring-like climate of this zone drew large numbers of indigenous people in the distant past and, more recently, has drawn Europeans. Here, such crops as corn, beans, squash, various green vegetables, wheat, and coffee are grown.

Between 6500 and 12,000 feet (2000 to 3600 meters) is the *tierra fria* ("cool land"). Many crops such as wheat, fruit trees, root vegetables, and cool-weather vegetables—cabbage and broccoli, for example—do very well at this altitude. Many animals—such as llamas, sheep, and guinea pigs—are raised for food and fiber. Sev-

eral modern population centers are in this zone, including Mexico City, Mexico, and Quito, Ecuador.

Above 12,000 feet (3600 meters) is the *tierra helada* ("frozen land"). In the highest reaches of this zone, vegetation is almost absent, and mountaintops emerge from under snow and glaciers. A remarkable feature of such tropical mountain zones is that in a single day of strenuous hiking, one can encounter many of the climate types found on earth.

Precipitation The pattern of precipitation throughout the region is influenced by the interaction of global wind patterns with

Climate Zones

Tropical Humid Climates
- Tropical wet
- Tropical wet/dry

Arid and Semiarid Climates
- Desert
- Steppe

Temperate Climates
- Midlatitude, moist all year
- Subtropical, winter dry
- Mediterranean, summer dry

Cool Humid Climates
- Continental, winter dry
- Continental, moist all year

Coldest Climates
- Arctic
- High altitude
- Winds
- Ocean currents

Northeast trade winds bring heavy seasonal rains.

Seasonal winds bring rains.

Peru Current brings cold surface waters. Air above is very dry. **El Niño** brings warm water instead of cold every few years.

Rain shadow The Andes block winds off the Atlantic.

Southeast trade winds bring rain.

Globe-encircling eastward-blowing winds bring steady cold rains.

Rain shadow The Andes block rains coming from the west.

A Tropical wet, Belém, Brazil

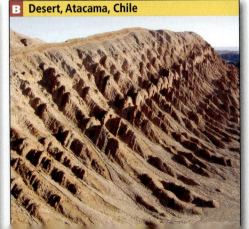

B Desert, Atacama, Chile

C Continental, moist all year, Tierra del Fuego

trade winds winds that blow from the northeast and the southeast toward the equator

mountains and ocean currents (see Photo Essay 1.6 on pages 54–55). The **trade winds** sweep off the Atlantic, bringing heavy seasonal rains to places roughly 23 degrees north and south of the equator (see Photo Essay 3.1 Map on page 123). Winds from the Pacific bring seasonal rain to the west coast of Central America, but mountains block that rain from reaching the Caribbean side, which receives heavy rainfall from the northeast trade winds.

The Andes are a major influence on precipitation in South America. They block the rains borne by the trade winds off the Atlantic into the Amazon Basin and farther south, creating a rain shadow on the western side of the Andes in northern Chile and southwestern Peru (see Photo Essay 3.1 Map). Southern Chile is in the path of eastward-trending winds that sweep out of the Southern Ocean, bringing steady, cold rains that support forests similar to those of the Pacific Northwest of North America. The Andes block this flow of wet, cool air and divert it to the north. They thereby create another extensive rain shadow on the eastern side of the mountains along the southeastern coast of Argentina (Patagonia).

The pattern of precipitation is also influenced by the adjacent oceans and their currents. Along the west coasts of Peru and Chile, the cold surface waters of the Peru Current bring cold air that cannot carry much moisture. The combined effects of the Peru Current and the central Andes rain shadow have created what is possibly the world's driest desert, the Atacama of northern Chile (see Photo Essay 3.1B on page 123).

El Niño periodic climate-altering changes, especially in the circulation of the Pacific Ocean, now understood to operate on a global scale

El Niño One aspect of the Peru Current that is only partly understood is its tendency to change direction every few years (on an irregular cycle, possibly linked to sunspot activity). When this happens, warm water flows eastward from the western Pacific, bringing warm water and torrential rains, instead of cold water and drought, to parts of the west coast of South America. The phenomenon was named **El Niño**, or "the Christ Child," by Peruvian fishermen, who noticed that when it does occur, it reaches its peak around Christmastime.

El Niño also has global effects, bringing cold air and drought to normally warm and humid western Oceania and unpredictable weather patterns to Mexico and the southwestern United States. The El Niño phenomenon in the western Pacific is discussed further in Chapter 11.

Hurricanes In this region, many coastal areas are threatened by powerful storms that can create extensive damage and loss of life. These form annually, primarily in the Atlantic Ocean north of the equator and close to Africa. A tropical storm begins as a group of thunderstorms. A few hurricanes also form in the southeastern Pacific and can affect the western coasts of Middle America before turning west toward Hawaii. When enough warming wet air comes together, the individual storms organize themselves into a swirling spiral of wind that moves across the earth's surface. The highest wind speeds are found at the edge of the eye, or center, of the storm. Once wind speeds reach 75 miles (120.7 kilometers) per hour, such a storm is officially called a

hurricane. Hurricanes usually last about 1 week; because they draw their energy from warm surface waters, they slow down and eventually dissipate as they move over cooler water or land. Human vulnerability to hurricanes is increasing as coastal populations increase. Some scientists also think that climate change is leading to an increase in the number and intensity of hurricanes (see Photo Essay 3.3B on page 127).

THINGS TO REMEMBER

1. The rain forests of the Amazon Basin are planetary treasures of biodiversity that also play a key role in regulating the earth's climate.

2. There are four temperature-altitude zones in the region that influence where and how people live and the crops they can grow.

3. Significant environmental hazards include earthquakes, volcanic eruptions, and hurricanes.

Environmental Issues

Environments in Middle and South America were among the first to inspire concern about the use and misuse of the earth's resources. Beginning in the 1970s, construction of the Trans-Amazon Highway provided migrant farmers with access to the Amazon Basin. They followed the new road into the rain forest and began clearing it to grow crops. Scholars warned of an impending crisis in the tropical rain forests of this region. We now know that even in prehistory, every human settlement has had consequences for environments across Middle and South America. Today's impacts are particularly severe, however, because not only has population density increased substantially, but per capita consumption has increased exponentially and local environments now supply global demands.

Tropical Forests, Climate Change, and Globalization

As explained in Chapter 1 (see pages 45–48), forests release oxygen and absorb carbon dioxide, the greenhouse gas most responsible for global warming. Hence, the loss of large forests, such as those in the Amazon Basin and Middle America, contributes to global warming by releasing carbon dioxide as a product of burning, and by reducing the amount of carbon dioxide that can be taken out of the atmosphere. A British study reported in the science journal *Nature* in 2009 showed that tropical rain forests absorb at least 18 percent of the 32 billion tons of CO_2 added to the earth's atmosphere yearly by human activity. Because 50 percent of the earth's remaining tropical rainforests are in South America, keeping these forests is crucial to controlling climate change.

Middle and South American rain forests are being diminished by multiple human impacts (Photo Essay 3.2). Primary among them is the clearing of land to raise cattle and grow crops, such as soybeans (for animal feed) and corn for ethanol, intended for the global market. Because many of the cleared trees are simply burned rather than used as building material (see Photo Essay 3.2C, D), not only is the forest's function as an absorber of

While there is a wide variety of human impacts on the environments of this region, here we focus on the processes of land cover and land use change that lead to the conversion of forests to grazing land or farmland. The forces guiding this process are complex, driven by the needs of poor people for livelihoods, the desire of governments to assert control over lightly populated areas, and the demands for wood, meat, and food that arise in distant urban centers and the global market. Together these forces have led to a rapid loss of forest cover throughout much of this region. Brazil loses more forest cover each year than any country on the planet.

A A newly built road that connects to the Trans-Amazon Highway in Brazil.

B A farmer and his children play with a monkey outside their home in the Amazonian state of Pará, Brazil.

C A field in Chiapas, Mexico, recently cleared for agriculture by burning.

D Cattle on recently burned land in Pará, Brazil.

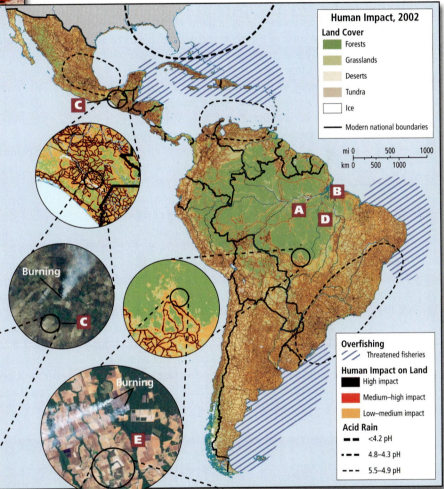

Human Impact, 2002

Land Cover
- Forests
- Grasslands
- Deserts
- Tundra
- Ice
- — Modern national boundaries

mi 0   500   1000
km 0   500   1000

Overfishing
- Threatened fisheries

Human Impact on Land
- High impact
- Medium–high impact
- Low–medium impact

Acid Rain
- – – <4.2 pH
- ·–·– 4.8–4.3 pH
- - - - 5.5–4.9 pH

Burning

Burning

E Soy fields recently cleared of forest in Mato Grosso, Brazil.

CO_2 lost, large amounts of carbon stored in the bodies of the trees are released into the atmosphere as CO_2. Brazil now ranks as the fourth-largest emitter of greenhouse gases in the world after the United States, China, and Indonesia (see Thematic Concepts Part A on page 118). But soybeans are an important source of revenue for Brazil, which exports one-third of the world's soybeans (2007), mostly to China as animal feed to fulfill that country's increasing appetite for meat. In Middle America, forests are cleared to create pastures for beef cattle grown primarily for the U.S. fast-food industry. The forests of Central America and the Caribbean have long been cleared for export crops, such as sugar, cotton, tobacco (in the past), and bananas (today). If deforestation continues in Middle America at the present rates, the natural forest cover will be entirely gone in 20 years. In some of the poorest countries, like Haiti and El Salvador, the nearly total removal of native forests has resulted in such severe soil loss that extreme poverty is the result.

Also contributing to deforestation are the logging of hardwoods and the extraction of underlying minerals, including oil, gas, and precious stones. Investment capital comes from Asian multinational companies that have turned to the Amazon forests after having logged up to 50 percent of the tropical forests in Southeast Asia. Moreover, the construction of access roads to support these activities continues to accelerate deforestation by opening new forest areas to migrants. The governments of Peru, Ecuador, and Brazil encourage impoverished urban people to occupy cheap land along the newly built roads (see Photo Essay 3.2A on page 125), and encourage deforestation by supplying chainsaws to the settlers. However, the settlers have a difficult time learning to cultivate the poor soils of the Amazon. After a few years of farming, they often abandon the land, now eroded and depleted of nutrients, and move on to new plots. Ranchers may then buy the worn-out land from these failed small farmers to use as cattle pastures.

Ironically, while Middle and South America are significantly contributing to global warming by adding CO_2 to the earth's

ecotourism nature-oriented vacations, often taken in endangered and remote landscapes, usually by travelers from industrialized nations

atmosphere—both by removing forest that absorbs carbon and by the large-scale burning of the removed trees—many in this very region are particularly vulnerable to the increasing threats of climate change. Photo Essay 3.3 illustrates several such cases: shortages of clean water brought on by intensifying droughts and by glacial melting, vulnerability to rising sea levels, and the effects of increasingly violent storms.

Environmental Protection and Economic Development

In the past, governments in the region argued that economic development was so desperately needed that environmental regulations were an unaffordable luxury. Now, there are increasing attempts to embrace economic development as necessary to raise standards of living while trying to minimize its negative effects on the environment. Extensive environmental regulations are being developed in all the countries of the Amazon, but as yet, enforcement is typically weak.

Ecotourism Other economic development efforts focus on earning money from the beauty of undegraded natural environments. Many countries are now promoting **ecotourism,** which encourages visitors from developed countries to appreciate ecosystems and wildlife that do not exist in their parts of the world. Travel experiences in unfamiliar natural and cultural environments can sensitize both travelers and hosts to the complexity of environmental issues. Sustainable use and conservation of resources can be achieved while providing a livelihood to local people and the broader host community. Ecotourism is now the most rapidly growing segment of the global tourism and travel industry, which itself is the world's largest industry, with $3.5 trillion spent annually. Many Middle and South American nations also have spectacular national parks that can provide a basis for ecotourism (Figure 3.6).

Ecotourism has its downsides, however. Mismanaged, it can be similar to other kinds of tourism that damage the environment

FIGURE 3.6 Ecotourism in the Amazon. (A) A tourist poses in front of a giant lupuna tree in the Peruvian Amazon. **(B)** An Amazon river dolphin being fed by a tourist. **(C)** An ecotourism river boat operating out of Manaus, Brazil, cruises a tributary of the Amazon.

Adapting to the multiple stresses that climate change is bringing to this region is proving to be quite a challenge. Drought, hurricanes, and glacial melting are combining with growing populations and persistent poverty to create a complex landscape of vulnerability to climate change.

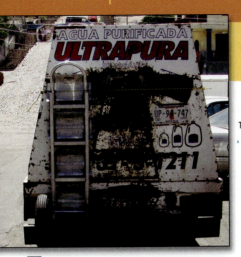

A A water truck in Nogales, Mexico, where much of the population has no other access to water. The higher temperatures that climate change is bringing could make this already water-scarce area even dryer. Hence the millions of Mexicans who have moved to work in factories along the U.S. border are highly vulnerable to climate change.

B Honduran rescue workers search for survivors of Hurricane Mitch, which killed 18,000 people. Hurricanes are likely to intensify as temperatures rise with climate change. Poor countries like Honduras are particularly vulnerable to the damage these storms bring.

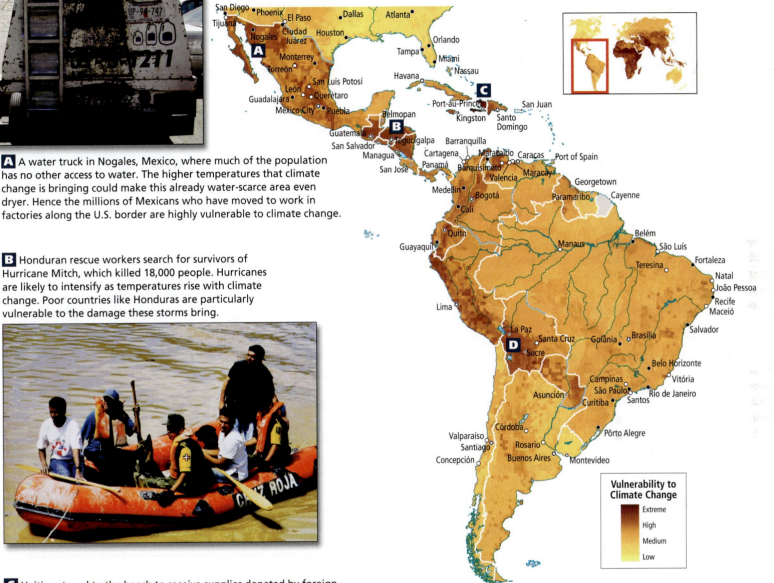

Vulnerability to Climate Change

- Extreme
- High
- Medium
- Low

C Haitians travel to the beach to receive supplies donated by foreign governments after flooding cut their village off from the rest of the country. So much of Haiti's forest cover has been removed that even mild tropical storms can cause catastrophic flooding. Much of Haiti's impoverished population is already dependent on foreign aid, a situation likely to get worse with climate change.

D La Paz and many smaller cities and towns in Bolivia's drought-prone highlands receive much of their drinking water from glaciers in the Andes. Higher temperatures are causing these glaciers to melt rapidly. Many have already disappeared and the rest could be gone in 15 years.

and return little to the surrounding community. While there is the potential to use the profits of ecotourism to benefit local communities and environments, the margins of profit may be small.

Personal Vignette Puerto Misahualli is a small river boomtown in the Ecuadorian Amazon that is currently enjoying significant economic growth. Its prosperity is due to the many European, North American, and other foreign travelers who come for experiences that will bring them closer to the now-legendary rain forests of the Amazon.

The array of ecotourism offerings can be perplexing. One indigenous man offers to be a visitor's guide for as long as desired, traveling by boat and on foot, camping out in "untouched forest teeming with wildlife." His guarantee that they will eat monkeys and birds does not seem to promise the nonintrusive, sustainable experience the visitor might be seeking. At a well-known "eco-lodge," visitors are offered a plush room with a river view, a chlorinated swimming pool, and a fancy restaurant serving "international cuisine."

By contrast, the solar-powered Yachana Lodge (see a similar establishment in **Thematic Concepts Part B** on page 118 has simple rooms and local cuisine. Its knowledgeable resident naturalist is a veteran of many campaigns to preserve Ecuador's wilderness. Profits from the lodge fund a local clinic and various programs that teach sustainable agricultural methods that protect the fragile Amazon soils while increasing farmers' earnings from surplus produce. The nonprofit group running the Yachana Lodge—the Foundation for Integrated Education and Development—earns just barely enough to sustain the clinic and agricultural programs.
[Source: Alex Pulsipher's field notes in Ecuador.] ∎

Climate Change and the Water Crisis

Although this world region receives more rainfall than any other and has three of the world's six largest rivers (in volume), it is experiencing a water crisis. Most of the factors causing the water crisis are induced by humans, but as a set these factors then expose many communities to yet more crucial water crises born of climate change that may bring increased drought or increased flooding.

commodification turning something not previously thought of as having economic value into a good, or commodity, which can be bought and sold

For example, the air pollution that hangs over so many cities in this region is contributing to global warming, which is thought to be the cause of rapid glacial melting in the Andes. These glaciers feed rivers that are the main source of water for millions of people. When the glaciers are gone, the rivers, at best, will run only seasonally, devastating communities and industries that depend on them.

Lax environmental policies may allow industries to pollute with few restraints. Waterways along the Mexican border with the United States are polluted by numerous factories set up to take advantage of trade with the United States and Canada (see the discussion of maquiladoras on page 138). Only a small percentage of the often highly toxic water discharges are treated and disposed of properly. For example, in Mexico City, it is estimated that as much as 90 percent of urban wastewater goes untreated.

Rapid and often unplanned urbanization has left many poor people without access to clean drinking water and sanitation. In the largest cities, the water infrastructure may be so inadequate that as much as 50 percent of freshwater is lost due to leaky pipes. Moreover, in some cities as much as 80 percent of the population has no access to decent sanitation; toilets may be entirely lacking, leaving people to relieve themselves on the city streets. This poses a major health hazard and so pollutes local water resources that cities must bring in drinking water from distant areas (see Photo Essay 3.3A on page 127).

Behind these immediate problems lie systemic failures. Political corruption diverts public funds that should be going to water infrastructure improvements. Some efforts have been made to increase access to clean water in cities by selling water utilities that were once government-run to private companies. In some cases, however, profit-driven private companies have increased water prices to levels that few urban residents can afford, while doing little to improve water supply or delivery systems. In Bolivia, for example, a subsidiary of the global engineering firm Bechtel even charged urban residents for water taken from their own wells and rainwater harvested off their roofs! Popular protests have forced many countries to take back control of water services from private companies. A Canadian nongovernmental organization (NGO), Blue Gold, is monitoring such **commodification** of water. The term refers to efforts by a firm to surreptitiously "marketize" a water supply long thought of as a public resource—for instance, a mountain stream near a rural village. The firm then markets the water for profit to outsiders in units—plastic bottles, for example—and takes the profits.

THINGS TO REMEMBER

1. Middle and South American rain forests are being diminished by multiple human impacts: the needs of poor people for livelihoods; the desire of governments to assert control over lightly populated areas; and the demands for wood, meat, and food in distant urban centers and in the global market.

2. There are increasing attempts to embrace economic development as necessary to raise standards of living while trying to minimize its negative effects on the environment.

Human Patterns over Time

The conquest of Middle and South America by Europeans set in motion a series of changes that helped create the ways of life found in this region today. The conquest wiped out much of indigenous civilization and set up colonial regimes in its place. It introduced many new cultural influences and led to the disparities in the distribution of power and wealth that continue to this day.

The Peopling of Middle and South America

Recent evidence suggests that between 25,000 and 14,000 years ago, groups of hunters and gatherers from northeastern Asia spread throughout North America after crossing the Bering land bridge on foot, or moving along shorelines in small boats, or both. Some

FIGURE 3.7 Incan building and agricultural infrastructure in Peru. (A) The Incas constructed architectural wonders, such as the city of Machu Picchu, which they built on a high ridge deep in the Andes mountains. **(B)** Living in an earthquake zone, the Incas mastered the art of carving tightly interlocked stone blocks without using mortar, creating walls that could withstand violent shaking. **(C)** The Incas also built elaborate stone terraces and irrigation canals to support agriculture in the mountains. In the sunken terraces shown here, plants were sheltered from cold winds.

of these groups remained in North America, while others ventured south across the Central American land bridge, reaching the tip of South America by about 13,000 years ago.

By 1492, there were 50 to 100 million indigenous people in Middle and South America. In some places, population densities were high enough to threaten sustainability. People altered the landscape in many ways. They modified drainage to irrigate crops, constructed raised fields in lowlands, terraced hillsides, and built paved walkways across swamps and mountains. They constructed cities with sewer systems and freshwater aqueducts, and raised huge earthen and stone ceremonial structures that rivaled the pyramids of Egypt.

The indigenous people also perfected the system of **shifting cultivation** that is still common in wet, hot regions in Central America and the Amazon Basin. In this system, small plots are cleared in forestlands, the brush is dried and burned to release nutrients into the soil, and the clearings are planted with multiple crop species. Each plot is used for only 2 or 3 years and then abandoned to allow the forest to regrow. If there is sufficient land, this system is highly productive per unit of land and labor. If population pressure increases to the point that a plot must be used before it has fully regrown and its fertility restored, yields will decrease drastically.

The **Aztecs** of the high central valley of Mexico had some technologies and social systems that rivaled or surpassed those of Asian and European civilizations of the time, although they lacked the wheel and gunpowder. Particularly well developed were urban water supplies, sewage systems, and elaborate marketing systems. Historians have concluded that, on the whole, Aztecs probably lived more comfortably than their contemporaries in Europe by 1500.

In 1492, the largest state in the region was that of the **Incas**, stretching from southern Colombia to northern Chile and Argentina. The main population clusters were in the Andes highlands, where the cooler temperatures at these high altitudes eliminated the diseases of the tropical lowlands, while proximity to the equator guaranteed mild winters and long growing seasons. For several hundred years, the Inca Empire was one of the most efficiently managed empires in the history of the world. Highly organized systems of labor were used to construct paved road systems, elaborate terraces and irrigation systems, and great stone cities in the Andean highlands. Incan agriculture was advanced, particularly in the development of crops such as numerous varieties of potatoes and grains (**Figure 3.7**).

European Conquest

The European conquest of Middle and South America was one of the most significant events in human history. It rapidly

> **shifting cultivation** a productive system of agriculture in which small plots are cleared in forestlands, the dried brush is burned to release nutrients, and the clearings are planted with multiple species; each plot is used for only 2 or 3 years and then abandoned for many years of regrowth
>
> **Aztecs** indigenous people of high-central Mexico noted for their advanced civilization before the Spanish conquest
>
> **Incas** indigenous people who ruled the largest pre-Columbian state in the Americas, with a domain stretching from southern Colombia to northern Chile and Argentina

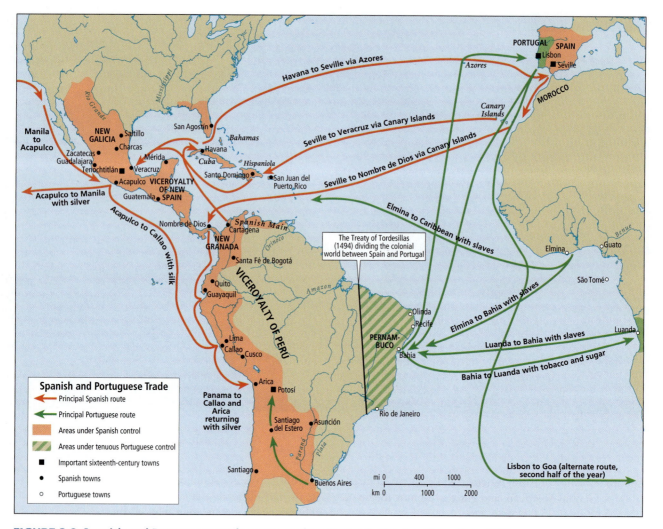

FIGURE 3.8 Spanish and Portuguese trade routes and territories in the Americas, circa 1600.
The major trade routes from Spain to its colonies led to the two major centers of its empire, Mexico and Peru. The Spanish colonies could trade only with Spain, not directly with one another. By contrast, there were direct trade routes from Portuguese colonies in Brazil to Portuguese outposts in Africa. Many millions of Africans were enslaved and traded to Brazilian plantation and mine owners (as well as to Spanish, British, French, and Dutch colonies in the Caribbean and Middle and South America). [Adapted from *Hammond Times Concise Atlas of World History* (Maplewood, N.J.: Hammond, 1994), pp. 66–67.]

altered landscapes and cultures and ended the lives of millions of indigenous people through disease and slavery.

Columbus established the first Spanish colony in 1492 on the Caribbean island of Hispaniola (presently occupied by Haiti and the Dominican Republic). This initial seat of the Spanish Empire expanded to include the rest of the Greater Antilles—Cuba, Puerto Rico, and Jamaica. After learning of Columbus's exploits, other Europeans, mainly from Spain and Portugal on Europe's Iberian Peninsula, conquered Middle and South America.

The first part of the mainland to be invaded was Mexico, home to several advanced indigenous civilizations, most notably the Aztecs. The Spanish were unsuccessful in their first attempt to capture the Aztec capital of Tenochtitlán, but they succeeded a few months later after a smallpox epidemic decimated the native population. The Spanish demolished the grand Aztec capital in 1521 and built Mexico City on its ruins.

The conquest of the Incas in South America was achieved by a tiny band of Spaniards, again aided by a smallpox epidemic. Out of the ruins of the Inca Empire, the Spanish created the Viceroyalty of Peru, which originally encompassed all of South America except Portuguese Brazil. The newly constructed capital of Lima flourished, in large part as a transshipment point for enormous quantities of silver extracted from mines in the highlands of what is now Bolivia.

Diplomacy by the Roman Catholic Church prevented conflict between Spain and Portugal over the lands of the region. The Treaty of Tordesillas of 1494 divided Middle and South America at approximately 46° W longitude (Figure 3.8). Portugal took all lands to the east and eventually acquired much of what is today Brazil; Spain took all lands to the west.

The conquest of Brazil by the Portuguese was unique in some key respects. Brazil had no highly organized urban cultures, although dense indigenous settlements based on sophisticated agriculture definitely existed. Most Atlantic coastal people were annihilated early on, and the population in the huge Amazon Basin declined sharply as contagious diseases spread through trading. Because it was difficult to penetrate the lowland tropical forests, the Portuguese focused on extracting gold and precious gems from the Brazilian Highlands and on establishing plantations along the Atlantic coast.

The superior military technology of the Spanish and Portuguese speeded the conquest of Middle and South America. A larger factor, however, was the vulnerability of the indigenous people to diseases carried by the Europeans. In the 150 years following 1492, the total population of Middle and South America was reduced by more than 90 percent to just 5.6 million. To obtain a new supply of labor to replace the dying indigenous people, the Spanish initiated the first shipments of enslaved Africans to the region in the early 1500s.

By the 1530s, a mere 40 years after Columbus's arrival, all major population centers of Middle and South America had been conquered and were rapidly being transformed by Iberian colonial policies. The colonies soon became part of extensive trade networks within the region and globally with Europe, Africa, and Asia (see Figure 3.8). One of Spain's most lucrative trading links was between Acapulco in Mexico and Manila in the Philippines. By 1570, two Spanish galleons crossed the Pacific each year, one sailing west carrying Mexican silver, one sailing east carrying Asian spices, porcelain, ivory, lacquerware, and silk cloth. Galleons crossing the Pacific and Atlantic laden with this wealth proved irresistible targets for pirates.

A Global Exchange of Crops and Animals

From the earliest days of the conquest, plants and animals were exchanged between Middle and South America, Europe, Africa, and Asia via the trade routes illustrated in Figure 3.8. Many plants essential to agriculture in Middle and South America today—rice, sugarcane, bananas, citrus, melons, onions, apples, wheat, barley, and oats, for example—were all originally imports from Europe, Africa, or Asia. When disease decimated the native populations of the region, the colonists turned the abandoned land into pasture for herd animals imported from Europe, including sheep, goats, oxen, cattle, donkeys, horses, and mules.

Plants first domesticated by indigenous people of Middle and South America have changed diets everywhere and have become essential components of agricultural economies around the globe. The potato, for example, had so improved the diet of the European poor by 1750 that it fueled a population explosion. Manioc (cassava) played a similar role in West Africa. Corn, peanuts, vanilla, and cacao (the source of chocolate) are globally important crops to this day, as are peppers, pineapples, and tomatoes (see Table 3.1 on page 132).

The Legacy of Underdevelopment

In the early nineteenth century, wars of independence left Spain with only a few colonies in the Caribbean. Figure 3.9 shows the European colonizing country and the dates of independence for the countries in the region. The supporters of the nineteenth-century

FIGURE 3.9 The colonial heritage of Middle and South America. Most of Middle and South America was colonized by Spain and Portugal, but important and influential small colonies were held by Britain, France, and the Netherlands. Nearly all the colonies had achieved independence by the late twentieth century. Those for which no date appears on the map are still linked in some way to the colonizing country. [Adapted from *Hammond Times Concise Atlas of World History* (Maplewood, N.J.: Hammond, 1994), p. 69.]

TABLE 3.1   Globally important domesticated plants that originated in the Americas

Type	Common name	Scientific name	Place of origin
Seeds Quinoa	Amaranth	*Amaranthus cruentus*	Southern Mexico, Guatemala
	Beans	*Phaseolus* (4 species)	Southern Mexico
	Maize (corn)	*Zea mays*	Valleys of Mexico
	Peanut	*Arachis hypogaea*	Central lowlands of South America
	Quinoa	*Chenopodium quinoa*	Andes of Chile and Peru
	Sunflower	*Helianthus annuus*	Southwestern and southeastern North America
Tubers Tannia (foreground); maize (background)	Manioc (cassava)	*Manihot esculenta*	Lowlands of Middle and South America
	Potato (numerous varieties)	*Solanum tuberosum*	Lake Titicaca region of Andes
	Sweet potato	*Ipomoea batatas*	South America
	Tannia	*Xanthosoma sagittifolium*	Lowland tropical America
Vegetables Chayotes	Chayote (christophene)	*Sechium edule*	Southern Mexico, Guatemala
	Peppers (sweet and hot)	*Capsicum* (various species)	Many parts of Middle and South America
	Squash (including pumpkin)	*Cucurbita* (4 species)	Tropical and subtropical America
	Tomatillo (husk tomato)	*Physalis ixocarpa*	Mexico, Guatemala
	Tomato (numerous varieties)	*Lycopersicon esculentum*	Highland South America
Fruit Cacao	Avocado	*Persea americana*	Southern Mexico, Guatemala
	Cacao (chocolate)	*Theobroma cacao*	Southern Mexico, Guatemala
	Papaya	*Carica papaya*	Southern Mexico, Guatemala
	Passion fruit	*Passiflora edulis*	Central South America
	Pineapple	*Ananas comosus*	Central South America
	Prickly pear cactus (tuna)	*Opuntia* (several species)	Tropical and subtropical America
	Strawberry (commercial berry)	*Fragaria* (various species)	Genetic cross of Chile berry + wild berry from North America
	Vanilla	*Vanilla planifolia*	Southern Mexico, Guatemala, perhaps Caribbean
Ceremonial and drug plants Coca	Coca (cocaine)	*Erythroxylon coca*	Eastern Andes of Ecuador, Peru, and Bolivia
	Tobacco	*Nicotiana tabacum*	Tropical America

Source: Brian M. Kermath, Bradley C. Bennett, and Lydia M. Pulsipher, "Food Plants in the Americas: A Comprehensive Survey," http://www.umfk.maine.edu/fileshare/Kermath.AmericanFoodPlants.Complete.Draft.23.Nov.09.pdf.

revolutions were primarily **Creoles** (people of mostly European descent born in the Americas) and relatively wealthy **mestizos** (people of mixed European, African, and indigenous descent). The Creoles' access to the profits of the colonial system had been restricted by mercantilism, and the mestizos were excluded by racist colonial policies. Once these groups gained power, however, they became a new elite that controlled the state and monopolized economic opportunity. They did little to expand economic development or access to political power for the majority of the population.

Today, the economies of Middle and South America are much more complex and technologically sophisticated than they once were. Nevertheless, disparities persist: around 30 percent of the population is poor and lacks access to land, adequate food, shelter, water and sanitation, and basic education. Meanwhile, a small elite class enjoys levels of affluence equivalent to those of the very wealthy in the United States. These conditions are in part the lingering results of colonial economic policies that favored the export of raw materials and fostered undemocratic privileges for elites and outside investors who often spend their profits elsewhere rather than reinvesting them within the region. These policies will be further discussed in the Economic and Political Issues section below.

> **Creoles** people mostly of European descent born in the Americas
>
> **mestizos** people of mixed European, African, and indigenous descent

THINGS TO REMEMBER

1. Shifting cultivation of small plots is a traditional and sustainable agricultural strategy still used by some throughout the region, but often underappreciated by society.

2. Spain and Portugal were the primary colonizing countries of Middle and South America, with smaller colonies held by Britain, France, Denmark, and the Netherlands. By the late twentieth century, nearly all the colonies had gained independence.

II CURRENT GEOGRAPHIC ISSUES

Historically, power and wealth in the countries of Middle and South America have been concentrated in the hands of a few, giving them the means for not only economic domination but also political control. Globalization, economic modernization, and the transformation from rural to urban societies have not changed that reality, although recent shifts toward democratization are beginning to give more political power to the majority. Culturally, this region retains diverse influences from Europe, Africa, and indigenous peoples. Meanwhile, shifting gender roles and new religious movements are powerful agents of change.

Economic and Political Issues

Although Middle and South America are not as poor on average as sub-Saharan Africa, South Asia, or Southeast Asia, they face serious economic challenges. The unique colonial and postcolonial history of this region that resulted in wide income or wealth disparity and persistent political disempowerment of the poor make it necessary to combine the discussions of economic and political issues.

Income Disparity

Many economic challenges in this region are related to **income disparity.** With the exception of a few small countries in Middle America, the gap between rich and poor is one of the largest on earth. According to the most recent UN figures, the richest 10 percent of the population was between 19 and 94 times richer than the poorest 10 percent, depending on which country was analyzed (see Table 3.2 on page 134). In 2009, although disparities were shrinking in Brazil, Chile, Mexico, and Venezuela over what they were in 2001, primarily because of new government policies, nearly one-third of the regional population was living in poverty. In Bolivia, Colombia, Guatemala, and Peru, disparities have increased (see Table 3.2). In Haiti, Nicaragua, Bolivia, Paraguay, Guatemala, and Peru, more than half the population now lives in poverty. The global economic downturn (2007 and after) is increasing poverty rates and disparities in wealth.

Phases of Economic Development

The current economic and political situation in Middle and South America derives from the region's history, which can be divided into three major phases: the early extractive phase, the import substitution industrialization phase, and the structural adjustment and marketization phase of the present. All three phases have helped entrench wide income disparities despite a consensus that more egalitarian development is desirable.

Early Extractive Phase The **early extractive phase** began with the European conquest and lasted until the early twentieth century in parts of the region. Economic development was guided by a policy of **mercantilism,** in which European rulers sought to increase the power and wealth of their realms by extracting resources from the colonies and managing all aspects of production, transport, and commerce to the benefit of the "mother" country. Nearly all

> **income disparity** the gap in wealth between the richest 10 percent and the poorest 10 percent of a country's population
>
> **early extractive phase** a phase in Central and South American history, beginning with the Spanish conquest and lasting until the early twentieth century, characterized by a dependence on the export of raw materials
>
> **mercantilism** the policy by which European rulers sought to increase the power and wealth of their realms by managing all aspects of production, transport, and commerce in their colonies

TABLE 3.2	Income disparities in selected countries[a]

	Ratio of wealth of richest 10% to poorest 10%[b] of the population					
Country	HDI rank, 2001	1987–1995[c] Richest 10% to poorest 10%	HDI rank, 2003	1998–2000[c] Richest 10% to poorest 10%	HDI rank, 2009	2004–2007[c,d] Richest 10% to poorest 10%
Middle and South America						
Bolivia	104	91:1	114	25:1	113	94:1
Brazil	69	49:1	65	66:1	75	41:1
Chile	39	34:1	43	43:1	44	26:1
Colombia	62	43:1	64	43:1	77	60:1
Guatemala	108	29:1	119	29:1	122	34:1
Mexico	51	26:1	55	35:1	53	21:1
Peru	73	23:1	82	22:1	78	26:1
Venezuela	61	24:1	69	44:1	58	19:1
Other selected countries						
China	87	13:1	104	13:1	92	13:1
France	13	9:1	17	9:1	8	9:1
Jordan	88	9:1	90	9:1	96	10:1
Philippines	70	16:1	85	17:1	105	14:1
South Africa	94	33:1	111	65:1	129	35:1
Thailand	66	12:1	64	13:1	87	13:1
Turkey	82	14:1	96	13:1	79	17:1
United States	6	17:1	7	17:1	13	16:1

[a] The UN used data from 1987–1995, 1998–2000, and 2004–2007 on either income or consumption to calculate an approximate representation of how much richer the wealthiest 10 percent of the population is than the poorest 10 percent. The lower the ratio, the more equitable the distribution of wealth in the country.

[b] Decimals rounded up or down.

[c] Survey years fall within this range.

[d] Ratios are from the United Nations Development Report 2009.

Source: United Nations Human Development Report 2001, Table 12; UNHDR 2003, Table 13; UNHDR 2009, Table M.

hacienda a large agricultural estate in Middle or South America, more common in the past; usually not specialized by crop and not focused on market production

commercial activity in Middle and South America was under colonial control and little effort was made to build stable local economies.

A small flow of foreign investment and manufactured goods entered the region, but a vast flow of raw materials left for Europe and beyond. The foreign money to fund the farms, plantations, mines, and transport systems that enabled the extraction of resources for export came first from Europeans and later from North Americans and other international sources. One example is the still highly lucrative Panama Canal. The French first attempted to build the canal in 1880; it was later completed and run by the United States and finally turned over to Panamanian control in 1999 when it was nearly obsolete. The profits from

these ventures were usually banked abroad, depriving the region of investment that could have made it more economically independent. Industries were slow to develop in the region, and hence even essential items, such as farm tools and household utensils, had to be purchased from Europe and North America at relatively high prices. Many people simply did without.

A number of economic institutions arose in Middle and South America to supply food and raw materials to Europe and North America. Large rural estates called **haciendas** were granted to colonists as a reward for conquering territories and people for Spain. For generations these estates were then passed down through the families of those colonists. Over time, the owners, who often lived in a distant city or in Europe, lost interest in the day-to-day operations of the haciendas. Hence, productivity was generally low and hacienda laborers remained extremely poor. Nevertheless, haciendas did produce a diverse array of products (cattle, cotton, sisal, rum, sugar) for local consumption and export.

Plantations were large factory farms, meaning that in addition to growing crops such as sugar, coffee, cotton, or (more recently) bananas, some processing for shipment was done on site. Plantation owners made larger investments in equipment, and these farms were more efficient and profitable than haciendas. Because most profits were either reinvested in the plantations themselves or invested abroad, few connections with local economies or ancillary industries developed. And instead of employing local populations, they imported slave labor from Africa.

First developed by the European colonizers of the Caribbean and northeastern Brazil in the 1600s, plantations became more common throughout Middle and South America by the late nineteenth century. Unlike haciendas, which were often established in the continental interior in a variety of climates, plantations were for the most part situated in tropical coastal areas with year-round growing seasons. Their coastal and island locations gave them easier access to global markets via ocean transport.

As markets for meat, hides, and wool grew in Europe and North America, the livestock ranch emerged, specializing in raising cattle and sheep. Today, commercial ranches serving such markets as the fast-food industry are found in the drier grasslands and savannas of South America, Central America, northern Mexico, and even in the wet tropics on freshly cleared rain forest lands.

Mining was another early extractive industry. Important mines (at first primarily gold and silver) were located on the island of Hispaniola (placer gold nuggets were already extracted there in 1493), in north-central Mexico, in the Andes, in the Brazilian Highlands, and in many other locations. Extremely inhumane labor practices were common in all these mines. Today, oil and gas have been added to the mineral extraction industry, but rich mines throughout the region continue to produce gold, silver, copper, tin, precious gems, titanium, bauxite, and tungsten (**Figure 3.10**).

> **plantation** a large factory farm that grows and partially processes a single cash crop

Profits from the region's mines, ranches, plantations, and haciendas continued to leave the region, even after the countries

FIGURE 3.10 The Escondida Mine in Chile. **(A)** Copper from Chile's Escondida mine, the largest copper mine in the world, accounts for 15 percent of the country's exports. The area shown in this satellite image covers roughly 220 sq mi (560 sq km). **(B)** Hundreds of millions of tons of copper ore are removed each year from the mine's open pits. **(C)** For this and other work, the mine is highly dependent on imported machinery, such as large Caterpillar™ dump trucks made in the United States.

import substitution industrialization (ISI) policies that encouraged local production of machinery and other items that previously had been imported at great expense from abroad

land reform a policy that breaks up large landholdings for redistribution among landless farmers

recession a slowing of economic activity

external debts debts a country owes to foreign banks or governments that are repayable only in foreign currency

structural adjustment policies (SAPs) policies that require economic reorganization toward less government involvement in industry, agriculture, and social services; sometimes imposed by the World Bank and the International Monetary Fund as conditions for receiving loans

privatization the selling of formerly government-owned industries and firms to private companies or individuals

marketization the development of a free market economy in support of *free trade*

gained independence in the nineteenth century. One of the main reasons for this was that wealthy European and North American private investors had purchased and retained control of many of the extractive enterprises.

Import Substitution Industrialization Phase In the 1950s, there were waves of protest against the continuing domination of the economy and society by local elites and foreign businesses. Many governments—Mexico and Argentina most prominent among them—proclaimed themselves socialist democracies. To keep money and resources within the region in order to foster economic self-sufficiency, they enacted a set of policies that became known as **import substitution industrialization (ISI)** policies. Import substitution policies encouraged local production of machinery and other commonly imported items. To some extent these ISI policies did help reduce consumer costs and retain profits at home, but generally they failed to lift the region out of poverty.

National governments seized the most profitable extractive industries from foreign owners (usually with some payment). The intention was to use any profits to create local manufacturing industries that could supply the goods once purchased from Europe and North America. The plan included exporting these goods to foreign markets to earn more money for home economies. To encourage local people to buy manufactured goods from local suppliers, governments placed high tariffs on imported manufactured goods. The money and resources kept within each country were expected to provide the basis for further industrial development. This would create well-paying jobs, raise living standards for the majority of people, and ultimately replace the extractive industries as the backbone of the economy.

The goal of lifting large numbers of people out of poverty through jobs and general economic growth was never realized. The state-owned manufacturing sectors on which the success of ISI depended were never able to produce high-quality goods that could compete with those produced in Asia, Europe, and North America. This was largely because they could not afford the technologies and lacked the managerial skills to run globally competitive factories. Further, local populations were not large or prosperous enough to support these industries as consumers. And because exports never took off, employment stagnated, tax revenues remained low, and social programs were not adequately funded.

Not all state-owned corporations were losing propositions, however. ISI still survives in some countries and is being considered for reimplementation in a few others, such as Bolivia and Venezuela. Brazil—with its aircraft, armament, and auto industries—and Mexico—with its oil and gas industries—both

had success with some ISI programs. Interest in state-supported manufacturing industries continues, although the general trend is now toward more market-oriented management and global competitiveness. Dependence on the export of raw materials persists.

Other policies enacted during the ISI phase also had some success. Income disparities were reduced and spending on public health, education, and infrastructure increased. Meanwhile, **land reform** broke up some large landholdings and distributed them among poor landless farmers.

Beginning in the early 1970s, a debt crisis helped diminish the ISI phase. Sharp increases in oil prices and decreases in global prices of raw materials ended a period of global prosperity that had begun in the early 1950s. Reluctant to let go of the prospect of rapid development, governments and private interests continued to pursue ambitious plans to modernize and industrialize their national economies, even as prices for raw material exports—a main source of income—fell. With false optimism they paid for these projects by borrowing millions of dollars from major international banks, most of which were in North America or Europe.

In 1980, a global **recession** put a halt to the development plans of many governments in the region. Because their economies could not meet their targets for growth, governments were unable to repay their loans. The damage to the region was worsened by the fact that the biggest borrowers—such as Mexico, Brazil, and Argentina—also had the largest economies in the region. Hence, huge **external debts** (debts owed to foreign banks or governments and repayable only in foreign currency) now burdened the very countries that had been the most likely to grow (**Figure 3.11**).

Structural Adjustment and the Marketization Phase In the 1980s, alarmed over the mounting debt of their clients, foreign banks that had made loans to governments in the region took action. The International Monetary Fund (IMF) developed and enforced policies that mandated profound changes in the organization of national economies. In order to ensure that sufficient money would be available to repay loans (from foreign banks taken out by governments to finance the now-discredited ISI phase), the IMF required **structural adjustment policies (SAPs)**. These SAPs were based on concepts of **privatization** (the selling of formerly government-owned industries and firms to private investors), **marketization** (the development of a *free market* economy in support of *free trade*), and *globalization* (see Chapter 1 on pages 32–37). At the time, these concepts were considered the soundest ways for countries to achieve economic expansion and hence to repay the debts to banks in North America, Europe, and Asia.

Often the investors to whom government firms were sold were multinational corporations located in Asia, North America, and Europe; hence, ironically, the SAP era resulted in industries across the region being returned to foreign ownership. The other main SAP policy, marketization, meant that in order to obtain further loans, governments were required to remove tariffs on imported goods of all types, thus putting local industries at risk.

Crucially, SAPs also reversed the ISI-era trend toward the expansion of government social programs and the building of infrastructure. To free up funds for debt repayment, governments were required to fire many civil servants and drastically reduce spending on public health, education, job training, day care,

FIGURE 3.11 Debt, maquiladoras, and trade blocs in Middle and South America. The high rate of debt in the region stems from countries borrowing money to finance industrial and agricultural development projects intended to replace imports and reduce poverty by providing jobs. Debt is presented as total debt service (repayment of public and private loans) as a percentage of the country's GDP. Maquiladoras have played a major role in efforts to reduce government debts. The creation of trade blocs such as NAFTA and Mercosur is another means countries in the region are using to reduce debt by lowering tariffs and reducing dependence on higher-cost imports from outside the region. [Adapted from *United Nations Human Development Report 2005* and *2007–2008* (New York: United Nations Development Programme), Tables 20 and 18, respectively.]

water systems, sanitation, and infrastructure building and maintenance. While severely cutting these badly needed government programs, SAPs encouraged the expansion of industries that were already earning profits by lowering taxes on their activities, thus further shrinking revenues to pay for government services. In the countries of Middle and South America, as in most developing countries, the most profitable industries remained those based on the extraction of raw materials for export.

📺 **61. MANY PERUVIANS STRUGGLE TO GAIN HEALTHCARE ACCESS**

export processing zones (EPZs) specially created legal spaces or industrial parks within a country where, to attract foreign-owned factories, duties and taxes are not charged

maquiladoras foreign-owned, tax-exempt factories, often located in Mexican towns just across the U.S. border from U.S. towns, that hire workers at low wages to assemble manufactured goods which are then exported for sale

Export Processing Zones A major component of SAPs was the expansion of manufacturing industries in **export processing zones (EPZs)**, also known as *free trade zones*—specially created areas within a country where, in order to attract foreign-owned factories, taxes on imports and exports are not charged. The main benefit to the host country is the employment of local labor, which eases unemployment and brings money into the economy. Products are often assembled strictly for export to foreign markets.

EPZs exist in nearly all countries on the Middle and South American mainland and on some Caribbean islands. However, the largest of the EPZs is the conglomeration of assembly factories, called **maquiladoras**, located along the Mexican side of the U.S.–Mexico border (Figure 3.11 on page 137 and Figure 3.12).

FIGURE 3.12 Living standards along the U.S.–Mexico border. An outhouse in Nogales, Mexico, that serves a family employed at the maquiladoras in the area. Many Mexicans have moved to the U.S.–Mexico border zone in search of a better life, only to find low wages and poor living conditions.

Although these factories do provide employment, they might not be as beneficial to the Mexican economy as once thought, as the following vignette illustrates.

Personal Vignette Orbalin Hernandez has just returned to his self-built shelter in the town of Mexicali on the border between Mexico and the United States. Recently fired for taking off his safety goggles while loading TV screens onto trucks at Thompson Electronics, he had just gone to the personnel office to ask for his job back. Because there were no previous problems with him, he was rehired at his old salary of $300 per month ($1.88 per hour). Thompson is a French-owned electronics firm that took advantage of NAFTA (page 140) when it moved to Mexicali from Scranton, Pennsylvania, in 2001. There, its 1100 workers had been paid an average of $20.00 per hour. Now 20 percent of the Scranton workers are unemployed, and many feel bitter toward the Mexicali workers.

Though Orbalin's salary at Thompson-Mexicali is barely enough for him to support his wife, Mariestelle, and four children, they are grateful for the job. They are originally from a farming community in the Mexican state of Tabasco, where, for people with no high school education, wages averaged $60 per month.

Nonetheless, in 2006 Orbalin and his fellow Mexicali workers were being told that their wages were too high. Maquiladora firms claimed that to compete globally they had to cut costs even further. Fourteen Mexicali plants closed and moved to Asia, where in 2005, workers with the same skills as Orbalin earned just $0.35 an hour. Those firms remaining in Mexicali cut wages and reduced benefits in response. By 2009, it appeared that the global recession would affect Chinese–Mexican–U.S. relations in new ways. Rising fuel costs and other external factors such as storage and minor quality differences meant that in aggregate it was making less and less sense to move a Mexican factory to China. In some cases even keeping a factory in the United States made sense when all was considered. It is not yet clear whether this trend is short-lived or the wave of the future. *[Source: NPR reports by John Idste, August 14, 2001, August 25, 2003, and August 16, 2003, and by Gary Hadden, August 27, 2003; David Bacon, "Anti-China Campaign Hides Maquiladora Wage Cuts," ZNet, February 2, 2003, at http://www.zcommunications.org/anti-china-campaign-hides-maquiladora-wage-cuts-by-david-bacon; April 2006, http://www.maquilaportal.com/cgi-bin/public/index.pl; "How Rising Wages Are Changing the Game in China," Business Week, March 27, 2006, at http://www.businessweek.com/magazine/content/06_13/b3977049.htm; Pete Engardio, Bloomberg Businessweek, June 4, 2009, "So Much for the Cheap 'China Price,'" at http://www.businessweek.com/magazine/content/09_24/b4135054963557.htm.]* ■

Outcomes of SAPs and EPZs The SAP/EPZ era did not produce the sustained economic growth that was expected to relieve the debt crisis and achieve broader prosperity. SAPs increased some kinds of economic activity in Middle and South America, resulting in modest rates of total national GDP growth, but overall the SAP era was a time of backsliding. During the last two decades of the ISI phase (1960 to 1980), per capita GDP grew by 82 percent for the region as a whole. When SAPs were implemented across the region (1980 to 2000), growth in per capita GDP slowed to just 12.6 percent, and between 2000 and 2005 it slowed to only 1 percent.

For a time, the lowering of trade barriers expanded opportunities for local entrepreneurs and EPZs encouraged investors from other countries, especially multinational corporations, to invest in manufacturing in the region, an activity called **foreign direct investment (FDI)**. U.S.-, European-, and Asian-owned companies invested in garment making, electronics assembling, mining, agriculture, logging, and fishing industries. Partnerships between foreign companies and local businesspeople operated the largely unregulated gold mining, forestry, and other extractive industries of the Amazon Basin. However, here, as elsewhere in the region, SAPs encouraged industries to expand in ways that raised concerns about environmental impacts, worker safety, sustainability, and most recently, vulnerability to global recession. A 2009 report by the Brookings Institute revised downward its economic outlook for the six largest economies in the region in light of the shrinking global economy, projecting that average growth for them would shrivel to 1.4 percent for that year, with Mexico, Argentina, and Venezuela in negative growth. Up until the global recession beginning in 2008, a number of countries had managed to significantly lower their debt as a percent of GDP (see Figure 3.11 on page 137), but as earnings decrease, debt is likely to increase.

A major reason that SAPs failed to stimulate economic growth was that they encouraged greater dependence on exports of raw materials just when prices for these items were falling in the global market. Ironically, prices fell partly because of SAPs! Since SAPs forced indebted countries around the globe to boost raw materials exports simultaneously, a global glut was created and prices were driven down. After 2008, prices fell also because demand for raw materials decreased during the global recession.

Voter Backlash Against SAPs In the late 1990s and since, millions of voters within this region registered their opposition to SAPs (and in some cases to EPZs). Since 1999, presidents who explicitly opposed SAPs were elected in eight countries in the region: Argentina, Bolivia, Brazil, Chile, Ecuador, Nicaragua, Uruguay, and Venezuela. Both Brazil and Argentina made considerable sacrifices to pay off their debts and be liberated from the restrictions of SAPs (see Figure 3.11 on page 137). Venezuela, under President Hugo Chávez, has emerged as a leader of the SAP backlash. In 2005, the Chávez administration **nationalized** foreign oil companies operating in Venezuela. Since then it has also used its own considerable oil wealth to help other countries, such as Argentina, Bolivia, and Nicaragua, pay off their debts. Bolivia, led by Evo Morales (often an ally of Chávez), reversed the SAP policy of privatization by nationalizing that country's natural gas industry. Loans taken from the IMF dropped from $48 billion in 2003 to just $1 billion in 2008 and during this same period IMF policy reversed itself from promoting SAPs to promoting policies that attempted to combine marketization with strong investment in social programs. This new IMF position is holding, and so it appears that voters had an important effect by electing officials who successfully challenged IMF policies.

63. NEW BOLIVIAN ENERGY POLICY CAUSES CONCERN

67. U.S.–VENEZUELA ENERGY TIES ENDURE DESPITE DETERIORATING POLITICAL RELATIONS

The Informal Economy

For centuries, low-profile businesspeople throughout Middle and South America have operated in the *informal economy*, meaning that they support their families through inventive entrepreneurship but are not officially recognized in the statistics and do not pay business, sales, or income taxes. Most are small-scale operators involved with street vending or recycling used items such as clothing, glass, or waste materials. In some instances, the informal economy can serve as an incubator for new businesses that may expand, eventually providing legitimate jobs for family and friends.

Throughout this region, nearly every citizen depends on the informal economy in some way as either a buyer or a seller. Critics argue that informal workers are just treading water, making too little to ever expand their businesses significantly. Moreover, the bribes they have to pay to avoid arrest or fines are much less beneficial to the economy as a whole than the taxes paid by legitimate businesses. Work in the informal economy is also risky because there is no protection of workers' health and safety and no sick leave, retirement, or disability benefits (also often lacking in the formal economy). Nevertheless, the informal economy can be a lifesaver during times of economic recession. For example, after recession hit Peru in 2000, 68 percent of urban workers were in the informal sector. They generated 42 percent of the country's total gross domestic product. Most of them worked as street vendors (**Figure 3.13**).

> **foreign direct investment (FDI)** the amount of money invested in a country's business by citizens, corporations, or governments of other countries
>
> **(to) nationalize** to seize private property and place it under government ownership, with some compensation

FIGURE 3.13 A street vendor selling religious items on a street corner in Lima, Peru.

Unfortunately, the informal economy goes hand in hand with corruption because entrepreneurs operating on the fringe of legality are vulnerable to unscrupulous agents who may demand protection money or extract high-priced favors for not turning vendors in to the authorities. On the other hand, those in the informal economy may actually have chosen the illegal route precisely because corrupt government officials demanded high bribes to dispense legal licenses to operate.

The Role of Remittances

Migration for the purpose of supporting family members back home with remittances is an extremely important economic activity throughout this region (see further discussion on pages 149–152). Most such migrations are within countries, and the amounts of cash sent home are small but economically significant for families that increasingly find farming and other subsistence activities no longer sufficient to cover the basics and school fees for their children. Migrants to North America, legal and undocumented, are almost always planning to support extended families back home.

One report (by geographer Dennis Conway and anthropologist Jeffrey H. Cohen) suggests that couples who migrate from indigenous villages in Mexico typically work at menial jobs in the United States, living frugally in order to save a substantial nest egg. Then they may return home for several years to build a house (usually a family self-help project) and buy furnishings. Because a few thousand dollars saved in the United States will accomplish a great deal in rural Mexico, the family may live off their nest egg for several years while one or both members of the couple renders volunteer community service to the home village. When the money runs out, the couple, or just the husband or wife, may migrate again to save another nest egg.

Among those who sent money from the United States to Mexico in the early 2000s, the average remittance was only about $350, but this amount was sent several times a year, amounting to well over $1000 annually. The global recession lasting through 2009 affected the amount that Mexicans remitted. From January through November of 2009, $19.6 billion was remitted, compared with $23.4 billion during the same period one year earlier. According to the Inter-American Development Bank, in 2004, total migrant remittances from the United States to Middle and South America amounted to $32 billion, which is roughly double the amount of U.S. foreign aid to the region in that year.

Regional Trade and Trade Agreements

The growth in international trade and foreign investment in the region, encouraged in part by SAPs, has been joined by growth in regional free trade agreements. Such agreements, inspired in part by the free trade initiatives of the IMF and other external finance institutions, reduce tariffs and other barriers to trade among a group of neighboring countries. The two largest free trade agreements within the region are NAFTA and Mercosur (see Figure 3.11 on page 137).

The **North American Free Trade Agreement (NAFTA)**, a free trade bloc consisting of the United States, Mexico, and Canada, and containing more than 450 million people, was created in 1994. The goals of NAFTA are to reduce barriers to trade; to create an expanded market for the goods and services produced in North America; to establish clear and mutually advantageous trade rules; to improve working conditions across the region; and to develop and expand world trade and international cooperation. Since 1994, the value of the economies of these three countries grew steadily and by 2008 was worth at least $17 trillion. But there have been problems, as further discussions will illustrate. The continuing success of NAFTA is far from certain, especially insofar as Mexico is concerned. A subsequent effort by the United States to create a NAFTA-like trade bloc for all of the Americas (FTAA) has stalled in recent years, largely because of dissatisfaction with effects of globalization and with the IMF and World Bank, perceived to be controlled by the United States.

Mercosur links the economies of Argentina, Brazil, Paraguay, and Uruguay to create a common market with 250 million potential consumers and an economy worth nearly $3 trillion per year (Bolivia, Chile, Colombia, Ecuador, and Peru are associate Mercosur members and are not included in these figures). Mercosur, which is smaller and older than NAFTA, having been created in 1991, is seen as a more equitable alternative to NAFTA, largely because the United States is not involved. Venezuela was invited to join in 2006, but by mid-2009 its bid was not yet ratified by the congresses of Brazil and Paraguay, apparently because of reservations about the intentions of Venezuelan president Hugo Chávez. Chávez, a chief opponent to SAPs, was by 2007 fashioning himself as an increasingly authoritarian leader (page 143) at home and an aggressive anti-U.S. spokesman for *neo-socialism*. Mercosur members were increasingly wary of his ambitions to gain regional power and influence because he was distributing oil-derived funds to causes in Argentina, Bolivia, Peru, Ecuador and Honduras (see page 139).

▶️ **62. MANY VENEZUELANS UNCERTAIN ABOUT CHÁVEZ'S 21ST CENTURY SOCIALISM**

The **World Trade Organization (WTO)** is ostensibly intended to operate at the global level to promote free trade, but in the eyes of many countries in Middle and South America it tends more to look after the interests of the more developed countries. The G22, a global coalition of 22 of the richest developing countries, led in part by Argentina, Brazil, Cuba, and Venezuela, has challenged the powerful developed countries of the G8 (the United States, Canada, France, Germany, Italy, Japan, Russia, and the United Kingdom) to stop the hypocrisy of promoting free trade for others while practicing protectionist policies themselves. The G22 protested being shut out of G8 markets by G8 rulings that amounted to tariffs. They also protested government subsidies paid to G8 farmers that made it possible to sell G8 farm products on the global market at less than cost, thus putting farmers from

North American Free Trade Agreement (NAFTA) a free trade agreement made in 1994 that added Mexico to the 1989 economic arrangement between the United States and Canada

Mercosur a free trade zone created in 1991 that links the economies of Brazil, Argentina, Uruguay, and Paraguay to create a common market

World Trade Organization (WTO) a global institution made up of member countries whose stated mission is to lower trade barriers and to establish ground rules for international trade

fair trade movement a global movement to distribute profits more fairly to producers by upgrading their knowledge of markets and hence to increase their competitiveness

poor countries out of business in their own homelands and rendering G22 citizens ever more reliant on expensive imported food.

The overall record of regional free trade agreements so far is mixed. While they have increased trade, the benefits of that trade usually are not spread evenly among regions or among all sectors of society. In Mexico, for instance, the benefits of NAFTA have gone to elites and have been concentrated in the northern states that border the United States. There, the agreement facilitated the growth of maquiladoras by easing cross-border finances and transit. But, significantly, as many as one-third of small-scale farmers throughout Mexico have lost their jobs as a result of increased competition from U.S. corporate agriculture, which now, with NAFTA, has unrestricted access to Mexican markets. Corn imports from the United States have undermined the market for locally produced corn, impoverishing small farmers or driving them out of business. These unemployed and often displaced farmers make up a significant number of the undocumented workers coming into the United States.

The Fair Trade Movement

Coffee used to be a primary commercial crop in the Pacific coastal uplands of Guatemala. Now, many of the coffee plantations (fincas) are abandoned. The workers have migrated to Mexico and the United States, seeking some other way to support their families. Too many poor tropical farmers on 10 or 12 acres trying to make a living in coffee produced a glut on the global coffee market. As a consequence, most are paid only a tiny amount per pound by the itinerant coffee trader (called a coyote) who is their only access to the market. When you buy a $1.00 cup of coffee, the grower gets 10 to 12 cents, the trader 3 cents, and the shipper 4 cents. The roaster—usually a large multinational company—gets 65 to 70 cents. The retailer typically gets 10 to 15 cents. The global **fair trade movement**, still limited in scope, seeks to address these inequities for the grower by upgrading their skills as producers.

THINGS TO REMEMBER

1. Three phases of economic development—extractive, import substitution, and structural adjustment—dominated the region well into the late twentieth century, creating huge wealth disparities and large populations of poor people.

2. In the 1990s, voters across the region rebelled and began electing governments on the side of the poor and opposed to SAPs.

3. Since the difficulties of the 1990s, free trade and fair trade policies were adopted in some industries, and regional trading blocs were formed, with mixed results.

Food Production and Contested Space

The agricultural lands of Middle and South America (Figure 3.14) are an example of what geographers call **contested space**, in which various groups are in conflict over the right to use a specific territory as they see fit. The conflicts take place at various scales, as the following vignettes demonstrate, and all are in one way or another related to inequities that have been in place at least since colonial times.

> **contested space** any area that several groups claim or want to use in different and often conflicting ways, such as the Amazon or Palestine

Local Scale: A Banana Worker in Costa Rica For centuries, while people labored for very low wages on haciendas, the owners would at least allow them a bit of land to grow a small garden or some cash crops. In recent years, however, SAPs have encouraged a shift to large-scale, mechanized, export-oriented farms and plantations. The rationale is that these operations can earn larger profits and hence can help countries pay off debts faster.

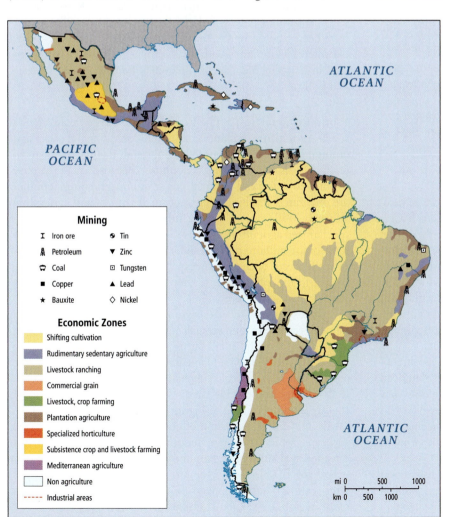

FIGURE 3.14 Agricultural and mineral zones in Middle and South America. [Adapted from *Goode's World Atlas*, 21st edition (Chicago: Rand McNally, 2005), p. 137, Minerals and Economic map.]

Mining

I	Iron ore	⚲	Tin
⚒	Petroleum	▼	Zinc
⌂	Coal	⊡	Tungsten
■	Copper	▲	Lead
★	Bauxite	◇	Nickel

Economic Zones

- Shifting cultivation
- Rudimentary sedentary agriculture
- Livestock ranching
- Commercial grain
- Livestock, crop farming
- Plantation agriculture
- Specialized horticulture
- Subsistence crop and livestock farming
- Mediterranean agriculture
- Non agriculture
- Industrial areas

ATLANTIC OCEAN

PACIFIC OCEAN

ATLANTIC OCEAN

mi 0   500   1000
km 0   500   1000

SAPs have made it easier for foreign multinational corporations, such as Del Monte, with their highly efficient, state-of-the-art operations, to dominate the most profitable agribusinesses. Meanwhile, throughout Middle and South America, increasing numbers of rural people find themselves displaced from lands they once cultivated. As a result, they now work as migrant laborers for low wages, as the following story illustrates.

Personal Vignette Aguilar Busto Rosalino used to work on a Costa Rican hacienda. He had a plot on which to grow his own food and in return worked 3 days a week for the hacienda. Since a banana plantation took over the hacienda, he rises well before dawn and works 5 days a week from 5:00 A.M. to 6:00 P.M., stopping only for a half-hour lunch break. Because he doesn't have time to farm, he now has to buy most of his food.

Aguilar places plastic bags containing pesticide around bunches of young bananas. He prefers this work to his last assignment of spraying a more powerful pesticide, which left him and 10,000 other plantation workers sterile. He works very hard because he is paid according to how many bananas he treats. Usually he earns between $5.00 and $14.50 a day.

Right now Aguilar is working for a plantation that supplies bananas to the Del Monte corporation, but he thinks that in a few months he will be working for another plantation nearby. It is common practice for these banana operations to fire their workers every 3 months so that they can avoid paying the employee benefits that Costa Rican law mandates. Although Aguilar makes barely enough to live on, he has no plans to press for higher wages because he knows that he would be put on a "black list" of people that the plantations agree not to hire. [*Source: Adapted from Andrew Wheat, "Toxic Bananas," Multinational Monitor 17 (9), September 1996, pp. 6–7; updated 2007, at http://www.magney.org/photofiles/CostaRica-Bananas1.htm.*] ∎

Provincial Scale: The Zapatista Rebellion In the southern Mexican state of Chiapas, agricultural activists and indigenous leaders have mobilized armed opposition to the economic and political systems that have left them poor and powerless. The Mexican government redistributed some hacienda lands to poor farmers early in the twentieth century, but most land in Chiapas is still held by a wealthy few who use mechanized methods to grow cash crops for export. The poor majority farm tiny plots on infertile hillsides. In 2000, about three-fourths of the rural population was malnourished, and one-third of the children did not attend school.

The Zapatista rebellion (named for the Mexican revolutionary hero Emiliano Zapata) began on the day the North American Free Trade Agreement took effect in 1994. The Zapatistas view NAFTA as a threat because it diverts the support of the Mexican federal government from land reform to large-scale mechanized export agriculture. The Mexican government used the army to suppress the rebellion.

In 2003, after 9 years of armed resistance, the Zapatista movement redirected part of its energies. It began a nonviolent political campaign to democratize local communities by setting up people's governing bodies parallel to local official governments. Before the national elections in 2006, the Zapatistas toured the country in an effort to turn the political climate against globalization and toward greater support for indigenous people and the poor. The election ended in a near tie, but in the end the Zapatista-favored candidate was not judged the winner by the election commission. The Zapatistas continue to be active in efforts to reform Mexican politics to include more grassroots participation.

National Scale: Brazil's Landless Movement The trends in agriculture described above for Middle America have sparked rural resistance in Brazil as well. Sixty-five percent of Brazil's arable and pasture land is owned by wealthy farmers who make up just 2 percent of the population. Since 1985, more than 2 million small-scale farmers have been pressed to sell their land to large-scale, mechanized farms. Because the larger farms specialize in major export items, such as cattle and soybeans, they have been favored by governments under pressure from SAPs to increase exports. As a result, many poor farmers have been forced to migrate.

To help these farmers, organizations such as the Movement of Landless Rural Workers (MST) began taking over unused portions of some large farms. Since the mid-1980s, the MST has coordinated the occupation of more than 21 million hectares (51 million acres) of Brazilian land (an area about the size of Kansas). Some 250,000 families have gained land titles, while the elite owners have been paid off by the Brazilian government and have moved elsewhere. Movements with goals similar to those of MST now exist in Ecuador, Venezuela, Colombia, Peru, Paraguay, Mexico, and Bolivia.

World Regional Scale: The Persistence of Large-Scale Mechanized Agriculture Instances of workers contesting the use of agricultural land are exceptions to overall trends that favor the growth of large-scale mechanized agriculture. Because of the money-making potential and political power of big agriculture, conflicts are only rarely resolved in favor of small farmers and agricultural workers. Moreover, in rapidly urbanizing countries, large-scale mechanized agriculture is seen as the only way to supply sufficient food for the millions of city dwellers. Ironically, many of these new urbanites were once farmers capable of feeding themselves; they moved to the cities because they were unable to compete with mechanized agriculture.

Zones of modern mixed farming, including the production of meat, vegetables, and specialty foods for sale in urban areas, are located on large and small farms around most major urban centers. Many areas have had large farms for centuries. For example, a wide belt of large-scale mechanized grain production, similar to that found in the midwestern United States, stretches through the Argentine pampas (see Figure 3.14 on page 141). And, as we saw in the earlier discussion of environmental issues (page 124), soybeans, which are often produced by mechanized farms on once-forested lands in and around the Amazon Basin, are a major export for Brazil. Hence, despite the recent political tensions, large-scale agriculture appears to have a solid place in the future of this region. ▶ **57. HAITI'S RISING COST OF FOOD WORRIES AID GROUPS**

THINGS TO REMEMBER

1. *Contested space* is an important geographical concept in Middle and South America.

2. The term often refers to competition over agricultural land that occurs with modernization, such as individual farmers losing access to land on which to produce their own food.

3. In these contests over land, large-scale agriculture, often aimed at the export market, usually wins because of economic clout.

Political Reform and Democratization

After decades of elite and military rule, almost all countries in the region now have multiparty political systems and democratically elected governments. In the last 25 years, there have been repeated peaceful and democratic transfers of power in countries once dominated by rulers who seized power by force. These **dictators** often claimed absolute authority, governing with little respect for the law or the rights of their citizens. Their authority was based on alliances between the military, wealthy rural landowners, wealthy urban entrepreneurs, foreign corporations, and even foreign governments such as that of the United States.

Fragile Democracies Although democracy has transformed the politics of the region, problems remain (**Photo Essay 3.4** on page 145). Elections are sometimes poorly or unfairly run, and their results are frequently contested. Elected governments are sometimes challenged by citizen protests or threatened with a **coup d'état**, in which the military takes control of the government by force. Such coups are usually a response to policies that are unpopular with large segments of the population, powerful elites, the military, or the United States. In the last decade, coups have threatened Venezuela, Honduras, Ecuador, Colombia, and Bolivia.

Change has been especially dramatic in Mexico. The entrenched political machine of the Institutional Revolutionary Party (PRI) ruled the country for almost three-quarters of a century. But in 2000, it was unseated by a coalition led by Vicente Fox, a wealthy Coca-Cola executive. Fox managed to pass a number of reform-oriented laws in his 6-year term. In the 2006 presidential election, the conservative Felipe Calderón (from Fox's political party) narrowly defeated the populist mayor of Mexico City, Andrés López Obrador. López Obrador, with significant public support, created a constitutional crisis by refusing to accept the election results. He organized a so-called parallel government, declaring himself president and appointing ministers. However, he failed to force President Calderón to resign. Now Mexico is facing a new threat to political stability (see **Photo Essay 3.4A** on page 145). The government (federal and local) seems unable to control powerful drug cartels that perpetrate corruption and death in urban and rural areas alike, with much of criminal activity connected to the North American drug market (discussion follows).

📹 **64. CONTROVERSY IN MEXICO'S CLOSE PRESIDENTIAL ELECTIONS**

📹 **65. MEXICO'S WINNING CANDIDATE FAVORS STRONG TIES TO U.S.**

In Venezuela, the landslide election of Hugo Chávez as president in 1998 appeared to advance the cause of democracy there. Elites had long dominated Venezuelan politics, and profits from the country's rich oil deposits had not trickled down to the poor. One-third of the population lived on $2.00 (PPP) a day, and per capita GDP (PPP) was lower in 1997 than it had been in 1977. Chávez campaigned as a champion of the poor, calling for the government to provide jobs, community health care, and subsidized food for the large underclass. These policies, all of which were enacted soon after his election, eroded Chávez's original support from the small middle class and elites, who feared a turn toward general government control. When the U.S. government expressed alarm, Chávez became a major critic of U.S. and other foreign dominance in the region. Chávez sought to lead the region-wide backlash against SAPs and condemned the long history of U.S. intervention in the internal affairs of countries across the region (page 139).

> **dictator** a ruler who claims absolute authority, governing with little respect for the law or the rights of citizens
>
> **coup d'état** a military- or civilian-led forceful takeover of a government

📹 **66. SOCIAL PROGRAMS AT ROOT OF CHÁVEZ'S POPULARITY**

Chávez was reelected in 2000, and then briefly deposed in a coup d'état in 2001, with the rumored approval of the Bush administration. He was quickly reinstated, however, by a groundswell of popular support among the large underclass, who stood to benefit from his policies. His presidency was sustained in a referendum in 2004 and again in the landslide election of 2006. In 2009, voters approved a constitutional amendment removing presidential term limits, thereby allowing Chávez to run for president indefinitely. Some saw this as a triumph of populism; others saw this as a loss for democracy (see **Photo Essay 3.4 Map** on page 145).

📹 **69. VENEZUELANS REJECT CONSTITUTIONAL CHANGES**

📹 **68. WILL CHÁVEZ INHERIT CASTRO'S REVOLUTIONARY MANTLE?**

The Drug Trade

The international illegal drug trade is a major factor contributing to corruption, violence, and subversion of democracy throughout the region (see **Photo Essay 3.4A,B** on page 145). Cocaine and marijuana are the primary drugs of trade (but cultivation of opium poppies is increasing), with most drugs produced in or passing through northwestern South America, Central America, and Mexico. **Figure 3.15** on the next page graphically illustrates the geographic distribution of cocaine seizures of more than 10 kilograms in 2006. (Note: Such seizures may not reveal activity in areas where law enforcement is lax or entirely co-opted by the drug traders.)

Most coca growers are small-scale farmers of indigenous or mestizo origin in remote locations who can make a better income for their families from these plants than from other cash crops. Production of addictive drugs is illegal in all of Middle and South America. However, public figures, from the local police on up to high officials, are paid to turn a blind eye to the industry. Those who oppose the cartels, whether farmers, journalists, or law enforcement officials, often end up kidnapped, tortured, and brutally murdered.

In Colombia, the illegal drug trade has financed all sides of an ongoing civil war that has threatened the country's democratic

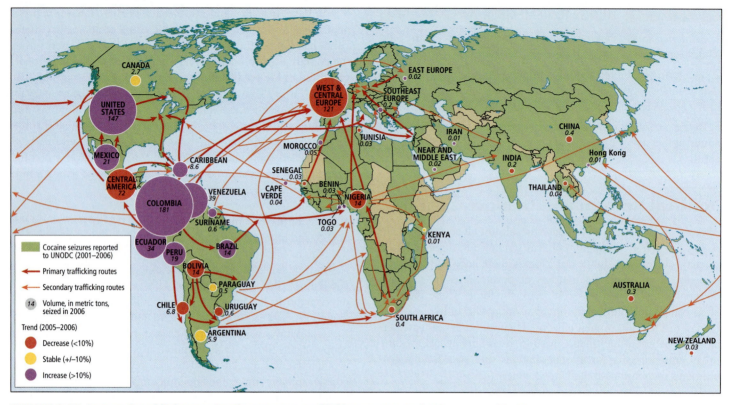

FIGURE 3.15 Interregional linkages: Cocaine sources, trafficking routes, and seizures worldwide. Colombia, Peru, and Bolivia are the most important sources for the cultivation and production of cocaine in the world; Colombia produces 55 percent of the world's cocaine; Peru produces the second most, and Bolivia, the third. The big cocaine markets are in the Americas; use has declined by 36 percent in the United States since 1998 and increased in South America. Cocaine use in western Europe and West Africa is on the rise, due in part to new trade routes through West Africa to Europe. (Hawaii is not included here because it is covered in Oceania, Chapter 11.) [Adapted from *2008 World Drug Report* (Vienna: United Nations Office on Drugs and Crime, 2005), pp. 13 and 15, at http://www.unodc.org/documents/wdr/WDR_2008/WDR_2008_eng_web.pdf.]

traditions and displaced more than 1.5 million people over the past several decades (see Photo Essay 3.4B). Meanwhile, competing drug-smuggling rings and paramilitaries bribe, intimidate, kidnap, and murder Colombian citizens and government officials who seek to reduce or eradicate the drug trade.

72. THOUSANDS KIDNAPPED IN COLOMBIA IN LAST DECADE

U.S. policy has emphasized stopping the production of illegal drugs in Middle and South America rather than curtailing the demand for drugs in the United States. As a result, the U.S. *war on drugs* has led to a major U.S. presence in the region that is focused on supplying intelligence, eradication chemicals, military equipment, and training to military forces in the region. A consequence is that U.S. military aid to Middle and South America is now about equal to U.S. aid for education and other social programs in the region. Meanwhile, drug production in Middle and South America is now greater than ever, exceeding demand in the U.S. market, where street prices for many drugs have fallen in recent years.

Foreign Involvement in the Region's Politics

Interventions in the region's politics by outside powers have frequently compromised democracy and human rights. Although the former Soviet Union, Britain, France, and other European countries have wielded much influence, by far the most active foreign power in this region has been the United States.

In 1823, the United States proclaimed the Monroe Doctrine to warn Europeans that the United States would allow no further colonization in the Americas. Subsequent U.S. administrations interpreted this policy more broadly to mean that the United States itself had the right to intervene in the affairs of the countries of Middle and South America, and it has done so many times. The official goal for such interventions was usually to make countries safe for democracy, but in most cases the driving motive was to protect U.S. political and economic interests.

During the past 150 years, U.S.-backed, unelected political leaders, many of them military dictators, have been installed at some point in many countries in the region (see **Photo Essay 3.4C**). In recent decades, the United States funded armed interventions in Cuba (1961), the Dominican Republic (1965), Nicaragua (1980s), Grenada (1983), and Panama (1989). Perhaps the most infamous intervention occurred in Chile in 1973. With U.S. aid, the elected socialist-oriented government of Salvador Allende was overthrown, Allende killed, and a military dictator, General Augusto Pinochet, installed in his place. Over the next

Democratization is well underway in Middle and South America, although many barriers remain. Decades of elite and military rule combined with rampant corruption have led to popular revolutionary movements aimed at taking control of the government by force. The result has been war and brutal repression by both governments and militants that denies citizens the freedoms essential to democracy. The drug trade has also emerged as a major obstacle to democracy, with huge amounts of illicit cash being used to pay off elected officials, civil servants, police forces, and the military. Nevertheless, recent decades have seen a dramatic expansion of democracy and a decline in violence for the region as a whole.

A Federal agents in Hermosillo, Mexico arrest drug traffickers who killed another federal agent. Mexico's ongoing drug war has led to more than 10,000 deaths over the past several years and corruption that has eroded local democratic institutions, especially near the U.S. border. Most of Mexico's drug production supplies the United States.

B Victims of Colombia's ongoing conflict march in protest in Bogotá. This complex conflict, the worst in the region, is driven by popular revolutionary movements and an ongoing drug war. Democratic institutions and the rule of law have been badly damaged, with many high-ranking officials linked to extreme abuses of human rights and civil liberties.

C School children in Leon, Nicaragua, play in front of a mural commemorating students killed by government troops in 1959 during a protest against then-president Anastasio Somoza. Sixty thousand Nicaraguans died during decades of conflict between Somoza's military dictatorship and the pro-democracy political opposition.

D Cuban revolutionaries Che Guevara (right) and Camillo Cienfuegos (left) are immortalized in wax in the Museum of the Revolution in Havana. Since coming to power in 1959, Cuba's communist government has been criticized for jailing and sometimes executing leaders of the political opposition.

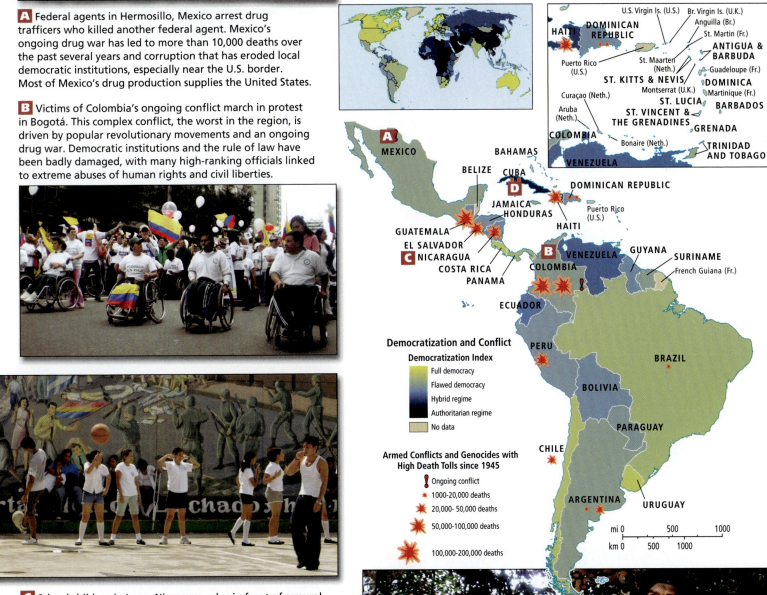

U.S. Virgin Is. (U.S.) — Br. Virgin Is. (U.K.)
HAITI — DOMINICAN REPUBLIC
Anguilla (Br.)
St. Martin (Fr.)
Puerto Rico (U.S.)
St. Maarten (Neth.)
ANTIGUA & BARBUDA
ST. KITTS & NEVIS
Guadeloupe (Fr.)
Curaçao (Neth.)
Montserrat (U.K.) — DOMINICA
Martinique (Fr.)
Aruba (Neth.)
ST. LUCIA
BARBADOS
ST. VINCENT & THE GRENADINES
GRENADA
COLOMBIA
Bonaire (Neth.)
TRINIDAD AND TOBAGO
VENEZUELA

A MEXICO
BAHAMAS
BELIZE — CUBA
D
JAMAICA
DOMINICAN REPUBLIC
HONDURAS
Puerto Rico (U.S.)
GUATEMALA
HAITI
EL SALVADOR
C NICARAGUA
COSTA RICA
PANAMA
B VENEZUELA
GUYANA
SURINAME
French Guiana (Fr.)
COLOMBIA
ECUADOR
PERU
BRAZIL
BOLIVIA
PARAGUAY
CHILE
ARGENTINA
URUGUAY

Democratization and Conflict

Democratization Index
- Full democracy
- Flawed democracy
- Hybrid regime
- Authoritarian regime
- No data

Armed Conflicts and Genocides with High Death Tolls since 1945
- Ongoing conflict
- 1000-20,000 deaths
- 20,000-50,000 deaths
- 50,000-100,000 deaths
- 100,000-200,000 deaths

mi 0 — 500 — 1000
km 0 — 500 — 1000

17 years, the Pinochet regime imprisoned and killed thousands of Chileans who protested the loss of democracy.

📹 74. EX-CHILEAN DICTATOR AUGUSTO PINOCHET DIES

📹 70. NORIEGA LAWYERS CONTINUE TO FIGHT
AGAINST EXTRADITION

The Cold War and Post–Cold War Eras Since 1959, the Caribbean island of Cuba has been governed by a radically socialist government led first by Fidel Castro and Che Guevara (see **Photo**

Essay 3.4D on page 145) and then, starting in 2006, by Fidel's brother Raúl. The revolution that brought Fidel Castro to power transformed a plantation and tourist economy once known for its extreme income disparities into one of the most egalitarian in the region, though troubled economically and in matters of human rights. Because Castro adopted socialism and allied Cuba with the Soviet Union, relations with the United States became extremely hostile. The United States has funded many efforts to destabilize Cuba's government, and it actively discourages other countries

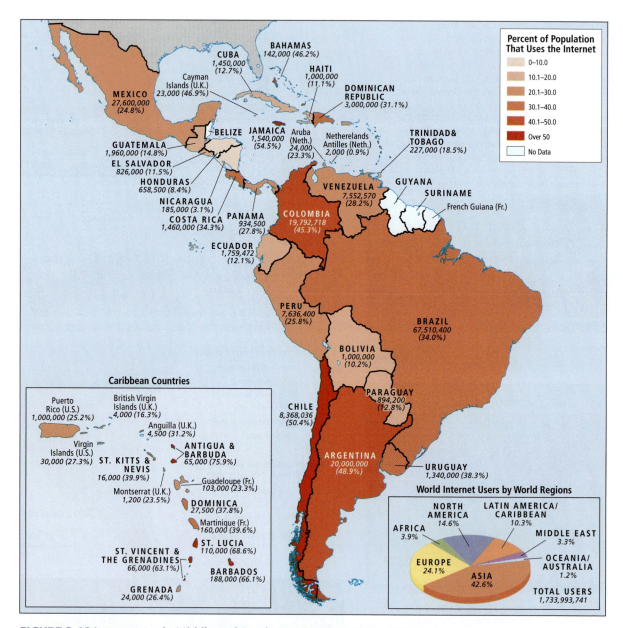

FIGURE 3.16 Internet use in Middle and South America, September 2009. The numbers on the map indicate the number of Internet users in each country; percents indicate the percentage of the country's population that uses the Internet. By September, 2009, over 184 million people (26.7 percent of the population in Middle and South America were using the Internet; the number of users more than doubled since 2005. Chile has the highest percentage of users (50.4) in South America; Antigua & Barbuda (75.9 percent) the highest in the Caribbean; and Costa Rica (34.3 percent) the highest in Middle America. [Source: "Internet Usage Statistics—The Big Picture" and "Internet Usage Statistics for the Americas," at http://www.internetworldstats.com/stats.htm, http://www.internetworldstats.com/stats10.htm, and http://www.internetworldstats.com/stats11.htm.]

from trading with Cuba. The United States maintains an ongoing embargo against Cuba, and so far efforts to lift it have failed.

73. EU SPLIT OVER DEVELOPMENT COMMISSIONER'S PROPOSAL TO NORMALIZE RELATIONS WITH CUBA

With help from the Soviet Union, Castro managed to dramatically improve the country's life expectancy, literacy, and infant mortality rates. Unfortunately, he also imprisoned or executed thousands of Cubans who disagreed with his policies. Much of the Cuban upper class fled to southern Florida, which remains a popular destination for exiled Cubans. Following the demise of the Soviet Union, Castro opened the country to foreign investment. Many countries responded, and Cuba is now a major European tourist destination. However, the Castro regime is intact, political repression in Cuba persists, and relations with the United States remain cool, although the Obama administration has moderated some policies against travel and fund transfers, but only for Cuban Americans.

The Political and Economic Impacts of Information Technology

The Internet already plays a role in the politics of Middle and South America by helping activists, such as the Zapatistas of Chiapas and the MST (page 142) of Brazil, get their message out to the rest of the world. However, too few poor have Internet access; meanwhile, the middle classes are gaining access quickly (Figure 3.16). In 2000, Internet users made up only 3.2 percent of the region's population; but that figure had risen to 26.7 percent by September, 2009. The highest rate of use was on the Caribbean islands of Antigua and Barbuda, where 75.9 percent of the population used the Internet in that year. Brazil had the largest number of Internet users at 67.5 million, which represented 34 percent of the population.

Overall, Middle and South America is more advanced in technological achievement than other developing regions of the world. In 2009, Brazil, with 34 percent of the population connected, ranked fifth in the world for total number of Internet users. It has two world-class technology hubs in the environs of São Paulo and is attempting to cover the entire country with broadband fiber-optic cable networks for telecommunications and Internet service. Mexico, with 21.7 percent of its population connected (see Figure 3.16), has recently launched a program to give its citizens access to training and higher education via the Internet. This is part of an effort to stem the flow of migration to the United States. On the other hand, the possibility of sending remittances electronically has facilitated migration and increased the use of bank accounts in Mexico.

The region tends to lag a bit in cell phone use, but that is changing rapidly as the realization grows that the cell phone transforms access to development for the poorest of the poor. As African villagers are demonstrating, it can be used to earn a living—selling calls; to get crop market prices, reach wider markets, or bank money; and as a substitute for travel. A recent report in *The Economist* (September 26, 2009) noted that adding just 10 mobile phones per 100 people boosts GDP per capita by 0.8 percent. In Brazil, cell phone use grew 22 percent in the year ending March 2009.

1. Democracy has transformed the politics of the region, yet problems remain. Elections are sometimes poorly or unfairly run, with results frequently contested. Elected governments are at times challenged by citizen protests or with a coup d'état, because policies are unpopular with large segments of the population, powerful elites, or the military.

2. The drug trade has emerged as a major obstacle to democracy, with huge amounts of illicit cash being used to pay off elected officials, civil servants, police forces, and the military.

3. The Internet plays an important role in the politics of the region, which is more technologically advanced than the world's other developing regions.

Sociocultural Issues

Under colonialism, a series of social structures evolved that guided daily life—standard ways of organizing the family, community, and economy. They included rules for gender roles, race relations, and religious observance. These social structures, combined with economic systems, influenced population distribution and growth. Traditional social structures, still widely accepted in the region, are nonetheless changing in response to urbanization, economic development, migration, and globalization. The results are varied. In the best cases, change is leading to a new sense of initiative on the part of women, men, and the poor. In the worst cases, the result is loss of economic viability and community cohesion and the breakdown of family life, with some of the most extreme impacts felt by indigenous communities.

Population Patterns

Today, populations in Middle and South America continue to grow, but at a slower pace, due to lower birth rates. At the same time, a major migration from rural to urban areas is transforming traditional ways of life. A second, international, migration trend is also growing as many people leave their home countries, temporarily or permanently, to seek opportunities in the United States and Europe. As of mid-2009, about 580 million people were living in Middle and South America, close to 10 times the population of the region in 1492. This is 239 million more than presently live in North America.

Population Distribution The population density map of this region reveals a very unequal distribution of people. If you compare Photo Essay 3.5 on the next page with the regional map (see Figure 3.1 on pages 114–115), you will find areas of high population density in a variety of environments. Some of the places with the highest densities, such as those around Mexico City and in Colombia and Ecuador, are in highland areas. But high concentrations are also found in lowland zones along the Pacific coast of Central America and especially along the Atlantic coast of South America. The cool uplands (*tierra templada*) were densely occupied even before the European conquest. Most coastal lowland

A Barbados has the highest population density in the region at 1683 people per square mile.

B Honduras. Central America has the highest population growth rates in the region.

C A couple strolls in Havana. Cuba has the lowest population growth rate in the region.

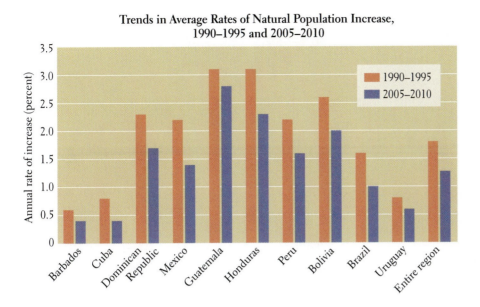

Trends in Average Rates of Natural Population Increase, 1990–1995 and 2005–2010

FIGURE 3.17 **Trends in natural population increases, 1990–2010.** The orange and blue columns show that rates of natural increase have declined steadily throughout the region and are projected to continue to do so into the future. Nevertheless, in many countries, natural population increase remains high enough to outstrip efforts at improving standards of living. Note that the rates of natural increase given here are for ranges of time, and the total percentage is for the 34 independent countries in the region. [Adapted from *United Nations Human Development Report 2009* (New York: United Nations Development Programme), Table 5.]

concentrations, in the *tierra caliente*, are near seaports with vibrant economies and a cosmopolitan social life that attract people.

Population Growth Rates of natural population increase were high in Middle and South America in the twentieth century. They are now declining, as shown in **Figure 3.17**; however, because 30 percent of the region's population is under age 15, the population will continue to grow as this large group reaches the age of reproduction, even if couples choose to have only one or two children. Population projections for the year 2050 were recently revised downward from 778 million to 724 million—an increase overall of 18 percent. Though significantly lower than previous forecasts, for a region struggling to increase living standards, supplying 18 percent more people with food, water, homes, schools, and hospitals will be a challenge.

During the twentieth century, cultural and economic factors combined with improvements in health care to create a population explosion. Most of the population growth was due to decreasing mortality, not to higher birth rates. Improved living conditions (food, shelter, sanitation habits) and improved education and medical care all contributed to longer life expectancy. Birth rates were sustained for a while because the Roman Catholic Church discourages systematic family planning, and cultural mores encouraged men and women to reproduce prolifically. In agricultural areas, children were seen as sources of wealth because they could do useful farm and household work at a young age and eventually would care for their aging elders. Worry over infant death rates persisted, so some parents had four or more children to be sure of raising at least a few to adulthood. By the 1930s, access to medical care was improving and incomes and education rates rose as more people moved to the cities. By 1975, death rates were one-third what they had been in 1900. Amazingly, by 2008 the overall death rate of 6 per 1000 in Middle and South America was lower than that of North America, where the rate is 8 per 1000.

By the 1980s, migration out of the region and the limiting of family sizes was curbing population growth. Now the region is undergoing a demographic transition (see Figure 1.12 on page 20). Between 1975 and 2009, the annual rate of natural increase for the entire region fell from about 1.9 percent to 1.4 percent—still a higher rate of growth than the world average of 1.2 percent. The low death rate continues to contribute to population growth in most of the region.

HIV-AIDS The global epidemic of HIV-AIDS is now taking a significant toll on populations throughout this region, accounting for 9.5 percent of all deaths in 2005. In 2009, about 2.9 million people were HIV-positive. Of these, at least 60,000 were children. Some Caribbean islands have the highest HIV-AIDS infection rates outside sub-Saharan Africa, where the rate is 5 percent of the population. On the impoverished island of Haiti, 2.2 percent of the population (ages 15–49) has HIV-AIDS; in the Bahamas, 3 percent; and in Guyana, 2.5 percent. In most of the Caribbean, aggressive education programs about HIV, along with high literacy rates and the relatively high status of women are limiting the number of infections.

The Venezuelan journalist Silvana Paternostro, in her book *In the Land of God and Man: Confronting Our Sexual Culture* (1998), argues that cultural practices contribute to HIV infection in this region. Cultural mores discourage the open discussion of sex. Men are rarely expected to be monogamous, and many visit prostitutes, making their wives and any children they might bear vulnerable to infection. Condom use is rare, and a wife would be loath to ask her husband to use one. Drug use is also a significant factor in the spread of HIV.

Human Well-Being The wide disparities in wealth that are a feature of life in Middle and South America have been discussed above. Too often, development efforts linked to urbanization and globalization have only increased the gap between rich and poor. Each of the three maps in **Figure 3.18** on page 150 shows one of the various indicators of well-being used by the United Nations to show how countries are doing in providing a decent life for their citizens.

Map A in Figure 3.18 is of average GDP per capita (PPP); it is immediately apparent that no country in this region falls in the highest category (purple), and just a few islands in the Caribbean

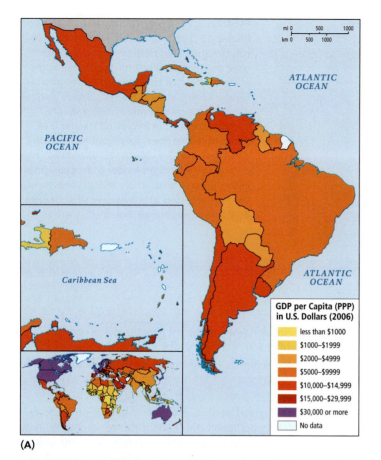

(A)

(B)

FIGURE 3.18 Maps of human well-being. (A) GDP per capita, adjusted for purchasing power parity (PPP). Notice that few countries in this region are either extremely rich or extremely poor. **(B)** HDI rank. Notice that some countries, such as Venezuela, are relatively lower in HDI than in GDP (PPP). This may reflect inadequate access to health care or to education for the majority of people. Meanwhile other countries, such as some islands in the Caribbean, are higher in HDI than GDP (PPP), suggesting that other nonmonetary factors influencing human development are higher here. **(C)** Female earned income as a percent of male earned income (F/MEI). Notice that women earn considerably less than men in Mexico and Central America than in the poorer Andean countries. [Maps adapted from data: "Human development indices," (Maps A and B): Table 2, pages 28–32; (Map C): Table 5, pages 41–44 at http://hdr.undp.org/en/media/HDI_2008_EN_Tables.pdf.]

(C)

(Trinidad and Tobago, Barbados, and Antigua and Barbuda, inset) fall into the second-highest category (deep red). Most of the Caribbean is in the global middle class, with one major exception, Haiti, which is the only country in the entire Middle and South American region to fall into either of the lowest GDP categories (yellow). Across the mainland, only Mexico, Panama, Venezuela, Uruguay, Argentina, and Chile have an average per capita GDP between U.S.$10,000 and U.S.$14,999. All other countries in the region, including the large country of Brazil, have average per capita GDPs that are less than $9999, many of them significantly less. Also, most of these countries have rather wide disparities in wealth (see Table 3.2 on page 134) that are masked on a map like this. In

Brazil for example, the most affluent 20 percent have incomes that are 21 times greater than the bottom 20 percent, and in actuality some Brazilians survive on less than $1000 a year while others have incomes well over $100,000 per year.

Map B, which charts the UNHDI rank for all countries in the region, shows that most countries rank in the high and medium ranges (dark red, medium red) in providing the basics for their citizens (education, health care, and income). The global inset map confirms this mid-range status for Middle and South America. Most of Africa, Central Asia, and South Asia rank lower, while North America, Europe, Australia, Japan, and South Korea rank higher. Much of East Asia, Russia, and some of the post-independent states of the former Soviet Union rank in the same range as Middle and South America. Of course the UNHDI is a country-wide index and does not give any indication of how well-being is distributed within a given country. On this map and the GDP map, the countries of the Central American isthmus stand out as poorer than Mexico, the Caribbean, and South America.

Map C, which shows the regional pattern for female earned income as a percentage of male income, confirms that this region does not yet begin to approach gender equality in pay. In fact, on this indication of well-being, the region ranks even lower than some parts of Africa. The fact that Paraguay, Bolivia, and Colombia rank higher on this index than Brazil and Argentina and much higher than Chile and Mexico and others with the yellow color is not an indication that pay for women is higher there (compare with Map A), but only that pay for men and women is somewhat closer to being equitable.

Cultural Diversity

The region of Middle and South America is culturally complex because many distinct indigenous groups were already present when the Europeans arrived, and many cultures were introduced during and after the colonial period. Due both to warfare and the introduction of new diseases, in the Caribbean, the Guianas (Guyana, Suriname, and French Guiana), and Brazil, the arrival of a relatively small number of European newcomers resulted in severe population reduction among the indigenous cultures. Various people from outside the region then populated these nearly empty areas.

From 1500 to the early 1800s, some 10 million Africans from many different parts of Africa were brought to plantations on the islands and in the coastal zones of Middle and South America. After the emancipation of African slaves in the British-controlled Caribbean islands in the 1830s, more than half a million Asians were brought there from India, Pakistan, and China as indentured agricultural workers. Their cultural impact remains most visible in Trinidad and Tobago, Jamaica, and the Guianas. In some parts of Mexico, Central America, the Amazon Basin,

and the Andean Highlands, indigenous people have remained numerous. To the unpracticed eye, they may appear little affected by colonization, but this is not the case.

Today, in the Caribbean and along the east coast of Central America as well as the Atlantic coast of Brazil, the population majority is a mixture of African and European ancestry. *Mestizos* are now the majority in Mexico, central America, and much of South America. In some areas, such as Argentina, Chile, and southern Brazil, people of central European descent are also numerous. The Japanese, though a tiny minority everywhere in the region, increasingly influence agriculture and industry, especially in Brazil (Figure 3.19), the Caribbean, and Peru. Alberto Fujimori, a Peruvian of Japanese descent, was the controversial president of Peru for 10 years.

In some ways, diversity is increasing as the media and trade introduce new influences. At the same time, the processes of **acculturation** and **assimilation** are also accelerating, meaning that diversity is being erased to some extent as people adopt twenty-first century ways. This is especially true in the biggest cities, where people of widely different backgrounds live in close proximity.

> **acculturation** adaptation of a minority culture to the host culture enough to function effectively and be self-supporting; cultural borrowing
>
> **assimilation** the loss of old ways of life and the adoption of the lifestyle of another culture

Race and the Social Significance of Skin Color

People from Middle and South America, especially those from Brazil, often proudly claim that race and color are of less

FIGURE 3.19 Cultural diversity in Brazil. Japanese-Brazilian country singer Maurício Miya sings in Portuguese, but his music sounds like it came from Nashville. Brazil is now home to 1.4 million people of Japanese ancestry, the largest such population outside of Japan. For some of Miya's music, see http://www.mauriciomiya.com/.

consequence in this region than in North America. They are right in certain ways. Skin color is less associated with status than in the colonial past. A person of any skin color, by acquiring an education, a good job, a substantial income, the right accent, and a high-status mate, may become recognized as upper class.

Nevertheless, the ability to erase the significance of skin color through one's actions is not quite the same as race having no significance at all. Overall, those who are poor, less educated, and of lower social standing tend to have darker skin than those who are educated and wealthy. And while there are poor light-skinned people throughout the region (often the descendants of migrants from Ireland, central Europe, and the eastern Mediterranean over the last century), most light-skinned people are middle and upper class. Indeed, race and skin color have not disappeared as social factors in the region. In some countries—Cuba, for example, where overt racist comments are socially unacceptable—it is common for a speaker to use a gesture (tapping his forearm with two fingers) to indicate that the person referred to in the conversation is African-Caribbean.

The Family and Gender Roles

The basic social institution in the region is the **extended family**—which may include cousins, aunts, uncles, grandparents, and more distant relatives. It is generally accepted that the individual should sacrifice many of his or her personal interests to those of the extended family and community and that individual well-being is best secured by doing so.

The arrangement of domestic spaces and patterns of socializing illustrate these strong family ties. Families of adult siblings, their mates and children, and their elderly parents frequently live together in domestic compounds of several houses surrounded by walls. Social groups in public spaces are most likely to be family members of several generations rather than unrelated groups of single young adults or married couples, as would be the case in Europe or the United States. A woman's best friends are likely to be her female relatives. A man's social or business circles will include male family members or long-standing family friends.

Gender roles in the region have been strongly influenced by the Roman Catholic Church. The Virgin Mary is held up as the model for women to follow through a set of values known as *marianismo. Marianismo* emphasizes chastity, motherhood, and service to the family. The ideal woman is the day-to-day manager of the house and of the family's well-being. She trains her sons to enter the wider world and her daughters to serve within the home. Over the course of her life, a woman's power increases as her skills and sacrifices for the good of all are recognized and enshrined in family lore.

Her husband, the official head of the family, is expected to work and to give most of his income to his family. Men have much more autonomy and freedom to shape their lives than women because they are expected to move about the larger community and establish relationships, both economic and personal. A man's social network is deemed just as essential to the family's prosperity and status in the community as his work.

In addition, there is an overt double sexual standard for males and females. While all expect strict fidelity from a wife to a husband in mind and body, a man is much freer to associate with the opposite sex. Males measure themselves by the model of *machismo*, which considers manliness to consist of honor, respectability, fatherhood, household leadership, attractiveness to women, and the ability to be a charming storyteller. Traditionally, the ability to acquire money was secondary to other symbols of maleness. Increasingly, however, a new market-oriented culture prizes visible affluence as a desirable male attribute.

Many factors are transforming these family and gender roles. With infant mortality declining steeply, couples are now having only two or three children instead of five or more. Moreover, because most people still marry young, parents are generally free of child-raising responsibilities by the time they are 40. Left with 30 or more years of active life to fill in other ways (life expectancies in the region average in the mid-70s), middle-aged mothers and grandmothers are increasingly working in urban factory or office jobs that put to use the organizational and problem-solving skills they perfected while managing a family. Employment outside the home is a way to gain a measure of independence as a woman and also contribute to the needs of the extended family. Some women are even moving into high-level jobs, such as management positions, professorships, and positions that traditionally went to men—for example, the presidencies of Chile and Argentina.

Migration and Urbanization

Since the early 1970s, Middle and South America have led the world in migration from rural to urban communities. Now more than 70 percent of the people in the region live in settlements of at least 2000. Increasingly, one city, known as a **primate city**, is vastly larger than all the others, accounting for a large percentage of the country's total population (Photo Essay 3.6). Examples in the region are Mexico City, Mexico (with 23 million people, a bit more than 21 percent of Mexico's total population); Managua, Nicaragua (33 percent of that country's population); Lima, Peru (29 percent); Santiago, Chile (30 percent); and Buenos Aires, Argentina (36 percent).

The concentration of people into just one or two large cities in a country leads to uneven spatial development and to government policies and social values that favor urban areas. Wealth and power are concentrated in one place, while distant rural areas, and even other towns and cities, have difficulty competing for talent, investment, industries, and government services. Many provincial cities languish as their most educated youth leave for the primate city.

Brain Drain It always takes some resourcefulness to move from one place to another. Those people who already have some years of education and strong ambition are the ones who migrate to cities. The loss of these resourceful young adults in which the

---

extended family a family that consists of related individuals beyond the nuclear family of parents and children

marianismo a set of values based on the life of the Virgin Mary, the mother of Jesus, that defines the proper social roles for women in Middle and South America

machismo a set of values that defines manliness in Middle and South America

primate city a city, plus its suburbs, that is vastly larger than all others in a country and in which economic and political activity is centered

Urbanization is at the heart of many transformations in this region. The last several decades has seen cities grow extremely fast, and 77 percent of the region now lives in cities. New opportunities have opened up for migrants from rural areas, but new stresses have emerged as well. The low-skilled jobs they can get often don't pay well, and many are forced to live in unplanned neighborhoods on the outskirts of huge cities.

A Santiago is one of the region's primate cities, home to 34 percent of Chile's population. It dominates Chile's economy, generating 40 percent of the country's GDP.

B Below, children in Lima, Peru search through a garbage can in search of food to eat. Some families disintegrate after the move to the city as parents are forced to work long hours at low-paying jobs. Even in those that remain intact, children are sometimes forced to beg for money and food instead of going to school.

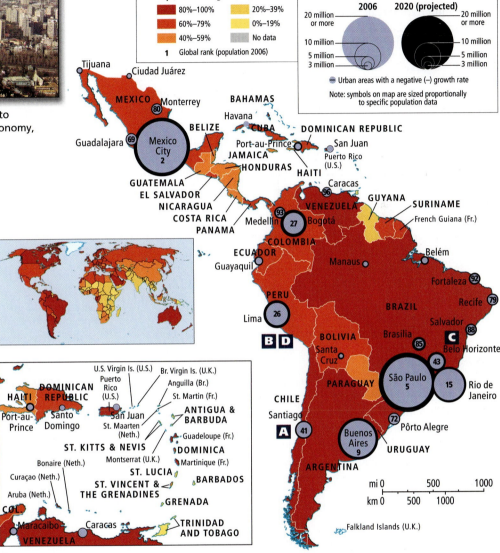

Population living in urban areas
- 80%–100%
- 60%–79%
- 40%–59%
- 20%–39%
- 0%–19%
- No data

1 Global rank (population 2006)

Population of urban areas
2006 | 2020 (projected)
- 20 million or more
- 10 million
- 5 million
- 3 million

Urban areas with a negative (−) growth rate

Note: symbols on map are sized proportionally to specific population data

C A woman works as a street vendor in Salvador, Brazil. Many new urban migrants work in similar "informal sector" jobs that generally pay less and are much less secure than "formal sector" jobs.

D An informal "squatter" settlement forming outside of Lima, Peru. Structures made out of woven grass mats are the first to go up, followed by wooden shacks and eventually cement or brick houses. Water and electricity may eventually be extended to such settlements, but only after years or decades of occupation.

community has invested years of nurturing and education is referred to as **brain drain.** Brain drain happens at several scales in Middle and South America: through rural-to-urban migration from villages to regional towns, and from towns and small cities to primate cities. There is also international brain drain when migrants move to North America and Europe; for example, one in four U.S. doctors are foreign born. Families often encourage their children to migrate so that they can benefit from the remittances, goods, and services that migrants send back to their home communities.

> **brain drain** the migration of educated and ambitious young adults to cities or foreign countries, depriving the communities from which the young people come of talented youth in whom they have invested years of nurturing and education
>
> **favelas** Brazilian urban slums and shantytowns built by the poor; called colonias, barrios, or barriados in other countries

Favelas: Unplanned Urban Neighborhoods A lack of planning to accommodate the massive rush to the cities has created urban landscapes that are very different from the common U.S. pattern. In the United States, a poor older inner city is usually surrounded by more affluent suburbs, with clear and planned spatial separation of wealthy and working-class residencies. Residential, industrial, and commercial areas are also separated. In Middle and South America, by contrast, both affluent and working-class areas have become unwilling neighbors to unplanned slums filled with poor migrants. Too destitute even to rent housing, these "squatters" occupy parks and small patches of vacant urban land wherever they can be found, as depicted in the diagram in Figure 3.20.

The best known of these unplanned communities are Brazil's **favelas** (Figure 3.21). In other countries they are known as *slums, shantytowns, colonias, barrios,* or *barriadas.* The settlements often spring up overnight, without city-supplied water or electricity and with housing made out of whatever is available (see Figure 3.21B, C, D). Once they are established, efforts to eject squatters usually fail. The impoverished are such a huge portion of the urban population that even those in positions of power will not challenge them directly. Hence, nearby wealthy neighborhoods simply barricade themselves with walls and security guards.

The squatters frequently are enterprising people who work hard to improve their communities. They often organize to press governments for social services. Some cities, such as Fortaleza in northeastern Brazil, even contribute building materials so that favela residents can build more permanent structures with basic indoor plumbing. Over time, as shacks and lean-tos are transformed through self-help into crude but livable suburbs, the economy of favelas can become quite vibrant with much activity in the informal sector. Housing may be intermingled with shops, factories, warehouses, and other commercial enterprises. Favelas and their counterparts can become centers of pride and support for their residents, where community work, folk belief systems, crafts, and music (for example, Favela Funk) flourish. Many of the best steel bands of Port of Spain, Trinidad, have their homes in the city's shantytowns.

Personal Vignette Favelas are everywhere in Fortaleza, Brazil. The city grew from 30,000 to 300,000 residents during the 1980s. By 2006, there were more than 3 million residents, most of whom had fled drought and rural

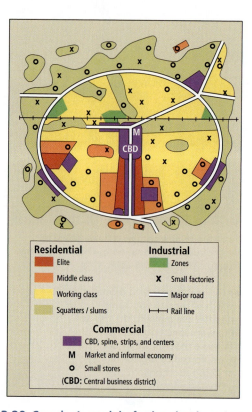

FIGURE 3.20 Crowley's model of urban land use in mainland Middle and South America. William Crowley, an urban geographer who specializes in Middle and South America, developed this model to depict how residential, industrial, and commercial uses are mixed together, with people of widely varying incomes living in close proximity to one another and to industries. Squatters and slum dwellers ring the city in an irregular pattern. [From *Yearbook of the Association of Pacific Coast Geographers* 57 (1995): 28; printed with permission.]

poverty in the interior. City parks of just a square block or two in middle-class residential areas were suddenly invaded by squatters. Within a year, 10,000 or more people were occupying crude, stacked, concrete dwellings in a single park, completely changing the ambience of the upscale neighborhood. In the early days of the migration, lack of water and sanitation often forced the migrants to relieve themselves on the street.

One day, while strolling on the Fortaleza waterfront, Lydia Pulsipher chanced to meet a resident of a beachfront favela who invited her to join him on his porch. There, he explained how one becomes an urban favela dweller. He and his wife had come to the city 5 years before, after being forced to leave the drought-plagued interior when a newly built irrigation reservoir flooded the rented land their families had cultivated for generations. With no way to make a living, they set out on foot for the city. In Fortaleza, they constructed the building they used for home and worked from objects they collected along the beach. Eventually they were able to purchase roofing tiles, which gave it an air of permanency. At the time of the visit, he maintained a small refreshment stand, and his wife a beauty parlor that catered to women from the beach settlements. *[Source: Lydia Pulsipher's field notes on Brazil, updated with the help of John Mueller, Fortaleza, Brazil, 2006.]* ■

FIGURE 3.21 A favela in Rio de Janeiro. (A) Favela Dona Marta, indicated by the white circle, clings to a once-vacant hillside that was considered too steep for apartment buildings or even for road access. Now home to 7000 people, it is still served only by footpaths and narrow alleyways. **(B)** Water is only available sporadically, so many residents store water in tanks on their roof. **(C)** Electrical connections consist of dangerous tangles of wires, such as those hanging over this man's head, illegally connected to a main power line. **(D)** Favelas are often full of children who spend much of the day under the supervision of a friend or relative while their parents work in distant parts of the city.

Urban Transport Issues Transportation in large, rapidly expanding urban areas with millions of poor migrants can be a special challenge. Favelas are often on the urban fringes, far from available low-skill jobs and are ill-served by roads and public transport (see Crowley's model, Figure 3.20). Workers must make lengthy, time-consuming, and expensive commutes to jobs that pay very little. The entire urban population gets caught up in endless traffic jams that cripple the economy and pollute the environment. Rapid growth seems to have outstripped the capacities of even the most enlightened urban planning.

Yet the southern Brazilian city of Curitiba (Figure 3.1 on pages 114–115) has carefully oriented its expansion around a master plan (dating from 1968) that includes an integrated transport system funded by a public-private collaboration. Eleven hundred minibuses making 12,000 trips a day bring 1.3 million passengers from their remote neighborhoods to terminals, where they meet express buses to all parts of the city (see **Figure 3.22** on the next page). The large pedestrian-only inner-city area is a boon

for merchants because minibus commuters spend time shopping rather than looking for parking. Being able to get to work quickly and cheaply has helped the poor find and keep jobs. The reduced use of cars means that emissions and congestion are lowered and urban living is more pleasant.

Gender and Urbanization Interestingly, rural women are just as likely as rural men to migrate to the city. This is especially true when employment is available in foreign-owned factories that produce goods for export. Companies prefer women for such jobs because they are a low-cost and usually passive labor force. Ironically, rural development projects may actually force female migration to urban areas. Due to the mechanization of formerly labor-intensive agricultural systems, these projects often end up decreasing the number of available jobs. Women are rarely considered for training as farm-equipment operators and mechanics. In urban areas, unskilled women migrants usually find work as street vendors (see Figure 3.13 on page 139 and

FIGURE 3.22 Public transportation in Curitiba, Brazil. Express buses and tube stations serve patrons with fast, safe, and efficient transport throughout the city.

Photo Essay 3.6C on page 153) or domestic servants. Low wages often force female servants to live in inadequate quarters in the households where they work. There, they are often subject to sexual predation by their employers, yet the servants are the ones blamed for the failure to abide by rigid standards of behavior.

Male urban migrants tend to depend on short-term, low-skilled day work in construction, maintenance, small-scale manufacturing, and petty commerce. Many work in the informal economy as street vendors, errand runners, car washers, and trash recyclers, and some turn to crime. The loss of family ties and village life is sorely felt by both men and women, and the chances for recreating the family life they once knew are extremely low in the urban context.

Families constructed in the cities tend to disintegrate because of extreme poverty and malnutrition, long commutes for both working parents, and poor quality of day care for children (see Figure 3.21D). At too young an age, children are left alone or sent into the streets to scavenge for food or to earn money for the family (see Photo Essay 3.6B on page 153 and Figure 3.21D on page 155).

Religion in Contemporary Life

populist movements
a political philosophy that supports the rights of the poor in their resistance to the power and privilege of elites

Judaism, Islam, Hinduism, and indigenous beliefs are found across the region, but the majority are at least nominal Christians, most of them Roman Catholic. While the Roman Catholic Church remains highly influential and relevant to the lives of believers (Figure 3.23), it has had to contend both with popular efforts to reform it and with increasing competition from other religious movements. From the beginning of the colonial era, the church was the major partner of the Spanish and Portuguese colonial governments. It received extensive lands and resources from those governments, and it sent thousands of missionary priests to convert indigenous people.

For centuries, the Roman Catholic Church encouraged working people to accept their low status, be obedient to authority, and postpone their rewards until heaven. Furthermore, the church ignored those teachings of Christ that admonish the privileged to share their wealth and attend to the needs of the poor. Nonetheless, poor people throughout Spanish and Portuguese America embraced the faith and still make up the majority of the church's members. Many are indigenous people who put their own spin on Catholicism, creating multiple folk versions of the Mass with music, participation of women in worship services, and interpretations of Scripture that vary greatly from European versions. A range of African-based belief systems (Candomblé, Umbanda, Santería, Obeah, and voodoo) combined with Catholic beliefs is found in Brazil, northern South America, and in Middle America and the Caribbean—wherever the descendants of Africans have settled. These African-based religions have attracted adherents of European or indigenous backgrounds as well, especially in urban areas.

The power of the Roman Catholic Church began to erode in the nineteenth century in places such as Mexico. **Populist movements** seized and redistributed church lands. They also canceled the high fees the clergy had been charging for simple rites of passage such as baptisms, weddings, and funerals. Over the years, the Catholic Church became less obviously connected to the elite and more attentive to the needs of poor and non-European people. By the mid-twentieth century, the church was abandoning many of its racist policies and ordaining indigenous and African-American clergy. Women were also given a greater role in religious ceremonies.

FIGURE 3.23 Catholics in Mexico City praying for the end of swine flu. The devotees are wearing masks to reduce the risk of contagion while praying in Mexico City's Metropolitan Cathedral.

evangelical Protestantism a Christian movement that focuses on personal salvation and empowerment of the individual through miraculous healing and transformation; some practitioners preach to the poor the "gospel of success"—that a life dedicated to Christ will result in prosperity for the believer

liberation theology a movement within the Roman Catholic Church that uses the teachings of Jesus to encourage the poor to organize to change their own lives and the rich to promote social and economic equality

In the 1970s, a radical Catholic movement known as **liberation theology** was begun by a small group of priests and activists. They sought to reform the church into an institution that could combat the extreme inequalities in wealth and power common in the region. The movement portrayed Jesus Christ as a social revolutionary who symbolically spoke out for the redistribution of wealth when he divided the loaves and fish among the multitude. The perpetuation of gross economic inequality and political repression was viewed as sinful, and social reform as liberation from evil.

At its height in the 1970s and early 1980s, liberation theology was the most articulate movement for region-wide social change. It had more than 3 million adherents in Brazil alone. But the Vatican objected to this popularized version of Catholicism and today its influence is diminished. In countries such as Guatemala and El Salvador, rightist regimes targeted liberation theology participants, vilifying them as communist collaborators; several priests and nuns were murdered. One of the best known was Archbishop Óscar Romero of El Salvador, murdered while celebrating mass. Liberation theology has also had to compete with newly emerging evangelical Protestant movements.

▶ **79. POPE OPENS TOUR OF LATIN AMERICA IN BRAZIL**

Evangelical Protestantism has diffused from North America into Middle and South America and is now the region's fastest-growing religious movement. About 10 percent of the population, or at least 50 million people, are adherents; and the movement is growing rapidly in Brazil and Chile, in the Caribbean, Mexico, and Middle America, especially among poor and middle classes in both rural and urban settings. It does not, however, share liberation theology's emphasis on combating the region's extreme inequalities in wealth and power. Some evangelical Protestants teach a "gospel of success," stressing that those who are true believers and give themselves to a new life of hard work and clean living will experience prosperity of the body (wealth) as well as of the soul.

The movement is *charismatic*, meaning that it focuses on personal salvation and empowerment of the individual through miraculous healing and psychological transformation. Evangelical Protestantism is not hierarchical in the same way as the Roman Catholic Church, and there is usually no central authority; rather, there are a host of small, independent congregations led by entrepreneurial individuals who may be either male or female.

THINGS TO REMEMBER

1. The pace of population growth in this region is declining in part because of improved health practices, emigration, and family-planning choices made by women with new education and work options.

2. Human well-being has improved steadily over several decades and provides a hopeful picture in the region as a whole, but disparities in wealth and health remain a problem for large segments of the people.

3. Gender roles in the region have been strongly influenced by the Catholic Church through the ideals of *marianismo* and *machismo*.

4. Middle-aged women are increasingly working in factory or office jobs, where they put to use the organizational and problem-solving skills they perfected while supervising their families.

5. Evangelical Protestantism came to the region from North America and is now the fastest-growing religious movement in the region.

Reflections on Middle and South America

European colonialism in Middle and South America launched the modern global economic system. It was in this region that large-scale extractive industries were inaugurated. Raw materials were shipped at low prices to distant locales, where they were turned into high-priced products, the profits of which went to Europe.

In most of the region, this pattern has persisted for 500 years. After the massive outflows of raw materials during the colonial era, the region tried a number of economic development strategies, with mixed results. In a few cases, import substitution industries (ISIs) worked to lower the rate of imports, but the industries failed to lift the region out of poverty. The huge debts that many governments accumulated while inefficiently pursuing these projects later led to the imposition of structural adjustment programs. SAPs also failed to produce economic growth as they reduced investment in human capital and relied heavily on raw materials exports just when prices for these exports were falling on global markets. Now deforestation, much of it driven by the need for export earnings, is contributing to both global climate change and the loss of the region's great biodiversity.

Massive rural-to-urban and international migrations resulted from the shift from small-scale agriculture for local consumption to large-scale, mechanized agriculture for export. Most cities were unprepared for the influx of migrants, and water crises developed as existing supply and delivery systems were overwhelmed. While the informal economy has been a lifesaver for many poor urbanites, its ability to lift people out of poverty is limited. Traditional gender roles have also changed with urbanization, giving new opportunities to some but also straining family ties.

However, there are many positive signs in the region. Environmental regulations and alternatives to deforestation hold the potential to reduce pressure on the region's forests. Increasingly, regional trade organizations are emerging that may keep more wealth within the region. Finally, although models of democracy may not fit North American and European norms, nearly all countries now hold regular peaceful elections, and when leaders make moves reminiscent of old-style dictators, they must eventually answer to an ever better informed public.

Critical Thinking Questions

1. If the European colonists had come to Middle and South America in a different frame of mind—say, they were simply looking for a new place to settle and live quietly—how do you think the human and physical geography of the region would be different today?

2. Explain two main ways in which tectonic processes account for the formation of mountains in Middle and South America.

3. Reflecting on the whole chapter, pick some locations where you were impressed with the ways in which people are presently dealing with either environmental issues or issues of income/wealth disparity or human well-being. Explain your selections.

4. Discuss the ways in which you see the historical circumstances of colonization affecting modern approaches to economic problems in Mexico, Bolivia, Brazil, Venezuela, or Cuba.

5. Describe the main patterns of migration in this region and discuss the effects of migration on both the sending and receiving societies.

6. Discuss how gender roles in this region have been affected by migration patterns.

7. Name three factors you see as important in increasing democratic participation in this region. In which countries would you say these factors are making the biggest difference?

8. Explain how the Amazon Basin and its resources constitute an example of contested space.

9. Argue for or against the proposition that free trade blocs, such as NAFTA and Mercosur, assist upward mobility for the lowest-paid workers.

10. How would you respond to someone who suggested that Middle and South America were helped toward development and modernization by the experience of European colonization?

Chapter Key Terms

acculturation, 151

assimilation, 151

Aztecs, 129

biodiversity, 121

brain drain, 154

commodification, 128

contested space, 141

coup d'état, 143

Creoles, 133

dictator, 143

early extractive phase, 133

ecotourism, 126

El Niño, 124

evangelical Protestantism, 157

export processing zones (EPZs), 138

extended family, 152

external debts, 136

fair trade movement, 140

favelas, 154

foreign direct investment (FDI), 139

haciendas, 134

import substitution industrialization (ISI), 136

Incas, 129

income disparity, 133

indigenous, 117

land reform, 136

liberation theology, 157

machismo, 152

maquiladoras, 138

marianismo, 152

marketization, 136

mercantilism, 133

Mercosur, 140

mestizos, 133

Middle America, 120

(to) nationalize, 139

North American Free Trade Agreement (NAFTA), 140

plantation, 135

populist movements, 156

primate city, 152

privatization, 136

recession, 136

shifting cultivation, 129

silt, 121

South America, 120

structural adjustment policies (SAPs), 136

subduction zone, 121

temperature-altitude zones, 121

trade winds, 124

World Trade Organization (WTO), 140

THEMATIC CONCEPTS

Population • Gender • Development
Food • Urbanization • Globalization
Democratization • Climate Change • Water

Europe

A **Alps, Kühtai, Austria**

B **Uplands, Saarschleife, Germany**

C **North European Plain, Schleswig-Holstein, Germany**

D **Rhine River, Heldsburg, Switzerland**

E **Danube River, Budapest, Hungary**

FIGURE 4.1 Regional map of Europe.

Global Patterns, Local Lives

Grigore Chivu (a pseudonym) wanders through the empty pig pens on his farm near Lugoj in western Romania. For generations his family has made a meager but rewarding living by raising and processing hogs and selling the meat. A few years ago, just before Romania joined the European Union (EU), he had more than 250 pigs. At Christmas, he and thousands of pig farmers across the country would slaughter a certain number of their pigs and preserve the meat, using age-old methods. Pig slaughtering was a time of high spirits, community cooperation, and celebration, as the farmers contemplated the coming feast and their profits. Flavorful sausages, which they smoked and hung high in the rafters of their kitchens for further drying, were slowly parceled out to select customers, providing a steady income well into summer.

When Romania entered the European Union in 2007, all farmers were required to conform to EU standards for processing meat. The old methods were no longer allowed, and the new standards were prohibitively expensive.

At first the farmers were uncertain just how to respond. Before they could organize a butchering cooperative that conformed to EU standards, and for which development funds were available, an American company stepped into the breach. Smithfield, a Fortune 500 meat company based in Virginia, was expanding operations into Middle America and Central Europe, where it planned to produce pork and pork products and market them globally. Eager to enter the EU market, Smithfield enlisted the help of Romanian politicians and got permission to establish a conglomerate that included feed production, pig breeding, modernized sanitary barns for fattening thousands of hogs, and slaughterhouses.

As the old picturesque Romanian agricultural landscape is transformed into one of huge factory farms, the number of pig farmers has been reduced by more than 90 percent (**Figure 4.2**). Unable to compete with the lower prices Smithfield can charge, Grigore Chivu, like thousands of his fellow pig farmers, is thinking of migrating to western Europe where, because of his traditional farming background, he will be eligible for only menial labor. [*Source: Conversations with geographer Margareta Lelea, a Romanian specialist, Visiting Assistant Professor, Bucknell University; Doreen Carvajal and Stephen Castle, "A U.S. Hog Giant Transforms Eastern Europe," New York Times, May 6, 2009, at http://www.nytimes.com/2009/05/06/business/global/06smithfield. html?ref=europe; http://www.nytimes.com/2009/05/06/business/global/06smithfield. html?ref=europe.]* ■

European Union (EU) a supranational organization that unites most of the countries of West, South, North, and Central Europe

cultural homogenization the tendency toward uniformity of ideas, values, technologies, and institutions among associated culture groups

Grigore Chivu is faced with disruptive change as his country adjusts to new circumstances in the **European Union (EU)**, but he is also part of a global revolution in food production and marketing that is changing food patterns across the world. For example, Smithfield pork trimmings, produced and processed at low cost in Romania, are now marketed in West Africa, where the low prices are putting African pig farmers out of business.

The European Union is a supranational organization that unites most of the countries of West, South, North, and Central

FIGURE 4.2 Pig farms in Romania and the United States.
(A) Shepherds on a farm in Romania that raises small numbers of cows, sheep, and pigs (shown in **B**). **(C)** An industrial hog farm in Georgia, U.S., capable of raising thousands of pigs at a time.

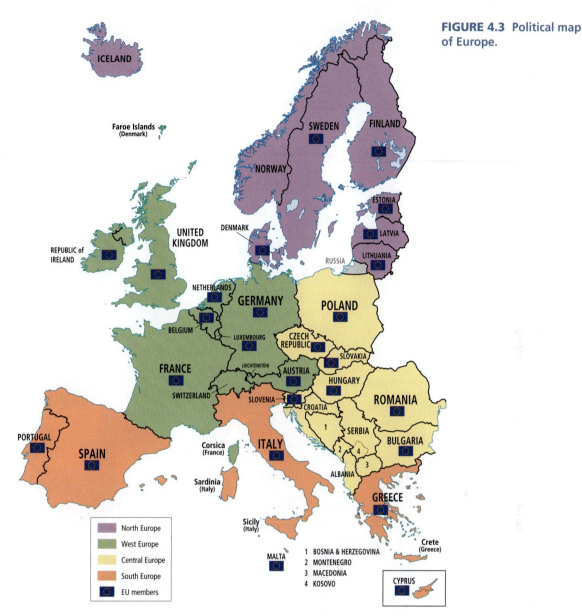

FIGURE 4.3 Political map of Europe.

ICELAND

Faroe Islands
(Denmark)

NORWAY

SWEDEN

FINLAND

ESTONIA

LATVIA

RUSSIA

LITHUANIA

REPUBLIC of
IRELAND

UNITED
KINGDOM

DENMARK

NETHERLANDS

GERMANY

POLAND

BELGIUM

LUXEMBOURG

CZECH
REPUBLIC

SLOVAKIA

FRANCE

LIECHTENSTEIN

AUSTRIA

HUNGARY

SWITZERLAND

SLOVENIA

CROATIA

ROMANIA

PORTUGAL

SPAIN

Corsica
(France)

ITALY

1

SERBIA

4

BULGARIA

2

3

ALBANIA

Sardinia
(Italy)

GREECE

Sicily
(Italy)

Crete
(Greece)

MALTA

	North Europe
	West Europe
	Central Europe
	South Europe
	EU members

1 BOSNIA & HERZEGOVINA
2 MONTENEGRO
3 MACEDONIA
4 KOSOVO

CYPRUS

Europe (**Figure 4.3**). In principle, throughout the European Union, people, goods, and money can move freely. Many people, like Grigore Chivu, decide to migrate only reluctantly, and often do so because the longstanding inequities that persist across Europe make life especially difficult for those from poorer regions. Adapting to the changes that have come in Europe is wrenching for many, especially for poor, rural people.

There are now 27 countries in the European Union, the most recent joining in 2007, with more hoping to join in the next several years. The new members (with the exceptions of Malta and Cyprus) are formerly Communist countries in eastern Europe, with lower standards of living and higher unemployment rates than western Europe. As their own economies have faltered, hundreds of thousands of workers from eastern Europe have taken advantage of the EU principle that (for the most part) citizens of all member states may move to any other EU state.

Some EU residents fear that the migration of workers in an expanding European Union is bringing very different people into close contact with each other, resulting in political tensions and increased costs for social and educational services. Others lament the **cultural homogenization** that is occurring. Across the European Union, as in Romania, distinctive local ways of life are disappearing as a result of new well-intentioned EU regulations. Meanwhile, economic planners and employers across Europe argue that without modernization and workers willing to migrate, economic growth and competitiveness in the global market is impossible.

Many of the points touched on in this introduction to Europe are elaborated on in the Thematic Concepts on pages 164–165. They are further expanded upon in the chapter.

THINGS TO REMEMBER

Throughout the European Union, smaller family-run farms are giving way to larger farms run by corporations. This move is strongest in the recently admitted EU states of Central Europe, resulting in many farmers and their families migrating to other EU countries in search of work.

Globalization, Development, and Colonialism: European powers conquered vast overseas empires, creating trade relationships that laid the foundation for the modern global economy. The profits from these empires transformed economic development in Europe by helping to fund the industrial revolution.

(A) A poster depicting Britain's conquest of Sudan in 1897.

(B) Sugar plantation workers in the 1890s in Puerto Rico, then a colony of Spain.

(C) A steel mill in the Netherlands in 1890, financed in part by profits from a large colonial empire.

Urbanization and Democratization: As Europe urbanized during the industrial revolution, widespread poverty, squalid living conditions, and unsafe work environments created pressure for change in Europe's political order. Decades of struggle eventually led to democratic reforms, including the establishment of strong social welfare institutions that transformed many European cities.

(D) Destitute Londoners applying for entrance to a homeless shelter are shown in this 1872 print, "Refuge: Applying for Admission," by Gustave Doré.

(E) This painting by Eugène Delacroix, "Liberty Leading the People," commemorates a pro-democracy uprising in Paris in 1830.

(F) Public housing in London. Decent housing for low-income people is a central achievement of Europe's democratic "welfare states."

Climate Change: Europe leads the world in responding to climate change. Europe has made more commitments to reduce greenhouse gas emissions, and taken more concrete steps to do so, than any other region. Although emissions are still high, Europe's dense cities and well-developed mass transit networks reduce per capita emissions to about half of North America's.

(G) New housing in London powered partially by micro wind turbines.

(H) Bicycles parked by commuters outside a train station in Lund, Sweden.

(I) These buildings in Amsterdam save energy because of their adjoining walls.

Population and Gender: Europeans are choosing to have fewer children, and as a result, the population as a whole is aging. Small families are in part a result of women pursuing demanding careers in occupations once dominated by men. Many women delay or avoid child bearing so that they can focus on work.

(J) A mother and her only child celebrate Cernavale in Viareggio, Italy.

(K) Two elderly women and their caregiver go for their daily walk in Cordoba, Spain.

(L) Ségolène Royal (in white), France's first female major party candidate for president. Royal was defeated in 2007.

Food: Throughout the European Union, smaller family-run farms are giving way to larger farms run by corporations. This move is strongest in the Central European states recently admitted to the EU. Smaller-scale organic food production is increasingly being promoted as an alternative throughout the EU.

(M) A rural landscape in Germany dominated by small family farms.

(N) A harvester making its way across a huge field on a corporate-run farm in France.

(O) An organic vineyard in Slovenia.

Water: Europe's many seas are increasingly threatened by water pollution from agriculture, industry, and cities. Several seas are nearly landlocked, and pollution here stays around longer because it is only slowly diluted by the Atlantic Ocean.

(P) The Amoco *Cadiz* sinks off the Atlantic coast of France in 1978, creating the worst oil spill in European history.

(Q) The narrow Strait of Gibraltar (circled), slows the Mediterranean's exchange of water with the Atlantic.

(R) Sewage dumped into the Mediterranean by the resort city of Benidorm, Spain, has recently forced beaches to close.

I THE GEOGRAPHIC SETTING

Terms in This Chapter

This book divides Europe into four subregions (see Figure 4.3 on page 163)—*North*, *West*, *South*, and *Central Europe*.

For convenience, we occasionally use the term *western Europe* to refer to all the countries that were not part of the experiment with communism in the Soviet sphere and Yugoslavia. That is, *western Europe* comprises the combined subregions of *North Europe* (except Estonia, Latvia, and Lithuania), *West Europe* (except the former East Germany), and *South Europe*. When we refer to the countries that were part of the Soviet sphere up to 1989, we use the pre-1989 label *eastern Europe*. When we refer to the group of countries that includes Albania, Bosnia and Herzegovina, Bulgaria, Croatia, Macedonia, Montenegro, Romania, Serbia, and Slovenia (known to some collectively as the *Balkans*), we use the term *southeastern Europe*. *Central Europe* is now the commonly used term for all those countries formerly in eastern Europe that are now in the European Union, plus the countries that were formerly in Yugoslavia, as well as Albania (see Figure 4.3).

Physical Patterns

Europe is a region of peninsulas upon peninsulas (see **Figure 4.1** on pages 160–161). The entire European region is one giant peninsula extending off the Eurasian continent. Its very long coastline has many peninsular appendages, large and small. Norway and Sweden share one of the larger appendages. The Iberian Peninsula (shared by Portugal and Spain), Italy, and Greece are other large peninsulas. Then there are small peninsulas along most coastlines, numbering in the thousands. One result of these many fingers jutting into oceans and seas is that much of Europe feels the climate-moderating effect of the large bodies of water that surround it.

Landforms

Although European landforms are fairly complex, the basic pattern is mountains, uplands, and lowlands, all stretching roughly west to east in wide bands. As you can see in Figure 4.1, Europe's largest mountain chain stretches west to east through the middle of the continent, from southern France through Switzerland and Austria. It extends into the Czech Republic and Slovakia, and curves southeast into Romania. The *Alps* are the highest and most central part of this formation. This network of mountains is mainly the result of pressure from the collision of the northward-moving African Plate with the southeasterly moving Eurasian Plate (see Figure 1.26 on page 51). Europe lies on the westernmost extension of the Eurasian Plate.

South of the main Alps formation, mountains extend into the peninsulas of Iberia and Italy, and along the Adriatic Sea through Greece to the east. The northernmost mountainous formation is shared by Scotland, Norway, and Sweden. These northern mountains are old (about the age of the Appalachians in North America) and have been worn down by glaciers and millions of years of erosion.

Extending northward from the central mountain zone is a band of low-lying hills and plateaus curving from Dijon (France) through Frankfurt (Germany) to Krakow (Poland). These uplands (see Figure 4.1B) form a transitional zone between the high mountains and lowlands of the *North European Plain*, the most extensive landform in Europe (see Figure 4.1C). The plain begins along the Atlantic coast in western France and stretches in a wide band around the northern flank of the main European peninsula, reaching across the English Channel and the North Sea to take in southern England, southern Sweden, and most of Finland. The plain continues east through Poland, then broadens to the south and north to include all the land east to the Ural Mountains (in Russia).

The coastal zones of the North European Plain are densely populated all the way east through Poland. Crossed by many rivers and holding considerable mineral deposits, this coastal lowland is an area of large industrial cities and densely occupied rural areas. Over the past thousand years, people have transformed the natural seaside marshes and vast river deltas into farmland, pastures, and urban areas by building dikes and draining the land with wind-powered pumps. This is especially true in te low-lying Netherlands, where concern over climate change and sea level rises is considerable.

The rivers of Europe link its interior to the surrounding seas. Several of these rivers are navigable well into the upland zone, and Europeans have built large industrial cities on their banks. The Rhine carries more traffic than any other European river, and the course it has cut through the Alps and uplands to the North Sea also serves as a route for railways and motorways (see Figure 4.1D). The area where the Rhine flows into the North Sea is considered the economic core of Europe. Here Rotterdam, Europe's largest port, is located. The larger and much longer Danube River flows southeast from Germany, connecting the center of Europe with the Black Sea. As the European Union expands to the east, the economic and environmental roles of the Danube River basin, including the Black Sea, are getting increased attention (Figure 4.1E).

Vegetation

Nearly all of Europe's original forests are gone, some for more than a thousand years, to make way for farmland, pasture, towns, and cities. Today, forests with very large and old trees exist only in scattered areas, especially on the more rugged mountain slopes (see **Photo Essay 4.1C**) and in the northernmost parts of *Scandinavia* (the area occupied by Denmark, Norway, Sweden, and Finland; see Photo Essay 4.1C). In parts of central and southeastern Europe, forests have been sustainably managed for generations—in Slovenia, for example, the owner of a woodlot may not cut one of her own trees for use as firewood or lumber without a special permit—and across Europe, forests are now regenerating where small farms have been abandoned. Today, although regenerating forests cover about one-third of Europe, the dominant vegetation is crops and pasture grass. Former forestlands are covered with industrial sites, railways, roadways, parking lots, canals, cities, suburbs, and parks.

Climate

Europe has three main climate types: temperate midlatitude, Mediterranean, and humid continental (**Photo Essay 4.1**). The

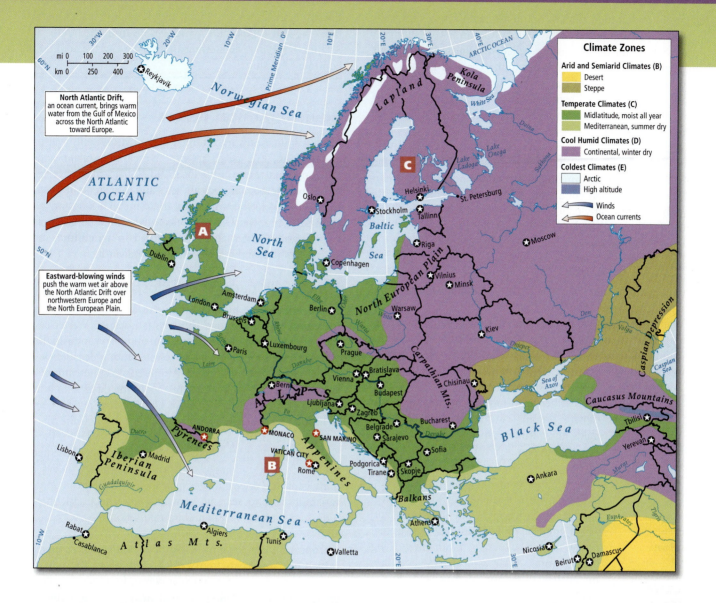

Climate Zones

Arid and Semiarid Climates (B)
- Desert
- Steppe

Temperate Climates (C)
- Midlatitude, moist all year
- Mediterranean, summer dry

Cool Humid Climates (D)
- Continental, winter dry

Coldest Climates (E)
- Arctic
- High altitude

- Winds
- Ocean currents

North Atlantic Drift, an ocean current, brings warm water from the Gulf of Mexico across the North Atlantic toward Europe.

Eastward-blowing winds push the warm wet air above the North Atlantic Drift over northwestern Europe and the North European Plain.

A Midlatitude, moist all year, Scotland

B Mediterranean, summer dry, Corsica

C Continental, winter dry, Finland

167

temperate midlatitude climate dominates in northwestern Europe, where the influence of the Atlantic Ocean is very strong. A broad warm-water ocean current called the **North Atlantic Drift** brings large amounts of warm water to the coasts of Europe. It is really just the easternmost end of the Gulf Stream, which carries water from the Gulf of Mexico north along the eastern coast of North America and across the North Atlantic to Europe (see Photo Essay 2.1 on page 71).

The air above the North Atlantic Drift is warm and wet. Eastward-blowing winds push it over northwestern Europe and the North European Plain, bringing moderate temperatures and rain deep into the Eurasian continent. These factors create a climate that, although still fairly cool, is much warmer than elsewhere in the world at similar latitudes. To minimize the effects of heavy precipitation runoff, people in these areas have developed elaborate drainage systems for their houses and communities. Forests are both evergreen and deciduous. For the most part, the food crops in the temperate zone are not native to Europe, coming instead from around the world (see Figure 1.29 on page 57).

There is some concern that global warming (see pages 45–46 in Chapter 1) could weaken the North Atlantic Drift, leading to a significantly cooler, drier Europe. However, in recent years Europe has actually experienced abnormally warm temperatures, but also less precipitation.

Farther to the south, the **Mediterranean climate** prevails—warm, dry summers and mild, rainy winters. In the summer, warm, dry air from North Africa shifts north over the Mediterranean Sea as far north as the Alps, bringing high temperatures and clear skies. Crops grown in this climate, such as olives and grapes, citrus, apple and other fruits, and wheat, must be drought-resistant or irrigated. Forests tend to be evergreen with low-growing hardy shrubs, such as rosemary and bay. In the fall, this warm, dry air shifts to the south and is replaced by cooler temperatures and rainstorms sweeping in off the Atlantic. Overall, the climate here is mild, and houses along the Mediterranean coast are often open and airy to afford comfort in the hot, sunny summers.

In eastern Europe, without the moderating influences of the Atlantic Ocean and the Mediterranean Sea, the climate is more extreme. In this region of **cool humid continental climate**, summers are fairly hot, and the winters become longer and colder the farther north or deeper into the interior of the continent one goes. Here, houses tend to be well insulated, with small windows, low ceilings, and steep roofs that can shed snow. Crops must be adapted to much shorter growing seasons and include corn and other grains; a wide variety of vegetables, including root crops and cabbages adapted to cold; and fruit trees.

temperate midlatitude climate as in south-central North America, China, and much of Europe, a climate that is moist all year with relatively mild winters and long, warm-to-hot summers

North Atlantic Drift the easternmost end of the Gulf Stream, a broad warm-water current that brings large amounts of warm water to Europe

Mediterranean climate a climate pattern of warm, dry summers and mild, rainy winters

cool humid continental climate a midlatitude climate pattern in which summers are fairly hot and moist, and winters become longer and colder the deeper into the interior of the continent one goes

Biosphere the parts of the earth and its atmosphere in which all living organisms exist

---

THINGS TO REMEMBER

Europe, a region of peninsulas upon peninsulas, has three main landforms—mountain chains, uplands in the north, and the vast North European Plain; and three principal climates—the temperate midlatitude, which results from the North Atlantic Drift; the Mediterranean climate; and the cool humid continental pattern.

Environmental Issues

Having dramatically transformed their environments over the past 10,000 years, Europeans are now increasingly taking action on environmental issues at the local and global scales.

Nevertheless, Europe's air, seas, and rivers remain some of the most polluted in the world, and there is still a long way to go to meet the European Union's stated environmental goals of clean air and water; sustainable development in agriculture, industry, and energy use; and maintenance of biodiversity.

Europe's Impact on the Biosphere

There is a geographic pattern to the ways that human activity over time has transformed Europe's landscapes. Western Europe shows the effects of dense population and heavy industrialization, and eastern Europe reveals the results of long decades of willful disregard for the environment. At present, despite Europe's activist role in alerting global citizenry about the dangers to humanity posed by pollution and climate change, this region continues to have a major impact on the **Biosphere** through the air, water, and sea pollution it generates (Photo Essay 4.2). Furthermore, Europe is itself especially vulnerable to a number of the potential effects of climate change, such as cooler, drier climates in the north, warmer, drier climates near the Mediterranean, and wider variability year to year, all of which will affect agriculture and industries such as tourism and transportation.

Europe's Energy Resources Europe's main energy sources have shifted over the years from coal to petroleum and natural gas, and in some countries to nuclear power. Increasingly, alternative energy sources are being pursued in response to rising energy costs and efforts to cut greenhouse gas emissions.

The 27 members of the European Union (often referred to as the EU-27) get a large portion of their fuel supplies from Russia—32 percent of their crude oil and 42 percent of their natural gas, as of 2008. Most of the gas now comes via pipelines through Belarus and Ukraine, but Turkey is negotiating to supply Russian gas to Europe via the Black Sea, and it already hosts a pipeline that carries gas from the Caspian Sea to Europe. Russia is negotiating for yet another trans-Turkey pipeline to carry oil and gas to the Mediterranean. Europeans fear that Russia will use this dependency against them, withholding and releasing flows at will as it did in 2008 with oil and gas flowing through Ukraine and Belarus. Another 30 percent of the gas and oil the European Union consumes comes from the Middle East. Large oil and gas deposits in the North Sea, most controlled by Norway (not an EU member), have alleviated Europe's dependence on "foreign" sources of energy, but the production of oil from the North Sea has already peaked and is expected to run out by 2018.

82. Winter on the Way

Most of Europe has been transformed by human activity. Western Europe has some of the most heavily impacted landscapes and ecosystems, but the former communist countries of Central Europe also have severe environmental problems. Meanwhile, agricultural intensification is creating new problems in South Europe. The map shows some of the impacts on Europe's land, sea, and air.

A The Garzweiler open pit coal mine in Germany, one of the largest in the world, covers 25.3 square miles (66 square km). Its operations have forced the abandonment of twelve villages and towns and threaten local groundwater resources. The coal produced by this mine is a highly polluting variety called lignite, which is responsible for much of Europe's acid rain problem because of its high sulfur content.

B A coal-fired power plant at an industrial facility in Poznan, Poland, in 2009. Central Europe has thousands of similar small but highly polluting facilities, many of them built during the era of Soviet domination or before, when environmental safeguards were rare.

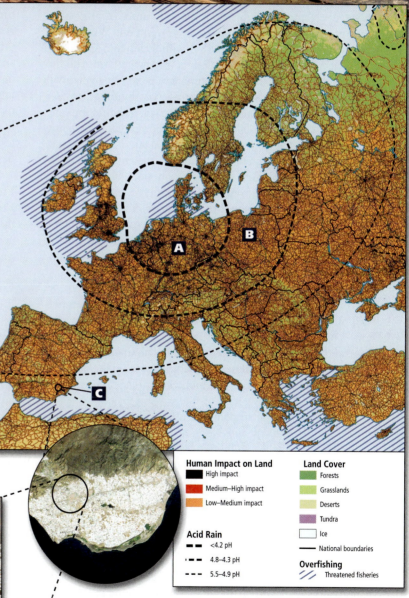

Human Impact on Land
- ⬛ High impact
- 🟥 Medium–High impact
- 🟧 Low–Medium impact

Acid Rain
- – – <4.2 pH
- –·– 4.8–4.3 pH
- – – – 5.5–4.9 pH

Land Cover
- 🟩 Forests
- 🟩 Grasslands
- 🟨 Deserts
- 🟪 Tundra
- ⬜ Ice
- —— National boundaries

Overfishing
- ⫽⫽ Threatened fisheries

C A sea of plastic greenhouses covers most of Campo de Dalía, a coastal plain in Spain's Almería province that has seen intense growth in year-round export-oriented vegetable production over the past several decades. Pesticides and fertilizers are heavily used, and health effects are beginning to be seen among the mainly Moroccan migrant workers and their families who work and live among the greenhouses. Meanwhile, water shortages are so severe that a desalination plant is being built nearby.

169

The use of nuclear power to generate electricity has been more common in Europe than in North America. The EU-27 depend on nuclear power for 30 percent of their total needs. In France, 78 percent of the electricity is generated by nuclear power (compared with only 20 percent in the United States). However, support for nuclear power has declined, partly in response to the disastrous 1986 explosion of a nuclear power plant in Chernobyl, Ukraine, and partly because of worries over how to safely dispose of nuclear waste products.

The European Union wishes to increase its use of renewable energy in order to reduce fuel imports and thereby increase energy security, stimulate the economy with new energy-related jobs, and combat climate change. The goal is to reduce CO_2 emissions by 20 percent, increase use of renewable energy by 20 percent, and power 60 percent of EU homes with renewably generated electricity, all by 2020. A wide array of alternative energy projects is now attracting massive investment. Wind power is generally the favored technology, with Denmark, Germany, and Spain in the lead. Germany has the most windfarms and the most windpower capacity; but Denmark is pioneering what will be the world's largest offshore windfarm—91 turbines spread over a 13.5-square-mile (35-square-kilometer) area. When complete, the farm will power 200,000 homes. Solar power plants are being built in Spain, some using panels developed and built in China; geothermal power generation abounds in Iceland; and energy from biomass and biofuels is increasingly popular throughout the region. In Central Europe, exceedingly efficient wood stoves make use of surplus timber from the region's well-managed forests. Super-energy-efficient homes are built in Germany and France; the Riko company in Slovenia builds individually designed energy-efficient homes and ships them throughout Europe.

Air Pollution At present, significant air pollution exists in much of Europe, but it is particularly heavy over the North European Plain. This is a region of heavy industry, dense transportation routes, and large and affluent populations. The intense fossil fuel use associated with such lifestyles results not only in the usual air pollution but also in *acid rain*, which can fall far from where it was generated.

The highest level of air pollution is found in the former communist states of Central and North Europe. Mines in Central Europe produce highly polluting soft coal (**Photo Essay 4.2B** on page 169) that is burned in out-of-date factories. This region produces the world's highest per capita emissions from burning oil and gas, and it also receives air pollution blown eastward from western Europe. In Upper Silesia, Poland's leading coal-producing area, acid rain has destroyed forests, contaminated soils and the crops grown on them, and raised water pollution to deadly levels. Residents have experienced birth defects, high rates of cancer, and lowered life expectancies. Industrial pollution was one of Poland's greatest obstacles to entry into the European Union, which it barely achieved.

Central Europe's severe environmental problems developed in part because the Marxist theories and policies promoted by the Soviet Union portrayed nature as existing only to serve human needs. During the Soviet era, there was little opportunity for public protest against pollution, but the recent shift to democracy has enabled greater action in places like Hungary, where popular protest has resulted in reductions in air pollution.

Recent investment from wealthier EU countries may help reduce environmental degradation in Central Europe through the use of new, cleaner technologies. One hopeful example is Poland's joint venture with the French multinational glass and high-tech building materials company, Saint-Gobain, to construct modern glass-making factories. The first factory, in Silesia, has emissions well below the standards set by Germany, the industry leader, and all glass waste is completely recycled. By 2008, Saint-Gobain had 19 factories and subsidiaries producing glass and high-tech materials in Poland.

The new market economies in the former Soviet bloc countries are improving energy efficiency and reducing emissions. Power plants, factories, and agriculture are polluting less, and the countries with the worst emissions records, such as Estonia and Latvia, have been making the most progress. All are working toward the 2010 reduction levels set by the Kyoto Protocol. Highly anticipated meetings in Copenhagen in December 2009, meant to upgrade the Kyoto Protocol and get the United States on board, disappointed nearly everyone. The delegates managed only to redefine and gain wider attention to the issues, and to institute new working groups among the super-polluters.

Freshwater and Seawater Pollution Sources of water pollution in Europe include insufficiently treated sewage, chemicals and silt in the runoff from agricultural plots and residential units, consumer packaging litter, petroleum residues, and industrial effluent. Most inland waters contain a variety of such pollutants, and any that enter Europe's wetlands, rivers, streams, and canals quickly reach Europe's surrounding coastal waters. Thus far, the Atlantic Ocean, the Arctic Ocean, and the North Sea are able to disperse most pollutants dumped into them because they are part of, or closely connected to, the circulating flow of the world ocean. In contrast, the Baltic, Mediterranean, and Black seas are nearly landlocked bodies of water that do not have the capacity to flush themselves out quickly; thus, all three, but the Mediterranean to a lesser extent, are prone to accumulating pollution (see red stains in **Figure 4.4**).

There are 34 countries with coastlines on Europe's many seas, all with different economies, politics, and cultural traditions. Such diversity often makes it difficult for them to cooperate to minimize pollution or even reduce the risk of severe pollution.

In the Mediterranean, the effect of the pollution that pours in from rivers, adjacent cities, hotel resorts, and farms is exacerbated by the fact that there is just one tiny opening to the Atlantic Ocean (Figure 4.4 and **Thematic Concepts Part Q** on page 165). At the surface, seawater flows in from the Atlantic through the narrow Strait of Gibraltar and moves eastward. At the bottom of the sea, water exits through the same narrow opening, but only after it has been in the Mediterranean for 80 years! The natural ecology of the Mediterranean is attuned to this lengthy cycle, but the nearly 460 million people now living in the countries surrounding the sea have upset the balance. Their pollution stays in the Mediterranean for decades.

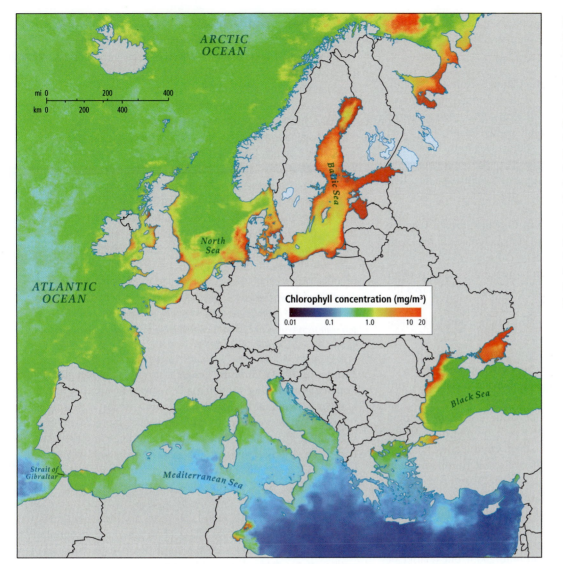

FIGURE 4.4 Pollution of the seas. Europe's exceptionally long and convoluted coast affords easy access to the world's oceans. However, pollution of the nearly landlocked Baltic, Black, and Mediterranean seas is causing increasing concern. Chlorophyll concentration, one measure of pollution levels, is significant in the Atlantic and Arctic oceans as well. [Adapted from a NASA image created by Jesse Allen, Earth Observatory, using data provided courtesy of the SeaWiFS Project, NASA/Goddard Space Flight Center and ORBIMAGE, available at http://earthobservatory.nasa. gov/Newsroom/NewImages/images. php3?img_id517332.]

Although most of the Mediterranean's pollution is generated by Europe (see Thematic Concepts Part R on page 165 and the green stains in Figure 4.4), rapidly growing populations and development on the North African and eastern Mediterranean coasts of the sea also pose an environmental threat—several large containership ports are planned or are under construction, for example. Since 1995, the European Union has been working to lessen pollution in the Mediterranean Basin. However, the North African countries still lack adequate urban sewage treatment and environmental regulations to control agricultural and industrial wastes. It may be that modernized ports will raise the profiles of countries along the southern and eastern Mediterranean shores, providing jobs, suggesting future expansion, and making conformity to pollution standards more attractive.

The Wide Reach of Europe's Environmental Impact In Chapter 1 on page 42, we discussed average per capita water needs per year (18.25 m³) and noted that to arrive at a person's annual total water footprint, one must add the water required for basic needs to a person's virtual water footprint (see Table 1.2 on page 42).

Virtual water per capita is the volume of water used to produce all that a person consumes in a year. Although Europeans do not import as much of their consumer goods as do Americans, they still consume one-fifth of the world's imports and many of these goods have a high *virtual water component* (meaning the water consumed in the production process). Nearly all EU countries import more virtual water than they export, which means that the countries that produce Europe's imported goods are losing in the water exchange. Many of these are countries that have very little water to begin with. As yet, the costs of the loss of the virtual water are not being adequately figured into the price of the products exported to Europe. In all fairness then, the environmental impacts of Europe's virtual water consumption should be counted against Europe's total impact on the Biosphere.

The term *virtual water* has become current largely because water is now recognized as being in short supply globally, and not endlessly renewable, as previously thought. However, other nonrenewable resources from outside Europe that go into Europe's per capita consumption of imports (lumber, fuel, minerals) could also be calculated as *virtual resources*. Were this done, it would be evident that the impact on the Biosphere of consumers in rich regions, such as Europe and North America, is much greater than previously thought.

Europe's Vulnerability to Climate Change Europe's wealth, technological sophistication, and well-developed emergency response systems make it more resilient to the consequences of climate change than most regions of the world. Nevertheless, some areas are much more vulnerable than others, due to their location, dwindling water resources, rising sea levels, or the consequences of poverty. Photo Essay 4.3 illustrates some of the consequences that have been predicted.

European Leadership in Response to Global Climate Change Europe leads the world in response to global climate change, with EU governments having agreed to cut greenhouse gas emissions by 20 percent by 2020. Europe has been more willing than any other region to act on climate change, largely because it recognizes the economic sense in doing so. Recent research suggests that investments in energy conservation, alternative energy, and other measures would cost EU governments 1 percent of their GDP. In contrast, doing nothing about global warming would *shrink* GDP by 20 percent. 📹 **83. German Investments in Clean Alternative Energy Pay Off**

Europe's increasing concern about global warming may also be influenced by public alarm at recent abnormal weather. The summer of 2003 broke all high-temperature records for Europe. Crops failed, freshwater levels sank, forests burned, and deaths soared. In France alone, 3000 people died. On the other hand, in 2002 and 2006, rainfall and snowfall in Central Europe reached record levels. In the spring of 2006, the rivers of Central Europe—the Elbe, the Danube, and the Morava—flooded for weeks. Then in the winter of 2006–2007, snowfall was unusually light.

Progress in Green Behavior By global standards, Europeans use large amounts of resources and contribute about one-quarter of the world's greenhouse gas emissions. However, one European resident averages only one-half the energy consumption of the average North American resident. Europeans live in smaller dwellings, which need less energy to heat, and air conditioning is rare. They drive smaller, more-fuel-efficient cars and use public transportation widely. Because communities are denser, many people walk or bicycle to their appointments.

Green environmentally conscious

These energy-saving practices are related in part to the high population densities and social customs of the region, but also to widespread explicit support for ecological principles. **Green** (environmentally activist) political parties influence national policies in all European countries, often serving in coalition governments. And Green policies are central to the agenda of the European Union. Results include strong regional advocacy for emission controls, community recycling programs now well entrenched across the European Union, and grassroots efforts on local environmental concerns.

📹 **84. Iceland: Energy to Spare**

📹 **85. Britain Requires Energy Inspections for Home Sales**

📹 **86. McDonald's Garbage Heats and Lights British City**

📹 **87. London Leads by Example to Curb Pollution, Climate Change**

Personal Vignette Under cover of darkness, gardeners—who in the daytime are bureaucrats, stock traders, and computer jockeys—sneak into Central London to plant colorful flowers and foliage in traffic islands and roundabouts. They are part of a movement called Guerrilla Gardeners, which has quickly grown to embrace more than 500 activists and a Web site, http://www.guerrillagardening.org/.

Bypassing the town councils (which tend to impose crippling rules), the Guerrilla Gardeners make quick assaults late at night, armed with trowels, spades, mulch, and watering cans. Authorities are unable to stop the guerrillas from covering even the most obscure bit of neglected urban land with blooming hyacinths, tulips, marigolds, shrubs, and even trees. Yet more radical are the Seed Bombers, a group that packs flower seeds, soil, and water into compact parcels and tosses them into derelict patches of public land, where they shatter on impact, spewing forth seeds that will produce plants capable of outcompeting the weeds. *[Adapted from Rob Gifford, "Gardeners Brighten London Under Cover of Dark," Morning Edition, National Public Radio, May 15, 2006, at "http://www.npr.org/templates/story/story.php?storyId=5404229" http://www.npr.org/templates/story/story.php?storyId=5404229.]* ∎

Progress in *Green Building* There is no generally accepted definition of *green building*, though the U.S.-based Leadership in Energy and Environmental Design (LEED) rating system is gaining some international acceptance. In the European Union, green building refers to the all-around sustainability of a building design: the use of recyclable materials and natural lighting, reduction of energy use and of CO_2 emissions, recycling and reusing energy lost through a building's operations, and the recycling and reuse of water, including storm water.

At present each EU country certifies a building as green, but there will soon be a general EU pact on green certification and a sharing of methods and materials across Europe. Techniques presently in use include glazing of balconies and stairwells to passively collect heat and use natural light; use of skylights to focus natural light into office spaces; using lost heat energy to run generators; capturing rainfall on special roofs that are densely planted to retain and recycle the water for other uses; and installing solar and wind panels on building roofs.

Thus far Europe has outpaced the United States in green building. For example, Denmark recycles 55 percent of the electricity used in buildings, the Netherlands and Finland recycle 35 percent, and the United States recycles only 8 percent.

Changes in Transportation Europeans have long favored fast rail networks for both passengers and cargo rather than private cars, trucks, and multilane highways. Yet, despite high and rising gasoline prices over the last two decades, and worries about the relationship between CO_2 emissions and climate change, there was a noticeable trend toward less-energy-efficient but more flexible motorized road transport. Now, however, rising fuel costs and CO_2 emissions are increasingly being considered in the design of *multimodal transport* that links high-speed rail to road, air, and water transport (Figure 4.5 and see Photo Essay 4.6C, D on page 194). An EU transportation report notes that 1 kilogram of

Europe's wealth, technological sophistication, and well developed emergency response systems make it more resilient to climate change than most regions. Nevertheless, some areas are much more vulnerable than others due to dwindling water resources or poverty.

A A once-submerged bridge and an old church re-emerge from a sinking reservoir in Catalonia, Spain. As temperatures rise, Spain's climate will get drier and scarce water resources will shrink further, threatening agriculture and drinking water, especially along the Mediterranean coast.

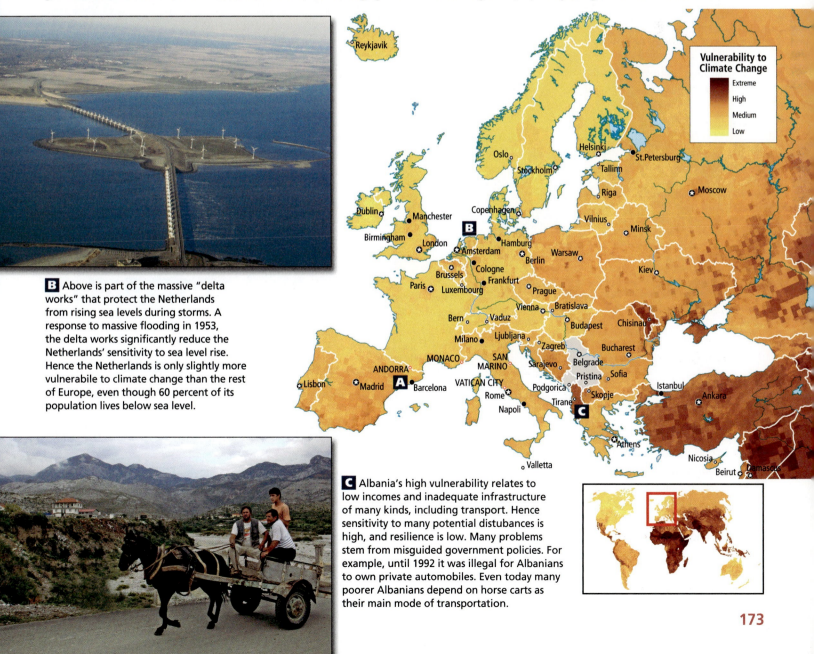

B Above is part of the massive "delta works" that protect the Netherlands from rising sea levels during storms. A response to massive flooding in 1953, the delta works significantly reduce the Netherlands' sensitivity to sea level rise. Hence the Netherlands is only slightly more vulnerabile to climate change than the rest of Europe, even though 60 percent of its population lives below sea level.

Vulnerability to Climate Change

Extreme
High
Medium
Low

C Albania's high vulnerability relates to low incomes and inadequate infrastructure of many kinds, including transport. Hence sensitivity to many potential distubances is high, and resilience is low. Many problems stem from misguided government policies. For example, until 1992 it was illegal for Albanians to own private automobiles. Even today many poorer Albanians depend on horse carts as their main mode of transportation.

173

FIGURE 4.5 The Trans-European transport network. Achievement of the Priority Projects, May 2008.

gasoline can move 50 tons of cargo a distance of 1 kilometer by truck, but the same amount of gasoline can move 90 tons of cargo 1 kilometer by rail and 127 tons by waterway.

The low cost of water transport and Europe's reliance on global trade have been a boon for the development of ocean ports that cater to container ships: Helsinki, Riga, Hamburg, Copenhagen, Antwerp, Rotterdam, Plymouth, Southampton, Le Havre, Marseille, Barcelona, and Koper are only a few of Europe's modern container ports. Soon newly built ports in Morocco, Algeria, Tunisia, Malta, and Egypt, all catering at least in part to European markets, will join these ports—one-third of the world's container traffic now goes through the Mediterranean. The container ship industry, attuned to the CO_2 and climate change debates, now carefully calculates the optimal (often slower) speed for container ships, in order to minimize fuel consumption and emissions while maximizing profits.

1. Europe's impact on the environment globally is large; the region's internal production and consumption patterns have had an adverse impact on the environment and increased its vulnerability to air, water, and land pollution.

2. Europeans produce about one-quarter of the world's greenhouse gas emissions, and the ecology of the Mediterranean is threatened by water pollution from the 320 million people now living in the countries surrounding this sea.

3. In response to these challenges, Europe has emerged as the world's leader on combating global warming. The European Union has a goal of cutting greenhouse gas emissions by 20 percent by 2020. EU citizens see changes in energy production, transportation, and most efforts at environmental protection as an incentive for innovation, not a hindrance to economic performance.

4. European residents average only one-half the energy consumption of the average North American, and fuel costs and CO_2 emissions are increasingly being considered in the design of multimodal transport that links high-speed rail to road, air, and water transportation.

Human Patterns over Time

Over the last 500 years, Europe has profoundly influenced how the world trades, fights, thinks, and governs itself. Attempts to explain this influence are wide-ranging. One argument is that Europeans are somehow a superior breed of humans. Another is that Europe's many bays, peninsulas, and navigable rivers have promoted commerce to a greater extent there than elsewhere. In fact, much of Europe's success is based on technologies and ideas it borrowed from elsewhere. For example, the concept of the peace treaty, so vital to current European and global stability, was first documented not in Europe but in ancient Egypt. In an effort to explain how Europe gained the leading global role it continues to play, it is worth taking a look at the broad history of this area.

Sources of European Culture

Starting about 10,000 years ago, the practice of agriculture and animal husbandry gradually spread into Europe from the Tigris and Euphrates river valleys in Southwest Asia and from farther east in Central Asia and beyond (see Figure 1.29 on page 57). Mining, metalworking, and mathematics also came to Europe from these places and from North Africa. All of these innovations opened the way for a wider range of economic activity—most notably trade—in Europe.

The first European civilizations were ancient Greece (800 to 86 B.C.E.) and Rome (100 B.C.E. to 450 C.E.). Located in southern Europe, both Greece and Rome (Rome was, and is, centered in what is today Italy) initially interacted more with the Mediterranean rim, Southwest Asia, and North Africa than with the rest of Europe, which then had only a small and relatively poor rural population. Later European traditions of science, art, and literature were heavily based on Greek ideas, which were themselves derived from yet earlier Egyptian and Southwest Asian sources.

The Romans, after first borrowing heavily from Greek culture, also left important legacies in Europe. Many Europeans today speak *Romance* languages, such as Spanish, Portuguese, Italian, French, and Romanian, all of which are largely derived from Latin, the language of the Roman Empire. Rome was the origin of European laws that determine how individuals own, buy, and sell land. These laws have been spread throughout the world by Europeans.

Roman practices used in colonizing new lands also shaped much of Europe. After a military conquest, the Romans secured control in rural areas by establishing large plantation-like farms. Politics and trade were centered on new Roman towns built on a grid pattern that facilitated both commercial activity and military repression of rebellions. These same systems for taking and holding territory were later used when Europeans colonized the Americas, Asia, and Africa.

The influence of Islamic civilization on Europe is often overlooked. After the fall of Rome, while Europe experienced a period of decline known as the *Early Middle Ages* (roughly 450–1000 C.E.), pre-Muslim (Arab and Persian) and then Muslim scholars preserved learning from Rome and Greece. Muslim Arabs originally from North Africa ruled Spain from 711 to 1492. From the 1400s through the early 1900s, the Ottoman Empire (based in what is now Turkey) dominated much of southeastern Europe and Greece. The Arabs, Persians, and Turks all brought new technologies, food crops, architectural principles, and textiles to Europe from Arabia, Persia, Anatolia, China, India, and Africa. Arabs also brought Europe its numbering system, mathematics, and significant advances in medicine and engineering, building on ideas they picked up in South Asia.

Beginning 500 years ago, Europe began to draw on a host of cultural features from the various colonies that were established in the Americas, Asia, and Africa. For example, the food crops now popular in Europe came mostly from the Americas (potatoes, corn, peppers, tomatoes, beans, and squash). Thousands of years earlier western Central Asia was the source of food plants (wheat, buckwheat, leafy greens, onions, various cousins to the cabbage, and hardy nut and fruit trees, including apples and apricots).

The Inequalities of Feudalism

As the Roman Empire declined, a social system known as *feudalism* evolved during the *medieval period* (450–1500 C.E.). This system originated from the need to defend rural areas against local bandits and raiders from Scandinavia and the Eurasian interior. The objective of feudalism was to have a sufficient number of heavily armed, professional fighting men, or knights, to defend a much larger group of *serfs*, who were legally bound to live on and cultivate plots of land for the knights. Over time, some of these knights became a wealthy class of warrior-aristocrats, called the *nobility*, who controlled certain territories. Some nobles gained so much power that they became centralized rulers—called "kings" or "monarchs"—of a certain area.

FIGURE 4.6 Podsreda Castle, Slovenia. Podsreda Castle in eastern Slovenia exemplifies the feudalism from which modern Europe eventually emerged. It was begun in the twelfth century, has been modified over the centuries since, and is now renovated and preserved as a national landmark in Kozjanski Park. [Mac Goodwin.]

The often lavish lifestyles and elaborate castles (Figure 4.6) of the wealthier nobility were supported by the labors of the serfs. Most serfs lived in poverty outside castle walls, and much like slaves, were legally barred from leaving the lands they cultivated for their protectors.

The Role of Urbanization in the Transformation of Europe

While rural life followed established feudal patterns, new political and economic institutions were developing in Europe's towns and cities. Here, thick walls provided defense against raiders, and commerce and crafts supplied livelihoods. Thus the people could be more independent from feudal knights and kings.

Located along trade routes, Europe's urban areas were exposed to new ideas, technologies, and institutions from Southwest Asia, India, and China. Some institutions, such as banks, insurance companies, and corporations, provided the foundations for Europe's modern economy. Over time, Europe's urban areas established a pace of social and technological change that left the feudal rural areas far behind.

Urban Europe flourished in part because of laws that granted basic rights to urban residents. With adequate knowledge of these laws, set forth in legal documents called *town charters*, people with few resources could protect their rights even if challenged by those who were more wealthy and powerful. Town charters provided a basis for European notions of *civil rights*, which have proved hugely

humanism a philosophy and value system that emphasizes the dignity and worth of the individual

mercantilism a strategy for increasing a country's power and wealth by acquiring resources cheaply and turning them into exportable products. Colonies aided this effort. All aspects of their production, transport, and trade were managed for the colonizer's benefit.

influential throughout the world. With strong protections for their civil rights, some of Europe's townsfolk grew into a small middle class whose prosperity moderated the feudal system's extreme divisions of status and wealth. A related outgrowth of urban Europe was a philosophy known as **humanism,** which emphasized the dignity and worth of the individual regardless of wealth or social status.

The liberating influences of European urban life transformed the practice of religion. Since late Roman times, the Catholic Church had dominated not just religion but also politics and daily life throughout much of Europe. In the 1500s, however, a movement known as the *Protestant Reformation* arose in the urban centers of the North European Plain. Reformers such as Martin Luther challenged Catholic practices that stifled public participation in religious discussions, such as holding church services in Latin, which only a tiny educated minority understood. Protestants also promoted individual responsibility and more open public debate of social issues, which altered the perception of the relationship between the individual and society. These ideas spread faster with the invention of the European version of the printing press, which enabled widespread literacy.

European Colonialism: An Acceleration of Globalization

A direct outgrowth of the greater openness of urban Europe was the exploration and subsequent colonization of much of the world by Europeans. The increased commerce and cultural exchange began a period of accelerated *globalization* that persists today (see the discussion in Chapter 1 on pages 32–37).

In the fifteenth and sixteenth centuries, Portugal took advantage of advances in navigation, shipbuilding, and commerce to set up a trading empire in Asia and a colony in Brazil. Spain soon followed, founding a vast and profitable empire in the Americas and the Philippines. By the seventeenth century, however, England and the Netherlands had seized the initiative from Spain and Portugal. They perfected **mercantilism,** a strategy for increasing a country's power and wealth by acquiring resources cheaply and turning them into exportable products. Colonies aided this effort. All aspects of their production, transport, and trade were managed for the colonizer's benefit. Mercantilism supported the Industrial Revolution in Europe by supplying cheap resources from around the globe for Europe's new factories. The colonies also supplied markets for European manufactured goods (Figure 4.7).

🎦 **88. Art Exhibit Encompasses the World of Portuguese Explorations**

By the nineteenth century, the English, French, and Dutch (the people who live in the Netherlands) overshadowed the Spanish and Portuguese colonial empires and extended their influence into Asia and Africa (see **Thematic Concepts Parts A, B** on page 164). By the twentieth century, European colonial systems had strongly influenced nearly every part of the world.

The overseas empires of England, the Netherlands, and eventually France were the beginnings of the modern global economy. The riches they provided shifted wealth and investment away from southern Europe and the Mediterranean and toward western Europe (**Figure 4.8A, B** on page 178). The regional trading

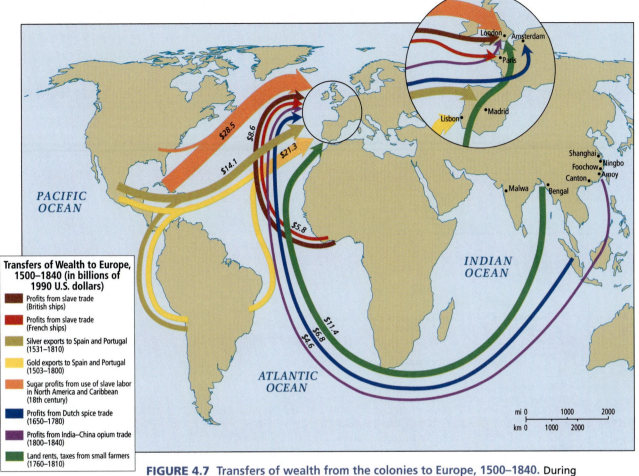

FIGURE 4.7 **Transfers of wealth from the colonies to Europe, 1500–1840.** During the period of mercantilism, Europe received billions of dollars of income from its overseas colonies. [Adapted from Alan Thomas, *Third World Atlas* (Washington, D.C.: Taylor & Francis, 1994), p. 29.]

networks that were developed by some Italian city-states during the medieval period (450–1500 C.E.) were overshadowed by the evolving global trading networks of cities such as Amsterdam, London, and Paris.

By the mid-1700s, wealthy merchants in these and other western European cities were investing in new industries. Workers from rural areas poured into urban centers in England, the Netherlands, Belgium, France, and Germany to work in new manufacturing industries and mining (Figure 4.8C). Wealth and raw materials flowed in from colonial ports in the Americas, Asia, and Africa. Some cities, such as Paris and London, were elaborately rebuilt in the 1800s to reflect their roles as centers of global empires.

By 1800, London and Paris, each of which had a million inhabitants, were Europe's largest cities, a status that eventually brought them to their present standing as *world cities* (cities of worldwide economic or cultural influence). London is a global center of finance, and Paris is a cultural center that has influence over global consumption patterns, from food to fashion to tourism.

Urban Revolutions in Industry and Democracy

The wealth derived from Europe's colonialism helped fund two of the most dramatic transformations in a region already characterized by rebirth and innovation: the industrial and democratic revolutions. Both first took place in urban Europe.

The Industrial Revolution Europe's Industrial Revolution—particularly Britain's ascendancy as the leading industrial power of the nineteenth century—was intimately connected with colonial expansion. In the seventeenth century, Britain developed a small but growing trading empire in the Caribbean, North America, and South Asia, which provided it with access to a wide range of raw materials and to markets for British goods.

Sugar, produced by British colonies in the Caribbean, was an especially important trade crop (see Figure 4.7). Sugar production was a complex process requiring major investments in equipment and slaves forcibly brought in from Africa. The skilled management and large-scale organization needed later provided a production model for the Industrial Revolution. The mass production of sugar also generated enormous wealth that helped fund industrialization.

By the late eighteenth century, Britain was introducing mechanization into its industries, first in textile weaving and then in the production of coal and steel. By the nineteenth century, Britain was the world's greatest economic power, with a huge and growing empire, expanding industrial capabilities, and the world's

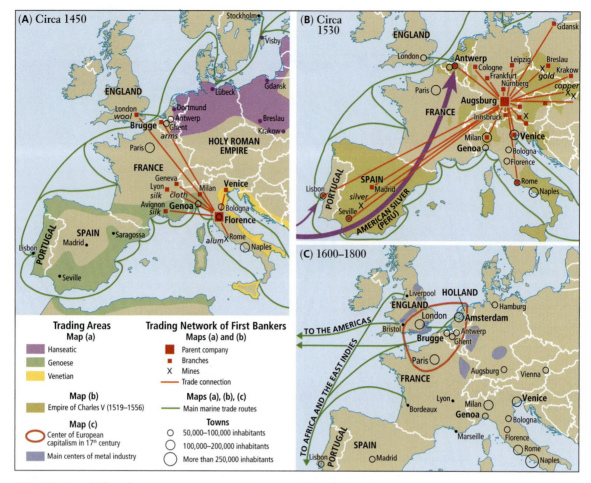

FIGURE 4.8 Shifts of power among urban centers, 1450–1800. [This map was prepared for this text with the assistance of geographer John Agnew.]

most powerful navy. Its industrial technologies spread throughout Europe, North America, and elsewhere, transforming millions of lives in the process.

Urbanization and Democratization Industrialization led to massive growth in urban areas in the eighteenth and nineteenth centuries. Extremely low living standards in Europe's cities created tremendous pressures for change in the political order. Most industrial jobs were dangerous and unhealthy (see Thematic Concepts Part C on page 164), demanding long hours and offering little pay. Poor people usually had to wait to get into shelters, where they were packed into tiny, airless spaces that they shared with many others (see Thematic Concepts Part D on page 164). Children were often weakened and ill due to poor nutrition, industrial pollution, inadequate sanitation and health care, and child labor. Water was often contaminated, and most sewage ended up in the streets. Opportunities for advancement through education were restricted to the tiny few who could afford it.

The ideas and information gained by the few key people who did learn to read gave them the incentive to organize and protest for change. After lengthy struggles, democracy was expanded to Europe's huge and growing working class, and power, wealth, and opportunity were distributed more evenly throughout society.

However, the road to democracy in Europe was rocky and violent, just as it is now in many parts of the world.

In 1789, the French Revolution led to the first major inclusion of common people in the political process in Europe. Angered by the extreme disparities of wealth in French society, and inspired by news of the popular revolution in North America, the poor rebelled against the monarchy and the elite-dominated power structure that still controlled Europe. As a result, the general populace, especially in urban areas, became involved in governing through democratically elected representatives. The democratic expansion created by the French Revolution ultimately proved short-lived as elite-dominated governments soon regained control in France. Nevertheless, the French Revolution (see Thematic Concepts Part E on page 164) provided crucial inspiration to later urban democratic political movements in many parts of the world.

The Impact of Communism During the struggles that resulted ultimately in the expansion of democracy, popular discontent erupted periodically in the form of new revolutionary political movements that threatened the established civic order. The political philosopher and social revolutionary Karl Marx framed the mounting social unrest in Europe's cities as a struggle between socioeconomic classes. His treatise *The Communist Manifesto* (1848) helped social reformers across Europe articulate ideas on how wealth could be more equitably distributed. East of Europe in Russia, Marx's ideas inspired the creation of a revolutionary communist state in 1917, the Union of Soviet Socialist Republics, often referred to as the USSR or the Soviet Union. Eventually the Soviet Union extended its ideology and state power throughout Central Europe.

Popular Democracy and Nationalism In the cities of western Europe, communism gained some followers, but because political movements among workers were more successful at expanding democracy, violent revolution was avoided and communism was less popular. Nonetheless, innumerable struggles between labor and the authorities continued throughout the nineteenth and early twentieth centuries until eventually workers gained the right to unionize and all adults got the right to vote.

Throughout the nineteenth and twentieth centuries, the development of democracy was also linked to the idea of **nationalism,** or allegiance to the state. The notion spread that all the people who lived in a certain area formed a nation, and that loyalty to that nation should supersede loyalties to family, clan, or individual monarchs. Eventually, the whole map of Europe was reconfigured, and the mosaic of kingdoms gave way to a collection of nations or countries. All of these new countries were slowly but relentlessly transformed into democracies by the political movements arising in Europe's industrial cities. Nationalism was a major component of World Wars I and II, so in the post-war era, and as the European Union was constructed, nationalism was deemphasized.

Democracy and the Welfare State Channeled through the democratic process, public pressure for improved living standards moved most European governments toward becoming **welfare states** by the mid-twentieth century. Such governments accept responsibility for the well-being of their people, guaranteeing basic necessities such as education, employment, and health care for all citizens. In time, government regulations on wages, hours, safety, and vacations established more harmonious relations between workers and employers. The gap between rich and poor declined somewhat, and overall civic peace increased. And although welfare states were funded primarily through taxes on industry, industrial productivity did not decline.

Modern Europe's welfare states have yielded generally adequate-to-high levels of well-being for all citizens. Just how much support the welfare state should provide is still a subject of debate, however, and one that has been resolved in different ways across Europe (see Thematic Concepts Part F on page 164, Figure 4.21 on page 199, and the discussion on pages 199–201).

Two World Wars and Their Aftermaths

Despite Europe's many advances in industry and politics, by the beginning of the twentieth century the region still lacked a system of collective security that could prevent war among its rival countries. Between 1914 and 1945, two horribly destructive world wars left Europe in ruins, no longer the dominant region of the world. At least 20 million people died in World War I (1914–1918) and 70 million in World War II (1939–1945). During World War II, Germany's Nazi government killed 15 million civilians during

its failed attempt to conquer the Soviet Union. Eleven million civilians died at the hands of the Nazis during the **Holocaust,** a massive execution of 6 million Jews and 5 million **Roma** (Gypsies); disabled and mentally ill people; gays, lesbians, and transgendered people; political dissidents, and ethnic Poles and other Slavs (**Figure 4.9**).

🎥 **89. World Remembers Victims of Holocaust**

The defeat of Germany, the country seen as the instigator of both world wars, resulted in a number of enduring changes in Europe. After World War II ended in 1945, Germany was divided into two parts. West Germany became an independent democracy allied with the rest of western Europe—especially Britain and France—and the United States. The Soviet Union controlled East Germany and the rest of what was then called eastern Europe (Latvia, Lithuania, Estonia, Poland, Czechoslovakia, Hungary, Romania, Bulgaria, Ukraine, Moldova, and Belarus). The line between East and West Germany was part of what was called the **iron curtain,** a

nationalism devotion to the interests or culture of a particular country, nation, or cultural group; the idea that a group of people living in a specific territory and sharing cultural traits should be united in a single country to which they are loyal and obedient

welfare state a government that accepts responsibility for the well-being of its people, guaranteeing basic necessities such as education, employment, and health care for all citizens

Holocaust during World War II, a massive execution by the Nazis of 6 million Jews and 5 million Roma (Gypsies); disabled and mentally ill people; gays, lesbians, and transgendered people; political dissidents; and ethnic Poles and other Slavs

Roma the now-preferred term in Europe for Gypsies

iron curtain a long, fortified border zone that separated western Europe from (then) eastern Europe during the Cold War

FIGURE 4.9 The Holocaust: The Warsaw Ghetto Uprising. The photo shows the end of the largest single revolt by Jewish people against the German army during World War II. For 5 months in 1943, more than 60,000 residents of the Jewish ghetto in Warsaw, Poland, resisted Nazi efforts to evict them from their homes and send them to their deaths in concentration camps. Most were eventually killed, either during the uprising or later at nearby concentration camps. However, it is possible that the boy in the picture is Tsvi Nussbaum, who survived the Holocaust and now lives in upstate New York.

Cold War a period of conflict, tension, and competition between the United States and the Soviet Union that lasted from 1945 to 1991

capitalism an economic system characterized by privately owned businesses and industrial firms that adjust prices and output to match the demands of the market

communism an ideology and economic system, based largely on the writings of the German revolutionary Karl Marx, in which, on behalf of the people, the state owns all farms, industry, land, and buildings (a version of socialism)

central planning a communist economic model in which a central bureaucracy dictates prices and output with the stated aim of allocating goods equitably across society according to need

long, fortified border zone that separated western Europe from (then) eastern Europe.

The Cold War The division of Europe created a period of conflict, tension, and competition between the United States and the Soviet Union known as the **Cold War**, which lasted from 1945 to 1991. During this time, once-dominant Europe, and indeed the entire world, became a stage on which the United States and the Soviet Union competed for dominance. The central issue was the competition between **capitalism**—characterized by privately owned businesses and industrial firms that adjusted prices and output to match the demands of the market—and **communism** (actually a version of socialism), in which the state owned all farms, industry, land, and buildings. After 1945, in most of what we now call Central Europe, the Soviet Union forcibly implemented a communist economic model known as **central planning**: a central bureaucracy dictated prices and output with the stated aim of allocating goods and services equitably across society according to need. This system was successful in alleviating long-standing poverty and illiteracy among the working classes, but it was rife with waste, corruption, and bureaucratic bungling. It ultimately collapsed in the 1990s due to inefficiency, insolvency, high levels of environmental pollution, and public demands for more democratic participation.

In the rest of Europe, especially western Europe, a successful demonstration of capitalism was mounted, underwritten by financial support from the United States under the *Marshall Plan*. Basic facilities, such as roads, housing, and schools were rebuilt, and governments were aided in reestablishing free market economic systems. Economic reconstruction proceeded rapidly in the following decades.

📹 91. Marshall Plan's 61st Anniversary

Decolonization, Democratization, and Conflict in Modern Europe Europe's decline during the two world wars also led to the loss of its colonial empires. Many European colonies had participated in the wars, with some (India, the Dutch East Indies, Burma, and Algeria) suffering extensive casualties, and almost all emerging economically devastated. After the war, Europe could not afford to help these areas rebuild, and demands for independence grew. By the 1960s, most former European colonies had gained independence, often after bloody wars fought against European powers and their local allies.

Despite democracy's long history in Europe (if not in Europe's colonies), it nearly disappeared during the two world wars when dictators in Germany (Hitler) and Italy (Mussolini) sought continental domination. Then post-war western Europe made remarkable progress in reorganizing itself around democratic principles and humanitarian ideals. Exceptions in the west were Portugal, which remained a dictatorship and colonial master in Africa until

Patterns of democratization and conflict in Europe and in former European colonies generally support the notion that countries with lower levels of democratization also suffer the most from violent conflict. Since World War II, the most violent conflicts in Europe have occurred in Central Europe, where repressive governments frequently denied their populations basic democratic freedoms. Outside Europe, many long and brutal wars were fought to gain independence from European powers. These wars stemmed in part from the inequitable extraction of resources from the "colonies" by Europeans, but another factor was the denial of basic democratic freedoms to the local non-European population. The map shows how low levels of Democratization persist even today in the countries that fought wars of independence, decades after the formal end of colonial rule. Other factors worked against Democratization in these countries, but European imperialism helped set many on an antidemocratic trajectory that has only recently started to change.

D Portuguese soldiers loading a bomb into a plane in Angola in the 1960s. Portugal, at the time an undemocratic authoritarian dictatorship, fought three major wars in Africa during the 1960s and 1970s. The rising human and financial costs of these wars eventually brought a revolution in Portugal itself in 1974, resulting in independence for the colonies and the eventual establishment of democracy in Portugal. None of Portugal's former colonies have achieved a high level of democratization, due in part to authoritarian governing structures put in place by the Portuguese as well as the lengthy civil wars that followed decolonization.

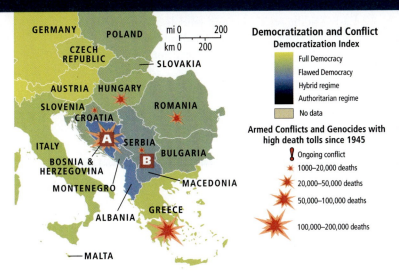

Democratization and Conflict

Democratization Index

- Full Democracy
- Flawed Democracy
- Hybrid regime
- Authoritarian regime
- No data

Armed Conflicts and Genocides with high death tolls since 1945

- Ongoing conflict
- 1000–20,000 deaths
- 20,000–50,000 deaths
- 50,000–100,000 deaths
- 100,000–200,000 deaths

GERMANY, POLAND, CZECH REPUBLIC, SLOVAKIA, AUSTRIA, HUNGARY, SLOVENIA, ROMANIA, CROATIA, ITALY, SERBIA, BULGARIA, BOSNIA & HERZEGOVINA, MONTENEGRO, MACEDONIA, ALBANIA, GREECE, MALTA

A The aftermath of Bosnia's civil war in a suburb of Sarajevo in 1996. Attempts by Serbia to limit democratic decision making within the former Yugoslavia resulted in several wars during the 1990s. In Bosnia, a civil war that pitted different ethnicities against one another resulted in 175,000 deaths.

B Members of the Kosovo Liberation Army turn in their weapons to NATO forces at the end of the war in 1999. Like other parts of the former Yugoslavia, Kosovo fought for independence after Serbia's anti-democratic policies denied basic political freedoms.

C Two murals in Derry, Northern Ireland marking an entrance to a "nationalist" neighborhood. The mural on the right depicts a boy wearing a gas mask and holding a homemade "petrol" bomb. Both murals were created during "the Troubles," a time of violent conflict lasting from 1969 to 1998. One aspect of the still-sensitive conflict is competing claims about Northern Ireland's democratic legitimacy. The "nationalists" (who are usually Catholic) claim that Northern Ireland was undemocratically separated from the rest of Ireland in 1920 by the government of the United Kingdom during Ireland's war of independence from the UK. The "loyalists" (usually Protestant) see Northern Ireland as an expression of the will of the voters in the state, most of whom are Protestant and want to remain part of the UK.

YOU ARE NOW ENTERING FREE DERRY

1974 (Photo Essay 4.4D) and Spain, which labored under a dictator until 1975 and then made a slow transition to what in Spain is called a *parliamentary monarchy*. Greece also had a military dictatorship from 1967 to 1974. Another exception was Northern Ireland, which until the late 1990s saw sectarian violence between Catholics and Protestants so severe that normal life could not proceed (Photo Essay 4.4C). In eastern Europe, democracy did not arrive until the Soviet Union dissolved after 1989, and its former satellites, one by one, declared independence and began to develop representative democracies. Yugoslavia, a large Communist Republic in southeastern Europe outside the Soviet bloc, also dissolved beginning in the early 1990s. There, with the exception of Slovenia, democracy again nearly lost out to a rising tide of ethnic xenophobia instigated by the dominant Serbs who fomented hostilities among Croats, Bosnians, Kosovars, and other smaller groups, based on ethnic hatred (Photo Essay 4.4A, B).

THINGS TO REMEMBER

1. To evolve into the current European Union, Europe went through several tumultuous stages over the past 2000 years— the Early Middle Ages; medieval feudalism; urbanization; mercantilism and colonialism; globalization and nationalism; two world wars; democratization, socialism, and communism; and the Cold War.

2. As European powers like Spain, the United Kingdom, and the Netherlands conquered vast overseas territories, they created trade relationships that laid the foundation for the modern global economy.

3. Most of the countries in the world have been ruled by a European colonial power at some point in their history.

4. Urban Europe flourished in part because of laws that granted basic rights to urban residents.

5. The road to democracy in Europe was rocky and violent, just as it is now in many democratizing parts of the world.

6. Patterns of democratization and conflict in Europe generally support the notion that countries with lower levels of democratization also suffer the most from violent conflict.

II CURRENT GEOGRAPHIC ISSUES

Europe today is in a state of transition as a result of two major changes that occurred during the 1990s: the demise of the Soviet Union (discussed above and in Chapter 5) and the rise of the European Union. These developments have already brought greater peace, prosperity, and world leadership to Europe. Nonetheless, problems and tensions remain.

Economic and Political Issues

At the end of World War II, European leaders felt that closer economic ties among the European nation-states would prevent the kind of hostilities that had led to two world wars. Over the next 50 years, a series of steps were taken that increasingly united Europe economically, socially, and to a certain extent politically. More recently, political unification has been tentatively attempted.

The European Union: A Rising Superpower

The original plan after the trauma of World War II was simply to work toward a level of economic and social integration that would make possible the free flow of goods and people across national borders. While this goal remains central, some Europeans believe that the European Union, which is already a global economic power, should become a global counterforce to the United States in political and military affairs.

Steps in Creating the European Union The first major step in achieving economic unity took place in 1958, when Belgium, Luxembourg, the Netherlands, France, Italy, and West Germany formed the *European Economic Community (EEC)*. The members of the EEC agreed to eliminate certain tariffs against one another and to promote mutual trade and cooperation. Denmark, Ireland, and the United Kingdom joined in 1973; Greece, Spain, and Portugal in the 1980s; and Austria, Finland, and Sweden in 1995, bringing the total number of member countries to 15. In 1992, the concept of the EEC was expanded to that of the European Union, which is concerned with more than just economic policy.

In the 1980s, the Solidarity movement—a Polish workers' rebellion that began to break Soviet control of Eastern Europe —helped strengthen the ongoing process of building a broader European Union, which would eventually include Central Europe. By 1990, the lengthy process of economic and political reunification of East and West Germany had begun. A united Germany facilitated the further expansion of the European Union into the former communist countries of Central Europe—Estonia, Latvia, Lithuania, Poland, the Czech Republic, Slovakia, Slovenia, and Hungary—in 2004 (Malta and Cyprus in the Mediterranean also joined in 2004), and then into southeastern Europe when Romania and Bulgaria joined in 2007. Figure 4.10 shows the current members of the European Union and the dates they joined, as well as new candidates for membership.

90. Poles Celebrate the 25th Anniversary of Solidarity.

Membership in the European Union became especially attractive to countries in Central Europe after the demise of the Soviet Union, when their economies and socialist safety nets began to deteriorate. Many workers had lost their jobs, and in some of the poorest countries, such as Romania, Bulgaria, and Serbia, social turmoil and organized crime threatened stability. Investments by wealthier EU member countries in West and North Europe have reduced many of these problems in the Central European EU economies.

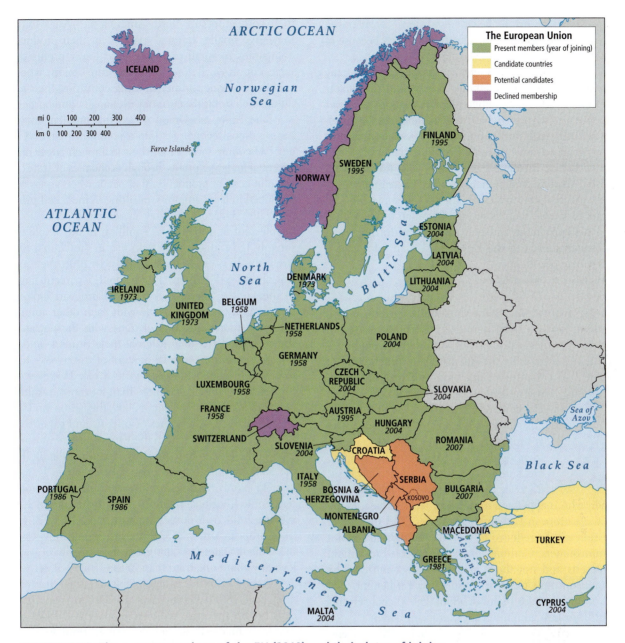

FIGURE 4.10 The current members of the EU (2010) and their dates of joining.
The European Union, formed as the EEC in 1958 with the initial goal of economic integration, became the EU in 1992, and has led the global movement toward greater regional cooperation. It is older and more deeply integrated than its closest competitor, the North American Free Trade Agreement (NAFTA), which was created largely in response to the EU.

According to the preamble of the EU Charter of Fundamental Rights, the European Union is "founded on the indivisible, universal values of human dignity, freedom, equality and solidarity; it is based on the principles of democracy and the rule of law." Standards for EU membership are rather specific. A country must achieve political stability and have a democratically elected government. Each country has to adjust its constitution to EU standards that guarantee the rule of law, human rights, and respect for minorities. Each must also have a functioning market economy that is open to investment by foreign-owned companies

and that has well-controlled banks. Finally, farms and industries must comply with strict regulations governing the finest details of their products and the health of environments.

A few countries have chosen not to join the European Union: Switzerland in the center of Europe, and Norway and Iceland on the northern periphery. These three countries have long treasured their neutral role in world politics. Moreover, as wealthy countries, they have been concerned about losing control over their domestic economic affairs. However, during the 2008–2009 recession, Iceland, much reduced in wealth and in debt to the

International Monetary Fund, began discussions to enter the European Union, but was not immediately welcomed because of its financial troubles.

Several countries on the perimeter of Europe may eventually join over the next few decades. Of these, Turkey is the most likely candidate. However, Turkey's strained relations with Greece over the island country of Cyprus (which was admitted to the European Union in 2004), its history of human rights violations against minorities (especially against its large Kurdish populations), and some issues regarding separation of religion and the state are strikes against it. Ukraine, and perhaps even the Caucasian republics (Armenia, Azerbaijan, and Georgia), may at some time be invited to join. However, there is strong opposition to this within Europe, and Europe's huge and potentially powerful neighbor, Russia, opposes expansion of the European Union into what it regards as its sphere of influence.

◻ 93. EU Tells Turkey to Deepen Reforms

EU Governing Institutions Somewhat similar to the United States, the European Union has one executive branch and two legislative bodies. A formal agreement for revising the details of the functioning of the European Union, the Treaty of Lisbon, was ratified by all members and took effect December 1, 2009. The *European Commission* acts like an executive branch of government, proposing new laws, implementing decisions, and generally running the European Union on a day-to-day basis. Each of the 27 member states gets one commissioner, who is appointed for a 5-year term, subject to the approval of the European Parliament; but after 2014 the size of the Commission will be reduced to 18, with the right to appoint rotating commissioners. Commissioners are expected to uphold common interests and not those of their own countries. The entire Commission must resign if censured by Parliament. The European Commission also includes about 25,000 civil servants who work in Brussels to administer the European Union.

The *European Parliament* is directly elected by EU citizens, with each country electing a proportion of seats based on its population, much like the U.S. House of Representatives. The Parliament elects the president of the European Commission, who serves for 5 years as a head of state and head of foreign policy. Laws must be passed in Parliament by 55 percent of the member states, which must contain 65 percent of the EU total population. In other words, a simple majority does not rule. The *Council of the European Union* is similar to the U.S. Senate in that it is the more powerful of the two legislative bodies. However, its members are not elected but consist of one minister of government from each EU country; which minister assigned to attend depends on the agenda: foreign affairs, agriculture, industry, the environment, and so forth. The Council of the European Union acts with Parliament to enact legislation.

Economic Integration and a Common Currency Economic integration has solved a number of problems in Europe. Individual European countries have far smaller populations than their competitors in North America and Asia. This means that they have smaller internal markets for their products. Typically, companies in small countries earn lower profits than those in large countries. Before the European Union, when businesses sold their products to neighboring countries, their earnings were diminished by tariffs, currency exchanges, and border regulations.

The European Union solved this problem by joining European national economies into a common market. Companies in any EU country now have access to a much larger market and the potential for larger profits through **economies of scale** (reductions in the unit costs of production that occur when goods or services are produced in large amounts, resulting in increased profits per unit).

The EU economy now encompasses close to 492 million people (out of a total of 525 million in the whole of Europe)—roughly 185 million more than live in the United States. Collectively, the EU countries are wealthy. In 2008, their joint economy was almost $15 trillion (PPP), about 5 percent larger than that of the United States, making the European Union the largest economy in the world. Trade with each other amounts to about twice the monetary value of trade with the outside world. And the combined EU total external trade (imports and exports with non-EU countries) was 19 percent of the world's total, equal to that of the United States (Figure 4.11). However, whereas the United States imports far more than it exports, resulting in a trade deficit of $26 billion in March 2009, the European Union enjoyed a trade balance—the values of imports and exports were roughly equal. Economic growth in the European Union is usually slower than in the United States due to careful regulation. Regulation of the banking industry and financial markets in Europe is greater than in North America, with the result that Europe was more resilient to the global financial crisis that began in 2008. Also, the average 2008 gross domestic product (GDP, PPP) per capita for the European Union ($33,400) was significantly less than that of the United States ($47,000); but the disparity of wealth was also lower by about one-third than in the United States.

Some EU countries are notably wealthier than others (Figure 4.12, page 186). EU funds are raised through an annual 1.27 percent tax on the gross national product (GNP) of all members. Although financial allotments change from year to year, most member countries in North and West Europe contribute more than they receive back in allocations of revenue from the central government. Most EU countries in South Europe, Central Europe, and Southeast Europe on the other hand, tend to receive more than they contribute.

Since 1993, the EU agenda has expanded to include the creation of a common European currency, the defense of Europe's interests in international forums, and negotiation of EU-wide agreements on human rights and social justice. The European Union is also experimenting with various forms of political unification, including the creation of a common European military.

The official currency of the European Union is the **euro (€)**. Sixteen EU countries now use the euro: Austria, Belgium, Cyprus, Finland, France, Germany, Greece, Ireland, Italy, Luxembourg, Malta, the Netherlands, Portugal, Slovakia, Slovenia, and Spain. Countries that use the euro have a greater voice in the creation of EU economic policies, and the use of a common currency greatly facilitates trade, travel, and migration within the European Union.

economies of scale reductions in the unit cost of production that occur when goods or services are efficiently mass produced, resulting in increased profits per unit

euro (€) the official (but not required) currency of the European Union as of January 1999

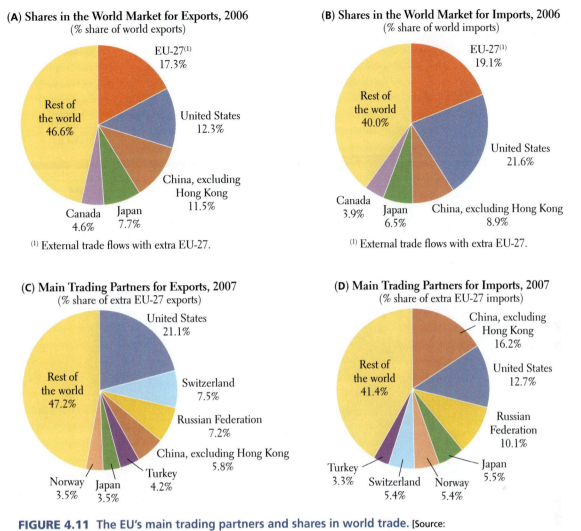

(A) Shares in the World Market for Exports, 2006
(% share of world exports)

- EU-27(1) 17.3%
- United States 12.3%
- China, excluding Hong Kong 11.5%
- Japan 7.7%
- Canada 4.6%
- Rest of the world 46.6%

(1) External trade flows with extra EU-27.

(B) Shares in the World Market for Imports, 2006
(% share of world imports)

- EU-27(1) 19.1%
- United States 21.6%
- China, excluding Hong Kong 8.9%
- Japan 6.5%
- Canada 3.9%
- Rest of the world 40.0%

(1) External trade flows with extra EU-27.

(C) Main Trading Partners for Exports, 2007
(% share of extra EU-27 exports)

- United States 21.1%
- Switzerland 7.5%
- Russian Federation 7.2%
- China, excluding Hong Kong 5.8%
- Turkey 4.2%
- Japan 3.5%
- Norway 3.5%
- Rest of the world 47.2%

(D) Main Trading Partners for Imports, 2007
(% share of extra EU-27 imports)

- China, excluding Hong Kong 16.2%
- United States 12.7%
- Russian Federation 10.1%
- Japan 5.5%
- Norway 5.4%
- Switzerland 5.4%
- Turkey 3.3%
- Rest of the world 41.4%

FIGURE 4.11 The EU's main trading partners and shares in world trade. [Source: *Europe in Figures: Eurostat Yearbook 2009* (Eurostat Statistical Books, Luxembourg: Office for Official Publications of the European Commission, 2009), pp. 388 and 392, Figures 10.5, 10.6, 10.9, and 10.10, at http://epp.eurostat.ec.europa.eu/cache/ITY_OFFPUB/KS-CD-09-001/EN/KS-CD-09-001-EN.PDF.]

All non-euro member states except Sweden and the United Kingdom have currencies whose value is determined by that of the euro. Depending on global financial conditions, either the euro or the U.S. dollar is the preferred currency of international trade and finance.

For the less affluent countries in South and Central Europe, the use of the euro has been a point of pride—proof that their economies are viable; but for unwary consumers it was a dangerous step. By 2005, both citizens and businesses had entered into credit arrangements with banks to buy luxuries or expand, and then, as the global recession hit and jobs were lost and markets shrank, the borrowers found themselves unable to keep up the payments. The result was bankruptcies, which are a bigger and more lingering disgrace in Europe than in the United States. Borrowers in Greece, Poland, Portugal, Italy, Spain, and Hungary were hardest hit. In 2010, EU financial authorities became aware that Greek officials had hidden a growing insolvency. The Greek debt was so large that the European Union was forced to bail out the country to the tune of nearly $1 trillion, in hopes that

this would save the value of the euro and prevent other heavily indebted members of the EU from defaulting. Aside from this very serious threat, overall the euro seems to have been a stabilizing influence during the recession. During the crisis in Greece, it temporarily sank in value relative to the U.S. dollar.

The European Union and Globalization The European Union is pursuing a number of strategies designed to ensure that it continues to be economically competitive with the United States, Japan, and the developing economies of Asia, Africa, and South America. A primary focus is on keeping exports a central component of national economies—Germany, France, and the United Kingdom are Europe's main exporters—but doing so in high-wage economies is difficult. One strategy is to relocate factories from the wealthiest EU countries to the relatively poorer, lower-wage member states. Until 2009, this strategy worked well, helping poorer European countries prosper while keeping the costs of doing business low enough to restrain European companies from moving outside Europe to developing countries where costs are

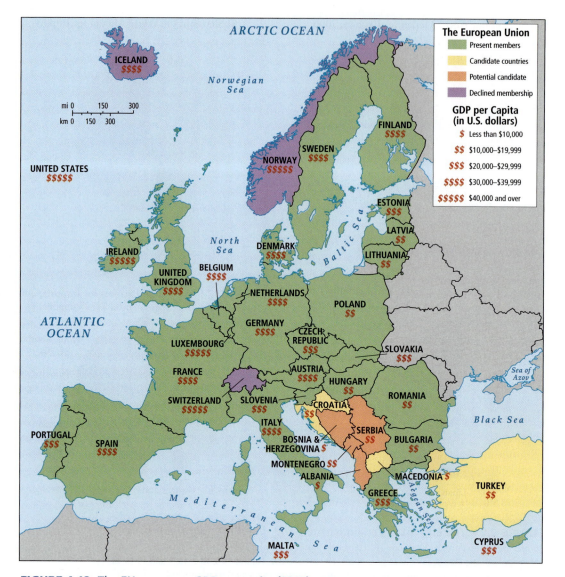

FIGURE 4.12 The EU economy: GDP per capita (2007). GDP per capita varies considerably among the 27 countries in the European Union. [Source: *Europe in Figures: Eurostat Yearbook 2009* (Eurostat Statistical Books, Luxembourg: Office for Official Publications of the European Commission, 2009), pp. 388 and 392, Figure 2.1 (page 75), at http://epp.eurostat.ec.europa.eu/cache/ITY_OFFPUB/KS-CD-09-001/EN/KS-CD-09-001-EN.PDF.]

lower still. However, the resulting reduction of industrial capacity (*deindustrialization*) in Europe's wealthiest countries led to higher unemployment for them. And the global recession that began in 2008 jeopardized export markets as world consumers slowed purchases.

Other strategies include holding down domestic wages and emphasizing the quality of European exports. Germany, the world's leading exporter (in value), accomplished both these feats well until the recession of 2009. Germany's exports, like those of all EU countries, fell by about 19 percent in 2009 and are likely to stay down into the future as competition increases from China and other developing economies. Germany, France, and the United Kingdom will try to encourage domestic consumption to take up the slack, but measures to encourage consumption, like tax cuts and government stimulus packages, are less likely due to fiscally conservative EU governments.

So how will EU countries maintain or increase their share of world trade, especially exports? Like the United States, the European Union presently has a large share of world trade (see Figure 4.11) and therefore exerts a powerful influence on the global trading system. The European Union often negotiates privileged access to world markets for European firms and farmers and for former European colonies, and some EU firms have moved abroad to cut costs. The European Union also employs protectionist measures that favor European producers by making goods from outside the European Union more expensive. These higher-priced goods create added expense for European consumers, but jobs and control over supplies are ensured. Of course, non-European producers shut out of EU markets are unhappy, and increasingly they have united to protest the European Union's failure to open its economies to foreign competition. So far such protests have met with little success.

NATO and the Rise of the European Union as a Global Peace-maker A new role for the European Union as a global peacemaker and peacekeeper is developing through the **North Atlantic Treaty Organization (NATO),** which is based in Europe. During the Cold War, European and North American countries cooperated militarily through NATO to counter the influence of the Soviet Union. NATO, established in 1949, included the United States, Canada, and ten countries of western Europe; it now includes almost all the EU countries as well.

Since the breakup of the Soviet Union, other nations came to assist the United States in the difficult task of addressing global security issues only after a major failure to avert a bloody ethnic conflict during the breakup of Yugoslavia in 1991. NATO has since focused mainly on providing the international security and cooperation needed to expand the European Union. When the United States invaded Iraq in 1993, most EU members opposed the war and as worldwide opposition to the United States built, the global status of the European Union was elevated. With the United States preoccupied with Iraq, NATO assumed more of a role as a global peacekeeper. NATO now commands the International Security Assistance Force (ISAF) from 46 countries, including all 28 NATO members in Afghanistan, but enthusiasm for this war is waning in Europe.

In addition to the major role they now play in Afghanistan, NATO and the European Union are also helping patrol the world ocean. In the spring of 2009 during the Somali pirate crisis off the northeast coast of Africa, NATO reported that both French and Portuguese naval vessels foiled attempts by pirates to seize merchant ships. Also in May of 2009, France announced that in accordance with its role in NATO, it had established a base in Abu Zaby (Abu Dhabi).

▶◀ **94. NATO's Future Role Debated**
▶◀ **95. NATO to Project Different Philosophy**
▶◀ **96. Concern over Common Values at the U.S.–EU Summit**
▶◀ **97. NATO Leaders, Putin Meet in Bucharest**

Food Production and the European Union

Europeans prefer food from European farms to imported food, and because of these sentiments they pay more for food than do people in the United States. Most food is now produced on large mechanized farms that are efficient: they require less labor and are more productive per acre. One result is that just over 2.35 percent of Europeans are now engaged in full-time farming, and one-fifth of these people are workers in Romania, where most holdings still rely on hand and animal power. Romania has over one-quarter of the farms in the European Union (France and Poland also have large agricultural sectors) and most are small. A second result of the mechanization of agriculture is that farmland has declined as a land use in Europe since the mid-1990s, while forestlands have increased.

The Common Agricultural Program (CAP) The drastic decline of labor and land in farming worries Europeans, who see it as endangering the goal of self-sufficiency in food. Toward this end, the European Union established its wide-ranging **Common Agricultural Program (CAP),** meant to guarantee secure and safe food supplies at affordable prices. The CAP aids farmers by placing tariffs on imported agricultural goods and by giving **subsidies**

(payments to farmers) to underwrite their costs of production. It was this provision that the Smithfield company (with its European partners) took advantage of in setting up its giant pig farms in Romania (recall the opening vignette). Subsidies are expensive—payments to farmers are the largest expense category in the EU budget. While these policies do ensure a safe and sufficient food supply and provide a decent living standard for farmers, they also effectively raise food costs for millions of consumers.

Protective agricultural policies, like tariffs and subsidies—also found in the United States, Canada, and Japan—are unpopular in the developing world. Tariffs lock farmers in poorer countries out of major markets. Subsidies encourage overproduction in rich countries (in order to collect more payments). The result is a glut of farm products that are then sold cheaply on the world market, as is the case with Smithfield pork. This practice, called *dumping,* lowers global prices and thus hurts farmers in developing countries while it aids those in developed countries by keeping their supplies low and prices high.

North Atlantic Treaty Organization (NATO) a military alliance between European and North American countries that was developed during the Cold War to counter the influence of the Soviet Union; since the breakup of the Soviet Union, NATO has expanded membership to include much of eastern Europe and Turkey, and is now focused mainly on providing the international security and cooperation needed to expand the European Union

Common Agricultural Program (CAP) a European Union program, meant to guarantee secure and safe food supplies at affordable prices, that places tariffs on imported agricultural goods and gives subsidies to EU farmers

subsidies monetary assistance granted by a government to an individual or group in support of an activity, such as farming, that is viewed as being in the public interest

Growth of Corporate Agriculture and Food Marketing As small family farms disappear in the European Union, just as they did several decades ago in the United States, the trend is toward consolidating smaller farms (see Thematic Concepts Part M, on page 165) into larger, more profitable operations run by European and foreign corporations (see Thematic Concepts Part N on page 165 and Photo Essay 4.2C on page 169). These farms tend to employ very few laborers and use more machinery and chemical inputs.

The move toward corporate agriculture is strongest in Central Europe. When communist governments gained power in the mid-twentieth century, they consolidated many small, privately owned farms into large collectives. After the breakup of the Soviet Union, these farms were rented to large corporations, which in turn further mechanized the farms and laid off all but a few laborers. Rural poverty rose and small towns shrank as farmworkers and young people left for the cities. Many fear that without the EU's generous subsidies to farmers, a similar pattern could sweep across the European Union.

Green Food Production During the communist era, Slovenia was unlike most of the rest of Central Europe in that the farms were not collectivized. Therefore, it has not been necessary to redistribute land. Instead, the problem is that farms are too small for efficient production: the average farm size is just 8.75 acres (3.5 hectares). Although Slovenia has plenty of rich farmland, as standards of living have risen it has become a net importer of food, mostly from EU countries such as Italy, Spain, and Austria. Nonetheless, Slovenia's new emphasis on private entrepreneurship, combined with a growing demand throughout Europe for organic foods, has encouraged some Slovene farmers to carve out a niche for themselves in local markets. The case of Vera Kuzmic is illustrative.

Personal Vignette Vera Kuzmic (a pseudonym) lives 2 hours by car south of Ljubljana, Slovenia's capital. Her family has farmed 12.5 acres (5 hectares) of fruit trees near the Croatian border for generations. After Slovenia became independent in 1991, Vera and her husband lost their government jobs due to economic restructuring. The Kuzmic family decided to try earning its living in vegetable market gardening because vegetable farming could be more responsive to market changes than fruit tree cultivation. By 2000, the adult children and Mr. Kuzmic were working on the land, and Vera was in charge of marketing their produce and that of neighbors whom she had also convinced to grow vegetables.

Vera secured market space in a suburban shopping center in Ljubljana, where she and one employee maintained a small vegetable and fruit stall (**Figure 4.13**). Her produce had to compete with much less expensive Italian-grown produce sold in the same shopping center—all of it produced on large corporate farms in northern Italy and trucked in daily. But Vera gained market share by bringing her customers special orders and by guaranteeing that only animal manure, no pesticides or herbicides, was used on the fields. For a while, her special customer services and her organically grown produce kept her in business. But when Slovenia joined the European Union in 2004, she had to do more to compete with produce growers and marketers across Europe.

Anticipating the challenges to come, the Kuzmics' daughter Lili completed a marketing degree at the University of Ljubljana. The family incorporated their business, and Lili is now its Ljubljana-based director, while Vera manages the farm. Lili's market research shows that it would be wisest to diversify. The Kuzmics continue to focus on Ljubljana's expanding professional population, who are willing to pay extra for fine organic vegetables and fruits. But now, in a recently built banquet facility on the farm, Vera also prepares

special dinners for bus-excursion groups interested in traditional Slovene dishes made from homegrown organic crops. *[Source: Lydia Pulsipher's conversations with Vera Kuzmic and Dusan Kramberger, 1993 through 2009.]* ■

Europe's Growing Service Economies

As industrial jobs have declined across the region, most Europeans (about 70 percent) have found jobs in the service economy. *Services* such as the provision of health care, education, finance, tourism, and information technology are now the engine of Europe's integrated economy, drawing hundreds of thousands of new employees to the main European cities. For example, financial services located in London and serving the entire world play a huge role in the British economy, and many transnational companies are headquartered in London.

A major component of Europe's service economy is *tourism*. Europe is the most popular tourist destination in the world, and one job in eight in the European Union is related to tourism. Tourism generates 13.5 percent of the EU's gross domestic product and 15 percent of its taxes, although this varies with global economic conditions. Europeans are themselves enthusiastic travelers, visiting one another's countries as well as many exotic world locations frequently. This travel is made possible by the long paid vacations—usually 4 to 6 weeks—that Europeans are granted by employers. The most popular holiday destinations among EU members in 2006 included Austria, Ireland, and the traditional Mediterranean destinations of Cyprus, Malta, Spain, and Italy (2005).

Service occupations increasingly involve the use of technology. Europe has lagged behind North America in the development and use of personal computers and the Internet (**Figure 4.14,** pie diagram inset). However, the information economy is advanced in Europe, with West Europe leading the way. In fact, Europe leads the world in cell phone use. In South Europe and Central Europe, where personal computer ownership is lowest, public computer facilities in cafés and libraries are common, and surfing the Internet is popular among schoolchildren and grandmothers alike.

FIGURE 4.13 Vera Kuzmic in her market stall in Ljubljana.

THINGS TO REMEMBER

1. The European Union's original plan was to reach a level of economic and social integration that would make possible the free flow of goods and people across national borders; for the most part, those goals have been achieved among the current 27 members.

2. The European Union has one executive branch—the European Commission—and two legislative branches—the European Parliament, directly elected by EU citizens, and the Council of the European Union, whose members consist of one minister from each EU country.

3. The European Union joined the members' national economies into a common market. By 2008 the EU's economy was almost $15 trillion (PPP), about 5 percent larger than that of the United States, making the European Union the largest economy in the world.

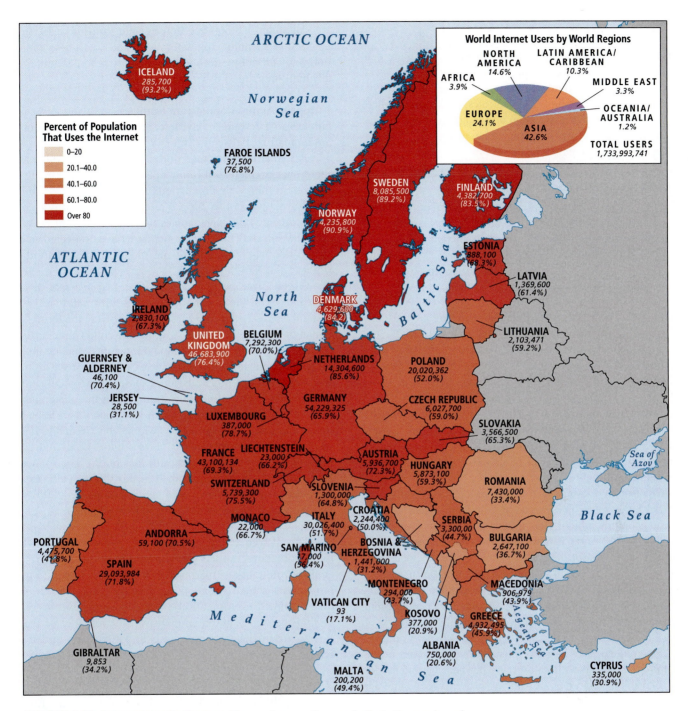

FIGURE 4.14 Internet use in Europe. The numbers on the map indicate the number of Internet users as of September 30, 2009. The percentages indicate the percent of the population that uses the Internet. Compare European Internet use with that in Middle and South America, which is shown in Figure 3.16. [Source: http://www.internetworldstats.com/stats4.htm.]

4. A new role for the European Union as a global peacemaker and peacekeeper is developing through the North Atlantic Treaty Organization (NATO), which is based in Europe.

5. Europeans prefer food from European farms to imported food, and because of this they pay more for food than do people in the United States.

6. Throughout the European Union, smaller family-run farms are giving way to larger corporate farms, especially in the recently admitted EU states in Central Europe.

7. As industrial jobs have declined across the region, most Europeans (about 70 percent) have found jobs in the service economy.

Sociocultural Issues

The European Union was conceived primarily to promote economic cooperation and free trade, but its programs have social implications as well. As population patterns change across Europe, attitudes toward immigration, gender roles, and social welfare programs are also evolving. Religion and language, once divisive issues in the region, are now fading as a focus of disputes. Immigration, however, is an increasing source of tension.

Population Patterns

Population patterns in Europe foretell processes that are emerging around the world. Europe's high population density and urbanization trends are long-standing. A newer phenomenon, the aging of the populace, is well advanced in Europe and affects all manner of social policies. Europe's issues with aging are common to Russia and its neighbors, and to Japan, China, and India.

Population Distribution and Urbanization There are currently about 525 million Europeans. Of these, 492 million live within the European Union. The highest population densities stretch in a discontinuous band from the United Kingdom south and east through the Netherlands and central Germany into northern Switzerland (Photo Essay 4.5 on page 193). Northern Italy is another zone of high density, along with pockets in many countries along the Mediterranean coast. Overall, Europe is one of the more densely occupied regions on earth, as shown on the population density map in Photo Essay 1.1 on pages 16–17. Most of this population now lives in cities.

Today, Europe is a region of cities surrounded by well-developed rural hinterlands. These cities are the focus of the modern European economy, which, though long grounded in agriculture, trade, and manufacturing, is now primarily service based. In West, North, and South Europe, more than 75 percent of the population lives in urban areas. Even in Central Europe, the least urbanized part of the region, around 70 percent of people live in cities. As noted earlier, many European cities began as trading centers more than a thousand years ago and still bear the architectural marks of medieval life in their historic centers (Photo Essay 4.6A on page 194). These old cities are located either on navigable rivers in the interior or along the coasts because water transport figured prominently (and still does) in Europe's trading patterns (see Figure 4.8 on page 178).

Since World War II, nearly all the cities in Europe have expanded around their perimeters in concentric circles of apartment blocks (Photo Essay 4.6B). Usually well-developed rail and bus lines link the blocks to one another and to the old central city. Land is scarce and expensive in Europe, so only a small percentage of Europeans live in single-family homes, although the number is growing. Even single-family homes tend to be attached or densely arranged on small lots. Except for public parks, which are common, one rarely sees the sweeping lawns that many North Americans are accustomed to. Publicly funded transportation is widely available, so many people live without cars in apartments near city centers (Photo Essay 4.6C, D). However, many others commute by car daily from suburbs or ancestral villages to work in nearby cities.

Although deteriorating housing and slums do exist, substantial public spending (on sanitation, water, utilities, education, housing, and public transportation) help most maintain a generally high standard of urban living.

Europe's Aging Population Europe's population is aging as families are choosing to have fewer children and life expectancies are increasing. Between 1960 and 2009, the proportion of those 14 years and under declined from 27 percent to 15 percent, while those over 65 increased from 9 percent to 16 percent. By comparison, in the global population, the overall share of young persons was 27 percent in 2009, while older generations accounted for 8 percent. Life expectancies now range close to 80 years in North, West, and South Europe, and this is reflected in urban landscapes, where the elderly are more common than children (Photo Essay 4.6C).

Overall, Europe is now close to a negative rate of natural increase (<0.0), the lowest in the world. Birth rates are lowest in Central Europe—the countries that were part of the former Soviet Union or Yugoslavia. Increasingly, the one-child family is common throughout Europe, except among immigrants, who are the major source of population growth. However, once they have assimilated to European life, immigrants, too, choose to have small families.

The declining birth rate is illustrated in the population pyramids of European countries, which look more like lumpy towers than pyramids. The population pyramid of Sweden and of the whole European Union are examples (Figure 4.15B and C). The pyramid's narrowing base indicates that for the last 35 years, far fewer babies have been born than in the 1950s and 1960s, when there was a post-war baby boom across Europe. By 2000, twenty-five percent of Europeans were choosing to have no children at all.

The reasons for these trends are complex. For one thing, more and more women want professional careers. This alone could account for late marriages and lower birth rates; 25 percent of German women are choosing to remain unmarried well into their thirties. Governments also make few provisions for working mothers beyond paid maternity leave. In Germany, for example, there is insufficient day care available for children under 3, and school days run from 7:30 A.M. to noon or 1:00 P.M. Many German women choose not to become mothers because they would have to settle for part-time jobs in order to be home by 1:00 P.M. To encourage higher birth rates, there is a move within the European Union to give one parent (mother or father) a full year off with reduced pay after a child is born or adopted, provide better pre-school care, lengthen school days, and serve lunch at school (see discussion of gender on pages 197–198).

As a result of the low birth rate and long life expectancy, the vast majority of Germans are older than 30 (see Figure 4.15A), and those above 80 are the fastest-growing age group. As older people die, the population may eventually settle into a stable age structure, as Sweden's population has already done (see Figure 4.15B). In this structure, each age category has roughly similar numbers of people, tapering only at the top after age 65.

A stable population with a low birth rate has several consequences. Because fewer consumers are being born, economies may contract over time. Demand for new workers, especially

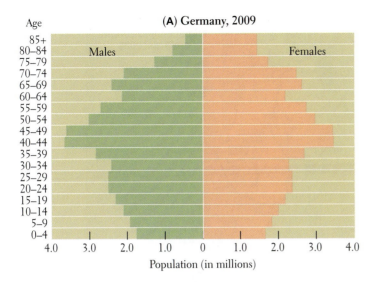

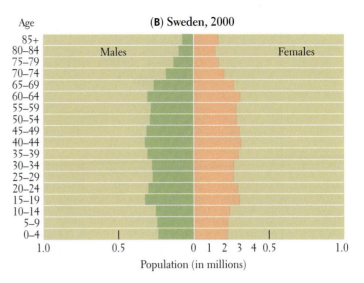

FIGURE 4.15 Population pyramids for Germany, Sweden, and the EU in 2009–2010. These population pyramids have quite different shapes, but all exhibit a narrow base, indicating low birth rates. The EU pyramid (**C**) shows the percent of the total population, not the numbers of people. [Adapted from "Population Pyramids of Germany" and "Population Pyramids of Sweden," U.S. Census Bureau International Data Base, 2009, at http://www.census.gov/ipc/www/idb/country.php; and from *Europe in Figures: Eurostat Yearbook 2009* (Luxembourg: Eurostat, 2009), p. 139, Figure 3.5, at http://epp.eurostat.ec.europa.eu/portal/page/portal/publications/eurostat_yearbook.]

highly skilled ones, may go unmet, though retraining the elderly in new skills and encouraging them to work until age 70 or 72 is being tried in some countries. Further, the number of younger people available to provide expensive and time-consuming health care for the elderly, either personally or through tax payments, will decline. Currently, for example, there are just two German workers for every retiree. Immigration provides one solution to the dwindling number of young people. However, Europeans are reluctant to absorb large numbers of immigrants, especially from distant parts of the world where cultural values are very different from those in Europe.

Immigration and Migration: Needs and Fears

Until the mid-1950s, the net flow of migrants was out of Europe, to the Americas, Australia, and elsewhere. By the 1990s the net flow was into Europe. In the 1990s, most of the European Union (plus Iceland, Norway, and Switzerland) implemented the **Schengen Accord**, an agreement that allows free movement of people and goods across common borders. The accord has facilitated trade, employment, tourism, and most controversially,

migration within the European Union. The Schengen Accord has also indirectly increased both the demand for immigrants from outside the European Union and their mobility once they are in the European Union (Figure 4.16).

📺 **92. After 50 years, Europe Still Coming Together**

Schengen Accord an agreement signed in the 1990s by the European Union and many of its neighbors that called for free movement across common borders

Attitudes Toward Internal and International Migrants and Citizenship Like citizens of the United States, Europeans have ambivalent attitudes toward migrants. The internal flow of migration is mostly from Central Europe into North, West, and South Europe. These Central European migrants are mostly treated fairly, although prejudices against the supposed backwardness of Central Europe are still evident, and despite the Schengen Accord, people can be made to feel unwelcome. Europeans are especially sensitive to the nuances of cultural difference—body language, dress, pronunciation—that can telegraph a person's origins; they modify their relationships accordingly. These tendencies toward exclusion are moderating as skilled and

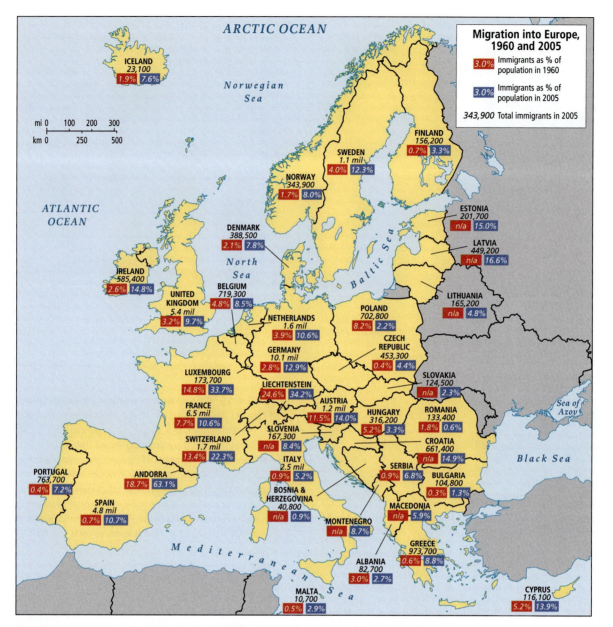

FIGURE 4.16 Migration into Europe, 1960 and 2005. Migration into Europe increased from 1960 through the 1990s and continued to increase into the twenty-first century for most countries. Migration continues to be a crucial issue in EU debates. The percentages of total population are shown in the two color boxes for each country: the red box for 1960 and the blue box for 2005. The number of migrants living in each country in 2005 is above the two boxes. [Source: *United Nations Human Development Report 2009* (New York: United Nations Development Programme), at http://hdrstats.undp.org/en/indicators/6.html.]

educated Europeans move into neighboring countries, often creating their own social circles of high-status outsiders. Immigrants from outside Europe, so-called *international immigrants*, meet with varying acceptance.

International immigrants often come legally and illegally from Europe's former colonies and protectorates across the globe. Many Turks and North Africans

guest workers legal workers from outside a country who help fulfill the need for temporary workers but who are expected to return home when they are no longer needed

come legally as **guest workers** who are expected to stay for only a few years, fulfilling Europe's need for temporary workers in certain sectors. Other immigrants are refugees from the world's trouble spots, such as Afghanistan, Iraq, Haiti, and Sudan. Many also come illegally from all these areas.

Some Europeans see international immigrants as important contributors to their economies, providing needed skills and making up for the low birth rates. Others, however, are alarmed by recent increases in immigration. Studies of public attitudes in

Europe's population is not growing, a fact that raises concerns about future economic conditions as the population ages. Many governments now encourage large families, with generous maternity and paternity leave (up to 10 months with full pay), free day care, and other incentives. However, few countries have seen much change in their population growth rate. A major reason is that more women now are working or want to work than ever before, so childbearing is being delayed and families remain small. As this trend shows no signs of decreasing, some officials are looking to immigration as a possible solution. While Europe has been attracting large numbers of immigrants for decades now, many Europeans are wary of hosting large populations of foreigners who might not share their values.

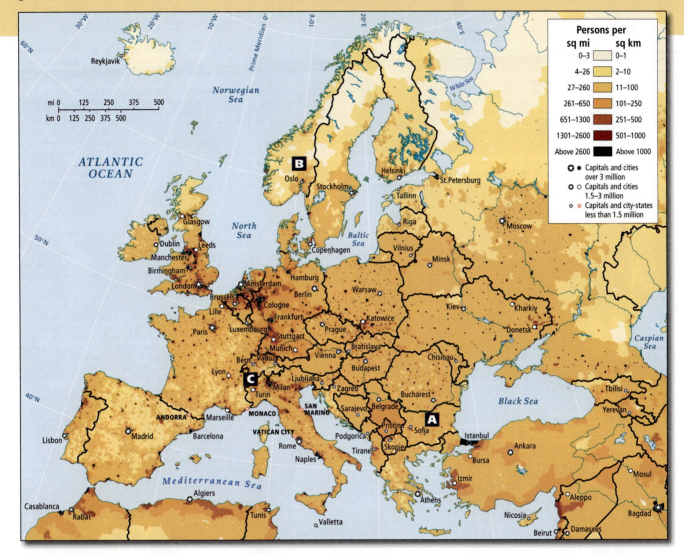

Persons per	
sq mi	sq km
0–3	0–1
4–26	2–10
27–260	11–100
261–650	101–250
651–1300	251–500
1301–2600	501–1000
Above 2600	Above 1000

✪ ● Capitals and cities over 3 million
✪ ○ Capitals and cities 1.5–3 million
○ ○ Capitals and city-states less than 1.5 million

A Children in Bulgaria, where despite government incentives for childbearing, the population is expected to shrink by 35 percent by 2050. Bulgaria's population decline is the most severe in the world.

B Norwegian politician, diplomat, physician, and international leader on sustainable development, Gro Harlem Brundtland. As prime minister of Norway, she enacted policies designed to make it easier for working women to have children.

C West African immigrants working as street vendors in Turin, Italy. Illegal immigration from Africa has become a major political issue in Italy.

Europe's cities are world famous for their architecture, economic dynamism, and cultural vitality. Many are quite ancient but have expanded in recent decades with large apartment blocks connected to the old city centers by public transport. Despite their global draw, many European cities will shrink in the next several decades, due to declining national populations (as in Italy and parts of Central Europe), as well as to deliberate efforts to shift urban growth to other urban centers, as in the case of London.

A Shown here is the old urban core of Prague, Czech Republic, founded in the 9th century c.e. on the banks of the Vltava river.

B Hi-rise apartment blocks dot the periphery of Berlin's old urban core.

C La Rambla, a tree-lined pedestrian mall in Barcelona, Spain, is a popular stroll among locals and tourists.

D A tram in Paris, France. The city's metropolitan transit system moves roughly 6 million people per day and is the second largest in Europe after London's.

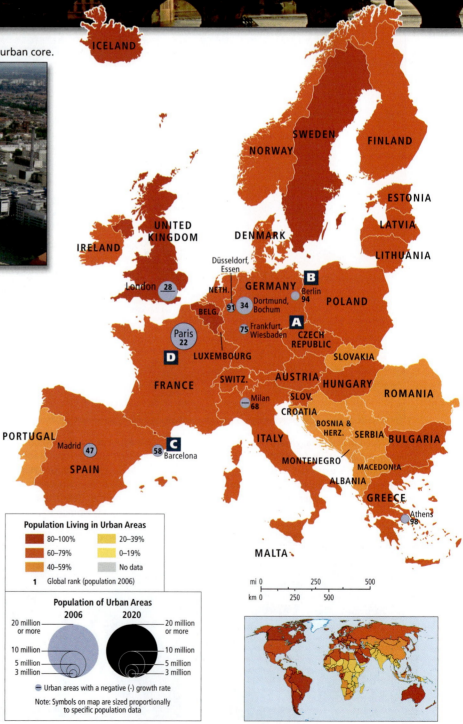

Population Living in Urban Areas

80–100%	20–39%
60–79%	0–19%
40–59%	No data

1 Global rank (population 2006)

Population of Urban Areas

2006	2020
20 million or more	20 million or more
10 million	10 million
5 million	5 million
3 million	3 million

● Urban areas with a negative (-) growth rate

Note: Symbols on map are sized proportionally to specific population data

mi 0   250   500
km 0   250   500

Europe show that immigration is least tolerated in areas where incomes and education are low, suggesting that many people fear the unknown and worry that an influx of migrants may drive down wages further or inhibit their own social mobility. Central and South Europe are the least tolerant of new immigrants. North and West Europe, with higher incomes and generally more stable economies, are the most tolerant.

Cultural issues also influence attitudes toward immigrants. Indeed, the presence of so many new cultural groups in Europe raises some thorny questions. Is Germany no longer a German place? Should Islam, the religion of many immigrants, have equal footing with Christianity? How will Europeans cope with unfamiliar value systems and family types and modes of dress?

Across Europe, anti-immigration views, especially toward non-Europeans, are becoming more common. Mainstream politicians increasingly support stricter controls. In Germany, Austria, Italy, and France, far-right political parties now openly advocate forcing non-European immigrants, legal or illegal, to leave. In 2009, the Swiss (Switzerland is not an EU member), in a tangential blow to religious freedom, voted to not allow the future construction of minarets on Muslim mosques. Some countries are also limiting the number of migrants from other EU countries, especially those from Central Europe. In response, the European Union is increasing its efforts to curb illegal immigration from outside Europe while at the same time helping EU citizens to be more tolerant.

An incident in France shows how humor can diffuse some of the tensions surrounding immigration. In 2005, French nationalists complained in the press that Polish plumbers were taking French jobs. (In fact, there was a significant shortage of plumbers in France.) The "Polish plumber" quickly became shorthand across Europe for issues related to immigration and jobs. Poland responded by featuring on its tourism posters a "hunky" male model posing as a plumber (Figure 4.17) and saying seductively, "I'm staying in Poland; won't you come over?"

Anti-foreigner sentiment has been a hindrance to acquiring citizenship across Europe, but as the advantages of a multicultural, multiracial, multigenerational Europe become more apparent, requirements for citizenship are gradually being relaxed. Until recently in Germany, the German-born children and grandchildren of Turkish or North African immigrant workers were not considered citizens. But as of January 2000, all children born in Germany since 1975 are citizens. The UK, probably the most multicultural of all European countries, recently granted full citizenship to several hundred thousand immigrants from former colonies in the Caribbean, South Asia, and Africa. In 2004, more than 650,000 became citizens of an EU country, with the largest numbers in Germany, France, and the UK. Citizenship usually requires an extended period of legal residency, evidence of a good work record, and proficiency in the country's main language. If an eligible migrant seeks European citizenship, this is viewed as evidence of formal assimilation.

Rules for Assimilation: Muslims in Europe In Europe, culture plays a larger role in defining differences between people than race and skin color. An immigrant from Asia or Africa may be accepted

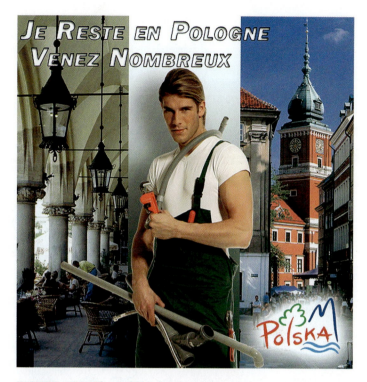

FIGURE 4.17 The Polish plumber. This poster, inviting the French to meet the plumber in Poland, became an emblematic reaction to anti-immigrant sentiment.

into the community if he or she has gone through a comprehensive change of lifestyle. **Assimilation** in Europe usually means giving up the home culture and adopting the ways of the new country. If minority groups—such as the Roma, who have been in Europe for thousands of years—maintain their traditional ways, it is nearly impossible for them to blend into mainstream society. And many in the larger society believe the Roma, for example, do not want to assimilate.

> **assimilation** the loss of old ways of life and the adoption of the lifestyle of another country

101. The Art of Integration in Germany

As a prime example of assimilation, consider the case of Londoner Iqbal Wahhab, the son of Bangladeshi immigrants, whose restaurant called *Roast* is aiming to revive interest in what he admits is rather bland eighteenth-century British cuisine. He created *Roast* as a reaction against *fast food* and the safety issues posed by the globalization of food. Speaking as an assimilated Brit, Wahhab talks of the value of using local ingredients and of the need to counteract the loss of British tradition as multiculturalism spreads through the country.

Muslims, presently the focus of assimilation issues in the European Union, have lived in Europe in small pockets for well over 1000 years. Like Mr. Wahhab, these Muslims have assimilated thoroughly to, and identify with, their home countries. For these long-standing Europeans, religion is not their main identity; rather they think of themselves as French or Swiss or Bosnian while occupation, gender, and social class are perhaps more important features of their identity.

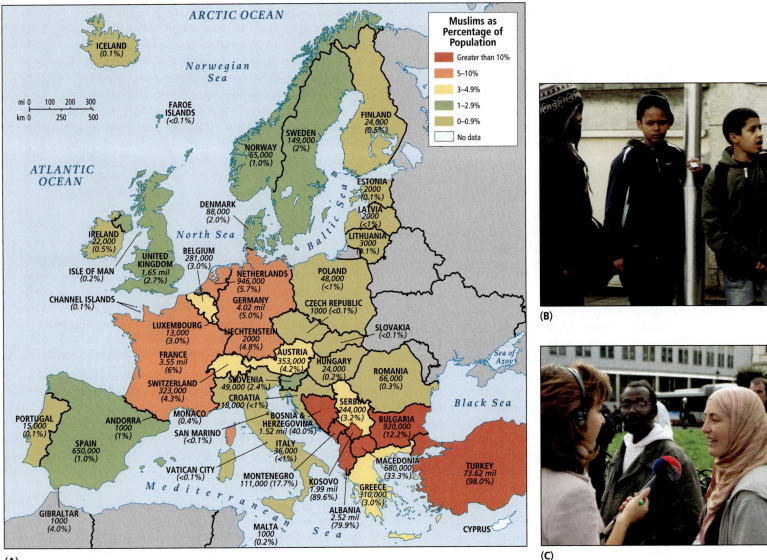

FIGURE 4.18 Muslims in Europe. (A) Muslims are a small minority in most European countries, but many intend to make the region their permanent home. Just how, or if, the assimilation of Muslims into largely secular Europe will proceed is a major topic of public debate. [Source: *Mapping the Global Muslim Population*, The Pew Forum on Religion & Public Life, Pew Research Center, October 2009, pp. 31–32, at http://pewforum.org/newassets/images/reports/Muslimpopulation/Muslimpopulation.pdf.] **(B)** Who is French? Youths shown here are waiting for a bus in a Paris suburb. There is a growing recognition that French society discriminates against immigrants and their children, especially when it comes to hiring practices. The unemployment rate is 9 percent for those of French ancestry, but 14 percent for those of foreign ancestry. NGOs like SOS Racisme have documented systematic discrimination in private sector hiring practices. The tensions periodically explode into violent episodes, such as the riots of 2005, which occurred throughout France, especially in the poorer suburbs that are home to most immigrants and children of immigrants. **(C)** A Muslim woman is interviewed during a demonstration in Brussels, Belgium, by a group calling itself Stop the Islamization of Europe. She is Belgian and has converted to Islam.

Europe's small but growing Muslim immigrant population (Figure 4.18) comes from a wide range of places and cultural traditions: North Africa, Turkey, Central Asia, Pakistan, Bangladesh, Indonesia, the Guianas. Some have clung to traditional dress, gender roles, and religious values, while others have accepted and practice European culture. The deepening alienation that has boiled over in recent years among some Muslim immigrants and their children—resulting in protests, riots, and occasional terrorism—relates primarily to the systematic exclusion of these less-assimilated Muslims from meaningful employment, from social services, and from higher education, which are seen correctly by immigrant youth as the paths to success.

The riots that broke out in Paris in the fall of 2005 illustrate this point. Young people of North African descent were not protesting on behalf of Islam, but rather because of their lack of access to higher education, jobs, and housing. Subsequent investigations by the French media revealed that their complaints were indeed legitimate. However, many Europeans, unaware of the context,

double day the longer workday of women with jobs outside the home who also work as caretakers, housekeepers, and/or cooks for their families

viewed the Paris riots as linked to deadly terrorist bombings in Madrid in 2004 and London in 2005. Both of those violent events were carried out by young Muslims, some of them born in Europe, protesting British and Spanish involvement in the war in Iraq.

In some cases conflicts have also arisen over European perceptions of cultural aspects of Islam, such as the *hijab* (one of several traditional coverings for women). For example, in France in 2004, wearing of the hijab by observant Muslim schoolgirls became the center of a national debate about civil liberties, religious freedom, and national identity. French authorities wanted to ban the hijab but were wary of charges of discrimination. Eventually the French declared all symbols of religious affiliation illegal in French schools (including crosses and yarmulkes, the Jewish head covering for men), a move that resulted in widespread protest among Muslims throughout the world. In 2006, movements to ban Muslim dress arose in the United Kingdom and the Netherlands but did not gain popular support. In France the issue persisted; in 2009 a French minister of government of Algerian descent openly declared that as a feminist, she viewed most Muslim religious garb for women as an affront to European values. In a related case, crucifixes were banned in Italian public schools in 2009.

📺 **102. Religious Tolerance Facing Test in Britain**

📺 **103. London's Mega Mosque Stirs Controversy**

Changing Gender Roles

Gender roles in Europe have changed significantly from the days when most women married young and worked in the home or on the family farm. Increasing numbers of European women are working outside the home (though the percentages vary considerably among different parts of the region), and the percentage of women in professional and technical fields is growing (Figure 4.19). Nevertheless, European public opinion among both women and men largely holds that women are less able than men to perform the types of work typically done by men, and that men are less skilled at domestic, caregiving, and nurturing duties. In most places, men have greater social status, hold more managerial positions, earn higher pay, and have greater autonomy in daily life than women—more freedom of movement, for example. These male advantages have a stronger hold in Central and South Europe today than they do in West and North Europe.

Although younger men are now assuming more domestic duties, women who work outside the home usually face what is called a **double day** in that they are expected to do most of the domestic work in the evening in addition to their job outside the home during the day. United Nations research shows that in most of Europe, women's workdays, including time spent in housework and child care, are 3 to 5 hours longer than men's. (Iceland and Sweden reported that women and men there share housework equally.) Women burdened by the double day generally operate with somewhat less efficiency in a paying job than do men. They also tend to choose employment that is closer to home and that offers more flexibility in the hours and skills required. Compared to typical male jobs, these more flexible jobs (often erroneously classified as part-time) almost always offer lower pay and less opportunity for advancement, though not necessarily fewer working hours.

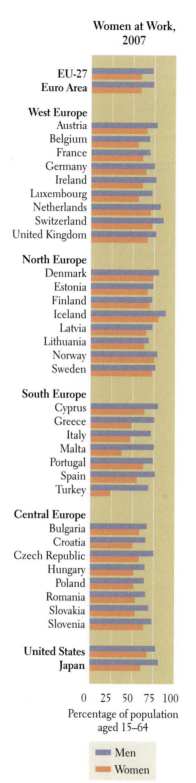

Women at Work, 2007

Percentage of population aged 15–64

■ Men
■ Women

FIGURE 4.19 Women at work in Europe (and selected other countries for comparison), 2007. A majority of women in Europe work outside the home, and their numbers are increasing. The exceptions are in Greece, Italy, and Malta. Included for comparison is Turkey, where slightly less than 25 percent of women work outside the home. A related trend is declining birth rates in these countries.
[Source: *Europe in Figures: Eurostat Yearbook 2009* (Luxembourg: Eurostat, 2009), p. 27, Figure 7.3: Employment Rate by Gender, 2007, at http://epp.Eurostat.ec.europa.eu/portal/page/portal/publications/eurostat_yearbook.]

Many EU policies encourage gender equality. Managers in the EU bureaucracy are increasingly female, and well over half the university graduates in Europe are now women. However, the political influence and economic well-being of European women lag behind those of European men. As **Figure 4.20** illustrates, in most European national parliaments, women make up less than a third of elected representatives. Only in North Europe do women come anywhere close to filling 50 percent of the seats in the legislature, and there several women have served as heads of government. In West Europe, the United Kingdom has had numerous women officials in high office, but elsewhere in the region this trend is only beginning. In 2009, Germany reelected Angela Merkel as its first woman chancellor (prime minister), and in France since the 2007 election of President Sarkozy (he defeated a female opponent), women have occupied a number of French cabinet-level positions. Although change is clearly underway in the European Union, women generally serve only in the lower ranks of government bureaucracies in positions that deprive them of a strong voice in the formation of national policies.

Because women are largely absent from policy-making positions, their progress has been slow on many fronts. For example, in 2006, female unemployment was higher than male unemployment in all but a few countries (the United Kingdom, Germany, the Baltic Republics, Ireland, Norway, and Romania), where the differences were slight, and 32 percent of women's jobs were part time, as opposed to only 7 percent of men's jobs. Throughout the European Union, women are paid less than men for equal work (see Figure 4.22C on page 201), despite the fact that young women tend to be more highly educated than young men.

Norway leads Europe in redefining gender in society. It directs much of its most innovative work on gender equality toward advancing men's rights in traditional women's arenas. For example, men now have the right to at least a month of paternity leave with a newborn or adopted child. This policy recognizes a father's responsibility in child rearing and provides a chance for father and child to bond early in life. Even when addressing male violence against women, Norway recognizes in its public discussions that this global problem stems in part from cultural customs that suffocate sensitivity in individual males. Male children's games that include derision of boys who are not violent are now openly discouraged in Norway.

At the same time, equal rights for women are closer to being a reality in Norway than in any other country. In 1986, Gro Harlem Brundtland, a physician, became the country's first female prime minister, and appointed women to 44 percent of all cabinet posts (see Photo Essay 4.5B on page 193). Norway became the first country in modern times to have such a high proportion of women in important government policy-making positions. By 2009, women made up 36 percent of Norway's parliament (Sweden had 47 percent; Finland had 42 percent; see Figure 4.20). They also held nearly 33 percent of elected positions in municipal government and more than 40 percent of posts in county government. The growing policy-making power of women is perhaps best seen in the fact that by the 1990s, women were already leaders of Norway's three major political parties.

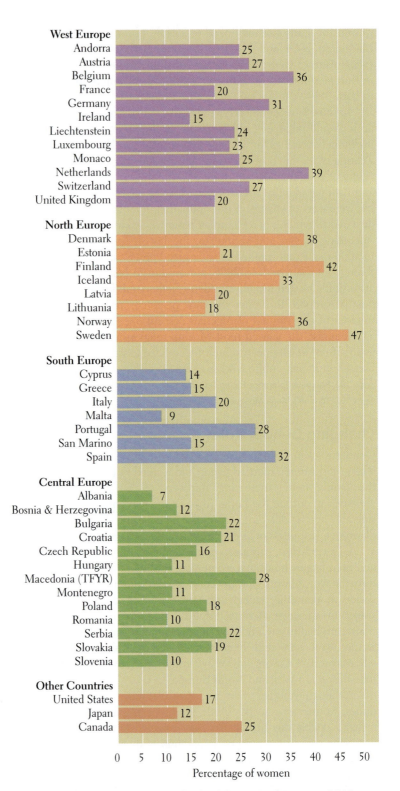

FIGURE 4.20 Women in the legislatures of Europe, 2009. Women comprise about half the adult population of the EU, but they have nowhere near their fair share of representation in European legislatures; hence, their influence on legislation is seriously restricted. Notice that some regions are closer to equity than others, and also notice how the United States compares. [Source: *United Nations Human Development Report 2009* (New York: United Nations Development Programme), pp. 186–189, Table K, "Gender Empowerment Measure and Its Components," at http://hdr.undp.org/en/statistics/data/hdi2009/.]

Politics is not the only area in which Norway seeks gender equality. It was the first country in the world to have an ombudsperson whose job is to enforce the 1979 Equal Status Act. This act has a "60–40 rule" that requires a minimum of 40 percent representation by each sex on all public boards and committees. Even though the 60–40 rule has not yet been achieved in all cases, female representation averages over 35 percent for state and municipal agencies. In addition, employment ads are required to be gender neutral, and all advertising must be nondiscriminatory.

Social Welfare Systems and Their Outcomes

In nearly all European countries, tax-supported systems of **social welfare** or **social protection** (the EU term) provide all citizens with basic health care; free or low-cost higher education; affordable housing; old age, survivor, and disability benefits; and generous unemployment and pension benefits. Europeans generally pay much higher taxes than North Americans (the rate for EU countries is about 40 percent of GDP; for the United States, 27 percent; and for Canada, 30 percent), and in return they expect more in services. Nonetheless, the European Union spends on average about $3000 per person for health care, while the United

States spends more than $7000. Surprisingly, the European Union has more doctors and acute-care hospital beds per citizen (but less than half as much expensive medical equipment) and better outcomes than the United States in terms of life expectancy and infant mortality.

Europeans do not agree on the goals of these welfare systems, or on just how generous they should be. Some argue that Europe can no longer afford high taxes if it is to remain competitive in the global market. Others maintain that Europe's economic success and high standards of living are the direct result of the social contract to take care of basic human needs for all. The debate has been resolved differently in different parts of Europe, and the resulting regional differences have become a source of concern in the European Union. With open borders, unequal benefits can encourage those in need to flock to a country with a generous welfare system and overburden the taxpayers there.

European welfare systems can be classified into four basic categories (Figure 4.21). *Social democratic welfare systems*, common

social welfare (in the European Union, **social protection**) in Europe, tax-supported systems that provide citizens with benefits such as health care, pensions, and child care

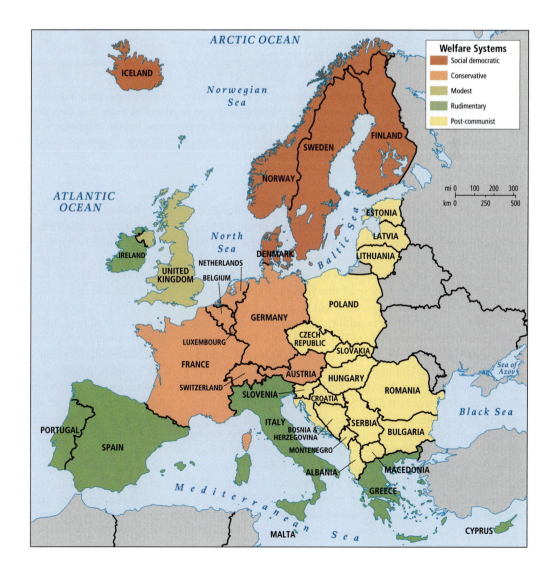

FIGURE 4.21 European social welfare/protection systems. The basic categories of social welfare systems shown here and described in the text should be taken as only an informed approximation of the existing patterns.

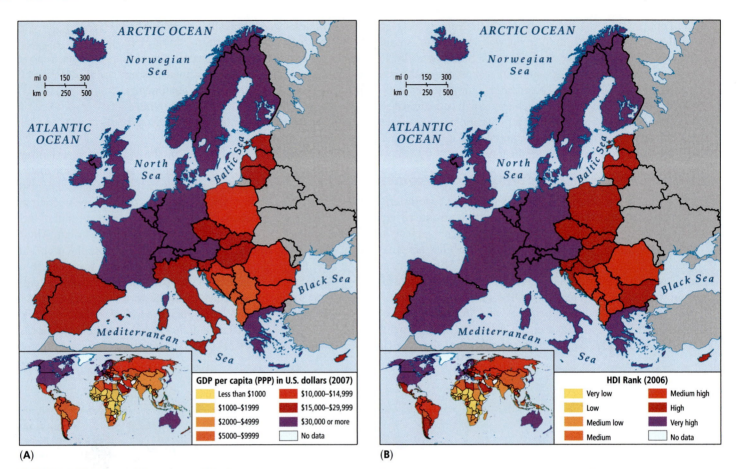

FIGURE 4.22 Maps of human well-being. (A) GDP per capita, adjusted for purchasing power parity. Notice that all the countries in this region have a GDP over U.S.$5000 and most considerably more than that. **(B)** HDI rank. All countries in the region have an HDI ranking of medium or above. **(C)** Female earned income as a percent of male earned income.
[Sources: Maps adapted from data: "Human development indices," (Maps A and B): Table 2, pages 28–32; (Map C): Table 5, pages 41–44 at http://hdr.undp.org/en/media/HDI_2008_en_Tables.pdf.]

in Scandinavia, are the most generous systems. They attempt to achieve equality across gender and class lines by providing extensive health care, education, housing, and child and elder care benefits to all citizens from cradle to grave. Child care is widely available, in part to help women enter the labor market. But early childhood training, a key feature of this system, is also meant to ensure that in adulthood every citizen will be able to contribute to the best of his or her highest capability, and that citizens will not develop criminal behavior or drug abuse. And while finding comparable data is very difficult, surveys of crime victims in Scandinavia and the European Union show that generally the rate of crime in Scandinavia is lower than in other parts of Europe. Finally, although gender equality is a stated goal, throughout this lifelong support system traditional gender roles are tacitly emphasized.

The goal of *conservative and modest welfare systems* is to provide a minimum standard of living for all citizens. These systems are common in the countries of West Europe. The state assists those in need but does not try to assist upward mobility. For example, college education is free or heavily subsidized for all, but strict entrance requirements in some disciplines can be hard

for the poor to meet. State-supported health care and retirement pensions are available to all, but although there are movements to change these systems, they still reinforce the traditional "housewife contract" by assuming that women will stay home and take care of children, the sick, and the elderly. The "modest" system in the United Kingdom is considered slightly less generous than the "conservative" systems found elsewhere in West Europe. The two are combined here but not in Figure 4.21.

Rudimentary welfare systems do not accept the idea that citizens have inherent rights to government-sponsored support. They are found primarily in South Europe and in Ireland. Here, local governments provide some services or income for those in need, but the availability of such services varies widely, even within a country. The state assumes that when people are in need, their relatives and friends will provide the necessary support. The state also assumes that women work only part time, and hence are available to provide child care and other social services free. Such ideas reinforce the custom of paying women lower wages than men.

Post-communist welfare systems prevail in the countries of Central Europe. During the communist era, these systems were

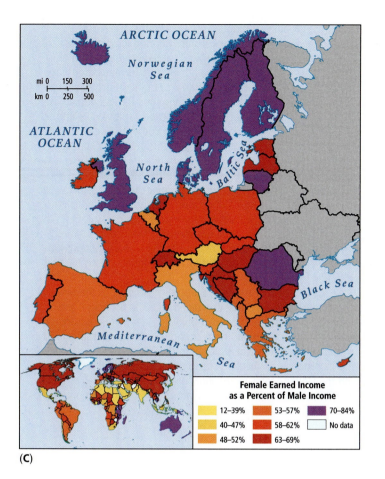

Female Earned Income as a Percent of Male Income

12–39%	53–57%	70–84%
40–47%	58–62%	No data
48–52%	63–69%	

(C)

Macedonia, and Albania—fall below $10,000 per capita. West and North Europe (except for Estonia, Latvia, and Lithuania) are the wealthiest regions. It is worth noting that Greece, which in the 1970s was still quite poor, joined the European Union in 1981 and began to realign its economy (industrializing, reducing its debt and rate of inflation) in order to meet EU specifications. While it was remarkable that in less than 30 years Greece appeared to have joined the ranks of Europe's wealthiest countries, in 2010, the extent to which Greece's prosperity was financed with expensive borrowing was revealed. Greece's huge debt obligations will undoubtedly lower its standard of living, at least for a while. The world map inset shows how Europe ranks in comparison to other parts of the world.

Figure 4.22B depicts the ranks of European countries in the United Nations Human Development Index (UNHDI), which is a calculation (based on adjusted real income, life expectancy, and educational attainment) of how well a country provides for the well-being of its citizens. It is possible to have a high GDP rank and a lower HDI rank if social services are inadequate or if there is a wide disparity of wealth. In Europe the two maps (Figure 4.22A, B) are very similar, but the few differences illustrate an important point. Notice that on Figure 4.22B (HDI), Spain and Italy are now in the highest-ranking category (purple), and Poland and Croatia also rank higher on the HDI (B) than they do on GDP (A), as do Bosnia and Herzegovina, Serbia, Montenegro, Macedonia, and Albania. This means that all of these countries, regardless of their actual wealth, provide for their people relatively well, and wealth disparity is under control. In most cases this is because the countries have or had socialist governments or were until recently in communist societies that emphasized meeting basic needs.

Figure 4.22C shows how well countries are doing in ensuring gender equality—in this case, earned income. Here, note many surprises. First, notice that even in this richest part of the world, nowhere are women paid equally to men. Second, only North Europe plus the UK and Romania are in the highest category (purple), in which females have an earned income that is 70 to 84 percent of male earned income. Romania ranks high even though everyone earns much less than in North Europe (see Figure 4.22A) because, due to the emphasis given to female education and employment under communism, earned income approaches gender equality. It is worth noting the countries where income for males and females is grossly unequal. In Luxembourg and Belgium—both very wealthy countries that are administrative centers for the European Union—and in Spain, women earn just over 50 percent of what men earn; in Italy, women earn less than half of what men earn; in Austria, women earn just 40 percent of what men earn.

comprehensive, resembling the cradle-to-grave social democratic system in Scandinavia, except that women were pressured to work outside the home. Benefits often extended to nearly free apartments, health care, state-supported pensions, and subsidized food and fuel. However, in the post-communist era, state funding has collapsed, forcing many to do without basic necessities. In many post-communist countries, welfare systems are being reviewed, but usually with an eye to reducing benefits and to extending work lives past age 65.

Geographic Patterns of Human Well-Being To a significant extent, the social welfare policies described above affect patterns of well-being, which, not surprisingly, have a geographical pattern. As we have observed, although Europe is one of the richest regions on Earth, there is still considerable disparity in wealth and well-being. GDP per capita is one measure of well-being because it shows in a general way how much people in a country have to spend on necessities, but it is an average figure and does not indicate how income is distributed. As a result, it provides no way to determine if a few are rich and most poor, or if income is more equally apportioned.

Figure 4.22A, showing GDP per capita (PPP) for all of Europe, illustrates two points. Nowhere in the region does GDP fall below $5000 per capita, and only five small countries in southeastern Europe—Bosnia and Herzegovina, Serbia, Montenegro,

---

![THINGS TO REMEMBER]

1. Europeans are choosing to have fewer children, and as a result the population as a whole is aging. Small families are in part a result of women pursuing careers.

2. Because Europe's population is not growing and consumers are dying faster than they are being born, economies may

contract over time. Many governments now encourage large families with generous maternity and paternity leave (up to 10 months with full pay), free day care, and other incentives.

3. Despite high levels of education, the political influence and economic well-being of European women lag behind those of European men; women are paid on average 15 percent less than men for equal work.

4. Europe's cities are world famous for their architecture, economic dynamism, and cultural vitality. Many are quite ancient but have expanded in recent decades with large apartment blocks connected to the old city centers by public transport. In West, North, and South Europe, around 80 percent of the population lives in urban areas.

5. Anti-foreigner sentiment has been a hindrance to acquiring citizenship across Europe, but as the advantages of a multicultural, multiracial, and multigenerational Europe become more apparent, requirements for citizenship are gradually being relaxed. Future trends are uncertain.

Reflections on Europe

Many of the issues confronting people in Europe are similar to those encountered elsewhere around the world. How can economic development be balanced with the need to address global warming and the problems of already-stressed environments? Will Europe's economies decline as aging populations strain the region's ability to maintain high productivity? Will gender roles and support for education have to change further to encourage optimal population growth? How can a country create and retain living-wage jobs for its people when there are qualified workers nearby, or even far away, willing to work for much less? And if immigrants come, how can they be absorbed fairly and with the least disruption? How can a society help the poor and unemployed to a better life while at the same time lessening their drain on public funds? How can poor areas be stimulated to develop without straining prosperous areas too much? How can countries in economic transition be enabled to maintain their newly acquired democratic political systems? No less complex are the cultural questions of how countries with very different traditions, mores, and social welfare systems can find sufficient common ground to cooperate economically and politically.

For all its current troubles, Europe's head start as the birthplace of the Industrial Revolution, combined with the economic and political success of the European Union, give it unique advantages. Nowhere else in the world have countries collaborated so extensively on such a broad range of issues as in the European Union. Moreover, with its historical connections to so many parts of the world, and with the United States increasingly preoccupied with Iraq and the so-called war on terror, Europe is in a position to maintain and extend its global influence.

Critical Thinking Questions

1. How do the issues of immigration in Europe compare with those in the United States? If you were a poor, undocumented immigrant searching for a way to support your family, would you choose the United States or Europe as a possible destination? Why?

2. How was Europe transformed by its colonial empires? How is Europe still linked to its former colonies?

3. How have urban areas been central to changes in Europe before and after the Industrial Revolution?

4. What are some of the potential consequences of negative population growth rates in Europe? Why might an understanding of a region's population age structure affect immigration policy?

5. How might the European Union's status as the world's largest economy, and the rising importance of NATO, lead to changes in global power relationships?

6. What does the evolution of democracy in Europe suggest about current efforts to establish democracy in Iraq?

7. Why is the European Union's generous support for European farmers criticized by developing countries?

8. Which European state-supported welfare systems have offered women the greatest opportunities for employment outside the home?

9. Cite the evidence that European ways of life contribute less to global warming than North American ways.

10. What could the European Union do to reduce its contribution to pollution in Europe's seas? What other efforts might be effective at addressing this problem?

Chapter Key Terms

assimilation, 195

Biosphere, 168

capitalism, 180

central planning, 180

Cold War, 180

Common Agricultural Program (CAP), 187

communism, 180

cool humid continental climate, 168

cultural homogenization, 162

double day, 196

economies of scale, 184

euro, 184

European Union (EU), 162

Green, 172

guest workers, 192

Holocaust, 179

humanism, 176

iron curtain, 179

Mediterranean climate, 168

mercantilism, 176

nationalism, 179

North Atlantic Drift, 168

North Atlantic Treaty Organization (NATO), 187

Roma, 179

Schengen Accord, 191

social welfare (in the EU, social protection), 199

subsidies, 187

temperate midlatitude climate, 168

welfare state, 179

A North European Plain, Belarus

E West Siberian Plain, Russia

B Caucusus Mountains, Georgia

C Volga River, Russia

Greenland

UNITED KINGDOM
Manchester
London
FRANCE
BELGIUM
Brussels
Amsterdam
NETHERLANDS
LUX.
Hamburg
GERMANY
Berlin
Copenhagen
PRAGUE
CZECH REP.
POLAND
Warsaw
SLOVAKIA
HUNGARY
Carpathian Mts.
Lviv
ROMANIA
MOLDOVA
Chisinau
Odesa
UKRAINE
Kiev
Kherson
Sevastopol

North Sea
DENMARK
SWEDEN
Stockholm
Baltic Sea
Kaliningrad
LITHUANIA
Vilnius
LATVIA
Riga
ESTONIA
Tallinn
Gulf of Bothnia
FINLAND
Helsinki
Gulf of Riga
St. Petersburg
Lake Ladoga
Lake Onega
Brest
BELARUS
Minst
Smolensk
Homyel
Pripyat
Chernobyl
Bryansk
Tver
Yaroslavl
Ivanovo
Moscow
Kursk
Tula
Lipetsk
Penza
Voronezh
Nizhniy Novgorod
Kirov

Norway
Murmansk
Kirovsk
Kola Peninsula
White Sea
Barents Sea
Arkhangelsk
Northern Dvina Canal
Divina
Pechora
North European Plain
Kazan
Perm
Kama
Ufa
Saratov
Volga-Don Canal
Samara
Tolyatti
Volgograd
Astrakhan

Svalbard
Franz Josef Land
Novaya Zemlya
Kara Strait
Kara Sea
ARCTIC OCEAN
North Land
Dikson
Laptev Sea
Lena R.
Taymyr Peninsula
North Siberian Lowland
Norilsk
Olenek
Central Siberian Plateau
Vilyuy Res.

Pechora
Vorkuta
Salekhard
Yamal Peninsula
Pechora Basin
Urengoy
West Siberian Plain
Nizhniy Tagil
Yekaterinburg
Tyumen
Surgut
Chelyabinsk
Magnitogorsk
Orenburg
Orsk

Ural Mountains
Steppes
Caspian Depression
Aral Sea
Caspian Sea

RUSSIAN FEDERATION
(RUSSIA)

Tunguska Basin
Seversk
Tomsk
Omsk
Novosibirsk
Novokuznetsk
Krasnoyarsk
Gladkaya
Tayshet
Bratsk
Bratsk Res.
E. Sayan Mts.
W. Sayan Mts.
Kyzyl
Angarsk
Irkutsk
Irkutsk Basin
Ulan-Ude
Lake Baikal

Black Sea
Sea of Azov
Rostov-na-Donu
Dnipropetrovsk
Donetsk
Mykolayiv
Caucasus Mts.
GEORGIA
Batumi
TURKEY
ARMENIA
Yerevan
AZERBAIJAN
Baku
Tbilisi
Groznyy
IRAQ
Lake Urmia
IRAN
Tehran
Esfahan
Mashhad
Elburz Mts.
Zagros Mts.
Lake Van

TURKMENISTAN
Ashkhabad
UZBEKISTAN
Amu Darya
Samarkand
Tashkent
Dushanbe
TAJIKISTAN
Pamirs
Hindu Kush
AFGHANISTAN
PAKISTAN
Islamabad
Faisalabad
Lahore
INDIA
New Delhi

KAZAKHSTAN
Aktogay
Astana
Qaraghandy
Leninsk
Syr Darya
Shymkent
Bishkek
KYRGYZSTAN
Almaty
Lake Balkhash
Tarim
Tarim Basin
Tien Shan
Junggar Basin
Urumqi
Altai Mts.
MONGOLIA
Ulan Bator
CHINA
Gobi
Baotou
Lanzhou
Xian

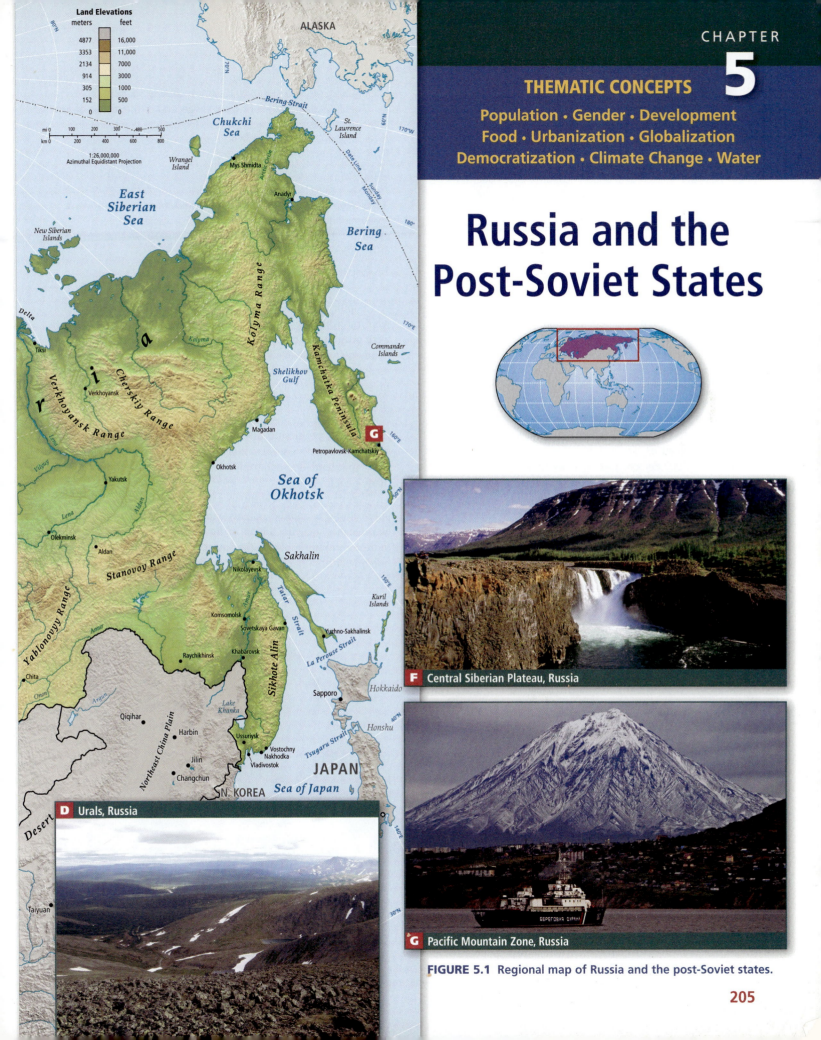

THEMATIC CONCEPTS

Population • Gender • Development
Food • Urbanization • Globalization
Democratization • Climate Change • Water

Russia and the Post-Soviet States

Land Elevations

meters	feet
4877	16,000
3353	11,000
2134	7000
914	3000
305	1000
152	500
0	0

mi 0   100   200   300   400   500
km 0   200   400   600   800

1:26,000,000
Azimuthal Equidistant Projection

F Central Siberian Plateau, Russia

G Pacific Mountain Zone, Russia

D Urals, Russia

FIGURE 5.1 Regional map of Russia and the post-Soviet states.

Global Patterns, Local Lives It's June 2009, and despite the gathering crowds the Rustaveli Cinema in Tbilisi, Georgia, is closed. Located just in front of the Parliament, the cinema usually runs 20 showings a day, but since April, protestors demanding the resignation of Georgian president Mikheil Saakashvili have blocked the cinema entrance. Surrounding businesses have shuttered, and local economic losses due to the demonstrations are mounting as foreign investors, once pleased with opportunities in Georgia, pull out.

The irony of President Saakashvili's predicament was striking. In 2001, then age 34, he had resigned as Minister of Justice and accused then-president Eduard Shevardnadze of corruption and human rights abuses. By 2003, Saakashvili was leading a grassroots pro-democracy and anticorruption political campaign featuring mass nonviolent demonstrations. What became known as the "Rose Revolution" ultimately brought Saakashvili, an American-educated lawyer, to the presidency (**Figure 5.2**). But by 2007, protestors were organizing *against* Saakashvili, and Georgian police were shutting down privately owned TV stations and using batons, rubber bullets, and water cannons to clear the streets of several thousand demonstrators. Observers speculated that Saakashvili was himself embarrassed by the similarity of the 2007 protests to the Rose Revolution—protests he himself had led against Shevardnadze.

After the 2007 demonstrations, eager to salvage his image as a popular democratic leader, Saakashvili called for elections within 2 months (January 2008). He won by a large margin, but his popularity did not last. Faced with the everyday difficulties of managing a democracy, Saakashvili turned once again to authoritarian measures. By April 2009, mass demonstrations were again demanding his resignation, calling him a bloodthirsty tyrant and an embarrassment to Georgia. This time, sections of the military were taking part. In May 2009, a Georgian army tank battalion attempted to organize a mutiny to remove Saakashvili from office.

The 2009 demonstrations grew in size, and by the middle of June 2009, the success story that had been Georgia was disintegrating *[Source: Compiled with the aid of the following authors: Nana Sajaia, Rustaveli Ave. Blues, June 16, 2009, Transitions Online, at http://www.tol.cz/look/TOL/article.tpl?IdLanguage=1&IdPublication=4&NrIssue=325&NrSection=1&NrArticle=20641&tpid=10; Sreeram Chaulia, "Democratisation, NGOs and 'Colour Revolutions,'" January 19, 2006, Open Democracy, at http://www.opendemocracy.net/globalization-institutions_government/colour_revolutions_3196.jsp; and Irakliy Khaburzaniya, "Neither White Knight nor Blackguard," June 2, 2009, Transitions Online, at http://www.tol.org/client/article/20620-neither-white-knight-nor-blackguard.html.]* ■

Popular discontent with ongoing corruption is undeniably what motivated the demonstrators, but the rambunctious protests were possible in large part because of work by NGOs that promote democracy in places like Georgia and Ukraine. Georgia is widely recognized as one of the most democratic post-Soviet states. Free speech and assembly are constitutionally guaranteed. Thus, the situation of leaders like Georgia's Saakashvili is rather ironic. They came to power leading political movements aimed at ending corruption and promoting public participation, yet now find themselves the target of similar movements. Their predicament shows both how hard democratization is to achieve and how badly so many people want to achieve it.

Georgia is a tiny country in a huge region that has changed its political and economic systems entirely in just a few years. Barely two decades ago, the **Union of Soviet Socialist Republics (USSR)**, more commonly known as the **Soviet Union**, was the largest political unit on earth, stretching from Central Europe to the Pacific Ocean. It covered one-sixth of the earth's land surface. In 1991, the Soviet

(A)

(B)

FIGURE 5.2 Globalization and democratization in Georgia. (A) Student activists chant their demands during Georgia's "Rose Revolution" of 2003. Like other "color revolutions" that have occurred in this region, the Rose Revolution was funded by international NGOs, such as the Open Society Institute of George Soros (which itself receives funding from the U.S. State Department), and it followed models codified by U.S. political scientist Gene Sharp. **(B)** A picture of Georgian president Mikheil Saakashvili, who was brought to power by the Rose Revolution, graces a bus in Tbilisi during the 2007–2008 presidential elections.

FIGURE 5.3 Political map of Russia and the post-Soviet states.

Union of Soviet Socialist Republics (USSR) the nation formed from the Russian empire in 1922 and dissolved in 1991

Soviet Union see Union of Soviet Socialist Republics

private businesses. The conversion has proven difficult.

Politically, the Soviet Union has been replaced by a loose alliance of Russia and 11 post-Soviet states—Ukraine; Belarus; Moldova; the Caucasian states of Georgia, Armenia, Azerbaijan; and the Central Asian states of Kazakhstan, Kyrgyzstan, Tajikistan, Turkmenistan, and Uzbekistan (Figure 5.3). Russia, which was always the core of the Soviet Union, remains predominant in this alliance and influential in the world because of its size, population, military, and huge oil and gas reserves.

Geopolitically, this region is looking to the west, south, and east to sort out future relationships. The Cold War between the Soviet Union and the United States and its allies is over. Many former Soviet allies in Central Europe have already joined the European Union, and some other countries in this region may eventually do the same. Trade with Europe, especially in oil and gas, is booming, if fraught with conflict. Meanwhile, the Central Asian states currently allied with Russia may align themselves with neighbors in Southwest Asia or South Asia. The far eastern parts of Russia and Russia as a whole are already finding common trading ground with East Asia and Oceania. The map in Figure 5.4 shows these potential regional realignments.

Union broke apart, ending a 70-year experiment with nearly complete government control of the economy, society, and politics. Over the course of a few years, attempts were made to replace economic control by government bureaucrats with capitalist systems similar to that of the United States, based on competition among

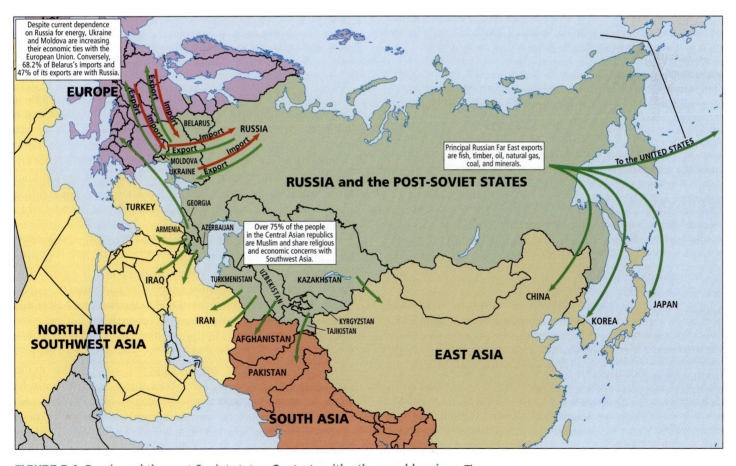

FIGURE 5.4 Russia and the post-Soviet states: Contacts with other world regions. The region of Russia and the post-Soviet states is in flux as all of the component countries rethink their geopolitical positions relative to one another and to adjacent regions. The arrows indicate various types of contact, ranging from economic to religious and cultural.

After 70 years of authoritarian rule, elections are becoming the norm throughout this region. Whether they are actually free and fair elections is debatable, as opposition candidates are increasingly marginalized by lack of access to both print and broadcast media. Economic and political instability have led many voters to support strong leaders who tolerate little criticism. An explosion of crime and corruption has many wondering if the new democracies in this region are strong enough to endure (see Thematic Concepts on pages 210–211).

THINGS TO REMEMBER

1. The former Soviet Union has been replaced by a loose alliance of Russia and 11 post-Soviet states.

2. Elections have become common in the region, although whether they are free and fair is uncertain.

3. Democracies in the region are threatened by crime and corruption.

I THE GEOGRAPHIC SETTING

Terms in This Chapter

There is no entirely satisfactory new name for the former Soviet Union. In this chapter we use *Russia and the post-Soviet States*. Russia is still closely associated with all of these states economically, but they are independent countries, and their governments are legally separate from Russia's. Russia itself is formally known as the **Russian Federation** because it includes more than 30 (mostly) ethnic *internal republics*—such places as Chechnya, Ossetia, Tatarstan, and Buryatya—which all together constitute about one-tenth of its territory and one-sixth of its population. *Federation* should not be taken to mean that the internal republics share power equally with the Russian government.

Physical Patterns

The physical features of Russia and the post-Soviet states vary greatly over the huge territory they encompass. The region bears some resemblance to North America in size, topography, climate, and vegetation. Russia is the largest country in the world, nearly twice the size of the second-largest country, Canada.

Landforms

Because the region is so complex physically, a brief summary of its landforms is useful (Figure 5.1A–G). Moving west to east, there is first the eastern extension of the North European Plain, including the alluvial plain of the Volga River (Figure 5.1A, C), then the Ural Mountains (Figure 5.1D), then the West Siberian Plain (Figure 5.1E), followed by an upland zone called the Central Siberian Plateau (Figure 5.1F), and finally, in the far east, a mountainous zone bordering the Pacific (Figure 5.1G). To the south of these territories from west to east is an irregular border of mountains (the Caucasus, Figure 5.1B), semiarid grasslands, or **steppes** (in western Central Asia, not pictured), and barren uplands and high mountains (in eastern Central Asia, not pictured). The adjacent regions of Southwest Asia, South Asia, and East Asia skirt this region to the south.

Russian Federation Russia and its political subunits, which include 30 internal republics and more than 10 so-called autonomous regions

steppes semiarid, grass-covered plains

permafrost permanently frozen soil just a few feet beneath the surface

tundra a treeless area, between the ice cap and the tree line of arctic regions, where the subsoil is permanently frozen

The eastern extension of the North European Plain rolls low and flat from the Carpathian Mountains in Ukraine and Romania 1200 miles (about 2000 kilometers) east to the Ural Mountains (see Figure 5.1A, C, D on pages 204–205). The part of Russia west of the Urals is often called *European Russia* because the Ural Mountains are traditionally considered part of the indistinct border between Europe and Asia. European Russia is the most densely settled part of the entire region and is its agricultural and industrial core. Its most important river is the Volga, which flows into the Caspian Sea. The Volga River—with its various tributaries and canals to adjacent rivers—is a major transportation route that connects many parts of the North European Plain to the Baltic and White seas in the north and to the Black and Caspian seas in the south (see Figure 5.1C).

The Ural Mountains extend in a fairly straight line south from the Arctic Ocean into Kazakhstan (see Figure 5.1). The Urals, a low-lying range similar in elevation to the Appalachians, are not much of a barrier to humans and only partially to nature (some European tree species do not extend east of the Urals). There are several easy passes across the mountains, and winds carry moisture all the way from the Atlantic and Baltic across the Urals and into Siberia. Much of the Urals' once-dense forest has been felled to build and fuel new industrial cities.

The West Siberian Plain, east of the Urals, is the largest wetlands plain in the world (see Figure 5.1E). A vast, mostly marshy lowland about the size of the eastern United States, it is drained by the Ob River and its tributaries, which flow north into the Arctic Ocean. Long, bitter winters mean that in the northern half of this area a layer of permanently frozen soil (**permafrost**) lies just a few feet beneath the surface. In the far north, the permafrost comes to within a few inches of the surface. Because water does not percolate through this dense layer, swamps and wetlands form in the summer months, providing habitats for many migratory birds. In the far north lies the **tundra**, where because of the extreme cold, the shallow soils, and the permafrost, only mosses and lichens can grow. The West Siberian Plain has some of the world's largest reserves of oil and natural gas, although their extraction is made difficult by the harsh climate and the permafrost.

The Central Siberian Plateau and the Pacific Mountain Zone farther to the east together equal the size of the United States (see

Figure 5.1F, G). Permafrost prevails except along the Pacific coast. There, the ocean moderates temperatures, and additional heat is supplied by the many active volcanoes created as the Pacific Plate sinks under the Eurasian Plate. Places warmed by these forces are the Kamchatka Peninsula, Sakhalin Island, and Sikhote-Alin on Russia's southeastern Pacific coast. Only lightly populated, these places are havens for wildlife, but unfortunately also for poachers.

To the south of the West Siberian Plain is an irregular band of steppes and deserts stretching from the Caspian Sea to the mountains bordering China. To the south and east of these grasslands is a wide, curving band of mountains, including the Caucasus (see Figure 5.1B), Elburz, Hindu Kush, Pamir, Tien Shan, and Altai. Their rugged terrain has not deterred people from crossing these mountains. In this Caucasus region and adjacent areas, for literally tens of thousands of years, people have exchanged plants (apples, onions, citrus, rhubarb, wheat), animals (horses, sheep, goats, cattle), technologies (cultivation, animal breeding, portable shelter construction, rug and tapestry weaving), and religious belief systems (principally Islam and Buddhism, but also Christianity and Judaism).

Climate and Vegetation

The climates and associated vegetation in this large region are varied but not as much as in other regions because so much territory is taken up by expanses of midlatitude grasslands, and northern taiga forest and tundra that are cold much of the year. No inhabited place on earth has as harsh a climate as the northern part of the Eurasian landmass occupied by Russia and particularly by Siberia (Photo Essay 5.1 Map and 5.1C on page 204). The winters are generally long and cold, with only brief hours of daylight. Summers are short and cool to hot, with long days. Precipitation is moderate, coming primarily from the west. Farthest north, the natural vegetation is tundra grasslands. The major economic activities here are the raising of reindeer and extraction of oil, gas, and some minerals. Just south of the tundra lies a vast cold-adapted coniferous forest known as **taiga** that stretches from northern European Russia to the Pacific. The largest portion of taiga lies east of the Urals, and here forestry—often unrestrained by ecological concerns—is a dominant economic activity. The short growing season and large areas of permafrost in this northern zone generally limit crop agriculture, except in the southern West Siberian Plain, where grain is grown.

Because massive mountain ranges in Central and East Asia to the south block access to warm wet air from the Indian Ocean, most rainfall in the entire region comes from storms that blow in from the Atlantic Ocean far to the west. By the time these initially rain-bearing air masses arrive, most of their moisture has been squeezed out over Europe. A fair amount of rain does reach Ukraine, Belarus, European Russia (see Photo Essay 5.1A), and the Caucasian states, and these regions are especially important in food production (vegetables, fruits, and grain). The natural vegetation in these western zones is open woodland and steppe (grassland), though in ancient times, forest was common.

East of the Caucasus Mountains, the lands of Central Asia have semiarid to arid climates (see Photo Essay 5.1B) influenced by their location in the middle of a very large continent. The summers are scorching and short, the winters intense. In the desert

zones, daytime-to-nighttime temperatures can vary by 50 degrees or more. Northern Kazakhstan produces grain and grazing animals. The more southern areas (southern Kazakhstan, Uzbekistan, and Turkmenistan), support grasslands (steppe), which are also used for herding. In modern times, these grasslands have been used for irrigated commercial agriculture that is dependent upon glacially fed rivers, but most land is not useful for farming (see Photo Essay 5.1). The climates in the area where Kazakhstan, Uzbekistan, Tajikistan, and Kyrgyzstan meet are varied, supporting a number of small-scale agricultural activities, some of them commercial.

taiga subarctic forests

Environmental Issues

Soviet ideology held that nature was the servant of industrial and agricultural progress, and that humans should dominate nature on a grand scale. Joseph Stalin, General Secretary of the Communist Party of the USSR from 1922 to 1953, and a major architect of Soviet policy, is famous for having said, *"We cannot expect charity from Nature. We must tear it from her."* Hence, during the Soviet years huge dams, factories, and other industrial facilities were built without regard for their effect on the environment or public health. Now Russia and the post-Soviet states have some of the worst environmental problems in the world. By 2000, more than 35 million people in the region (15 percent of the population) were living in areas where the soil was poisoned and the air was dangerous to breathe (Photo Essay 5.2A, B, C on page 215).

Since the collapse of the Soviet Union, the region's governments, beset with myriad problems, have been reluctant to address environmental issues. As one Russian environmentalist put it, "When people become more involved with their stomachs, they forget about ecology." Pollution controls are complicated by a lack of funds and by an official unwillingness to correct even a few of the past environmental abuses. Photo Essay 5.2 shows just a few examples of human impact on the region's environment, most of which are related to ongoing reckless industrial pollution (as in Norilsk, in Photo Essay 5.2A) or failure to properly shut down factories made obsolete when a more market-based economy was introduced (as in Donetsk, Ukraine, in Photo Essay 5.2B). Now, oil and gas extraction and sale on the global market are a major source of income for Russia and several of the post-Soviet states, and still there is little attention to environmental impacts (see Photo Essay 5.2C).

Urban and Industrial Pollution

Urban and industrial pollution was ignored during Soviet times as cities expanded quickly to accommodate new industries. As workers flooded into the cities from the countryside, lethal levels of pollutants were generated. Today, this region has some of the most polluted cities on the planet.

It is often difficult to link urban pollution directly to health problems because the sources of contamination are multiple and difficult to trace. Such **nonpoint sources of pollution** include

nonpoint sources of pollution diffuse sources of environmental contamination, such as untreated automobile exhaust, raw sewage, and agricultural chemicals that drain from fields into water supplies

Environment, Development, and Urbanization:
Seventy years of Soviet-style centrally planned economic development created some of the worst pollution in the world, in both rural and urban areas. Much of the population was moved into cities that were built rapidly with little concern for protecting inhabitants from pollution. While some improvements have been made, many problems remain.

(A) Deforested hills surround a small industrial city in Russia that is generating significant air pollution.

(B) Bumper cars in Pripyat, Ukraine. This city was abandoned after the nuclear disaster at Chernobyl in 1986.

(C) A power plant adds to a thick haze of smog that hangs over Moscow.

Climate Change, Food and Water:
Russia is now emerging as a major producer of greenhouse gases. Meanwhile, parts of Central Asia are highly vulnerable to climate change, with irrigated agricultural systems facing massive water shortages.

(D) Natural gas being flared off at an oil well in Siberia. Gas flaring accounts for much of Russia's greenhouse gas emissions.

(E) Melting glaciers in Kyrgyzstan. Much of Central Asia depends on water from melting glaciers that will be gone in 50 years.

(F) A man in Tajikistan gets water from a public spigot. The water infrastructure is underdeveloped throughout much of Central Asia.

Globalization and Development:
The post-Soviet era has brought economic reform, new development, and accelerated globalization to the region. Some have become rich, but millions of others have lost jobs and access to health care.

(G) Disney characters now vie for attention with a Soviet-era statue of Lenin.

(H) The yachts of Moscow's new fabulously wealthy business class.

(I) An elderly veteran forced to beg because of the declining value of his pension.

Democratization: After 70 years of authoritarian rule, elections are becoming the norm throughout this region. However, economic and political instability has led many voters to support strong leaders who tolerate little criticism. Meanwhile, an explosion of crime and corruption has many wondering if the new democracies in this region are strong enough to endure.

(J) A Ukrainian stamp commemorating the "Orange Revolution" of 2004–2005, a pro-democracy movement that brought new leadership to power.

(K) A tribute to slain Russian journalist Anna Politskaya, who was known for criticizing Russian policies in Chechnya.

(L) Russian president Dimitri Medvedev (left) was picked for the job by his predecessor, Vladimir Putin (right).

Population: Russia and the countries bordering Europe have the most rapidly declining populations on earth. Meanwhile, the Central Asian and Caucasian countries are growing, some rapidly. Cultural and economic differences may account for some of the disparity, as well as different experiences with recent economic transitions.

(M) An abandoned house in Ukraine, where the population decline is intense.

(N) An only child in Moscow. Russia offers $12,000 to couples having a second child.

(O) Children in Tajikistan, where the population will double by 2050.

Gender, Development, and Democratization: While the Soviet Union gave strong incentives for women to work and become involved in politics, the post-Soviet era has seen growing gaps between men and women in employment and political representation. New threats to women have also emerged, such as international prostitution rings that are now active in some countries.

(P) A stamp commemorating the first woman in space, Valentina Tereshkova.

(Q) Tatyana Yakovleva, one of a handful of female parliamentarians to have been elected since the end of the Soviet Union.

(R) A sign in Kiev, Ukraine, warns women about prostitution rings that pose as foreign employment agencies.

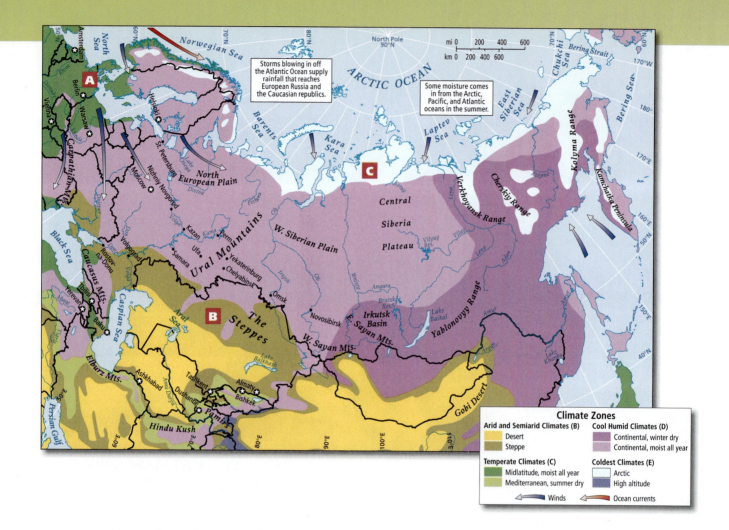

Storms blowing in off the Atlantic Ocean supply rainfall that reaches European Russia and the Caucasian republics.

Some moisture comes in from the Arctic, Pacific, and Atlantic oceans in the summer.

Climate Zones

Arid and Semiarid Climates (B)
Desert
Steppe

Temperate Climates (C)
Midlatitude, moist all year
Mediterranean, summer dry

Cool Humid Climates (D)
Continental, winter dry
Continental, moist all year

Coldest Climates (E)
Arctic
High altitude

Winds
Ocean currents

A Midlatitude, moist all year, Russia

B Steppe, Kazakhstan

C Arctic, Russia

untreated automobile exhaust, raw sewage, and agricultural chemicals that drain from fields into urban water supplies. Moscow, for example, is located at the center of a large industrial area, where infant mortality and birth defects are particularly high. Researchers are convinced that these effects result from a complex mixture of pollutants of diffuse origins. In all urban areas of the region, air pollution resulting from the burning of fossil fuels is skyrocketing as more people purchase cars and as the industrial and transport sectors of the economy continue to grow (see Thematic Concepts, Parts A, C on page 210 and Photo Essay 5.2A on page 215).

Some cities were built around industries that produce harmful by-products. The former chemical weapons manufacturing center of Dzerzhinsk is listed in the *Guinness Book of World Records* as the most chemically polluted city in the world. Men here have a life expectancy of 42 and women, 47. In the city of Norilsk, not a single living tree exists within 30 miles of the world's largest metal smelting complex. Male life expectancy is only 50 years, despite the city's general prosperity and availability of free health care and sports clubs (Photo Essay 5.2B on page 215).

Ukraine serves as an example of overall lack of public involvement with the environment. Ukrainian public officials ignore the knotty problems created by polluters because there is little public pressure on the government to take action. Yet this country has some of the worst pollution on earth, including underground burning garbage and industrial waste dumps, chemical pollution of soils and of fresh water, and the lingering effects of nuclear disasters (see the following section). Offending firms simply pay assessed fines or bribes and continue to pollute.

The Globalization of Nuclear Pollution

Russia and the post-Soviet states are also home to extensive nuclear pollution, and its effects have spread globally. The world's worst nuclear disaster occurred in Ukraine in 1986, when the Chernobyl nuclear power plant exploded (see Thematic Concepts, Part B). The explosion severely contaminated a vast area in northern Ukraine, southern Belarus, and Russia. It spread a cloud of radiation over much of eastern Europe, Scandinavia, and eventually the entire planet. As a direct result of this incident, 5000 people died, 30,000 were disabled, and 100,000 were evacuated from their homes.

▆ 278. Safety at the Center of Nuclear Power Operations Worldwide

The Arctic Ocean and the Sea of Okhotsk in the northwestern Pacific (see map in Figure 5.1 on pages 204–205, and see Photo Essay 5.2C on page 215) are also polluted with nuclear waste dumped at sea. Although the Soviet government signed an international antidumping treaty, it later sank 14 nuclear reactors and dumped thousands of barrels of radioactive waste in these seas that flow into the world ocean.

From 1949 to 1989, remote areas of eastern Kazakhstan served as a testing ground for Soviet nuclear devices (including more than 100 nuclear bombs). The residents were sparsely distributed and no one took the trouble to evacuate them. A museum in Kazakhstan now displays hundreds of deformed human fetuses and newborns. According to the Nuclear Threat Initiative (an international NGO), Kazakhstan has 237.2 metric tons of radioactive waste

waiting for disposal. The disposal and reclamation of the poisoned land will cost $1.1 billion, but only $1 million has been budgeted. Incredibly, one proposal is for Kazakhstan to earn the money by taking in the nuclear waste of other countries eager to be rid of it (France is interested). Estimates are that the country could earn from $30 billion to $40 billion by becoming a commercial receptacle for radioactive waste and could use the profits to dispose of its own waste. But no safe system has yet been found for storing nuclear waste until it is no longer radioactive, and strong opposition on moral grounds has been raised by environmental and human rights NGOs.

The Globalization of Resource Extraction and Environmental Degradation

Russia and the post-Soviet states have considerable natural and mineral resources (see Figure 5.14 on page 226). Russia itself has the world's largest natural gas reserves, major oil deposits, and forests that stretch across the northern reaches of the continent. Russia also has major deposits of coal and industrial minerals such as iron ore and nickel. The Central Asian states share substantial deposits of oil and gas, which are centered on the Caspian Sea and extend east toward China. The value of all these resources is determined in the global marketplace and their extraction affects the global environment.

After the demise of the Soviet Union, general internal pollution levels fell for a while simply because the economy slowed. But now, as economies rebound, demand for resources is once again growing. Greenhouse gas emissions and water pollution levels are on the rise. In Russia, for example, the government is issuing contracts to foreign timber companies for rapid and unsustainable clear-cutting, which not only depletes a valuable resource but removes carbon-absorbing trees and increases erosion and water pollution.

Despite obvious environmental problems, cities like Norilsk (see Photo Essay 5.2A on page 215), which sits on huge mineral deposits, continue to attract foreign investment crucial to the new Russian economy. Thirty-five percent of the world's nickel supply, 10 percent of its copper, and 40 percent of its platinum are in the Norilsk area. By 2006, Norilsk Nickel, the privatized company that runs the smelter, was producing 2 percent of the Russian GDP and had attracted major investment from U.S. and European banks. The success of Norilsk Nickel led the company to globalize despite its shockingly poor environmental record. The company has purchased mining operations in Australia, Botswana, Finland, the United States, and South Africa (Figure 5.5).

Russia's booming oil and gas industries are also creating some of the world's worst oil spills. In the Arctic and Siberia, inland oil spills have contaminated the ocean and coastal wetlands, lakes, rivers, and wildlife (see Photo Essay 5.2D). Recently Lake Baikal, one of the world's largest and least damaged freshwater lakes, was threatened by the construction of a 2500-mile (4000-kilometer) pipeline to carry Russian oil to Asian Pacific markets. After local and international protests, President Vladimir Putin unexpectedly agreed to divert the pipeline a safe distance around the lake. Still, the future success of public environmental protests is uncertain because public activism is so new.

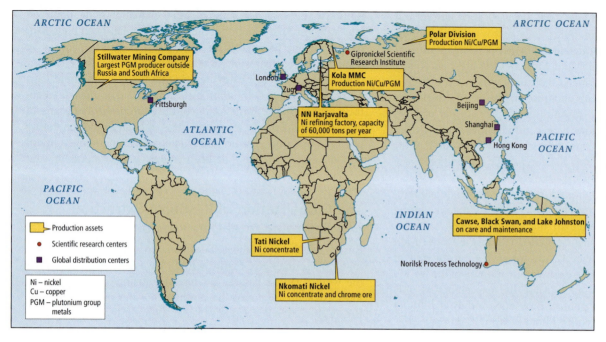

FIGURE 5.5 Norilsk Nickel's worldwide operations. The map depicts the main assets of the company. To see more detailed information and maps of specific mines, plants, and other elements, go to http://www.nornik.ru/en/our_products/ and click on Operations. [Adapted from "About Norilsk Nickel," at http://www.nornik.ru/en/about/.]

In Central Asia, a particularly troubling legacy of mechanized agriculture, mining, and heavy industry persists in the complex mountainous borderlands of Tajikistan and Kyrgyzstan, where Central Asia touches China, Pakistan, and Afghanistan. Soviet efforts to develop commercial agriculture and industry and to supply a labor force for this development resulted in high levels of pollution in the soil and water, including uranium-related contamination (**Figure 5.6**).

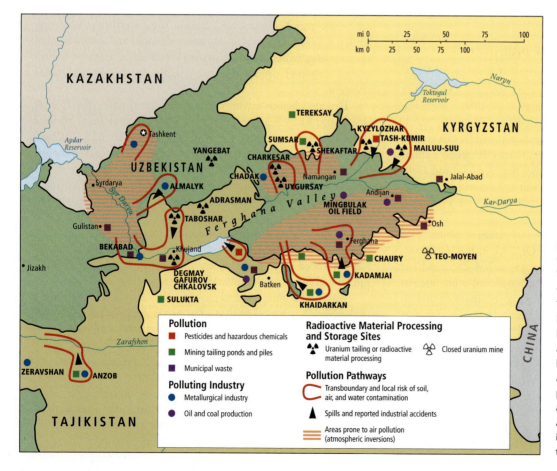

FIGURE 5.6 Industrial pollution and waste hot spots in the Ferghana Valley. In the remote Ferghana Valley, which touches both Kyrgyzstan and Tajikistan, mining and heavy industry have contaminated the soil with toxic heavy metals. Because these industries are essential to the local economies, however, government officials have taken a callous attitude toward controlling pollution. [Adapted from Viktor Novikov and Philippe Rekacewicz, UNEP/GRID-Arendal, April 2005, at http://maps.grida.no/go/graphic/industrial_pollution_and_waste_hotspots_in_the_ferghana_valley.]

This region is home to some of the worst pollution on the planet. Seventy years of industrial development with few environmental safeguards have wreaked havoc on ecosystems. Industries are attempting to become cleaner, but decades may pass before significant improvements are realized.

A A cloud of pollution from a nearby nickel smelter hangs over the city of Norilsk. The smelter emits around 1 percent of the entire global emissions of sulfur dioxide, which helps form acid rain. Enough heavy metals have accumulated in the soils around Norilsk that they can now be mined commercially. The company that runs the smelter has promised improvements, but it estimates that no significant changes in emissions will occur until 2015–2020.

B This false color satellite image shows Norilsk. Purple indicates areas where vegetation has been killed by pollution.

Human Impact, 2002

Land cover
- Forests
- Grasslands
- Deserts
- Tundra
- Ice

Human impact on land
- High impact
- Medium–high impact
- Low–medium impact
- —— Modern national boundaries

C An abandoned factory in Ukraine's Donetsk basin, one of the region's most important industrial areas. Many abandoned facilities still contain many dangerous substances, but money is rarely available to clean them up.

D Oil production in the Russian arctic. With such a short growing season, many arctic environments are extremely fragile. Some of the largest oil spills in history have occurred here, causing severe damage to ecosystems and the indigenous people who depend on them.

Water Issues

Water, once assumed to be inexhaustible, usable for endless purposes, yet in need of little maintenance, is now recognized by most countries in this region as a vital resource.

River Systems The river systems of this region are particularly interesting because they both help and hinder development and because of the environmental issues their use raises. In the west, the Dnieper is the principal river of Ukraine. It has long served as a transport conduit, flowing through the center of Ukraine's industrial and agricultural heartland to the Black Sea. From there cargo reaches the world ocean through the Bosporus and then the Mediterranean. The Dnieper has numerous hydroelectric earthen-dammed reservoirs with power stations. These reservoirs increased in importance after nuclear-generated power lost credibility because of the Chernobyl nuclear power disaster in the 1980s. The durability of the hydroelectric dams is of some concern, as is urban and agricultural pollution spilling into the Dnieper and its tributaries.

The Volga River is the chief waterway of European Russia, serving as a transport artery through the central part of this region. However, the river is not naturally well connected to the Baltic Sea and it flows into the inland Caspian Sea with no outlet to the world ocean. As part of the effort to connect the Volga to the outside world, canals have been built from its upper reaches to the Baltic and from its lower reaches to rivers flowing southwest to the Black Sea (see Figure 5.1 on pages 204–205). A great deal of pollution is added to the river by agricultural effluent and by Moscow and other urban settlements which discharge sewage that has been only partially treated into the river. The Volga is also the site of numerous hydroelectric power plants.

The major rivers east of the Urals—the Ob, the Yenisy, and the Lena—all flow from south to north, meandering to the Arctic Sea through lightly settled frigid, flat land. As such, these rivers are not of much economic use to Russia, and there has long been talk of doing something to change the direction of their flow. Fortunately, before such plans could be implemented, their ecological role was recognized. There is now considerable concern about the extensive marshy plains through which these rivers flow and the role the plains play in global physical systems, especially since the marshy plains are turning into swamps due to global warming and killing trees .

Rivers, Irrigation, and the Loss of the Aral Sea To the south of Russia in the Central Asian states, the glacially fed rivers of the Syr Darya and Amu Darya have long served an irrigation role, supporting commercial cotton agriculture (see the discussion of the effects of global climate change on glaciers on page 217). For millions of years, the landlocked Aral Sea, once the fourth-largest lake in the world, was fed by the Syr Darya and Amu Darya rivers, which bring snowmelt and glacial melt from mountains in Kyrgyzstan (see **Thematic Concepts, Part E** on page 210). Water was first diverted from the two rivers in 1918 when the Soviet leadership ordered its use to irrigate millions of acres of cotton in Kazakhstan and Uzbekistan.

So much water was consumed by these projects that within a few years, the Aral Sea had shrunk measurably (**Figure 5.7**). By the early 1980s, no water at all from the two rivers was reaching the Aral Sea, and by early 2001, the sea had lost 75 percent of its volume and had shrunk into three smaller lakes. The huge fishing industry, on which the entire USSR population depended for some 20 percent of their protein, died out due to increasing water salinity. Once-active port cities were marooned many kilometers from the water. The decline and disappearance of the Aral Sea has been described as the largest man-made ecological disaster on Earth.

The shrinkage of the Aral Sea caused changes in the local climate and human health. The country around the sea has become drier; summers are hotter, and winters cooler and longer. Winds sweep across the newly exposed seabed, picking up salt and chemical residues and creating poisonous dust storms. At the southern end of the Aral Sea in Uzbekistan, 69 of every 100 people report chronic illness due to air pollution and underdeveloped water and sanitation. In some villages, life expectancy is just 38 years (the national average is 66 years).

Efforts to increase water flows to the sea have been confronted with continuing high demands for irrigation and political tangles, because the Aral Sea is shared by the now-independent countries of Uzbekistan and Kazakhstan. Uzbekistan is the world's fifth-largest cotton grower; cotton accounts for one-third of its total exports and employs 40 percent of its labor force. Nevertheless, some actions have been taken to restore the Aral Sea. Kazakhstan, which now relies on oil more than cotton for income, has built a dam to keep fresh water in the northern Aral Sea (see Figure 5.7A). By 2006, the water had risen 10 feet, and the fish catch was improving. Kazakh fishers note that there is now more open public debate about what to do next regarding the Aral Sea. 📹 **107. Dying Sea Makes Comeback**

Climate Change

Climate change is an issue in this region not only because high levels of CO_2 and other greenhouse gases are produced here, but also because Russia's forests play a role in absorbing global carbon and because portions of the region are particularly vulnerable to the negative effects of a changing climate. It is possible that melting in the Arctic zone will improve Russia's accessibility to the world ocean, but vulnerability to the negative effects of climate change is at medium levels throughout much of this region (see the map in **Photo Essay 5.3** on page 218), and at high and extreme levels in Central Asia, which is naturally prone to drought and water scarcity (see Figure 5.7A).

Although Russia has recently begun to reduce its CO_2 and other greenhouse gas emissions, it is still a major contributor of such emissions (see Figure 1.25 on page 47). Many wasteful practices date from the Soviet era, such as burning off or "flaring" natural gas that comes to the surface when oil wells are drilled. Still, in some ways, Russia has shown more willingness to limit its own emissions than other big polluters have: Russia has signed international treaties, such as the Kyoto Protocol and the Copenhagen Accord, to reduce greenhouse gas emissions.

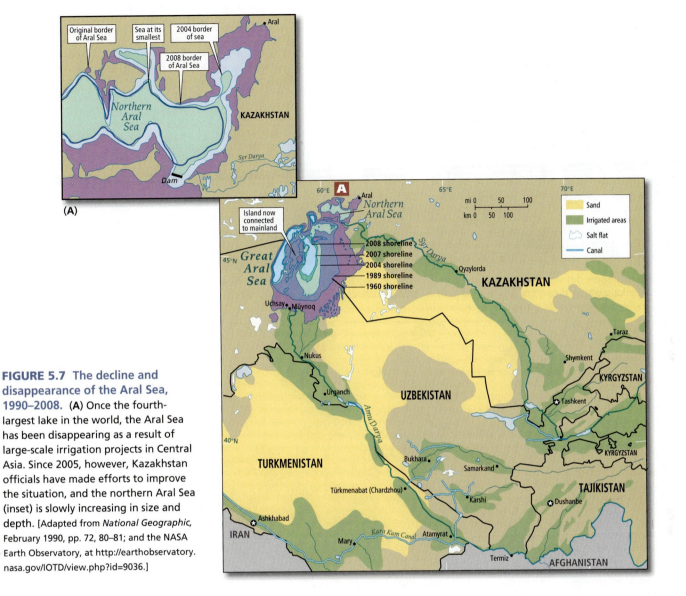

FIGURE 5.7 The decline and disappearance of the Aral Sea, 1990–2008. (A) Once the fourth-largest lake in the world, the Aral Sea has been disappearing as a result of large-scale irrigation projects in Central Asia. Since 2005, however, Kazakhstan officials have made efforts to improve the situation, and the northern Aral Sea (inset) is slowly increasing in size and depth. [Adapted from *National Geographic,* February 1990, pp. 72, 80–81; and the NASA Earth Observatory, at http://earthobservatory. nasa.gov/IOTD/view.php?id=9036.]

Central Asian dependence on irrigated agriculture, using water from rivers that are fed by glaciers in the Pamirs and other mountains in Kyrgyzstan and Tajikistan (see Photo Essay 5.3B, C) leaves the region vulnerable to global climate change. Irrigated agriculture was already unsustainable before the threat of climate change became apparent. Now climate change is bringing more extreme temperature ranges and decreased rain- and snowfall. The glaciers are melting and shrinking. If the glaciers no longer act as water storage systems, supplying melted ice flow to the rivers, the rivers may run dry during the summer, when irrigation is most needed for growing crops. Because most rain falls in the mountains in the winter and spring, either Central Asia's agricultural systems would have to adapt to a winter/spring growing season, or farmers would have to store water for use in the summer. Either proposition demands complex and expensive changes on a massive scale.

THINGS TO REMEMBER

1. The largest country in the world, Russia, is nearly twice the size of the second-largest country, Canada. The region encompasses numerous landforms and associated climates.

2. Soviet ideology held that nature was the servant of industrial and agricultural progress, and that humans should dominate nature—an ideology that wreaked havoc on ecosystems.

3. Industries are attempting to become cleaner, but decades may pass before significant improvements are realized.

4. Water and air pollution, especially in urban and industrial areas, are some of the worst in the world.

5. Global climate change is also a rising concern, with Russia a major contributor of greenhouse gases.

Vulnerability to climate change is at medium levels throughout much of this region, but at high and extreme levels in Central Asia that are highly exposed to drought and water scarcity. Both problems will get worse as temperatures rise and glaciers in the Pamirs and other mountains melt. The dependence of so many Central Asians on agriculture makes them highly sensitive to these problems, and widespread poverty combined with poorly developed disaster response systems reduces overall resilience to these and other disturbances.

A Semi-nomadic people get water at a nearly dry river in Tajikistan. Much of Central Asia depends on rivers for drinking water. If rivers receive less water from glaciers, underground aquifers may not be able to supply adequate water, especially in rural areas.

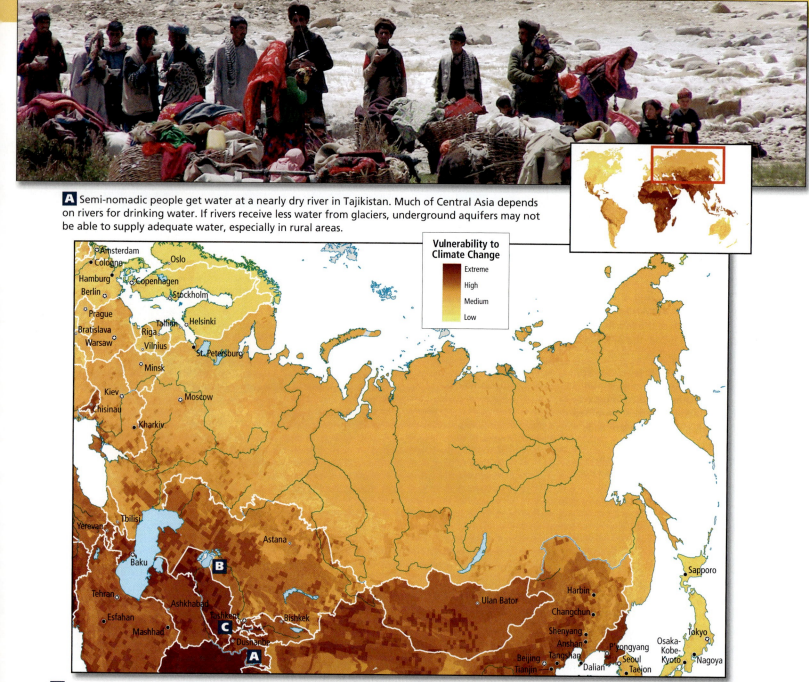

Vulnerability to Climate Change

Extreme
High
Medium
Low

B A fishing boat stranded in what used to be part of the Aral Sea, which has shrunk dramatically due to overuse of the river waters that feed the sea for irrigation of cotton and food crops. Even less water will reach the Aral Sea as glaciers continue melting.

C Workers in a cotton field in Uzbekistan. Forty-four percent of Uzbeks work in agriculture, and cotton is the country's leading export. If irrigation water becomes less available, the economy would be hurt and millions of Uzbeks, 33 percent of whom live below the poverty line, would be out of work.

Human Patterns over Time

The core of the entire region has long been European Russia, the most densely populated area and the homeland of the ethnic Russians. Expanding gradually from this center, the Russians conquered a large area inhabited by a variety of other ethnic groups. These conquered territories remained under Russian control as part of the Soviet Union (1917–1991), which attempted to create an integrated social and economic unit out of the disparate territories. The breakup of the Soviet Union has reversed this gradual process of Russian expansion for the first time in centuries.

The Rise of the Russian Empire

For thousands of years, the militarily and politically dominant people in the region were **nomadic pastoralists** who lived on the meat and milk provided by their herds of sheep, horses, and other grazing animals, and used animal fiber to make yurts, rugs, and clothing. These are the people who are thought to have first domesticated the horse from wild herds. Their movements followed the changing seasons across the wide grasslands that stretch from the Black Sea to the Central Siberian Plateau. The nomads would often take advantage of their superior horsemanship and hunting skills to plunder settled communities. To defend themselves, permanently settled peoples gathered in fortified towns.

Towns arose in two main areas: the dry lands of greater Central Asia and the moister forests of the Caucasus, Ukraine, and Russia. As early as 5000 years ago, Central Asia had settled communities that were supported by irrigated croplands (various grains, vegetables, and apples were first domesticated in Central Asia). The communities were enriched by trade along what became known as the Silk Road, that vast ancient interwoven ribbon of major

and minor trading routes between the Pacific coast of China and the Mediterranean, with lesser connections to southern China, Russia, Southeast Asia, the Indian Ocean, the Black Sea, the Arabian Peninsula, sub-Saharan Africa, Turkey, and Europe (**Figure 5.8**).

About 1500 years ago, the **Slavs**, a group of farmers including those known as the Rus (of possible Scandinavian origin), emerged in what is now Poland, Ukraine, and Belarus. They moved east, founding numerous settlements, including the towns of Kiev in about 480 and Moscow in 1100. By 600, Slavic trading towns were located along all the rivers west of the Ural Mountains. The Slavs prospered from a lucrative trade route over land and along the Volga River that connected Scandinavia (North Europe) and Central and Southwest Asia (via Constantinople, modern-day Istanbul. Refer to Figure 5.8 to locate these developments. Powerful kingdoms developed in Ukraine and European Russia. In the mid-800s, Greek missionaries introduced both Christianity and the Cyrillic alphabet, still used in most of the region's countries.

In the twelfth century, the Mongol armies of Genghis Khan conquered the forested lands of Ukraine and Russia. The **Mongols** were a loose confederation of nomadic pastoral people centered in East and Central Asia. Moscow's rulers became tax gatherers for the Mongols, dominating neighboring kingdoms and eventually growing powerful enough to challenge local Mongol rule. In 1552, the Slavic ruler Ivan IV ("Ivan the Terrible") conquered the Mongols, marking the beginning of the Russian empire.

nomadic pastoralists people whose way of life and economy are centered on the tending of grazing animals who are moved seasonally to gain access to the best grasses

Slavs a group of farmers who originated between the Dnieper and Vistula rivers in modern-day Poland, Ukraine, and Belarus

Mongols a loose confederation of nomadic pastoral people centered in East and Central Asia, who by the thirteenth century had established by conquest an empire stretching from Europe to the Pacific

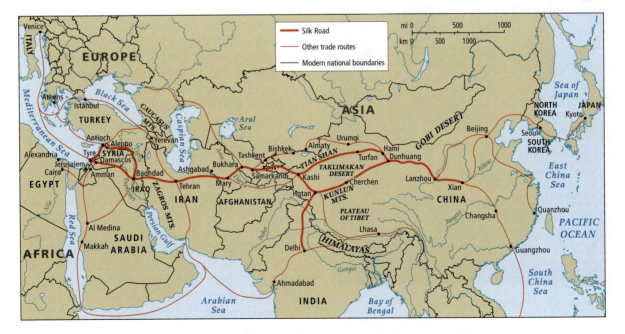

FIGURE 5.8 The ancient Silk Road and related trade routes. Merchants who worked the Silk Road rarely traversed the entire distance. Instead, they moved back and forth along part of the road, trading with merchants to the east or west. [Adapted from "The Silk Road and Related Trade Routes" map, the Silk Road Project and the Stanford Program on International and Cross-Cultural Education, at http://www.silkroadproject.org/tabid/177/default.aspx.]

czar title of the ruler of the Russian empire; derived from the word "caesar," the title of the Roman emperors

Bolsheviks a faction of communists that came to power during the Russian Revolution

communism an ideology, based largely on the writings of the German revolutionary Karl Marx, that calls on workers to unite to overthrow capitalists and establish an egalitarian society where workers share what they produce

capitalists usually a wealthy minority that owns the majority of factories, farms, businesses, and other means of production

Communist Party the political organization that ruled the USSR from 1917 to 1991; other communist countries, such as China, Mongolia, North Korea, and Cuba, also have communist parties

St. Basil's Cathedral, a major landmark in Moscow, commemorates the victory (Figure 5.9).

By 1600, Russians centered in Moscow had conquered many former Mongol territories, integrating them into their growing empire and extending it to the west as shown in Figure 5.10. The first major non-Russian area to be annexed was western Siberia (1598–1689). Russian expansion into Siberia (and even into North America) resembled the spread of European colonial powers throughout Asia and the Americas. Russian colonists forcibly took Siberian resources from the local populations and treated the

people of those cultures as inferiors. Moreover, massive migrations of laborers from Russia to Siberia meant that indigenous Siberians were vastly outnumbered by the eighteenth century. By the mid-nineteenth century, Russia had also expanded its control to the south in order to gain control of Central Asia's cotton crop, its major export.

The Russian empire was ruled by a powerful leader, the **czar**, who lived in splendor (along with a tiny aristocracy) while the vast majority of the people lived short brutal lives in poverty (Figure 5.11). Many Russians were *serfs* who were legally bound to live on and farm land owned by an aristocrat. If the land was sold, the serfs were transferred with it. Serfdom was ended legally in the mid-nineteenth century. However, the brutal inequities of Russian society persisted into the twentieth century, fueling opposition to the czar. By the early twentieth century, a number of violent uprisings were underway.

The Communist Revolution and Its Aftermath

Periodic rebellions against the czars and elite classes took place over the centuries; by the mid-nineteenth century, these rebellions were being led not only by the serfs but by those who could read and write. Resistance to repression grew in the pre–World War I era. Then, in 1917, at the height of Russian suffering during World War I, Czar Nicholas II was overthrown in a revolution. What followed was a civil war between rival factions with different ideas about how the revolution should proceed. Eventually the revolution brought a complete restructuring of the Russian economy and society according to what at first promised to be a more egalitarian model—a model that was extended to adjacent countries.

Of the disparate groups that coalesced to launch the Russian revolution, the **Bolsheviks** succeeded in gaining control during the post-revolution civil war. The Bolsheviks were inspired by the principles of **communism** as explained by the German revolutionary philosopher Karl Marx. Marx criticized the societies of Europe as inherently flawed because they were dominated by **capitalists**—the wealthy minority who owned the factories, farms, businesses, and other means of production. Marx pointed out that the wealth of capitalists was actually created in large part by workers. The workers' labor was so undervalued and underpaid that this propertyless majority was usually extremely poverty-stricken. Under communism, workers were called upon to unite to overthrow the capitalists, take over the means of production (farms, factories, services), and establish a completely egalitarian society without government or currency. The philosophy held that people would work out of a commitment to the common good, sharing whatever they produced so that each had their basic needs met.

One Bolshevik leader, Vladimir Lenin, declared that the people of the former Russian empire needed a transition period in which to realize the ideals of communism. Accordingly, Lenin's Bolsheviks formed the **Communist Party**, which set up a powerful government based in Moscow. Lenin believed that the government should run the economy, taking land and resources from the wealthy and using them to benefit the poor majority. Lenin's emphasis on centralized power (rather than power lodged with the workers) struck some as contradicting the move to an egalitarian

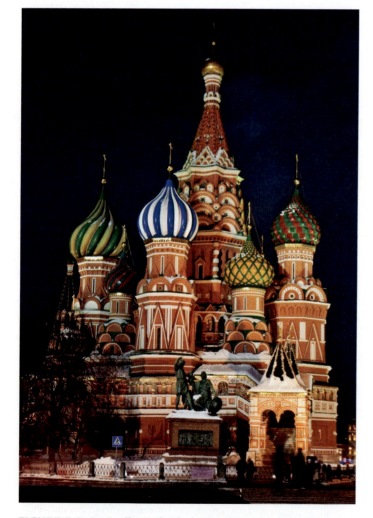

FIGURE 5.9 St. Basil's Cathedral, Moscow. Now the most recognized building in Russia, St. Basil's Cathedral was built between 1555 and 1561 to commemorate the defeat of the Mongols by Ivan the Terrible, the first Russian czar. A central chapel, capped by a pyramidal tower, stands amid eight smaller chapels, each with a colorful "onion dome." Each chapel commemorates a saint on whose feast day Czar Ivan had won a battle.

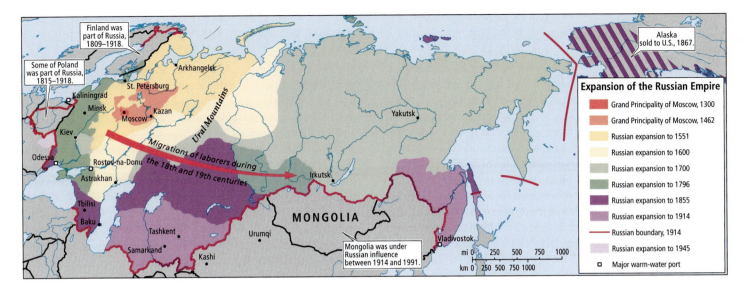

FIGURE 5.10 Russian imperial expansion, 1300–1945. A long series of powerful entities expanded Russia's holdings across Eurasia to the west and east. Expansion was particularly vigorous after 1700 when Siberia, the largest area, was acquired. [Adapted from Robin Milner-Gulland with Nikolai Dejevsky, *Cultural Atlas of Russia and the Former Soviet Union*, rev. ed. (New York: Checkmark Books, 1998), pp. 56, 74, 128–129, 177.]

society. A colleague of Lenin's and an important revolutionary theorist, Leon Trotsky, disagreed with Lenin. Trotsky had been responsible for strengthening the Red (Bolshevik) Army early in the revolution, which had been critical in eliminating any resistance to the revolution. Once the revolution was solidified, Trotsky favored carrying communism to the world in a global *permanent revolution* of the **proletariat** (the working class). He also appears to have favored more democratic principles than Lenin or Lenin's successor, Joseph Stalin. Trotsky had ever more serious differences with Stalin's brutal tactics and was eventually forced to leave Russia. For a time Trotsky lived in Istanbul, in France, and eventually in Mexico, where a Soviet agent assassinated him in 1940.

The Stalin Era After Lenin suffered a series of strokes in 1922, Joseph Stalin stepped in to reorganize the society and economy in a much more autocratic style than Lenin had developed. He began his 31-year rule of the Soviet Union as party chairman and premier, remaining in office until his death in 1953. Stalin brought to the country a mixture of brutality and revolutionary change that set the course for the rest of the Soviet Union's history. He sought to cure production shortfalls through rapid industrialization made possible by a **centrally planned**, or **socialist, economy**. The state would own all real estate and means of production, while government bureaucrats in Moscow would direct all economic activity. This system became known as the *command economy*. In this economy, central control was used to locate all factories, apartment blocks, and transport infrastructure, and to manage all production, distribution, and pricing of products. The idea was that under a socialist system the economy would grow more quickly, which would hasten the transition to the idealized communist state in which everyone shared equally. This notion was reflected in the new name chosen for the country: the Union of Soviet Socialist Republics.

> **proletariat** the working class; the lowest social or economic class
>
> **centrally planned**, or **socialist, economy** an economic system in which the state owns all real estate and means of production, while government bureaucrats direct all economic activity, including the locating of factories, residences, and transportation infrastructure

FIGURE 5.11 "Burlaks on the Volga." This painting depicts the destitution of the Burlaks, poor men and women hired to haul freight upstream along the Volga River between the sixteenth and twentieth centuries. Created in 1873 during the reign of Czar Alexander II, the painting illustrates the poverty that would inspire revolution against the czars in the early twentieth century.

Stalin used the powers of the command economy with fervor and cruelty. To increase agricultural production, he forced farmers to join large government-run collectives. Those who resisted were relocated to Siberia, indentured to work in factories, or executed.

Cold War the contest that pitted the United States and Western Europe, espousing free market capitalism and democracy, against the USSR and its allies, promoting a centrally planned economy and a socialist state

For Stalin, the true key to achieving economic growth was increasing industrial production, for which a quick secure labor force was necessary. Accordingly, he ordered massive government investments in gigantic development projects, such as factories, dams, and chemical plants, some of which are still the largest of their kind in the world. The millions of resisters sent to Siberia became part of the forced labor system known as the Gulags, labor camps where the inmates mined for minerals and built and then worked in the new industrial cities. These new cities included Norilsk and those clustered along the southern rail line that ran through Chelyabinsk, Novosibirsk, and Krasnoyarsk, to Angarsk and Irkutsk, near Lake Baikal.

This strategy of government control met many expectations of the central planners. Eventually, government-controlled companies monopolized every sector of the economy, from agriculture to mining to industry to clothing design and production. The millions who were forced to work in the Gulags of Siberia and those who more willingly worked in urban industries helped develop the country and support a large military; their labor eventually brought higher standards of living to those who survived (millions did not). Schools were provided for previously uneducated poor and rural children and contributed to major technological and social advances. During the Great Depression of the 1930s, the Soviet Union's industrial productivity grew steadily even while the economies of other countries stagnated.

However, beyond its cruelty, Stalin's model had some serious flaws. One problem was that production was geared largely toward heavy industry (the manufacture of machines, transport equipment, military vehicles, and armaments). Less attention was paid to the demand for consumer goods and services that could have dramatically improved daily life for the Russian people. The most destructive aspect of Stalin's rule, however, was his ruthless use of the secret police, starvation, and mass executions to silence anyone who dared to oppose him. The lucky ones were those merely sentenced to labor camps in remote Siberia. Between 34 and 60 million others were killed as a result of Stalin's policies.

World War II and the Cold War

During World War II, the Soviet Union did more than any other country to defeat the armies of Nazi Germany. A failed attempt to conquer the Soviet Union exhausted Hitler's war machine. In the process of defeating Germany on the Eastern Front, 23 million Soviets were killed, more than all the other European combatants combined. After the war, Stalin was determined to erect a buffer of allied communist states in eastern Europe that would be the battleground of any future war with Europe. Perhaps not realizing Stalin's intent, or because of inattention, Western leaders (Churchill and Roosevelt) ceded control of much of Central Europe to Russia while they busied themselves with reconstructing postwar

economies. When Russia erected the Iron Curtain (so named by Churchill in 1946), it became clear that the USSR intended to utterly dominate Central Europe. In response, the United States and its allies organized to stop the Soviets from extending their power even farther into Europe and perhaps elsewhere. The result was the **Cold War**, a nearly 50-year-long global geopolitical rivalry that pitted the Soviet Union and its allies against the United States and its allies (**Figure 5.12**).

In an attempt to match the global military power of the United States, the Soviets diverted resources to their military and away from much-needed economic and social development. Internationally, the Soviets promoted communism far and wide, with major efforts in China, Mongolia, North Korea, Afghanistan, Cuba, Vietnam, Nicaragua, and various African nations. Closer to home, in Central Europe and Central Asia, the Soviet Union maintained a pervasive influence through political, economic, and military coercion. By co-opting local leaders, Soviet officials were able to manipulate the economies and resources of individual countries to the benefit of the Soviet Union.

Afghanistan provides an example of how Cold War politics brought on political change in the Soviet Union. In the 1970s in Afghanistan, which lies just to the south of the Central Asian states, there was contention between a small group who favored a Western-style democratic government, those who favored communism, and those who favored a theocratic state based on Islamic law. Russia feared that the strongly anti-communist Muslim movement would influence Central Asian countries to unite under Islam and rebel against Soviet control. Therefore, seeking a foothold in Afghanistan, Russia built military bases and backed an unpopular pro-communist government there. This brought strong resistance from the Muslim fundamentalist Mujahedeen who were at that time viewed by the wider world as merely Afghan anti-communist freedom fighters. Soviet finances and morale were heavily invested in propping up this shaky government, and in 1979 Russia launched a war to support the government against the Mujahedeen resistance.

Over the next decade, the Soviets spent large amounts of money and many lives on what Russia's general population saw as a terrible waste. It is now clear that the Afghan war played an important role in bringing down the Soviet Union (see Photo Essay 5.4A on page 232). Russian forces were badly beaten by the highly motivated Afghan guerilla fighters, who, interestingly, were financed, armed, and trained by the United States, with Pakistan as the conduit for $10 billion in weapons and aid. Once the Russians were defeated (1988), the U.S. forces left and the more radical elements of the Mujahedeen, known as the Taliban, took over. The Soviet–Afghanistan conflict and the aftermath are discussed further in Chapter 8.

THINGS TO REMEMBER

1. As early as 5000 years ago, Central Asia had settled communities that were supported by irrigated croplands and enriched by trade along what became known as the Silk Road.

2. The Bolsheviks, a group inspired by the principles of communism, were the dominant leaders of the Russian Revolution of 1917. Their goal, never achieved, was an

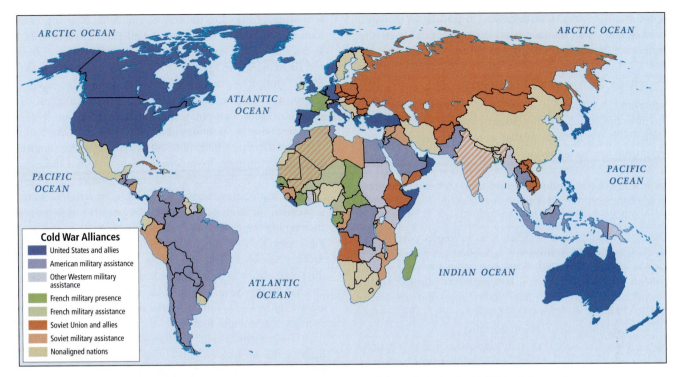

FIGURE 5.12 The Cold War in 1980. In the post–World War II contest between the Soviet Union and the United States, both sides enticed allies through economic and military aid. The group of countries militarily allied with the Soviet Union was known as the Warsaw Pact. Some countries remained unaligned. [Adapted from Clevelander, at http://en.wikipedia.org/wiki/Image:New_Cold_War_Map_1980.png.]

egalitarian society without government or currency, and their philosophy held that people would work out of a commitment to the common good, sharing whatever they produced.

3. Taking control in 1922 while Lenin was ill, Joseph Stalin brought to the country a mixture of brutality and revolutionary change that set the course for the rest of the Soviet Union's history. He established a centrally planned, or socialist, economy in which the state owned all real estate and means of production, while government bureaucrats in Moscow directed all economic activity. This system became known as the command economy.

4. In the aftermath of World War II, a nearly 50-year-long global geopolitical rivalry known as the Cold War pitted the Soviet Union and its allies against the United States and its allies.

II CURRENT GEOGRAPHIC ISSUES

The goals of the Soviet experiment begun in 1917 were unique in human history: to reform quickly and totally an entire human society for the benefit of the people. The Soviet Union's collapse brought an even more rapid shift from centrally planned economies to market economies. The transition remains incomplete and for many, life is proceeding amid economic, political, and social uncertainty.

Economic and Political Issues

When the Soviet Union disbanded and the leaders of the post-Soviet states embraced various versions of a market economy, everyone had to learn how capitalism works. That early period was characterized by a pell-mell tumble into a market economy that gave the advantage to a few high-level Soviet bureaucrats who had run government industries in the old command economy. They often acquired previously publicly held assets at fire-sale prices, thereby becoming fabulously rich very quickly, and assumed the role of **oligarchs**—individuals who are so wealthy that they wield enormous, often clandestine, political control.

Oligarchs continue to exercise power in government and private enterprise. Recently, an attempt was made to harness this power for the good by creating a master of business administration (MBA) program on a lavish new campus near Moscow. With a *who's who* of oligarchs serving as funders, board members, and even teachers, the Moscow School of Management, Skolkovo, opened in 2009. The curriculum addresses the fact that many of the early self-made millionaires of the 1990s

oligarchs in Russia, those who acquired great wealth during the privatization of Russia's resources and who use that wealth to exercise power

lacked the management skills necessary to lead firms that were solid and durable. Its emphasis is on entrepreneurial leadership honed in actual business situations.

The Former Command Economy

To understand the major economic and political issues of the decades after the collapse of the Soviet Union, it is important to be familiar with the institutions that were developed and implemented during the Soviet period.

Characteristics of the Command Economy Although the long-term goal of the command economy was to achieve a communist society, the shorter-term goal was to end the severe deprivation suffered under the czars. To a large extent, the Soviets met this goal. The Soviet economy grew rapidly until the 1960s, and abject poverty was largely eradicated. Basic necessities such as housing, food, health care, education, and transportation (the still beautiful Moscow subway was built under Stalin, some of it with forced labor, in the 1930s) were all provided for free or at very low cost—a remarkable accomplishment for any country.

Nonetheless, the Soviet centrally planned economy was notably less efficient than market economies in allocating resources. Because of the lack of competition in the economy, producers had no incentive to use more efficient production methods or to come up with products that were more effective or attractive. Quality also suffered because hard work and innovation in the workplace were rarely rewarded. Promotions generally went to those with connections in the Communist Party. As a result, most consumer goods, such as cars or washing machines, were of poor quality and available only at high cost to a privileged few.

Until the 1980s, punishment was severe for privately producing and selling (on the *black market*) food, liquor, or consumer products. Then, under Soviet leader Mikhail Gorbachev, a private market economy began to develop. Private producers of food, especially meat, were allowed to sell legally, but their prices were beyond the means of most.

In science and military technology, however, the Soviet Union achieved feats that had the United States and its allies scrambling to keep up. Soviet scientists and engineers launched the world's first satellite in 1957 and the first manned spacecraft in 1961. The scientists and engineers of the region are still making significant breakthroughs in physics, biology, metallurgy, computer technology, and a wide variety of other technical fields. However, research in the social sciences lagged considerably, mostly because of an official fear that such researchers would undermine the communist system by asking difficult questions and discovering unwelcome trends. This bias against the social sciences continues across the post-Soviet region.

📺 **104. Former Soviet Union Launched Space Age 50 Years Ago**

Soviet Strategic Regional and Transport Planning One enduring legacy of Soviet central planning was the strategic location of huge industrial projects in the farthest reaches of Soviet territory. Leaders did this to buttress Russia's claims to distant territories by bringing the higher standards of living enjoyed in industrialized European Russia to all parts of the Soviet Union. Economic development projects also continued the pre-communist Russian empire's policy of **Russification**—forcing non-Russian ethnic groups to conform to the state's goals by swamping the groups with large numbers of ethnic Russian migrants. Russian migrants (other than political prisoners) were given the best jobs and most powerful positions in regional governments. This strategy minimized the possibility of revolt by remote ethnic minorities. Yet another concern was to disperse industrial centers throughout the country to make them safer from enemy attack.

These ambitious regional development schemes never really succeeded, in large part because of transport problems related to the region's huge size and challenging physical geography. While the region's rivers run mainly south to north flowing into the Arctic (the Volga River is the only major exception flowing from the north, south to the Caspian Sea), the primary transport needs created by the Soviet regional development schemes were east-west (**Figure 5.13**). This created a need for land transport systems such as railroads and highways.

The construction of land transport systems has long been held back by the region's climate. Long, harsh winters, during which it is difficult to build, gave way to a period called the *rasputitsa* or "quagmire season," when melting permafrost turns many roads and construction sites into impassable mud pits. Huge distances and complex topography, especially in Siberia, add to the problem and as a result, few roads or railroads were ever built outside European Russia.

Even today this region—more than two and a half times the size of the United States—has less than one-sixth the number of hard-surface roads and virtually no multilane highways. No road, and only one main rail line—the Trans-Siberian Railroad (Figure 5.13)—runs the full east-west length, connecting Moscow with Vladivostok, the main port city on the Russian Pacific coast. By contrast, the road and rail network is fairly dense within European Russia, Belarus, and Ukraine, with Moscow as the main hub.

Oil and Gas Development: Fueling Globalization

Crude oil and natural gas are the region's most lucrative exports and are dominating economies more than ever before. The struggle over price and allocation of the region's oil and gas resources has become global and could be a source of conflict for years to come (**Figure 5.14**). However, the dire need for reliable oil and gas trading relationships that will secure profits so badly needed for development and economic diversification is likely to bring greater stability to the region's energy markets.

Energy-Based Economic Volatility in the Region In 1998, after severe economic disruptions, Russia's debt was staggeringly high—90 percent of GDP—and a major impediment to the country's economic development. It was especially difficult to attract foreign investment because the economy was so shaky. But suddenly rising revenues from oil and gas began to finance Russia's economic

Russification the assimilation of all minorities to Russian (Slavic) ways

FIGURE 5.13 Principal industrial areas and land transport routes of Russia and the Post-Soviet states.
The industrial, mining, and transport infrastructure is concentrated in European Russia and adjacent areas. The main trunk of the Trans-Siberian Railroad and its spurs link industrial and mining centers all the way to the Pacific, but the frequency of these centers decreases with distance from the borders of European Russia. [Adapted from Robin Milner-Gulland with Nikolai Dejevsky, *Cultural Atlas of Russia and the Former Soviet Union,* rev. ed. (New York: *Checkmark Books,* 1998), pp. 186–187, 198–199, 204–205, 216–217; and http://www.russia-ukraine-travel.com/map-trans-siberian.html.]

recovery. By 2000, Russia was running balance-of-payments surpluses and by 2008 had foreign exchange reserves of $600 billion. Today, Russia is the world's largest exporter of natural gas (Figure 5.14). The state-owned energy company **Gazprom** is the tenth-largest oil and gas entity in the world. But over time, stabilizing Russia's oil and gas industry is likely to be a critical economic issue.

Many foreign investors are also interested in the energy resources of Central Asia. A tug-of-war has evolved between Russia and multinational energy corporations over the rights to develop, transport, and sell these resources. As holders of these considerable oil reserves, the new Central Asian states are not powerful or experienced enough to stave off the covetous advances of world powers (**Figure 5.15**). Nor do they have the capital to develop the resources themselves, yet individuals and special interests in the Central Asian countries are reaping enormous financial rewards as deals are made.

Contention over Central Asian oil and gas hinges primarily on pipeline routes (see Figure 5.14). Russia, the United States, the European Union, Turkey, China, and India have all developed or proposed various routes, and by 2009 China was building a pipeline from the east side of the Caspian Sea to China's Xinjiang Uygur Autonomous region, a distance of 1,900 miles (3000 km). By 2006, the United States had installed military bases to protect its interests in Georgia and Uzbekistan, both of which have pipelines to the West.

Gazprom in Russia, the state-owned energy company; it is the tenth-largest oil and gas entity in the world

📺 **108. Energy Revenues and Corruption Increase in Russia**

Relations with the European Union and the Global Economy

Russia's trade in oil and gas with the European Union has been especially crisis ridden. Gazprom, for example, earns 65 percent of its revenue from sales to the European Union, which buys

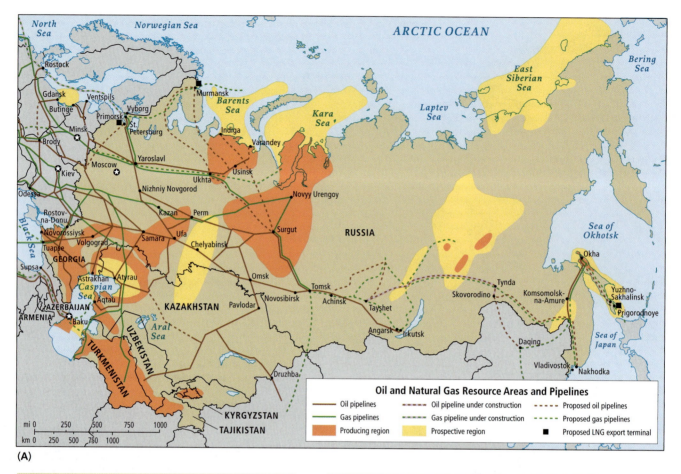

(A)

Oil, 2006	
Oil reserves	**80 billion barrels**
Oil reserves, as percentage of world	7 percent
Saudi Arabian reserves	264 billion barrels
U.S. reserves	30 billion barrels
Oil production	**10 million barrels per day**
Oil production, as percentage of world	12 percent
U.S. oil production	7 million barrels per day
Oil exports	**7 million barrels per day**
Oil exporter, rank	2
Oil exports, to U.S.	370,000 barrels per day

(B)

Natural Gas, 2006	
Gas reserves	**48 trillion cubic meters**
Gas reserves, as percentage of world	26 percent
Iranian reserves	28 trillion cubic meters
U.S. reserves	6 trillion cubic meters
Gas production	**612 billion cubic meters**
Gas production, as percentage of world	21 percent
U.S. gas production	524 billion cubic meters
Gas exports	**263 billion cubic meters**
Gas exporter, rank	1
Gas exports, to Europe	151 billion cubic meters

(C)

FIGURE 5.14 Oil and natural gas: Russia and the post-Soviet states' resources and pipelines.
For additional information and maps see, http://www.eia.doe.gov/emeu/cabs/ and click on Newly
Independent States on the map. [Map adapted from U.S. Department of Energy, Energy Information Administration,
"Russia Country Analysis Brief," May 2008, at http://www.eia.doe.gov/emeu/cabs/Russia/images/Russian%20Energy%20
at%20a%20Glance%202007.pdf. [Anatoly Ustinenko].

25 percent of its natural gas from Gazprom. In fact, the European Union is Russia's primary trading partner and investor; about 80 percent of Russia's direct foreign investment comes from EU countries. Meanwhile, Russia is the EU's third largest trading partner. From 2000 to 2008, trade between the EU and Russia tripled in value, but then it contracted due to the global economic crisis.

This trading relationship remains volatile because in 2007, before the recession, Russia attempted to use Europe's significant dependence on Russia's oil and gas wealth as a pawn in global geopolitics. Often the threats were veiled and hard to prove, but on a number of occasions Russia intimated that it would be necessary to cut off supplies of oil to Europe, especially that oil that flows through Ukraine and Belarus. These threats caused much anxiety among poorer Europeans, especially those in Central and Eastern Europe, who often faced energy shortages during the Soviet period. Many went back to cutting wood to burn as fuel. Then

FIGURE 5.15 Pipeline construction in Kazakhstan. Large oil and gas reserves are driving rapid economic growth in Kazakhstan. GDP per capita (PPP) nearly tripled between 2000 and 2008, and the government plans similar growth in the future, based largely on pipeline deals signed with companies from China, Russia, the EU, and the United States.

during the global economic crisis beginning in 2008, when oil prices fell sharply, Russia was suddenly less affluent and became more conciliatory as it was reminded of its dependence on income from Europe.

Both the European Union and the United States want to ensure that whatever the price of oil or gas on the world market, Russia's oil and gas wealth does not finance a return to the hostile relations of the Cold War. For this reason, Russia was invited to join the **Group of Eight (G8),** an organization of the wealthiest and most highly industrialized nations. All parties also agree that Russia should gain membership in global trading institutions, such as the World Trade Organization.

Meanwhile, Russia, still casting about for a clear leadership role in the global economy, is courting Brazil, India, and China to form a trade consortium known as BRIC. All four countries have experienced spectacular economic growth over the last decade. In June 2009, Russia hosted the first meeting of BRIC. The idea was to form an economic counterweight to the United States and the

European Union, but, in fact, Russia may be the weakest and least diverse economy in the grouping.

> **Group of Eight (G8)** an organization of highly industrialized countries: France, the United States, Britain, Germany, Japan, Italy, Canada, and Russia

Food Production in the Post-Soviet Era

Across most of the region, agriculture is precarious at best, either hampered by a short growing season and boggy soils or requiring expensive inputs of labor, water, and fertilizer. Only 10 percent of Russia's vast expanse is suitable for agriculture due to harsh climates and rugged landforms. The Caucasian mountain zones are, on the other hand, areas in the region where rainfall adequate for agriculture coincides with a relatively warm climate and long growing seasons. Together with Ukraine and European Russia, this area is the agricultural backbone of the region (**Figure 5.16**). The best soils are in an area stretching from Moscow south toward the Black and Caspian seas, and extending west to include much of Ukraine and Moldova. In Central Asia, irrigated agriculture on collective farms was extensive, especially where long growing seasons supported cotton, fruit, and vegetables.

Changing Agricultural Production During the 1990s, agriculture went into a general, if temporary, decline across the entire region. In most of the former Soviet Union, yields dropped by 20 to 30 percent compared to previous production levels. This was due mostly to the collapse of the subsidies and trade arrangements of the Soviet era. Many large and highly mechanized collective farms suddenly found themselves without access to equipment, fuel, or fertilizers. Since that time, the massive collective farms of the Soviet era have slowly been privatized, and better management has resulted in efficiency gains. Thousands of collective farms have been broken up into small tracts and sold to the people who worked on them. These independent farmers now have a much greater stake in their productivity and are making better decisions about what to grow, how to grow it, and where to sell it.

In Central Asia, agriculture was reorganized with less emphasis on collective farming of export crops and more on smaller food-producing family farms. Grain and livestock farming (for meat, eggs, and wool) are increasing and production levels per acre and per worker are increasing. China, through the Asian Development Bank, has provided assistance in reducing the use of agricultural chemicals.

Georgian farmers, blessed with warm temperatures and abundant moisture from the Black and Caspian seas, can grow citrus fruits and even bananas; they do so primarily on family, not collective, farms (**Figure 5.17**). Before 1991, most of the Soviet Union's citrus and tea came from Georgia, as did most of its grapes and wine. In 2005, seventy-five percent of Georgian wine (60 million bottles) went to Russia. But Georgia and Russia have long had a contentious relationship. In May 2006, Russia unexpectedly banned the importation of Georgian wine, claiming it was polluted with heavy metals. Trade experts suspect this was punishment for Georgia's resistance to Russian efforts to control the marketing of Georgia's wine and other products in the European Union. Although Georgia would like to reestablish its wine

FIGURE 5.16 Agriculture in Russia and the post-Soviet states. Agriculture in this part of the world has always been a difficult proposition, partly because of the cold climate and short growing seasons, and partly because soil fertility and lack of rainfall are problems in all but a few places (Ukraine, Moldova, and Caucasia). [Adapted from Robin Milner-Gulland with Nikolai Dejevsky, *Cultural Atlas of Russia and the Former Soviet Union,* rev. ed. (New York: Checkmark Books, 1998), pp. 186–187, 198–199, 204–205, 216–217.]

market in Russia, Georgian wine now enjoys a market in 42 other countries around the world.

Family Food Gardens in Urban Space Urban family gardens have long been an important source of nourishment in the region. In Soviet times, cities were surrounded by small private garden plots on government-held land. Urbanites had a place to relax on weekends and to grow food that was cheaper and often less contaminated with pesticides than purchased food.

Since the fall of the Soviet Union, family gardens have proven essential to the diets of millions of people across the region. During the economic crises of the 1990s, the cost of food suddenly took up as much as half of family budgets. Even those with access to gardens were eating fewer vegetables; less meat, fewer eggs and dairy products; and even less bread, long a staple of the region's diet. Regional nutrition levels fell and life expectancies sank. Today, entrepreneurial farmers (especially in Ukraine and Caucasia) supply attractive fresh food to the best city markets. However, even with the recent economic improvements, few ordinary people can afford these luxuries and many still grow much of

FIGURE 5.17 Cabbage-planting season for Georgian family farmers.

their own produce. These gardens are highly productive, accounting for 20 to 30 percent of agricultural produce in some countries on as little as 1 percent of the agricultural land area.

As urban economies expand, the growing market for real estate is displacing family gardens. For example, since the transition to a market economy, middle-class and wealthy suburbs are beginning to ring the city of Moscow, surrounding the long-cultivated garden plots of inner-city residents (**Figure 5.18**). However, given the instability of jobs in Moscow and other large cities, garden plots are likely to remain important supplements to family nutrition that urban residents will fight hard to keep.

Economic Reforms in the Post-Soviet Era

Russia's recent economic reforms have been ambitious but haphazard. Recently, rising oil prices put much more money into government coffers, especially in oil- and gas-rich Russia and the Central Asian states. Many economic sectors improved and, for some, life became luxurious. Then oil prices fell and ventures had to be curtailed.

Privatization and the Lifting of Price Controls The Soviet economy consisted almost entirely of industries owned and operated by the government. These have now been sold to private companies or individuals in a process called **privatization.** The hope is that they would be run more efficiently in a competitive free market setting. By 2000, approximately 70 percent of Russia's economy was in private hands, an earthshaking change from the 100 percent state-owned economy of 1991.

Another transforming reform is that prices are no longer kept artificially low by the government to make goods affordable to all. Instead, they are determined by supply and demand. The lifting of price controls led to skyrocketing prices in the 1990s for the many goods that were in high demand but also in short supply. While a tiny few grew rich, many people had to use their savings to pay for basic necessities such as food. Eventually, as opportunities opened and competition developed, the supply of goods increased and prices fell. But in the interim, many people suffered.

Unemployment and the Loss of Social Services Since becoming privatized, most formerly state-owned industries have cut many jobs in an attempt to compete against more efficient companies in the global free-market economy. Losing a job is especially devastating in a former Soviet country because one also loses the subsidized housing, health care, and other social services that were usually provided along with the job. Work was also the center of community life, so when a job ended, so too did the social support group.

The new companies that have emerged rarely offer benefits to employees. There is little job security because most small private firms appear quickly and often fail. Discrimination is also a problem given the absence of equal opportunity laws. Job ads often contain wording such as "only attractive, skilled people under 30 need apply."

Unemployment figures for the region vary widely but tend toward the high end, as has been common since the end of the Soviet era. By early 2010, official unemployment rates were just 1.9 percent in Belarus and 9.2 percent in Russia, but 27 percent in parts of Caucasia and Central Asia and in some recent years in

FIGURE 5.18 Gardening by city dwellers. Russians, like many Europeans, often journey on weekends to the urban fringe, where they cultivate small garden plots. Here family members harvest strawberries from their plot outside Moscow.

Turkmenistan as high as 60 percent. Actual unemployment may be higher because many remaining state-owned firms cannot pay employees still listed as workers. It is estimated that in the mid-1990s, three-fifths of the Russian labor force was not being paid in full and on schedule. The rate of **underemployment,** which measures the number of people who are working too few hours to make a decent living or who are highly trained yet working at menial jobs, is even higher in all these countries.

Small Businesses Face Challenges of Inflation and Institutionalized Corruption In the fall of 2009, Russia's president, Dmitri Medvedev, recognizing the drag placed on progress by unemployment and the lingering state-owned firms, called for more diversification and privatization of the economy. As discussed above, the Russian economy is overly dependent on oil and gas, even more so than it was 10 years ago. To encourage diversification, Medvedev urged that more take up entrepreneurialism. But to make this happen, the banks will have to give better credit terms to small businesses, which up to now they have been loath to do. For many small businesses, because of inflation, the cost of credit has risen from 8 percent to 25 percent.

Moreover, crippling bureaucratic tangles and bribery make starting a business outlandishly expensive. A business license officially costs under $100, but bribes push actual costs to as much as $10,000. And finding space is nearly impossible. The government still owns many buildings, but efforts to fairly allot space also end in bribery. Even those who make it through the labyrinth of setting up a business must then face protection racketeers. Small businesses may indeed eventually be the salvation of the Russian economy, as they are in so many developed countries where they are the chief engine of job growth; but for this to happen, high officials will themselves have to destroy the culture of corruption that is so pervasive in Russia.

privatization the sale of industries that were formerly owned and operated by the government to private companies or individuals

underemployment the condition in which people are working too few hours to make a decent living or are working at menial jobs even though highly trained

FIGURE 5.19 The informal economy.
Women inspect shirts for sale by a street vendor in Moscow before the winter holidays. Like almost all businesses in Moscow, street vendors pay "protection" money to gangsters. A much smaller number pay taxes.

The Growing Informal Economy To some extent, the new informal economy in the region is an extension of the old one that flourished under communism. The *black market* of that time was based on currency exchange and the sale of hard-to-find luxuries. In the 1970s, for example, savvy Western tourists could enjoy a vacation on the Black Sea paid for by a pair or two of smuggled Levi's blue jeans and some Swiss chocolate bars. Today, many people who have lost stable jobs due to privatization now depend on the informal economy for their livelihood (Figure 5.19).

Workers in the informal economy tend to be very young unskilled adults, retirees on minuscule pensions, or those with only a low-level education. They may simply be employed "off the books" in a formal sector enterprise. Some may have formal sector jobs by day and informal second jobs at night. The majority, however, are unregistered full-time clandestine entrepreneurs who operate out of their homes, selling cooked foods, vodka made in their bathtubs, electronics they have assembled, or smuggled Asian-made T-shirts, to name just a few products. Such a large percentage of the economy is now in the informal sector that in many countries of the region, people may actually be better off financially than official gross domestic product (GDP per capita) figures suggest (see Figure 5.24A on page 240).

Despite the fact that the informal economy helps people survive and keeps the lid on social discontent, it is not popular with governments. Unregistered enterprises do not pay business and sales taxes and usually are so underfinanced that they do not grow into job-creating formal sector businesses. In many cases, informal businesspeople must pay protection money to local gangsters (the so-called *mafia tax*) to keep from being reported to the authorities.

Personal Vignette Natasha is an engineer in Moscow. She has managed to keep her job and the benefits it carries, but in order to better provide for her family, she sells secondhand clothes in a street bazaar on the weekends. "Everyone is learning the ropes of this capitalism business," she laughs. "But it can get to be a heavy load. I've never worked so hard before!" Asked about her customers, Natasha says, "Many are former officials and high-level bureaucrats who just can't afford the basics for their families any longer." Some are older retired people whose pensions are so low that they resort to begging on Moscow's elegant shopping streets close to the parked Rolls-Royces of Moscow's rich. These often highly educated and only recently poor people buy used sweaters from Natasha and eat in nearby soup kitchens. *(Source: This composite story is based on work by Alessandra Stanley, David Remnick, David Lempert, and Gregory Feifer.)* ■

THINGS TO REMEMBER

1. The Soviet centrally planned economy was less efficient than market economies in allocating resources. An enduring legacy of Soviet central planning is the location of huge industrial projects in the farthest reaches of Soviet territory.

2. Command economy strategy created a need for modern, efficient land transportation systems such as railroads and highways, but the construction of these systems has long been held back by the climate and vastness of the region.

3. The struggle to control this region's huge crude oil and natural gas resources is increasing international tensions, while the need to profit from them is creating new incentives for international cooperation.

4. Across most of the region, agriculture is precarious, either hampered by a short growing season and boggy soils or requiring expensive inputs of labor, water, and fertilizer. Yet global corporations have shown interest in agricultural processing in Russia and some of the post-Soviet states. Further, family urban gardens have long been an important source of nourishment across the region, especially in the Russian Federation.

5. The benefits of economic reforms in the post-Soviet era have been dampened by widespread corruption, lower incomes for many, and the loss of social services.

Democratization in the Post-Soviet Years

The beginnings of democratization came to the USSR somewhat unexpectedly. In 1985, the Soviet Union's reform-minded president, Mikhail Gorbachev, responded to the various pressures for change. He opened up public discussions of social and economic problems, an innovation known as **glasnost**. He also attempted to revitalize the Soviet economy through **perestroika**, or restructuring, though these efforts resulted in little real change. Gorbachev is credited with effectively ending the Cold War by opening dialogue with the West (Europe and the United States).

In August of 1991, in reaction to Gorbachev's liberalizing policies, a group of Communist Party and military officials led an attempt to seize control of the Soviet government. Their failure led to the abrupt dissolution of the Soviet Union. Since 1991, democratization and the newly introduced market economy have proceeded down a bumpy road (see the map in Photo Essay 5.4 on page 232).

Do Elections Lead to a Democratic Society? In Russia, Vladimir Putin, a former high official in the *KGB* (Russia's intelligence agency, during the Cold War, now known as the Federal Security Service or FSB), took over as acting president in 1999. It was hoped that he would rescue Russia after a tumultuous period led by Boris Yeltsin, who, as historian Peter Rutland says, was given to "drunken pranks and erratic policy shifts." During Yeltsin's presidency, corruption thrived, foreign investors fled, unemployment was rampant, and living standards fell.

In 2000, Putin was elected as president and served the two terms allowed by law, during which he exercised tight control over political and economic policy and consolidated political power in Moscow. He was very popular for bringing relative peace and prosperity and for restoring Russia's image of itself as a world power, despite disallowing meaningful democratic reforms at the local, state, and national levels, and extending government control over the press and the media (see page 234). Ostensibly, criticism of the government is now allowed, but behind the scenes a new autocracy, known as the *securocrats* or *siloviki* (former FSB functionaries) governs Russia, and critics fear retribution. Putin is thought to have at least tacitly allowed the repression and possibly even the execution of rivals, dissidents, and critics of his policies.

The graph in Figure 5.20, developed by the political analyst Olga Kryshtanovskaya, gives insight into the role that the military and security personnel (*siloviki*) played as advisors to Putin. She assessed the percent of *siloviki* in government and elite groups during the Gorbachev, Yeltsin, and Putin administrations, and found that there has been a striking increase over time. Other analysts corroborate these findings and assert that this does not bode well for democracy because the *siloviki* see themselves as an elite, superior to ordinary citizens and therefore the rightful "bosses." By 2006, Kryshtanovskaya found that 77 percent of the people in the top 1016 government positions had backgrounds in security agencies.

In 2008, Putin tapped a former aide, Dmitri Medvedev (see Thematic Concepts, Part L on page 211), as the heir apparent to the presidency. Medvedev was then elected president with no opposition and Putin engineered his own appointment to the

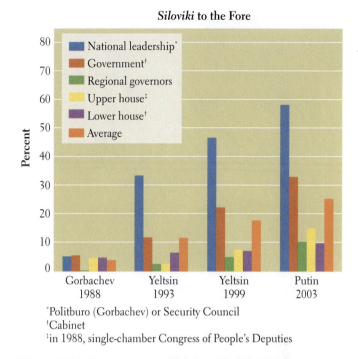

Siloviki to the Fore

* Politburo (Gorbachev) or Security Council
† Cabinet
‡ in 1988, single-chamber Congress of People's Deputies

Figure 5.20 Government officials with *Siloviki* ties. Percent of people with military/security (*Siloviki*) backgrounds in elite groups during the administrations of Gorbachev, Yeltsin, and Putin. [Source: Olga Kryshtanovskaya, "The Making of a Neo-KGB State," The *Economist*, August 23, 2007, at http://www.economist.com/world/displaystory.cfm?story_id=E1_JGRNGNT.]

office of prime minister, which has no term limits. Many see these moves as an effort by Putin to hold on to power indefinitely.

117. Putin Confirmation

123. New Russian Leader

Elsewhere in the region, progress toward true democracy is similarly limited. In Belarus, Moldova, Russian Caucasia (Chechnya), and Central Asia, authoritarian leaders unaccustomed to criticism or to sharing power are still the norm (see the map in Photo Essay 5.4). Elections often result in the systematic intimidation of voters and arrests of the political opposition

110. Kazakhstan's President to Meet with George W. Bush

The Color Revolutions and Democracy In a few of the post-Soviet states, there have been a number of relatively peaceful transitions to somewhat greater democracy—the so-called *Color Revolutions*, including the Rose Revolution in Georgia (November 2003–2004, discussed in the opening vignette), the Orange Revolution in Ukraine (2004–2005; see Photo Essay 5.4B), and the Tulip Revolution in Kyrgyzstan (April 2005).

The democratization movements were named the Color Revolutions because they were largely led by coalitions of educated young adults who rallied under various symbolic colors meant to unite them despite differences. For a time it appeared that the movements would remain peaceful and result in significant democratic participation by the citizenry.

glasnost literally, "openness"; the opening up of public discussion of social and economic problems that occurred in the Soviet Union under Mikhail Gorbachev in the late 1980s

perestroika literally, "restructuring"; the restructuring of the Soviet economic system in the late 1980s in an attempt to revitalize the economy

Since World War II, most conflicts in this region occurred in Caucasia and Central Asia. Most occurred just before or shortly after the breakup of the Soviet Union. Since then, pro-democracy movements have emerged in several countries, where they hold the promise of political change without violent conflict. However, their long-term effectiveness remains to be seen.

A A fighter plane lies in Afghanistan among other pieces of wreckage dating from the Soviet Union's 10-year war in that country. The Soviets intervened in 1979 to prop up an authoritarian Afghan government that had initiated radical reforms and responded to widespread dissent by killing off the political opposition. The war resulted in at least 1.5 million deaths and a humiliating defeat for the USSR. Fourteen thousand Soviet soldiers died and billions of dollars were spent and practically nothing gained. Public outrage over the Afghan war played a major role in the breakup of the Soviet Union. Since the fall of the USSR, democratization has proceeded at a much more rapid pace throughout the region.

Democratization and Conflict

Armed Conflicts and Genocides with High Death Tolls Since 1945

- Ongoing conflict
- 1000–5000 deaths
- 5000–50,000 deaths
- 50,000–300,000 deaths
- 300,000–1,000,000 deaths

Democratization Index

- Full democracy
- Flawed democracy
- Hybrid regime
- Authoritarian regime
- No data

B Protesters pushing for a transition from authoritarian to more democratic government in Ukraine decorated police barricades with balloons and flowers during the Orange Revolution of 2004–2005. The movement followed a model pioneered in Serbia and Georgia, where student groups, funded by NGOs and governments in Europe and the United States, spearheaded grassroots political campaigns featuring massive nonviolent demonstrations. A similar "color revolution" occurred in Kyrgyzstan, and one is developing in Belarus.

C A bombed-out apartment building in Grozny, Chechnya, where separatist rebels have fought two unsuccessful wars and conducted ongoing terror campaigns against Russia since the fall of the Soviet Union. Political instability in Caucasia is one of the greatest forces working against democratization in the region.

However, they have not resulted in durable reforms. Although elections have taken place, many forms of authoritarian control and practices that disrupt orderly legislative and judicial procedures remain. Elected representative assemblies often act as mere rubber stamps for very strong presidents and exercise only limited influence on policy. In **Caucasia**, democracy is the victim of continuing political instability. In Ukraine, the Orange Revolution began in 2004 (see Photo Essay 5.4B). Before the 2004 election, candidate Victor Yuschenko, who favored closer ties with the West and Ukrainian membership in NATO, was poisoned with dioxin. He recovered but was horribly disfigured and his health was undermined. After the obviously rigged 2004 election, supposedly won by Viktor Yanukovych (who favored closer ties with Russia), widespread protests went on for days. The persistent protests and international disapproval of the rigged election and of the apparent poisoning of Yuschenko led to Yuschenko's being declared president in 2005. His term was characterized by years of conflict between two deeply divided, equally powerful factions. Yuschenko's faction favored closer ties with the European Union, while the opposition, buttressed by a large Russian-speaking minority, favored close ties with Russia. In the 2009 election, Yuschenko finished far back in the ranks of candidates for president. Yanukovych, the pro-Russian original winner in 2004, came in first and was in a runoff in 2010 against Yulia Tymoshenko, a co-leader of the Orange Revolution. Yanukovych won the runoff and Tymoshenko eventually conceded. The Orange Revolution was apparently over, and democracy succeeded, assuming the election was fair.

Cultural Diversity and Democracy Russia's long history of expansion into neighboring lands has left it with an exceptionally complex internal political geography. As the Russian czars and the Soviets pushed the borders of Russia eastward toward the Pacific Ocean over the past 500 years, they conquered a number of small non-Russian areas. These 30 republics and 10 autonomous regions now constitute 25 percent of the Russian Federation's land area (**Figure 5.21**). Many have significant ethnic minority populations that are descendants of indigenous people or trace their origins to long-ago migrations from Germany, Turkey, or Persia. Even during the Soviet period, but especially after, minorities organized to resist Russification and to enhance local ethnic identities.

Shortly after the breakup of the Soviet Union, several internal republics demanded greater autonomy, and two of them, Tatarstan and Chechnya, declared outright independence. Tatarstan has since been placated with greater economic and political autonomy. However, Chechnya's stronger resistance to Moscow's authority has led to the worst bloodshed of the post-Soviet era.

Chechnya, located on the fertile northern flanks of the Caucasus Mountains, is home to 800,000 people (see Figure 5.21 inset map). The Chechens converted to the Sunni branch of Islam in the 1700s, largely in response to Russian oppression. Since then, Islam has served as an important symbol of the Chechen identity and an emblem of resistance against the Orthodox Christian Russians, who annexed Chechnya in the nineteenth century. Stalin, born an Orthodox Christian in nearby Georgia, saved some of his most vicious wrath for the Chechens. In 1924, he forced the

Chechens onto collective farms, which supplied produce to distant Soviet cities. Chechen resistance against the Russians continued. In 1942, when the Germans invaded Russia, they did so through the Caucasus and encountered some Chechen separatists who briefly considered the possibility of finding protection under the Germans. Stalin seized the opportunity to accuse the Chechens of collaborating with the Germans and drew up plans to deport all Chechens. In 1944, close to 400,000 Chechen villagers were surrounded, loaded at gunpoint onto trains and sent to Kazakhstan and the Russian Far East (Siberia), where they were held in concentration camps, with many dying of starvation. The Chechens were finally allowed to return to their villages in 1957, but a heavy propaganda campaign portrayed them as traitors to Russia, preparing the way for later official demonizing.

> **Caucasia** the mountainous region between the Black Sea and the Caspian Sea

In 1991, as the Soviet Union was dissolving, Chechnya declared itself an independent state, which to Russians was a dangerous precedent that could spark similar demands by other cultural enclaves throughout Russia. Wishing to retain the agricultural and oil resources of the Caucasus, and planning to build pipelines across Chechnya to move oil and gas to Europe from Central Asia, Russia once again sought to squelch any thoughts of independence. Since 1991, Chechen guerrillas have used terrorism to repeatedly challenge Russian authority, and the Russians have responded with bombing raids and other military operations that have killed tens of thousands and created 250,000 refugees (see Photo Essay 5.4C).

Russia's brutality throughout the conflict has raised not only doubts about its commitment to human rights but about its overall ability to address internal political dissent. Since 2008, Russia has invested substantially in rebuilding Chechnya and on April 16, 2009, Russia's counterterrorism effort officially ended.

116. Chechnya Hangs On to Uneasy Peace

Just south of Chechnya, conflicts between Russia and Georgia over the ethnic republics of South Ossetia and Abkhazia have worsened in the post-Soviet era. Both republics became part of Georgia at the behest of Joseph Stalin (a native Georgian). He then moved Georgians and Russians into Ossetia and Abkhazia so that the native people became a minority in their own place. More recently, Russia has strategically supported the ethnic Ossetian and Abkhazian populations' agitation for secession from Georgia, even granting Russian citizenship to between 60 to 70 percent of the non-Georgian population. Russia may have done this in retaliation for Georgia's increasingly close relations with the United States and the European Union, as evidenced by its candidacy for NATO membership. A pipeline that links the oil fields of Azerbaijan with the Black Sea via Georgia is another source of contention. Russia would like to enhance its control over the oil resources of the Caspian Sea region by routing pipelines through Russian-controlled territory.

In August 2008, Georgia's military, attempting to gain control over rebelling parts of South Ossetia, engaged with the Russian army on Russian territory under circumstances that are in contention. South Ossetia had become a major smuggling center between Russia and Georgia, including the smuggling of nuclear

FIGURE 5.21 Ethnic character of Russia and percentage of Russians in the post-Soviet states. Russia, with all of its internal republics and autonomous regions, is formally called the Russian Federation. The ethnic character of many of the more than 30 internal republics was changed by the policy of central planning, so Russians now form significant minorities. As of 2002, the ethnic makeup of Russia was 79.8 percent Russian, 3.8 percent Tatar, 2 percent Ukrainian, 1.2 percent Bashkir, 1.1 percent Chuvash, and 12.1 percent other. [Adapted from James H. Bater, *Russia and the Post-Soviet Scene* (London: Arnold, 1996), pp. 280–281; Graham Smith, *The Post Soviet States* (London: Arnold, 2000), p. 75; and *The World Factbook 2009,* at https://www.cia.gov/library/publications/the-world-factbook/.]

materials. Russia responded to the Georgian attack with its much larger military, driving Georgia's forces out of South Ossetia and following them back into Georgia, heading for the capital. At the same time, fighters in Abkhazia drove Georgia's military out of that province, and Russian planes bombed a town near a Georgian pipeline. At present, it is not clear if Abkhazia and South Ossetia will break away and become independent countries, become provinces within the Russian Federation, or be satisfied by offers of greater autonomy within Georgia's federal structure.

The Media and Political Reform In the Soviet era, all communication media were under government control. There was no free press, and public criticism of the government was a punishable offense. Yet it was journalists, risking retribution for criticizing public officials and policies, who were instrumental in bringing an end to the Soviet Union. Between 1991 and the early 2000s, the communications industry was a center of privatization, and several media tycoons emerged to challenge the Russian government. Privately owned newspapers and television stations regularly criticized

the policies of various leaders of Russia and the other states. It appeared that a free press was developing.

Vladimir Putin's rise to power was a turning point, after which the most critical newspapers and TV stations in Russia were shut down. Since then, critical analysis of the government has become rare throughout Russia. Journalists openly critical of Putin's policies have been treated to various forms of censorship, exile, or violence. From 2000 to 2008, more than 80 were killed, most notably Ana Politkovskaya, a well-known investigative journalist who covered Russia's repression of Chechens (see Thematic Concepts, Part K on page 211 and the discussion of Chechnya on page 233).

Closely related to the struggle to develop a free press is the availability of communication technology, including the Internet, to the general public (Table 5.1). Television sets are widely available, and many people receive European and other stations via satellite dishes. Access to telephone landlines is still limited, but mobile phone use has increased dramatically across the region and is essential to many new enterprises. Russia now has more

	TABLE 5.1 Increase in Internet use, 2000–2009		
Country	Internet users as percentage of population, 2007	Internet users as percentage of population December, 2009	Percentage increase in internet users since 2000
Armenia	5.1	6.4	536.7
Azerbaijan	8.0	29.7	20,271.1
Belarus	35.0	32.2	1626.1
Georgia	4.0	22.2	5020.0
Kazakhstan	2.7	14.9	3185.7
Kyrgyzstan	5.2	15.6	1547.3
Moldova	10.9	19.7	3300.0
Russia	16.5	32.3	1359.7
Tajikistan	0.1	8.2	29,900
Turkmenistan	0.5	1.5	3650.0
Ukraine	11.5	22.7	5077.0
Uzbekistan	3.3	8.9	32,820.0
United States	69.4	76.3	145.8

Sources: Internet World Stats, "Internet Usage in Asia," at http://www.internetworldstats.com/stats3.htm; Internet World Stats, "Internet Usage in Europe," at http://www.internetworldstats.com/stats4.htm; and Internet World Stats, "Internet Usage and Population in North America," at http://www.internetworldstats.com/stats14.htm.

mobile phones per person than does the United States, where landlines are still widely used. Mobile phone service, however, remains expensive and unreliable, and it is not available to many people (Figure 5.22). The same is true of personal computers.

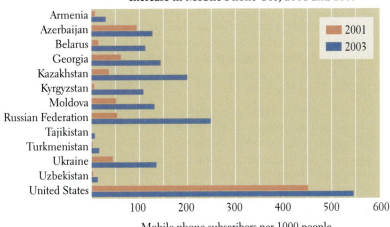

Increase in Mobile Phone Use, 2001 and 2003

Mobile phone subscribers per 1000 people

FIGURE 5.22 Access to mobile phones in Russia and the post-Soviet states. Access to information throughout the region is low compared with that in more developed countries. Since 2001, however, people's access to mobile phones has increased significantly. Note that in Russia, mobile phone use is higher than it is in the United States. [*United Nations Human Development Report 2003* and *2005* and *2007–2008* (New York: United Nations Development Programme), Tables 11 and 13.]

Corruption and Organized Crime

Any potential benefits of political and economic reforms in the post-Soviet era have been dampened by widespread corruption and the growth of organized crime. Many oligarchs became closely connected to the so-called *Russian Mafia*, a highly organized criminal network dominated by former KGB (the Soviet intelligence agency and counterpart to the CIA) and military personnel who control the thugs on the streets. The Russian Mafia extended its influence into nearly every corner of the post-Soviet economy, especially in illegal activities and the arms trade.

In the post–9/11 world, concern spread that military corruption in this region could put nuclear weapons from the former Soviet arsenal in the hands of terrorists.

After huge military funding cuts, weapons, uniforms, and even military rations were routinely sold on the black market. In 2007, Russian smugglers were caught trying to sell nuclear materials on the black market. Recent developments are more encouraging. In an effort to fight corruption and the temptation to sell nuclear materials on the black market, the Russian government has given impoverished military personnel long-delayed pay raises or termination compensation. Moreover, all countries in the region are now cooperating with the International Atomic Energy Agency in controlling nuclear material. For years, the United States bought weapons-grade uranium from Russia as a way to remove it from future use in nuclear weapons, but in May 2009, for the first time the United States agreed to buy virgin uranium from Russia, thus opening up the possibility for legal uranium commerce in Russia, which, it is hoped, will keep it off the black market.

THINGS TO REMEMBER

1. Democratization came to the USSR somewhat unexpectedly in 1985. Since 1991, democratization and the newly introduced market economy have proceeded down a bumpy road.

2. Since World War II, most conflicts in this region occurred in Caucasia and Central Asia, mainly just before or shortly after the breakup of the Soviet Union. In the years after the breakup, pro-democracy movements have emerged in several countries, where they hold the promise of political change without violent conflict. However, their long-term effectiveness remains unclear.

3. There continue to be limitations on the free press and related media across the region, but people who have experienced glimmers of democracy are seeking increased access to information through wireless sources: mobile phones, computers, and the Internet.

4. The potential benefits of political and economic reforms in the post-Soviet era have been dampened by official corruption and the growth of organized crime.

Sociocultural Issues

When the winds of change began to blow through the Soviet Union in the 1980s, few anticipated the rapidity and depth of the transformations or the social instability that resulted. On the one hand, new freedoms have encouraged self-expression, individual initiative, and cultural and religious revival; on the other hand, the post-Soviet era has brought very hard times to many as their jobs and social safety net were obliterated.

Population Patterns

This region shares some of the same population characteristics as the United States and Europe—low birth rates and an aging population—both usually features of wealthy and developed societies. But in Russia and the post-Soviet states, low birth rates and the rate of aging in the population are more extreme, and, generally speaking, human well-being is much lower than in the United States, Europe, and other developed areas of the world. Several circumstances contribute to this situation.

Population Distribution and Urbanization The region of Russia and the post-Soviet states is one of the largest on earth but is the least densely populated, with only about 249 million people. European Russia is the most heavily settled zone (Photo Essay 5.5), but even this subregion, with an average density of 22 people per square mile (8.5 per square kilometer), is still much less dense than the United States (80 per square mile; 30 per square kilometer). A broad area of moderately dense population forms an irregular triangle that stretches from Ukraine on the Black Sea north to St. Petersburg on the Baltic Sea and east to Novosibirsk, the largest city in Siberia. In this triangle, settlement is highly urbanized, but the cities are widely dispersed; the basic urban infrastructures were centrally planned for industrial workers with an emphasis on giant apartment blocks. The capital city of Moscow, with 11 million people, is a primate city in Russia (see Photo Essay 5.6A on page 239), as is Kiev in Ukraine, Minsk in Belarus, and Chisinau in Moldova.

Beyond Novosibirsk on the West Siberian Plain, settlement follows an irregular and sparse pattern of industrial and mining development in widely spaced cities, stretching east across Siberia. These economic activities and the cities they support are linked primarily to the route of the Trans-Siberian Railroad (see Figure 5.13 on page 225). Although Siberia is a desolate, lonely place, nearly 90 percent of its people are concentrated in a few relatively large urban areas. The costs of maintaining these settlements in this remote and harsh environment are considerable.

A secondary spur of dense settlement extends south from European Russia into Caucasia, the mountainous region between the Black Sea and the Caspian Sea, where there are several primate cities of well over 1 million people each. In Central Asia, another patch of relatively dense settlement is centered on the cities of Tashkent and Almaty (both over 1 million in population) and the new capital of Kazakhstan, Astana, which has 750,000 people (see Photo Essay 5.6C). Along major Central Asian rivers during the Soviet period, the development of irrigated cotton farming and mineral extraction resulted in patches of high rural density, fueled partly by ethnic Russian immigration.

Many of the major cities in this region grew rapidly after 1991 with the closing of collective farms and rural industries, but more recently some cities have begun to shrink in size (see below). Urban life is hampered by the residual effects of grossly inadequate central planning during the Communist era. Housing shortages and cramped drab apartments with shared crude kitchens and bathrooms mean that even middle-class urban families live in slum-like conditions. Inadequate sewage, garbage, and industrial waste management pose serious long-term health and environmental threats. Crumbling infrastructures and the decline of once-beautiful urban amenities like parks (see Photo Essay 5.6B) add to the depressing atmosphere, especially in winter. Time-wasting and polluting traffic jams are so extreme that it is said in Moscow, the average car speed is only 9–10 miles (15–18 kilometers) per hour.

Shrinking Populations: High Death Rates, Low Birth Rates This region is experiencing a unique variant of the demographic transition (see Chapter 1 on page 20). In all but the Caucasian and Central Asian states, populations are shrinking faster than in any other world region. During the Soviet era, the increase in women's opportunities to become educated and work outside the home curtailed population growth. Severe housing shortages were an additional disincentive to procreation. Many chose to have only one or two children. Free health care and adequate retirement pensions also helped lower incentives for large families. Population numbers grew very slowly, but during the economic crisis brought on by the breakup of the Soviet Union, Russia's population has shrunk more than 5 percent, to 142 million. The United Nations predicts that Russia will drop to just 116 million people by 2050. Populations are also shrinking in Belarus, Moldova, and Ukraine. In Central Asia, where birth rates are higher, all

Population trends in this region are highly uneven. While Central Asia and some countries in Caucasia are growing, populations in the rest of the region are shrinking. Some of this may be related to cultural differences or to varying levels of dependence on social welfare institutions that collapsed with the end of the Soviet Union.

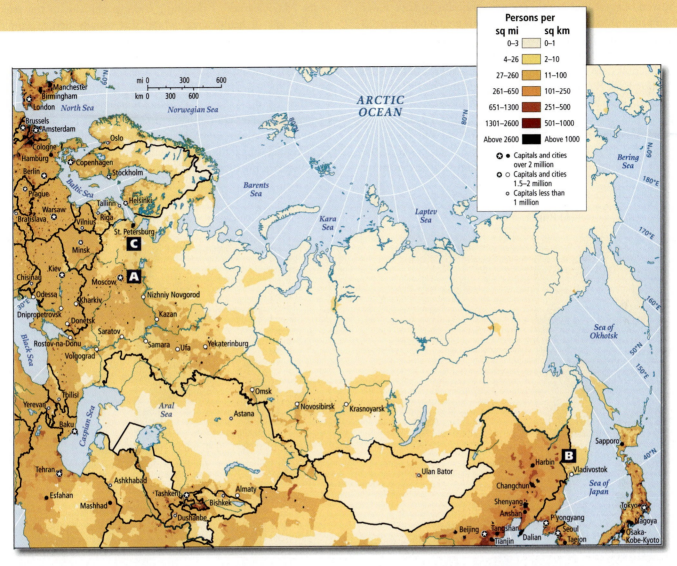

Persons per

sq mi	sq km
0–3	0–1
4–26	2–10
27–260	11–100
261–650	101–250
651–1300	251–500
1301–2600	501–1000
Above 2600	Above 1000

⊕ ● Capitals and cities over 2 million
⊕ ○ Capitals and cities 1.5–2 million
○ Capitals less than 1 million

A A child in Moscow. Fertility rates declined dramatically after the collapse of the Soviet Union, but they have increased in recent years with better economic conditions.

B Unemployed men fishing in Vladivostok, Russia. The loss of jobs and associated health-care benefits has driven death rates to record levels, especially for men.

C Men on a drinking binge in St. Petersburg. Alcohol-related deaths are high in Russia and tend to rise during periods of economic stress, such as that which followed the collapse of the Soviet Union.

countries are losing population due to emigration, as is also happening in the Caucasian states, where Georgia is expected to have 28 percent fewer people by 2050.

Surveys suggest that people are now choosing to have fewer children primarily out of concern over gloomy economic prospects for the near future, but also out of the desire to make money and have fun. Russia is attempting to reverse population loss by paying couples to contribute the time and effort to raise more children and by attracting back Russians and their dependents who live abroad. In 2007 and 2008, Russia spent $300 million to send emissaries to the far corners of the earth (Brazil, Egypt, Germany, and all the post-Soviet states) to lure "returnees." Only 10,300 were recruited.

In addition to negative birth and migration rates, much of the reason for the population decline is the declining life expectancy.

In Russia, for example, between 1990 and 2008, male life expectancy dropped from 63.9 to 60 years, the shortest in any industrialized country. Female life expectancy dropped from 74.4 to 73 years. Major causes of declining life expectancies in the region are the loss of health care, which was usually tied to employment, and the physical and mental distress caused by lost jobs and social disruption. The high male death rate is explained in large part by alcohol abuse and related suicides (women are less prone to both but tend to smoke in excess). In Russia, approximately 7 million deaths per year are alcohol related (see Photo Essay 5.5C on page 237). This is almost 100 times the number of similar deaths in the United States, and Russia's population is only half as large!

Population pyramids for several countries (**Figure 5.23**) show the overall population trends in the region and reflect geographic

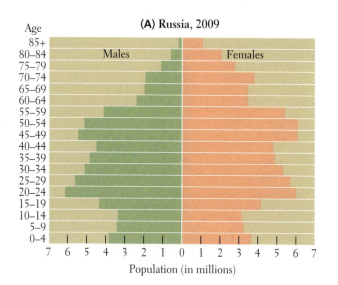

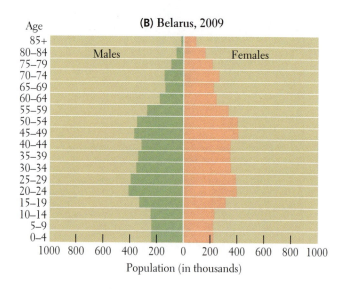

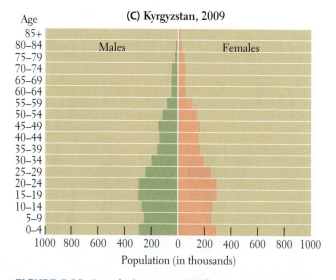

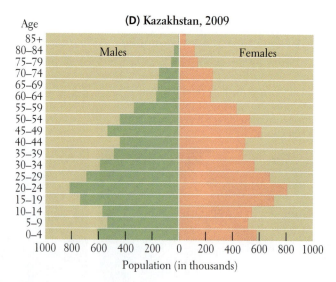

FIGURE 5.23 Population pyramids for Russia, Belarus, Kyrgyzstan, and Kazakhstan. Note that the pyramid for Russia is at a different scale (millions) than the other three pyramids (thousands) because of Russia's larger total population. This difference does not significantly affect the pyramid's shape. [Adapted from U.S. Census Bureau, Population Division, "Population Pyramids of Russia," "Population Pyramids of Belarus," "Population Pyramids of Kyrgyzstan," and "Population Pyramids of Kazakhstan," International Data Base, 2009, at http://www.census.gov/ipc/www/idb/pyramids.html.]

An uneven pattern of urbanization is developing in this region. In several countries, the largest cities are growing rapidly, as they are centers of new development and globalization. Meanwhile, in Russia and elsewhere, many cities are shrinking as the overall population declines.

A Moscow is already the region's primate city; its population of almost 11 million is larger than the next three largest cities combined. Moscow produces 20 percent of Russia's GDP, and this number is growing, fueled by the city's status as the headquarters of Russia's globalizing economy. By 2020 Moscow will gain 900,000 inhabitants, five times what any other city in the region will gain.

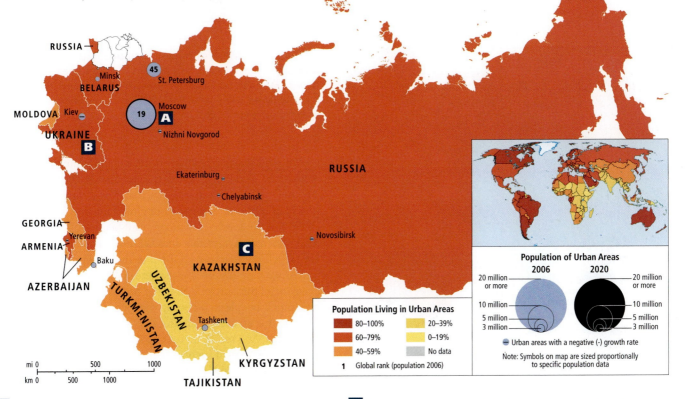

Population Living in Urban Areas

- 80–100%
- 60–79%
- 40–59%
- 20–39%
- 0–19%
- No data

1 Global rank (population 2006)

Population of Urban Areas

2006 — 2020

20 million or more
10 million
5 million
3 million

⊖ Urban areas with a negative (-) growth rate

Note: Symbols on map are sized proportionally to specific population data

B A public park in Kharkov, Ukraine suffers from neglect, a symptom of the city's decline. Kharkov has lost 10 percent of its population since 1989, and it will lose another 10 percent by 2020. The population decrease is related both to Ukraine's overall population decline and to the closing of many of Kharkov's defense-related industries after the collapse of the USSR.

C The presidential palace and other public buildings in Astana, Kazakhstan, which doubled its population in 10 years after being designated the new capital. Astana's growth has been funded by Kazakhstan's new oil and gas wealth.

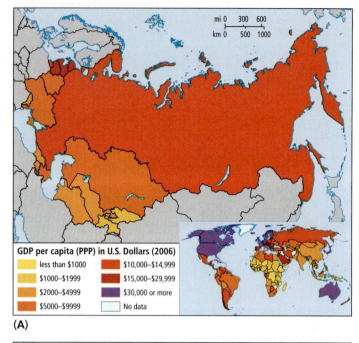

(A)

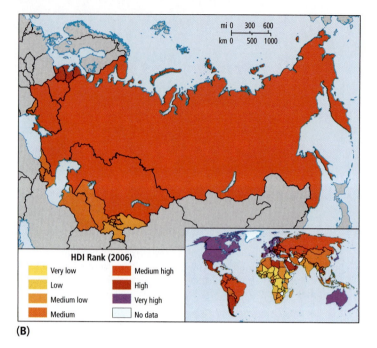

(B)

FIGURE 5.24 Maps of human well-being. (A) GDP. **(B)** HDI. **(C)** Female earned income as a percent of male earned income. [Sources: Maps adapted from data: "Human development indices," (Maps A and B): Table 2, pages 28–32; (Map C): Table 5, pages 41–44 at http://hdr.undp.org/en/media/HDI_2008_en_Tables.pdf.]

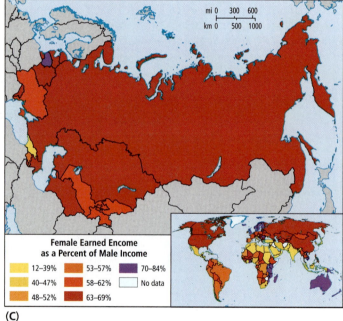

(C)

differences in patterns of family structure and fertility. The pyramids for Belarus and Russia resemble those of European countries (for example, Germany, as shown in Figure 4.15A on page 191). They are significantly narrower at the bottom, indicating that birth rates in the last several decades have declined sharply. The narrower point at the top for males in all four pyramids depicts their much shorter average life span. The base of the pyramid for Kazakhstan in Central Asia is also narrowing as a drop in birth rates accompanies urbanization; however, in Kyrgyzstan births appear to have recently rebounded to levels of a decade ago.

Geographic Patterns of Human Well-Being During the Communist and post-Communist eras, local conditions and changing

government involvement in social welfare policies have shaped patterns of well-being across this region. In Maps A and B of Figure 5.24, the shading shows that Russia and the post-Soviet states fall in the middle range of global income (GDP per capita PPP) and in levels of human well-being (UNHDI). Only in Tajikistan and Kyrgyzstan does GDP per capita (PPP) fall below U.S.$2000 per year and HDI fall in the medium-low category.

As we have observed elsewhere, GDP per capita can be used as a general measure of well-being because it shows in a broad way how much people in a country have to spend on necessities, but because it is an average figure it does not indicate how income is distributed. As a result, there is no way of knowing if a few are rich and most poor, or if income is more equally apportioned. In this particular region, the relatively high HDI ranks are in large part a holdover from the Soviet era when socialism kept wages relatively equal (though usually lower on average outside European Russia) and the strong social safety net provided basic health care, food, and housing for all. Since 1991, disparities in wealth have steadily increased within the Russian Federation and each of the post-Soviet states. We have observed that in Russia and elsewhere in the region, a few have become fabulously wealthy; their wealth has undoubtedly skewed the average figure. GDP per capita (PPP) is actually well below the U.S.$10,000 indicated on Map A for many Russians and well below the stated GDP per capita figures for the other states, as well.

Figure 5.24B depicts the ranks on the United Nations Human Development Index (UNHDI), which is a calculation (based on adjusted real income, life expectancy, and educational attainment) of how well a country provides for the well-being of its citizens. It is very possible to have a high GDP rank and a lower HDI rank if social services are inadequate or if there is a wide disparity of wealth; but in this case Russia again falls in the medium-high category, and Kazakhstan, Ukraine, and Belarus actually improve their rankings, joining Russia. Their higher HDI rank relative to GDP is a holdover result of the Soviet emphasis on broad education and basic social services that greatly improved human well-being over most of the twentieth century. The lower HDI rankings in comparison to Russia for Central Asia (other than Kazakhstan), the Caucasian states, and Moldova reflect long-standing disparities, but conditions in these states have also worsened for many since 1991.

Figure 5.24C shows how well countries are doing in ensuring gender equality in earned income. First, as is true across the globe, nowhere in the region are women paid equally to men. Second, with only a few exceptions, most of the entire region ranks in the second category, paying women on average between 63 and 69 percent of what males are paid. Again, the Soviet system, which encouraged education and careers for women, is partly responsible for this record. In Ukraine, Armenia, Uzbekistan, Tajikistan, Kyrgyzstan, and especially Georgia, average female earned income as compared to males falls significantly lower. Still, as the global insert map shows, this region ranks higher in gender income equity than do Brazil, China, India, and nearly all of Africa.

Gender: Challenges and Opportunities in the Post-Soviet Era

Soviet policy that encouraged all women to work for wages outside the home was effective. By the 1970s, ninety percent of able-bodied women were working full time, giving the Soviet Union the highest rate of female paid employment in the world. However, the traditional attitude that women are the keepers of the home persisted. The result was the *double day* for women. Unlike men, most women worked in a factory or office or on a farm for 8 or more hours and then returned home to cook, care for children, shop daily for food, and do the housework (without the aid of household appliances). Because of shortages (the result of central-planning miscalculations), they also had to stand in long lines to procure necessities for their families.

The Female Labor Force When the first market reforms in the 1980s reduced the number of jobs available to all citizens, President Gorbachev espoused surprisingly outdated ideas on gender equality when he encouraged women to go home and leave the increasingly scarce jobs to men. Many women lost their jobs involuntarily, and by the late 1990s, 70 percent of the registered unemployed were women, despite the fact that due to illness, death, or divorce, many if not most were the sole support of their families. Consequently, many had to find new jobs.

On average, the female labor force in Russia is now better educated than the male labor force. The same pattern is emerging in Belarus, Ukraine, Moldova, and parts of Muslim Central Asia. In Russia, the best-educated women commonly hold jobs as economists, accountants, scientists, family-care physicians, and technicians, but they are unlikely to hold senior supervisory positions. As recently as 2005, the wages of women professional workers averaged 36 percent less than those of men.

The Trade in Women During the economic boom stimulated by marketization and oil and gas wealth, the "marketing" of women had become one of the less savory entrepreneurial activities. One part of this market is the Internet-based mail-order bride services aimed at men in Western countries. Any viewer can easily find such services, which pop up on virtually all Web sites related to Russia or other countries in the region. A woman in her late teens or early twenties, usually seeking to escape economic hardship, pays about $20 to be included in an agency's catalog of pictures and descriptions. (One Internet agency advertises 30,000 such women.) She is then assessed via email or Facebook by the prospective groom, who then travels—usually to Russia or Ukraine—to choose from the women he has selected from the catalog.

In the recent boom years, there has been a large increase in sex work. Very young women are often kidnapped from inside or outside the region for this work. Precise numbers are hard to come by. However, in 2000, the *Economist* estimated that 300,000 such women were smuggled each year into the European Union, where the sex trade then generated $9 billion annually (**Figure 5.25**).

FIGURE 5.25 The trade in women: Ukrainian sex workers in Amsterdam, Netherlands. Over 100,000 Ukrainian women work in the sex industry outside Ukraine, many of them in western Europe. Most were lured abroad with the promise of a good job in an office only to find themselves sold into prostitution.

The business of supplying sex workers is dominated by members of the Russian Mafia, who have been known to kidnap schoolgirls or deceive women desperately seeking jobs as domestic servants or waitresses in Europe and then force them to work as prostitutes or strippers.

The Political Status of Women The most effective way for women to address institutionalized discrimination is to achieve positions where they can affect wide-ranging policies—usually as elected officials. Although women were granted equal rights in the Soviet constitution, they never held much power. In 1990, women accounted for 30 percent of Communist Party membership, but just 6 percent of the governing Central Committee. The very few in party leadership often held these positions at the behest of male relatives. Ironically, since the fall of the Soviet Union, the political empowerment of women has advanced the least where democracy has developed the most, perhaps because of long-standing cultural bias against women in positions of power. Where governments are less democratic and more authoritarian, as in Belarus, Moldova, and Kyrgyzstan, women actually hold more legislative positions (**Figure 5.26**). However, in those countries, leaders who wanted to appear more progressive in the eyes of international donors may have promoted them undemocratically, choosing women who were least likely to work for change. Ironically, support among women for women's political movements is not widespread, as many fear being seen as anti-male or against traditional feminine roles.

Religious Revival in the Post-Soviet Era

The official Soviet ideology was atheism, and religious practice and beliefs were seen as obstacles to revolutionary change. The Orthodox Church was tolerated, but few went to church, in part because the open practice of religion could be harmful to one's career. Now, religion is a major component of the general cultural revival across the former Soviet Union. Throughout the Russian Federation, especially among indigenous ethnic minorities, people are turning back to ancient religious traditions. For example, the Buryats, from east of Lake Baikal in Siberia, who are related to the Mongols, are relearning the prayers and healing ceremonies of the Buryat version of Tibetan Buddhism, which they adopted in the eighteenth century. The shamans who lead them have organized into a guild to give official legitimacy to their spiritual work. They now pay taxes on their clergy income.

In European Russia (as well as in Georgia and Armenia), most people have some ancestral connection to Orthodox Christianity. Those with Jewish heritage form an ancient minority in European Russia and Caucasia, where they trace their heritage back to 600 B.C.E. Religious observance by both groups increased markedly in the 1990s, and many sanctuaries that had been destroyed or used for non-religious purposes by the Soviets were rebuilt and restored.

A countertrend to the robust revival of Orthodox Christianity is the spread of evangelical Christian sects from the United States (Southern Baptists, Adventists, and Pentecostals). Evangelical Christianity first came to Russia in the eighteenth century, but after 1991, American missionary activity increased markedly. The notion often promoted by this movement—that with faith comes economic success—may be particularly comforting both to those struggling with hardship and to those adjusting to new prosperity. **Figure 5.27** is a sculptor's humorous attempt to show the jarring cultural change that has transpired over the last few decades.

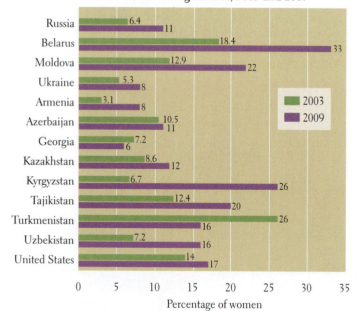

Women in Legislatures, 2003 and 2009

Legend: 2003, 2009

Country	2003	2009
Russia	6.4	11
Belarus	18.4	33
Moldova	12.9	22
Ukraine	5.3	8
Armenia	3.1	8
Azerbaijan	10.5	11
Georgia	7.2	6
Kazakhstan	8.6	12
Kyrgyzstan	6.7	26
Tajikistan	12.4	20
Turkmenistan	26	16
Uzbekistan	7.2	16
United States	14	17

Percentage of women

FIGURE 5.26 Women legislators. This graph shows the percentage of legislators in Russia and the post-Soviet states (with the United States for comparison) who are women. Belarus and Kyrgyzstan stand out as having the most women lawmakers, but both are authoritarian societies in which true democratic participation is rare. [Adapted from *United Nations Human Development Report 2009* (New York: United Nations Development Programme), Gender Empowerment Measure, Table K.]

Personal Vignette Valerii, age 35, once a government research scientist, now makes a comfortable living importing and exporting goods in the informal economy of Moscow. Although he has to bribe officials and pay protection money to mobsters, his income has made his family much wealthier than their longtime friends. Valerii's wife, Nina, is the only woman among them who does not work outside the home. In search of values that will guide them in these new circumstances, both have recently been baptized in an evangelical Christian sect. They say they chose this particular religious group because it promotes modesty, honesty, and commitment to hard work. *(Source: Adapted from Timo Piirainen,* Towards a New Social Order in Russia: Transforming Structures and Everyday Life *(Aldershot, UK, and Brookfield, Vermont: Dartmouth Press, 1997), pp. 171–179; updated in 2007.)* ∎

Chapter Key Terms

Bolsheviks, 220
capitalists, 220
Caucasia, 233
centrally planned or socialist economy, 221
Cold War, 222
communism, 220
Communist Party, 220
czar, 220
Gazprom, 225
glasnost, 231

Group of Eight (G8), 227
Mongols, 219
nomadic pastoralists, 219
nonpoint sources of pollution, 209
oligarchs, 223
perestroika, 231
permafrost, 208
privatization, 229
proletariat, 221
Russian Federation, 208

Russification, 224
Slavs, 219
Soviet Union, 207
steppes, 208
taiga, 209
tundra, 208
underemployment, 229
Union of Soviet Socialist Republics (USSR), 207

Land Elevations

meters	feet
4877	16,000
3353	11,000
2134	7000
914	3000
305	1000
152	500
0	0

mi 0   100   200   300   400   500
km 0   200   400   600   800

1:24,500,000
Lambert Azimuthal Equal Area Projection

ATLANTIC OCEAN

A Sahara, Morocco

B Atlas Mountains, Algeria

C Red Sea, Gulf of Aqaba

D Rub' al Khali dunes as seen from space

North Africa and Southwest Asia

E **Tigris River, Iraq**

F **Mt. Ararat, Turkey**

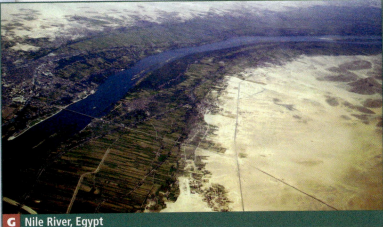

G **Nile River, Egypt**

H **A Wadi, Morocco**

FIGURE 6.1 Regional map of North Africa and Southwest Asia.

247

Global Patterns, Local Lives

Amos Oz, a well-known Israeli author (**Figure 6.2**), has written a novel about the founding of the state of Israel. It has no heroes, but rather tells an honest tale about Jewish settlers and the Palestinians they displaced. The book, *A Tale of Love and Darkness,* has gained wide acclaim for depicting both sides of the Israeli–Palestinian story with compassion and insight. Oz is frequently interviewed on the Arab network Al Jazeera, the most popular broadcast system in North Africa and Southwest Asia.

Elias Khoury is a prominent Palestinian lawyer. His son George, a student at Hebrew University, was jogging on the West Bank and was shot and killed by fellow Palestinians, who mistook him for a Jew. In his grief, Elias Khoury searched for a fitting memorial to his son, who had been noted for his open, multicultural views and for his friendships with both Jews and Arabs. Elias Khoury had been touched by Amos Oz's book and proposed to pay for translating it into Arabic and distributing it so that sensitive, open-minded Arabs across the region could read it.

Amos Oz was touched by this proposal and agreed. Oz strongly argues for partition and a two-state solution to the Israeli–Palestinian dispute—a space for the Israeli state and a separate space for a Palestinian state.

Islam a monotheistic religion that emerged in the seventh century c.e. when, according to tradition, the archangel Gabriel revealed the tenets of the religion to the Prophet Muhammad

Islamism a grassroots religious revival in Islam that seeks political power to curb what are seen as dangerous non-Muslim influences; also seeks to replace secular governments and civil laws with governments and laws guided by Islamic principles

fossil fuel a source of energy formed from the remains of dead plants and animals

FIGURE 6.2 Nily and Amos Oz.

Al Jazeera's coverage of Oz and the Khoury affair is itself highly controversial. Based in Doha, Qatar (with offices in Washington, D.C., Kuala Lumpur, and London), and supported both by advertising and by an income from the Qatari emir, Sheikh Hamad bin Khalifa, Al Jazeera tries to remain neutral on the Israeli–Palestinian conflict. Many broadcasts reflect the sentiments of its mainly Arab viewers, who tend to side with the Palestinians. When Al Jazeera tries to cover the views of Israelis, even those with views as sensitive to Palestinian

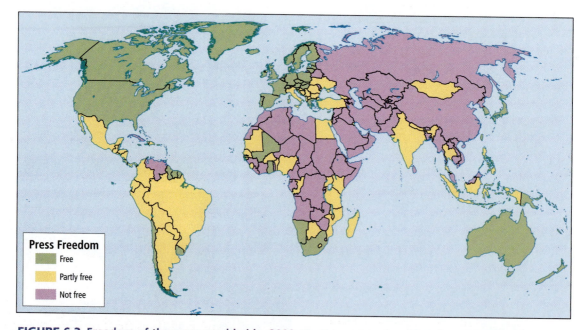

FIGURE 6.3 Freedom of the press worldwide, 2009. The three categories of freedom of the press shown on the map are based on several factors: the legal environment in which the media operate; the degree of independence from government ownership and influence, economic pressures on news content; and violations of press freedom ranging from murder of journalists to extralegal abuse and harassment. Israel, Italy, and Hong Kong slipped from being "free" to being "partly free" in 2008. It is troubling that there have been steady declines in press freedom over the past 3–5 years in some emerging democracies: Argentina, Mexico, Peru, the Philippines, Senegal, Sri Lanka, and Thailand. [Adapted from Freedom House, "Map of Press Freedom 2009," at http://www.freedomhouse.org/template.cfm?page=251&year=2009.]

FIGURE 6.4 Political map of North Africa and Southwest Asia. The occupied Palestinian Territories are not shown here because of the scale of the map.

issues as Amos Oz, they risk the sharp criticism of Arabs who accuse them of having a "pro-Israeli" bias. [*Sources: Adapted from broadcasts available at Al Jazeera.com, February 19, 2009: http://english.aljazeera.net/programmes/rizkhan/2009/02/200921865142701932.html; and CNN.com: Christiane Amanpour, "The Role of Literature in the Path to Peace," March 11, 2010, at http://transcripts.cnn.com/TRANSCRIPTS/1003/11/ampr.01.html.]* ■

Freedom of the press, which is essential for democracy, is a fragile concept in any part of the world, and in North Africa and Southwest Asia it is only beginning to develop (Figure 6.3). In part, this is because the 21 countries in the region (Figure 6.4) did not achieve political independence until the twentieth century. Most countries in the region now have parliaments and elections, but the parliaments have few powers and elections are controlled to ensure that a ruler or a particular party is reseated.

North Africa and Southwest Asia is a region of striking continuities but equally striking contrasts. To begin with, while most countries in this area share an arid climate, there is much variation in degrees of aridity. Also, although the vast majority of people practice **Islam**, a monotheistic religion that emerged between 601 and 632 C.E., Islam is a faith with multiple aspects. Only a minority is drawn to ultra-fundamentalist Islam or Islamism. Many practitioners (and probably a majority) are moderate in their thinking, accepting the validity of other beliefs (especially Christianity—Mary and Jesus play important roles in Islam), and are eager for interaction with outsiders.

The term **Islamism** refers to grassroots religious revivals that seek political power to curb what they see as dangerous secular influences that are spreading because of globalization. Some such movements characterize Western influence as corrupt and destructively self-serving. But Islamist movements also vary greatly, are unevenly distributed across the region, and have waxed and waned in influence, often in sync with economic recessions and booms.

Access to oil wealth is thought to be common in this region, but it is not. **Fossil fuel** reserves, formed from the fossilized remains of dead plants and animals, are abundant in some countries but totally absent in others. In this region, oil and gas reserves are found mainly around the Persian Gulf. These fossil fuels are extracted and exported throughout the world for tremendous profits, but these profits are not equitably distributed, especially within oil-rich countries (see the following discussion of wealth disparities in Photo Essay 6.5A, B on page 274).

Nor is this an entirely Arab region. The Turks, Iranians, Kurds, and Israelis do not consider themselves Arabs and do not use Arabic as their first language. There are many other sources of variety in this region. In some countries, nearly everyone lives in a city, while in others life is still primarily rural. Even in a forward-looking country like Turkey, women may be highly educated, outspoken, and active in commerce, public life, and government, or they may lead secluded domestic lives with few educational opportunities. Under all of these cultural umbrellas, local ethnicity can add another variable, and migration, within and from without the region, brings yet another component of diversity.

The nine Thematic Concepts covered in this chapter are illustrated here with photo essays that show examples of how each concept is experienced in this region (see pages 250–251). In all cases, interactions between two or more of these concepts are illustrated and explained in the captions.

THINGS TO REMEMBER

1. While most countries have some elected bodies of government, democratization has been slow.

2. Al Jazeera, the most popular television channel in the region, covers sensitive political issues, advocates free speech and political reform, and exposes political corruption—practices for which many governments in the region have banned it. Many of its broadcasts favor the Arab perspective, but it aspires to remain neutral. When Al Jazeera tries to cover Israeli views, it risks criticism from Arabs who accuse it of having a pro-Israeli bias.

3. The region has striking continuities as well as contrasts: the area is arid, but there is great variability in the degree of aridity; most people of the region follow Islam, but they range in religious observance from ultra-conservative to moderate; the countries of the Persian Gulf have rich oil deposits, but the other countries of the region have little or none. There are several primarily Arab countries, but there also are countries that are not Arab.

Water, Food, and Climate Change:
As the population of this dry region increases, obtaining enough water for agriculture will become more difficult. Many countries are highly dependent on imported food, which may cause problems if climate changes reduce global food output. Some low-lying agricultural areas are also highly vulnerable to climate change. New technologies may offer solutions, but not all can afford them.

(A) Flood irrigation in Iraq. Many agricultural areas depend on scarce water supplies that cannot be easily expanded.

(B) Nile delta flooding. Climate change could bring more flooding and salt-water intrusion into this low, densely populated area.

(C) Drip irrigation, pioneered in Israel, could dramatically reduce water use in agriculture, but it is too expensive for many farmers.

Development, Globalization, and Fossil Fuels:
This region's vast petroleum energy resources have brought wealth to a privileged few and shaped development in many countries, linking economies to a global system of investment, labor, and expertise. Politics have also become globalized, with Europe and the United States strongly influencing many governments.

(D) The yacht of Saudi Arabia's King Abdullah, docked in Malta. About 40 percent of Saudi Arabia's budget goes to the royal family.

(E) U.S. and European consultants dine with businessmen in Riyadh, Saudi Arabia.

(F) A U.S. war plane over an oil rig in the Persian Gulf. Foreign political influence in the region is increasingly controversial.

Urbanization and Globalization:
Two highly globalized patterns of urbanization have emerged in the region. In the oil-rich countries, spectacular new luxury-oriented urban development is tied to global flows of money, goods, and people. Elsewhere, economic reforms aimed at improving global competitiveness have brought massive migration from rural areas, creating crowding and slums.

(G) Dubai's skyline now boasts the tallest building on earth. Luxury villas and apartments line its coast.

(H) Indian workers in front of the apartments they are building in Dubai. Foreign workers make up 71 percent of Dubai's population.

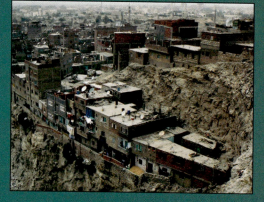

(I) A slum on the outskirts of Cairo that has been taking in rural migrants for decades now hosts 800,000 people in makeshift housing.

Population and Gender: This region has the second-highest population growth rate in the world after sub-Saharan Africa. In part, this is because women are generally less educated than men and tend not to work outside the home. Hence, childbearing remains crucial to a woman's status, a situation that tends to encourage large families.

(J) A woman with her child attends an adult literacy class in Cairo. With few girls going to school, female literacy rates are 30 percent lower than for men in Egypt.

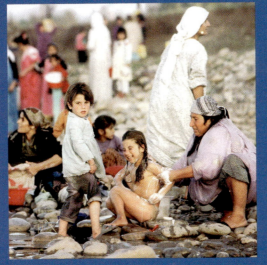

(K) An Iraqi woman bathes her child. For the many women who work only within the home, motherhood provides the main source of status within the community.

(L) A family in rural Turkey. Families are often large within this region. With so many incentives for childbearing, relatively few women practice contraception.

Democratization: While most countries have some elected bodies of government, democratization remains low as real power remains with the wealthy, the politically connected, the clerics, and often the military. Governments frequently repress popular political movements that could lead to greater public participation in decision making. This repression leads some political movements to turn to terrorism.

(M) A woman takes part in days of massive protests that swept through Iran after its June 2009 election, which was widely criticized as having been "rigged."

(N) Sheikh Mohammed, prime minister of the United Arab Emirates and one of several unelected monarchs in the region, speaks with a U.S. Air Force officer in Dubai.

(O) The aftermath of an attack on a building in Saudi Arabia housing U.S. military personnel. Islamic militants who oppose the Saudi monarchy claimed responsibility.

I THE GEOGRAPHIC SETTING

Terms in This Chapter

The common term *Middle East* is not used in this chapter because it describes the region from a European perspective. To someone in Japan, the region lies to the far west, and to a Russian, it lies to the south. The *Arab World* is also not used because not all people in the region are of Arab ethnicity. In this book, the term used for this region is *North Africa and Southwest Asia*.

occupied Palestinian Territories (OPT) Palestinian lands occupied by Israel in 1967

We use the term **occupied Palestinian Territories (OPT)** to refer to Gaza and the West Bank, since that is how the United Nations currently refers to those areas where Israel still exerts control despite treaty agreements; the word "occupied" is not capitalized to show the supposed temporary quality of the occupation. The U.S. Department of State uses the term *Palestinian Territories*.

Physical Patterns

Landforms and climates are particularly closely related in this region. The climate is dry and hot in the vast stretches of relatively low flat land; it is somewhat moister where mountains are able to capture rainfall. The lack of vegetation perpetuates aridity. Without plants to absorb and hold moisture, the rare but occasionally copious rainfall simply runs off, evaporates, or sinks rapidly into underground aquifers.

Climate

No other region in the world is as extensively dry as North Africa and Southwest Asia (Photo Essay 6.1). A belt of dry air that circles the planet between roughly 20° and 30° N creates desert climates in the Sahara of North Africa, the Arabian Peninsula, Iraq, and Iran.

The Sahara's size and location under this high-pressure belt of dry air make it a particularly hot desert region. In some places, temperatures can reach 130°F (54°C) in the shade at midday. With little water or moisture-holding vegetation to retain heat, nighttime temperatures can drop quickly to below freezing. Nevertheless, in even the driest zones, humans survive at scattered oases, where they maintain groves of drought-resistant plants such as date palms. Desert inhabitants often wear light-colored, loose, flowing robes that reflect the sunlight and retain body moisture during the day and provide warmth at night.

In the uplands and desert margins, enough rain falls to nurture grass, some trees, and limited agriculture. Such is the case in western Morocco, northern Algeria and Tunisia, Turkey, and the northern parts of Iraq and Iran. The rest of the region, generally too dry for cultivation, has long been the prime herding lands for nomads, such as the Kurds of Southwest Asia, the Berbers in North Africa, and the Bedouin of the steppes and deserts on the Arabian Peninsula. Recently, most nomads have been required to settle down and some of their lands are now irrigated for commercial agriculture, but the general aridity of the region means that sources of irrigation water are scarce.

Landforms and Vegetation

The rolling landscapes of rocky and gravelly deserts and steppes cover most of North Africa and Southwest Asia (Figure 6.1 on pages 246–247 and see Photo Essay 6.1A, B, C, D). In a few places, mountains capture moisture, allowing plants, animals, and humans to flourish. In northwestern Africa, the Atlas Mountains stretch from Morocco on the Atlantic coast to Tunisia on the Mediterranean coast. They block and lift damp winds from the Atlantic Ocean, creating rainfall of more than 50 inches (127 centimeters) per year on windward slopes. In some Atlas Mountain locations, snowfall is sufficient to support a skiing industry.

Africa and Southwest Asia are separated by a rift formed between two tectonic plates—the African Plate and the Arabian Plate—that are moving away from each other (see Figure 1.26 on page 51). The rift, which began to form about 12 million years ago, is now filled by the Red Sea. The Arabian Peninsula lies to the east of this rift. There, mountains bordering the rift in the southwestern corner rise to 12,000 feet (3,658 meters).

Behind these mountains to the east lies the great desert region of the Rub'al Khali. Like the Sahara, it has virtually no vegetation. The sand dunes of the Rub'al Khali, which are constantly moved by strong winds, are among the world's largest, some reaching more than 2000 feet (610 meters).

The landforms of Southwest Asia are more complex than those of North Africa. The Arabian Plate is colliding with the Eurasian Plate and pushing up the mountains and plateaus of Turkey and Iran (see Figure 1.26 on page 51). Turkey's mountains lift damp air passing over Europe and the Mediterranean from the Atlantic, resulting in considerable rainfall. Only a little rain makes it over the mountains to the interior of Iran, which is very dry. The tectonic movements that create mountains also create earthquakes, a common hazard in Southwest Asia.

There are only three major river systems in the entire region, and all have attracted human settlement for thousands of years. The Nile flows north from the moist central East African highlands. It crosses arid Sudan and desert Egypt and forms a large delta on the Mediterranean. The Euphrates and Tigris rivers both begin with the rain that falls in the mountains of Turkey; the rivers flow southeast to the Persian Gulf. A fourth and much smaller river, the Jordan, starts as snowmelt in the uplands of southern Lebanon and flows through the Sea of Galilee to the Dead Sea. Most other streams are dry riverbeds, or wadis, most of the year, carrying water only after generally light rains that fall between November and April.

North Africa and Southwest Asia were home to some of the very earliest agricultural societies. Today, however, agriculture is possible only in a few places. In spots along the Mediterranean coast, rain is sufficient to grow citrus fruits, grapes, olives, and many vegetables, though supplemental irrigation is often needed.

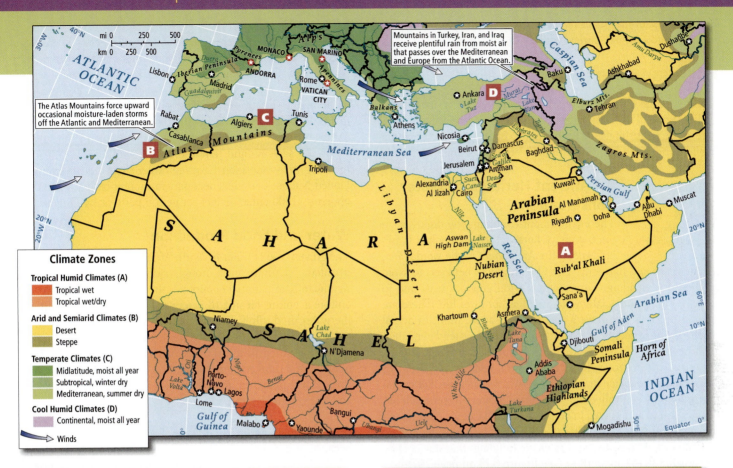

Mountains in Turkey, Iran, and Iraq receive plentiful rain from moist air that passes over the Mediterranean and Europe from the Atlantic Ocean.

The Atlas Mountains force upward occasional moisture-laden storms off the Atlantic and Mediterranean.

Climate Zones

Tropical Humid Climates (A)
- Tropical wet
- Tropical wet/dry

Arid and Semiarid Climates (B)
- Desert
- Steppe

Temperate Climates (C)
- Midlatitude, moist all year
- Subtropical, winter dry
- Mediterranean, summer dry

Cool Humid Climates (D)
- Continental, moist all year

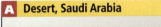 Winds

A Desert, Saudi Arabia

B Steppe, Morocco

C Mediterranean, summer dry, Algeria

D Continental, moist all year, Turkey

In the valleys of the major rivers, seasonal flooding and irrigation support cotton, wheat, barley, vegetables, and fruit trees.

Environmental Issues

Environmental concerns are only beginning to be an overt focus in this region. This is in part because in many ways the residents here have consistently focused on the challenges their natural habitat poses, with a special emphasis on water scarcity.

An Ancient Heritage of Water Conservation

Qur'an (or Koran) the holy book of Islam, believed by Muslims to contain the words Allah revealed to Muhammad through the archangel Gabriel

The environment is not directly discussed in the **Qur'an** (or **Koran**), the holy book of Islam, but *Salam*, the Arabic root of the word *Islam*, means peace and harmony. The Qur'an guides believers to avoid spoiling or degrading human and natural environments and to share resources, especially water, with all forms of life. These are ancient precepts that have a very modern application. In actual practice, the residents of this region conserve water better than most people in the world do and many of their conserving practices have ancient roots. Daily bathing is a religious requirement and it often takes place in public baths where water use and sanitation are tightly controlled. For millennia, mountain snowmelt has been captured and moved to dry fields and villages via constructed underground water conduits called *qanats*. Likewise, traditional architectural designs are used to create buildings that stay cool by maximizing shade and airflow (see Figure 6.13 on page 267).

Despite their long history of water-conserving technologies and practices, however, this region's 450 million residents now have such limited water resources that even clever combinations of ancient and modern measures are no longer sufficient to ensure an adequate supply. Rapidly growing populations and unwise modern usages of water virtually guarantee that water shortages will be more extreme in the future (Photo Essay 6.2B, C, D). According to the map in Figure 6.5, Turkey is the only country in the region that is not vulnerable to losing the availability of its fresh water. All of the countries in North Africa and on the Arabian Peninsula face freshwater scarcity (and Iran and Iraq are vulnerable to scarcity), meaning they may soon have less water than the minimum the United Nations considers necessary to support basic human development—1000 cubic meters per person per year (see Figure 6.5).

Water and Food Production

The greatest use of water in North Africa and Southwest Asia is for irrigated agriculture, even though agriculture does not contribute significantly to national economies. In Tunisia, for example, agriculture accounts for just 10 percent of GDP but 86 percent of all the water used. Only 13 percent of water is used in homes, and only 1 percent by industry. Nonetheless, when irrigated agriculture is measured in terms of its value to local diets, jobs, family budgets, and rural economies, it emerges as essential in this water-stressed region; hence, most countries subsidize irrigation for food and fiber crops in some way.

Imported Food and Virtual Water Despite agricultural subsidies, because of growing populations, rising living standards, and the relative unproductivity of much agricultural land, almost all people consume imported food. The water used to produce this imported food must be added to the *virtual water* consumption of

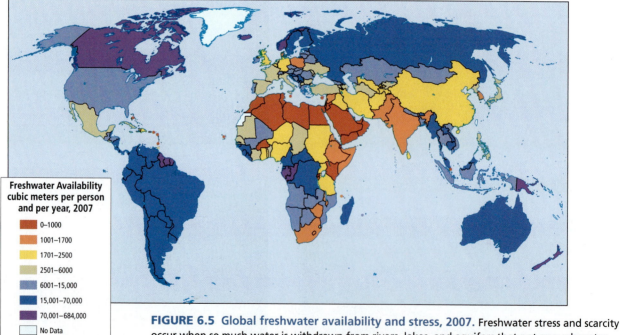

FIGURE 6.5 Global freshwater availability and stress, 2007. Freshwater stress and scarcity occur when so much water is withdrawn from rivers, lakes, and aquifers that not enough water remains to meet human and ecosystem requirements for sustainability. [Adapted from United Nations Environmental Programme, *Vital Water Graphics—An Overview of the State of the World's Fresh and Marine Waters*, 2nd ed., 2008, available at http://www.unep.org/dewa/vitalwater/article69.html.]

Freshwater Availability
cubic meters per person and per year, 2007

- 0–1000
- 1001–1700
- 1701–2500
- 2501–6000
- 6001–15,000
- 15,001–70,000
- 70,001–684,000
- No Data

War and water scarcity have resulted in major impacts on the biosphere in this region. Growing populations promise to increase these stresses on ecosystems.

A An attack on an oil pipeline near Kirkuk, Iraq, resulted in a large spill that quickly caught fire. The largest oil spill in history occurred during the first Gulf War (1991) when the Iraqi government deliberately spilled 300 million gallons of oil (30 times the amount spilled by the *Exxon Valdez*) into the Persian Gulf in order to thwart a land invasion by the United States. Lebanon suffered a smaller, though similarly devastating spill in 2006 as a result of Israeli bombing raids.

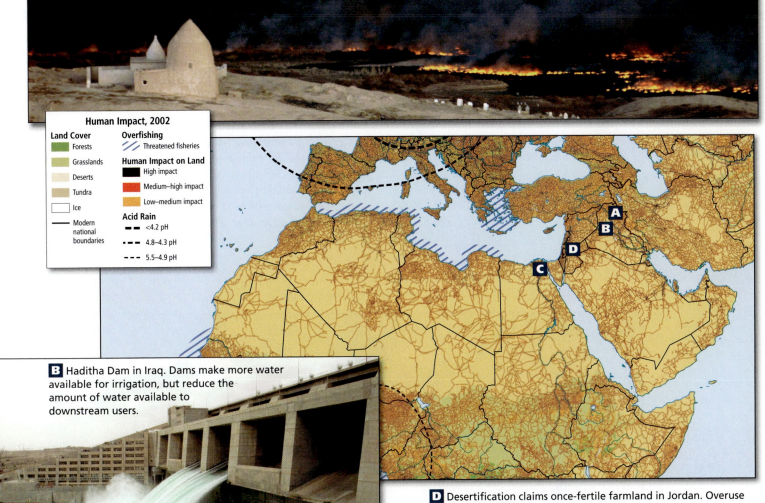

Human Impact, 2002

Land Cover
- Forests
- Grasslands
- Deserts
- Tundra
- Ice
- Modern national boundaries

Overfishing
- Threatened fisheries

Human Impact on Land
- High impact
- Medium–high impact
- Low–medium impact

Acid Rain
- <4.2 pH
- 4.8–4.3 pH
- 5.5–4.9 pH

B Haditha Dam in Iraq. Dams make more water available for irrigation, but reduce the amount of water available to downstream users.

D Desertification claims once-fertile farmland in Jordan. Overuse of water for irrigation upstream from Jordan has reduced flows of water in the Jordan and Yarmouk rivers. As soil moisture has decreased, many plants have died off.

C In Egypt, a gas-powered engine pumps water from a ditch onto a field. This type of irrigation has increased soil salinity and reduced soil fertility.

the citizens of this region. Virtual water, now a widely accepted term in water scarcity discussions, is the volume of water used to produce all that a person consumes in a year (see Chapter 1 on page 42). For example 1 kilogram (2.2 pounds) of beef requires 15,500 liters (4,094 gallons) of water to produce; 1 kilogram of goat meat requires 4000 liters (1,056 gallons); 1 kilogram of corn requires 900 liters (238 gallons). Beef and corn are common imports in the region, while goat meat is produced locally and has a somewhat less significant water component than beef.

Until the twentieth century, agriculture was confined to a few coastal and upland zones where rain could support agriculture, and to river valleys (such as the Nile, Tigris, and Euphrates valleys) where farms could be irrigated with simple gravity-flow technology (see Thematic Concepts Part A on page 250). However, to accommodate population growth and development, Libya, Egypt, Saudi Arabia, Tunisia, Turkey, Israel, and Iraq all now have ambitious mechanized irrigation schemes that have expanded agriculture deep into formerly uncultivated desert environments (see Thematic Concepts Part A, C on page 250; Photo Essay 6.2B, C, D on page 255; and Figure 6.6).

salinization a process that occurs when large quantities of water are used to irrigate areas where evaporation rates are high, leaving behind dissolved salts and other minerals

seawater desalination the removal of salt from seawater, usually accomplished through the use of expensive and energy-intensive technologies, making the water suitable for drinking or irrigating

Over time, irrigation projects damage soil fertility through **salinization.** When irrigation is used in hot, dry environments, the water evaporates, leaving behind a salty residue of minerals or other contaminants. Over time, so much residue accumulates that the plants are unable to grow or even survive. This human-made process is one of the largest environmental water issues that the world faces today in arid and semiarid regions.

Israel has developed relatively efficient techniques of *drip irrigation* that dramatically reduce the amount of water used, thereby limiting salinization and freeing up water for other uses (see Thematic Concepts Part C on page 250). Until very recently, however, poorer states have been unable to afford this somewhat complex technology, which requires an extensive network of hoses and pipes to deliver water to each plant. Other countries are wary of depending on a technology developed by Israel, a country they deeply distrust.

Strategies for Increasing Access to Water Some strategies have been developed for increasing supplies of fresh water, but each presents a set of difficulties. All of them are expensive, some have enormous potential to cause increased wasting of water, and most do not include guarantees of equal access to the increased supply.

Seawater desalination The fossil fuel–rich countries of the Persian Gulf have invested heavily in **seawater desalination**

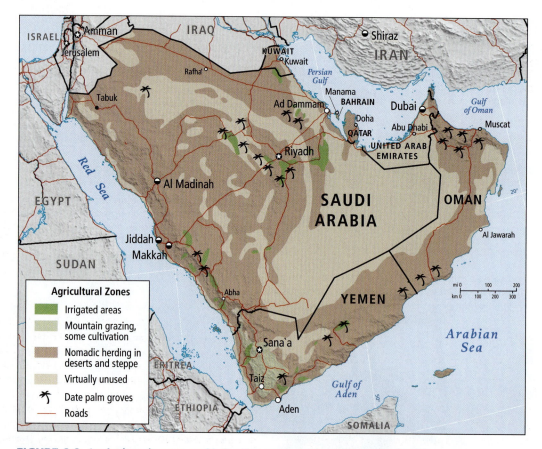

FIGURE 6.6 Agricultural zones and irrigation areas in the Arabian Peninsula.

technologies that remove the salt from seawater, making it suitable for drinking or irrigating. Seventy percent of Saudi Arabia's drinking water is supplied by desalination plants, and some of its wheat fields are irrigated with desalinated water. However, the process of desalination uses huge amounts of energy, thus contributing to global warming. Furthermore, if all the costs of producing food with desalinated water were counted—the energy to desalinate, the irrigation equipment, the fact that irrigated soil inevitably loses productivity due to soil salinization—the wheat so produced would be exorbitantly expensive.

Groundwater pumping Many countries pump groundwater from underground aquifers to the surface for irrigation or drinking water. Libya has invested some of its earnings from fossil fuels into one of the world's largest groundwater pumping projects, known as the *Great Man-Made River*. This project draws on ancient *fossil water*, deposited at least 14,000 years ago, to supply almost 2 billion gallons of water per day to Libya's coastal cities, and to 600 square miles of agricultural fields. With this irrigated agriculture, Libya hopes to grow enough food to end its current food imports, and even supply food to the EU. However, the aquifer that the project depends on is not being replenished, making this use of groundwater unsustainable. By 1982, hydrologists were reporting fissures in the land surface above the aquifer of widths up to 16 inches (100 cm) that are associated with land sinking, or *subsidence*, as the water is withdrawn. Another problem is seawater intruding as fresh water is withdrawn, so that aquifers and wells in coastal zones are rendered brackish and useless.

Dams and reservoirs Dams and reservoirs have been built on the regions' major river systems to increase water supplies, but these projects have created new problems (see Photo Essay 6.2B on page 255). In Egypt, for example, the natural cycles of the Nile River have been altered by the construction of the *Aswan Dams*. Downstream of these two dams, water flows have been reduced and floods no longer deposit fertility-enhancing silt on the land. As a result, expensive fertilizers are now necessary. Moreover, with less water and silt coming downstream, parts of the Nile delta are sinking into the sea. Upstream, the artificial reservoir created by the dams has flooded villages, fields, wildlife habitats, and historic sites, and created still-water pools that harbor parasites. Dams can also cause hardship across borders. The Aswan dam sits on Egypt's border with Sudan, long an area of contention.

Water-acquisition strategies strain cross-border relations The need to share the water of rivers that flow across or close to national boundaries often impedes national development efforts. Turkey's Southeastern Anatolia Project, which involves the construction of several large dams on the Euphrates River for hydropower and irrigation purposes, has reduced the flow of water to the downstream countries of Syria and Iraq. In negotiations over who should get Euphrates water, for example, Turkey argues that it should be allowed to keep more water behind its dams because the river starts in Turkey and most of its water originates there as mountain rainfall (Figure 6.7). Meanwhile, Iraq points out that the Euphrates travels the longest distance in Iraq.

285. The Jordan River Is Dying

Virtual water and land A very new strategy for acquiring sufficient food in this region of scarce water combines the concept of

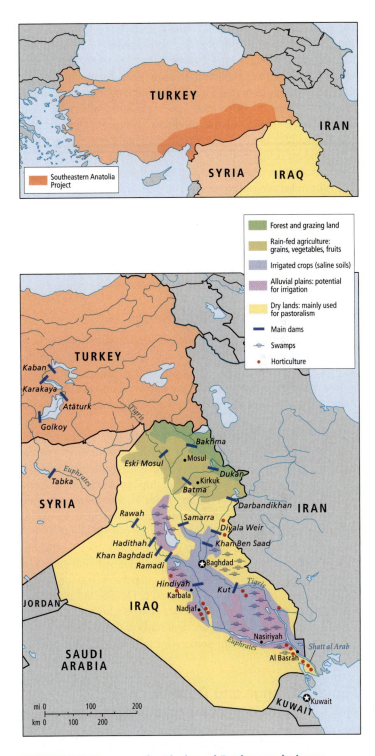

FIGURE 6.7 Dams on the Tigris and Euphrates drainage basins. Turkey's projects to manage the Tigris and Euphrates river basins through dam construction have international implications. Water that is retained in Turkey will not reach its neighbors. The main map shows Turkey's dams on the headwaters of the two rivers, as well as dams built in Syria and Iraq, all of which will have environmental effects, especially on the lower reaches of both rivers in Iraq. The smaller map shows the full extent of Turkey's Southeastern Anatolia Project. [Adapted from United Nations Environment Programme, *Vital Water Graphics: Problems Related to Freshwater Resources*, "Turning the Tides" map, at http://www.unep.org/dewa/assessments/ecosystems/water/vitalwater/22.htm.]

virtual water with a new idea, *virtual land.* As we have observed, imported food has a component of virtual water—that which was used to grow, produce, and transport the food. Now virtual land is being added to the equation. To fairly assess the environmental footprint of such imported food, the virtual land must be counted. Dependence on imported food leaves a country vulnerable to all sorts of uncertainties in the producer countries. Food security can be easily imperiled when political conditions change in or with food-exporting countries. The Gulf States are using a new approach that involves spending some of their oil and gas wealth to buy up productive well-watered farmland in Africa, Asia, or Latin America, where they can run their own food-producing farms and bypass local producers. By doing this, they are acquiring virtual land as well as the virtual water it contains. In a March 10, 2010, interview on Al Jazeera, Kofi Annan, the former secretary general of the United Nations, questioned whether this practice might be a new and insidious version of colonialism.

Vulnerability to Climate Change

North Africa and Southwest Asia are especially vulnerable to both the effects of climate change (especially global warming) and the world's attempts to reduce greenhouse gas emissions, which will ultimately reduce revenues from the sale of oil and gas. As the climate warms, a sea-level rise of a few feet could severely impact the Mediterranean coast, especially the Nile delta, one of the poorest and most densely populated areas in the world (Photo Essay 6.3A). Elsewhere, shifting rainfall patterns resulting from changes in the regional climate could drastically reduce water availability where it is already scarce and where people are impoverished (Photo Essay 6.3B, C), causing more nondesert lands to become transformed into deserts. Conversely, under some climate-change scenarios, unusually intense rainfall could increase, causing flooding.

Independent of any environmental changes, global efforts to reduce fossil fuel consumption could devastate oil- and gas-based economies and transform the region's geopolitics, leaving it with much less income and much less power. Despite the enormous wealth from fossil fuel sales that has flowed into some of these countries over the past 40 years, few have undertaken significant economic diversification to prepare for reduced global consumption of fossil fuels.

Climate Change and Desertification

The changes that characterize the conversion of nondesert lands into deserts are referred to as **desertification** (see Photo Essay 6.2D on page 255). As soil moisture and plant cover are reduced for any reason, those plants better adapted to dry conditions spread but often provide only poor cover for the arid soil. Bare patches of soil can become badly eroded by wind, and eventually sand dunes can blow onto formerly vegetated land.

desertification a set of ecological changes that converts nondesert lands into deserts

In the grasslands (steppes) bordering the Sahara or in Anatolia in Turkey, a wide array of land-use changes can contribute to the general drying. For example, as groundwater levels fall due to pumping for cities and irrigated agriculture, plant roots can no longer reach sources of moisture. In other cases, international development agencies have encouraged nomadic herders to take up settled cattle ranching of the type practiced in western North America. The thinking is that nomadism is dysfunctional in the modern world where record keeping, taxing, and schooling for children are valued. However, ranching on fragile grasslands has led to overgrazing and excessive water use both by the cattle and by the people. Historically, nomadic herders were accustomed to using very little water; but when encouraged to settle, take up "modern" ways of life, and raise cattle to sell as meat, their water use can increase dramatically. If new sustainable sources of water are not found, these efforts at development can actually create new desert environments.

Urbanization and the Environment

Several of the region's environmental issues are either caused by urbanization or especially affect urban residents. Fossil fuel use and the air pollution it creates have decreased the quality of life and threaten human health in all cities. Cairo (page 275) is an exemplary case.

Urban water issues also abound. The scarcity of fresh water and the poor management of wastewater complicate and endanger the lives of people in the old medieval inner cities and in urban fringe shantytowns (see Thematic Concepts Part I on page 250 and Photo Essay 6.5C on page 274).

Commercial and residential structures hastily built with few amenities and aesthetically incompatible with ancient architectural masterpieces impose a utilitarian bleakness on urban landscapes. Baghdad may be the city most affected by a grim constructed environment, much of it installed as protection against suicide bombings during the U.S. occupation, but Algiers, Beirut, Cairo, Istanbul, Izmir, and Jerusalem are also affected by the blight of too much unadorned concrete.

THINGS TO REMEMBER

1. Landforms and climates are particularly closely related in this region, and it is the driest region in the world. Rolling deserts and steppes cover most of the region. In a few places, mountains capture moisture, allowing plants, animals, and humans to flourish—but sustainability is becoming more difficult.

2. As the population of such a dry region increases, obtaining enough cultivable land and water for agriculture will become harder. Many countries are dependent on imported food, which causes food insecurity when global prices rise. Climate change could reduce food output both locally and globally. New technologies may offer solutions, but are expensive.

Human Patterns over Time

Important developments in agriculture, societal organization, and urbanization took place long ago in this part of the world. Three of the world's great religions were born here: Judaism, Christianity, and Islam.

Water scarcity, sea-level rise, poverty, and political instability are some of the factors that make parts of this region highly vulnerable to climate change.

A Alexandria, Egypt, located on the Mediterranean coast of the Nile delta region, is a low-lying city highly exposed to rising sea levels. Efforts to improve the sea wall around the city may be outpaced by rising sea levels. The entire Nile delta is struggling to adapt to salt water that is moving up rivers and permiating soils, making agriculture extremely difficult and polluting freshwater resources on which cities depend. Half of Egypt's population lives in the Nile delta, and 80 percent of the country's imports and exports run through Alexandria.

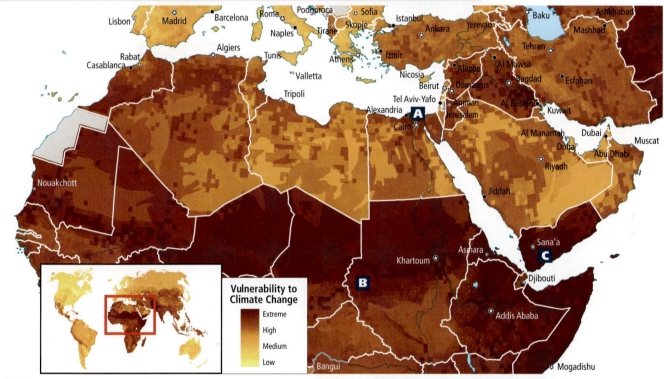

Vulnerability to Climate Change

- Extreme
- High
- Medium
- Low

B Refugees line up for food and water in Darfur, Sudan. This water-scarce area is highly exposed to drought. Meanwhile widespread poverty leaves the population highly sensitive to any disruption in food or water supplies. Political instability complicates emergency response and longer-term planning efforts that might lend greater resilience.

C Residents on the outskirts of Sana'a, Yemen, ride out a sandstorm at the side of the road. The same factors that make Darfur, Sudan, highly vulnerable to climate change also apply to Yemen. However, Yemen also receives large numbers of refugees from Somalia and other parts of Africa, further stretching resources and frustrating planning efforts.

Agriculture and the Development of Civilization

Between 10,000 and 8000 years ago, formerly nomadic peoples founded some of the earliest known agricultural communities in the world. These communities were located in an arc formed by the uplands of the Tigris and Euphrates river systems (in modern Turkey and Iraq) and the Zagros Mountains of modern Iran. This zone is often called the **Fertile Crescent** (Figure 6.8) because of its plentiful freshwater, open forests and grasslands, abundant wild grains, and fish, goats, sheep, wild cattle, and other large animals.

Fertile Crescent an arc of lush, fertile land formed by the uplands of the Tigris and Euphrates river systems and the Zagros Mountains, where nomadic peoples began the earliest known agricultural communities

The skills of these early people in domesticating plants and animals allowed them to build ever more elaborate settlements. The settlements eventually grew into societies based on widespread irrigated agriculture along the base of the mountains and in river basins, especially along the Tigris and Euphrates. Nomadic herders living in adjacent grasslands traded animal products for the grain and manufactured goods produced in the settled areas.

Over the next several thousand years, agriculture spread to the Nile Valley, west across North Africa, east to the mountains of Persia (modern Iran), and ultimately influenced other cultivation systems worldwide. Eventually, the agricultural settlements took on urban qualities: dense populations, specialized occupations, concentrations of wealth, and centralized government and bureaucracies. For example, the agricultural villages of Sumer (in

FIGURE 6.8 The Fertile Crescent, one of the earliest known agricultural sites. About 10,000 years ago, people in the Fertile Crescent began domesticating cereal grains, legumes, and animals, especially sheep and goats. The uses of domesticated animals spread into Europe and Africa as agricultural peoples traded their surpluses for other goods or moved into other regions. Three major empires developed successively in the eastern part of the Fertile Crescent: the Sumerian, the Babylonian, and the Assyrian. [Map adapted from Bruce Smith, *The Emergence of Agriculture* (New York: Scientific American Library, 1995), p. 50.] **(A)** Sheep graze near the ruins of Apamea, founded about 300 B.C.E. by one of Alexander the Great's generals in Syria.

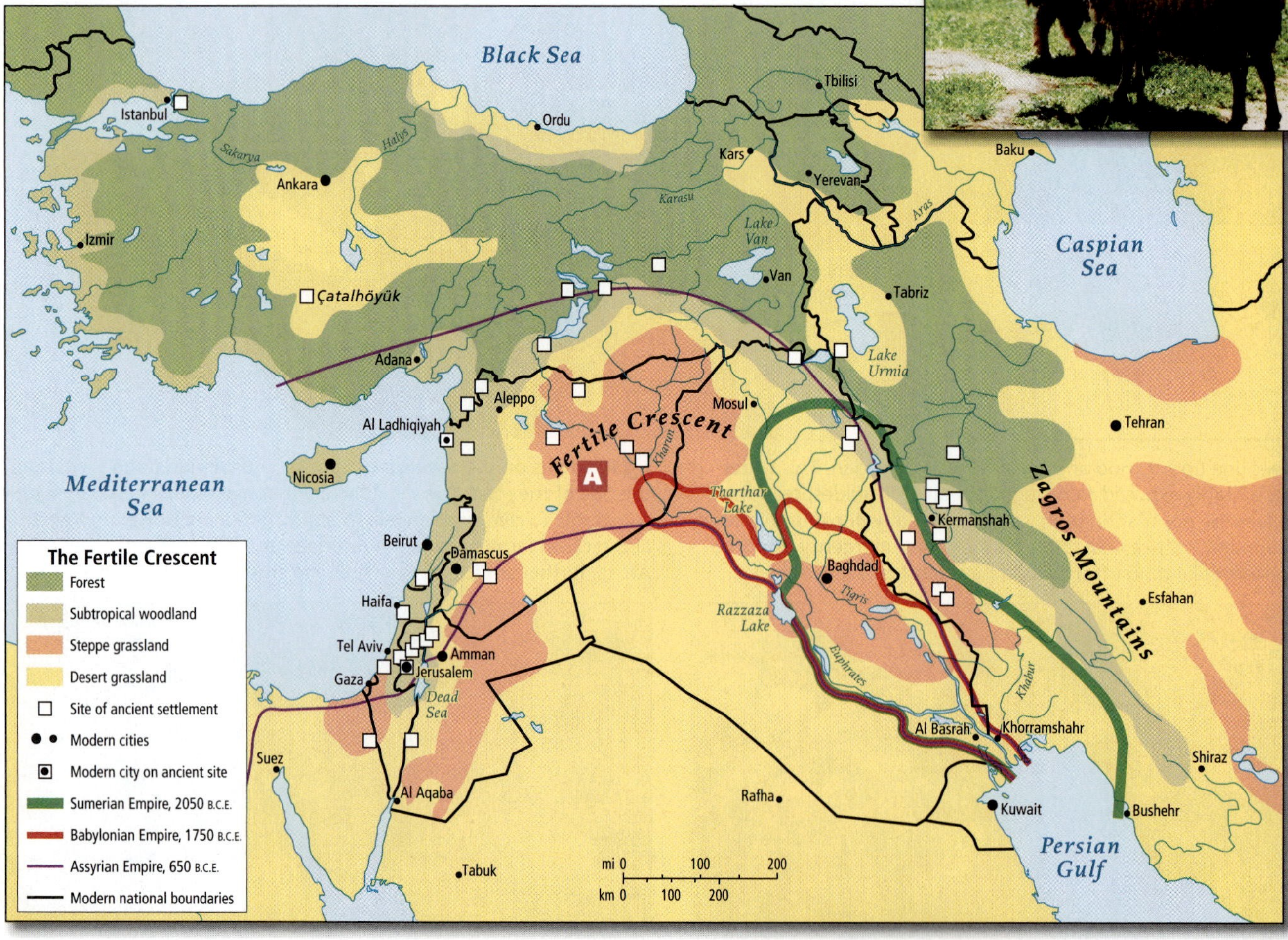

The Fertile Crescent
- Forest
- Subtropical woodland
- Steppe grassland
- Desert grassland
- □ Site of ancient settlement
- ●● Modern cities
- ◉ Modern city on ancient site
- — Sumerian Empire, 2050 B.C.E.
- — Babylonian Empire, 1750 B.C.E.
- — Assyrian Empire, 650 B.C.E.
- — Modern national boundaries

modern southern Iraq), which existed 5000 years ago, gradually turned into city-states that extended their influence over the surrounding territory. The Sumerians developed wheeled vehicles, oar-driven ships, and irrigation technology.

At times, nomadic tribes who had adopted the horse as a means of conquest banded together and, with devastating cavalry raids, swept over settlements. They then set themselves up as a ruling class, but soon they adopted the settled ways and cultures of the peoples they conquered and thus themselves would become vulnerable to attack.

Agriculture and Gender Roles

Increasing research evidence suggests that the dawning of agriculture may have marked the transition to markedly distinct roles for men and women. Archaeologist Ian Hodder reports that at the 9000-year-old site of Çatalhöyük, near Konya in south-central Turkey (see Figure 6.8), where the economy was primarily hunting and gathering, there is little evidence of gender differences. Families were small and men and women performed similar chores in daily life. Both had comparable status and power, and both played key roles in social and religious life.

Scholars think that after the development of agriculture, as wealth and property became more important in human society, a concern with family lines of descent and inheritance emerged. This led in turn to the idea that women's bodies needed to be controlled so that a woman could not become pregnant by a man other than her mate and thus confuse lines of inheritance.

The Coming of Monotheism: Judaism, Christianity, and Islam

The very early religions of this region were based on a belief in many gods who controlled natural phenomena; such was the case through the Greek era and into the Roman period. Several thousand years ago, **monotheistic** belief systems—those based on one god—began to emerge. The three major monotheistic world religions—Judaism, Christianity, and Islam—all have connections to the eastern Mediterranean—Jerusalem is sacred to all three (Figure 6.9). Muslims also revere Makkah (Mecca) and Al Madinah (Medina) in Saudi Arabia. All three religions have a connection to a sacred text: the Old Testament of the Bible for Jews; the Old and New Testaments

> **monotheistic** pertaining to the belief that there is only one god

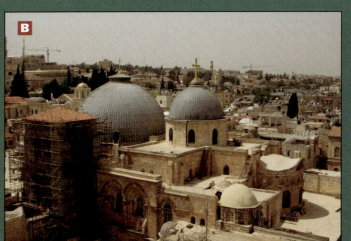

FIGURE 6.9 Holy sites of the three major religions of the region. (A) The Temple Mount in Jerusalem is the holiest site for many Jews, who revere it as the location of the first Jewish temple. The wall on the right side of the photo is held sacred as the last remnant of the temple. Today the Temple Mount is partially occupied by Al Aqsa Mosque, the golden dome of which can be seen in the left of the photo. Muslims consider this their second holiest site because the Prophet Mohammed is said to have ascended into heaven from here. **(B)** Many Christians revere the Church of the Holy Sepulchre in Jerusalem, marking it as the site where Jesus died and was resurrected. **(C)** Located in al-Haram Mosque in Makkah, Saudi Arabia, the Kaaba—the black cube in the photo—is the most sacred site for Muslims, who must visit it at least once during their lifetime. All mosques point toward the Kaaba, and Muslims are encouraged to pray toward it five times a day.

Judaism a monotheistic religion characterized by the belief in one god, Yahweh, a strong ethical code summarized in the Ten Commandments, and an enduring ethnic identity

diaspora the dispersion of Jews around the globe after they were expelled from the eastern Mediterranean by the Roman Empire beginning in 73 C.E.; can now refer to other dispersed cultural groups

Christianity a monotheistic religion based on the belief in the teachings of Jesus of Nazareth, a Jew, who described God's relationship to humans as primarily one of love and support, as exemplified by the Ten Commandments

Muslims followers of Islam

for Christians; and for Muslims both the Bible and the Qu'ran are sacred.

Judaism was founded approximately 4000 years ago. According to tradition, the patriarch Abraham led his followers from Mesopotamia (modern Iraq) to the shores of the eastern Mediterranean (modern Israel and the occupied Palestinian Territories) where he founded Judaism. Jewish religious history is recorded in the Torah (the first five books of the Bible's Old Testament). Judaism is characterized by the belief in one God, a strong ethical code summarized in the Ten Commandments, and an enduring ethnic identity reinforced by dietary and religious laws.

After the Jews rebelled against the Roman Empire, which culminated in their expulsion in 73 C.E. from the eastern Mediterranean, some were enslaved by the Romans and most migrated to other lands in a movement known as the **diaspora** (the dispersion of an originally localized people). Many Jews dispersed across North Africa and Europe, and others

went to various parts of Asia. After 1500, Jews were among the earliest European settlers in all parts of the Americas.

Christianity is based on the teachings of Jesus of Nazareth, a Jew who, claiming to be the son of God, gathered followers in the area of Palestine about 2000 years ago. Jesus, who became known as Christ (meaning *anointed one* or *Messiah*), taught that there is one God, who primarily loves and supports humans, but who will judge those who do evil. This philosophy grew popular, and both Jewish religious authorities and Roman imperial authorities of the time saw Jesus as a dangerous challenge to their power.

After Jesus' execution in Jerusalem in about 32 C.E., his teachings were written down (the Gospels) by those who followed him, and his ideas spread and became known as Christianity. Centuries of persecution ensued, but by 400 C.E., Christianity had become the official religion of the Roman Empire. However, following the spread of Islam after 622 C.E., only remnants of Christianity remained in Southwest Asia and North Africa.

Islam is now the overwhelmingly dominant religion in the region. Islam emerged in the seventh century C.E., after the Prophet Muhammad transmitted the Qur'an to his followers by writing down what was conveyed to him by Allah. Born in about 570 C.E., Muhammad was a merchant and caravan manager in the small trading town of Makkah (Mecca) on the Arabian Peninsula

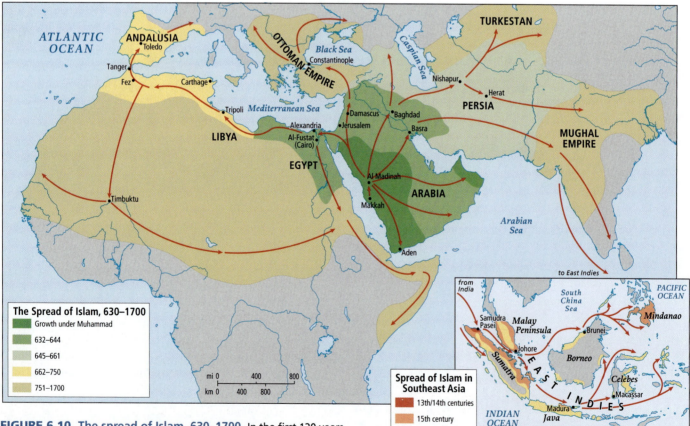

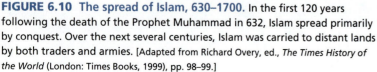

FIGURE 6.10 The spread of Islam, 630–1700. In the first 120 years following the death of the Prophet Muhammad in 632, Islam spread primarily by conquest. Over the next several centuries, Islam was carried to distant lands by both traders and armies. [Adapted from Richard Overy, ed., *The Times History of the World* (London: Times Books, 1999), pp. 98–99.]

near the Red Sea. Followers of Islam, called **Muslims**, believe that Muhammad was the final and most important in a long series of revered prophets which includes Abraham, Moses, and Jesus.

The religion of Islam has virtually no central administration and only informal religious hierarchy (this is somewhat less true of the Shi'ite version of Islam; see the discussion on page 266). The world's 1 billion Muslims may communicate directly with God (Allah). A clerical intermediary is not necessary, though there are numerous mullahs (clerical leaders) who help their followers interpret the Qur'an. An important result of the lack of a central authority is that the interpretation of Islam varies widely within and among countries and from individual to individual.

The Spread of Islam

Among the first converts to Islam were the Bedouin—nomads of the Arabian Peninsula. By the time of Muhammad's death in 632 C.E., they were already spreading the faith and creating a vast Islamic sphere of influence. Over the next century, Muslim armies built an Arab–Islamic empire over most of Southwest Asia, North Africa, and the Iberian Peninsula of Europe (Figure 6.10).

While most of Europe was stagnating during the medieval period (450–1500), the Arab–Islamic empire nurtured learning and economic development. Muslim scholars traveled throughout Asia and Africa, advancing the fields of architecture, history, mathematics, geography, and medicine. Centers of learning flourished from Baghdad (Iraq) to Toledo (Spain). During the early Arab–Islamic era, the development of banks, trusts, checks, receipts, and bookkeeping fostered vibrant economies and wide-ranging trade. The traders founded settlements and introduced new forms of living spaces. The architectural legacy of Arabs and Muslims lives on in Spain, India, Central Asia (Figure 6.11), the Americas, and in countless buildings across the world.

By the end of the tenth century, the Arab–Islamic empire had begun to break apart. From the eleventh to the fifteenth centuries, Mongols from eastern Central Asia (converted to Islam by 1330) conquered parts of the Arab-controlled territory, forming the Muslim Mughal Empire centered in what is now north India. Meanwhile, beginning in the 1200s, nomadic Turkic herders from Central Asia began to converge in western Anatolia (Turkey) where they eventually forged the Ottoman Empire, which became the greatest Islamic empire the world has ever known.

By the 1300s, the Ottomans had become Muslims and by the 1400s they had defeated the Christian Byzantine Empire, which

(A)

(B)

(C)

FIGURE 6.11 Islamic architecture outside the region. The architectural legacy of Islamic empires lives on today in Spain, India, Central Asia, and beyond. **(A)** The "Court of the Lions" at the Alhambra in Granada, Spain. Built between 1238 and 1391 by Moroccan architects and craftsmen, Alhambra was occupied by the last of several Moroccan dynasties to rule parts of Spain. **(B)** Iranian architects built the Taj Mahal in Agra, India, between 1632 and 1653 for Mughal Emperor Shah Jahan as a mausoleum for his favorite wife, who died in childbirth. **(C)** The Registan, a public square in Samarkand, Uzbekistan, is flanked by three Islamic schools, or madrassas, built between 1417 and 1660. Like the Taj Mahal, the buildings and plan of the Registan were heavily influenced by Iranian architecture.

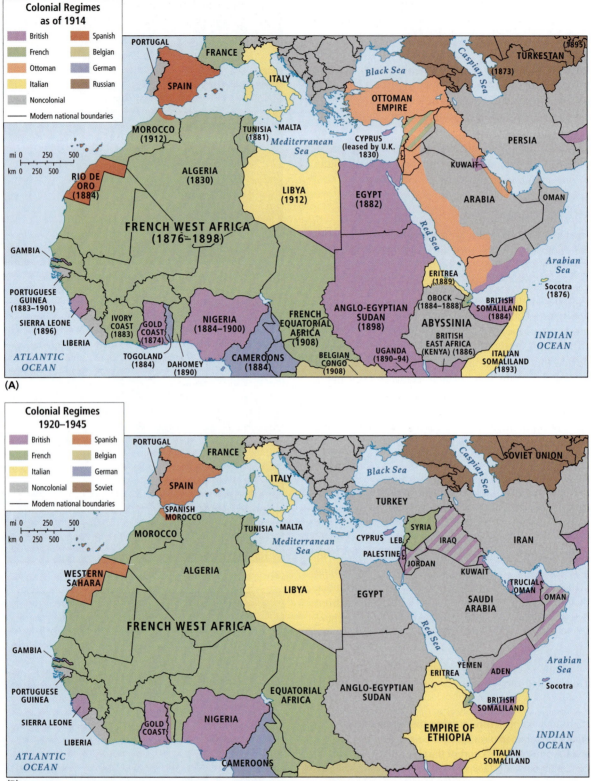

FIGURE 6.12 Colonial regimes in North Africa and Southwest Asia. **(A)** European powers began influencing the affairs of the region in the nineteenth century and expanded their control by 1914 at the beginning of World War I. The dates on the map indicate when the Europeans took control of each country. [Adapted from *Hammond Times Concise Atlas of World History* (Maplewood, N.J.: Hammond, 1994), pp. 100–101.] **(B)** Between 1920 and 1945, what was left of the Ottoman Empire in the eastern Mediterranean became protectorates administered by the British and French. The striped areas reflect the British colonial practice of allowing local rulers to govern while controlling many of their policies and actions. [Adapted from *Rand McNally Historical Atlas of the World* (Chicago: Rand McNally, 1965), pp. 36–37; and *Cultural Atlas of Africa* (New York: Checkmark Books, 1988), p. 59.]

was the successor to the Roman Empire. The Ottomans took over the Byzantine capital, Constantinople, renamed it Istanbul, and soon controlled most of the eastern Mediterranean and Egypt and Mesopotamia. By the late 1400s, they also controlled much of southeastern and central Europe. At about the same time, the Arab Muslims lost their control of the Iberian Peninsula to Christian kingdoms. Today, Islam still dominates in a huge area that stretches from Morocco to western China and includes northern India, and Malaysia and Indonesia in Southeast Asia.

Once a location was completely conquered, the Ottoman Empire, like the Arab–Islamic empire before it, encouraged religious tolerance toward the conquered peoples so long as they adhered to a religion with a sacred text. Jews, Christians, Buddhists, and Zoroastrians were allowed to practice their religions, although there were attractive economic and social advantages to converting to Islam. Multicultural urban life in Ottoman cities facilitated vast trading networks spanning the known world, and Istanbul became a cosmopolitan capital, outshining European cities until the nineteenth century.

Western Domination and State Formation

The Ottoman Empire ultimately withered in the face of a Europe made powerful by colonialism and the Industrial Revolution. Throughout the nineteenth century, North Africa provided raw materials for Europe in a trading relationship dominated by European merchants. In 1830, France became the first European country to exercise direct control over a North African territory (Algeria). France took control of Tunisia in 1881 and Morocco in 1912; Britain gained control of Egypt in 1882 and Sudan in 1898; and Italy took control of Libya in 1912 (Figure 6.12A).

World War I (1914–1918) brought the fall of the Ottoman Empire, which in a strategic error had linked itself with Germany. At the end of the war, the victorious Allied powers dismantled the Ottoman Empire and all of the former Ottoman territories; only Turkey was recognized as an independent country. The rest of the formerly Ottoman-controlled territory was allotted to France and Britain as protectorates (see Figure 6.12B). On the Arabian Peninsula, Bedouin tribes were consolidated under Sheikh Ibn Saud in 1932, and Saudi Arabia began to emerge as an independent country.

The aftermath of World War II further affected the political development of North Africa and Southwest Asia. Most famously, in the aftermath of the Holocaust in Europe, the Jewish state of Israel was created in the eastern Mediterranean on land occupied by unaffiliated groups of Arab small farmers and nomadic herders as well as some Jews (see the discussion on pages 286–287).

By the 1950s, European and U.S. energy and textile companies played a key role in deciding who ruled Iran and Saudi Arabia—where vast oil deposits were to become especially lucrative—and Egypt, which produced cotton. The governments of these countries showed their loyalty to the foreign companies with low taxes on oil and cotton exports and easy access to land. While a tiny ruling elite in these countries grew fabulously wealthy, the oil-tax revenues were not invested in creating opportunities for poor or middle-class people. Over time, ever more political power accrued to the foreign energy companies. The United States and Western Europe supported those autocratic local leaders who would keep order and were most sympathetic to their Cold War policies—as opposed to those of the Soviet Union—and were most likely to maintain a friendly attitude toward U.S. and European business interests. As a result of these concerns with commercial interests, both the Europeans and the Americans supported nondemocratic governments and stood in the way of reforms that would have resulted in a more educated populace able to participate in democracy.

THINGS TO REMEMBER

1. About 10,000 years ago in the Fertile Crescent, formerly nomadic peoples founded some of the world's earliest known agricultural communities. The domestication of plants and animals allowed them to build ever more elaborate settlements that eventually grew into societies based on widespread irrigated agriculture.

2. The three major monotheistic world religions—Judaism, Christianity, and Islam—all have their origins in this region in the eastern Mediterranean. Islam is by far the largest in numbers of adherents in the region, and it is the principal faith in all of the region's countries except Israel.

3. Beginning in the nineteenth century and continuing through the end of World War II, European colonial powers ruled or controlled most countries in the region.

4. Following World War II, the state of Israel was created, and in countries with oil deposits, the United States and Western Europe supported those autocratic local leaders most likely to maintain a friendly attitude toward U.S. and European business interests.

II CURRENT GEOGRAPHIC ISSUES

To outside observers, social and political change in this region seems to be lagging behind economic development. Why is the wealth generated by oil not resulting in an opening up of public discourse, a spreading of opportunities, such as broad public education, and democratization? Most analysts say that actually social change is underway and just beginning to gain momentum. On the other hand, political reform is much less apparent, but it too is bound to speed up as social change, such as the education of women, leads the way and economic development continues.

Sociocultural Issues

This section will explore the basics of Islam and examine the broad social changes occurring in the region with regard to family values, gender roles and gendered spaces, demographic change, urbanization and migration, and patterns of human well-being.

Religion in Daily Life

Ninety-three percent of the people in the region are followers of Islam; for them, the Five Pillars of Islamic Practice embody the central teachings of Islam. Some Muslims are fully observant, some are not, but the Pillars have an impact on daily life for all. The call to prayer, broadcast five times a day in all parts of the region, is a constant reminder to all people to reflect on their beliefs.

The Pillars of Muslim Practice

1. A testimony of belief in Allah as the only God and in Muhammad as his messenger (prophet).

2. Daily prayer at five designated times (daybreak, noon, midafternoon, sunset, and evening). Although prayer is an individual activity, Muslims are encouraged to pray in groups and in mosques.

3. Obligatory fasting (no food, drink, or smoking) during the daylight hours of the month of Ramadan, followed by a light celebratory family meal after sundown.

4. Obligatory almsgiving (*zakat*) in the form of a "tax" of at least 2.5 percent. The alms are given to Muslims in need. *Zakat* is based on the recognition of the injustice of economic inequity. Although it is usually an individual act, the practice of government-enforced *zakat* is returning in certain Islamic republics.

5. Pilgrimage (**hajj**) at least once in a lifetime to the Islamic holy places, especially Makkah (Mecca), during the twelfth month of the Islamic calendar (see **Figure 6.9C** on page 261).

Source: Carolyn Fluehr-Lobban, Islamic Society in Practice *(Gainesville: University of Florida Press, 1994).* ■

hajj the pilgrimage to the city of Makkah (Mecca) that all Muslims should make at least once in their lifetime

shari`a literally, "the correct path"; Islamic religious law that guides daily life according to the interpretations of the Qur'an

Sunni the larger of two major groups of Muslims, with different interpretations of shari`a

Shi`ite (or Shi`a) the smaller of two major groups of Muslims, with different interpretations of shari`a; Shi`ites are found primarily in Iran and southern Iraq

patriarchal relating to a social organization in which the father is supreme in the clan or family

Saudi Arabia occupies a prestigious position in Islam, as it is the site of two of Islam's three holy shrines: Makkah, the birthplace of the Prophet Muhammad and of Islam; and Al Madinah (Medina), the site of the Prophet's mosque and his burial place. (The third holy shrine is in Jerusalem.) The fifth pillar of Islam has placed Makkah and Al Madinah at the heart of Muslim religious geography. Each year, a large private-sector service industry, owned and managed by members of the huge Saud family, organizes and oversees the 5- to 7-day hajj for more than 2.5 million foreign visitors (see Figure 6.9C on page 261).

Islamic Religious Law and Variable Interpretations Beyond the Five Pillars, Islamic religious law, called **shari`a**, "the correct path," guides daily life according to the principles of the Qur'an. Some Muslims believe that no other legal code is necessary in an Islamic society, as the shari`a provides guidance in all matters of life, including worship, finance, politics, marriage, diet, hygiene, war, and crime. Insofar as the interpretation of shari`a is concerned, the Muslim community is split into two major groups: **Sunni** Muslims, who today account for 85 percent of the world community of Islam, and **Shi`ite** (or **Shi`a**) Muslims, who live primarily in Iran but also in southern Iraq and southern Lebanon. In fact, there are many interpretations of the Qur'an and a wide variety of versions of the observant Muslim life.

The Sunni–Shi`ite split dates from shortly after the death of Muhammad, when divisions arose over who should succeed the Prophet and have the right to interpret the Qur'an for all Muslims. This division continues today. The original disagreements have been exacerbated by countless local disputes over land, resources, and philosophies. In Iraq, for example, conflict between Sunnis and Shi`ites has been intensified by the rivalry over political power and fossil fuel resources that followed the U.S. invasion in 2003. Shi`ites recognize an authoritative priestly class whom they call *mullahs*.

Family Values and Gender

Perhaps because Islam has so little religious hierarchy, the family is the most important institution in this region. Although the role of the family is changing, a person's family is still such a large component of personal identity that the idea of individuality is almost a foreign concept. Each person is first and foremost part of a family, and the defining parameter of one's role in the family is gender identity.

The head of the family is nearly always a man; even when a woman is widowed or divorced, she will come under the tacit supervision of a male, perhaps her father or her son. An educated unmarried woman with a career outside the home is likely to live in the home of her parents or a brother and defer to them in decision making. Traditionally, men are considered more capable of making decisions and thus, it is thought, they should be in charge. These ideas, often labeled **patriarchal**, are now changing.

Gender Roles and Gendered Spaces

Carefully specified gender roles are common in many cultures, and there is often a spatial component to these roles. In the region of North Africa and Southwest Asia, in both rural and urban settings, the ideal is for men and boys to go forth into *public spaces*—the town square, shops, the market. Women are expected to inhabit primarily *private spaces*. But this ideal is an abstract concept; there are many exceptions.

To facilitate this ideal, traditional family compounds included a courtyard that was usually a private, female space within the home (**Figure 6.13A, C**); the only men who could enter were relatives. For the urban upper classes, female space was an upstairs set of rooms with latticework or shutters at the windows, which increased the interior ventilation and from which it was possible to look out at street life without being seen (**Figure 6.13B**). Today, the majority of people in the region live in urban apartments, yet even here there is a demarcation of public and private space. One

FIGURE 6.13 Domestic spaces. Many of the older cities and buildings in this region reflect the division between public space and private or domestic space. **(A)** As seen from above, nearly every house in this old part of Baghdad has a courtyard with a garden in it. **(B)** Projecting windows covered with lattice or louvers, known as a *mashrabiya,* in Aleppo, Syria. *Mashrabiyas* allow women to look out on the street below and catch breezes without being seen. **(C)** An interior view of a courtyard in Cairo. Courtyards are places where women can do chores and manage children while remaining secluded from the outside world.

or two formally furnished reception areas are reserved for nonfamily visitors, and rooms deeper into the dwelling are for family-only activities. When guests are present, women in the family are absent or present only briefly. Today, many women as well as men go out into public spaces, but how women enter these spaces remains an issue. Customs vary not only from country to country, but also from rural to urban settings and by social class.

In some parts of the region, particularly in Saudi Arabia, the requirement that women stay out of public view (also known as **female seclusion**) is strictly enforced. Women should not be in public spaces except when on important business and accompanied by a male relative. Elsewhere (in Egypt, for example), affluent urban women may observe seclusion either very little (especially if they are highly educated), or even more strictly than do rural women. Although rural women are traditional in their outlook, they have many tasks that they must perform outside the home: agricultural work, carrying water, gathering firewood,

female seclusion the requirement that women stay out of public view

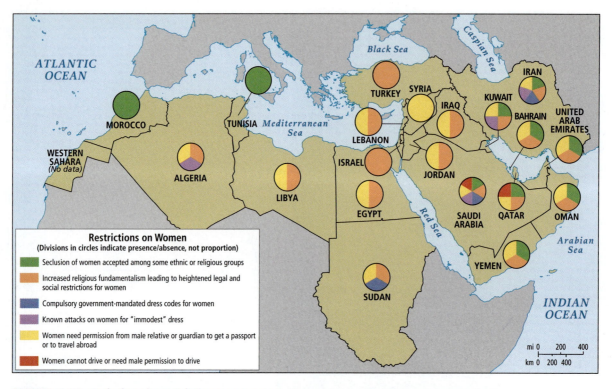

FIGURE 6.14 **Variations in restrictions on women.** The restrictions placed on women vary from country to country. Women's rights are perhaps most strongly protected in Turkey and Israel, where equality for women is constitutionally guaranteed, though religious fundamentalists are working to repeal these guarantees in both countries. [Adapted from Joni Seager, *The Penguin Atlas of Women in the World*, 3rd ed. (New York: Viking Penguin, 2003), pp. 14–15.]

veil the custom of covering the body with a loose dress and/or of covering the head—and in some places the face—with a scarf

Gulf States Saudi Arabia, Kuwait, Bahrain, Oman, Qatar, and the United Arab Emirates

polygyny the taking by a man of more than one wife at a time

and marketing, while upper-class women can afford servants to perform daily tasks in public spaces.

In the more secular Islamic countries—Morocco, Tunisia, Libya, Egypt, Turkey, Lebanon, and Iraq—women regularly engage in activities that place them in public spaces. Some wear conservative religious clothing; others dress in Western styles. Increasingly, female doctors, lawyers, teachers, and businesspeople are found in even the most conservative societies. Figure 6.14 compares the various levels of restrictions on women across the region.

Many women in this region use clothing as a way to create private space (Figure 6.15). This is done with the many varieties of the **veil**, which may be a garment that totally covers the person's body and face, or just a scarf that covers the person's hair. In some cultures, even prepubescent girls wear the veil; in others they go unveiled until their transition into adulthood is observed. The veil allows a devout Muslim woman to preserve a measure of seclusion when she enters a public space, thus increasing the space she may occupy with her honor preserved. A young woman may choose to wear a headscarf with jeans and a T-shirt in order to signal to the public that she is both a modern woman and an observant Muslim.

There is considerable debate about the origin and validity of female seclusion and veiling as specifically Muslim customs. Scholars of Islam say that these ideas predate Islam by thousands of years and do not derive from the teachings of the Prophet Muhammad. In fact, Muhammad may have been reacting against such customs when he advocated equal treatment of males and females. Muhammad's first wife, Khadija, did not practice seclusion, and worked as an independent businesswoman whose counsel Muhammad often sought.

📺 **146. Education, Economic Empowerment Are Keys to Better Life for Muslim Women**

The Rights of Women in Islam

The most conservative area in the region is the area known as the **Gulf States** (Saudi Arabia, Kuwait, Bahrain, Oman, Qatar, and the United Arab Emirates), where women cannot travel independently or even drive a car or shop without male supervision. Yet even here changes have recently been made.

In the decade of the 2000s, women in the Gulf States became noticeably more active in public life, education, and business, even staging mini-demonstrations now and then (one Saudi woman posted a video on YouTube of herself driving a car). In Qatar and the United Arab Emirates (UAE), progress in legal rights is the result of the will of some men and a few activist women, such as Sheikha Moza, a wife of the emir of Qatar, and other female lawyers and journalists who have challenged persistent patriarchal attitudes. Saudi Arabia remains the most restrictive, but even there it is now possible for a woman to register a business without first proving that she has hired a male manager.

Personal Vignette What is a well-to-do, highly educated young woman in Saudi Arabia to do if she wants to avoid an arranged marriage, and in fact might like to find a compatible mate in a society that allows virtually no contact between the sexes during adolescence and young adulthood? In a daring novel, Rajaa al-Sanea, herself such a girl, chronicles the daily lives of four premed and dentistry college women in *The Girls of Riyadh*. Freed by the Internet to make at least "virtual" contact with men, because they have little experience, the women make the expected naïve assumptions about the men they meet, only to be dumped. After a series of disappointing encounters, they rashly conclude that all men are pathetic pawns in a society that gives them power but no responsibilities. The stories are interesting but the main value of the novel is the way in which it discloses the ins and outs of life as a modern cloistered Saudi woman. For example, while abroad and liberated from the burka (a loose garment that covers the entire body except for the eyes), the girls of Riyadh enjoy their freedom but avoid all contact with other Saudis, lest their behavior and dress be reported back home. When flying home from trips to Europe or the United States, they endure long lines outside the airplane bathrooms as they make the last-minute change from Western clothes back to the burka. [*Sources: Claudia Rot Pierpont, "Found in Translation: The Contemporary Arabic Novel,"* The New Yorker, *January 18, 2010, pp. 74–80; and* The Complete Review, *at http://www.completereview.com/reviews/arab/alsanea.htm#ours.*] ∎

Polygyny—the taking by a man of more than one wife at a time—is a family practice that is a source of contention within this region and of censure from abroad. Although the Qur'an allows polygyny under certain conditions, it is not encouraged, nor is it a common modern practice. In some countries in the region, however, polygyny is legally permitted for Muslim men. It is estimated, for example, that less than 4 percent of males in North Africa have more than one wife. The Qur'an limits the number of wives to four and requires that all be treated as equals, which usually means they must have their own separate living quarters. This imposes a financial limit on the practice of polygyny.

Urbanization and modernization are also important limiting factors on polygynous practice. When agriculture was the main economic activity, multiple wives with several children each may have been economically productive. Urban life, with its small living spaces and cash requirements, favors smaller families. Democratization has led to bans against polygyny, first in Turkey (1926) and later in Tunisia, Lebanon, and Palestine. The custom does persist, especially in rural areas.

The Lives of Children

Three sweeping statements can be made about the lives of children in the Islamic cultures of North Africa and Southwest Asia.

FIGURE 6.15 Variations on the veil. There is an almost infinite variety of interpretations of the veil. **(A)** An Iraqi woman wears a scarf through which her hair can be seen. **(B)** A Turkish woman wears a headscarf that covers her hair completely. **(C)** Schoolchildren in Iran wear a uniform that covers their hair completely and a suit that covers most of their bodies. **(D)** An Egyptian woman covers all but her eyes and her hands.

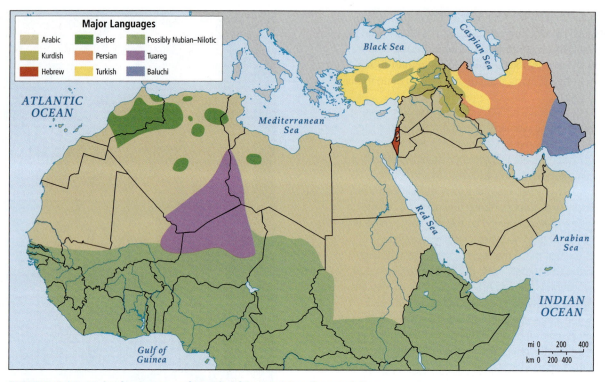

FIGURE 6.16 Major languages of North Africa and Southwest Asia. [Adapted from Charles Lindholm, *This Islamic Middle East: An Historical Anthropology* (Oxford: Blackwell, 1996), p. 9.]

First, in most families children contribute to the welfare of the family starting at a very young age. In cities, they run errands, clean the family compound, and care for younger siblings. In rural areas, they do all these chores and also tend grazing animals, fetch water, and tend gardens. Second, their daily lives take place overwhelmingly within the family circle. Both girls and boys spend their time within the family compound or in urban areas in adjacent family apartments. Their companions are adult female relatives and siblings and cousins of both sexes. Even teenage boys in most parts of the region identify more with family than with age peers.

In rural areas, prebertal girls often have considerable spatial freedom—the ability to move about in public space—as they go about their chores in the village. After puberty they may be restricted to the family compound and required to wear the veil. The U.S. geographer Cindi Katz found that until puberty, rural Sudanese Muslim girls have considerably more spatial freedom than do girls of similar ages in the United States, who are rarely allowed to range alone through their own neighborhoods.

The third observation is that school, television, and the Internet increasingly influence and broaden the lives of children. Most children go to school; many boys go for a decade or more, and increasingly girls go for more than a few years, and in some countries (Algeria, Israel, Jordan, Lebanon, Libya, the occupied Palestinian Territories, Oman, Qatar, Tunisia, and the UAE), a larger percentage of girls than boys go to school. Like educated women everywhere, these girls will make life choices different from those of their mothers.

Even in rural areas, it is fairly common for the poorest families to have access to a television, and it is often on all day, in part because it provides a window on the world for secluded women. Television may serve either to reinforce traditional cultural values or as a vehicle for secular values, depending on which channels are watched.

Language and Diversity

The eastern Mediterranean is the ancient home of people who spoke Semitic languages, including the people of Assyria, Babylonia, and Phoenicia. Modern speakers of Semitic languages include the Arabs and Jews.

Arabic is now the official language in all countries of North Africa and Southwest Asia except Iran, Turkey, and Israel (where the official languages are Farsi, Turkish, and Hebrew, respectively). This uniformity of language masks considerable cultural diversity, however. The many dialects of Arabic indicate deep cultural variation across the region. Also, numerous minorities within the region, such as Berbers, Tuaregs, Nubians, Kurds, and Turkomans, have their own non-Arabic languages (**Figure 6.16**). This cultural diversity of language is being counteracted by the mass media, where Arabic is the dominant language. French and English are also important second languages, especially in urban areas. The dominance of English on the Internet has contributed to its spread in recent years.

🦚🦚 **THINGS TO REMEMBER** 🦚🦚

1. For Muslims (93 percent of the population in this region), the Five Pillars of Islamic Practice embody the central teachings of Islam. Some Muslims are fully observant, some are not; but the

Pillars have an impact on daily life for all. The call to prayer, broadcast five times a day in all parts of the region, is a reminder to all people to reflect on their beliefs.

2. Given the wide diversity of thought, beliefs, and practices of Islam, the family is probably the most important societal institution for Muslims in the region. Most Muslim families here are patriarchal; and the role of women, the spaces they occupy, and in some cases, the clothes they wear, are carefully defined and rigidly enforced.

Changing Population Patterns

Although the region as a whole is nearly twice as large as the United States, most of the population is concentrated in the few areas that are useful for agriculture. Vast tracts of desert are virtually uninhabited, while the region's 477 million people are packed into coastal zones, river valleys, and mountainous areas that capture orographic rainfall (compare Photo Essay 6.4 on page 272 with Photo Essay 6.1 on page 253). Population densities in these areas can be quite high. For example, some of Egypt's urban neighborhoods have over 260,000 people per square mile (100,000 per square kilometer), a density four times higher than that of New York City, the densest city in the United States.

Although fertility rates have dropped significantly since the 1960s, at 3.1 children per woman in 2009, they are still higher than the world average of 2.6. Only sub-Saharan Africa, at 5.4 children per woman, is growing faster. At present growth rates, the population of the region will reach 540 million by 2025. Such population growth is severely straining supplies of freshwater, food, housing, jobs, and medical care, and requires the building of millions more schools. Also, high birth rates mean that the region's population is very young (see Figure 6.18 on page 273); the majority of people are under 25 years of age and all of these young people are in need of some sort of educational services to compete globally.

Population Growth and Gender Status

This region's high population growth rate can be explained in part by the persistent low status of women. As noted in many world regions, population growth rates are higher in societies where women are not accorded basic human rights, are less educated, and work primarily inside the home. In places where women have opportunities to work or study outside the home, they usually choose to have fewer children (see Photo Essay 6.4A, B, C on page 272). Figure 6.17 shows that as of 2005, considerably less than 50 percent of women across the region (except in Israel and Kuwait) worked outside the home at jobs other than farming. Moreover, on average only 74.8 percent of adult females can read, whereas 88 percent of adult males can.

For uneducated women who work only at home or in family agricultural plots, children remain the most important source of personal status, family involvement, and power (see Thematic Concepts Parts J, K, L on page 251). This may partially explain

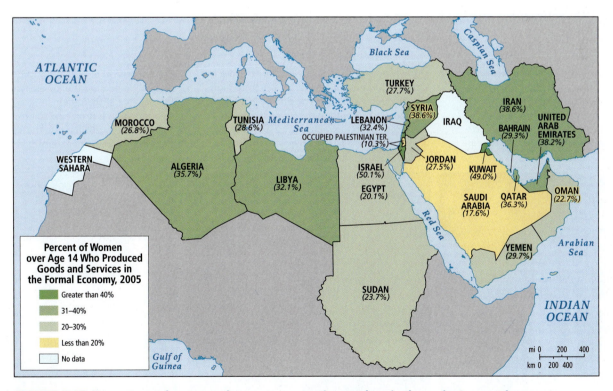

FIGURE 6.17 Percentage of women who are wage-earning workers in the region's countries. [Adapted from Joni Seager, *Women in the World: An International Atlas* (New York: Viking Penguin, 1997), pp. 66–67; and *United Nations Human Development Report 2007–2008* (New York: United Nations Development Programme), Table 31.]

This region has the second-highest population growth rate in the world and, at the present fertility rate, the population could reach 540 million by 2025, putting severe strains on water, food, and housing supplies. The high growth and fertility rates are due in part to the persistently low status of women in nearly all countries in the region: women are generally less educated than men and tend not to work outside the home.

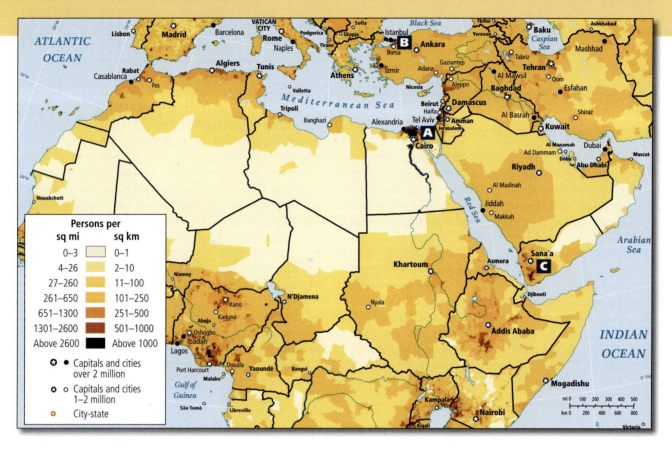

Persons per

sq mi	sq km
0–3	0–1
4–26	2–10
27–260	11–100
261–650	101–250
651–1300	251–500
1301–2600	501–1000
Above 2600	Above 1000

✪ ● Capitals and cities over 2 million
✪ ○ Capitals and cities 1–2 million
✪ City-state

A Israeli military police in Tel Aviv. Female literacy is high in Israel, and many women work outside the home. Female income is 67 percent of male income, closer than in any other country in the region. Population growth is moderate, with a projected increase of 49 percent by 2050.

B Female and male students take an exam in Istanbul, Turkey. Female literacy rates are high for the region at 80 percent, and by 2009 more women worked outside the home than in any other country in the region (51.2 percent). Population growth is also the slowest in the region, with a 19 percent increase projected by 2050.

C Children in a junkyard outside Sana`a, Yemen. Yemen has the lowest female literacy rate in the region at 38.6 percent. Its population is projected to grow by 151 percent by 2050, a rate that is higher than that of any other country in the region.

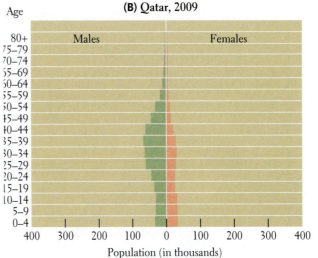

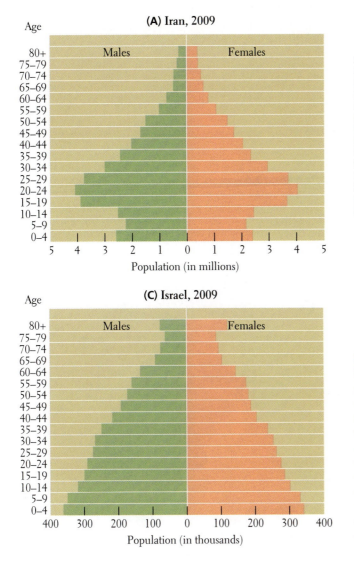

FIGURE 6.18 Population pyramids for Iran, Israel, and Qatar. The population pyramid for Iran is at a different scale (millions) from those for Israel and Qatar (thousands). The imbalance of Qatar's pyramid in the 25–54 age groups is caused by the presence of numerous male guest workers. Note, too, that pyramids A and C show missing females in the younger age groups. (This is most easily observed by drawing lines from the ends of the male and female age bars to the scale at the bottom of the pyramid and comparing the numbers.) [Adapted from U.S. Census Bureau, Population Division, "Population Pyramids for Iran," "Population Pyramids for Qatar," and "Population Pyramids for Israel,", International Data Base, May 2009, at http://www.census.gov/ipc/www/idb/country.php.]

why in 2009 only about 35 percent of women in this region were using modern methods of contraception, and only 54 percent were using any method of contraception at all. Both of these numbers are well below the world average of 55 and 62 percent, respectively. Other factors in the low use of contraception include the unavailability or high cost of effective birth control products and, most important, the unwillingness of males to consider the wishes of women in reproductive decisions.

The deeply entrenched cultural preference for sons in this region is both a cause and a result of women's lower social and economic standing. Families sometimes continue having children until they have a desired number of sons. Moreover, some young females may not survive because of malnutrition and associated illnesses or even because female fetuses are sometimes aborted. The result is that males slightly outnumber females in several age cohorts of the population pyramids, even those above age 5 in Figure 6.18 (see also the discussion in Chapter 1 on pages 18–19). In Qatar and the UAE, gender imbalance is extreme (see

Figure 6.18B) The unusually large number of males in Qatar over the age of 15 is the result of the presence of a large number of male guest workers. These men are kept separate from the rest of society and the UAE media has recognized the social pathology that such a concentration of men can bring, including the victimization of trafficked young women who are brought in to the male camps as sex workers (Justin Thomas, "The Biggest Gender Imbalance? It's Not in China," *The National*, February 2, 2010, at http://www.thenational.ae/apps/pbcs.dll/article?AID=/20100203/OPINION/702029924/1080).

Urbanization, Globalization, and Migration

Until recently, most people lived in small settlements. Since the late 1970s, significant migration from rural villages to urban areas occurred in response to economic forces driven by oil wealth and globalization, and by agricultural modernization that displaced small farmers. By 2008, more than 70 percent of the region's people lived in urban areas (the definition of urban varies by

Globalization has brought two distinct patterns of urbanization to this region. In fossil fuel–rich countries, populations are more urbanized and cities have undertaken lavish building booms, drawing in laborers and highly skilled workers from across the globe. In countries without fossil fuel, cities are receiving massive flows of poor rural migrants. These inflows are partially a result of economic reforms aimed at making rural and urban economies more globally competitive.

A Palm Jumeirah is an artificial palm-shaped island in Dubai, UAE. Built by more than 40,000 workers, mostly low-wage migrants from South Asia, Palm Jumeirah will eventually house 65,000 people, mostly in villas and condos that cost millions of dollars each. Palm Jumeirah and several similar developments are central to Dubai's efforts to build a globalized tourism-based economy that will prosper long after the region's fossil fuel resources are exhausted.

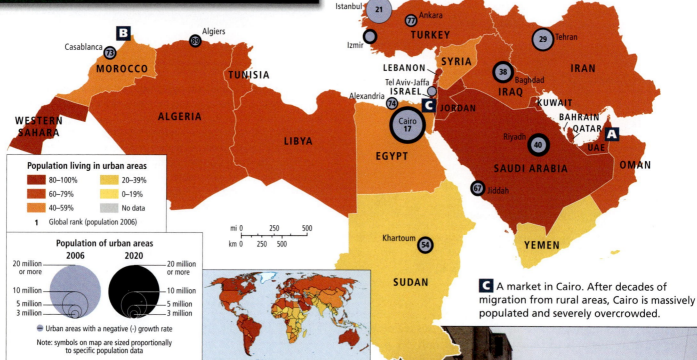

Population living in urban areas

- 80–100%
- 60–79%
- 40–59%
- 20–39%
- 0–19%
- No data

1 Global rank (population 2006)

mi 0   250   500
km 0   250   500

Population of urban areas

2006 | 2020

20 million or more
10 million
5 million
3 million

Urban areas with a negative (-) growth rate

Note: symbols on map are sized proportionally to specific population data

C A market in Cairo. After decades of migration from rural areas, Cairo is massively populated and severely overcrowded.

B Tensions run high at the fishing docks in Casablanca, Morocco. The city's population is growing quickly due largely to migration from Morocco's drought-stricken interior, but a lack of jobs is resulting in a higher crime rate.

country; see Photo Essay 6.5). By 2009, there were more than 434 cities with populations of at least 100,000, and 37 cities with more than 1 million people. The largest city in the region, and one of the largest in the world is Cairo, with 16.3 million residents.

The fossil fuel–rich Gulf States are now highly urbanized, and their sleek, modern cities draw money and workers from all over the world (see Thematic Concepts Parts D, E, G on page 250). In the Persian Gulf states, between 70 and 100 percent of the population now live in urban areas, which are extravagant in design. For example, Dubai (one of the United Arab Emirates), has built elaborate (and as yet largely unoccupied) new condominiums for the rich on Palm Jameirah, a fanciful palm tree–shaped island and peninsula off of its Persian Gulf coastline (see Photo Essay 6.5A). The economic boom being enjoyed by the Gulf States has slowed drastically since the recent recession of 2008. The wealthier Emirates subsidize the living standards of those that do not have oil wealth and for the most part poverty is apparent only among foreign contract laborers (see Thematic Concepts Part H on page 250).

In the ancient cities of the region—Istanbul, Baghdad, Cairo, and Casablanca, for example—growth has been more haphazard, driven by rapid rural-to-urban migration (see Photo Essay 6.5B). For example, in 1950 Baghdad had only 580,000 residents, while today it is home to 5 million. Cairo has had to provide for millions of new residents who live in huge makeshift slums (see Thematic Concepts Part I on page 250). Cairo's middle class occupies the medieval interiors of the old city, where streets are narrow pedestrian pathways (see Photo Essay 6.5C).

Of those who remain rural residents, some live and work on farms, but increasingly, in central Turkey, for example, even rural people live in blocks of government-subsidized medium-rise apartment buildings. A few still work on mechanized farms, but most work in factories and firms that were purposely located in the countryside to stem the flow of migrants to the cities.

Migration is an active force in urbanization. Rural-to-urban migration is *pushed* by agricultural modernization and global commodity price structures that are disadvantageous to small producers. To some extent, rural people are also *pulled* into cities by the prospect of jobs.

Immigrants come from all over the world to be temporary guest workers in the Arabian Peninsula. They work in construction and other industries that help make cities function. In the Arabian Peninsula, where foreign immigrants make up 88 percent of the labor force, most employers prefer Muslim guest workers, and over the last two decades, several hundred thousand Muslim workers have arrived from Palestinian refugee camps in Lebanon and Syria. Other countries that supply labor are Egypt, Pakistan, and India (see Thematic Concepts Part H on page 250). Some female domestic and clerical workers come from Muslim countries in South and Southeast Asia, including the Philippines. As discussed above these immigrant workers are overwhelmingly temporary residents with no job security and no rights to earn citizenship. They remit most of their income to their families at home and often live in stark conditions alongside the opulent lifestyles of those enriched by oil and gas.

A final category of migrants are the refugees: this region has the largest number of refugees in the world. Usually they are escaping human conflict, but environmental disasters such as earthquakes or long-term drought also displace many people. When Israel was created in 1949, many Palestinians were placed in refugee camps in Lebanon, Syria, Jordan, the West Bank, and the Gaza Strip. Palestinians still constitute the world's oldest and largest refugee population, numbering at least 5 million. Elsewhere, Iran is sheltering more than a million Afghans and Iraqis because of continuing violence and instability in their home countries. Across the region, even more people are refugees within their own countries: 1.7 million Iraqis are internal refugees, and in Sudan, between 5 and 6 million Sudanese are living in refugee camps.

Refugee camps often become semipermanent communities of stateless people in which whole generations are born, mature, and die. Although residents of these camps can show enormous ingenuity in creating a community and an informal economy, the cost in social disorder is high. Tension and crowding create health problems. Because birth control is generally unavailable, refugee women have an average of 5.78 children each. Disillusionment is widespread. Years of hopelessness, extreme hardship, and lack of employment take their toll on youth and adults alike, leading some to adopt extremist attitudes against those they see as responsible for their suffering. Moreover, even though international organizations provide basic services for refugees, the refugees constitute a huge drain on the resources of their host countries. In Jordan, for example, native Jordanians are a minority in their own country because Palestinian refugees and their children account for well over half the total population of the country, and their presence has changed life for all Jordanians. Since the beginning of the Iraq war in 2003, over 750,000 Iraqis have also fled to Jordan.

Geographic Patterns of Human Well-Being

As we have observed in previous chapters, GDP per capita can be used as a general measure of well-being because it shows in a broad way how much people in a country have to spend on necessities, but it is only an average figure and does not indicate how income is distributed. As a result, there is no way to know if a few are rich and most poor, or if income is more equally apportioned. (The PPP designation means that all figures are adjusted for cost of living and so are comparable.)

Figure 6.19A on the next page points to three situations regarding GDP in this region. First, from country to country there is wide variation—roughly U.S.$2000 per person per year in Morocco and more than U.S.$30,000 per person per year in Kuwait, Bahrain, Qatar, and the UAE. Second, if wealth is more or less evenly distributed within each country, most people could be in at least the lower reaches of the global middle class (with the exception of Sudan). Unfortunately the UN data on how wealth is distributed is mostly missing (there is no data for Kuwait, Qatar, UAE, Bahrain, Libya, Oman, Saudi Arabia, Lebanon, the occupied Palestinian Territories, Syria, Sudan, and Iraq), which usually means that disparity is high and that the governments do not want to reveal the information. For those countries that do provide data (Israel, Turkey, Jordan, Tunisia, Iran, Algeria, Egypt, Morocco, and Yemen), wealth disparity is moderate, averaging about the same as in the United States. Third, oil and gas wealth does not necessarily

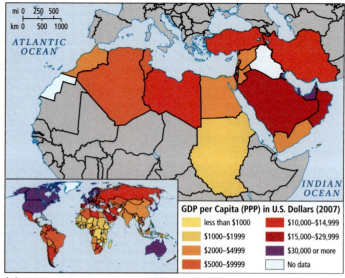

(A)

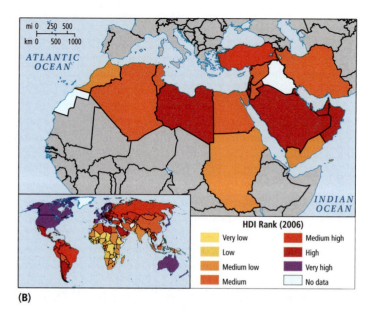

(B)

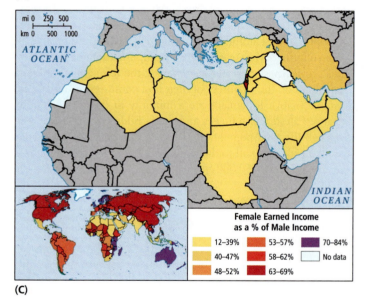

(C)

FIGURE 6.19 **Maps of human well-being.** **(A)** GDP. **(B)** HDI. **(C)** Female earned income as a percent of male earned income. [Maps adapted from data: "Human development indices," (Maps A and B): Table 2, pages 28–32; (Map C): Table 5, pages 41–44 at http://hdr.undp.org/en/media/HDI_2008_EN_Tables.pdf.]

translate into high GDP per capita—compare Figure 6.19A with Figure 6.20. Algeria has quite a bit of oil, but has a rather low GDP per capita. Saudi Arabia, for all its oil wealth, is high, but not in the top GDP per capita category; only Bahrain, Kuwait, Qatar, and the UAE are. Turkey, which has a diversified economy and very little oil, nonetheless has a GDP rank equal to that of Iran (in the middle range), which has significant oil and gas resources.

The Human Development Index (HDI) (Figure 6.19B). ranks countries on how well they provide the basic necessities for their people (income, education, and health). In this region, HDI ranks stretch from the medium-low to the high ranges. If on the HDI scale a country drops below its approximate position on the GDP scale, one might suspect that the lower HDI ranking is due to poor distribution of wealth and poor social services: a critical mass of people are simply not getting sufficient income to acquire good health and an education. Iran, Kuwait, Bahrain, and the UAE all rank lower on HDI than they do on the GDP scale; hence, one could say that these countries are not managing their relative wealth well enough to ensure the well-being of

their populations—disparities are high. Libya, on the other hand, ranks high on the HDI scale and only in the mid-range on GDP per capita. This probably indicates that Libya is investing quite heavily in social services that enhance human well-being. The fact that Sudan ranks higher on HDI than GDP is a bit of a mystery, because west central Sudan (Darfur) has suffered so much from drought, war, and genocide in recent years. It could be that despite grinding poverty, human well-being in Sudan has been raised a bit by services for refugees from emergency donor countries, or that available figures are based on faulty data.

Figure 6.19C shows the extent to which countries are ensuring gender equality in earned income. First, as is true across the globe, nowhere in this region are women paid equally to men. Second, with the sole exception of Israel, the entire region ranks uniformly low, paying women on average not even 40 percent of what men earn. Only Iran ranks slightly higher, paying women at best 47 percent of what men earn. As shown on the world inset map in Figure 6.19C, this region has a poorer gender equity record than most other parts of the world, with the exceptions of Pakistan, India, Indonesia, and Mexico.

FIGURE 6.20 **Economic issues: Oil and gas resources in North Africa and Southwest Asia.** [Adapted from U.S. Department of Energy, Country Analysis Briefs, at http://www.eia.doe.gov/emeu/cabs/Region_me.html; U.S. Department of Energy, "Selected Oil and Gas Pipeline Infrastructure in the Middle East," at http://www.eia.doe.gov/cabs/Saudi_Arabia/images/Oil%20and%20Gas%20Infrastructue%20Persian%20Gulf%20(large)%20(2).gif; and "Who Has the Oil," available at http://gcaptain.com/maritime/blog/who-has-the-oil-a-map-of-world-oil-reserves/.]

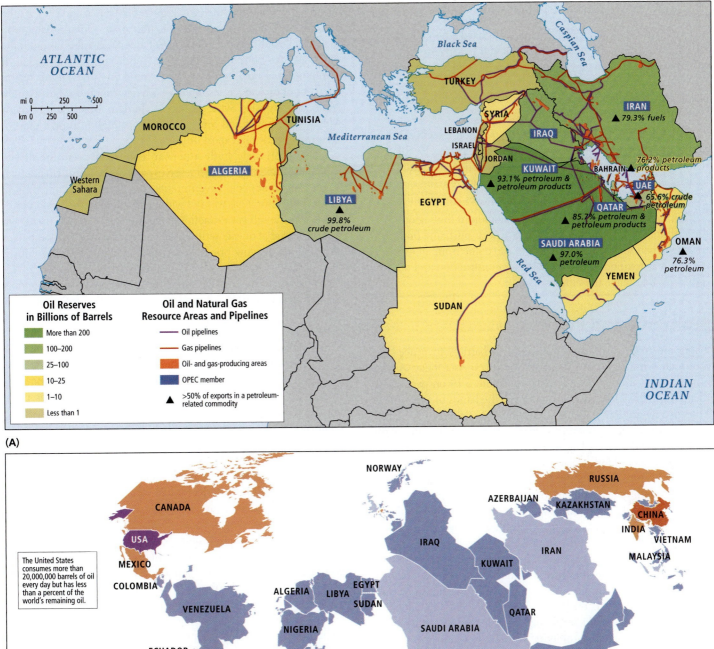

Oil Reserves in Billions of Barrels
- More than 200
- 100–200
- 25–100
- 10–25
- 1–10
- Less than 1

Oil and Natural Gas Resource Areas and Pipelines
- Oil pipelines
- Gas pipelines
- Oil- and gas-producing areas
- OPEC member
- ▲ >50% of exports in a petroleum-related commodity

IRAN ▲ 79.3% fuels

▲ 76.2% petroleum products

KUWAIT 93.1% petroleum & petroleum products

UAE ▲ 65.6% crude petroleum

QATAR ▲ 85.7% petroleum & petroleum products

LIBYA ▲ 99.8% crude petroleum

SAUDI ARABIA ▲ 97.0% petroleum

OMAN ▲ 76.3% petroleum

(A)

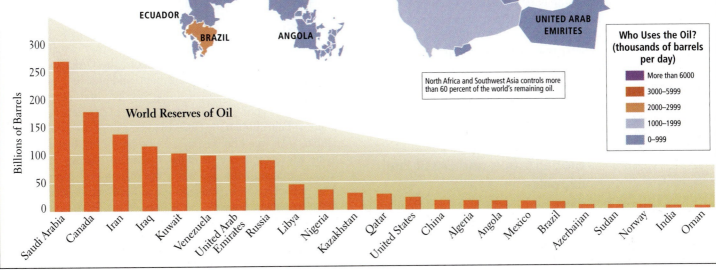

The United States consumes more than 20,000,000 barrels of oil every day but has less than a percent of the world's remaining oil.

North Africa and Southwest Asia controls more than 60 percent of the world's remaining oil.

Who Uses the Oil? (thousands of barrels per day)
- More than 6000
- 3000–5999
- 2000–2999
- 1000–1999
- 0–999

World Reserves of Oil

Billions of Barrels

Saudi Arabia, Canada, Iran, Iraq, Kuwait, Venezuela, United Arab Emirates, Russia, Libya, Nigeria, Kazakhstan, Qatar, United States, China, Algeria, Angola, Mexico, Brazil, Azerbaijan, Sudan, Norway, India, Oman

(B)

1. Two globalized patterns of urbanization have emerged in the region. In the oil-rich countries, spectacular luxury-oriented urban development is tied to flows of money, goods, and people. Elsewhere, economic reforms aimed at improving global competitiveness have brought massive migration of the rural poor into cities, creating crowding and slums.

2. This region has the second-highest population growth rate in the world, after sub-Saharan Africa. Part of the reason for this is that women are generally less educated than men and tend not to work outside the home. Hence, childbearing remains crucial to a woman's status, a situation that encourages large families.

3. Migration is an active force in urbanization in the region. Agricultural modernization and job prospects push and pull rural-to-urban migration. Immigrants also come from all over the world as temporary guest workers in construction and other industries that make the cities work.

Economic and Political Issues

There are major economic and political barriers to peace and prosperity within North Africa and Southwest Asia today. As the human well-being maps indicate (see Figure 6.19 on page 276), wealth from fossil fuel exports remains in the hands of a few elites. Most people are low-wage urban workers or relatively poor farmers or herders. The economic base is unstable because the main resources, fossil fuels and agricultural commodities, are subject to wide price fluctuations on world markets. Meanwhile, in many poorer countries, large national debts are forcing governments to streamline production and cut jobs and social services.

Political and economic cooperation in the region has been thwarted by a complex tangle of hostilities between neighboring countries. Many of these hostilities are the legacy of outside interference by Europe and the United States in regional politics, including earlier colonial intrusions and more recent neo-colonial activities by global corporations, including oil, gas, industrial, and agricultural enterprises. The Israeli–Palestinian conflict, which has profoundly affected politics throughout the region, is the result of the massive migrations of Jews from Europe and the former Soviet Union in the aftermath of the Holocaust and World War II. The Iran–Iraq war of 1980–1988 and the Gulf War of 1990–1991

cartel a group of producers strong enough to control production and set prices for its products

OPEC (Organization of Petroleum Exporting Countries) a cartel of oil-producing countries—including Algeria, Angola, Iran, Iraq, Kuwait, Libya, Nigeria, Qatar, Saudi Arabia, the United Arab Emirates, and Venezuela—that was established to regulate the production, and hence the price, of oil and natural gas

were instigated at least in part by pressures from outside the region. The long siege of violence in Iraq, though begun by a home-grown dictator, did not come to a head until the U.S. invasion and occupation of Iraq, which began in the spring of 2003.

Globalization and Fossil Fuel Exports

The area of North Africa and Southwest Asia is the major supplier to the world of fossil fuels; two-thirds of the world's known reserves of oil and natural gas are found here. Most oil and gas reserves are located around the Persian Gulf (see Figure 6.20 on page 277) in the countries of Saudi Arabia, Kuwait, Iran, Iraq, Oman, Qatar, and the United Arab Emirates (UAE). Oil and gas are also found in the North African countries of Algeria, Tunisia, Libya, and Sudan. It is important to note, however, that several countries in this region are without oil and gas resources or are net importers of fossil fuels: Morocco, Tunisia, Syria, Lebanon, Jordan, Israel, and Turkey. Meanwhile, Egypt and Sudan produce enough to supply most of their own needs but export little.

European and North American companies were the first to exploit the region's fossil fuel reserves early in the twentieth century. These companies paid governments a small royalty for the right to extract oil (natural gas was not widely exploited in this region until the 1960s, though gas extraction has grown rapidly since then). Oil was processed at on-site refineries owned by the foreign companies and sold at very low prices, primarily to Europe and the United States and eventually to other countries, such as Japan.

The current era of oil and gas price fluctuations on the global market began in 1973 when governments of the region abruptly declared all fossil fuel resources and industries to be the property of the state and raised crude oil prices dramatically. Oil income in Saudi Arabia alone shot up from U.S.$2.7 billion in 1971 to U.S.$110 billion in 1981. Even before this, however, in the 1960s, the oil-producing countries organized a **cartel**, a group of producers strong enough to control production and set prices for its products. The cartel is **OPEC—the Organization of Petroleum Exporting Countries**. OPEC now includes all of the states indicated in Figure 6.21 plus Ecuador, Venezuela, and Nigeria. OPEC members cooperate to periodically restrict or increase oil production, thereby significantly influencing the price of oil on world markets. A move to create an OPEC-like cartel for natural gas is currently underway.

OPEC countries, many of which were exceedingly poor just 40 years ago, were slow to invest much of their fossil fuel earnings at home in basic human resources, and they were slow to explore other economic strategies in case oil and gas ran out. Like their poorer neighbors, they have remained highly dependent on the industrialized world for much of their technology, manufactured goods, skilled labor, and expertise.

Throughout the 1990s and up to 2008, increasing global demand caused oil and gas prices to rise despite occasional dips in demand, and most OPEC producers are swimming in cash. Fossil fuel wealth has brought significant benefits to the region as a whole. The Gulf States with the largest reserves and highest petroleum-based incomes (Saudi Arabia, the UAE, and Kuwait) made the largest capital improvements and now have good roads, airports, new cities, irrigated agriculture, and petrochemical industries. A good example of the scale of projects in the Gulf States is Palm Jumeirah (Palm Island), a huge artificial island in Dubai, UAE (see Photo Essay 6.5A on page 274), that is intended to be the centerpiece of Dubai's planned tourism economy—insurance against the day when the regional oil economy fails. This project is now in financial jeopardy, primarily because of the global recession; tourism, an expensive luxury, is vulnerable to economic downturns.

Unfortunately, far too little attention has been given to education, social services, public housing, and health care. This is true across North Africa (with the possible exception of Libya) and

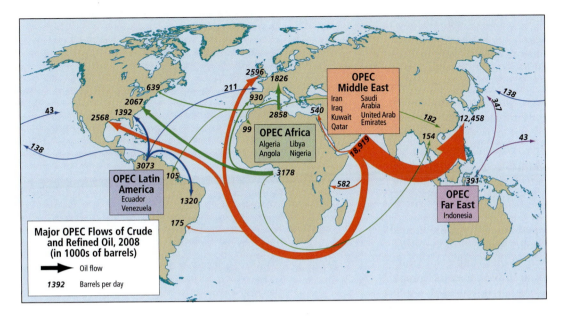

FIGURE 6.21 Major OPEC oil flows in 2008. The map shows the average number of barrels (in thousands) of crude and refined oil distributed per day by OPEC countries to the world. [Adapted from Organization of the Petroleum Exporting Countries, *Annual Statistical Bulletin 2008*, 2009.]

throughout most of the Gulf States. The exceptions in the Gulf States are Kuwait and Oman, which have developed ambitious plans for addressing all social service needs, with especially heavy investment in higher education. Meanwhile, the non–fossil fuel producing (non-OPEC) countries do not share directly in the wealth of the Persian Gulf states. For the most part, the oil-rich countries have not helped their poorer neighbors develop.

The major exceptions to this pattern of underinvestment in social services are Turkey, which has a sharply better record, and probably Iran, as the government tries to maintain public support, although data on Iran are hard to obtain.

Many factors outside of OPEC's control strongly influence world oil prices. Consumers outside the region affect prices by reducing or expanding demand (Figure 6.21). Geopolitical events, such as the September 11, 2001 attack on the United States, may cause an immediate spike in prices but ultimately depress prices, as the frightened and uncertain public decides not to travel and to consume less. On the other hand, recent rapid industrialization in China and India has increased demand and been a major factor in oil price rises. And many experts say that an era of diminished oil reserves has already begun.

As the global recession that began in 2008 gained momentum, consumption of petroleum products dropped steadily and precipitously. The decrease in demand caused crude oil prices to fall from U.S.$145 per barrel in July 2008 to U.S.$37 in January 2009. By July of 2009, the price had rebounded to U.S.$65—less than half of what it had been the previous July—and by January of 2010, demand was light for deep winter and the price was U.S.$78 per barrel. All oil- and gas-producing countries suffered during the recession. For a while the Gulf States appeared to be weathering the crisis well by drawing on savings to keep capital spending high. But as the crisis spread to real estate, GDP growth fell to half of what had been expected, investor confidence waned, and several of the massive building projects underway across the Gulf States screeched to a halt, including Palm Island and other projects in the UAE.

The effects of the delayed but nonetheless significant downturn in the economies of the Gulf States have reverberated globally. One of the most widespread effects has been the sharp decrease in remittances to all the countries that send immigrants to work in the Gulf State economies. Indonesia, southern Pakistan, the Philippine Islands, and the Malabar Coast of India have been severely affected.

Economic Diversification and Growth

Greater **economic diversification**—expansion of an economy to include a wider array of activities—could have a significant impact on the region. It could bring economic growth and broader prosperity and limit the damage caused by a drop in the price of oil, gas, or other commodities on the world market. Except for a few countries, however, actual diversification has been meager, despite—in some cases—wildly excessive spending.

economic diversification the expansion of an economy to include a wider array of activities

By far the most diverse economy is that of Israel, which has a large knowledge-based service economy and a particularly solid manufacturing base. Israel's goods and services and the products of its modern agricultural sector are exported worldwide. Turkey is the next most diversified, in large part because—like Israel—it has never had oil income to fall back on. Egypt, Morocco, Libya, and Tunisia are also starting to move into many new economic activities. Some fossil fuel–rich countries have tried to diversify into other industries. In the UAE, for example, only 25 percent of GDP is based directly on fossil fuels; trade, tourism, manufacturing, and financial services are now dominant. Even so, most Persian Gulf countries are still highly dependent on fossil fuel exports and the spin-off industries they generate.

Diversification has also been limited by economic development policies that favored *import substitution* (see Chapter 3 on pages 136–137). Beginning in the 1950s, many governments, such as those of Turkey, Egypt, Iraq, Israel, Syria, Jordan, Tunisia, and

sovereign-wealth funds in the Gulf States, state-owned savings from oil and gas income, which are then invested globally in a range of income-producing ventures

Libya, established state-owned enterprises to produce goods for local consumption. Among the major products were machinery and metal items, textiles, cement, processed food, paper, and printing. These enterprises were protected from foreign competition by tariffs and other trade barriers. With only small local markets to cater to, profitability was low and the goods were relatively expensive. Without competitors, the products tended to be shoddy and unsuitable for sale in the global marketplace. Meanwhile, the extension of government control into so many parts of the economy nurtured corruption and bribery and squelched entrepreneurialism.

Economic diversification and export growth were also limited by a lack of financing, private and public, from within the region (Figure 6.22). For example, members of the Saudi royal family generally prefer more profitable private investments in Europe, North America, or Southeast Asia. Much private and public investment has recently gone into lavish development in the UAE, particularly in Dubai. Only recently have governments recognized that they need to plan for a time when oil and gas run out. The Gulf States have established **sovereign-wealth funds**, state-owned savings from oil and gas income, which are then invested globally in a range of income-producing ventures. Such funds lost much of their value in the global recession beginning in 2008.

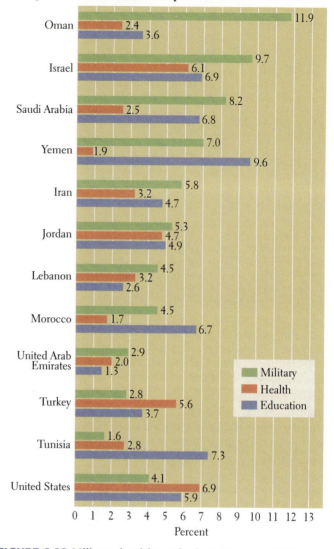

Military, Health and Education Expenditures as Percent of GDP

FIGURE 6.23 Military, health, and education expenditures as a percentage of GDP. These charts show only those countries in the region that have data for each variable. The United States is included for comparative purposes. [Adapted from *United Nations Human Development Report 2007–2008* (New York: United Nations Development Programme), Table 19, at http://hdr.undp.org/en/media/HDR_20072008_EN_Complete.pdf.]

FIGURE 6.22 Economic diversification and globalization. The textile industry in Syria has grown rapidly in recent years, but is now struggling to compete in the global market with cheaper products from China and India. Moreover, the industry has proven unstable during periods of global economic downturn, with many Syrian textile manufacturers having to close, then reopen when times are good, only to close again during the next crisis.

Finally, both international and domestic military conflicts and the ensuing political tensions have stymied economic diversification because they have resulted in some of the highest levels (proportionate to GDP) of military spending in the world. Military spending diverts funds from other types of development, such as health care and education (Figure 6.23). The top four spenders—Oman, Israel, Saudi Arabia, and Yemen—lead the world in military spending as a percentage of GDP. And all countries in the region, except Tunisia, are above the global average of 2 percent.

THINGS TO REMEMBER

1. This region's vast fossil fuel resources have long attracted interest from European and North American companies eager to make profits and from countries that need energy.

2. Profits to the oil- and gas-producing countries were low until OPEC was organized. Wealth has accrued to a privileged few,

and shaped development in many countries, linking economies to a global system of investment, labor, and expertise. Politics have also become globalized, with Europe and the United States strongly influencing many governments.

3. The vast wealth from energy resources, slow social reform, and the resultant political tensions have inhibited economic diversification and blinded elites to the consequences of declines in resources and demand. Governments tend to spend on the military rather than on other types of development.

Democracy and Islamism in a Globalizing World

Since the attacks of September 11, 2001, Islamism (see page 249) has been viewed by outsiders as a great threat to the region's political stability and economic development. But in fact, Islamist movements play a varied and nuanced role in the region. Some are simply grassroots religious revivals that are intended to counteract the secular influences that have become widespread, thanks to Western influence and globalization. In some cases, Islamists have been severely repressed by secular governments and hence have become more militant, even trying to undermine their own governments (Photo Essay 6.6E on page 282). In yet other instances, however, Islamists have been able to work productively within democratic processes, performing badly needed social services that governments have failed to provide.

The roots of Islamism are complex. For many decades, and into the present, virtually all governments in this region have tended to be authoritarian and undemocratic (see Thematic Concepts Part N on page 251). The violent conflicts that have occurred in nearly every country in the region (see Photo Essay 6.6 Map on page 283) may be primarily caused by the lack of options for general democratic participation. Free speech and the right to hold public meetings have been severely limited. Minorities have been denied the right to engage in their own cultural practices and speak their own languages. Journalists and private citizens have similarly not been able to criticize governments or expose discrimination and corruption or even visit neighboring countries (see Photo Essay 6.6A, B, D on pages 282–283). In many countries, the only public spaces in which people have been allowed to gather are mosques, and the only public discussions not subject to censorship by the government have been obscure religious discourses. Hence, political discussions and the political movements that have developed from them have been strongly shaped by the views of the region's Islamic leaders.

Islamism is also rooted in the mixture of governmental and religious authority that has characterized most of the region's empires and countries throughout their history. In **theocratic states**—Saudi Arabia, Yemen, the UAE, Oman, and Iran— Islam is the official religion; political leaders must be observant Muslims, and they are considered divinely guided. The government enforces the religious principles of Islam and the legal system is based on conservative interpretations of shari'a. In declared **secular states**—Algeria, Egypt, Libya, Morocco, Iraq, Turkey, and Tunisia—theoretically the state is neutral on matters of religion, treating all citizens equally regardless of religion, but protecting

the freedom of religion. There is no direct influence of religion on affairs of state, nor is the state atheistic or opposed to religion. In practice in this region, however, religion plays a public role even in secular states because most political leaders are at least nominally Muslim and Islamic ideas influence government affairs.

Globalization is a particular problem for Islamists, who object to what they view as immoral secular influences emanating from Europe and North America that are bound to undermine Islamic values. Some object to the liberalization of women's roles, especially to women being educated, employed, and otherwise active outside the home (see Figure 6.17 on page 271). Many also lament what they see as the global spread of open sexuality, consumerism, and hedonism, transmitted in part by TV, movies, and popular music. While seeing the advantages of bringing people into the computer age, Islamists abhor Internet pornography and free access to political ideas. The proper role of the press is also hotly debated. These concerns are similar to those of other devoutly religious people around the world.

> **theocratic states** countries that require all government leaders to subscribe to a state religion and all citizens to follow rules decreed by that religion
>
> **secular states** countries that have no state religion and in which religion has no direct influence on affairs of state or civil law

140. New Poll of Islamic World Says Most Muslims Reject Terrorism

Islamist factions vary greatly in the perspectives and fervor they bring to their causes. While some groups, such as Al Qaeda, advocate random violent resistance and drastic reforms once they gain control, most simply want to moderate secular influences in the hope that so doing will ameliorate the hardships of their lives. Islamist movements and the social services they provide have a strong popular base in the slums of the largest cities, where the economic situation of millions has been worsened by globalization and, more recently, by the global recession. Islamism also appeals to refugees from the region's military conflicts, whose suffering has led them to question the basic philosophy of the governments under which they live.

144. Jihadist Ideology and the War on Terror

139. What Motivates a Terrorist?

Repression of Islamists by nondemocratic secular governments has only increased their fervor. After decades of elitist and highly corrupt government in Iran by Shah Mohammad Reza Pahlavi, an Islamist revolution led by Ayatollah Ruhollah Khomeini overthrew the Shah's government in 1979. Fearing a repeat of this scenario, secular governments in several countries—Egypt and Algeria, for example—attempted to forcibly repress Islamist movements. As a result, both of these countries suffered from years of Islamist-sponsored terrorist campaigns. In Algeria, violence has diminished only in the last decade following free and fair elections that incorporated Islamists into the government. In Egypt, where repression persists and elections are plagued by government-sponsored intimidation and fraud, Islamist terror and violence occasionally re-erupts.

143. Slow Pace of Political Reform in Egypt Strains U.S.–Egyptian Relations

Turkey has been the most successful at using the democratic process to constructively engage Islamist political parties. In fact, Turkey's government has been controlled by Islamists on and off for decades. Nevertheless, there have been no significant departures from the country's safeguards for freedom of religion or its steady improvement on women's rights, education, and freedom of the press. Nor has the country's long-term goal of becoming a modern, prosperous member of the European Union been irreparably compromised by theocratic politics, though it has certainly been jeopardized.

📺 **153. Turkey Votes for Stability**

Perhaps, then, democratization is actually a path to peace with Islamists. In important cases, governments that have allowed free elections in which Islamists can run for office have enjoyed greater stability. This may partially explain the slight region-wide drift toward democracy in recent years. Elections have brought moderate Islamists to power in Turkey, Jordan, and Algeria. This has resulted in a decrease in discontent and a stoppage of violence. In other cases (primarily Iran and Gaza), elected Islamists have caused internal and external relations and interactions with the Muslim and non-Muslim world to worsen. In Iran, a government run both by appointed religious leaders and by elected officials repressed a large minority of political dissenters in 2009. Iran apparently has also sponsored Islamist terrorism in other countries, such as Israel, Iraq, and Lebanon.

While many countries are still ruled by kings and royal families, almost all countries have elected components to their national legislatures and municipal councils.

Democratization: The Role of the Press and Media

In some countries—Turkey, Israel, and Morocco, for example— the press is reasonably free and opposition newspapers are aggressive in their criticism. But in much of the region, journalism can be a risky career, often leading to prison. Egypt has a checkered history where the press is concerned. For years, Hisham Kassem was editor of the *Cairo Times*, an independent English-language weekly. When he became too critical of the government, he lost his license to publish in Egypt. For a time he took great pains to write and print his paper outside the country and smuggle it into Egypt, always risking arrest; but now with the backing of business leaders in Cairo, he publishes the *Al Masry Al Youm*, specializing in domestic issues of corruption, election fraud, and the need for an independent judiciary.

In Saudi Arabia, a dozen newspapers are on the newsstands every morning, but all are owned or controlled by the royal Saud family, and all journalists are constrained by the fact that they may not print anything critical of Islam or of the Saud family, which numbers in the tens of thousands. When accidents happen or some malfeasance by a public official is revealed, the story is blandly reported, with little effort to explore the causes of events or their effects or to hold responsible officials accountable.

The television broadcasting network Al Jazeera, introduced in the opening vignette, is credited with changing the climate for public discourse across the region and even with changing public opinion outside of the region. However, because it operates at the pleasure of the fairly liberal *emir* (Muslim ruler) of Qatar, it has apparently agreed not to cover sensitive issues in Qatar. A competing network owned by the Saud family, Al Arabiya, is far more constrained by censorship, though it was able to interview President Obama at some length early in his administration.

📺 **147. Al Jazeera Launches Global Broadcast Operation**

This region is one of the least democratized in the world. It is no coincidence that it also suffers more than most from violent conflict. Wars and terrorism here have reduced the potential for democratization by giving authoritarian regimes an excuse to hold on to power and forcibly repress legitimate political opposition groups. Meanwhile, the absence of democratic freedoms has prolonged conflicts that might have been resolved peacefully through free elections.

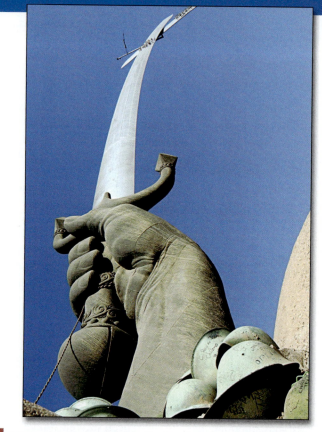

A A monument in Baghdad to the Iran–Iraq War (1980–1988), in which over one million died. When Iraq's then-dictator, Saddam Hussein, invaded Iran, one motivating factor was Iraq's majority Shia population who had been denied democratic freedoms that might have brought their leaders to power. Saddam feared that Iran's Shia-dominated government would back a Shia rebellion in Iraq that would remove him from power.

E An Iraqi medic and a U.S. soldier move a victim of a terrorist attack to an ambulance. Democratization was one stated goal of the U.S.-led war in Iraq. While some progress is being made, a complex and violent insurgency now limits the expansion of democracy.

LEBANON

SYRIA

PALESTINE
Golan Heights

PALESTINE
West Bank

PALESTINE
Gaza Strip

ISRAEL

JORDAN

EGYPT

SAUDI
ARABIA

B A painting in Cairo commemorating the 1973 war between Israel and Egypt. Ongoing tensions between Israel and its Arab neighbors has worked against democratization throughout the region. Many Arab governments use the conflict to justify repression of legitimate political parties. While historically far more democratic, Israel recently banned anyone who has visited "enemy countries" (Lebanon, Syria, Iraq, Iran, Saudi Arabia, Sudan, Libya, and Yemen) from running for national office for seven years.

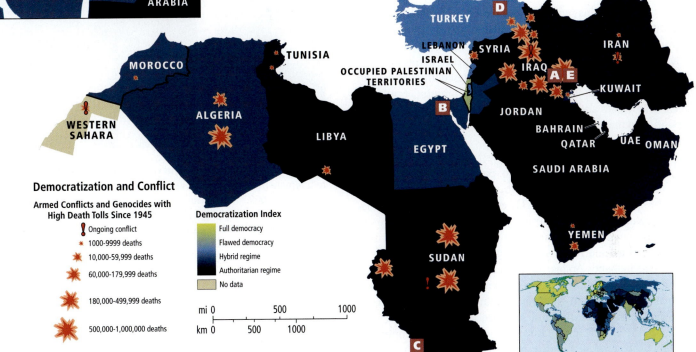

TURKEY **D**

LEBANON SYRIA

ISRAEL

OCCUPIED PALESTINIAN TERRITORIES

B

IRAN

IRAQ

A E

KUWAIT

JORDAN

BAHRAIN

QATAR UAE OMAN

SAUDI ARABIA

MOROCCO

TUNISIA

WESTERN SAHARA

ALGERIA

LIBYA

EGYPT

YEMEN

SUDAN

C

Democratization and Conflict

Armed Conflicts and Genocides with High Death Tolls Since 1945

! Ongoing conflict

✹ 1000-9999 deaths

✹ 10,000-59,999 deaths

✹ 60,000-179,999 deaths

✹ 180,000-499,999 deaths

✹ 500,000-1,000,000 deaths

Democratization Index

Full democracy

Flawed democracy

Hybrid regime

Authoritarian regime

No data

mi 0   500   1000

km 0   500   1000

D A Kurdish rebel in southeastern Turkey. Kurds form a majority of the population here and in neighboring parts of Iraq and Iran. Kurds have long been denied democratic freedoms and in some cases even the right to speak their language. Though Kurdish militias still exist, democratization has recently reduced public support for violence, especially in northern Iraq.

C A rebel army mobilizes in southern Sudan. Cultural tensions between the north and the south have led to two bloody civil wars and ongoing violence. The absence of basic democratic freedoms in Sudan has frustrated many efforts at reconciliation.

The role of the Internet is emerging, as illustrated by the following vignette about women bloggers in Yemen, and together with other electronic communication options, it is creating the climate for democratic reforms. These technologies also promise to open up the spaces for dissent, as was recently demonstrated in Iran after the contested election in June 2009. The fact that so many Iranians had video-capable cell phones meant that the world instantly saw via the Internet the brutality of the Iranian police and army. Young educated women have proven particularly adept at using the Internet to stage protests about discrimination. And in Egypt, the youthful dissident *April 6th Movement* was launched via a Facebook page. But the power of governments to censor is still strong. Syria blocked 225 Internet sites in 2008, shutting down Amazon, Facebook, and YouTube. In the UAE, new laws tighten Internet access and media freedom.

Democratization and Women

Most countries in this region now allow women to vote, and two countries—Israel and Turkey—have elected female heads of state (prime ministers) (Figure 6.24). Nevertheless, women who want to actively participate in politics still face many barriers. In the Iranian elections of 2009 and subsequent protest demonstrations in the streets of Tehran, women were especially prominent, wearing green headscarves and clothing as symbols of their support of the opposition to the government (see Thematic Concepts Part M on page 251).

Women's political status is the lowest in the Gulf States, but circumstances vary from country to country and change is in the wind. In Kuwait in 2009, four highly educated women were elected to parliament. They quickly energized the pace of legislation. They demanded seats on every major committee. They did their homework, refused to engage in the mutual flattery that men had engaged in, and reported the absences of male colleagues to the media. In Saudi Arabia, by contrast, women are denied the right to vote and only one woman serves as a public official. Oman, the first Gulf country to grant some women the vote in 1994, by 2003 had established universal suffrage. Several women were elected to the emir's Consultative Council. Qatar is attempting to broaden all types of democratic participation for women through suffrage rights and appointments to high office in government and educational institutions. Nonetheless, these are all only partial steps, and across the region there is still a tendency—even among women—to not support women candidates. Such was the case in Bahrain's elections of 2002, when women were a majority of the electorate but elected no female candidates. Across the region in 2009, women made up less than 10 percent of national legislatures, the lowest of any world region and half the world average.

In North Africa, women are entering politics in small but significant numbers. In June 2009 in Morocco, for example, women were elected to local councils for the first time, with 3406 elected, or about 12 percent of the total council membership.

An important impetus for change is the fact that more women are becoming educated and employed outside the home. In a number of countries, women outnumber men in universities; most notably, in Saudi Arabia women make up 70 percent of university students (but only 5 percent of the workforce). Such a large cadre of educated women is bound to have an effect on this country

FIGURE 6.24 Democracy and gender. Israel's Golda Meir. One of two female prime ministers ever elected in this region, Golda Meir led Israel from 1969 to 1974. During this time she led Israel through two wars with its Arab neighbors. Tansu Çiller served as prime minister of Turkey between 1993–1995, and was only the third female leader of a Muslim country, behind Benazir Bhutto of Pakistan (1988–1990 and 1993–1996) and Khaleda Zia in Bangladesh (1991–1996).

that is so conservative about gender. Saudi women activists have characterized their country as practicing a sort of apartheid with its own women, circumscribing or forbidding all manner of public activities. Yemen is another country that sees its female citizens as creatures to be set apart, yet some Yemeni women and men are leading change, as the following vignette illustrates.

Personal Vignette Raufa Hassan al-Sharki is in her 40s and holds a Ph.D. in social communications from the University of Paris. In 1996, she founded the Empirical Research and Women's Studies (ERWS) Center at San'a University in Yemen. As Yemen's most outspoken feminist, her overarching goal is to help women learn how to vote independently. Yemen's Islamist party, Islah, supports her efforts.

Despite having the right to vote, Yemeni women usually do not participate in the political process. Nearly 70 percent of the population live in rural areas, where only 4 in 10 women can read. Typically, few girls go to school. Most work at home, herding cattle, grinding wheat, and carrying water. The average adult woman bears six children.

Reception to change is mixed, and Dr. al-Sharki always gets permission from the local sheikh, a patriarchal tribal leader, before she talks to the women in a village. But lately the sheikhs, too, seem to be changing their views. Some recognize that if they support women's right to vote and encourage them to do so, the sheikhs' own sons may profit in future elections when female voters are more numerous. Other men support women's education but not necessarily their political empowerment.

For all her caution, Dr. al-Sharki eventually moved too fast for Yemen. In 1999, after she organized a successful international conference on women at ERWS, the government reorganized the center with an all-male staff and board of directors and removed al-Sharki as its director. Now she works as an independent consultant explaining gender issues on the Arabian Peninsula to audiences in American universities and is active in the international NGO, Women's Learning Partnership (WLP), which works to empower women in Muslim-majority societies to transform their families, communities, and societies.

Dr. al-Sharki's departure, however, was not the end of efforts to empower Yemeni women. In October 2009, a male trainer funded by the Hand in Hand NGO held a networking and social media workshop for 10 women gender activists in Yemen. Several attended wearing the full burka, yet they quickly learned how to use Facebook and how to blog, and a number began to do so immediately, protesting both the role and status of women in Yemen and the global stereotyping of their homeland as a failed state and supporter of Al Qaeda. *[Source: Daniel Pearl, "Yemen Steers a Path Toward Democracy with Some Surprises," Wall Street Journal, March 28, 1997, A1, A11; Beloit College Web site: "Noted Human Rights Activist Named 2006 Weissberg Distinguished Professor in International Studies," at http://www.beloit.edu/~pubaff/releases/05-06/0506weissberg_profile.htm; Women's Learning Partnership for Rights Development and Peace, 2009, at http://www.learningpartnership.org/en/about. Rising Voices: Helping the Global Population Join the Global Conversation, "Empowerment of Women Activists in Media Techniques—Yemen," at http://rising.globalvoicesonline.org/projects/empowerment-of-women-activists-in-media-techniques-yemen/.]* ■

Democratization and the Iraq War (2003–2010)

The U.S.-led invasion and occupation of Iraq has transformed the politics of this region. Anti-U.S. sentiment has exploded, together with criticism of leaders and governments, such as Israel's, who support the goals of the United States. Meanwhile, the U.S. attempt to create a democratic Iraq led to many unexpected consequences.

The origins of the Iraq war are complex. The United States had a long-standing amicable relationship with Iraq, driven in large part by Iraq's considerable oil reserves, the second-largest in the world after those of Saudi Arabia. The United States publicly supported Iraq in its 1980–1988 war with Iran (see Photo Essay 6.6A on pages 282), but also secretly supplied Iran with weapons when it appeared that Iraq might become more troublesome if it won the war. Ever since the overthrow of Shah Reza Pahlavi in 1980 (see discussion on page 281), Iran's revolutionary Islamist government has had a hostile relationship with the United States; but during the 1980s, the United States was more worried about Iraq than Iran. Iraq, with an advanced industrialized economy and well-educated populace but dictatorial government, had the

potential to develop nuclear weaponry and to use oil as part of a hostile geopolitical strategy. Relations with Iraq took a dramatic turn for the worse in 1990 when Saddam Hussein, Iraq's dictator, invaded Kuwait. The United States forced Iraq's military out of Kuwait in the Gulf War of 1990–1991, and afterward placed Iraq under crippling economic sanctions.

The U.S. administration of George W. Bush had launched the war on Afghanistan in the fall of 2001 (see Chapter 2 on page 84) as it looked for an explanation for the terrorist attacks of September 11, 2001. By 2003, the U.S. shifted its focus to Iraq and its president, Saddam Hussein. The Bush administration claimed that Iraq had or was creating an arsenal of weapons of mass destruction, and used this pretext to declare war on March 20, 2003, with the goals of confiscating Iraq's weapons of mass destruction, removing Saddam from power, and turning Iraq into a democracy. Such preemptive military action, which is counter to the UN charter, was out of character for the United States, as was the idea that democracy could be imposed. Great Britain and a few other countries, convinced that Iraq posed an immediate threat, joined the United States as allies, but most of the world objected to the war.

Contrary to official predictions, the Iraqis did not welcome the U.S. invasion. Terrorist attacks against the British and American (and other allies') troops erupted almost immediately. On May 1, 2003, President Bush declared the war won. However, terrorist bombs and insurgent attacks continued to take the lives of U.S. and allied soldiers at the rate of more than two per day. By February 2010, over 4380 U.S. troops had been killed and 31,500 wounded (see Photo Essay 6.6E on page 282). Furthermore, over 150,000 veterans of the conflict had been diagnosed with some form of serious mental disorder, including post-traumatic stress (PTSD). The death toll for Iraqis, including civilians, has been much higher, estimated as at least 103,535 by late December 2009. Some statistics place the death toll as high as 1,340,000, counting Iraqi deaths due to poor health and public safety conditions created by the war in addition to direct contact with foreign troops or insurgents.

📹 **136. Iraq War Enters Sixth Year**

📹 **152. Five Years After 'Mission Accomplished' in Iraq, War Continues**

The war has had many unintended consequences. While Saddam has been removed from power and executed, weapons of mass destruction were never found, nor were any links between the 9/11 attacks and Saddam's government uncovered. Democratization was the Bush administration's main stated justification for continuing the war, but the failure to understand and account for the long-simmering tensions between Iraq's major religious and ethnic groups crippled the democratizing process and brought added danger to the U.S.-led troops. Sunnis in the central northwest had dominated the country under Saddam, although they constituted only 32 percent of the population. Shi'ites in the south, with 60 percent of the population, have dominated politics since the fall of Saddam. This has given greater influence to Iran, whose population is mostly Shi'ite. Meanwhile, Kurds in the northeast, once brutally suppressed under Saddam Hussein (he was ultimately convicted of mass murder and subsequently executed in 2006 for his mistreatment of Kurds), have joined with Kurdish populations in Turkey and Syria and resist cooperating with the Iraqi national government in Baghdad. The end results of the many miscalculations by the Bush administration are an incompetently executed war, a huge increase in violence and loss of life within Iraq and among the allied troops, the creation of a major recruiting cause for Muslim extremists, and the most serious loss of international standing ever suffered by the United States.

134. Kurdish Nationalists in Iraq, Turkey Seek Land of Their Own

Regarding the still-incomplete move toward democracy in Iraq, 2010 elections were completed with much less violence than expected, but the results were disputed and six months later a government was still not formed. Recent studies of Iraqi public opinion indicate a number of important trends that will shape a more democratic Iraq. A majority of Iraqis want a strong central government that can protect them from violent insurgents and maintain control of the country's large fossil fuel reserves. Polls also show that the vast majority of Iraqis want all fighting to stop and all U.S. and allied military forces to leave. When elected in 2008, U.S. president Barack Obama promised to remove all troops by January 2012; many had left by summer, 2010. The timed withdrawal proceeded relatively calmly, despite continuing, though greatly diminished, suicide bombings aimed at Iraqi civilians and police.

THINGS TO REMEMBER

1. This region is one of the least democratized in the world. Power remains with the wealthy, the politically connected, clerics, and often with the military. Wars and terrorism have reduced the potential for democratization by giving authoritarian regimes an excuse to hold power and forcibly repress legitimate political opposition groups.

2. Generally, governments that have allowed free elections in which Islamists can run for office have enjoyed greater stability. This may partially explain the region-wide shift toward democracy in recent years.

3. Most countries now allow women to vote, and two countries have elected female heads of state. Even though women who seek active participation in politics still face barriers, patterns are changing as more women become educated and employed outside the home.

4. Demand for greater political freedoms is increasing, and public spaces for debate, other than mosques, are emerging. The press is reasonably free in some countries and severely curtailed in others. Al Jazeera is credited with changing the climate for public discourse across the region and with changing public opinion outside the region.

5. The situation in Iraq remains fluid as the United States begins withdrawing its troops and turns over power and policing operations to U.S.-trained Iraqi forces.

The State of Israel and the "Question of Palestine"

"The Palestinian people still yearn for the freedom and dignity denied them for decades. The Israeli people yearn for long-term security. Neither can achieve their legitimate demands without a settlement of the conflict."

These are the words of United Nations Secretary-General Ban Ki-moon, describing in February 2007 the Israeli–Palestinian conflict. This conflict, which has lasted longer than 60 years and has included several major wars and innumerable skirmishes, is a persistent obstacle to political and economic cooperation in the region.

Of the two entities (Palestine has not yet been recognized as a state), Israel has by far the most modern and diversified economy; indeed, Israel leads the region in this regard. Since the 1950s, Israel's development has been facilitated by the immigration of relatively well-educated middle-class Jews from the United States, Europe, Russia, and South America. Israel's excellent technical and educational infrastructure, its diverse and prospering economy, and large aid contributions from the United States and others, public and private, have made the country one of the region's wealthiest, most technologically advanced, and militarily powerful.

154. 60 Years After Israel's Founding, Palestinians Are Still Refugees

The Palestinian people, by contrast, are severely impoverished and undereducated after years of conflict, inadequate government, and meager lives (Table 6.1), often in refugee camps. Through a series of events over the past 60 years, Palestinians have lost most of the lands on which they used to live. They now live in two main areas—the West Bank and the Gaza Strip, both of which are highly dependent on Israel's economy. Israel often takes military action in these two zones in retaliation for Palestinian suicide bombings and rocket fire launched primarily from Gaza. The West Bank Palestinian territories continue to be encircled by security walls built by Israel to defend against violence and curtail Palestinian access (see Figure 6.27A on page 289).

131. Issues from 1967 Arab–Israeli War Remain Unresolved

The Creation of the State of Israel The idea of Israel actually began in Europe in the late nineteenth century. In response to centuries of discrimination and persecution in Europe and Russia, a small group of European Jews, known as **Zionists**, began to purchase land in an area of the Damascus province of the Ottoman empire that was known informally at the time as Palestine. Most sellers were wealthy non-Palestinian Arabs and Ottoman Turks living outside of Palestine. These absentee landowners had long leased their lands to Palestinian tenant farmers and herders or granted them the right to use the land. It should be noted that Bedouin nomads also had traditional rights to pass across these lands seasonally. Such rights were negated by the sales to Zionists, and historians still debate whether or not the Zionists or the former landlords adequately compensated the displaced indigenous inhabitants, including the Bedouin. In the aftermath of World War I (see page 265), the United Kingdom in 1922 received a Mandate from the League of Nations to administer the territory of Palestine (Figure 6.25A).

Earlier, in 1917, the British government had adopted the Balfour Declaration, which favored the establishment of a national home for Jews in the Palestine territory, and which explicitly stated that "nothing shall be done which may prejudice the civil and religious rights of existing non-Jewish communities in Palestine, or the rights and political status enjoyed by Jews in any other

TABLE 6.1	Circumstances and state of human well-being among Palestinians and Israelis			
	Population (in millions), 2009	Infant mortality (per 1000 live births) 2009	Percentage of unemployed, 2005	Percentage of population in poverty, 2007–2009
Palestinians	3.9	25	26.7	39
Israelis	7.6	3.6	9	23.6*

*Israel's poverty line is $7.30 per person per day (2005).

Source: United Nations Human Development Report 2005 (New York: United Nations Development Programme.)

country." It is at this time that the word *Palestine*, with roots far back in history, began to be used officially.

On this newly purchased land, the Zionists established communal settlements called *kibbutzim*, into which a small flow of Jews came from Europe and Russia. While Jewish and Palestinian populations intermingled in the early years, tensions emerged as Zionist land purchases displaced more and more Palestinians, in a seeming breach of the Balfour Declaration.

Efforts to secure a Jewish homeland in Palestine continued. By 1946, following the death of 6 million Jews in Nazi death camps during World War II, strong sentiment had grown throughout the world in favor of a Jewish homeland. Although a few other homeland sites were suggested, the only place seriously considered was the space in the eastern Mediterranean that Jews had shared in ancient time with Palestinians and other Arab groups and that was mentioned in a Biblical promise to Moses. Hundreds of thousands of Jews migrated to Palestine, and against British policy, many took up arms to support their goal of a Jewish state.

In November, 1947, after intense debate at the second session of the United Nations General Assembly, the UN adopted the "Plan of Partition with an Economic Union" that ended the earlier Mandate and called for the creation of both Arab and Jewish states and for an international regime for the city of Jerusalem (**Figure 6.25B**).

> **Zionists** those who have worked, and continue to work, to create a Jewish homeland (Zion) in Palestine

Sixty Years of Conflict The Palestinians and neighboring Arab countries fiercely objected to the establishment of a Jewish state and feared that they would continue to lose land and other resources upon which they were dependent. Then as now, the conflict between Jews and Palestinian Arabs was less about religion than control of land, settlements, and access to water.

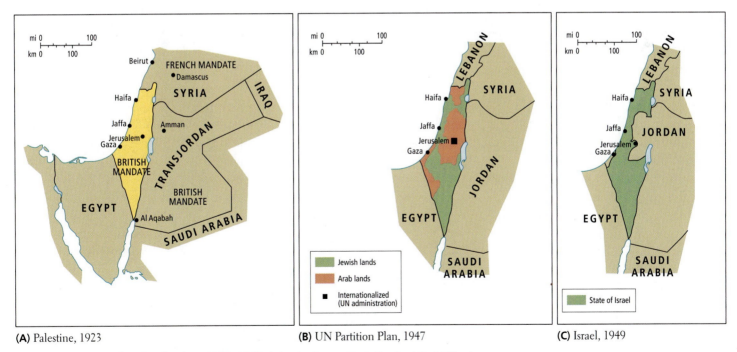

(A) Palestine, 1923   **(B)** UN Partition Plan, 1947   **(C)** Israel, 1949

FIGURE 6.25 Israel and Palestine, 1923–1949. (A) Palestine, 1923. Following World War I, Britain controlled what was called Palestine and is now Israel and Jordan (Transjordan was the precursor to Jordan). **(B)** UN Partition Plan, 1947. After World War II, the United Nations developed a plan for separate Jewish and Palestinian (Arab) states. **(C)** Israel, 1949. The Jewish settlers did not agree to that plan; instead, they fought and won a war, creating the state of Israel. [Adapted from Colbert C. Held, *Middle East Patterns—Places, Peoples, and Politics* (Boulder, CO: Westview Press, 1994), p. 184.]

On May 14, 1948, the British ended the Mandate and withdrew their forces. On that same day, the Jewish Agency for Palestine declared the state of Israel on land designated to them by the Plan of Partition. Warfare between the Jews and Palestinians began immediately. Neighboring Arab countries—Lebanon, Syria, Iraq, Egypt, and Jordan—supporting the Palestinian Arabs, invaded Israel the next day. Fighting continued for several months, during which Israel prevailed militarily. An armistice was reached in 1949, and as a result the Palestinians' land shrank still further, with the remnants incorporated into Jordan and Egypt (see Figure 6.25C on page 287).

intifada a prolonged Palestinian uprising against Israel

In the repeated conflicts over the next decades—such as the Six-Day War in 1967 and the Yom Kippur War in 1973—Israel again defeated its larger Arab neighbors, expanding into territories formerly controlled by Egypt (Sinai), Syria (Golan Heights), and Jordan (West Bank of the Jordan River, Figure 6.26A). Since 1948, hundreds of thousands of Palestinians have fled the war zones, with many removed to refugee camps in nearby countries. Some Palestinians stayed inside Israel and became Israeli citizens, but they have not been treated by the state as equal to Jewish Israelis.

📹 **128. Hezbollah—Serving Muslims with God and Guns**
📹 **133. Lebanese Oil Spill—Collateral Damage of the Bombings**

In 1987, the Palestinians mounted the first of two prolonged protest uprisings, known as the **intifada**, characterized by escalating violence. The first ran until 1993 when the Oslo Peace Accords were concluded (Figure 6.26B). The Oslo Accords included: Israeli withdrawal from parts of Gaza Strip and West Bank, and creation of Palestinian Authority for self-government. The second intifada began in 2000 and continues into the present, primarily fueled by the issue of Israeli settlements expanding, in breach of the Oslo Accords, into Palestinian territories in Gaza, the West Bank, and the Golan Heights (Figure 6.26C). Between 2000 and 2008, Palestinian suicide bombers targeted Israeli civilians, killing 1000, maiming thousands, and psychologically impacting all Israelis. In response, the Israeli military has used deadly force to quell demonstrations and to punish the families and communities of the suicide bombers. According to B'Tselem, an Israeli group that educates the Israeli public about human rights violations in the occupied Palestinian Territories, more than 5000 Palestinians and 64 foreigners died in the same period.

Territorial Disputes When Israel occupied Palestinian lands in 1967, the United Nations Security Council passed a resolution requiring Israel to return those lands, known as the *Occupied Palestinian Territories (OPT)*, in exchange for peaceful relations between it and neighboring Arab states. This *land-for-peace*

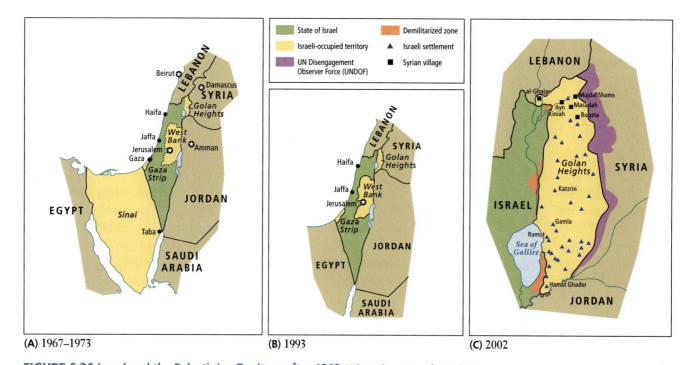

(A) 1967–1973   **(B)** 1993   **(C)** 2002

FIGURE 6.26 Israel and the Palestinian Territory after 1949. When the state of Israel was created in 1949, its Arab neighbors were opposed to a Jewish state. **(A)** In 1967, Israel soundly defeated combined Arab forces and took control over Sinai, the Gaza Strip, the Golan Heights, and the West Bank. **(B)** In subsequent peace accords, Sinai was returned to Egypt, but Israel maintained control over the Golan Heights and the West Bank, claiming that they were essential to Israeli security. **(C)** Although the Palestinians were granted some autonomy in the Gaza Strip and the West Bank, during the 1990s the Israelis, contrary to verbal agreements in the Oslo Accords, began building Jewish settlements in the West Bank and Golan Heights. [Adapted from Colbert C. Held, *Middle East Patterns—Places, Peoples, and Politics* (Boulder, CO: Westview Press, 1994), p. 184; and the Israeli Settlements in the Occupied Territories, 2002, Foundation for Middle East Peace, March, 2002, http://www.firstpr.com.au/nations.]

FIGURE 6.27 Conflict and cooperation between Israelis and Palestinians. Both conflict and cooperation between Israelis and Palestinians has increased in recent years. **(A)** The West Bank barrier and checkpoint (in the left) near Ramallah, West Bank. The barrier has hindered the flow of people and goods between Israel and the West Bank, severely damaging the latter's economy. **(B)** Palestinian residents of Gaza struggle to cross a border checkpoint with Egypt. Gaza is periodically plagued by shortages of food, water, and other necessities as Israel seals its borders in retaliation for rocket attacks into Israel from Gaza. **(C)** Palestinians harvest tomatoes from a farm in the West Bank. Food-processing industries have a large potential for growth throughout the West Bank, where much food is imported. **(D)** An Israeli food-processing company advertises its product at a holiday festival in northern Israel. Israeli food-processing companies have access to the expertise, equipment, and investment capital that could help food-processing industries in the West Bank and Gaza grow. While many partnerships are developing between Israeli and Palestinian companies, these suffer when violence forces border closures.

formula, which sets the stage for an independent Palestinian state, has been only partially fulfilled.

Despite the land-for-peace agreement, between 1967 and 2005 Israel secured ever more control over the land and water resources of the occupied territories, placing hundreds of Jewish settlements in the West Bank and the Gaza Strip (Figures 6.27B, C). Israel took a major step toward peace in 2005 when it removed all Jewish settlements from the Gaza Strip. However, in the eyes of Palestinians, this progress was negated by the blockade of Gaza's economy, the construction of the **West Bank barrier**, which began in 2003 (Figure 6.27A), and by continuing policies that constrain Palestinians' access to jobs and freedom of movement (see Figure 6.27B).

The West Bank barrier (see Figure 6.27A) encircles many of the remaining Jewish settlements on the West Bank and separates approximately 30,000 Palestinian farmers from their fields. It also blocks roads that once were busy with small businesses, and effectively annexes 6 to 8 percent of the West Bank to Israel (see Figure 6.26C). It also severely limits Palestinian access to much of the city of Jerusalem, most of which is now on the Israeli side of the barrier. The barrier, declared illegal

West Bank barrier a 25-foot-high concrete wall in some places and a fence in others that now surrounds much of the West Bank and encompasses many of the remaining Jewish settlements there

by the World Court and the United Nations, and opposed by the United States, is very popular among Israelis, however, because it has reduced the number of Palestinian suicide bombings.

132. West Bank Barrier, New Divide in Palestinian-Israeli Conflict

In 2006, Hezbollah, an anti-Israeli militia based in southern Lebanon and financed partly by Iran, kidnapped two Israeli soldiers. Israel responded with a 34-day counterattack that destroyed most of Lebanon's infrastructure. Israel failed to defeat Hezbollah, which launched thousands of rockets into northern Israel. Hezbollah, strengthened by this encounter, remains a powerful military and political force.

Once Israel had vacated its settlements in Gaza in partial fulfillment of the land-for-peace agreements, Israel and Egypt instituted a blockade of trade and movement that crippled the Palestinian economy (see Figure 6.27B on page 289). In retaliation, the Palestinians launched a series of rocket attacks from Gaza on southern Israel, to which Israel retaliated with a ground invasion in January 2009.

In March 2010, John Holmes, the top UN humanitarian official, "called for 'radical' changes in Israeli policies" that would allow Palestinians to have "normal and dignified lives" in the occupied territories. Evictions of Palestinians from their houses and subsequent demolitions in the West Bank should cease and border crossings to and from Gaza should be reopened. Palestinians, too, were to stop violence against Israelis. In seeming defiance a few days later, Israel announced plans to build another 1600 houses in the West Bank occupied Palestinian Territory in East Jerusalem.

Local Peace Efforts Although the peace efforts of Amos Oz and Elias Khoury (see page 248) get occasional international recognition, ordinary citizens, both Israeli and Palestinian, have separately and collaboratively been quietly designing ways to end the conflict.

Examples of these bottom-up peace initiatives include B'Tselem (mentioned on page 288); joint Israeli–Palestinian peace demonstrations and women's groups that have tried to end the Israeli occupation of the West Bank and Gaza Strip. Palestinian and Israeli members of Physicians for Human Rights have joined to address the medical problems of the overwhelmingly poor Palestinians. Groups from both sides hold youth camps so that Israeli and Palestinian children can create personal friendships that break the cycle of hatred. Israeli human rights groups regularly attempt to assist Palestinian communities that are coping with refugee status and forced culture change.

Can Economic Interdependence Be a Force for Peace? Economic relations with Israel are essential to Palestinian survival because Israel has long been the largest trading partner of the West Bank and Gaza Strip and provides their currency (the Israeli shekel), electricity, and most imports. Israelis employ tens of thousands of Palestinian workers in fields, factories, and homes within Israel (see Figure 6.27C, D on page 289). Increased economic cooperation could lead to an independent Palestinian state and then wider cooperation across the region.

Also rarely covered in the headlines are several cooperative industrial parks established jointly in recent years by the Palestinian Authority and the Israeli government in the West Bank and Gaza Strip. These developments have been dubbed "peace parks" because of their goal of using economic development to overcome conflict. Some critics regard this relationship as exploitative because Israel tightly controls Palestinian access to the world market and historically has discouraged industrial development in the occupied territories.

In an ideal world, economic cooperation between Israel and Palestine would be only the beginning of what could become a transformation across the region—a so-called *peace dividend*. The Gulf States, Turkey, Egypt, Iran, and all the other countries in the region are developing their human capital through social services and education. Business-level cooperation with Israel, which has long had the most developed economy in the region, could facilitate rapid improvements everywhere.

THINGS TO REMEMBER

1. While the modern conflict between Israel and the Palestinians is decades old, the situation remains dynamic and complex and is a persistent obstacle to political and economic cooperation in the region.

2. Citizens of both Israel and the occupied Palestinian Territories have organized grassroots peace initiatives.

3. The entire region would benefit from the *peace dividend* that would result from economic cooperation.

Reflections on North Africa and Southwest Asia

North Africa and Southwest Asia are frequently referred to as hopelessly mired in a morass of unsolvable problems. Many point to a gloomy future: violence, religious fundamentalism, and internal and external terrorism leading eventually to global disaster. If not that, then autocratic governments; economic inequities; water shortages; extreme gender discrimination; expanding populations; and for many countries, overdependence on energy resources that the world may soon declare as obsolete. Yet many of the changes currently underway could provide a route through this daunting maze of issues.

Some evidence suggests that shifting toward democracy and away from repressive authoritarian rule may diffuse the threat posed by fundamentalist Islam to the region's security and to relations with the rest of the world. Pressure for democracy is growing in the fossil fuel–rich countries, where an increasingly educated, informed, and globally aware citizenry is demanding greater freedom of speech, a larger voice in politics, and more equality for women. Al Jazeera television has introduced a new era of open debate and hard-nosed analysis. Across the region, urbanization and education for women promises to curb population growth as it has in much of the rest of the world. But if urbanization is to be a force for rising standards of living, then economic growth, increased employment, and upgrades to educational systems are essential.

Democracy and greater economic opportunities could calm sectarian and ethnic conflicts in places like Iraq, Israel, and Palestine. The most hopeful signs are the bottom-up peace initiatives and grassroots entrepreneurialism and constructive activism to be found across the region.

Improvements in education and economic growth could also aid the region's ability to provide its growing population with enough food and water as the climate changes. Ancient and modern water-saving agricultural technologies could stretch water supplies much further. But any significant shift in this direction would require farmers to become more educated and better able to invest in sophisticated and expensive technologies. If educational and employment opportunities were extended to women, then women's latent potential as innovators could be tapped and the region's high population growth rate would fall quickly, relieving some of the pressure on food and water resources.

Global warming and depleting fossil fuel reserves may provide additional motivation to make these changes. Water-saving technologies may become more necessary if rainfall patterns shift even slightly. Better-educated populations may find it easier to adapt to the complex economic disruptions created both by changing regional climates and a global shift away from fossil fuels. Similarly, citizens faced with the wide range of uncertainties that global warming may bring could demand a level of responsiveness and accountability on the part of governments that only democracies can deliver.

Critical Thinking Questions

1. Which social forces in North Africa and Southwest Asia modify the power of religion?

2. Discuss how people in this region have affected world diets (including especially the Americas), first through domestication of plants and animals and then through trade.

3. To what extent is the present-day map of North Africa and Southwest Asia related to the dismantling of the Ottoman Empire after World War I?

4. Consider the various factors that encourage relatively high fertility in this region and design themes for a public education program that would effectively encourage lower birth rates. Which population groups would you target? How would you incorporate cultural sensitivity into your project?

5. Consider the new forces that are affecting urban landscapes: immigration and globalization. Identify some of the expected effects on ordinary people of these abrupt changes in traditional living spaces.

6. Compare and contrast the public debate over the proper role of religion in public life in your country and in one country in this region (for example, Turkey, Morocco, Egypt, or Saudi Arabia). Contrast the roles of religious fundamentalists in the debates in your country and in the country chosen.

7. Gender is a complex subject in this region. Choose a rural location and an urban location and make a list of the forces in each that would affect the future of a 20-year-old woman. Describe those hypothetical futures objectively; that is, without using any judgmental terminology.

8. Why is it important to know that in some countries of this region agriculture may produce only a small amount of the GDP yet employ 40 percent or more of the people? What are some of the things such a relationship would indicate about the state of development in that country? What public policies would be appropriate in these circumstances—for example, should agriculture be deemphasized?

9. Describe the circumstances that led to support in Europe and the United States for the formation of the state of Israel. Why did the West overlook the Palestinian people in this political undertaking?

10. Discuss the possibilities that scarcity of water is or will become a cause of violence in the region. What is the evidence against this happening?

Chapter Key Terms

cartel, 278
Christianity, 262
desertification, 258
diaspora, 262
economic diversification, 279
female seclusion, 267
Fertile Crescent, 260
fossil fuel, 248
Gulf States, 268
hajj, 266
intifada, 288
Islam, 248

Islamism, 248
Judaism, 262
monotheistic, 261
Muslims, 262
Occupied Palestinian Territories (OPT), 252
OPEC (Organization of Petroleum Exporting Countries), 278
patriarchal, 266
polygyny, 268
Qur'an (or Koran), 254
salinization, 256
seawater desalination, 256

secular states, 281
shari'a, 266
Shi'ite (or Shi'a), 266
sovereign-wealth funds, 280
Sunni, 266
theocratic states, 281
veil, 268
West Bank barrier, 289
Zionists, 287

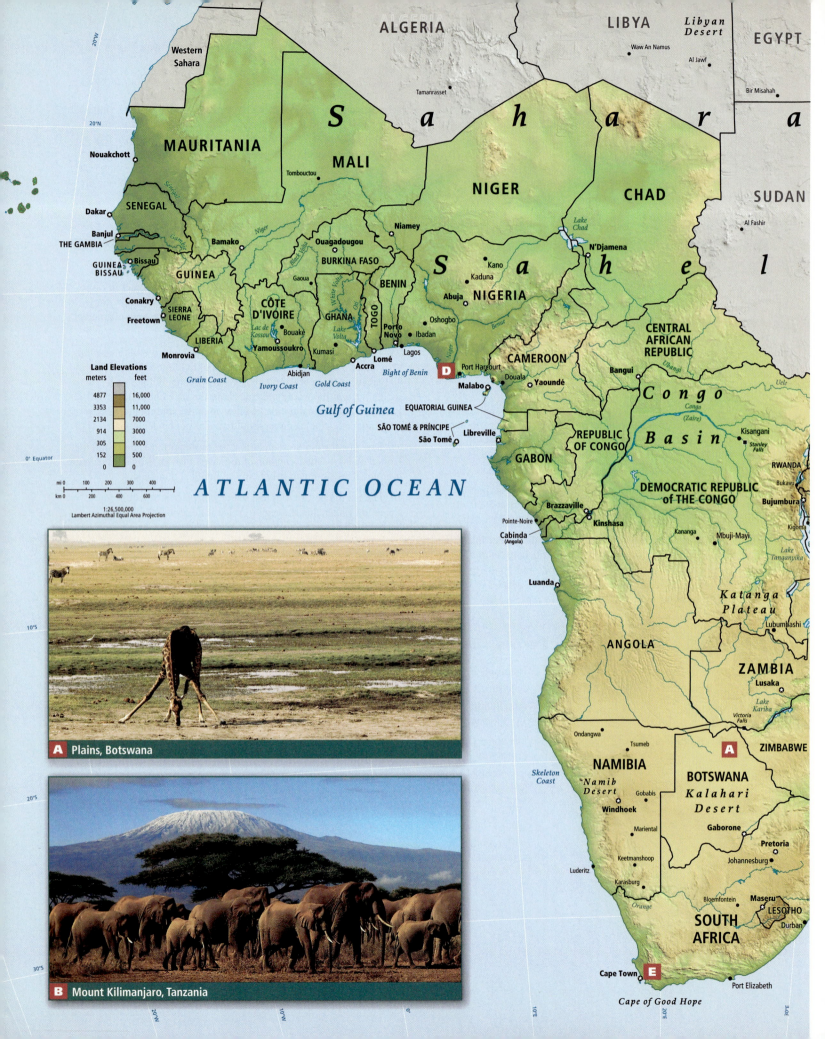

ALGERIA
LIBYA
Libyan Desert
EGYPT

Waw An Namus
Al Jawf
Bir Misahah

Western Sahara

S a h a r a

MAURITANIA

Nouakchott

Tombouctou

MALI

NIGER

CHAD

SUDAN

Tamanrasset

Al Fashir

MALI

Dakar
SENEGAL

Banjul
THE GAMBIA

GUINEA BISSAU

Bissau

GUINEA

Conakry

SIERRA LEONE

Freetown

LIBERIA

Monrovia

Bamako

Gaoua

BURKINA FASO

Ouagadougou

CÔTE D'IVOIRE

Lac de Kossou

Bouaké

Yamoussoukro

Kumasi

GHANA

Lake Volta

Abidjan

TOGO

BENIN

Niamey

S a h e l

NIGERIA

Kano

Kaduna

Abuja

Oshogbo

Ibadan

Porto Novo

Lomé

Lagos

Accra

Grain Coast

Ivory Coast

Gold Coast

Bight of Benin

Gulf of Guinea

Lake Chad

N'Djamena

CENTRAL AFRICAN REPUBLIC

Bangui

CAMEROON

Port Harcourt

Douala

Yaoundé

Malabo

EQUATORIAL GUINEA

SÃO TOMÉ & PRÍNCIPE

São Tomé

Libreville

GABON

C o n g o

B a s i n

Uele

Kisangani

Stanley Falls

RWANDA

Bukavu

Bujumbura

REPUBLIC OF CONGO

Brazzaville

Pointe-Noire

Kinshasa

Cabinda (Angola)

DEMOCRATIC REPUBLIC of THE CONGO

Kananga

Mbuji-Mayi

Kigoma

Lake Tanganyika

Luanda

ATLANTIC OCEAN

Land Elevations

meters	feet
4877	16,000
3353	11,000
2134	7000
914	3000
305	1000
152	500
0	0

mi 0 100 200 300 400
km 0 200 400 600

1:26,500,000
Lambert Azimuthal Equal Area Projection

0° Equator

10°S

20°S

30°S

Katanga Plateau

Lubumbashi

ANGOLA

ZAMBIA

Lusaka

Lake Kariba

Victoria Falls

Skeleton Coast

Ondangwa

Tsumeb

NAMIBIA

Namib Desert

Windhoek

Gobabis

Mariental

Keetmanshoop

Lüderitz

Karasburg

ZIMBABWE

BOTSWANA

Kalahari Desert

Gaborone

Pretoria

Johannesburg

Maseru

LESOTHO

SOUTH AFRICA

Bloemfontein

Durban

Orange

Cape Town

Cape of Good Hope

Port Elizabeth

A Plains, Botswana

B Mount Kilimanjaro, Tanzania

THEMATIC CONCEPTS

Population • Gender • Development
Food • Urbanization • Globalization
Democratization • Climate Change • Water

Sub-Saharan Africa

C Great Rift Valley, Tanzania

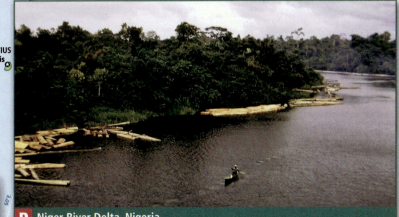

D Niger River Delta, Nigeria

FIGURE 7.1 Regional map of sub-Saharan Africa.

E Escarpment at Cape Town, South Africa

Global Patterns, Local Lives

Liberian environmental activist Silas Siakor is an affable and unassuming fellow. But his casual style conceals a fierce dedication to his homeland and a remarkable sleuthing ability. At great personal risk, Siakor uncovered evidence that 17 international logging companies were bribing Liberia's then-president, Charles Taylor, with cash and guns. Taylor looked the other way while the companies illegally logged Liberia's forests, which are home to many endangered species, such as forest elephants and chimpanzees. In return, the companies paid cash and weapons that Taylor used to equip his own personal armies. Made up largely of kidnapped and tragically abused children, Taylor's armies fought those of other Liberian warlords in a 14-year civil war that took the lives of 150,000 civilians. The war also spilled over into neighboring Sierra Leone, where another 75,000 people died. Meanwhile, the logging companies—based in Europe, China, and the Middle East—reaped huge fortunes from the tropical forests.

📹 175. 'Ezra,' Tragic Tale of Child Soldiers in Africa

Silas Siakor pulled together publicly available information that had been previously ignored by the international community and information provided by strategically placed informants in ports, villages, and lumber companies to prepare a clear, well-documented report substantiating the massive logging fraud (**Figure 7.2**). In response to Siakor's report, the United Nations Security Council voted to impose sanctions to stop the timber trade (**Figure 7.3**) and prosecute some of the people involved. Charles Taylor fled to

FIGURE 7.2 Silas Siakor at work. Siakor is documenting illegally taken logs that are being salvaged in Liberia's River Cess County.

Nigeria, but in 2006 he was turned over to face a war crimes tribunal in The Hague, Netherlands. His trial began in 2007 and is expected to end in 2011.

Democratic elections followed in Liberia, and in 2006 Ellen Johnson-Sirleaf took office, Africa's first elected woman president. In a move that was bold—given the poverty and political instability

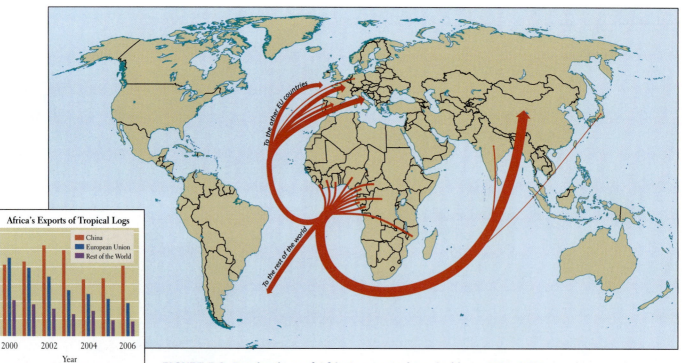

FIGURE 7.3 Destinations of Africa's exported tropical logs, 2000–2006. Most endpoint consumers of Africa's tropical woods are in North America and Europe. Though China is the largest single recipient of exported logs, much of that wood ends up in furniture sold by firms like Ikea in the United States and Europe. As of 2008, Ikea published standards that their wood suppliers must meet. [Export data is from GlobalTimber.org.uk, "Exports by, and Imports from, Africa," at http://www.globaltimber.org.uk/africa.htm; Ikea standards are available at http://www.ikea.com/ms/en_CA/about_ikea/pdf/IWAY_marketing_standard.pdf.]

of her country—she cancelled all contracts with timber companies pending a revision of Liberian forestry law. *[Sources: Adapted from "Silas Kpanan'Ayoung Siakor: A Voice for the Forest and Its People," The Goldman Environmental Prize, at http://www.goldmanprize.org/node/442; Scott Simon, "Reflections of a Liberian Environmental Activist," on Weekend Edition Saturday, National Public Radio, April 29, 2006; http://www.npr.org/templates/story/story.php?storyId=5370987; and authors' personal communication with Siakor, September 2006.]* ∎

The story of Silas Siakor and Liberia illustrates some of sub-Saharan Africa's challenges and relates to at least three of the nine **Thematic Concepts** we cover in this book: development, globalization, and democratization (see pages 296–297). Liberia, like much of this region, is still in the process of developing strong democratic institutions that can direct the sustainable use of the country's resources to the benefit of its citizens. The story of Liberia also exemplifies the ways in which sub-Saharan Africa has failed to benefit substantially from economic globalization. Liberia has rich timber and mineral (diamond) resources, but most of the profits from these industries have gone to Liberian elites or to foreign companies. And yet, Silas Siakor's success illustrates how Africans can find solutions to these challenges. It shows as well that African efforts toward achieving peace and prosperity can be both aided and thwarted by the rest of the world.

Sub-Saharan Africa (**Figure 7.1** on pages 292–293) is home to about 809 million people, a population that is growing at the rate of 2.5 percent per year. At this rate the population will double in 28 years. The region contains several of the fastest-growing economies in the world, as well as some of the world's richest deposits of oil, gold, platinum, copper, and other strategic minerals. During the era of European colonialism (1850s–1950s), this massive wealth flowed out of Africa, providing little benefit to its people. Even after African countries achieved political independence in the 1950s, 1960s, and 1970s, wealth continued to flow out of Africa. It is not surprising, then, that much of Africa is impoverished and often at war with itself. But amid the turmoil, many hopeful signs suggest that the conflicts can be resolved and the well-being of the majority enhanced.

THINGS TO REMEMBER

1. Europe's colonization of sub-Saharan Africa set up patterns of economic development that still leave many countries in a weak position in the global economy.

2. A shift toward democracy is occurring throughout Africa, but powerful forces continue to work against this trend in many countries.

❙ THE GEOGRAPHIC SETTING

Terms in This Chapter

We refer to this region as sub-Saharan Africa to recognize that North Africa is not included. Only occasionally do we refer to the whole continent and then we refer to it simply as Africa. Sub-Saharan Africa is defined by the countries shown in **Figure 7.4**. Notice that Sudan is not included. Sudan's location on the Nile River and the strong Arab influence in the capital of Khartoum historically brought that country into closer association with North Africa and Southwest Asia, and it is discussed in Chapter 6.

The naming of African countries can often be confusing. For example, there are two neighboring countries called Congo—the Democratic Republic of the Congo and the Republic of Congo. Because these designations are both lengthy and easily confused, we abbreviate them in this text. The Democratic Republic of the Congo (formerly Zaire) carries the name of its capital in parentheses: Congo (Kinshasa). The Republic of Congo is called Congo (Brazzaville). Check the regional map (see Figure 7.1) to see the locations of these countries and capitals.

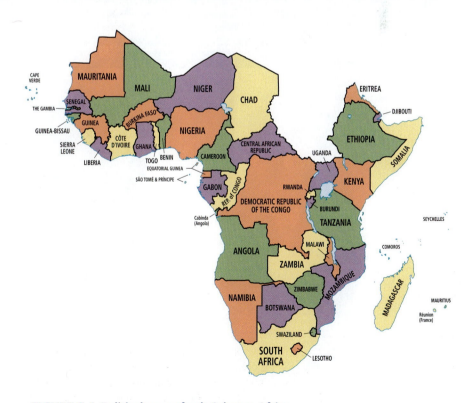

FIGURE 7.4 Political map of sub-Saharan Africa.

Climate Change, Food, and Water: This region is particularly vulnerable to climate change given the large number of poor people who depend on agriculture and herding for their livelihood. These occupations are sensitive to even slight changes in temperature, rainfall, and water availability that climate change is likely to bring. Moreover, poverty, political instability, and little access to cash reduce the resilience of rural populations, making them more likely to become uprooted by climate change.

(A) A child cultivates a field in Uganda, where 80 percent of the population is employed in agriculture.

(B) Floodwaters of the Tana River rip through a village in rural Kenya, uprooting people and crops.

(C) A woman and her grandchild in southern Ethiopia, where drought, conflict, and poverty produce chronic food insecurity.

Globalization and Development: Europe's colonization of this region set up highly globalized patterns of economic development that still leave many countries in a weak position in the global economy. Many national budgets rely heavily on low-value raw materials whose price can rise or fall dramatically from year to year. Meanwhile, production systems depend on expensive imported equipment whose prices are more steady. A few countries are beginning to prosper by exporting higher-value manufactued goods.

(D) Charles de Gaulle, a French general and politician, gives orders to Governor General Félix Éboué of Chad during World War II.

(E) A diamond miner in Sierra Leone. Global prices for the rough diamonds that Sierra Leone exports are extremely volatile.

(F) Textiles are made into clothing in Mauritius, one of the most prosperous countries in the region.

Democratization: A shift toward democracy is occurring throughout Africa, but powerful forces work against this trend in many countries. Free and fair elections have brought dramatic change to some countries, while in others elections have resulted in violence and sometimes civil war.

(G) Democratization has been transforming South Africa since 1994, when nonwhite South Africans were first allowed to vote.

(H) Demonstrators in Congo (Kinshasa) during post-election riots that left a million people homeless.

(I) Troops guard ballot boxes in Kenya, where as many as 1000 people died in rioting during a disputed election in 2008.

Urbanization and Population: Urbanization is slowing down population growth in sub-Saharan Africa, but not as quickly as in other regions. Rapid and unplanned urbanization has resulted in many slums where living conditions are poor and access to water and health care is uncertain. As in rural areas, these factors encourage couples to have many children to ensure that some survive to adulthood.

(J) Kibera slum in Nairobi, Kenya, home to over a million people. Access to water and health care is very poor.

(K) A police officer confronts a resident of Kibera slum during a protest for better access to water.

(L) A family looks out of their makeshift dwelling in Mathare slum, Nairobi, Kenya.

Gender: In some countries, recent efforts at political reform and controlling population growth have involved greater recognition of the important role of women in society. Education and the economic and political empowerment of women have been linked to lower population growth rates. In several countries in the region, women earn almost as much as men, and female political leaders are common. However, in many other countries, women remain undereducated, poorly paid, and politically disempowered.

(M) A village leader in Senegal celebrates the passage of a local resolution to end discrimination against women.

(N) Schoolgirls in Ghana, where female earned income is 71 percent of male earned income, a high number for any country.

(O) A family in Somalia, where female literacy and empowerment are low and population growth is high.

Physical Patterns

The African continent is big—the second largest after Asia and about 2 million square miles bigger than North America. But Africa's great size is not matched by its surface complexity. More than one-fourth of the continent is covered by the Sahara, many thousands of square miles of what, to the unpracticed eye, appears to be a homogeneous desert landscape (see Chapter 6). Africa has no major mountain ranges, but it does have several high peaks, including Mount Kilimanjaro (19,324 feet [5890 meters] high) and Mount Kenya (17,057 feet [5199 meters] high). At their peaks, both have permanent snow and ice, though these features have shrunk dramatically due to global climate change and associated local environmental changes.

Landforms

The surface of the continent of Africa can be envisioned as a raised platform, or *plateau*, bordered by fairly narrow and uniform coastal lowlands. The platform slopes downward to the north; it has an upland region with several high peaks in the southeast and lower areas in northwest. The steep *escarpments* (long cliffs) between plateau and coast have obstructed transportation and hindered connections to the outside world. Africa's long, uniform coastlines also provide few natural harbors.

Geologists usually place Africa at the center of the ancient supercontinent Pangaea (see Figure 1.26 on page 51). As landmasses broke off from Africa and moved away—North America to the northwest, South America to the west, and India to the northeast—Africa readjusted its position only slightly. Hence, it did not pile up long linear mountain ranges as did the other continents when their plates collided with one another (see page 50).

Today, Africa continues to break apart along its eastern flank. There the Arabian Plate has already split away and drifted to the northeast, leaving the Red Sea, which separates Africa and Asia. Another split, known as the Great Rift Valley, extends south from the Red Sea more than 2000 miles (3200 kilometers) (see Figure 7.1C on page 293). In the future, Africa is expected to split again along these rifts.

Climate and Vegetation

Most of sub-Saharan Africa has a tropical climate (Photo Essay 7.1). Average temperatures generally stay above 64°F (18°C) year-round everywhere except at the more temperate southern tip of the continent and in cooler upland zones (hills, mountains, high plateaus). Seasonal climates in Africa differ more by the amount of rainfall than by temperature.

Most rainfall comes to Africa by way of the **intertropical convergence zone (ITCZ)**, a band of atmospheric currents that circle the globe roughly around the equator (see inset, Photo Essay 7.1). At the ITCZ, warm winds converge from both the north

and south and push against each other. This causes the air to rise, cool, and release moisture in the form of rain. The rainfall produced by the ITCZ is most abundant in Africa near the equator (see Photo Essay 7.1D). Here, dense tropical rainforests flourish in places such as the Congo Basin (see Photo Essay 7.1A).

The ITCZ shifts north and south seasonally, generally following the area of Earth's surface that has the highest average temperature at any given time. Thus, during the height of summer in the Southern Hemisphere in January, the ITCZ might bring rain far enough south to water the dry grasslands, or steppes, of Botswana. During the height of summer in the Northern Hemisphere in August, the ITCZ brings rain as far north as the southern fringes of the Sahara—an area called the **Sahel**, where steppe and savanna grasses grow. Poleward of both of these extremes, the belt of air that has dumped its moisture while rising and cooling is now drier. At roughly 30° N latitude (the Sahara) and 30° S latitude (the Namib and Kalahari deserts) the drier air descends, forming a subtropical high-pressure zone that shuts out lighter, warmer, moister air. As a result of this system (which is in no way precise), deserts tend to be found in Africa (and on other continents) in these zones about 30 degrees north and south of the Equator.

The tropical wet climates that support equatorial rain forests are bordered on the north, east, and south by seasonally wet/dry tropical woodlands (see Photo Essay 7.1B). These give way to moist tropical savannas or steppes, where tall grasses and trees intermingle in a semiarid environment. These tropical wet and wet/dry climates have provided suitable land for agriculture for thousands of years. Farther to the north and south lie the true desert zones of the Sahara and the Namib and Kalahari (see Photo Essay 7.1C). This banded pattern of African ecosystems is modified in many areas by elevation and wind patterns.

Without mountain ranges to block them, wind patterns can have a strong effect on climate in Africa. Winds blowing north along the east coast keep ITCZ-related rainfall away from the **Horn of Africa**, the triangular peninsula that juts out from northeastern Africa below the Red Sea. As a consequence, the Horn of Africa is one of the driest parts of the continent (see Photo Essay 7.1 Map). Along the west coast of the Namib Desert, moist air from the Atlantic is blocked from moving over the desert by cold air above the northward-flowing Benguela Current. Like the Peru Current off South America, the Benguela is chilled by its passage past Antarctica. Rich in nutrients, it supports a major fishery along the west coast of Africa.

Environmental Issues

While Africans have generally contributed very little to the build-up of greenhouse gases in the atmosphere, deforestation in Africa, much of it to reap exportable hardwoods, is making significant contributions to global climate change. Because of Africa's poverty, its people are much less able to adapt to climate change than those in other regions. Yet in the face of these problems, many Africans are developing strategies to cope with uncertainties related to the region's changing climate.

🎦 **158. Disappearing Glaciers on Mt. Kilimanjaro Raise Environmental Concerns**

---

intertropical convergence zone (ITCZ) a band of atmospheric currents that circle the globe roughly around the equator; warm winds from both north and south converge at the ITCZ, pushing air upward and causing copious rainfall

Sahel a band of arid grassland, where steppe and savanna grasses grow, that runs east-west along the southern edge of the Sahara

Horn of Africa the triangular peninsula that juts out from northeastern Africa below the Red Sea and wraps around the Arabian Peninsula

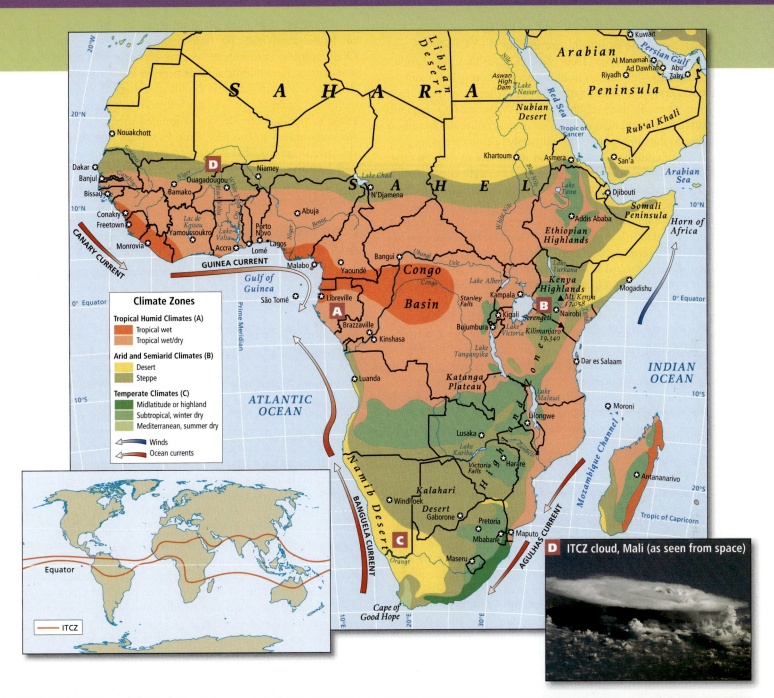

Climate Zones

Tropical Humid Climates (A)
- Tropical wet
- Tropical wet/dry

Arid and Semiarid Climates (B)
- Desert
- Steppe

Temperate Climates (C)
- Midlatitude or highland
- Subtropical, winter dry
- Mediterranean, summer dry

→ Winds
→ Ocean currents

ITCZ

D ITCZ cloud, Mali (as seen from space)

A Tropical wet, Gabon

B Subtropical, winter dry, Kenya

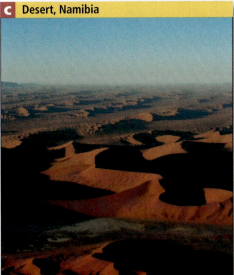

C Desert, Namibia

Deforestation and Climate Change

Deforestation is Africa's main contribution to CO_2 emissions and potential climate change. Trees absorb CO_2 as they photosynthesize, thus removing carbon from the air and storing it as biomass, a process known as **carbon sequestration.** When trees die, as they decompose or are burned, they release the stored CO_2 into the atmosphere. Hence, deforestation is a major contributor to carbon buildup in the atmosphere and ultimately to climate change.

African countries lead the world in the *rate* of deforestation, the percentage of total forest area lost. Of the eight countries that had the world's highest rates of deforestation between 1990 and 2005, six are African (Burundi, Togo, Nigeria, Benin, Uganda, and Ghana). This constitutes a major human environmental impact across the region (Photo Essay 7.2). In terms of *emissions* from deforestation, the leaders were Brazil and Indonesia, but Nigeria and Congo (Kinshasa), ranked third and fourth.

📺 **168. Plan to Clear-Cut Ugandan Forest Reserve for Growing Sugar Cane Sparks Controversy**

Most of Africa's deforestation is driven by the growing demand for farmland and fuelwood, although logging by international timber companies is also increasing. Africans use wood or charcoal to supply nearly all their domestic energy (see Photo Essay 7.2 A, B). Wood and charcoal remain the cheapest fuels available, in part because of African traditions that consider forests to be a free resource held in common. Even in Nigeria, a major oil producer, most people use fuelwood because they cannot afford petroleum products.

For a decade or more, many in sub-Saharan Africa have recognized the need to use fuelwood more sustainably. Solar ovens and fuel-efficient wood stoves have been widely promoted. While many users find them less convenient than old-fashioned cooking fires, the growing scarcity of fuelwood is driving greater acceptance of these technologies.

Some governments are also encouraging **agroforestry**—the raising of economically useful trees—to take the pressure off forests for fuelwood and construction materials. By practicing agroforestry on the fringes of the Sahel, a family in Mali, for example, can produce fuelwood, building and fencing materials, medicinal products, and food—all on the same piece of land. This would double what could be earned from single-crop agriculture or animal herding. However, critics of agroforestry note that sometimes the trees used are *invasive species*, non-native plants that threaten many African ecosystems by outcompeting indigenous species.

A strong promoter of agroforestry is the Green Belt Movement, founded by Dr. Wangari Maathai of Kenya. The Green Belt Movement helps rural women plant trees for use as fuelwood and to help reduce soil erosion. The movement grew from Maathai's belief that a healthy environment is essential for democracy to flourish. Thirty years and 30 million trees after she began, Maathai was awarded the 2004 Nobel Peace Prize for her contributions to sustainable development, democracy, and peace.

carbon sequestration the removal and storage of carbon taken from the atmosphere

agroforestry the raising of economically useful trees

subsistence agriculture farming that provides food for only the farmer's family and is usually done on small farms

mixed agriculture the raising of a variety of crops and animals on a single farm, often to take advantage of several environmental riches

Multiple other efforts are underway to reduce the extent of deforestation resulting from logging in sub-Saharan Africa. One approach is to require logging companies to use methods that reduce clear-cutting and damage to the surrounding forest. Another is to minimize the building of logging roads, because after the logging companies have left, poor farmers often use these roads to permanently occupy deforested areas. Under pressure from consumers in the EU, some European logging companies now use more sustainable practices, such as fewer logging roads and less clear-cutting. However, the Asian logging companies, which are expanding rapidly in sub-Saharan Africa, are generally not adopting such practices.

Food Production, Water Resources, and Vulnerability to Climate Change

Africa's food production systems have both advantages and disadvantages in helping the region adapt to climate change (Photo Essay 7.3 on page 302).

Agricultural Systems Most sub-Saharan Africans practice **subsistence agriculture,** which provides food for only the farmer's family and is usually done on small farms of about 2 to 10 acres (1 to 4 hectares). Most African farmers also practice **mixed agriculture,** raising a diverse array of crops and a few animals as livestock (see Photo Essay 7.3C). Many also fish, hunt, and gather some of their food from forest or grassland areas. In a wide variety of environments, animals are losing their habitat due to human pressure and climate change (see Photo Essay 7.3B). As awareness of endangered wildlife species increases, so does awareness that hunting practices also put pressure on wildlife and require modification.

These widely practiced food acquisition techniques have advantages and disadvantages in a world of climate change. Mixed agriculture provides a diverse array of strategies for coping with the changes in temperature and rainfall that climate change may bring. For example, traditional Nigerian farmers grow complex tropical gardens, often with 50 or more species of plants at one time. Some of the plants can handle drought, while others can withstand intense rain or heat.

However, the subsistence nature of most African farming can also leave families without much cash. If harvests are too low to provide saleable surpluses and hunting and gathering fail to provide enough food to feed the family, there is insufficient money to buy food. While such situations can lead to *famine,* it is important to note that the most serious famines in Africa have occurred not because of low harvests but rather because of political instability that disrupts economies and food growing and distribution systems.

📺 **162. Fellowship Program Aims to Open Doors for African Women in Agricultural Research**

Much of Africa is now shifting over to commercial agriculture, in which crops are grown for cash rather than solely as food for the farm family. This type of production also has advantages and disadvantages with respect to climate change. If harvests are good and prices for crops are adequate, farmers can earn enough cash to get them through a year or two of poor harvests (see Photo Essay 7.3D on page 302). Having some cash income can also allow poor families to invest in other means of earning a living,

Deforestation, desertification, and increasing water use have had major impacts on Sub-Saharan Africa's ecosystems.

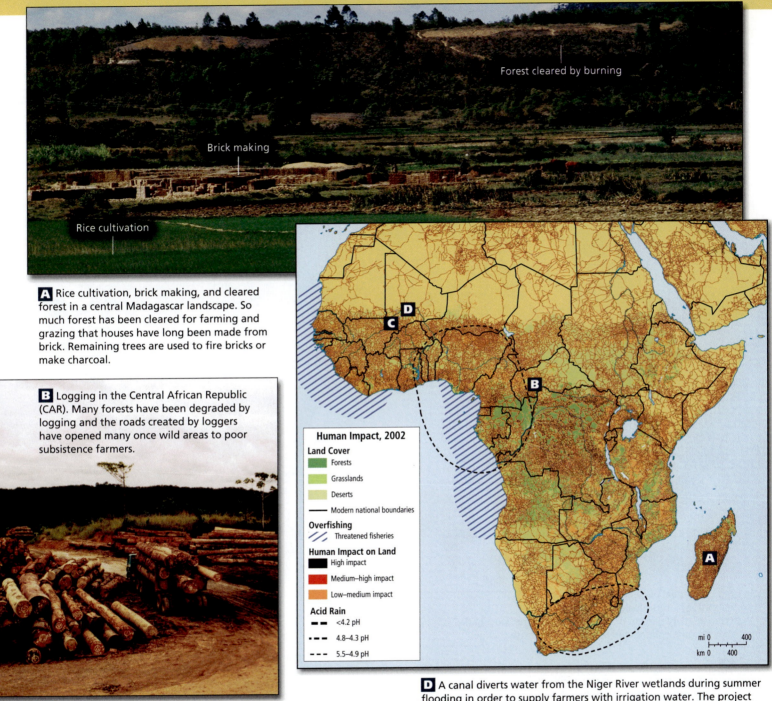

Forest cleared by burning

Brick making

Rice cultivation

A Rice cultivation, brick making, and cleared forest in a central Madagascar landscape. So much forest has been cleared for farming and grazing that houses have long been made from brick. Remaining trees are used to fire bricks or make charcoal.

B Logging in the Central African Republic (CAR). Many forests have been degraded by logging and the roads created by loggers have opened many once wild areas to poor subsistence farmers.

Human Impact, 2002

Land Cover
- Forests
- Grasslands
- Deserts
- Modern national boundaries

Overfishing
- Threatened fisheries

Human Impact on Land
- High impact
- Medium–high impact
- Low–medium impact

Acid Rain
- <4.2 pH
- 4.8–4.3 pH
- 5.5–4.9 pH

mi 0   400
km 0   400

C A fisher casts his net in the Niger River wetlands, where carefully synchronized resource use patterns have developed over millennia.

D A canal diverts water from the Niger River wetlands during summer flooding in order to supply farmers with irrigation water. The project increases food security for farmers elsewhere in Mali, while threatening the livelihoods of fishers, farmers, and pastoralists in the Niger wetlands.

Much of this region is highly exposed to drought, flooding, and other climatic disturbances. Poverty and low access to cash increase sensitivity to these events, while political instability makes it harder for governments to boost resilience with effective disaster management.

A Children bring water back to their home in Somalia. Drought, poverty, and political instability come together in Somalia to create extremely high vulnerability.

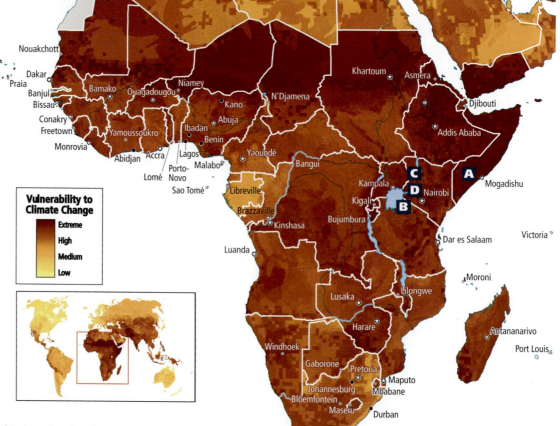

Vulnerability to Climate Change

- Extreme
- High
- Medium
- Low

B The annual wildebeest migration in Tanzania and Kenya. From 25 to 40 percent of species in Africa's national parks may become endangered as a result of climate change.

C A landscape of subsistence and mixed agriculture in Kenya. These diversified systems lend farmers some resilience to climatic disturbances, but they provide little cash to help farmers buy food if necessary.

D Tea harvested for export in Kenya. Commercial crops like tea can increase the resilience of farmers to climate change by providing cash to buy food if necessary. But tea, like many commercial crops, requires expensive and polluting fertilizers, and prices for tea are unstable.

such as opening a grinding mill to help process other farmers' harvests.

However, some of the most common commercial crops, such as peanuts, cacao beans (used for making chocolate), rice, or tea and coffee, are often less well adapted to environments outside their native range. This makes reduced yields more likely if temperatures increase or water becomes scarce. Moreover, to maximize profits, these crops are often grown in large fields of only a single plant species. This can leave crops vulnerable to pests and if crops fail, farmers have no other garden foods to rely on.

The potential for commercial agriculture to help farmers adapt to the uncertainties of global climate change is also limited by the instability of prices for commercial crops. Prices can rise or fall dramatically from year to year because of overproduction or crop failures both in Africa and abroad. Another downside of commercial production is that it is usually based on the permanent conversion of forests to fields. This can put a greater strain on soil and water resources. Like tropical soils everywhere (see Chapter 3, page 124), African soils rapidly lose their fertility when cultivated. This is especially true in the warmer, wetter areas, where the nutrient-carrying organic matter in the soil decays rapidly. To maintain soil quality in such a climate, subsistence farmers have long used **shifting cultivation**, in which patches of forest are cleared and cultivated for just 3 years or so and then left to regrow. After a few decades, the soil naturally replenishes its organic matter and nutrients and is ready to cultivate again.

When fields are cleared permanently, as many commercial agriculture systems demand, soil fertility declines and crops become more likely to fail even under ideal climatic conditions. Chemical fertilizers can compensate for this loss, but are often unaffordable for many farmers and after repeated use may lose their effectiveness. Moreover, rains almost always wash much of the fertilizer into nearby waterways, thus polluting them. This may ultimately hurt farmers and fishers by reducing the quantity of available fish, which are important sources of protein for rural people.

These aspects of commercial agricultural systems make them poorly adapted even to ideal climatic conditions in Africa, let alone the hotter and more drought-prone environments that climate change may bring. Foreign agricultural "experts" who often push commercial systems can be woefully ignorant of local conditions and the value of local cultivation techniques. For example, in Nigeria, it is women who grow most of the food for family consumption and who are guardians of the knowledge that makes Nigeria's complex farming systems possible. However, women were rarely included or even consulted by the British experts who promoted commercial agricultural development projects. At best, women were employed as field laborers. Consequently, diverse subsistence agricultural systems based on numerous plant species and deep agricultural knowledge were replaced by less-stable commercial systems based on a single plant species.

Agricultural scientists now recognize past mistakes and are trying to incorporate the knowledge of African farmers, female and male, into more diverse commercial agriculture systems. For example, scientists at Nigeria's International Institute for Tropical Agriculture are developing cultivation systems that, like traditional systems, use many species of plants that help each other cope with varying climatic conditions. Most of these systems are designed for both subsistence and commercial agriculture, and thus can give families both a stable food supply and cash to help them ride out crop failures and pay school fees for their children.

> **shifting cultivation** a productive system of agriculture in which small plots are cleared in forestlands, the dried brush is burned to release nutrients, and the clearings are planted with multiple species; each plot is used for only 2 or 3 years and then abandoned for many years of regrowth

Irrigation and Water Management Alternatives Because climate change is likely to result in changing rainfall patterns, many are looking to irrigation to provide more stability to agricultural systems. However, the massive ecological changes created by large-scale irrigation projects have made some Africans appreciate irrigation on a smaller scale.

Africa has a number of major rivers—the Nile, the Congo, the Zambezi, and the Limpopo—and climate change will affect all of them. Here we focus on the Niger, one of Africa's most important rivers, which flows through very different ecological zones. The Niger rises in the tropical wet Guinea Highlands and carries summer floodwaters northeast into the normally arid lowlands of Mali and Niger (see **Photo Essay 7.2 C, D**). There the waters spread out into lakes and streams that nourish wetlands. For a few months of the year (June through September), the wetlands ensure the livelihoods of millions of fishers, farmers, and pastoralists. These people share the territory in carefully synchronized patterns of land use that have survived for millennia. Wetlands along the Niger produce eight times more plant matter per acre than the average wheat field, provide seasonal pasture for millions of domesticated animals, and are an important habitat for wildlife.

The governments of Mali and Niger now want to dam the Niger River and channel its water into irrigated agriculture projects that will help feed the more than 26 million people in the two countries. Rising food prices in global markets, mounting population pressure, and the threat of changing rainfall patterns due to climate change are driving the two governments to undertake the massive project. However, the dams will forever change the seasonal rise and fall of the river. The irrigation systems may also pose a threat to human health. Systems that rely on surface storage lose a great deal of water through leakage, and the standing pools of water they create often breed mosquitoes that spread tropical diseases such as malaria and harbor the snails that host schistosomiasis.

Many smaller-scale alternatives are available. In some parts of Senegal, for example, farmers are using hand- or foot-powered pumps to bring water from rivers or ponds directly to the individual plants that need it (**Figure 7.5B**). This is in some ways a more modern version of traditional African irrigation practices whereby water is delivered directly to the roots of the plants by human water brigades (**Figure 7.5A**). Smaller-scale projects provide the same protection against drought that larger systems offer, but are much cheaper and simpler to operate for small farmers. They also avoid the large-scale ecological disruption of larger projects. Already in successful use for 15 years, these pumps will help farmers adjust to drier conditions that may come with climate change.

Herding and Desertification Herding, or **pastoralism,** is practiced by millions of Africans, primarily in savannas, on desert margins, or in the mixture of grass and shrubs called *open bush*. Herders use the milk, meat, and hides of their animals, and they typically circulate seasonally through wide areas, taking their animals to available pasturelands. They trade with settled farmers for grain, vegetables, and other necessities.

Many traditional herding areas in Africa are now experiencing **desertification**, the process by which arid conditions spread to areas that were previously moist. Traditional herding may be partially to blame, but economic development schemes that call for raising cattle to sell as meat can place greater stress on native grasslands than raising camels and goats. Also, long-term natural cycles may bring on long dry spells that are then made worse by human activities, leading to desertification.

The effects of desertification are most dramatic in the region called the Sahel (Arabic for "shore" of the desert). This band of arid grassland, 200 to 400 miles (320 to 640 kilometers) wide, runs east-west along the southern edge of the Sahara (see Photo Essay 7.1 Map on page 299). Over the last century, desertification has shifted the Sahel to the south. For example, the *World Geographic Atlas* in 1953 showed Lake Chad situated in a forest well south of the southern edge of the Sahel. By 1998, the Sahara itself was encroaching on Lake Chad (see Photo Essay 7.1) and the lake had shrunk to a tenth of the area it occupied in 1953.

The Sahel and other dry ecosystems in Africa are fragile environments where water and soil resources are barely sufficient for the needs of native grasses. Rainfall is already low in the Sahel, at 10 to 20 inches (25 to 50 centimeters) per year, and only low levels of organic matter exist in the soil to provide nutrients. Consequently, any further stress, such as fire, plowing, or intensive grazing, may cause grasses to die off. Rain evaporates more quickly from the barren land, and soon the dry soil is blown away by the wind. The remaining sand piles up into dunes as the grassland becomes more like a desert. Over the last 50 years in the Ethiopian highlands—an area that is home to 85 percent of Ethiopia's population and 75 percent of its livestock—desertification has damaged 80 percent of the land (compare the maps in Figures 7.1 on page 293 and Photo Essay 7.1 on page 299). Some areas are so severely damaged that food can no longer be produced there.

Scientists are now exploring the possibility that reducing the intensity of herding and agriculture in the Sahel may not only reduce desertification, but also help combat climate change. This would happen as a result of carbon sequestration. The vast extent of the Sahel and other dry grassland zones raises the possibility that these areas, if properly managed, could remove enough carbon from the atmosphere to at least partially counteract global climate change.

Climate Change and Land Grabs in Africa One other effect in Africa of global climate change that bears mentioning is a market economy phenomenon that is just beginning to be observed and bears watching (http://farmlandgrab.org). In non-sub-Saharan

(A)

(B)

FIGURE 7.5 Small-scale irrigation. (A) A boy in Senegal waters plants in his family's garden by bringing water up in a bucket from the nearby Senegal River. **(B)** A more modern variation on this technique is demonstrated by another farmer in Senegal who uses a foot-powered pump to deliver irrigation water to his crops via a hose.

African countries where farmland is particularly vulnerable to climate change or limited in size—in Saudi Arabia, the UAE, and Libya; or where populations are growing rapidly, as in China, and India; or where people are particularly concerned about having food security, as in Germany, the United Kingdom, and Scandinavia—governments and private investors are buying up or leasing large tracts of land in sub-Saharan Africa for large-scale food production for their homelands. Vast tracts of land—15 to 20 million acres—are involved, in places such as Kenya, Nigeria, Tanzania, Malawi, Congo (Kinshasa), Zambia, Uganda, Madagascar, and even Ethiopia, where millions of people receive food aid. These arrangements provide infrastructure development, jobs, and money for cash-strapped sub-Saharan African governments, but if the interests of local people are ignored, as appears to be the case, the land deals may spell economic disaster for displaced small farmers, severely erode food security, and may preclude future expansion of food production for sub-Saharan Africans.

Wildlife and Climate Change

Africa's world-renowned wildlife faces multiple threats from both human and natural forces, all of which could become more severe due to global climate change. The Intergovernmental Panel on Climate Change estimates that 25 to 40 percent of the species in Africa's national parks may become endangered as a result of climate change.

Wildlife managers need to develop new management techniques to help animals survive. For example, one of the greatest wildlife spectacles on the planet took a tragic turn in 2007. The annual 1800-mile-long natural migration of more than a million wildebeest, zebras, and gazelles in Kenya's Masai Mara game reserve requires animals to traverse the Mara River (see Photo Essay 7.3B on page 302). In the best of times, this is a difficult migration that usually results in a thousand or so animals drowning. In the past, the park's managers have taken a hands-off approach to the migration, considering the losses normal. However, in 2007 extremely heavy rains, which some scientists think are related to global climate change, swelled the Mara River to record levels. When the animals tried to cross, 15,000 drowned. Park managers are now considering taking a more active role in helping the migrating animals cope with unusual climatic conditions that may worsen with climate change. This may involve stopping animals from attempting a river crossing where many have already drowned and directing them to a safer crossing.

▶️ **169. Protecting Nature in Guinea Collides with Human Needs**

The dependence of farmers on hunting wild game (*bushmeat*) for part of their food and income is already a major threat to wildlife in much of Africa. For example, farmers who need food or extra income are killing endangered species such as gorillas and chimpanzees in record numbers in the Congo Basin. If crop harvests are diminished by global climate change, many farmers will become even more dependent on bushmeat. The threat to wild populations of various species has led to calls to expand protected areas for wildlife and establish new ones. However, Africa's existing parks are already struggling to deal with poaching (illegal hunting) within the parks by members of surrounding communities. Poaching is also fueled by demand for exotic animal parts (tusks, hooves, penises) as medicines and aphrodisiacs, especially in Asia. Some park managers have reduced poaching by promoting alternative livelihoods in the poachers' home communities.

Ecotourism already makes money for Africa's national parks, which constitute one-third of the world's preserved national park land. Some parks are now using profits from ecotourism to sponsor economic development in nearby communities that once depended economically in part on poaching. For example, in 1985 Kasanka National Park in Zambia was plagued by poaching that threatened its wildlife populations. Park managers decided to generate employment for the villagers through tourism-related activities. They built tourist lodges and wildlife-viewing infrastructure and started cottage industries to make products to sell to the tourists. Funding is also provided to local schools and clinics, and students are included in research projects within the park. Local farming has expanded into alternative livelihoods, such as beekeeping and agroforestry. Today, poaching in Kasanka is very low, its wildlife populations are booming, tourism is growing, and local communities have an ongoing stake in the park's success.

THINGS TO REMEMBER

1. The surface of the continent of Africa can be envisioned as a raised platform, or plateau, bordered by fairly narrow and uniform coastal lowlands. Most rainfall comes to Africa by way of the intertropical convergence zone (ITCZ), a band of atmospheric currents that circle the globe roughly around the equator.

2. Deforestation is Africa's main contribution to CO_2 emissions and potential climate change, and most of Africa's deforestation is driven by the growing demand for farmland and fuelwood, although logging by international timber companies is also increasing.

3. Long-standing physical challenges, such as deforestation, desertification, and increasing water scarcity, have had major impacts on sub-Saharan Africa's ecosystems. These challenges are likely to increase with climate change. The effects of desertification are most dramatic in the region called the Sahel.

4. Africa's food production systems, still largely subsistence based, but with some mixed agriculture, are threatened by climate change. Commercial food production for Africa's cities is needed, but commercial agriculture has many ecological drawbacks in Africa.

Human Patterns over Time

Africa's rich past has often been misunderstood and dismissed by people from outside the region. European slave traders and colonizers called Africa the "Dark Continent" and assumed it was a place where little of significance in human history had occurred. The substantial and elegantly planned cities of Benin in western Africa, Djenné in the Niger River basin (**Figure 7.6A**), and Loango in the Congo basin, which European explorers encountered in the 1500s, never became part of Europe's image of Africa. Even today, most people outside the continent are unaware of Africa's internal history or its contributions to world civilization.

(A)

(B)

(C)

FIGURE 7.6 Some historic sites of pre-colonial Africa.
(A) Djenné in the Niger River wetlands of Mali has been inhabited since 200 b.c.e. Rising behind the marketplace is the Great Mosque of Djenné, the largest adobe building in the world. **(B)** An aerial view of the Great Zimbabwe complex, which covered some 1800 acres at its peak between 1200 and 1300 c.e., when it was home to as many as 40,000 people. **(C)** Lamu town, Kenya, is a thousand-year-old Swahili settlement that was once a center of trade along the East African coast. It prospered as a slave port and protectorate of Oman from about 1652 until 1807, when the British abolished the slave trade.

The Peopling of Africa and Beyond

Africa is the original home of humans. It was in eastern Africa (in what is today the Ethiopian, Kenyan, and Ugandan highlands) that the first human species evolved more than 2 million years ago, although they differed anatomically from humans today. These early, tool-making humans (*Homo erectus*) ventured out of Africa, reaching north of the Caspian Sea and beyond as early as 1.8 million years ago. Anatomically, modern humans (*Homo sapiens*) evolved from earlier hominoids in eastern Africa about 200,000 years ago, and by about 90,000 years ago, they had reached the eastern Mediterranean. Like earlier humans, modern human migrations radiated out of Africa, spreading into Europe and across mainland and island Eurasia.

Early Agriculture, Industry, and Trade in Africa

In Africa, people began to cultivate plants as early as 7000 years ago in the Sahel and the highlands of present-day Sudan and Ethiopia (see Figure 1.29 on page 57 for a global map of the multiple sites where agriculture developed). Agriculture was brought south to equatorial Africa 2500 years ago and to southern Africa about 1500 years ago. Trade routes spanned the African continent, extending north to Egypt and Rome, and east to India and China. Gold, elephant tusks, and timber from tropical Africa were exchanged for a wide variety of goods.

About 3400 years ago, people in northeastern Africa learned how to smelt iron and make steel. By 700 c.e., when Europe was still recovering from the collapse of the Roman Empire, a remarkable civilization with advanced agriculture, iron production, and gold-mining technology had developed in the highlands of southeastern Africa in what is now Zimbabwe. This empire, now known as the Great Zimbabwe Empire (**Figure 7.6B**), traded with merchants from Arabia, India, Southeast Asia, and China, exchanging the products of its mines and foundries for silk, fine porcelain, and exotic jewelry. The Great Zimbabwe Empire collapsed around 1500 for reasons not yet understood.

Complex and varied social and economic systems existed in many parts of Africa well before the modern era. Several influential centers made up of dozens of linked communities developed in the forest and the savanna of the western Sahel. There, powerful kingdoms and empires rose and fell, such as that of Ghana (700–1000 c.e.), centered in what is now Mauritania and southwestern Mali, and the Mali Empire (1250–1600 c.e.), centered a bit to the

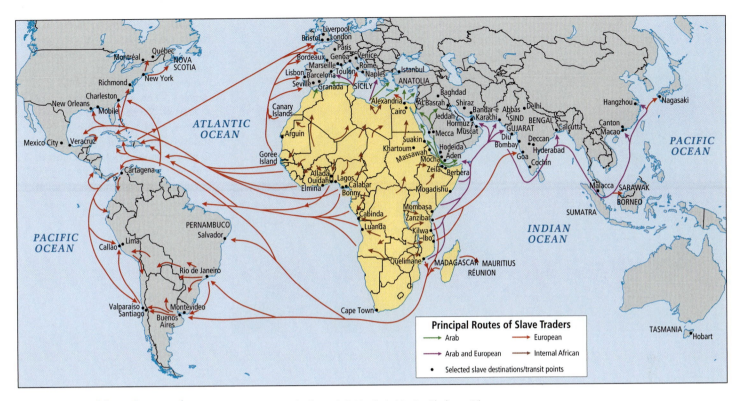

FIGURE 7.7 African slave trade. [Adapted from the work of Joseph E. Harris, in Monica Blackmun Visona et al., *A History of Art in Africa* (New York: Harry N. Abrams, 2001), pp. 502–503.]

east on the Niger River in the famous Muslim trading and religious center of Tombouctou (Timbuktu). Some rulers periodically sent large entourages on pilgrimages through Tombouctou to Makkah (Mecca), where their opulence was a source of wonder.

Africans also traded slaves. Long-standing customs of enslaving people captured during war fueled this trade. The treatment of slaves within Africa was sometimes brutal and sometimes reasonably humane. Long before the beginning of Islam, a slave trade developed with Arab and Asian lands to the east (**Figures 7.6C** and **7.7**). After the spread of Islam began around 700, the slave trade continued and close to 9 million African slaves were exported to parts of Southwest, South, and Southeast Asia by Muslim traders. When slaves were traded to non-Africans, indigenous checks on brutality were lost. For example, to ensure passivity and sterility, Muslim traders often preferred to buy castrated male slaves.

Europeans and the African Slave Trade

The course of African history shifted dramatically in the mid-1400s, when Portuguese sailing ships began to appear off Africa's west coast. The names given to stretches of this coast by the Portuguese and other early European maritime powers reflected their interest in Africa's resources: the Gold Coast, the Ivory Coast, the Pepper Coast, and the Slave Coast.

By the 1530s, the Portuguese had organized a slave trade with the Americas. The trading of slaves by the Portuguese, and then by the British, Dutch, and French, was more widespread and brutal than any trade of African slaves that preceded it. African slaves became part of the elaborate production systems supplying the raw materials and money that fueled Europe's Industrial Revolution.

To acquire slaves, the Europeans established forts on Africa's west coast and paid nearby African kingdoms with weapons, trade goods, and money to make slave raids into the interior. Some slaves were taken from enemy kingdoms in battle. Many more were kidnapped from their homes and villages in the forests and savannas. Most slaves traded to Europeans were male because they brought the highest prices and the raiding kingdoms preferred to keep captured women for their reproductive capacities. Between 1600 and 1865, about 12 million captives were packed aboard cramped and filthy ships and sent to the Americas. One-quarter or more of them died at sea. Of those who arrived in the Americas, about 90 percent went to plantations in South America and the Caribbean. Between 6 and 10 percent were sent to North America (see Figure 7.7).

The European slave trade severely drained the African interior of human resources and set in motion a host of damaging social responses within Africa that are not completely understood even today. The trade enriched those African kingdoms that could successfully conquer their neighbors and force some if not all of them into slavery. It also encouraged the slave-trading kingdoms to be dependent on European trade goods and technologies, especially guns and other weaponry. The slave trade impoverished the many smaller, less powerful kingdoms and other communities that were repeatedly terrorized and robbed of men, women, and children.

FIGURE 7.8 Modern slavery in Niger. Here, slaves collect water at a well in western Niger. Though slavery was "abolished" in Niger in 1960, and made a criminal offense in 2003, it remains deeply embedded in society. At least 43,000 people are enslaved in Niger, most of them born into the status. Slaves have no rights, little access to education, and will spend most of their lives herding cattle, tending fields, or working in their "master's" house.

Slavery persists in modern Africa and is a growing problem that some argue exceeds the trans-Atlantic slave trade of the past. Today, slavery is most common in the Sahel region, where several countries have made the practice officially illegal only in the past few years. People may become enslaved during war; or are sold by their parents or relatives to pay off debts; or are forced into slavery when trying to migrate or find a job in a city (**Figure 7.8**) or even in Europe. Slaves often work as domestic servants or as prostitutes, and they are increasingly used in commercial agriculture, mining, and war. It is hard to know exactly how many Africans are currently enslaved, but estimates range from several million to more than 10 million.

The Scramble to Colonize Africa

The European slave trade wound down by about the mid-nineteenth century, as Europeans found it more profitable to use African labor within Africa to extract raw materials for Europe's growing industries.

European colonial powers competed avidly for territory and resources, and by World War I, only two areas in Africa were still independent (**Figure 7.9**). Liberia, on the west coast, was populated by former slaves from the United States. Ethiopia (then called Abyssinia), in East Africa, managed to defeat early Italian attempts to colonize it. Otto von Bismarck, the German chancellor who convened the 1884 Berlin Conference at which the competing European powers formalized the partitioning of Africa, declared, "My map of Africa lies in Europe." With some notable exceptions, the boundaries of most African countries today derive from the colonial boundaries set up between 1840 and 1916 by Euro-

apartheid a system of laws mandating racial segregation in South Africa, in effect from 1948 until 1994

pean treaties. These territorial divisions lie at the root of many of Africa's current problems.

In some cases the boundaries were purposely drawn to divide tribal groups and thus weaken them. In other cases groups who used a wide range of environments over the course of a year were forced into smaller, less diverse lands or forced to settle down entirely, thus losing their livelihoods. As access to resources shrank, hostilities between competing tribal groups developed. Colonial officials often encouraged these hostilities, purposely removing tribal leaders with strong leadership skills, replacing them with leaders who could be manipulated. Food production, which was the mainstay of most African economies until the colonial period, was discouraged in favor of activities that would support European industries: cash-crop production (cotton, rubber, palm oil) and mineral and wood extraction. Eventually, formerly prosperous Africans were hungry and poor.

One of the main objectives of European colonial administrations in Africa, in addition to extracting as many raw materials as possible, was to create markets in Africa for European manufactured goods. The case of South Africa provides insights into European expropriation of African lands and the subjugation of African peoples. In this case, these aims ultimately led to the infamous system of racial segregation known as **apartheid.**

Case Study: The Colonization of South Africa

The Cape of Good Hope is a rocky peninsula sheltering a harbor on the southwestern coast of South Africa. Portuguese navigators seeking a sea route to Asia first rounded the Cape in 1488. The Portuguese remained in nominal control of the Cape until the 1650s, when the Dutch took possession with

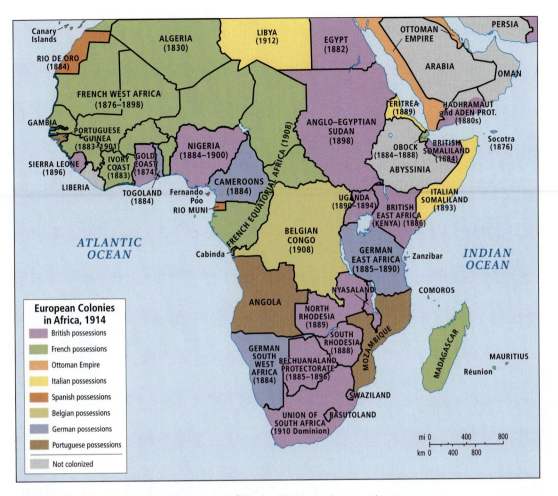

FIGURE 7.9 The European colonies in Africa in 1914. The dates on the map indicate the beginning of officially recognized control by the European colonizing powers. Countries without dates were informally occupied by colonial powers for a few centuries.
[Adapted from Alan Thomas, *Third World Atlas* (Washington, D.C.: Taylor & Francis, 1994), p. 43.]

the intention of establishing settlements. Dutch farmers, called Boers, expanded into the interior, bringing with them herding and farming techniques that used large tracts of land and depended on the labor of enslaved Africans. The British were also interested in the wealth of South Africa, and in 1795 they seized control of areas around the Cape of Good Hope. When slavery was outlawed throughout the British Empire in 1834, large numbers of slave-owning Boers migrated to the northeast. There, in what became known as the Orange Free State and the Transvaal, the Boers often came into intense and violent conflict with African populations.

In the 1860s, extremely rich deposits of diamonds and gold were unearthed in these areas, securing the Boers' economic future. Africans were forced to work in the diamond and gold mines under extreme hardship and unsafe conditions and for minimal wages. They lived in unsanitary compounds that travelers of the time compared to large cages.

Britain, eager to claim the wealth of the mines, invaded the Orange Free State and the Transvaal in 1899, waging the Boer War (Figure 7.10A on page 310). This brutal war gave

the British control of the mines briefly, until resistance by Boer nationalists forced the British to grant independence to South Africa in 1910. This independence, however, applied to only a small minority of whites: the Boers (who have since been known as Afrikaners) and some British who chose to remain. Black South Africans, who made up more than 80 percent of the population, lacked legal political rights until 1994.

In 1948, the long-standing segregation of South African society was reinforced by apartheid, a system of laws that required everyone except whites to carry identification papers at all times and to live in racially segregated areas (Figure 7.10B). Eighty percent of the land was reserved for the use of white South Africans, who at that time made up just 15 percent of the population. Blacks were assigned to ethnically based "homelands." The homelands were considered by the South African government to be independent enclaves within the borders of, but not legally part of, South Africa. Nevertheless, the South African government exerted strong influence in them. Democracy theoretically existed throughout South Africa, but blacks were allowed to vote only in the homelands.

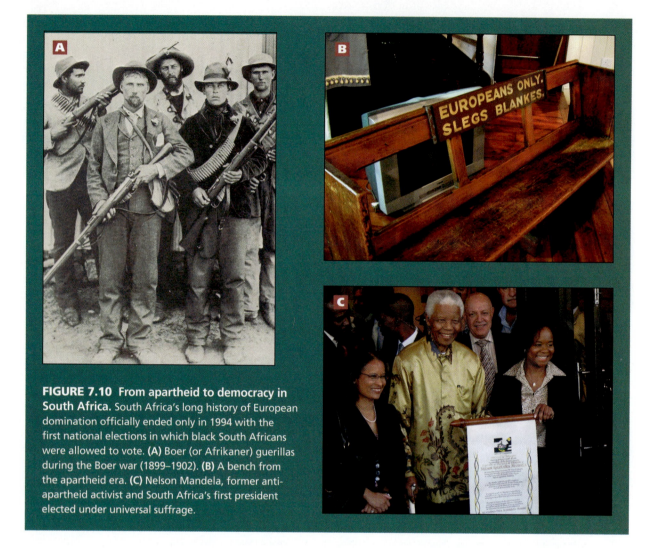

FIGURE 7.10 From apartheid to democracy in South Africa. South Africa's long history of European domination officially ended only in 1994 with the first national elections in which black South Africans were allowed to vote. **(A)** Boer (or Afrikaner) guerillas during the Boer war (1899–1902). **(B)** A bench from the apartheid era. **(C)** Nelson Mandela, former anti-apartheid activist and South Africa's first president elected under universal suffrage.

The African National Congress (ANC) was the first and most important organization fighting to end racial discrimination in South Africa. Formed in 1912 to work nonviolently for civil rights for all South Africans, the ANC grew into a movement with millions of supporters, most of them black, but some of them white. Its members endured decades of brutal repression by the white minority. One of the most famous member to be imprisoned was Nelson Mandela, a prominent ANC leader, who was jailed for 27 years and was finally released in 1990.

Violence increased throughout South Africa until the late 1980s, when it threatened to engulf the country in civil war. The difficulties of maintaining order, combined with international pressure, forced the white-dominated South African government to initiate reforms that would end apartheid. A key reform was the dismantling of the homelands and a range of laws that dictated where people could live and

civil society the totality of voluntary civic and social organizations and institutions that form the basis of a functioning society, as opposed to the structures of a state that are backed by force (regardless of that state's political system) and commercial institutions of the market

vote. Finally, in 1994, the first national elections in which black South Africans could participate took place. Nelson Mandela, the long-jailed ANC leader, was elected the country's president (**Figure 7.10C**). He was awarded the Nobel Peace Prize in 1993. Today in South Africa, the difficult process of dismantling systems of racial discrimination still continues. ∎

The Aftermath of Independence

The era of formal European colonialism in Africa was relatively short. In most places, it lasted for about 80 years, from roughly the 1880s to the 1960s. In 1957, Ghana became the first sub-Saharan African colonial state to achieve its independence. The last sub-Saharan African country to gain independence was Eritrea in 1993, although Eritrea won its independence not from a European power but from its neighbor Ethiopia after a 3-year civil war.

Africa entered the twenty-first century with a complex mixture of enduring legacies from the past and looming challenges for the future. Although it has been liberated from colonial domination, most old colonial borders remain intact (compare the colonial borders in Figure 7.9 on page 309 with the modern country borders in Figure 7.4 on page 295). Often these borders exacerbate conflict between incompatible groups by joining them into one resource-poor political entity; other borders divide potentially powerful ethnic groups, thus diminishing their influence, or cut off nomadic people from resources used on a seasonal basis.

With few exceptions, governments continue to mimic colonial bureaucratic structures and policies that distance them from their citizens. Corruption and abuse of power by bureaucrats, politicians, and wealthy elites stifle individual initiative, **civil society,** and entrepreneurialism, creating instead frustration and suspicion. Too often, coups d'état have been used to change governments. Democracy, where it exists, is often weakly connected only to elections and not to true participation in policy making at the local and national levels.

Another relic of the colonial era is a deemphasis on food production for local consumption and heavy dependence on commercial agricultural and mineral exports, for which profit margins are low and prices on the global market highly unstable. In addition, because Africa now must import food and still produces few of its manufactured needs, most African economies are dependent on relatively expensive imported food and manufactured goods. Hence, as an exporter and importer, sub-Saharan Africa finds itself at a disadvantage in a world of increasing globalization. It is not surprising, then, that sub-Saharan African countries remain economically entwined with their former European colonizers for trade and aid despite the fact that these entanglements rarely work to their advantage.

THINGS TO REMEMBER

1. The borders and administrative units of the African colonies were designed so that strong groups would be divided, groups hostile to one another would be under the same jurisdiction, and leaders would be weakened.

2. The main objectives of European colonial administrations in Africa were to extract as many raw materials as possible and to create markets in Africa for European manufactured goods and food exports.

3. The European slave trade severely drained the African interior of human resources and set in motion a host of damaging social responses within Africa with dimensions that are difficult to quantify even today. Slavery persists in modern Africa and is a growing problem that some argue exceeds the trans-Atlantic slave trade of the past.

4. Now-independent governments continue to mimic colonial policies that distance them from their citizens. Abuse of power stifles individual initiative, civil society, and entrepreneurialism. Democracy, where it exists, is often weak.

II CURRENT GEOGRAPHIC ISSUES

Most countries of sub-Saharan Africa have been independent of colonial rule for about 50 years. While many countries are still struggling with the lingering effects of the colonial era, others have moved forward.

Economic and Political Issues

Sub-Saharan Africa emerged from the exploitation and dependence of the colonial era just in time to get sucked into the vortex of globalization, where small or weak countries are often left with little control over their fates.

Commodities and Globalization in Africa

Sub-Saharan Africa's centuries-long role in the global economy as a producer of human labor and raw materials meant that the profits from turning these human resources and raw materials into higher-value manufactured and processed products went to wealthier countries in Europe and elsewhere. When traded, raw materials are called **commodities.** Two related conditions limit the potential of commodities to lift poor countries out of poverty: when there are many competing producers of the same commodity, competition for markets is stiff; such stiff competition can mean that the prices commodities bring on the world market are relatively low and unstable.

Commodity prices are usually low because, regardless of where they are produced, commodities are generally of more or less uniform quality and are traded as such on a global basis. For example, on the major commodities exchanges, a pound of copper may sell for U.S.$3 to $5, regardless of whether it is mined in South Africa, Chile, or Papua New Guinea. Price instability occurs because the global scale of this trade makes the price of any commodity highly responsive to changes anywhere in the world that might influence the supply or demand.

Similarly, the unpredictability of the prices of globally traded agricultural commodities can wreak havoc with farmers' incomes. When prices are high, farmers may overcommit to a certain crop and spend anticipated earnings unwisely, leaving them vulnerable if prices fall. Governments may overestimate revenues and commit to expensive infrastructure projects. Should commodity prices drop, farmers may have to take their children out of school (fees are charged for most schooling in Africa), and governments may need to cut essential services, such as electricity, road maintenance, or health care.

commodities raw materials that are traded, often to other countries, for processing or manufacturing into more valuable goods

Succeeding Eras of Globalization

At least three eras of globalization have transformed Africa over the past several centuries. This section examines the economic and political impacts of these three eras and describes the current factors moving some African countries away from raw materials exports and toward new types of economic development and regional integration.

The Era of Colonialism and Early Independence Most early European colonial administrations in Africa (Britain, France, Germany, Belgium) evolved directly out of private resource-extracting corporations, such as the German East Africa Company, which was organized with the backing of Otto von Bismarck in 1885. The welfare of Africans and their future development was a relatively low priority for most colonial administrators. Education and health care for Africans were generally neglected. Colonial administrations operated as *mercantilist* operations (see Chapter 3, page 133). The sub-Saharan Africans who worked in European-owned mines or on plantations were poorly paid, as were those who grew cash crops on their own lands. Moreover, the colonial governments of Africa strongly discouraged any other economic activities that might compete with Europe. Hence, it was impossible for sub-Saharan African economies to make a transition to the more profitable manufacturing-based industries that were transforming Europe and North America.

Not all European colonial powers had the same policies. The British in Nigeria, for example, tended to use education and Christianizing as tools to make Africans comply with British extractive economic policies. The Belgians in the Congo River basin, on the other hand, held their colonial subjects at arm's length and used extreme brutality to control them, giving very little attention to education.

When sub-Saharan African nations started gaining independence in the 1950s and 1960s, some of the most forward-thinking African leaders recognized that it would be essential for Africans to regain control of their own resources. Kwame Nkrumah, the U.S.-educated father of Ghanaian independence from Britain (1957) envisaged a United States of Africa where resources of the continent would be exploited for the benefit of Africans. Jomo Kenyatta, the founder of Kenya and its first president (1964–1978), held pan-African economic and political views similar to those of Nkrumah. Several other early post-independence leaders were less foresighted and tended to replicate the government institutions and development theories of European colonial officials, or like Hubert Maga of Benin, used force and assassination to retain power. For the most part, postindependence economies remained focused on the production of commodities—until recently just a single commodity—for export (**Figure 7.11**). Profits generally were not fairly distributed or wisely invested. Like their colonial predecessors, some leaders and bureaucrats considered it their right to enrich themselves with government tax revenues derived from commodity exports. Government jobs and contracts were handed out to cronies or tribal allies as a form of political patronage. Often, enormous sums of borrowed money went to fund government-run development projects that were plagued by corruption and poor planning. Meanwhile, farmers and miners remained so poorly paid that they had little capacity to consume. As a result, widespread poverty and the lack of

dual economy an economy in which the population is divided by economic disparities into two groups, one prosperous and the other near or below the poverty level

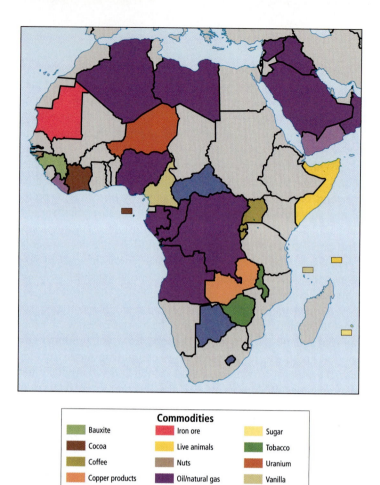

FIGURE 7.11 One-commodity countries, 2000. African countries depended on only one commodity for more than 50 percent of their export earnings in 2000. By 2010, apparently all countries in Africa had diversified their economies sufficiently to no longer depend on only one commodity, but commodity dependence persists. [Adapted from George Kurian, ed., *Atlas of the Third World* (New York: Facts on File, 1992), and the Central Intelligence Agency, *The World Factbook*, at https://www.cia.gov/library/publications/the-world-factbook/fields/2090.html?countryName=Angola&countryCode=ao&regionCode=af&#ao.]

a market for local products and services still characterize even sub-Saharan Africa's largest, most prosperous economies.

South Africa's economy is an exception to commodity dependence, but its success contains a warning. It is the only sub-Saharan African country with a strong manufacturing base. Early on, profits from its commodity exports (mainly minerals) were reinvested in related manufacturing industries that served the mining sector. Today, South Africa is a world leader in the manufacturing of mining and railway equipment. The country also has a well-developed service sector, with particular strengths in finance and communications that developed in part to support the mining and manufacturing industries. With only 6 percent of sub-Saharan Africa's population, South Africa today produces 30 percent of the region's economic output.

However, South Africa is a classic example of a **dual economy**. For centuries, it has had a well-off minority white population, whose skills and external connections fostered economic development within South Africa. After independence from Britain in 1910, the Dutch (Afrikaners or Boers) and British who remained in the country continued to dominate the economy. Even though the labor of black South Africans was essential to the country's prosperity, most black South Africans remained poor. Under the apartheid system, 84 percent of black South Africans lived at a bare subsistence level. This pattern began to change at the end of apartheid, as black South Africans benefited from government-backed loans and other programs designed to encourage entrepreneurship. In 2009, however, 50 percent of black South Africans still lived below the poverty line (compared with just 7 percent of white South Africans), and only 22 percent of black South Africans had finished high school (compared with 70 percent of white South Africans). The average income for a white South African is still more than 5 times that of a black South African. These inequalities contribute to high rates of crime and violence that have encouraged many skilled South Africans of all races to emigrate to Europe or North America.

The Era of Structural Adjustment By the 1980s, most African countries remained poor and dependent on their volatile and relatively low-value commodity exports (**Figure 7.12**). Generally, attempts at investing in manufacturing industries had failed due

FIGURE 7.12 Commodity dependence. Roads are in disrepair immediately outside the bauxite refinery in Fria, Guinea. Despite the country's rich bauxite resources (aluminum is made from bauxite), most Guineans lack access to safe water, adequate sanitation, or electricity.

to a number of factors, which are discussed below. Unfortunately, sub-Saharan African countries had taken out massive loans for these projects and were struggling to repay them. A breaking point came in the early 1980s when an economic crisis swept through sub-Saharan Africa and much of the rest of the developing world, leaving most countries unable to repay their debts. In response, the IMF and the World Bank designed *structural adjustment programs* (SAPs), also called belt-tightening programs (see Chapter 3, pages 136–139) to help countries repay their loans. SAPs did accomplish some useful results. They tightened bookkeeping procedures and thereby curtailed corruption and waste in bureaucracies. They closed some corrupt state-owned monopolies in industries and services, and opened some sectors of the economy to medium- and small-scale business entrepreneurs. And they made tax collection more efficient. But overall, SAPs and their successors had many unintended consequences (discussed below), and surprisingly, they failed at their primary objective—reducing debt (**Figure 7.13** on page 314).

Investments in manufacturing failed in the 1980s partly because of corruption and civil war, but tariffs in developed countries also played a crucial role. The story of a shoe factory in Tanzania provides a case in point. In the 1980s, the World Bank gave a loan to the government of Tanzania to develop an export-oriented shoe factory. Tanzania's large supply of animal hides was to be used to manufacture high-fashion shoes for the European market, using expensive imported machinery. However, due to EU tariffs that protected shoemakers in Europe, the Tanzanian factory never managed to export shoes to Europe. Ideally, the factory would have stayed in business by producing shoes for the large African shoe market. However, its machinery could produce only fancy dress shoes, not the practical working shoes most Africans needed. Years went by and, as the unused factory deteriorated and produced no income, Tanzania struggled to repay its loan to the World Bank. Such experiences, repeated hundreds of times over, left sub-Saharan African countries impoverished, highly indebted, and unable to repay their loans by the 1980s.

Structural adjustment programs had further negative effects across sub-Saharan Africa. In order to force countries to repay their loans, SAPs required governments to reduce their involvement in the economy by selling off government-owned enterprises, often at bargain-basement prices. Also, government payrolls, along with many social service, education, health, and agricultural programs, were slashed so that tax revenues could be devoted to loan repayment. If countries refused to implement SAP requirements, the international banks cut off any future lending for economic development.

Furthermore, prospective Western investors in sub-Saharan Africa were discouraged by problems that SAPs either ignored or made worse. Loss of public funds for schools perpetuated an underskilled workforce. As unemployment rose, so did political instability. Deteriorating infrastructure reduced the quality of remaining social services, transportation, and financial services, all of which scared away investors. Thus, SAPs made it harder for the poor majority to make a decent living and stay healthy.

Some of this negativity began to turn around during the global recession of 2008 and later as Singapore, India, Japan, China, Taiwan, and Korea all began investing in projects aimed

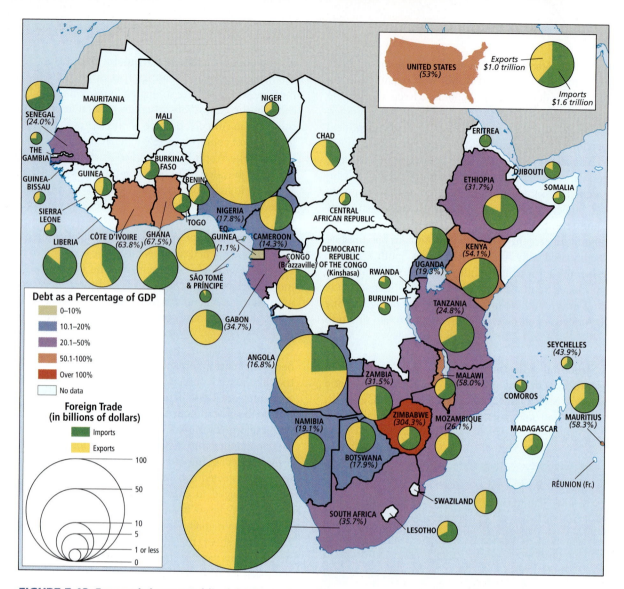

FIGURE 7.13 Economic issues: Public debt, imports, and exports. In most sub-Saharan African countries, imports exceed exports—even in the large diverse economy of South Africa. As these countries borrow money for development, their public debt increases. Even in Côte d'Ivoire, where exports are greater than imports, public debt is over 50 percent of the GDP. Oil producers Angola, Guinea, and Nigeria appear to have the most well-balanced trade, coupled with low debt. [Debt and trade data from the Central Intelligence Agency, *The World Factbook 2010,* at https://www.cia.gov/cia/publications/factbook/index.html.]

at developing sub-Saharan African resources in preparation for the time when global markets rebound (**Figure 7.14**). This increased investment by industrialized Asian economies—still only 3.8 percent of the global total FDI—could spell a real change for sub-Saharan Africans, but only if they are included in the planning and if average workers receive wage increases and other benefits as a result of the investments.

SAPs also reduced the availability of food for consumption in sub-Saharan Africa because agricultural resources were shifted toward the production of cash crops for export. Between 1961 and 2005, per capita food production in sub-Saharan Africa actually decreased by 14 percent, making it the only region on earth where people are eating less well now than in the past (see Figure 1.16 on page 30).

currency devaluation the lowering of a currency's value relative to the U.S. dollar, the Japanese yen, the European euro, or another currency of global trade

The litany of hardships imposed by structural adjustment policies is long. In order to sell more export crops, sub-Saharan African countries were encouraged to devalue their currency relative to currencies of other countries selling similar commodities. But **currency devaluation** also makes all imports more expensive. Thus farmers growing food for the sub-Saharan African market spend more on seeds, fertilizers, pesticides, and farm equipment (not yet produced within the region) and so food costs have risen. The shift to export crops also left farmers with less time and space to grow food for local markets. Increasingly, sub-Saharan Africans must pay for expensive imported food.

SAPs actually worked against the creation of export-oriented manufacturing industries. Consider sub-Saharan Africa's ancient textile industries: with some investment, these textiles could have gained a global market because of the artistic distinctiveness of the cloth. Instead, SAPs forced sub-Saharan African countries to

(A) The world, 2008

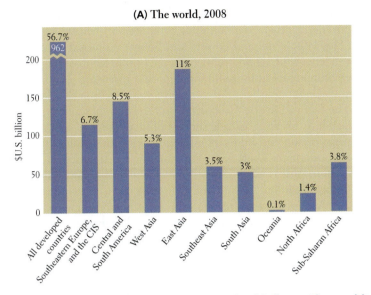

(B) Africa

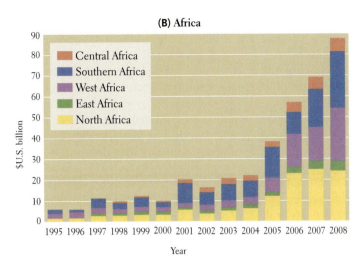

FIGURE 7.14 **Foreign direct investment (FDI) inflows: The world and Africa, 2008.** FDI inflows into developed countries declined in 2008, while those to developing countries and transition economies continued to increase and reached the second-highest percentage ever at 43.3 percent of the global total. FDI inflows into sub-Saharan Africa rose to a record level of $88 billion in 2008, and intra-African FDI is estimated to be about 13 percent, or $11.4 billion, of that total. [Source: UN Conference on Trade and Development, *Economic Development in Africa Report 2009: Strengthening Regional Economic Integration for Africa's Development* (New York and Geneva: United Nations, 2009), pp. 41–42, 76, at http://www.unctad.org/en/docs/aldcafrica2009_en.pdf.]

remove their own tariffs on textiles from other countries, resulting in a flood of imported cloth from China and, to a lesser degree, from India. Local weavers lost their local markets and were impeded from developing global markets. The overall result was factory closings and job losses in just-developing sub-Saharan African textile industries.

Ultimately, Africa's **informal economies** provided relief from the hardships created by SAPs. Informal economies in Africa are ancient and wide-ranging, providing employment and useful services and products. People may grow and sell garden produce, prepare food, make craft items and utensils, or tend to the sick. Others, however, earn a living doing less socially redeeming things, like distilling liquor (**Figure 7.15**) or smuggling scarce or illegal items, such as drugs, weapons, endangered animals, animal parts, or ivory. Because

informal economy all aspects of the economy that take place outside official channels

FIGURE 7.15 **An informal economy of illegal liquor.** Lucy Mugure is a single mother of five living in Kenya's Mathare slum where she brews an illegal and sometimes deadly liquor known as *chang'aa* in makeshift barrels. "I sell a glass at 10 shillings [U.S.$0.15]—less than half the price of legal beer. Although most of my customers drink on credit, paying at the end of the week or the month, some refuse to pay. I spend most of my time trying to avoid arrest by the police who will pour my liquor away before taking me to court where I am fined or jailed. During good months, I make a profit of between 500 and 1000 shillings [U.S.$7–$14], which is still not enough because I pay rent of 800 shillings [U.S.$12]. Sometimes I only make 50 shillings [U.S.$0.70] in a day, yet the children need to eat and we need to buy fuel. I usually end up sending the children to an eating place to spend 10 shillings [U.S.$0.15] each on a small plate of rice or a doughnut and beans to survive. The uniforms for my three school-going children are also expensive, and I have to pay examination fees. Sometimes the children are sent home from school and I cannot afford to take them back to school; most of the time, at least one is out of school." [Source: Quote from IRIN, *Lucy Mugure: "I Do Not Want My Children to Bring Up Their Children Here,* July 2007, at http://www.irinnews.org/HOVReport.aspx?ReportId=73094.]

most activities take place "under the radar," informal jobs may involve criminal acts, wildly unsafe procedures, and hazardous substances.

In most African cities, informal trade once supplied one-third to one-half of all employment; now it often provides more than two-thirds. This creates problems for governments because the informal economy typically goes untaxed, so less money is available to pay for government services or to repay debts. Moreover, the profits of many informal businesses have declined over time as more people compete to sell goods and services to people with less disposable income. And although women typically dominate informal economies, when large numbers of men lost their jobs in factories or the civil service, they crowded into the streets and bazaars as vendors, displacing the women and young people.

Many of the displaced women and girls have turned to, or been forced into, sex work—a growing informal sector—putting themselves at high risk of social rejection, emotional problems, and HIV-AIDS. In addition, more children from disintegrating families must fend for themselves on the streets. Cities such as Nairobi, Kenya, which had very few children living on the streets before SAPs, now have thousands.

In response to the now widely recognized failures of SAPs, and the overemphasis on the power of markets to guide development, the IMF and the World Bank in 2000 replaced SAPs with *Poverty Reduction Strategy Papers*, or *PRSPs*. While these policies are similar to SAPs in that they push market-based solutions, are intended to reduce the role of government in the economy, and are highly bureaucratic, they differ in several ways. They focus on reducing poverty, rather than on just "development" per se; they promote the participation of civil society; and they include the possibility that a country may have all or most of its debt "forgiven" (paid off by the IMF, the World Bank, or the African Development Bank) if the country follows the PRSP rules. Thirty sub-Saharan countries had qualified for and been approved to receive debt relief as of July, 2010. This debt relief program is designed to accelerate progress toward the United Nations Millennium

Development Goals (MDGs). The primary MDG is to reduce poverty in developing countries by at least 50 percent by 2015. (The MDGs are: end poverty and hunger, universal education, gender equality, child health, maternal health, combat HIV-AIDS, environmental sustainability, and global partnership.)

The Era of Diverse Globalization The current wave of globalization is resulting in new sources of investment as well as new pressures on the prices of Africa's export commodities. While Europe and the United States are still the largest sources of investment in Africa, Asia's influence on African economies is increasing significantly.

Perhaps the most hopeful sign for African economies is that Africans working and living abroad, primarily in Europe and North America, are sending home more in *remittances* (money sent to family members) than Africa is receiving from all other sources of foreign investment. According to World Bank estimates for 2007, sub-Saharan countries received 4.5 percent (or U.S.$15.2 billion) of all the global remittances, which amounted to U.S.$337 billion. This money was used to build houses, start small businesses, fund education for children, and help the needy. Remittances are a more stable source of investment than foreign direct investment, in that they tend to come regularly from committed donors who will continue their support for years. They are also much more likely to reach poorer communities. Yet this increasing reliance on international migrants presents its own form of vulnerability: the remittances tend to come in amounts too small to start anything but small projects, and they can abruptly cease in an economic recession, should the remitters lose their jobs.

In recent years, Africa as a whole, and especially sub-Saharan Africa, has become a new frontier for Asia's large and growing economies, especially those of China and India. Through their demand for Africa's export commodities, through direct investment in sub-Saharan Africa, and through the sale of their manufactured goods in sub-Saharan Africa, China and India now exert a more powerful influence on sub-Saharan Africa than ever before. Of the two, China's influence is by far the greater (**Figure 7.16**).

Together, China and India consume about 15 percent of Africa's exports, but their share is growing twice as fast as is that of any of Africa's other trading partners. However, the increasing Chinese and Indian demand for resources is a double-edged sword. While it has brought higher and more stable prices for sub-Saharan Africa's export commodities in recent years, the rising demand for food in China, India, and other parts of Asia has also helped drive up global food prices. This

FIGURE 7.16 China in Angola. The man in blue is one of the estimated 20,000 Chinese workers in Angola. China has given loans and aid to Angola in excess of U.S.$4 billion since 2004, and in return, China has been guaranteed a large portion of Angola's future oil production. In addition, 70 percent of Angola's development projects have been given to Chinese companies, most of which import workers from China.

has been particularly painful for the many poor African countries in which expanding urban populations, located where there is no access to subsistence plots, are highly dependent on food imports.

Will Asian investments in sub-Saharan Africa ultimately prove beneficial or hurtful to sub-Saharan African economies? Much like Europe during the colonial era, China has invested mainly in commodity exports, especially minerals, timber, and oil, and in the infrastructure—roads, airports—to remove these raw materials from sub-Saharan Africa. The improved infrastructure could ultimately facilitate intra-African trade, but the construction phase has not provided jobs for Africans because the agreements the governments have made require that all work be done by Chinese companies using Chinese labor. Most controversial has been the willingness of Chinese companies to deal with brutal and corrupt local leaders, such as Liberia's Charles Taylor (see the opening vignette) or Zimbabwe's Robert Mugabe, who barter away their country's nonrenewable resources at bargain prices and use the profits to enrich themselves and to wage war against their own citizens.

Some pro-marketization economists say that trade relationships with both China and India could have positive influences for Africa over the long term because they could facilitate intra-African trade, which has been poorly developed. China's involvement in road building, for example, could increase regional economic integration. China's investment in agriculture could result in more efficient production for export (and thus higher earnings) and greater food supplies for African internal markets, which would increase overall food security.

📹 **161. African Union Appeals to Diaspora to Aid Homelands**

📹 **160. Somaliland Expatriates Return Home to Help Native Land Develop**

Regional and Local Economic Development

Seeking alternatives to past development strategies, many African governments are focusing on regional economic integration similar to that of the European Union. Local agencies and public and private donors are pursuing grassroots development designed to foster very basic innovation.

Regional Integration Less than 20 percent of the total trade of sub-Saharan Africa is conducted between African countries. This is true partly because so many countries produce the same raw materials for export. And everywhere (except South Africa) industrial capacity is so low that the raw materials cannot be absorbed within the continent, so African countries compete with each other and all global producers to sell to Europe, Asia, and America (only 4 percent of U.S. imports came from sub-Saharan Africa in 2008, most of it oil).

Although overall intra-Africa trade is low in comparison with other developing regions, it does exist, and for many countries it is increasingly important because a wide variety of manufactured products (not just a few raw materials) dominate that trade. Much of this internal trade takes place within the regional trading blocs (discussed below) formed over the last several decades.

Thus far in each trade bloc only a few countries—so-called *trade poles*—are especially active and likely to become leaders of regional development. Particularly hampered in developing internal markets are countries that are land-locked and surrounded by poor neighbors with little capacity to cooperate in developing trade infrastructure (roads, airports, technical communications). For example, roads linking adjacent countries are often lacking and flying from one interior country to another can require connections through South Africa or even Europe.

Regional integration, which combines the markets of several countries (as do NAFTA and the EU), is beginning to achieve a market size sufficient to foster industrialization and entrepreneurialism within Africa. Nine regional trade associations (listed below) have evolved and share several goals: reducing tariffs between members, forming common currencies, reestablishing peace in war-torn areas, upgrading transportation and communication infrastructure, and building regional industrial capacity. Full-scale continent-wide economic union along the lines of the European Union is a long-term goal. **Figure 7.17** shows that various trade associations are already linking with one another.

The Major African Regional Economic Communities (RECs) by date of founding (oldest to youngest)

- **UMA (Arab Maghreb Union)**, created in 1989, is a free trade, full economic union group that includes Algeria, Libya, Mauritania, Morocco, and Tunisia.

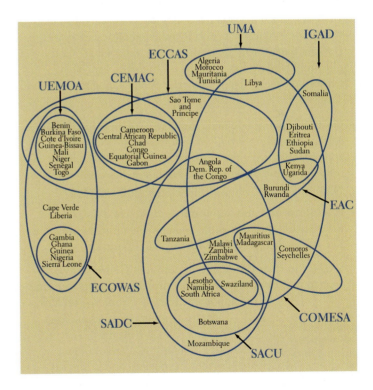

FIGURE 7.17 Principal trade organizations in sub-Saharan Africa. There is a lot of overlap between the countries and the regional trade organizations, but often one country is dominant in each group. South Africa, for example, is dominant in three regional trade organizations and Nigeria in two. [Adapted from UN Conference on Trade and Development, *Economic Development in Africa Report 2009: Strengthening Regional Economic Integration for Africa's Development* (New York and Geneva: United Nations, 2009), p. 12, Figure 1, at http://www.unctad.org/en/docs/aldcafrica2009_en.pdf.]

- **ECOWAS (Economic Community of West African States)**, created in 1993, is dominated by Nigeria and focuses on trade, currency, infrastructure, and political stability. The member countries are Benin, Burkina Faso, Cape Verde, Côte d'Ivoire, Gambia, Ghana, Guinea, Guinea-Bissau, Liberia, Mali, Niger, Nigeria, Senegal, Sierra Leone, and Togo.

- **UEMOA (West African Economic and Monetary Union)**, created in 1994, promotes economic integration among countries that share a common currency. Its member countries are Benin, Burkina Faso, Côte d'Ivoire, Guinea-Bissau, Mali, Niger, Senegal, and Togo.

- **COMESA (Common Market for Eastern and Southern Africa)**, created in 1994, is designed to create a free trade area among its member countries. Its member countries are Angola, Burundi, Comoros, Congo (Kinshasa), Djibouti, Egypt, Eritrea, Ethiopia, Kenya, Madagascar, Malawi, Mauritius, Namibia, Rwanda, Seychelles, Sudan, Swaziland, Uganda, Zambia, and Zimbabwe.

- **IGAD (Intergovernmental Authority on Development)** was created in 1996 as a free trade area for goods, services, investment, and migration. Its member countries are Djibouti, Eritrea, Ethiopia, Kenya, Somalia, Sudan, and Uganda.

- **CENSAD (Community of Sahel-Saharan States)**, created in 1998, is a free trade area that includes economic integration within some of its sectors. Its member countries are Benin, Burkina Faso, the Central African Republic, Chad, Côte d'Ivoire, Djibouti, Egypt, Eritrea, Gambia, Libya, Mali, Morocco, Niger, Nigeria, Senegal, Somalia, Sudan, Togo, and Tunisia.

- **CEMAC (Economic and Monetary Community of Central Africa)**, created in 1999, is a regional economic cooperation zone that has been hampered by wars and violence in its member countries, which are Cameroon, the Central African Republic, Chad, Congo, Equatorial Guinea, and Gabon.

- **EAC (East African Community)**, created in 2000, works toward a EU-style economic and political union between Kenya, Tanzania, Uganda, Rwanda, and Burundi.

- **SADC (Southern African Development Community)**, created in 2000, works on trade, currency, infrastructure, and political stability in Southern and East Africa. Its member countries are Angola, Botswana, Congo (Kinshasa), Lesotho, Malawi, Mauritius, Mozambique, Namibia, Seychelles, South Africa, Swaziland, United Republic of Tanzania, Zambia, and Zimbabwe.

- **SACU (Southern African Customs Union)**, created in 2004, is designed to facililtate cooperation with respect to goods, services, investment, and migration. Its member countries are Botswana, Lesotho, Namibia, South Africa, and Swaziland.

grassroots economic development economic development projects designed to help individuals and their families achieve sustainable livelihoods

self-reliant development small-scale development schemes in rural areas that focus on developing local skills, creating local jobs, producing products or services for local consumption, and maintaining local control so that participants retain a sense of ownership

- **ECCAS (Economic Community of Central African States)**, created in 2007, is a free trade, full economic union group. Its member countries are Angola, Burundi, Cameroon, the Central African Republic, Chad, Congo (Brazzaville), Congo (Kinshasa), Equatorial Guinea, Gabon, São Tomé and Príncipe, and Rwanda.

Local Development In addition to the structural changes that regional trade blocs are being used to create, regional integration is enhanced by **grassroots economic development** strategies. Those that are geared toward providing sustainable livelihoods in rural and urban areas using simple familiar technology offer the most hope. One approach is **self-reliant development**, which consists of small-scale self-help projects that use local skills to create products or services for local consumption. Crucially, local control is maintained so that participants retain a sense of ownership and commitment in difficult economic times. One district in Kenya has more than 500 such self-reliant groups. Most members are women who terrace land, build water tanks, and plant trees. They also build schools and form credit societies.

The issue of improving rural transportation illustrates how a focus on local African needs can generate unique solutions. When non-Africans learn that transportation facilities in Africa are in need of development, they usually imagine building and repairing roads for cars and trucks. But a recent study that analyzed village transportation on a local level found that 87 percent of the goods moved are carried via narrow footpaths *on the heads of women!* Women "head up" (their term) firewood from the forests, crops from the fields, and water from wells (**Figure 7.18**). An average adult woman spends about 1.5 hours each day moving

FIGURE 7.18 Rural water transport. A girl uses a footpath to bring home water in the Democratic Republic of Congo (Kinshasa). Some grassroots development efforts aimed at improving transportation in Africa are focusing on improving footpaths because so much material is transported along them.

the equivalent of 44 pounds (20 kilograms) more than 1.25 miles (2 kilometers).

Unfortunately, the often-dilapidated footpaths trod by Africa's load-bearing women have been virtually ignored by African governments and international development agencies, which tend to focus solely on roads for motorized vehicles (which are also badly needed). Grassroots-oriented nongovernmental organizations (NGOs) are now making much less expensive but equally necessary improvements to Africa's footpaths. Some women have been provided with bicycles, donkeys, and even motorcycles that can travel on the footpaths. This saves time and energy for women who can give more of their effort to education and generating income.

Africa's energy needs, which are currently unmet even at the most basic level of home electricity, can also be addressed by local solutions, as the following two vignettes illustrate.

Personal Vignette Myeka High School lies deep in the hinterland of South Africa and far from electric power lines. For years, lack of educational materials and general demoralization meant that few students graduated from the destitute school. Then the U.S.-based SELF foundation invested in solar power for the school, Dell contributed computers, and INFOSAT Telecommunications set up Internet service. Graduation rates soared, and students left with employable skills.

Intrigued by the idea of alternative power sources, a Myeka science teacher and several students devised a biogas system to use waste from the school's toilets to create methane gas, which now powers the school's electrical system (**Figure 7.19**). The spinoffs are impressive: sanitation problems are solved, and the manure fertilizes the school's gardens, which feed the many AIDS orphans in attendance. Teachers and students have mastered science, math, and technical skills that will serve them for years to come. http://www.solarengineering.co.za/myeka2html.htm.2004. ■

Personal Vignette In Malawi, 14-year-old William Kamkwamba was forced to drop out of school when a famine struck his country in 2001 and his family could no longer afford the $80 school fee. Depressed at the prospect of having no future, he went to a local library when he could. There he found a book in English called *Using Energy* that described an electricity-generating windmill. With an old bicycle frame, PVC pipes, and scraps of wood, he built a windmill that generated enough power to light his home, run a radio, and charge neighborhood cell phones.

Now known as "the boy who harnessed the wind," in 2009 William appeared on Jon Stewart's *Daily Show* in the United States to explain how he plans to start his own windmill company and other ventures that will bring power to remote places across Africa. He has returned to school, this time in the first pan-African prep school in South Africa. William's Web page is http://williamkamkwamba.typepad.com/about.html. ■

THINGS TO REMEMBER

1. Prospective investors in sub-Saharan Africa have been discouraged by problems that structural adjustment programs either ignored or worsened; yet in recent years, Africa has become a new frontier for Asia's large and growing economies, especially those of China and India.

FIGURE 7.19 A biogas digester being installed at Myeka High School. After earlier experiments with solar power, students and teachers at Myeka High School built a biogas plant as an alternative energy source. The students were amused by the idea of cooking and powering their computers with gas generated from cattle manure and human waste, but they have been pleased to find that it works very well.

2. Many African governments are focusing on regional economic integration along the lines of the European Union. Local-scale and locally designed grassroots development is also being pursued to foster very basic innovation.

3. Africa's informal economies have provided some relief from the hardships created by SAPs. These economies in Africa are ancient and wide-ranging, providing employment and useful services and products. The people in these economies do not pay taxes and therefore do not provide government revenues.

Democratization and Conflict

After years of rule by corrupt elites and the military, signs of a shift toward democracy and free elections are now visible across Africa. Yet progress is blocked by violent conflicts on the continent, and even when democratic reforms are enacted and free elections established, physical conflict often accompanies these elections.

Ethnic Rivalry Africa's democratic and economic progress has been held back by frequent civil wars that are in many ways the legacy of colonial era policies of **divide and rule**. Divisions and conflicts between ethnic or religious groups were deliberately intensified by European colonial powers. To make it hard for Africans to unite and overthrow their foreign rulers, the borders and administrative units of the African colonies were designed so that different and sometimes hostile groups would be put together under the same jurisdiction (**Figure 7.20** on page 320).

Colonial officials cast themselves as impartial governors with no ethnic loyalties, capable of benevolent intervention to settle disputes. After independence, however, rule by Africans was more difficult.

divide and rule the deliberate intensification of divisions and conflicts by European colonial powers

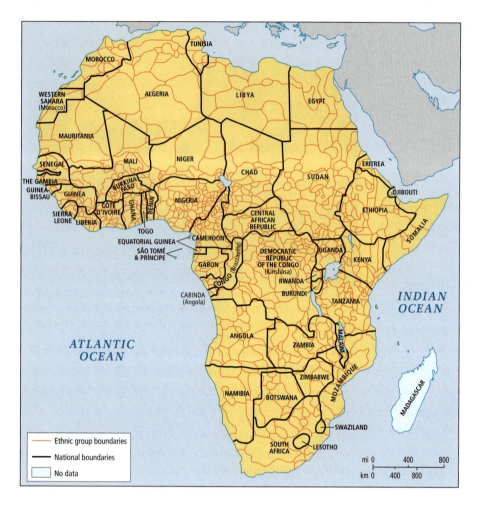

FIGURE 7.20 Ethnic groups in sub-Saharan Africa. This map indicates in simplified form the large number of ethnic groups spread across the continent of Africa. Nigeria, alone, has nearly 400 languages. Superimposed on this are the present national boundaries, which were, for the most part, imposed by European colonizers. Very rarely do ethnic group and national boundaries match. [Adapted from James M. Rubenstein, *An Introduction to Human Geography* (Upper Saddle River, N.J.: Prentice Hall, 1999), p. 246.]

Because African officials inevitably belonged to one local ethnic group or another, they could not be seen as impartial. Older traditions for ensuring ethical behavior, for resolving ethnic conflict, and for punishing greed on the part of leaders had been erased during the colonial era. The result has been years of carnage as ethnic and other hostilities, in the absence of democratic systems of conflict resolution, developed into civil wars.

Case Study: Conflict in Nigeria

Long-standing troubles in Nigeria illustrate the roots of conflict. Nigeria was and remains a creation of British divide-and-rule imperialism. Many disparate groups—speaking 395 indigenous languages—have been joined into one unusually diverse country (**Figure 7.21**).

In present-day Nigeria, there are four main ethnic groups. The Hausa (21 percent of the population) and Fulani (9 percent) are both predominantly Muslim. Until recently, they lived mostly as herders in the northern grasslands and semidesert. Hausa and Fulani elites collaborated extensively with the British during the colonial period, and since independence (1960), they have dominated the high levels of Nigeria's government and military. The Yoruba (20 percent), who practice animism (see *animist*

definition, page 333) and Christianity, live in southwestern Nigeria. At one time, nearly all Yoruba were settled farmers, but many now live in urban areas as laborers, tradespeople, and professionals. The Igbo (17 percent) are primarily farmers, centered in the southeast. Originally animists, many Igbo are now also Christian. Hundreds of other ethnic groups live among these larger groups.

The colonial administration reinforced a north-south dichotomy that mirrored the physical north (dry)-south (wet) patterns. Among the Hausa and Fulani in the north, the British ruled via local Muslim leaders who did not encourage public education. In the south, among the Yoruba and Igbo, Christian missionary schools open to the public were common. At independence, the south had more than ten times as many primary and secondary school students as the north. The south was more prosperous, and southerners also held most government civil service positions. Yet the northern Hausa dominated the top political posts. Over the years, bitter and often violent disputes have erupted between the southern Yoruba–Igbo and northern Hausa–Fulani regarding the distribution of economic development funds, jobs, and oil revenues, as well as over severe environmental damage from oil extraction and access to increasingly scarce clean water.

178. New Nigerian President Inherits Turbulent Niger Delta

FIGURE 7.21 Conflict in Nigeria. Merchants in the grain market of Laranto, Nigeria, after it was burned during interethnic riots in November 2008. Since 1999, twelve thousand Nigerians have died and many more have been made homeless as a result of interethnic violence.

The politics of oil have complicated the troubles in Nigeria. Nigeria is a major oil producer and exporter, and much of Nigeria's oil is located on lands occupied by the Ogoni people, which lie in the south along the edges of the Niger River delta. However, virtually none of the profits from oil production, and very little of the oil itself, go to the states in which this land lies, much less to the Ogoni's homeland, Ogoniland. While receiving few benefits from oil extraction, Ogoniland has suffered gravely from the resulting pollution (see the photo in Figure 1.6 on page 8). Oil pipelines crisscross Ogoniland, and spills and blowouts are frequent; between 1985 and 2009 there were hundreds of spills, many larger than that of the *Exxon Valdez* disaster in Alaska. Natural gas, a by-product of oil drilling, is burned off, even though it could be used to generate electricity—something many Ogoni lack. By its own admission, Royal Dutch Shell netted $200 million in profits yearly from Nigeria, but in 40 years it paid only $2 million total to the Ogoni community whose oil it had appropriated.

Geographic strategies have often been used to reduce tensions in Nigeria. One approach has been to create more political states (Nigeria now has 30) and thereby reallocate power to smaller local units with fewer ethnic and religious divisions. Recently, large, wealthy states have been subdivided, reputedly to spread oil profits more evenly, but the subdivision has had the secondary effect of muting the voice and power of protesters. ■

The Role of Cold War Geopolitics Cold War geopolitics between the United States and the former Soviet Union deepened and prolonged African conflicts that grew out of divide-and-rule policies. After independence, some sub-Saharan African governments turned to socialist models of economic development, often receiving economic and military aid from the Soviet Union. Other governments became allies of the United States, receiving equally generous aid (see Figure 5.12, page 223). Both the United States and the USSR tried to undermine each other's African allies by arming and financing rebel groups.

In the 1970s and 1980s, southern sub-Saharan Africa became a major area of tension. The United States aided South Africa's apartheid government in military interventions against Soviet-allied governments in Namibia, Angola, and Mozambique. Another area of Cold War tension was the Horn of Africa, where Ethiopia and Somalia fought intermittently throughout the 1960s, 1970s, and 1980s. At different times, the Soviets and Americans funded one side or the other.

Conflict and the Problem of Refugees Conflicts create refugees, and refugees are commonplace across sub-Saharan Africa. With only 11 percent of the world's population, this region contains about 19 percent of the world's refugees (**Figure 7.22**), and if people displaced within their home countries are also counted, this region has about 28 percent of the world's refugee population. Women and children constitute three-fourths of Africa's refugees because adult men who would be refugees are either combatants, jailed, or dead.

Throughout the last decade of the twentieth century, refugees from Somalia, Ethiopia, Uganda, Liberia, Sierra Leone, Congo (Kinshasa), Congo (Brazzaville), Rwanda, and Mozambique poured back and forth across borders and were displaced within their own countries. Often they were trying to escape **genocide**, efforts to murder an

genocide the deliberate destruction of an ethnic, racial, or political group

FIGURE 7.22 Refugees in Africa. These Bantu people from Somalia are waiting in Kenya to participate in a resettlement program that will bring them to the United States.

entire ethnic group. Such campaigns of ethnic cleansing and/ or genocide are often instigated by a corrupt government with its eye on resources, which then goads one ethnic or political faction into conflict against another and in the resulting fracas, seizes the resources. (For a discussion of ethnic cleansing and genocide, see Karyn Becker's article on the subject, at http://www.munfw.org/ archive/50th/4th1.htm).

As difficult as life is for these refugees, the burden on the areas that host them is also severe. Even with help from international agencies, the host areas find their own development plans complicated by the arrival of so many distressed people, who must be fed, sheltered, and given health care. Large portions of economic aid to Africa have been diverted to deal with the emergency needs of refugees.

Successes and Failures of Democratization Democratization in sub-Saharan Africa has produced mixed results. The number of elections held in the region has increased dramatically. In 1970, only 11 states had held elections since independence. By 2006, twenty-five out of 44 sub-Saharan African states had held open, multiparty, secret-ballot elections with universal suffrage. Although this trend toward democracy could lead to governments that are more responsive to the needs of their people, the implementation of democratic reforms has progressed unevenly.

Since 2006, according to the *Economist* magazine Democratization Index, the list of sub-Saharan democracies has shrunk by three: Kenya, Nigeria, and the Central African Republic. In Kenya, which had had several peaceful election cycles, an election in 2008 was so flawed by corruption that deadly riots broke out (**Photo Essay 7.4A**). More than 1000 people died and 600,000 were displaced by mobs of enraged voters. By 2010, a new Kenyan constitution gave some hope that democracy would return. In 2008 in Nigeria, local elections sparked deadly violence; and because of conflict across the country, by 2009 Nigeria was no longer officially considered an electoral democracy. In 2010, acting president Goodluck Jonathan asserted that Nigeria, like all sub-Saharan countries, must become a true democracy. The Central African Republic was demoted because standards to qualify as a democracy are now higher (**Photo Essay 7.4 Map**). The Congo (Kinshasa) has never managed to hold fair elections. In 2006, the first elections held there in 46 years resulted in violence that left more than 1 million Congolese refugees within their own country (see **Photo Essay 7.4C**).

Zimbabwe may represent the worst case of the failure of democratization. In the 1960s and 1970s, Robert Mugabe became a hero to many for his successful Maoist-type guerilla campaign in what was then called Rhodesia (now Zimbabwe). The highly discriminatory white minority government was allied with apartheid South Africa but was not formally recognized by any other country because of its extreme racist policies. Mugabe was elected president in 1980, following relatively free, multiparty elections. Over the years, however, his authoritarian policies impoverished and alienated more and more Zimbabweans. In the 1990s, he implemented a highly controversial land-redistribution program that resulted in his supporters gaining control of the country's best farmland, much of which had been in the hands of white Zimbabweans. This move decimated agricultural production and

contributed to a chronic food shortage and a massive economic crisis that left 80 percent of Zimbabweans unemployed (see **Photo Essay 7.4D**). The resulting political violence has created 3 to 4 million refugees, most of whom have fled to neighboring South Africa and Botswana. Mugabe held on to his office through the rigged elections of 2002 and 2008, but was then forced to share power in 2009 with Morgan Tsvangirai, who had won a plurality in 2008. The country is now in such dire straits that it is endangering the stability of its neighbors. Human rights and opposition activists are being beaten and arrested, civil servants are on strike, and food security is nonexistent.

📹 159. Zimbabwe's Robert Mugabe—A Profile

In some places, however (for example, Rwanda and South Africa), elections have helped end civil wars, as the possibility of becoming respected elected leaders has induced former combatants to lay down their arms. In Sierra Leone and Liberia, public outrage against corrupt ruling elites has resulted in elections that brought a fortuitous change of leadership.

One of the bright spots in the democratization process has been the increase in the number of women across Africa who are assuming positions of power (**Photo Essay 7.4B**). This chapter opened with a description of the democratic and environmental reforms that President Ellen Johnson-Sirleaf of Liberia has been making. Rwanda, devastated by genocide and mass rapes in the 1990s, is now the first country in the world where women constitute a majority of the national legislature. And Rwandan women are also leaders at the local level, where they make up 40 percent of the mayors. In Mozambique, 34.8 percent of the parliament is female, and in South Africa, 32.8 percent. All together, there are 11 African countries where the percentage of women in national legislatures is above the world average of 18.2 percent (the U.S. figure is 17 percent).

📹 170. Women Have Strong Voice in Rwandan Parliament

In many cases, these statistics reflect government policies as much as the willingness of Africans to be led by women. All of the countries cited above have lately enacted quotas that guarantee a certain percentage of legislative seats for women, and fewer women are elected in the countries that do not have quotas. The quotas are often a response to national crises, especially civil wars, in which women fought alongside men in battle or suffered disproportionately from the chaos and destruction of war. Quotas are usually a reflection of a larger post-conflict commitment toward the empowerment of women, sometimes written into a country's constitution, that encourages female participation in politics and civil society. It is important to note, however, that many female leaders in Africa, such as Ellen Johnson-Sirleaf and at least half of Rwanda's female parliamentarians, were elected without quotas.

🎏   THINGS TO REMEMBER   🎏

1. Africa's democratic and economic progress has been held back by frequent civil wars that are in many ways the legacy of colonial era policies of divide and rule.

2. While a shift toward democracy is occurring throughout Africa, powerful forces still work against democracy in many countries.

Democratization is helping to reduce the potential for civil war throughout this region, and the most democratized countries have had noticeably fewer violent conflicts since 1945. However, in many countries elections are still plagued by violence.

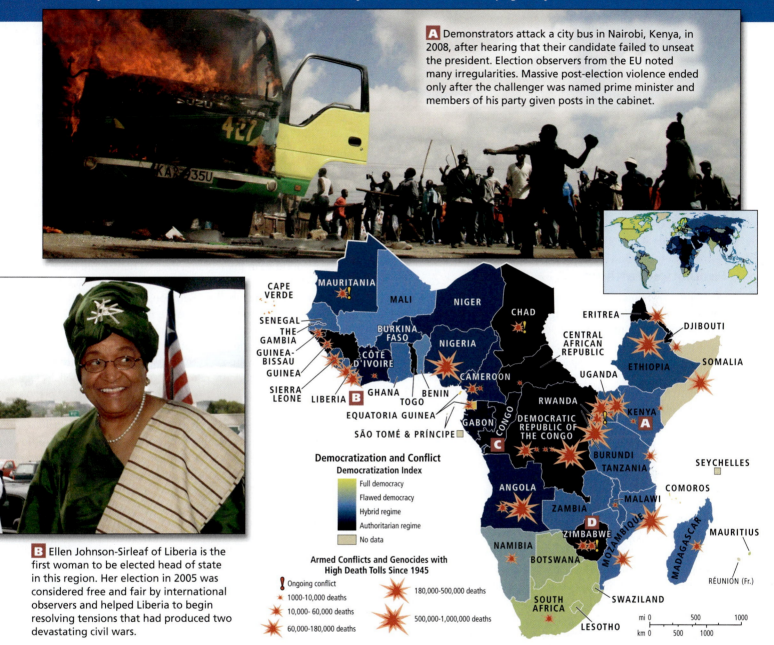

A Demonstrators attack a city bus in Nairobi, Kenya, in 2008, after hearing that their candidate failed to unseat the president. Election observers from the EU noted many irregularities. Massive post-election violence ended only after the challenger was named prime minister and members of his party given posts in the cabinet.

Democratization and Conflict

Democratization Index

- Full democracy
- Flawed democracy
- Hybrid regime
- Authoritarian regime
- No data

Armed Conflicts and Genocides with High Death Tolls Since 1945

- ! Ongoing conflict
- 1000–10,000 deaths
- 10,000–60,000 deaths
- 60,000–180,000 deaths
- 180,000–500,000 deaths
- 500,000–1,000,000 deaths

B Ellen Johnson-Sirleaf of Liberia is the first woman to be elected head of state in this region. Her election in 2005 was considered free and fair by international observers and helped Liberia to begin resolving tensions that had produced two devastating civil wars.

C Supporters of Joseph Kabila in Kinshasa, Congo, during the 2006 elections. Widespread irregularities combined with deep mistrust of the government to create post-election violence that left over 1 million homeless.

D Shelves go empty in Harare, Zimbabwe, during an economic crisis created by years of authoritarian, anti-democratic rule. Fraudulent elections in 2008 discouraged foreign-aid donors from helping, as there was no legitimate or trusted government to deal with. Widespread hunger and disease resulted.

3. Democratization is helping reduce the potential for civil conflict, and the most democratized countries have had noticeably fewer violent conflicts since 1945. However, in many countries elections are still plagued by violence.

4. The trend toward democracy in sub-Saharan Africa could lead to governments that are more responsive to the needs of their people, but since democratic reforms are not necessarily durable, a country can lose its standing as a democracy.

Sociocultural Issues

To the casual observer, it may appear that a majority of sub-Saharan Africans live traditional lives in rural villages. It is true that at just 35 percent urban, Africa is the world's most rural region. A closer look, however, reveals that migration and urbanization are occurring so rapidly that the impact on African life is monumental.

Population Patterns

Population dynamics in sub-Saharan Africa are particularly complex and confusing, and surprisingly different from general public perception. For example, when compared to the whole world (see Photo Essay 1.1 on pages 16–17), the region is not yet a particularly densely populated space. Also, while population growth rates are the highest in the world, the rates have slowed rather drastically over the last decades as a result of urbanization. They remain highest in rural areas, but are declining there too. On the other hand, in some of the most developed countries, life expectancy figures have deteriorated due to HIV-AIDS. Also, despite successful efforts to lower infant mortality rates, they still remain the highest on earth. How can all of this apparently conflicting information be explained?

Population Growth, Density, and the Demographic Transition
The overall low population density figures in sub-Saharan Africa—34.8 people per square kilometer, compared to the global average of 49 people per square kilometer—are misleading. First, people are distributed very unevenly across the region (Photo Essay 7.5 Map on page 326). Densities are extremely high in some places and very low over less habitable areas. While some areas have low density because of aridity, others have low density because of the dangers of disease or because natural factors, like impenetrable (yet valuable) forests or inadequate soils make dense occupation unwise. With so much of sub-Saharan African territory uninhabitable for one reason or another, some rural areas can be quite crowded—too crowded for the resource base.

Despite generally declining birth rates, sub-Saharan African populations are growing faster than in any other region in the world. In fewer than 50 years, sub-Saharan Africa's population has more than tripled, growing from around 200 million in 1960 to 828 million in 2009. By 2050, the population of this region is projected to be just under 1.7 billion. Hence, many places that are relatively uncrowded now may change dramatically over the next few decades. How can this be so if population growth rates are lower now than in the past and birth rates are shrinking? The short answer is that while families are now much smaller than in the past, people are still choosing to have more children than would be necessary to maintain population numbers, and more people now survive long enough to reproduce than did in the past.

■ 167. Africa's Expected Population Bulge Threatens Future Sustainability
Birth rates are as high as they are because many Africans are not yet choosing to have just the two children necessary to replace the parents. They view children as both an economic advantage and a spiritual link between the past and the future. Childlessness is considered a tragedy. Not only do children ensure a family's genetic and spiritual survival, they are viewed as having economic value. Children still do much of the work on family farms and in family-scale industries. Moreover, in this region of high infant mortality, parents have extra children in the hope of raising a few to maturity (Photo Essay 7.5B, C on page 326). In all but a few countries, the *demographic transition*—the sharp decline in births and deaths that usually accompanies economic development (see Figure 1.12 on page 20)—has barely begun.

Five countries in the region—the two richest and three of the smallest—have gone through the demographic transition. In South Africa, Botswana, Seychelles, Réunion, and Mauritius (the last three being small island countries off of Africa's east coast), circumstances have changed sufficiently to make smaller families desirable. In all five countries, per capita incomes are five to ten times the sub-Saharan average of U.S.$2000. Advances in health care have cut the infant mortality rate to about half the regional average of 88 infant deaths per 1000 live births; because of this, parents can have only a few children and expect most to live to adulthood. The circumstances of women have also improved, as reflected in female literacy rates of around 80 to 90 percent, compared to the regional average of 54.4 percent. Research also shows that opportunities for women to work outside the home at decent-paying jobs are greater in these countries than in the rest of the region. Thus, many women are choosing to use contraception because they have life options beyond motherhood. Indeed, the percent of married women using contraception in these five countries is double or even triple the rate for sub-Saharan Africa as a whole, which is only 22 percent (about one-third of the world average).

The population pyramids (Figure 7.23) demonstrate the contrast between countries that are growing rapidly (such as Nigeria, Africa's most populous country, with 152.6 million people) and countries that are already going through the demographic transition, such as South Africa, which has a population of just 50.7 million. Nigeria's pyramid is very wide at the bottom because over half the population is under the age of 20. In just 15 years, this entire cohort will be of reproductive age. Only 12 percent of Nigerian women use any sort of birth control. But even if each manages to have only 2 children, the population will grow exponentially.

In contrast to Nigeria, South Africa's pyramid has contracted at the bottom because its birth rate has dropped from 35 per 1000 to 23 per 1000 over the last 20 years. This decrease, primarily the consequence of economic and educational improvements and social changes that have come about since the end of apartheid in the early 1990s, is likely to persist as the advantages of smaller families, especially to women, become clear. The birth rate decrease is all the more remarkable because contraception is used by only 60 percent of South African women. Part of the birth rate

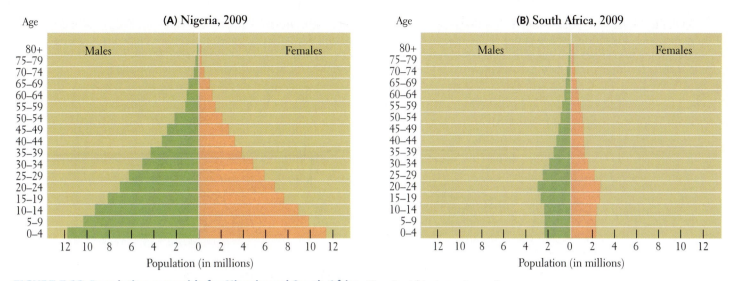

FIGURE 7.23 Population pyramids for Nigeria and South Africa. Nigeria, Africa's most populous country with a total population of 152.6 million, has a population growth rate of 2.6 percent, and has nearly twice the population of Ethiopia, Africa's second most populous country. South Africa, with a population of 50.7 million, has a growth rate of 0.9 percent and has the fifth largest number of people on the continent. [Adapted from U.S. Census Bureau, Population Division, "Population Pyramids of Nigeria" and "Population Pyramids of South Africa," International Data Base, 2009, at http://www.census.gov/ipc/www/idb/pyramids.html, and Population Reference Bureau, *2009 World Population Data Sheet.*]

decrease in South Africa is probably a consequence of the spread of HIV among young adults there (see Figure 7.24B on page 327). HIV-AIDS is now a major cause of slowing population growth rates and lowering life expectancies in Africa. Nigeria and South Africa represent the two extremes of growth rates in Africa.

Urbanization, Migration, and Population Growth In Africa, as in all parts of the world, the definition of "urban" varies from country to country. The great majority of sub-Saharan African urbanites live in towns and villages of several thousand people, not in large cities. Sub-Saharan Africa is experiencing a massive shift in population from tiny rural settlements of a few houses to substandard residential facilities in urban areas, and there are major implications for population growth. On average, rural sub-Saharan African women give birth to about 6.6 children, while urban African women give birth to 4.7. Town life strengthens all the factors that influence the demographic transition—increased economic development, better health care, and more educational opportunities. While sub-Saharan Africa's urban fertility rate of 4.7 children per woman is still almost double the world average and the demographic transition has only just begun, sub-Saharan Africa's birth rates will continue to decline.

Part of the reason that urban fertility rates remain as high as they are is that compared to cities in other regions, sub-Saharan Africa's cities have delivered fewer improvements in living standards, especially access to clean water and sanitation (Photo Essay 7.6 A, C on page 328). With poverty and disease still widespread in sub-Saharan Africa's cities, and for a variety of other reasons—lack of access to birth control, religious beliefs, low levels of education for women—urban families continue to have relatively large families.

Migration, not the birth rate, is the major factor behind sub-Saharan cities' growth rate of 5 percent per year, the highest rate in the world. In the 1960s, only 15 percent of sub-Saharan Africans lived in cities; now about 45 percent do (see Photo Essay 7.6). By 2030, the urban population of this region will have doubled to about 530 million and will account for about half of all Africans. In 1960, just 1 sub-Saharan African city—Johannesburg, South Africa—had more than 1 million people; in 2009, 52 do. The largest sub-Saharan African city is Lagos, Nigeria, where various estimates put the population at between 11 and 13 million; by 2020, Lagos is projected to have 20 million people (Photo Essay 7.6 Map). Much of this growth is taking place in *primate cities* (see Chapter 3, page 152). For example, Kampala, Uganda, with 1.8 million people, is almost ten times the size of Uganda's next largest city, Gulu.

Because governments and private investors have paid little attention to the need for affordable housing, most migrants have to construct their own dwellings using found materials (see Photo Essay 7.6). The vast unplanned one-story slums that result surround the older urban centers and house 72 percent of Africa's urban population. Transportation in these huge and shapeless settlements is a jumble of government buses and private vehicles. People often have to travel long hours through extremely congested traffic to reach distant jobs, getting most of their sleep while sitting on a crowded bus.

Public health is also a major concern, as many water distribution systems are contaminated with harmful bacteria from untreated sewage (see Photo Essay 7.6C). Only the largest sub-Saharan African cities have sewage treatment plants, and few of these extend to the slums that surround them. The result is frequent outbreaks of waterborne diseases such as cholera, dysentery, and typhoid.

The migration of sub-Saharan people looking for work in North Africa, Europe, Turkey, or the United States is now a familiar phenomenon, but the vast majority of African migrants

Of all the world regions, populations are growing fastest in sub-Saharan Africa. The demographic transition is taking hold in a few more prosperous countries where better health care and more economic and educational opportunities for women are allowing women to pursue careers and put off childbearing, resulting in smaller family size. [Data courtesy of Deborah Balk, Gregory Yetman, et al., Center for International Earth Science Information Network, Columbia University, at http://www.ciesin.columbia.edu.]

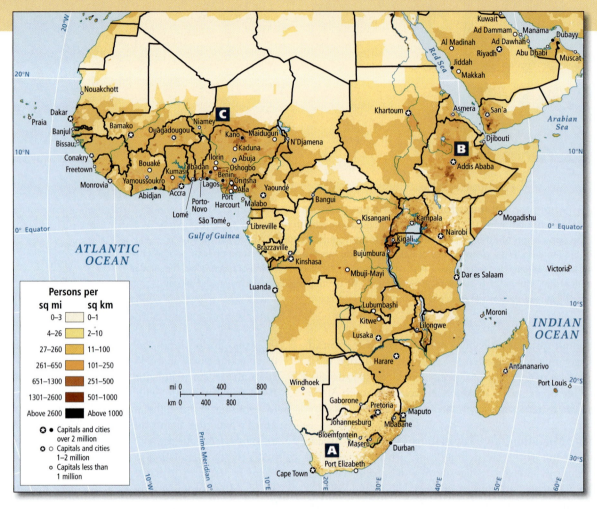

Persons per

sq mi	sq km
0–3	0–1
4–26	2–10
27–260	11–100
261–650	101–250
651–1300	251–500
1301–2600	501–1000
Above 2600	Above 1000

✪ ● Capitals and cities over 2 million
✪ ○ Capitals and cities 1–2 million
○ Capitals less than 1 million

A HIV-AIDS activist Zinny Thabetha of South Africa introduces a new project to international donors. Many development projects open up new opportunities for women.

B A woman and her several children await medical treatment in rural Ethiopia. Rural areas tend to have the highest population growth.

C A child at an emergency feeding center in Niger, where poverty and hunger are so widespread that families often have many children to ensure that some survive into adulthood.

seeking work or escaping violence move within the continent. In 2005, the United Nations estimated that there were 17 million such internal migrants. Most migrants go to West Africa and Southern Africa because jobs are more numerous. The five sub-Saharan African countries with the largest number of immigrants in 2005 were Côte d'Ivoire (2.4 million), Ghana (1.7 million), South Africa (1.1 million), Nigeria (1.0 million), and Tanzania (0.8 million). Together they have 40 percent of the migrants in Africa. Gabon, Gambia, and Côte d'Ivoire have the largest number of immigrants relative to their own populations (Gabon, 17.7 percent; Gambia, 15.3 percent; and Côte d'Ivoire, 13.1 percent). Even though the migrants typically make few demands, the impact on receiving countries that are unable to provide adequately for their own citizens is substantial. Like migrants who leave the continent, internal migrants live frugally and send much of their earnings home to families.

Population and Public Health Infectious diseases, including HIV-AIDS, are by far the largest killers in sub-Saharan Africa, responsible for about 50 percent of all deaths (**Figure 7.24**). Some diseases are linked to particular ecological zones. For example, people living between the 15th parallels north and south of the equator are most likely to be exposed to sleeping sickness (*trypanosomiasis*), which is spread among people and cattle by the bites of tsetse flies. The disease attacks the central nervous system and, if untreated, results in death. Several hundred thousand Africans suffer from sleeping sickness, and most of them are not treated because they cannot afford the expensive drug therapy.

Africa's most common chronic tropical diseases, *schistosomiasis* and *malaria*, are linked to standing fresh water. Thus, their incidence has increased with the construction of dams and irrigation projects. Schistosomiasis is a debilitating, though rarely fatal, disease that affects about 170 million sub-Saharan Africans. It develops when a parasite carried by a particular freshwater snail

enters the skin of a person standing in water. Malaria, spread by the anopheles mosquito (which lays its eggs in standing water), is more deadly. The disease kills at least 1 million sub-Saharan Africans annually, most of them children under the age of 5. Malaria also infects millions of adult Africans who are left feverish, lethargic, and unable to work efficiently because of the disease.

Until recently, relatively little funding was devoted to controlling the most common chronic tropical diseases. Now however, major international donors are funding research in Africa and elsewhere. More than 60 research groups in Africa are working on a vaccine that will prevent malaria in most people. The distribution of simple, low-cost, and effective mosquito nets is also reducing the transmission rates of malaria.

HIV-AIDS in Sub-Saharan Africa The leading cause of death in Africa is acquired immunodeficiency syndrome (AIDS), caused by the human immunodeficiency virus (HIV) (**Figure 7.25** on page 329). As of 2008, more than 15 million sub-Saharan Africans had died of the disease, and as many as 80 million more AIDS-related deaths are expected by 2025. In 2008, sub-Saharan Africa had two-thirds of the estimated worldwide total of 33 million people living with HIV. In Botswana, one of the richest countries in southern Africa, nearly 25 percent of the adult population is infected. The southern Africa subregion alone accounted for 38 percent of global HIV-AIDS deaths in 2007, and there the epidemic has significantly constrained economic development.

Worldwide, women account for half of all people living with HIV, but in Africa, HIV-AIDS affects women more than men: 59 percent of the region's HIV-infected adults are women. The reasons for this pattern are related to the social status of women in sub-Saharan Africa and elsewhere. Women are often infected by their mates, who may visit sex workers when they travel for work or business. Women have little power to insist that their partners use condoms.

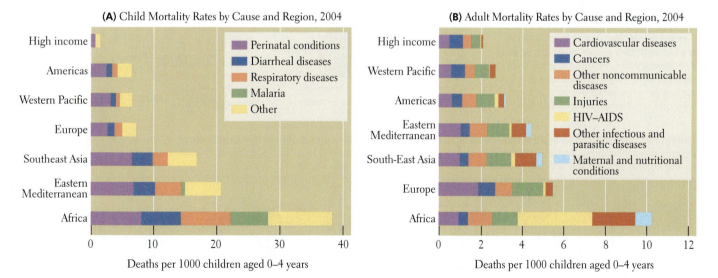

FIGURE 7.24 Mortality rates by major cause and region 2004. (**A**) Deaths per 1000 children aged 0–4. (**B**) Deaths per 1000 adults aged 15–59. [Adapted from The World Health Organization, *The Global Burden of Disease 2004 Update* (Geneva: World Health Organization Press, 2008), available at http://www.who.int/healthinfo/global_burden_disease/2004_report_update/en/index.html.]

Urban populations in sub-Saharan Africa are exploding due to migration from rural areas and high birth rates in cities. By 2030 half of the region will be living in cities, most in slums plagued by violence and inadequte access to water, sanitation, and education. Largely ignored by most governments, slum dwellers survive by helping themselves.

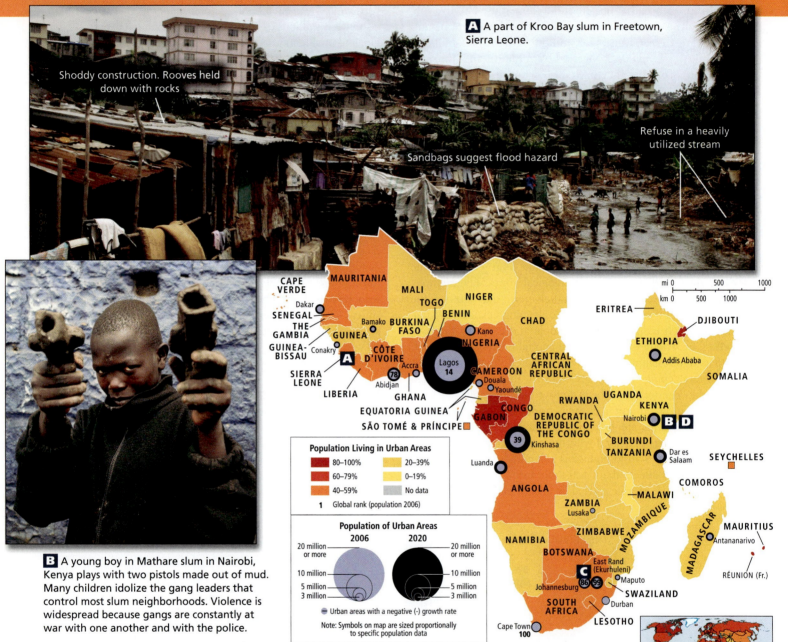

A A part of Kroo Bay slum in Freetown, Sierra Leone.

Shoddy construction. Rooves held down with rocks

Sandbags suggest flood hazard

Refuse in a heavily utilized stream

B A young boy in Mathare slum in Nairobi, Kenya plays with two pistols made out of mud. Many children idolize the gang leaders that control most slum neighborhoods. Violence is widespread because gangs are constantly at war with one another and with the police.

Population Living in Urban Areas

■ 80–100%	■ 20–39%
■ 60–79%	■ 0–19%
■ 40–59%	▨ No data

1 Global rank (population 2006)

Population of Urban Areas

2006 — 2020

20 million or more / 10 million / 5 million / 3 million

⊖ Urban areas with a negative (-) growth rate

Note: Symbols on map are sized proportionally to specific population data

C Children fetching water in Soweto, South Africa. Slum houses rarely have indoor plumbing. Water is obtained from a public spigot, and latrines provide the only sanitation.

D An organic garden in Kibera, one of many "self-help" projects found throughout Nairobi's slums.

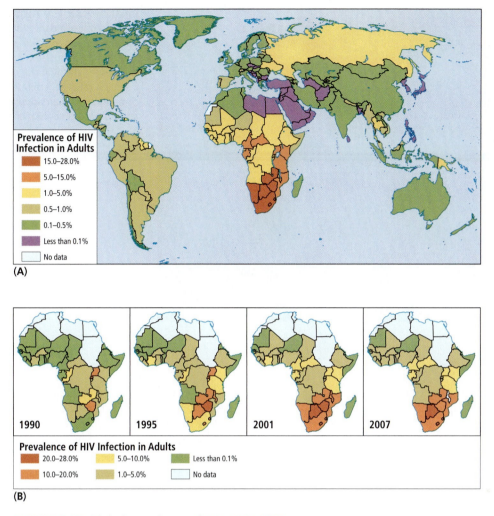

FIGURE 7.25 Global prevalence of HIV-AIDS, 2007. [Adapted from UNAIDS, *2008 Report on the Global AIDS Epidemic,* at http://www.searo.who.int/LinkFiles/Facts_and_Figures_global_HIV_epidemiology.pdf.]

The rapid urbanization of Africa has hastened the spread of HIV-AIDS, which is much more prevalent in urban areas. Many poor new urban migrants, removed from their village support systems, become involved in the sex industry. For some poor women, occasional sex work is part of what they do to survive economically. In some cities, virtually all sex workers are infected. Meanwhile, many urban men, especially those with families back in rural villages, visit prostitutes on a regular basis. These men often bring HIV-AIDS back to their rural homes. As transportation between cities and the countryside has improved, bus and truck drivers have also become major carriers of the disease to rural villages.

A number of myths and social taboos surround HIV-AIDS. This has exacerbated the problem of controlling HIV-AIDS and means that education is a key component to combating the epidemic. Many men think that only sex with a mature woman can result in infections, so very young girls are increasingly sought as sex partners. (This is sometimes referred to as "the virgin 'cure.'") Elsewhere, infection is considered such a disgrace that even those who are severely ill refuse to get tested.

Massive education programs are credited with stabilizing the HIV-AIDS epidemic by lowering rates of infection among those who can read and understand explanations of how HIV is spread (**Figure 7.26**). For example, Senegal started HIV-AIDS education in the 1980s, and levels of infection there have so far remained low. Major education campaigns in Uganda lowered the incidence of new HIV infections from 15 percent to just 4.1 percent between 1990 and 2004. By contrast, in areas such as South Africa, where a few top politicians have put forth untenable theories about the causes of HIV infection or have denied that HIV-AIDS was a problem, infection rates have soared.

Across the continent, the consequences of the HIV-AIDS epidemic are enormous. Millions of parents, teachers, skilled farmers, craftspeople, and trained professionals have been lost. More than 12 million children have been orphaned, many without any family left to care for them or to pass on vital knowledge and skills. The disease has severely strained the health-care systems of most countries. Demand for treatment is exploding, drugs are prohibitively expensive, and many health-care workers themselves

FIGURE 7.26 HIV-AIDS in the military.
A group of Rwandan soldiers perform a song they have composed for their anti–HIV-AIDS meeting. The UN has identified this region's armed forces as an especially high risk group for HIV-AIDS, with infection rates generally 2 to 5 times higher than in comparable civilian populations during peacetime, and up to 50 times higher during times of conflict.

are infected. Because so many young people are dying of AIDS, decades of progress in improving the life expectancy of Africans have been erased. For example, in 1990 adult life expectancy in South Africa was 63, in 2009 it was 52.

THINGS TO REMEMBER

1. The overall low population density figures in sub-Saharan Africa—34.8 people per square kilometer, compared to the global average of 49 people per square kilometer—are misleading. Densities are extremely high in some places and very low in less habitable areas.

2. Of all the world regions, populations are growing fastest in sub-Saharan Africa. The demographic transition is taking hold in a few more prosperous countries where better health care and child survival, along with more economic and educational opportunities for women, are encouraging women to pursue careers and put off childbearing, resulting in smaller family sizes.

3. Urban populations are exploding due to migration from rural areas and high birth rates in cities. By 2030, half of the region will be living in cities, most in slums with inadequate access to water, sanitation, and education. Largely ignored by most governments, slum dwellers survive by providing for themselves.

4. Sub-Saharan Africa suffers more than any other world region from infectious diseases, with the world's worst epidemics of malaria, schistosomiasis, and HIV-AIDS.

Geographic Patterns of Human Well-Being

As we have observed in previous chapters, GDP per capita PPP can be used as a general measure of well-being because it shows in a broad way how much income people in a country have to spend on necessities. However, it is only an average figure and does not indicate how income is distributed. As a result, there is no way of knowing if a few people are rich and most are poor, or if income is more equally apportioned. The PPP designation (purchasing power parity) means that all figures are adjusted for the cost of living in each place and so are comparable to figures for other regions.

The map in Figure 7.27A indicates three things regarding GDP in this region. First, most countries are yellow, indicating that they are in the two lowest GDP ranks. None are purple, the color that represents the highest GDP per capita rank, and only a very few are dark red. A look at the world map inset for GDP shows that of all regions on earth, sub-Saharan Africa has the lowest figures. In most of the countries of sub-Saharan Africa, per capita GDP is less than U.S.$2000 per year. Second, a band of countries with somewhat higher GDPs runs along the west side of the continent from Cameroon through South Africa. In these countries, GDP per capita ranges from U.S.$2000 to U.S.$29,999, but only tiny Equatorial Guinea is in the higher category; Gabon, Botswana, and the tiny island countries of Mauritius and Seychelles, far to the east in the Indian Ocean, are the next richest countries, averaging U.S.10,000 to U.S.$14,999 per year. Gabon and Botswana depend on extractive industries—selling timber, oil, manganese, and diamonds, and in both the income disparity is rather stark. South Africa, which is actually considered Africa's most developed country because of its industry, has a GDP that averages less than U.S.$9999. There is no data for Zimbabwe and Somalia; but as discussed elsewhere, both are dangerously troubled countries with seriously incompetent leaders.

The maps in Figure 7.27B show how countries rank on the human development index (HDI)—how well they provide the basics of income, education, and health for their people. In this

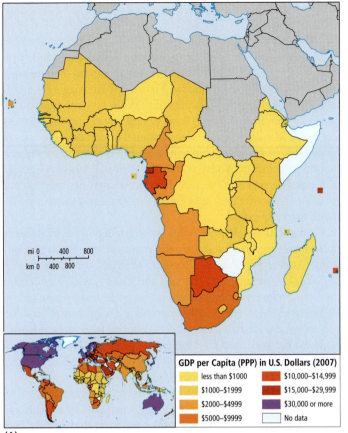

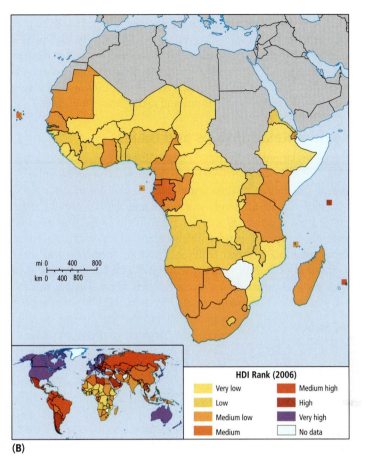

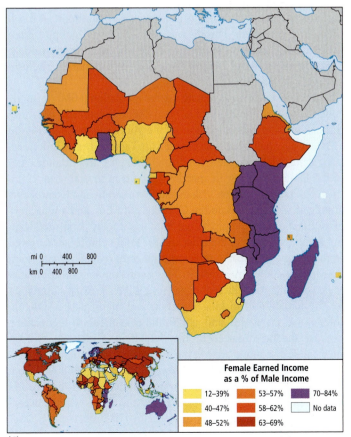

FIGURE 7.27 Maps of human well-being. (**A**) Gross domestic product (GDP). (**B**) Human development index (HDI). (**C**) Female earned income as a percent of male earned income (F/MEI). [Maps adapted from data: "Human development indices," (Maps A and B): Table 2, pages 28–32; (Map C): Table 5, pages 41–44 at http://hdr.undp.org/en/media/HDI_2008_EN_Tables.pdf.]

region, HDI ranks stretch from the very low to medium low, with only Equatorial Guinea and Gabon ranking medium. The island countries of Mauritius and Seychelles are alone in the medium-high category. As noted in Chapter 6 on page 276, if on the HDI scale a country drops below its approximate position on the GDP scale, one might suspect that the lower HDI ranking is due to the poor distribution of wealth and social services. Mali, Burkina Faso, Equatorial Guinea, Gabon, Angola, Botswana, and South Africa all rank lower on HDI than they do on GDP. It is fair to say that in these countries, a majority are simply not getting sufficient income to acquire an education and good health. A few countries rank marginally higher (medium low) on HDI than GDP: Mauritania, Ghana, Madagascar, Tanzania, and Kenya. These countries are beginning to invest in social services that enhance human well-being, but because of their relative poverty, their accomplishments are still slim.

Figure 7.27C shows how well countries are ensuring gender equality in earned income. First, as is true across the globe, nowhere in this region are women paid equally to men; but notice on the world inset map that gender disparity in pay is less extreme in sub-Saharan Africa than in North Africa and Southwest Asia.

Certainly it is important to remember that GDP per capita figures (a surrogate for average income) are generally very low for this region, but the greater equality of income does mean that women are economically active, and other evidence on gender (discussed previously and next) indicates that women in Africa are improving their social and economic status and taking leadership roles.

Gender Relationships and Issues

Gender relationships in sub-Saharan Africa are complex and variable, but in this region as in others, women are generally subordinate. Nonetheless, change is underway.

Ideas in Africa about how males and females should relate come from a variety of sources. Even in the ancient past, social controls tempered gender relationships. Most marriages were social alliances between families; therefore, husbands and wives spent most of their time doing their tasks with family members of their own sex rather than with each other. Then as today, women were influenced primarily by other women, men by other men. It is important to note that having multiple wives—the practice of **polygyny** (Figure 7.28)—is more common in sub-Saharan Africa (where it has ancient pre-Muslim, pre-Christian roots) than in Muslim North Africa (as discussed in Chapter 6 on page 269). Nevertheless, as Figure 7.28 indicates,

polygyny the taking by a man of more than one wife at a time

female genital mutilation (FGM) removing the labia and the clitoris and sometimes stitching the vulva nearly shut

today polygyny is prominent in only a few countries in western Sub-Saharan Africa.

The introduction of Islam in the 700s and Christianity (first in Ethiopia and adjacent areas just after Jesus's time and then elsewhere in sub-Saharan Africa during European colonialism; see maps on religions in Africa in Figure 7.30 on page 334) modified traditional gender relationships. With the introduction of both Christianity and Islam, men gained power and women lost freedoms. Muslim women were restricted to domestic spaces, no longer able to move about at will, trade in the markets, or engage in public activities. Today, although Christian women operate in the public sphere more than Muslim women do, they too are socialized to restrict their activities to the home and to subordinate themselves to men.

Long-standing African traditions dictate a fairly strict division of labor and responsibilities between men and women. In general, women are responsible for domestic activities, including raising the children, attending to the sick and elderly, and maintaining the house. Women collect water and firewood and prepare nearly all the food. Men are usually responsible for preparing land for cultivation. In the fields intended to produce food for family use, women sow, weed, and tend the crops as well as process them for storage. In the fields where cash crops are grown, men perform most of the work and retain control of the money earned.

When husbands in search of cash income migrate to work in the mines or in urban jobs, women take over nearly all agricultural work. They usually work with simple hand tools in the fields, and in the home, they often labor with a child or grandchild strapped on their back. When there are small agricultural surpluses or handcrafted items to trade, it is women who transport and sell them in the market. Throughout Africa, married couples often keep separate accounts and manage their earnings as individuals. When a wife sells her husband's produce at the market, she usually gives the proceeds to him.

A practice known as **female genital mutilation (FGM)** (formerly called female circumcision) has been documented in 27 countries throughout the central portion of the African continent (Figure 7.29). It also has been documented in Yemen, India, Indonesia, Iraq, Israel, Malaysia, the United Arab Emirates, and in some cases in the United States, with anecdotal reports in several other countries. The practice predates Islam and Christianity, and today occurs among all social classes and in Christian, Muslim, and animist religious traditions. In the procedure, which is usually performed without anesthesia, parts of the labia and the entire clitoris are removed from a young girl. In the most extreme cases (called *infibulation*), the vulva is stitched nearly shut. This mutilation far exceeds that of male circumcision, eliminating any possibility of sexual stimulation for the female and making urination and menstruation difficult. Intercourse is painful and childbirth is particularly devastating because the flesh scarred by the mutilation is inelastic. A 2006 medical study conducted with the help of 28,000 women in six African countries showed that women who had undergone FGM were 50 percent more likely to die during childbirth, and their babies were at similarly high risk. The practice also leaves women exceptionally susceptible to infection, especially HIV infection.

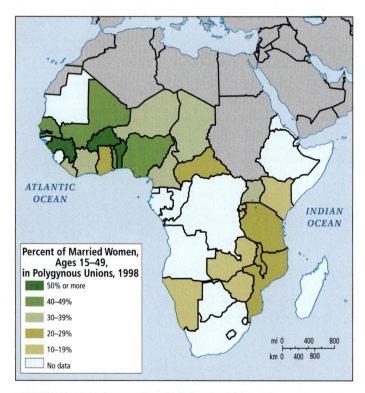

ATLANTIC OCEAN

INDIAN OCEAN

Percent of Married Women, Ages 15–49, in Polygynous Unions, 1998
- 50% or more
- 40–49%
- 30–39%
- 20–29%
- 10–19%
- No data

mi 0   400   800
km 0   400   800

FIGURE 7.28 Polygyny in sub-Saharan Africa. The practice of polygyny (men having multiple wives) is more common in sub-Saharan Africa than anywhere else in the world. Nevertheless, only a minority of Africans are in polygynous marriages today. [Adapted from United Nations, Department of Economic and Social Affairs, *The World's Women, 2000: Trends and Statistics* (New York: United Nations, 2000), pp. 27–28.]

▶ **171. Female Genital Mutilation Still Common in Somaliland**

In our traditional culture you have to make a sacrifice to powerful forces if you want to get results. It is the same here."

Generous gifts to churches are promoted as a way to bring divine intervention to alleviate miseries, whether physical or spiritual. Practitioners donate food, television sets, clothing, and money. One woman gave 3 months' salary in the hope that God would find her a new husband.

Like all religious belief systems, the gospel of success is best understood within its cultural context. Many of the believers, new to the city, feel isolated and are looking for a supportive community to replace the one they left behind. People view their material contributions to the church as similar to the labor and goods they previously donated to maintain standing in their home village. In return for these dues and volunteer services, members receive social acceptance and community assistance in times of need. ■

Ethnicity and Language *Ethnicity*, as we have seen throughout this book, refers to the shared language, cultural traditions, and political and economic institutions of a group. The map of Africa in Figure 7.31 shows a rich and complex mosaic of languages. Yet despite its complexity, this map does not adequately depict Africa's cultural diversity (compare it with Figure 7.20 on page 320).

Most ethnic groups have a core territory in which they have traditionally lived, but very rarely do groups occupy discrete and exclusive spaces. Often several groups share a space, practicing different but complementary ways of life and using different resources. For example, one ethnic group might be subsistence cultivators, another might herd animals on adjacent grasslands, and a third might be craft specialists working as weavers or metalsmiths.

People may also be very similar culturally and occupy overlapping spaces but identify themselves as being from different ethnic groups. Hutu farmers and Tutsi cattle raisers in Rwanda share similar occupations, languages, and ways of life. However, European colonial policies purposely exaggerated ethnic differences as a method of control, assigning a higher status to the Tutsis. Hutu and Tutsi now think of themselves as having very different ethnicities and abilities. In the 1990s and again in 2004, the Hutu-run government encouraged attacks on the minority Tutsi people. Hutus were also killed, but those killed tended to be people who opposed the genocide. The best estimates indicate that more than 1 million people, most of them Tutsis, died during the two genocide catastrophes.

Some African countries have only a few ethnic groups; others have hundreds. Cameroon, sometimes referred to as a microcosm of Africa because of its ethnic complexity, has 250 different ethnic groups. Different groups often have extremely different values and practices, making the development of cohesive national policies difficult. Nonetheless, the vast majority of African ethnic groups

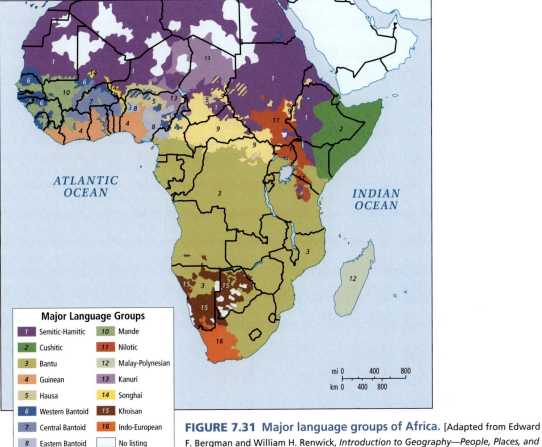

FIGURE 7.31 Major language groups of Africa. [Adapted from Edward F. Bergman and William H. Renwick, *Introduction to Geography—People, Places, and Environment* (Englewood Cliffs, N.J.: Prentice-Hall, 1999), p. 256; and Jost Gippert, *TITUS Didactica*, at http://titus.uni-frankfurt.de/didact/karten/afr/afrikam.htm.]

lingua franca a common language used to communicate by people who do not speak one another's native languages; often a language of trade

have peaceful and supportive relationships with one another.

To a large extent, language correlates with ethnicity. More than a thousand languages, falling into more than a hundred language groups, are spoken in Africa; Bantu is the largest such group (see Figure 7.31). Most Africans speak their native tongue and a **lingua franca** (language of trade). Some languages are spoken by only a few dozen people, while others, such as Hausa, are spoken by millions of people from Côte d'Ivoire to Cameroon. As linguas francas—such as Hausa, Arabic, and Swahili—take over because they better suit people's needs or have become politically dominant, some African languages die out. Former colonial languages such as English, French, and Portuguese (all classed as Indo-European languages in Figure 7.31) are also widely used in commerce, politics, education, and on the Internet.

THINGS TO REMEMBER

1. Gender relationships in sub-Saharan Africa are complex and variable, but in this region as in others, women are generally subordinate. Nonetheless, change is underway.

2. Female genital mutilation has been recognized as discrimination based on sex because it is rooted in gender inequalities and power imbalances between men and women.

3. Traditional African religions are among the most ancient on earth and are found in every part of the continent. Today, however, about one-third of sub-Saharan Africans are Muslim and about half are Christian, with many in the evangelical movement, a subset that promotes the Gospel of Success.

4. More than a thousand languages, falling into more than a hundred language groups, are spoken in Africa.

Reflections on Sub-Saharan Africa

Late one evening in a restaurant in Central Europe, after a lengthy conversation that touched on some of the world's perplexing problems, a friend leaned across the elegant white tablecloth and asked, "But don't you think that Africa is, after all, better off for having been colonized by Europeans?"

How does one politely reply to such a question? Of all regions on earth, sub-Saharan Africa, the ancient home of the human species, is the one in greatest need of attention. Sub-Saharan Africa is the poorest region of the world, yet the reasons for sub-Saharan

Africa's poverty are not immediately apparent to the casual observer and do not necessarily lie in Africa. Sub-Saharan Africa is blessed with many kinds of resources—agricultural, mineral, and forest—but most of the market value is added to these resources only after they leave Africa. Sub-Saharan Africa is not densely occupied, and birth rates are dropping, but population growth, especially in urban areas (due to inmigration), is hindering efforts to improve standards of living.

The deep reasons for sub-Saharan Africa's poverty and social and political instability emerge only through an exploration of its history over the last several centuries. Colonialism employed unimaginable brutality to methodically remove sub-Saharan Africans from control of their own societies and lives and forced sub-Saharan Africa's people and resources into the service of Europe. Even today, with colonialism officially dead for more than three decades, many sub-Saharan Africans are struggling with borders developed by Europeans and economies that are still focused on cheap and unstable raw material exports to Europe and elsewhere. In view of all this, it is hard to believe that anyone would think Africa is better off for having been colonized.

Sub-Saharan Africa is changing, however. Its people are beginning to devise new economic development strategies and political institutions to replace the ones imposed by outsiders. Sub-Saharan African leaders, an increasing number of whom are democratically elected, are articulating wider visions. In particular, the contributions of women to society, and the importance of gender issues in slowing population growth are increasingly being recognized.

Sometimes good-willed people suggest that the rest of the world should just "leave Africa to the Africans," but that view is unrealistic. Three more likely strategies are currently emerging as a consensus among sub-Saharan Africa specialists. First, SAPs should be completely abandoned and remaining debts to foreign governments and international lending agencies cancelled. Tax revenues could then once again support schools, health care, and social services. As they get relief from SAP-related debts, African leaders should take care that they do not become beholden to Asian investors now extending tempting deals in their direction. Second, the developed world should lower tariffs against sub-Saharan African manufactured products to foster the development of African industries.

A third suggestion is that future aid to sub-Saharan Africa should be designed to take advantage of indigenous skills and knowledge, and that development planning and aid money should address local needs as defined and managed by local African experts. This perspective may prove essential to efforts to reduce Africa's high vulnerability to climate change. Intimate knowledge of African environments and food production systems will be central to this region's effort to adapt to growing climatic uncertainty. Much of this knowledge lies with the rural Africans who have arguably gained the least from Africa's "development" so far. Perhaps efforts to adapt to global climate change will bring greater benefits to this group.

Critical Thinking Questions

1. If you were to investigate the origins of lumber and lumber products sold where you live, where would you start? What would make you think that you are or are not participating in the African timber trade? How is this trade related to the concept of neocolonialism?

2. Why does being an exporter of raw materials place a country at a disadvantage in the global economy? What would be some ways to amend the situation?

3. Explain the environmental links to the spread of the chronic communicable diseases of Africa. Why might some scientists argue that malaria is a worse threat to Africa's development than HIV-AIDS?

4. Why is rapid population growth still an issue in Africa? What are the factors that contribute to this growth? How might the developed countries help the situation?

5. Reflect on the many passages throughout this chapter regarding gender roles in Africa, then develop a statement of what you think are the most salient points to remember.

6. Describe the most important ways that women are exerting an influence on African political issues.

7. Technology is changing life in Africa. Explain what you think are the most crucial ways in which technology will modify the power of ordinary people to better their lives and to participate in civil society.

8. What are the various circumstances that make young heterosexual women so susceptible to HIV-AIDS in sub-Saharan Africa? How does the social and economic status of women (compared to men) influence the spread of HIV in women?

9. Looking at the various African locations where food is in short supply, explain the main factors that account for this situation. How would you respond to those who say that drought and climate are the principal causes of famine?

10. If you were asked to make a speech to the Rotary Club in your town about hopeful signs out of Africa, what would you include in your talk? Which pictures would you show?

Chapter Key Terms

agroforestry, 300
animist, 333
apartheid, 308
carbon sequestration, 300
civil society, 310
commodities, 311
currency devaluation, 314
desertification, 304

divide and rule, 319
dual economy, 312
female genital mutilation (FGM), 332
genocide, 321
grassroots economic development, 318
Horn of Africa, 298
informal economy, 315
intertropical convergence zone (ITCZ), 298

lingua franca, 336
mixed agriculture, 300
pastoralism, 304
polygyny, 332
Sahel, 298
self-reliant development, 318
shifting cultivation, 303
subsistence agriculture, 300

TURKMENISTAN

Mashhad

NORTHERN AFGHAN PROVINCES
1 JOWZJAN
2 BALKH
3 SAMANGAN
4 KONDUZ
5 BAGHLAN
6 PARVAN
7 VARDAK
8 TAKHAR
9 KAPISA
10 KABUL
11 LOWGAR
12 BADAKHSHAN
13 LAGHMAN
14 KONARHA
15 NANGARHAR

Mazār-e-Sharif

FARYAB

BADGHIS

Herat

HERAT

GHOWR

BAMIAN

Kabul

ORUZGAN

FARAH

GHAZNI

PAKTIA

NIMRUZ

ZABOL

PAKTIKA

Kandahar

KANDAHAR

HELMAND

AFGHANISTAN

Quetta

Pamirs

Hindū Kush

NORTHERN AREAS

NORTH-WEST FRONTIER

Khyber Pass

Peshawar

Islamabad

Rawalpindi

FED. ADM. TRIBAL AREAS

BALOCHISTAN

PAKISTAN

Central Makran Range

SINDH

Multan

Bahawalpur

Faisalabad

Lahore

Gujranwala

PUNJAB

AZAD KASHMIR

Srinagar

Jammu

JAMMU & KASHMIR

Ladakh

Karakoram Mts.

Mustagh (7,752)

Taklimakan Desert

CHIN

Plateau

Himal

Amritsar

HIMACHAL PRADESH

Chandigarh

Ludhiana

PUNJAB

HARYANA

Delhi

Meerut

New Delhi

DELHI

UTTARANCHAL

Ganga (Ganges)

NEPAL

Kathmandu

Ganges

Gangetic Plain

Lucknow

Ayodhya

Ghaghra

Patna

UTTAR PRADESH

Agra

Jaipur

RAJASTHAN

Jodhpur

Kanpur

Allahabad

Varanasi (Benares)

Ganga

Kochi

Zagros Mts.

IRAN

Gulf of Oman

Muscat

OMAN

Karachi

Indus River Delta

Rann of Kutch

Bhuj

Gulf of Kutch

Hyderabad

Thar Desert

INDIA

JHARKHAND

20° N

GUJARAT

Rajkot

Kathiawar Peninsula

Palitana

Ahmadabad

Vadodara

Surat

DAMAN & DIU

Gulf of Cambay

DADRA & NAGAR HAVELI

Nashik

Mumbai (Bombay)

Pune

Indore

Bhopal

MADHYA PRADESH

Jabalpur

Vindhya Range

Narmada

Satpura Range

Aurangabad

Nagpur

Jamshedpur

CHHATTISGARH

ORISSA

Arabian Sea

Bhima

MAHARASHTRA

Godavari

Hyderabad

ANDHRA PRADESH

Visakhapatnam

A Coastal Lowlands, South India

Krishna

Western Ghats

Deccan Plateau

Vijayawada

Yanam

Eastern Ghats

Panaji

GOA

Karwar

KARNATAKA

Bangalore

Mysore

Laccadive Sea

Nilgiri Hills

Mahe

Coimbatore

Chennai (Madras)

Pondicherry

Karikal

TAMIL NADU

Madurai

Palk Strait

Jaffna

Kavaratti

10° N

LAKSHADWEEP

KERALA

Cochin

Alleppey

Malabar Coast

Trivandrum

Gulf of Mannar

SRI LANKA

Colombo

Central Highlands

B Himalaya Mountains, Nepal

INDIAN

MALDIVES

60° E

70° E

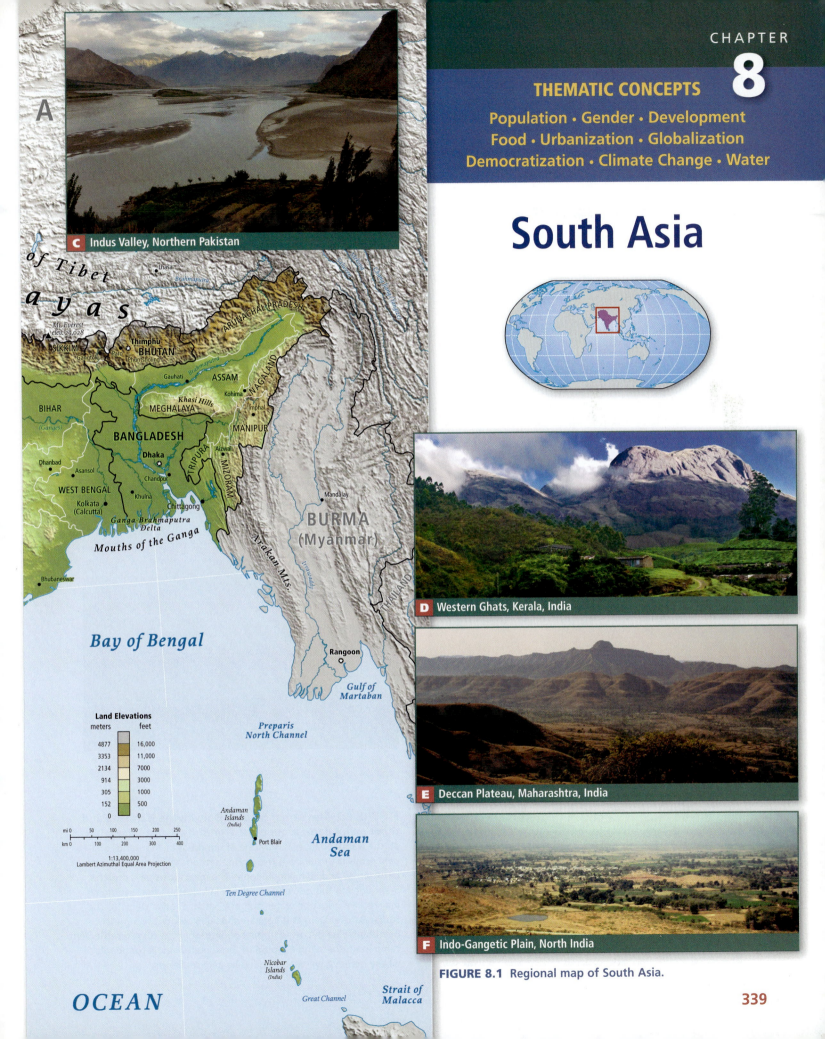

THEMATIC CONCEPTS

Population • Gender • Development
Food • Urbanization • Globalization
Democratization • Climate Change • Water

South Asia

C Indus Valley, Northern Pakistan

D Western Ghats, Kerala, India

E Deccan Plateau, Maharashtra, India

F Indo-Gangetic Plain, North India

FIGURE 8.1 Regional map of South Asia.

Land Elevations

meters	feet
4877	16,000
3353	11,000
2134	7000
914	3000
305	1000
152	500
0	0

mi 0   50   100   150   200   250
km 0   100   200   300   400

1:13,400,000
Lambert Azimuthal Equal Area Projection

Global Patterns, Local Lives On April 16, 2006, Narendra Modi, the governor of the state of Gujarat in India, draped in garlands by well-wishers, was embarking on a hunger strike. He was protesting a decision by India's national government in New Delhi to limit the height of the Sardar Sarovar Dam on the Narmada River in the neighboring state of Madhya Pradesh. Damming the river in neighboring Madhya Pradesh has become the centerpiece of Gujarat's efforts to deal with the periodic droughts that affect as many as 50 million of its citizens. Hence, Governor Modi saw limiting the height of the dam in the neighboring state as threatening his own state's many irrigation programs (**Figure 8.2B** and **Map**).

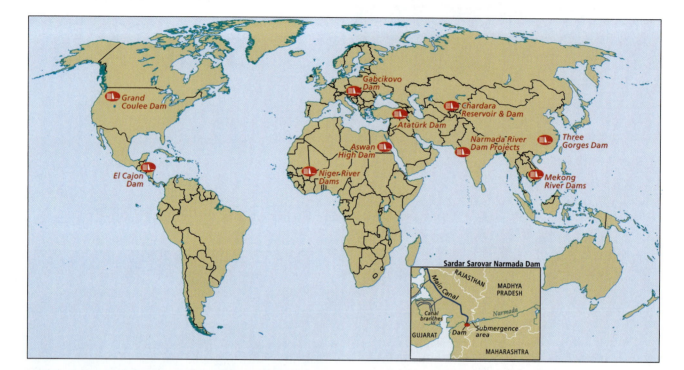

(A)

(B)

(C)

FIGURE 8.2 The Sardar Sarovar Dam on the Narmada and other major dams around the world. The Sardar Sarovar Dam on the Narmada River is only one of hundreds of thousands of dam projects throughout the world that together have displaced between 40 and 80 million people. The map shows dam projects mentioned in this book. **(A)** A village in Rajasthan that eventually may receive irrigation water from the Sardar Sarovar Dam. **(B)** The Sardar Sarovar Dam as it appeared in 2006 before the 5-meter height increase discussed in the opening vignette. **(C)** A resettlement camp in southern India typical of those built in India for people whose homes have been flooded by dam reservoirs.

Far away in New Delhi, Medha Patkar, the leader of the "Save the Narmada River" movement, was in day 18 of her hunger strike in protest against the same dam. As water rose in the dam's reservoir, 320,000 farmers and fishers were being forced to relocate. Although Indian law required that these people be given land or cash to compensate them for what they had lost, so far only a fraction had received any compensation. Some of the farmers demonstrated their objection by forfeiting their right to compensation and refusing to move, even as the rising waters of the reservoir consumed their homes. Forcibly removed by Indian police, many have since relocated to crowded urban slums.

Environmental problems have been at the heart of the Sardar Sarovar controversy since the project began in 1961. The Narmada River was once a placid, slow-moving river—and one of India's most sacred. Now the project has disturbed the river's natural cycles, causing massive die-offs of aquatic life and high unemployment among fishers.

The benefits of the dam have also been called into question. A major justification for the dam was that it would provide irrigation waters to drought-prone areas in the adjacent states of Gujarat and Rajasthan (see **Figure 8.2A** and **Map**). However, 80 percent of the areas in Gujarat most vulnerable to drought would not receive water from the project. Concerned that the economic benefits would be small and easily negated by the environmental costs, the World Bank withdrew its funding of the dam in the mid-1990s. Ecologists say that far less costly water-management strategies, such as rainwater harvesting, groundwater recharge (water that naturally replenishes the aquifer), and watershed management, would be better options for the drought-stricken farmers of Gujarat.

Less than a day after Gujarat's governor began his hunger strike, the Indian Supreme Court ruled that the Sardar Sarovar Dam could be raised higher, so, satisfied that he had won, Governor Modi ended the hunger strike. The following day, Medha Patkar ended her fast as well, because in the same decision the Supreme Court ruled that all people displaced by the dam must be adequately relocated (see **Figure 8.2C**). Furthermore, the decision confirmed that human impact studies are required for dam projects. Up to then, no such study had been done for the Sardar Sarovar Dam. By July 2009, the dam project was 5 years behind schedule and had stalled again in court. Both sides were using the Internet to promote their positions.

[Adapted from "Modi Goes on Fast over Narmada Dam," India eNews, April 16, 2006, http://www.indiaenews.com/politics/20060416/4531.htm; Rahul Kumar, "Medha Patkar Ends Fast After Court Order on Rehabilitation," One World South Asia; "Narmada's Revenge," Frontline 22(9), 2005, at http://www.hinduonnet.com/fline/fl2209/stories/20050506002913300.htm; "Water Harvesting, Addressing the Problem of Drinking Water," at http://www.narmada.org/ALTERNATIVES/water.harvesting.html; Friends of River Narmada, August/September, 2009, at http://www.narmada.org/; "Dam Delay," Indian Express.com, July 15, 2009, at http://www.indianexpress.com/news/dam-delay-sardar-sarovar-project-5-years-of/489337/.] ∎

The recent history of water management in the Narmada River valley highlights some key issues now facing South Asia and other regions that have developing economies. Across the world, large and poor populations are depending on increasingly overtaxed

FIGURE 8.3 Political map of South Asia.

environments, and this dependency is only increasing with climate change. Improving their standard of living nearly always requires more water and energy. Efforts to meet these urgent needs often make neither economic nor environmental sense, but are driven to completion by political and social pressures. In the case of the Sardar Sarovar Dam, the wealthier, more numerous, and more politically influential farmers of Gujarat have tipped the scales in favor of a project that is creating more problems than it is solving.

The countries that make up the South Asia region are Afghanistan and Pakistan in the northwest; the Himalayan states of Nepal and Bhutan; Bangladesh in the northeast; India (including the union territories of the Laccadives, Andaman, and Nicobar Islands); and the island countries of Sri Lanka and the Maldives (**Figures 8.1** and **8.3**).

The nine **Thematic Concepts** covered in each chapter of this book are illustrated here with photo essays that show examples of how each concept is experienced in this region. In all cases, interactions between two or more of these concepts are illustrated. The captions explain the pictures and how some of these interactions work in specific places.

THINGS TO REMEMBER

1. The Sardar Sarovar Dam and similar dams throughout the world are created primarily for irrigation, but they often create more problems than they solve.

2. Improving the standard of living of poor populations in many parts of the world nearly always requires more water and energy.

Food and Urbanization: Changes in South Asia's food production systems are contributing to urbanization. Farming is being made both more productive and more expensive with the introduction of new seeds, fertilizers, pesticides, and equipment. While some farmers have become wealthier, many have found themselves unable to compete and have moved to cities, where they often can find only low-paying jobs and inadequate housing.

(A) A man uses a tractor to work his rice field near Dhaka, Bangladesh.

(B) Recent rural migrants work in Dhaka as low-paid laborers pushing carts loaded with freight.

(C) A slum in Dhaka that houses many recent migrants from rural areas.

Population and Gender: A strong preference for sons has produced a gender imbalance throughout most of South Asia. Cultural norms enable sons to bring greater wealth and status to families. Hence, many wealthier families choose to abort female fetuses, while poorer families may commit "female infanticide." The result is an adult population where men significantly outnumber women.

(D) An Indian wedding. Dowries and other expenses traditionally borne by the bride's family make male children especially sought after.

(E) A sign at an Indian hospital advertising a law designed to curtail the abortion of female fetuses.

(F) Men in Mumbai, India, where the gender imbalance means that many men go unmarried or remain bachelors late into life.

Globalization and Development: Globalization has transformed economic development in this region, providing jobs for skilled workers in export-oriented industries such as software, manufacturing, call centers, and other types of "offshore outsourcing" enterprises. However, many low-skilled workers, especially in rural areas, have not benefited from these types of development.

(G) A worker at a "call center" where customers anywhere in the world can call for technical or other assistance.

(H) A swimming pool at Infosys, an information technology company based in India that employs over 100,000 people worldwide.

(I) A porter in rural Nepal carries a heavy drainpipe up a mountain road using a head strap.

Democratization and Conflict: Many conflicts in this region have been made worse by an unwillingness on the part of governments and warring parties to recognize the results of elections, or even to let people vote. Meanwhile, some conflicts have been defused, at least in the short run, by holding elections and letting former combatants run for office.

(J) An Afghan fighting against the Soviets in 1988. The conflict was sparked in part by the Soviet-supported Afghan government's antidemocratic policies.

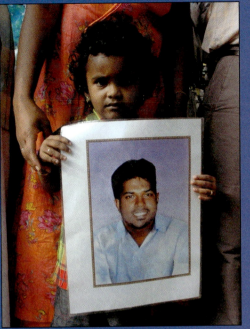

(K) A child in Sri Lanka holds up a photo of her missing uncle. The conflict intensified when ethnic Tamil plantation workers were denied the right to vote.

(L) A banner in a Nepalese town controlled by Maoists, who once waged a war against the government but have since won an election and become a relatively peaceful, if still unsettled, democratic political party.

Water and Climate Change: Glaciers in the Himalayas are melting so fast that they could eventually disappear entirely. Melting glaciers provide crucial domestic and irrigation water to hundreds of millions of South Asians, so their disappearance would be a disaster. Over the short term, flooding may intensify as rivers carry more glacial meltwater. Meanwhile, coastal areas are highly exposed to sea level rise as polar ice melts.

(M) One of the thousands of Himalayan glaciers that is melting faster than normal. The Indus, Ganga, and Brahmaputra rivers are all fed partially by meltwater from glaciers.

(N) Flooding along a river in southern Bangladesh. Increased flows in rivers as well as more severe cyclones and hurricanes have led to widespread flooding in recent years.

(O) Classes in northwestern Bangladesh are held in boats equipped with solar power and computers. The boats can provide schooling even during flooding.

343

I THE GEOGRAPHIC SETTING

Terms in This Chapter

Because its clear physical boundaries set it apart from the rest of the Asian continent, the term **subcontinent** is often used to refer to the entire Indian peninsula, including Nepal, Bhutan, India, Pakistan, and Bangladesh.

South Asians have recently adopted new place names to replace the names given them during British colonial rule. The city of Bombay, for example, is now officially *Mumbai*, Madras is *Chennai*, Calcutta is *Kolkata*, Benares is *Varanasi*, and the Ganges River is the *Ganga River*.

> **subcontinent** a term often used to refer to the entire Indian peninsula, including Nepal, Bhutan, India, Pakistan, and Bangladesh

Physical Patterns

Many of the landforms, and even the climates, of South Asia are the result of huge tectonic forces. These forces have positioned the Indian subcontinent along the southern edge of the Eurasian continent, where the warm Indian Ocean surrounds it and the massive mountains of the Himalayas shield it from cold airflows from the north (see Figure 8.1 on pages 338–339).

Landforms

The Indian subcontinent and the territory surrounding it dramatically illustrate what can happen when two tectonic plates collide. Millions of years ago, the Indian-Australian Plate, which carries India, broke free from the eastern edge of the African continent and drifted to the northeast (see Figure 1.26 on page 51). As it began to collide with the Eurasian Plate about 60 million years ago, India became a giant peninsula jutting into the Indian Ocean. As the relentless pushing from the south continued, both the leading (northern) edge of South Asia and the southern edge of Eurasia crumpled and buckled. The result is the world's highest mountains—the Himalayas, which rise more than 29,000 feet (8800 meters)—as well as other very high mountain ranges to the east and west that curve away from the central impact zone (see Figure 8.1). The continuous compression also lifted up the Plateau of Tibet, which rose up behind the Himalayas to an elevation of more than 15,000 feet (4500 meters) in some places. The compression and mountain-building process continues into the present.

South and southwest of the Himalayas are the Indus and Ganga river basins, also called the Indo-Gangetic Plain. Still farther south is the Deccan Plateau, an area of modest uplands 1000–2000 feet (300–600 meters) in elevation interspersed with river valleys. This upland region is bounded on the east and west by two moderately high mountain ranges, the Eastern and Western Ghats. These mountains descend to long but narrow coastlines interrupted by extensive river deltas and floodplains. The river valleys and coastal zones are densely occupied; the uplands only slightly less so.

Because of its high degree of tectonic activity and deep crustal fractures, South Asia is prone to devastating earthquakes, such as the magnitude 7.7 quake that shook the state of Gujarat in western India in 2001, the 7.6 quake that hit the India–Pakistan border region in 2005, and the 6.4 quake in Quetta Province, Pakistan, in 2008 that left 120,000 people homeless. Coastal areas are also vulnerable to tidal waves or *tsunamis* that are caused by undersea earthquakes. A massive tsunami originating off Sumatra in Southeast Asia wrecked much of coastal Sri Lanka and southern India in 2004, killing tens of thousands of people there.

🖳 **179. Town of Hambantota, Sri Lanka, Regains Some Normalcy After Tsunami a Year Ago**

Climate and Vegetation

The end of the dry [winter] season [April and May] is cruel in South Asia. It marks the beginning of a brief lull that is soon overtaken by the annual monsoon rains. In the lowlands of eastern India and Bangladesh, temperatures in the shade are routinely above a hundred degrees; the heat causes dirt roads to become so parched that they are soon covered in several inches of loose dirt and sand. Tornadoes wreak havoc, killing hundreds and flattening entire villages. . . .

This is also a time of hunger, as with each passing day thousands of rural families consume the last of their household stock of grain from the previous harvest and join the millions of others who must buy their food. Each new entrant into the market nudges the price of grain up a little more, pushing millions from two meals a day to one.

(ALEX COUNTS, GIVE US CREDIT (NEW YORK: TIMES BOOKS, 1996), P. 69

From mid-June to the end of October [summer] is the time of the river. Not only are the rivers full to bursting, but the rains pour down so relentlessly and the clouds are so close to village roofs that all the earth smells damp and mildewed, and green and yellow moss creeps up every wall and tree. . . . Cattle and goats become aquatic, chickens are placed in baskets on roofs, and boats are loaded with valuables and tied to houses. . . . As the floods rise villages become tiny islands . . . self-sustaining outpost[s] cut off from civilization . . . for most of three months of the year.

(JAMES NOVAK, BANGLADESH: REFLECTIONS ON THE WATER (BLOOMINGTON: INDIANA UNIVERSITY PRESS, 1993), PP. 24–25

These two passages highlight the contrasts between South Asia's dominant winter and summer wind patterns, known as **monsoons** (**Figure 8.4**). In winter, cool, dry air flows from the Eurasian continent to the ocean. In summer, warm, moisture-laden air flows from the Indian Ocean over the Indian subcontinent, bringing with it heavy rains. The abundance of this rainfall is amplified by the *intertropical convergence zone (ITCZ)*. Air masses moving south from the Northern Hemisphere converge near the equator with those moving north from the Southern Hemisphere. As the air rises and cools, copious precipitation is produced. As described in Chapter 7 (see page 298), the ITCZ shifts north and south seasonally. The intense rains of South Asia's summer monsoon

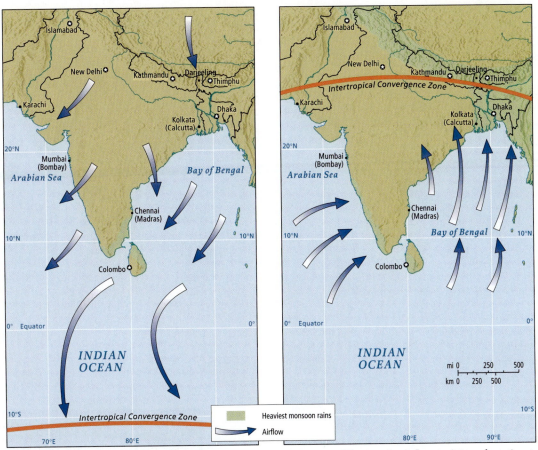

(A) Winter. Cool, dry air flows from Asian subcontinent.

(B) Summer. Warm, moist air flows to Asian subcontinent.

FIGURE 8.4 Winter and summer monsoons in South Asia. (A) In the winter, cool, dry air blows from the Eurasian continent south across India toward the ITCZ, which in winter lies far to the south. **(B)** In the summer, the ITCZ moves north across India, picking up huge amounts of moisture from the ocean, which are then deposited over India and Bangladesh.

are likely caused by the ITCZ being sucked onto the land by a vacuum created when huge volumes of air over the Eurasian landmass heat up and rise into the upper troposphere.

The monsoons are major influences on South Asia's climate. In early June, the warm, moist ITCZ air of the summer monsoon first reaches the mountainous Western Ghats. The rising air mass cools as it moves over the mountains, releasing rain that nurtures patches of dense tropical rain forests and tropical crops on the upper slopes of the Western and Eastern Ghats and in the central uplands. Once on the other side of India, the monsoon gathers additional moisture and power in its northward sweep up the Bay of Bengal, sometimes turning into tropical cyclones.

As the monsoon system reaches the hot plains of the Indian provinces of West Bengal and Bangladesh in late June, columns of warm rising air create massive, thunderous cumulonimbus clouds that drench the parched countryside. Monsoon rains run in a variable band parallel to the Himalayas that reaches across northern India all the way to northern Pakistan by July. Rainfall is especially intense in the east, north of the Bay of Bengal, where the town of Darjeeling holds the world record for annual rainfall—about 35 feet, even though no rain falls for half the year.

These patterns of rainfall are reflected in the varying climate zones (see Photo Essay 8.1 on page 346) and agricultural zones (see Figure 8.24 on page 374) of South Asia. Although sufficient rain falls in central India to support forests, most land has long been cleared of forest (see the discussion on page 348) and planted in crops. Patches of forest are now so fragmented they no longer provide suitable habitat for India's wildlife.

Periodically, the monsoon seasonal pattern is interrupted and serious drought ensues. This happened in July and August of 2009, when the worst drought in 40 years struck much of South Asia (see the discussion below). Then in September, heavy rains came to central south India, causing crop-damaging floods that killed hundreds of people. Scientists increasingly conclude that the extreme droughts and floods of recent years are not an anomaly but instead are part of the general global climate change.

By November, the cooling Eurasian landmass sends cooler, drier air over South Asia. This heavier air from the north pushes the warm, wet air back south to the Indian Ocean. Very little rain

monsoon a wind pattern in which, in summer months, warm, wet air coming from the ocean brings copious rainfall, and in winter months, cool, dry air moves from the continental interior toward the ocean

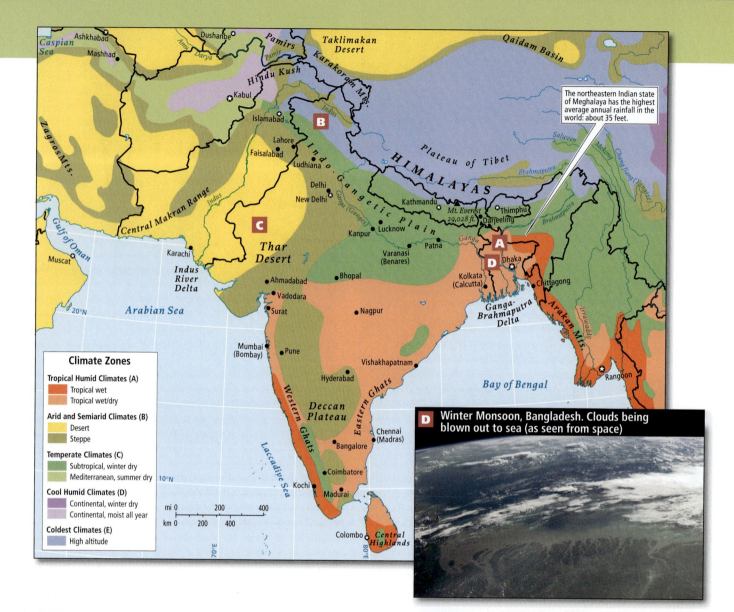

The northeastern Indian state of Meghalaya has the highest average annual rainfall in the world: about 35 feet.

Climate Zones

Tropical Humid Climates (A)
Tropical wet
Tropical wet/dry

Arid and Semiarid Climates (B)
Desert
Steppe

Temperate Climates (C)
Subtropical, winter dry
Mediterranean, summer dry

Cool Humid Climates (D)
Continental, winter dry
Continental, moist all year

Coldest Climates (E)
High altitude

mi 0   200   400
km 0   200   400

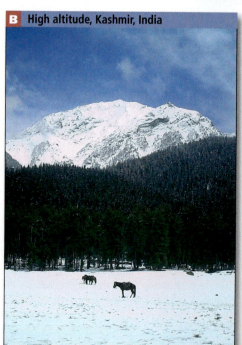

D Winter Monsoon, Bangladesh. Clouds being blown out to sea (as seen from space)

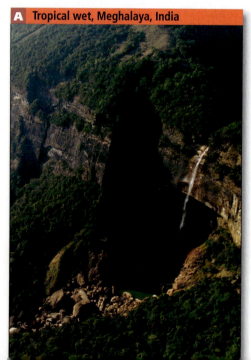

A Tropical wet, Meghalaya, India

C Desert, Jaisalmer, India

B High altitude, Kashmir, India

(A) Pre-monsoon Stage. The river flows in multiple channels across the flat plain.

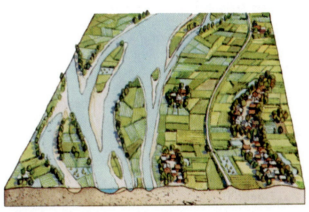

(C) Post-monsoon Stage. The river returns to its banks, but some of the new channels persist, changing the lay of the land. As the river recedes, it leaves behind silt and algae that nourish the soil. New ponds and lakes form and fill with fish.

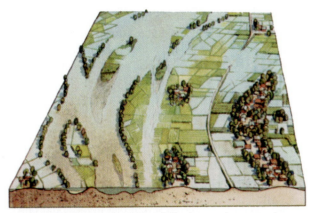

(B) Peak Flood Stage. During peak flood stage, the great volume of water overflows the banks and spreads across fields, towns, and roads. It carves new channels, leaving some places cut off from the mainland.

FIGURE 8.5 The Brahmaputra River in Bangladesh at various seasonal stages. People who live along the river have learned to adapt their farms to a changing landscape, and along much of the river farmers are able to produce rice and vegetables nearly year-round. [Adapted from *National Geographic,* June 1993, p. 125.]

falls in most of the region during this winter monsoon. However, as the ITCZ retreats southward across the Bay of Bengal, it picks up moisture, which is then released as early winter rains over parts of southeastern India and Sri Lanka.

The monsoon rains deposit large amounts of moisture over the Himalayas, much of it in the form of snow and ice that add to the mass of huge glaciers. Meltwater from these glaciers feeds the headwaters of the three river systems that figure prominently in the region: the Indus, the Ganga, and the Brahmaputra. All three rivers begin within 100 miles (160 kilometers) of one another in the Himalayan highlands near the Tibet–Nepal–India borders (see Figure 8.1 on pages 338–339 and the Map in Photo Essay 8.1). Forest vegetation is more common in the Himalayan highlands and in the foothills, but in many places human pressure has resulted in a widespread deforestation and an increasingly patchy forest.

These rivers, and many of the tributaries that feed them, are actively wearing down the surface of the Himalayas. They carry enormous loads of sediment, especially during the rainy season. When the rivers reach the lowlands, their velocity slows and much of the sediment settles out as silt. It is then repeatedly picked up and deposited by successive floods. As illustrated in the diagram of the Brahmaputra River in **Figure 8.5**, the movement of silt

constantly rearranges the floodplain landscape, and this complicates human settlement and agricultural efforts. However, the seasonal replenishment of silt nourishes much of the agricultural production in the densely occupied plains of Bangladesh. The same is true on the Ganga and Indus plains.

THINGS TO REMEMBER

1. Because of its high degree of tectonic activity and deep crustal fractures, South Asia is prone to devastating earthquakes.

2. Precipitation from monsoons is especially intense in the foothills of the Himalayas; an area north of the Bay of Bengal holds the world record for annual rainfall—about 35 feet, even though there is no rain at all for half the year.

Environmental Issues

South Asia has been occupied by people for millennia, but as recently as 1700 (just prior to British colonization), population density and human environmental impacts were far lower than they are today. By the beginning of the twenty-first century, population density and human impacts had grown exponentially. As a result, today South Asia has a range of serious environmental problems.

South Asia's Vulnerability to Climate Change

In South Asia, climate change—with the attendant sea level rise, water shortage, and crop failure—puts more lives at risk than in any other region in the world.

Sea Level Rise Bangladesh, where tens of millions of poor farmers and fishers live near sea level, has more people vulnerable to *sea level rise* (see pages 48–49 in Chapter 1) than any other country on earth (Photo Essay 8.2A). With 162 million people already squeezed into a country the size of Iowa, as many as 17 million people might have to find new homes if sea levels rise by 5 feet (Thematic Concepts Part N on page 343). The greatest economic impacts of sea level rise, however, could come from the submergence of parts of South Asia's wealthiest city, Mumbai, and the many other large, coastal urban centers (see Photo Essay 8.2C).

Also threatened by continuing sea level rise are the Maldives islands in the Indian Ocean, 80 percent of which lie 1 yard or less above the sea. Beach erosion is so severe that homes built only a few years ago in this richest of South Asia's countries are falling into the sea.

Water Shortage South Asia's three largest rivers are fed by glaciers high in the Himalayas that are now melting rapidly because of climate change. Smaller glacial-melt streams serve Afghanistan (see Photo Essay 8.2B). As many as 703 million people, almost half of the region's population, depend on these glacially fed rivers for irrigation and drinking water (see Thematic Concepts Part M on page 343). The immediate effect of glacial melting may be flooding. However, as glaciers shrink and provide less and less water to rivers each year during the dry season, the long-term effect will be severe water shortage. Unless dramatic action against climate change is successful, most Himalayan glaciers could eventually disappear. As a result, South Asia's largest rivers may run nearly dry during the winter monsoon when cool, dry air is flowing off the Eurasian continent (see Figure 8.4 on page 345).

Shifting rainfall patterns and rising temperatures are already resulting in drier conditions in much of South Asia, especially Afghanistan, Pakistan, and northwest India. High Himalayan communities in Ladakh, on the India/China border, are attempting to retain autumn glacial melt in stone catchments, where the melt refreezes over the winter and is available for spring irrigation.

Responses to Global Climate Change South Asia is pioneering some innovative responses to the multiple threats posed by global climate change. In the case of the present drought that could leave millions without sufficient food, India, by far the largest country in the region, has for some time had a system for storing grain and rotating it to retain its usability. There is also a federal emergency employment program that will give temporary jobs to those who lose their means of livelihood because of drought. These programs have a short history and their efficacy is not yet proven.

Public and private entities are taking measures to reduce contributions of CO_2 to the atmosphere. Although the use of fossil fuels in this region contributes significantly to global greenhouse gas emissions (see Figure 1.25 on page 47) and deforestation has removed most of the potential to sequester CO_2, some technological methods for reducing carbon emissions are being tried. For example, India is home to the world's largest producer of plug-in electric cars. The Reva company sells its cars in India and the UK for about U.S.$8000. Even factoring in emissions from the plants that generate the electricity used to charge the cars, electric cars

contribute substantially less to climate change than do gasoline- or diesel-powered cars. Also, several South Asian countries are investing private and public funds in solar energy, which could work better there than in many other regions because of the many cloudless days and the proximity to the equator. Wind power is another area of potential growth. In 2010, wind supplied only about 3 percent of India's electricity, yet this was enough to make India the world's fourth-largest user of wind power.

The melting of Himalayan glaciers will likely require both water conservation and increased water storage in order for supplies to last through the dry winter monsoon. Fortunately, South Asia has a lot of experience with such technologies. The Indus Valley in Pakistan was home to innovative underground water management even in ancient times. More recently, India and Pakistan have both pioneered methods of increasing the rate at which water deposited during the summer monsoon percolates through the soil and into underground aquifers (see Photo Essay 8.2D). More water is thus available for irrigation during the dry season. Because agriculture uses the most water of any human activity, drip irrigation technology, now thought to be the most efficient way to irrigate, would help conserve water. All South Asian countries have experience with drip irrigation, although the relatively high cost of modern efficient equipment has hampered widespread implementation of drip systems.

Deforestation

Deforestation has been occurring in South Asia since the first agricultural civilizations developed more than 5000 years ago. Ecological historians have shown that, as the forests vanished, the western regions of the subcontinent (from India to Afghanistan) became increasingly drier. The pace of deforestation increased dramatically over the past 200 years. By the mid-nineteenth century, perhaps a million trees a year were felled for use in building railroads alone.

Causes of Deforestation In the twenty-first century, South Asia's forests are still shrinking due to commercial logging and expanding village populations that use wood for building and for fuel. Many of South Asia's remaining forests are in mountainous or hilly areas, where forest clearing dramatically increases erosion during the rainy season. One result is massive landslides that can destroy villages and close roads. With fewer trees and less soil to retain the water, rivers and streams become clogged with runoff, mud, and debris. The effects can reach so far downstream that increased flooding in the plains of Bangladesh is linked to deforestation in the Himalayas.

Environmental Activism and Resistance Unlike China and many other nations facing similar problems, the countries of South Asia have a healthy and vibrant culture of environmental activism that has alerted the public to the consequences of deforestation. In 1973, for example, in the Himalayan district of Uttar Pradesh, India, a sporting-goods manufacturer planned to cut down a grove of ash trees so that the factory, in the distant city of Allahabad, could use the wood to make tennis racquets. The trees were sacred to nearby villagers, however, and when their protests were ignored,

Climate change is putting more lives at risk in South Asia than in any other region. However, many responses to climate change are emerging that could also increase resilience to climate-related hazards.

A A farmer in Bangladesh plows his field. Bangladesh is highly exposed to sea level rise, increased storm intensity, and flooding related to glacial melting. Bangladesh's large poor population makes it highly sensitive to these climate hazards.

B In Afghanistan, a crew improves irrigation infrastructure that may help reduce the sensitivity of nearby areas to water shortage. Widespread poverty and political instability contribute to Afghanistan's extremely high vulnerability to climate hazards.

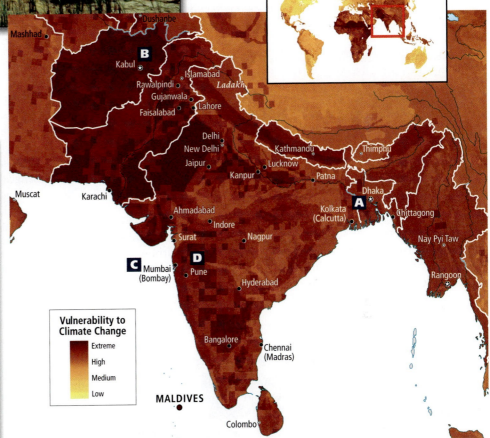

Mashhad, Dushanbe, **B** Kabul, Islamabad, Rawalpindi, Gujranwala, Lahore, Faisalabad, *Ladakh*, Delhi, New Delhi, Jaipur, Kathmandu, Lucknow, Kanpur, Thimphu, Patna, Muscat, Karachi, Ahmadabad, **A** Dhaka, Kolkata (Calcutta), Chittagong, Indore, Surat, Nagpur, Nay Pyi Taw, **C** Mumbai (Bombay), **D** Pune, Hyderabad, Rangoon, Bangalore, Chennai (Madras), **MALDIVES**, Colombo

Vulnerability to Climate Change
Extreme
High
Medium
Low

C Monsoon rains during a high tide in Mumbai, India. Low-lying Mumbai is highly exposed to sea level rise, and while its wealth and well-developed emergency response systems increase resilience to climate hazards, large slum areas remain highly sensitive.

D Fields in Maharashtra, India, are equipped with earthen berms that help capture rainwater during the summer monsoon, allowing water to filter down into the ground and replenish local aquifers. More water can then be pumped to the surface for irrigation during the dry season, reducing sensitivity to water shortage.

a group of local women took dramatic action. When the loggers came, they found the women hugging the trees and refusing to let go until the manufacturer decided to find another grove.

The women's action grew into the *Chipko* (literally, "hugging"), or *Social Forestry Movement*. The movement has spread to other forest areas, slowing deforestation and increasing ecological awareness. Proponents of the movement argue that the management of forest resources should be turned over to local communities. They say that people living at the edges of forests possess complex local knowledge of those ecosystems gained over generations—knowledge about which plants are useful as building materials, for food, medicinal uses, and fuel. They are more likely to manage the forests carefully because they want their descendants to benefit from forests for generations to come.

📹 **180. Tribal People in India Want to Protect Indigenous Way of Life**

Case Study: Nature Preserves in the Nilgiri Hills

The Mudumalai Wildlife Sanctuary and neighboring national parks in the Nilgiri Hills (part of the Western Ghats) harbor some of the last remaining forests in southern India. Here, in an area of about 600 square miles, live a few of India's last wild tigers (**Figure 8.6**) and a dozen or more other rare species, such as sloth bears and barking deer.

📹 **181. India's National Symbol Becoming More Difficult to Spot**

Even much smaller forest reserves play important roles in conservation. At 287 acres, Longwood Shola is a tiny remnant of the ancient tropical forests that once covered the Nilgiris.

Phillip Mulley, a naturalist, Christian minister, and leader of the Badaga ethnic group, points out that the indigenous peoples of the Nilgiris must now compete for space with

FIGURE 8.6 Endangered wildlife. This photo was taken by the authors during an unexpected encounter with a Bengal tiger in the Mudumalai Wildlife Sanctuary in the Nilgiri Hills.

a growing tourist industry (1.7 million visitors in 2005). In addition, huge tea plantations were cut out of forestlands by the state government to provide employment for Tamil refugees from the conflict in Sri Lanka (**Photo Essay 8.3D**). So while the forestry department and citizen naturalists are trying to preserve forestlands, the social welfare department, faced with a huge refugee population, is cutting them down. [*Sources: Lydia and Alex Pulsipher's field notes, Nilgiri Hills, June 2000; Tamil Nadu Human Development Report, 2003, at http://data.undp.org.in/shdr/tn/TN%20HDR%20 final.pdf.*] ∎

Water

One of the most controversial environmental issues in South Asia today is the use of water. South Asia has more than 20 percent of the world's population, but only 4 percent of its fresh water. It is not surprising, then, that disputes over access to water and the purity of water are common.

Conflicts Over the Use of Ganga River Water In recent years during the dry season, India has diverted 60 percent of the Ganga's flow to Kolkata to flush out channels where silt is accumulating and hampering river traffic (see the Ganga-Brahmaputra delta in Figure 8.1, and Photo Essay 8.1 Map). However, the diversions deprive Bangladesh of normal freshwater flow. Less water in the Ganga-Brahmaputra delta allows salt water from the Bay of Bengal to penetrate inland, ruining agricultural fields. The diversion has also caused major alterations in Bangladesh's coastline, damaging its small-scale fishing industry. Thus, to serve the needs of Kolkata's 15 million people, the livelihoods of 40 million rural Bangladeshis have been put at risk, triggering protests in Bangladesh.

When two or more countries share a water basin, conflicts can become geopolitical in scale. In the late 1990s, India signed a treaty with Bangladesh promising a fairer distribution of water, but as of 2008, Bangladesh was still receiving a considerably reduced flow.

Similar water use conflicts occur between states within India and Pakistan and between the wealthier and poorer sectors of the populations. For example, diversion of water from the Ganga in the state of Haryana deprives farmers downstream in Uttar Pradesh of the means to irrigate their crops. And in Delhi, just 17 five-star hotels use about 210,000 gallons (800,000 liters) of water daily, which would be enough to serve the needs of 1.3 million slum dwellers. The mining of water to bottle for elite consumption is yet another source of conflict.

📹 **182. Coca-Cola Blamed for India's Water Problems**

Water Purity of the Ganga River Water purity has become an issue in historic religious pilgrimage towns such as Varanasi, where each year millions of Hindus come to die, be cremated, and have their ashes scattered over the Ganga River. The number of such final pilgrimages has increased with population growth and affluence; hence, wood for cremation fires has become scarce. As a result, incompletely cremated bodies are being dumped into the river, where they pollute water used for drinking, cooking, and ceremonial bathing (see **Photo Essay 8.3B**). In an attempt to deal with this problem, the government recently installed an electric crematorium on the riverbank. It is attracting considerable

South Asia's huge population and growing industries have had major impacts on water and air quality, as well as on the extent and health of remaining ecosystems, such as forests.

A One of 230 leather-tanning facilities in Kasur, Pakistan, pumps effluent into a small river, where toxins are now 40 times the limit recommended by the World Health Organization. Local residents suffer dysentery, respiratory disorders, and skin diseases as a result. Still more tanneries are opening up in Kasur as companies try to avoid stricter environmental regulations on leather tanning in the European Union and China.

B Hindus flock to the highly polluted Ganga River in Varanasi, India, for ritual purification.

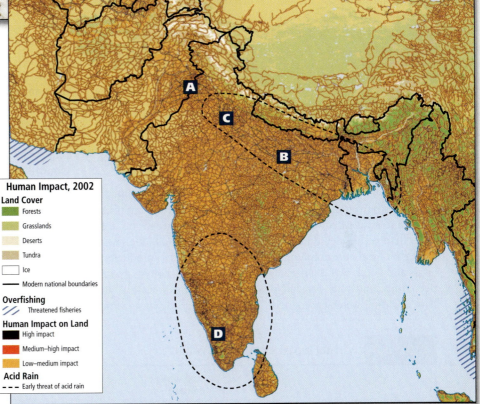

Human Impact, 2002

Land Cover
- Forests
- Grasslands
- Deserts
- Tundra
- Ice
- —— Modern national boundaries

Overfishing
- //// Threatened fisheries

Human Impact on Land
- High impact
- Medium–high impact
- Low–medium impact

Acid Rain
- – – – Early threat of acid rain

C New Delhi, India, where air pollution now rivals that of Beijing, China for the title of "World's worst." Sixty-seven percent of pollution comes from vehicles. Despite a shift to cleaner-burning compressed natural gas (CNG) and the banning of older vehicles, air quality continues to worsen as more people drive cars.

D A tea plantation in the Nilgiri Hills in Kerala, India. The expansion of tea plantations has resulted in habitat loss for many endangered species as well as degraded water resources.

business, as a cremation in this facility costs 30 times less than a traditional funeral pyre.

Of greater concern now is the large amount of industrial waste and sewage dumped into the river. Most sewage enters the river in raw form because city sewage systems (most built by the British early in the twentieth century) have long ago exceeded capacity. In Varanasi, pumps have been installed to move the sewage up to a new and expensive processing plant, but the plant is so overwhelmed by the volume of water during the rainy season that it can process only a small fraction of the city's sewage.

Veer Bhadra Mishra, a Brahmin priest (see page 361) and professor of hydraulic engineering at Banaras Hindu University in Varanasi, is on a mission to clean up the Ganga using unconventional methods. He is working with engineers from the United States to build a series of processing ponds that will use India's heat and monsoon rains to clean the river at half the cost of other methods. In addition, he preaches a contemporary religious message to the thousands who visit his temple on a bank of the sacred Ganga. The belief that the Ganga is a goddess who purifies all she touches leads many Hindus to think that it is impossible to damage this magnificent river. Mishra reminds them that because the Ganga is their symbolic mother, it would be a travesty to smear her with sewage and industrial waste.

Virtual Water The concept of *virtual water*, introduced in Chapter 1 (page 42) is especially useful in assessing the sustainability of water resources in the drier regions of South Asia. Afghanistan, Pakistan, and northwest India are all naturally dry regions that have been made more so by thousands of years of human use.

The water that goes into the production of all agricultural and industrial products of these regions is drawn from supplies that are so scarce that ordinary citizens often must survive on less water than is considered healthy. In the past when production was mostly for local consumption, the virtual water used in producing crops and goods was retained locally. Now, crops—cotton, grain, rice, fruits, and nuts—are all produced with irrigation; thus all have a virtual water component (see Table 8.1). Because these

crops are exported and the virtual water is consumed somewhere else in the global marketplace, these water-scarce regions may be subsidizing the water consumption of wealthy places with sufficient water. The same is true for manufactured goods, such as textiles, garments, hides, leather goods, and sporting goods that also have high virtual water inputs. Furthermore, in these dry regions, water for crops and industry is frequently drawn down faster than it is naturally replenished, so neither agriculture nor industry are not sustainable. Most importantly, the costs of depleted water used in production are not sufficiently accounted for in the pricing of goods exported from these regions.

Industrial Air Pollution

In many parts of South Asia, the air as well as the water is endangered by industrial activity. Emissions from vehicles and coal-burning power plants are so bad that breathing Delhi's air is equivalent to smoking 20 cigarettes a day (see Photo Essay 8.3C). The acid rain caused by industries up and down the Yamuna and Ganga rivers is destroying good farmland and such great monuments as the Taj Mahal.

M. C. Mehta, a Delhi-based lawyer, became an environmental activist partly in response to the condition of the Taj Mahal. For more than 20 years, he has successfully promoted environmental legislation that has removed hundreds of the most polluting factories from India's river valleys. His efforts are also a response to a disastrous event that took place in central India in 1984. At that time, an explosion at a pesticide plant in Bhopal produced a gas cloud that killed at least 15,000 people and severely damaged the lungs of 50,000 more. The explosion was largely the result of negligence on the part of the U.S.-based Union Carbide Corporation (which owned the plant) and the local Indian employees who ran the plant. In response to the tragedy, the Indian government launched an ambitious campaign to clean up poorly regulated factories.

THINGS TO REMEMBER

1. Low-lying, densely populated areas such as coastal Bangladesh are particularly vulnerable to sea level rise. However, even more people will be impacted by the melting of glaciers in the Himalayas that feed the region's major rivers during the dry season.

2. The *Chipko* ("tree hugging") *Social Forestry Movement* has spread to other forest areas, slowing deforestation and increasing ecological awareness.

3. India has water-allocation problems that impact farmers, towns, and cities, especially along the Ganga; water management in India also affects millions of people in Bangladesh.

4. Virtual water locked up in agricultural production in dry areas of the northwest is not included in the costs of products.

5. Pakistan, India, and Bangladesh have major air and water pollution problems related to industrial waste, lack of sewage treatment facilities, and end-of-life religious pilgrimages.

TABLE 8.1	**Virtual water component table**
Produce	**Virtual water content (in liters)**
1 apple	70
1 kg wheat	1300
1 kg rice	3400
1 kg cotton textile	11,000
1 kg leather	16,600
1 pair of bovine leather shoes	8000
1 U.S. dollar worth of industrial goods	80

Sources: Water Footprint Network, at http://www.waterfootprint.org/?page=files/productgallery&product=wheat

Human Patterns over Time

Over time, a variety of groups have migrated into South Asia, many of them as invaders who conquered peoples already there. Despite much blending over the millennia, the continued coexistence and interaction of many of these groups make South Asia both a richly diverse and an extremely contentious place.

The Indus Valley Civilization

There are indications of early humans in South Asia as far back as 200,000 years ago, but the first real evidence of modern humans in the region is about 38,000 years old. The first large agricultural communities, known as the **Indus Valley civilization** (or **Harappa culture**), appeared about 4500 years ago along the Indus River in modern-day Pakistan and northwest India. The architecture and urban design of this civilization were quite advanced for the time. Homes featured piped water and sewage disposal. Towns were well planned, with wide, tree-lined boulevards laid out in a grid. Evidence of a trade network that extended to Mesopotamia and eastern Africa has also been found.

Vestiges of the Indus Valley civilization's agricultural system survive to this day in parts of the valley, including techniques for storing monsoon rainfall to be used for irrigation in dry times. Remnants of language, and possibly superficial biological traits such as skin color, survive today among the Dravidian peoples of southern India, who originally migrated from the Indus region.

The reasons for the decline of the Indus Valley civilization after about 800 years (3700 years ago) are debated. Some scholars believe that complex geologic (seismic) and ecological changes (drier conditions) brought about a gradual demise. Others argue that foreign invaders brought a swift collapse. In any case, most agree that aspects of Harappa culture blended with subsequent foreign influences to form the foundation of many modern South Asian cultural traditions.

A Series of Invasions

The first recorded invaders to join the indigenous people of South Asia came from Central and Southwest Asia into the rich Indus Valley and Punjab about 3500 years ago. Many scholars believe that these people, referred to as Indo-European (a linguistic term, see page 359), in conjunction with the Harappa and other indigenous cultures, instituted some of the early elements of classical Hinduism, the major religion of India today. One of those elements was the still-influential caste system (discussed on pages 360–361), which divides society into hereditary hierarchical categories.

Other invaders from greater Central Asia included the Persians, the armies of the Greek general Alexander the Great, and numerous Turkic and Mongolian peoples. Defensive structures against these invaders can still be found across northwest South Asia (**Figure 8.7**). Jews came to the Malabar Coast of Southwest India more than 2500 years ago, Christians shortly after the time of Jesus. Arab traders came by land and sea to India long before the emergence of Islam; and then starting about 1000 years ago, Arab traders and religious mystics introduced Islam to what are now Afghanistan, Pakistan, and northwest India, and by sea to the coasts of southwestern India and Sri Lanka. In 1526, the invasion by the **Mughals**, a group of Turkic Persian people from Central Asia, intensified the spread of Islam. The Mughals reached the

> **Indus Valley civilization** the first substantial settled agricultural communities, which appeared about 4500 years ago along the Indus River in modern-day Pakistan and northwest India
>
> **Harappa culture** see Indus Valley civilization
>
> **Mughals** a dynasty of Central Asian origin that ruled India from the sixteenth to the nineteenth century

FIGURE 8.7 Mehrangahr Fort, Jodhpur, India. Started in 1459 by a king of the Rathore clan, Mehrengahr Fort stands 400 feet above the city of Jodhpur in western Rajasthan. Tens of thousands of forts like Mehrangahr can be found throughout South Asia, bearing testimony to millennia of invasions by outsiders and bitter rivalry between various South Asian dynasties.

height of their power and influence in the seventeenth century, controlling the north-central plains of South Asia. The last great Mughal ruler (Aurangzeb) died in 1707, but the legacy of the Mughals has remained as the power and extent of the dynasty declined. One aspect of that legacy is the more than 420 million Muslims now living in South Asia. The Mughals also left a unique heritage of architecture, art, and literature that includes the Taj Mahal, the fortress at Agra, miniature painting, and the tradition of lyric poetry. The Mughals contributed to the evolution of the Hindi language, which became the language of trade of the northern subcontinent and which is still used by more than 400 million people.

As Mughal rule declined, a number of regional states and kingdoms rose and competed with one another (Figure 8.8). The absence of one strong power created an opening for yet another invasion. By the late 1700s several European trading companies were competing to gain a foothold in the region. Of these, Britain's East India Company was the most successful. By 1857, the East India Company, acting as an extension of the British government, put down a rebellion against European intrusion and assumed dominant power in the region.

The Legacies of Colonial Rule

The British controlled most of South Asia from the 1830s through 1947 (Figure 8.9), profoundly influencing the region politically, socially, and economically. Even areas not directly ruled by the British felt the influence of colonialism. Afghanistan repelled British attempts at military conquest, but the British continued to intervene there, attempting to make Afghanistan a "buffer state" between British India and Russia's expanding empire. Nepal remained only nominally independent during the colonial period, and Bhutan became a protectorate of the British Indian government.

The Deindustrialization of South Asia As in their other colonies, the British used South Asia's resources primarily for their own benefit, often with disastrous results for South Asians. One example was the fate of the textile industry in Bengal (modern-day Bangladesh and the Indian state of West Bengal).

Bengali weavers, long known for their high-quality muslin cotton cloth, initially benefited from the greater access British traders gave them to overseas markets in Asia, the Americas, and Europe. By 1750, South Asia had an advanced manufacturing economy that produced 12 to 14 times more cotton cloth than Britain alone and more than all of Europe combined. However, during the second half of the eighteenth century, Britain's own highly mechanized textile industry, based on cotton grown in India, various other colonies, and the American South, developed cheaper cloth that then replaced Bengali muslin, first in the British colonies in the Americas, then in Europe, and eventually

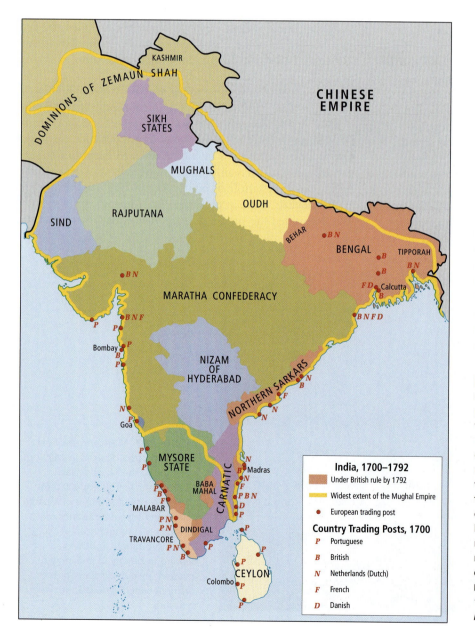

FIGURE 8.8 Precolonial South Asia. By 1700, several European nations had established trading posts along the coast of India and Ceylon. After the death of the Mughal ruler Aurangzeb in 1707, the ability of the Mughals to assert strong central rule throughout South Asia declined. A number of emergent regional states competed with one another for territory and power. Among the strongest was the Maratha Confederacy, composed of a number of small states dominated by the Maratha peoples, who were known for their martial skills. Weakness at the center and constant rivalry paved the way for British conquest by the end of the eighteenth century. [Adapted from William R. Shepherd, *The Historical Atlas* (New York: Henry Holt, 1923–1926), p. 137; and Gordon Johnson, *Cultural Atlas of India* (New York: Facts on File, 1996), p. 111.]

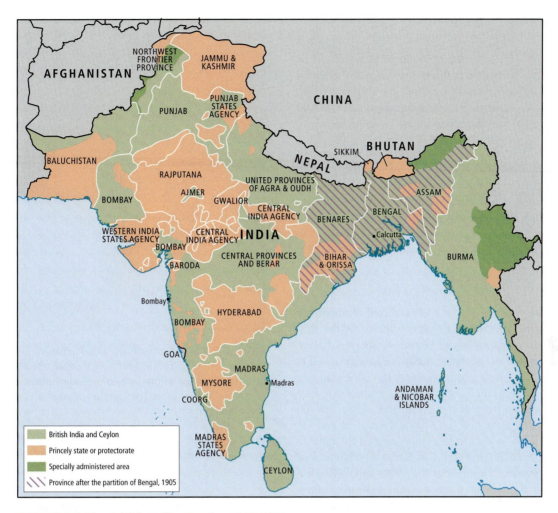

FIGURE 8.9 The British Indian Empire, 1860–1920. After winning control of much of South Asia, Britain controlled lands from Baluchistan to Burma, including Ceylon and the islands between India and Burma. [Adapted from Gordon Johnson, *Cultural Atlas of India* (New York: Facts on File, 1996), p. 158.]

throughout South Asia. The British textile industry was further aided by the British East India Company, which began severely punishing Bengalis who continued to run their own looms.

Many people who were pushed out of their traditional livelihood in textile manufacturing were compelled to find work as landless laborers. But rural South Asia already had an abundance of agricultural labor, so many migrated to emerging urban centers. In the 1830s, a drought worsened an already difficult situation, and more than 10 million people starved to death. It was during these trying times that South Asian workers were pressed into joining the stream of indentured laborers emigrating to other British colonies in the Americas, Africa, Asia, and the Pacific.

183. Caribbean Beat Pulses with Indian Accents

Economic development in South Asia was tightly controlled to benefit Britain. The production of tropical agricultural raw materials, such as cotton, tea, sugar, and indigo (a widely used blue dye), was encouraged in order to supply Britain and its other colonies. Industrial development, which might have competed with Britain's own industries, was discouraged.

Nonetheless, British rule did bring some beneficial changes. Trade with the rest of Britain's empire brought prosperity to a few areas, especially the large British-built cities on the coast, such as Bombay (now Mumbai), Calcutta (Kolkata), and Madras (Chennai). The British built a railroad system that boosted trade within South Asia and greatly eased the burden of personal transportation. In addition, English became a common language for South Asians of widely differing backgrounds, assisting both trade and cross-cultural understanding.

Democratic and Administrative Legacies Contemporary South Asian governments retain institutions put in place by the British to administer their vast empire. These governments inherited many of the shortcomings of their colonial forebears, such as highly bureaucratic procedures, resistance to change, and a tendency to remain aloof from the people they govern. Nonetheless, the governments have proved functional over time, but not without occasional major civil disturbances in virtually every country. Democratic government was not instituted on a large scale until

civil disobedience the breaking of discriminatory laws by peaceful means

Partition the breakup following Indian independence that resulted in the establishment of Hindu India and Muslim Pakistan

the final days of the empire, but since independence in 1947 (see below), it has given people an outlet for voicing their concerns and has enabled many peaceful transitions of elected governments. Still, the struggle to retain and build democratic institutions continues.

Independence and Partition The tremendous changes brought by the British inspired many resistance movements among South Asians. Some of these were militant movements intent on pushing the British out by force, while others focused on political agitation for greater democracy as a route to South Asian political independence. Although both types of action were brutally repressed, the democracy movements achieved worldwide attention and, after decades of struggle, success.

In the early twentieth century Mohandas Gandhi, a young lawyer from Gujarat, emerged as a central political leader of South Asia's independence movement. His tactics of **civil disobedience** focused on the nonviolent violation of laws imposed by the British that discriminated against South Asians. Gathering a large group of peaceful protesters, he would notify the government that the group was about to break a discriminatory law—for example, the law that made it illegal for South Asians to produce salt, thus requiring its importation from Britain. If the authorities ignored the act, the demonstrators would have made their point and the law would be rendered moot. On the other hand, if the government used force against the peaceful demonstrators, it would lose the respect of the masses. Throughout the 1930s and 1940s, this technique was used to slowly but surely undermine British authority across South Asia. The hunger strikes described in the chapter-opening vignette on the Sardar Sarovar Dam on the Narmada reflect Gandhi's legacy of nonviolent political protest.

In 1947, independence was granted to British India, which was divided into two independent countries: predominantly Hindu India and Muslim West and East Pakistan (Figure 8.10) (Afghanistan, Bhutan, and Nepal were never officially British colonies; Ceylon [now Sri Lanka] became independent in 1948). This division, labeled the **Partition,** which Gandhi greatly lamented, was perhaps the most enduring and damaging outcome of colonial rule.

The idea of two nations was first suggested by Muslim political leaders concerned about the fate of a minority Muslim population in a united India with a Hindu majority. Though Partition was highly controversial, it became part of the independence agreement between the British and the Indian National Congress (India's principal nationalist party). Northwestern and northeastern India, where the population was predominantly Muslim, became a single country consisting of two parts, known as West and East Pakistan, separated by northern India. Although both India and Pakistan maintained secular constitutions, with no official religious affiliation, the general understanding was that Pakistan would have a Muslim majority and India a Hindu majority. Fearing that they would be persecuted if they did not move, more than 7 million Hindus and Sikhs migrated to India from their ancestral homes in what had become Pakistan. A similar number of Muslims left their homes in India for Pakistan. In the process, families and com-

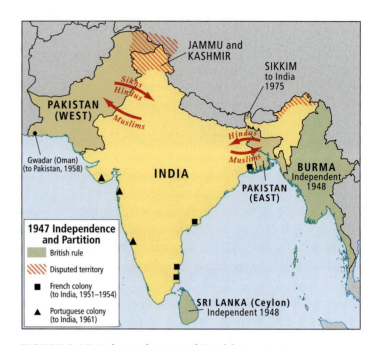

FIGURE 8.10 Independence and Partition. India became independent of Britain in 1947, and by 1948, the old territory of British India was partitioned into the independent states of India and East and West Pakistan. The Jammu and Kashmir region was contested space, and remains so today. Sikkim went to India, and both Burma and Sri Lanka became independent. Following additional civil strife, East Pakistan became the independent country of Bangladesh in 1971. [Adapted from *National Geographic,* May 1997, p. 18.]

munities were divided, looting and rape were widespread, and between 1 and 3.4 million people were killed in numerous outbreaks of violence. In 1971, a bloody civil war led to the division of Pakistan into Bangladesh (formerly East Pakistan) and Pakistan (formerly West Pakistan).

📺 **184. 60th Anniversary of India-Pakistan Partition on August 15th**

Partition was the tragic culmination of "divide-and-rule" tactics the British used throughout the colonial era (see Chapter 7, pages 320–321). This approach heightened tensions between South Asian Muslims and Hindus, thus creating a role for the British as seemingly indispensable and benevolent mediators. The legacy of these tactics includes not only Partition, but also the repeated wars and skirmishes, strained relations, and ongoing arms race between India and Pakistan.

The Postindependence Period In the more than 60 years since the departure of the British, South Asians have experienced both progress and setbacks. Democracy has expanded steadily, albeit somewhat slowly. India is now the world's most populous democracy and is gradually dismantling age-old traditions that hold back poor, low-caste Hindus, women, and other disadvantaged groups. Pakistan has had repeated elections and some governments have been modestly effective, but militaristic authoritarian and corrupt regimes have been more the norm, and Pakistan's long border feud with India (discussed on page 380) has sapped resources and national spirit. Bangladesh, long thought to be the most fragile since its union with Pakistan was violently dissolved in 1971, has

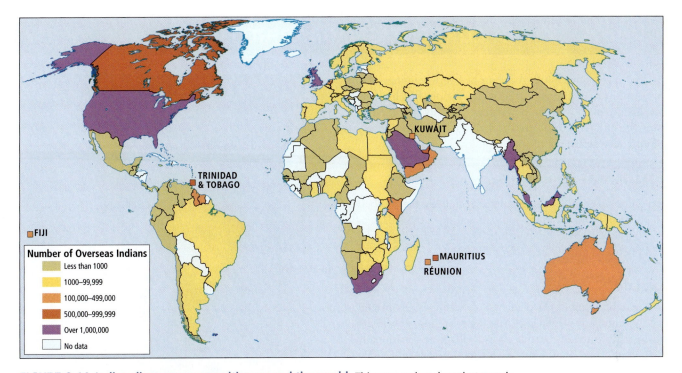

FIGURE 8.11 Indian diaspora communities around the world. This map makes clear that people of South Asian origin now live virtually everywhere on earth. They engage in many occupations, but are found particularly in the professions, technology development, commerce, and agriculture. [Data from the Ministry of External Affairs, Non Resident Indians & Persons of Indian Origin Division, *Report of the High Level Committee on The Indian Diaspora,* at http://indiandiaspora.nic.in/contents.htm; specifically, "Estimated Size of Overseas Community: Country-Wise," at http://indiandiaspora.nic.in/diasporapdf/part1-est.pdf.]

actually had a more stable, responsive and less militaristic government than Pakistan.

Industrialization became a main goal after independence, partly in response to the dismantling of industry during the colonial period. The emphasis on technical training has produced several generations of highly skilled engineers and technocrats whose talents are in demand around the world. Now, in most countries in the region, urban-based industrial and service economies constitute a far larger share of GDP than agriculture, with the information technology (IT) sector growing especially fast in India.

Under British rule, agricultural modernization lagged, and during World War II the Bengal Famine took an estimated 4 million lives. After independence, progress in agricultural production was slow until the late 1960s, when the green revolution brought marked improvements (discussed on page 373). The move to modernized farming on large tracts of land with far fewer agricultural workers than before brought prosperity to some South Asians and made food exports possible, but also forced millions to migrate to the cities.

These postindependence changes in South Asia brought economic growth throughout the region and especially to India. Various government subsidies for food and housing have reduced poverty significantly. In 1985, ninety-three percent of the Indian population lived on the equivalent of U.S.$1 a day per person; by 2007, with population growth factored in, just 54 percent were that poor. India's middle class now includes about 50 million people, but if present trends continue, by 2025 the Indian middle class will number 583 million, close to a third of the population. Nonetheless, taken as a whole, the South Asian countries had a collective average annual GDP per capita in 2009 of just over U.S.$3000 (adjusted for PPP); only sub-Saharan Africa's figures were lower.

Over the last three decades many highly skilled people have left the region to find better jobs elsewhere (**Figure 8.11**), contributing to an ongoing *brain drain* (see Chapter 3, pages 152–154). However, even before the global recession that began in 2007, many of these emigrants were returning to South Asia, anticipating a chance to participate in a new era of economic growth in this region.

THINGS TO REMEMBER

1. Following some 300 years of Mughal rule, the British controlled most of South Asia from the 1830s through 1947, profoundly influencing the region politically, socially, and economically.

2. In the early twentieth century, Mohandas Gandhi emerged as a central political leader of India's independence movement, using nonviolent civil disobedience that eventually led to independence.

3. Britain continued its "divide-and-rule" approach as it granted independence, separating the area into two countries, India and Pakistan. The tension between the two countries continues to this day.

II CURRENT GEOGRAPHIC ISSUES

The size and diversity of South Asia can be overwhelming. This section will provide a glimpse of the whole region, first by looking at the texture of daily life in both village and city, and then by examining the cultural characteristics that touch all lives yet vary greatly in practice across the region.

Sociocultural Issues

Travelers and natives to South Asia alike say that life there is best understood if it is observed in the intimate setting of the village or city neighborhood, where relationships among individuals, and within and between groups are easier to discern.

Village Life

The vast majority—about 70 percent—of South Asians live in hundreds of thousands of villages. Even many of those now living in South Asia's giant cities were born in a village or occasionally visit an ancestral rural community, so for most people village life is a lived experience (Figure 8.12).

 Personal Vignette The writer Richard Critchfield, who has studied village life in more than a dozen countries, writes that the village of Joypur (Bangladesh) in the Ganga-Brahmaputra delta is set in "an unexpectedly beautiful land, with a soft languor and gentle rhythm of its own." In the heat of the day, the village is sleepy: naked children play in the dust, women meet to talk softly in the seclusion of courtyards, and chickens peck for seeds.

In the early evening, mist rises above the rice paddies and hangs there "like steam over a vat." It is then that the village comes to life, at least for the men. The men and boys return from the fields, and after a meal in their home courtyards, the men come "to settle in groups before one of the open pavilions in the village center and talk—rich, warm Bengali talk, argumentative and humorous, fervent and excited in gossip, protest, and indignation" as they discuss their crops, an upcoming marriage, or national politics. ■

 Personal Vignette The anthropologist Faith D'Aluisio and her colleague Peter Menzel give another peek into village life as night falls in Ahraura, a village in the state of Uttar Pradesh in north-central India. In the enclosed women's quarters of a walled compound, Mishri is finishing her day by the dying cooking fire as her 1-year-old son tunnels his way into her sari to nurse himself to sleep. Mishri, who is 27, lives in a tiny world bounded by the walls of the courtyard she shares with her husband, five children, and several of her husband's kin. Like many villages in northern India, her village observes the practice of **purdah,** in which women keep themselves apart from men (see discussion, page 365).

That Mishri can observe purdah is a mark of status because it shows she need not help her husband in the fields. Within the compound, she works from sunup to sundown, only chatting

purdah the practice of concealing women from the eyes of non-family men

FIGURE 8.12 Life in a South Asian village. Although not Bangladeshi, here several generations of a northern Indian family from the village of Dharamkot in Himachal Pradesh province talk and laugh in their courtyard.

momentarily with two women who cover their faces and scurry from their own courtyards to hers for the short visit. Mishri is devoted to her husband, who was chosen for her by her family when she was 10; out of respect, she never says his name aloud. *[Source: Adapted from Faith D'Aluisio and Peter Menzel,* Women in the Maternal World, *1996.]* ■

City Life

South Asian cities are increasingly depicted as having split identities. On the one hand are sleek modern skyscrapers bearing the logos of powerful global high-tech firms; and on the other hand are chaotic, crowded, and violent urban environments, with over-stressed infrastructures and dilapidated low-rise housing. Mumbai exemplifies this dichotomy.

Bombay is the name by which most Westerners know South Asia's wealthiest city, now officially called Mumbai, after the Hindu goddess Mumbadevi. Mumbai is the largest deepwater harbor on India's west coast. Its metropolitan area, with over 21 million people, hosts India's largest stock exchange and the nation's central bank. It pays about a third of the taxes collected in the entire country and brings in nearly 40 percent of India's trade revenue. Its annual per capita GDP is three times that of India's next wealthiest city, the capital Delhi. Mumbai's wealth also extends into the realm of culture through the city's flourishing creative arts industries, including the internationally known film industry.

In India, Mumbai is known popularly as "Bollywood" because it produces popular Hindi movies portraying love, betrayal, and family conflicts. The stories, played out on lavish sets and accompanied by popular music and dance, serve to temporarily distract their huge audiences from the physical difficulties of daily life.

Mumbai's wealth is most evident when one looks up at the elegant high-rise condominiums built for the city's rapidly growing middle class. But at street level, the urban landscape is dominated by the large numbers of people living on the sidewalks, in narrow spaces between buildings, or in large, rambling shantytowns. The largest of the shantytowns is Dharavi, which houses up to a million people on less than 1 square mile and rivals Orangi Township in Karachi, Pakistan, for the title of the world's largest slum.

Implausibly, Dharavi is known for its inventive entrepreneurs. It is said to contain 15,000 one-room factories turning out thousands of products that make it into the global market, many of them from India's recycled plastic and metal. One young man, for example, collects and sells aluminum cans that once held ghee, a popular cooking oil in India. He says he makes about 15,000 rupees a month (U.S.$480), nearly twice the salary of the average teacher in India and much more than he made as a truck driver.

There is yet a third aspect to urban life in South Asia. Were one to carefully observe places as widely separated and culturally different as Mumbai, Chennai, Kathmandu, and Peshawar, one would discover that beyond the main avenues and shantytowns, these cities also contain thousands of tightly compacted, reconstituted villages where standards of living are relatively high and daily life is intimate and familiar, not anonymous as in Western cities. Such a place is Koli.

Personal Vignette Koli is an ancient fishing village that predates the city of Mumbai and is now squeezed between Mumbai's elegant coastal high-rises and the Bay of Mumbai (**Figure 8.13**), where some villagers still fish everyday. Koli is a labyrinth of low-slung, tightly packed homes ringed by fishing boats. The screeching of taxis and buses is soon lost in quiet calm as one ducks into a narrow covered passageway that winds through the village and branches in multiple directions. At first Koli appears impoverished, but inside, well-appointed homes, some with marble floors, TVs, and computers, open onto the dimly lit but pleasant alleys. The visitor soon learns that this is no warren of destitute shanties, no slum, but rather a community of educated bureaucrats, tradespeople, and artisans who constitute South Asia's rising urban middle class. ■

FIGURE 8.13 The village of Koli. Now surrounded by the vibrant city of Mumbai, Koli was once a small fishing settlement on a beautiful bay. There are still some fishers in Koli, but most residents are educated and work in the city.

Language and Ethnicity

Within the life of one South Asian village or urban neighborhood, there can be considerable cultural variety. Differences based on caste, economic class, ethnic background, religion, and even language are usually accommodated peacefully by long-standing customs that guide cross-cultural interaction. That everyone in South Asia is, in one way or another, a minority is to some extent an equalizing factor, in part because this fact of ubiquitous minority status has become part of the national dialogue about difference.

There are many distinct ethnic groups in South Asia, each with its own language or dialect. In India alone, 18 languages are officially recognized, but hundreds of distinct languages actually exist. This complexity results partly from the region's history of multiple invasions from outside. However, some groups were isolated for long periods of time, which also contributed to this complexity. In Figure 8.14, Number 21 indicates some of the most ancient culture groups in the region. These are descendants of Austro-Asiatic groups that were once more widely distributed but were left as isolated pockets when invaders brought with them sweeping cultural changes. The languages are distantly akin to others found farther east in Southeast Asia. The languages represented by Numbers 1–12 are linked to various groups of Indo-European people, who entered South Asia from Central Asia during prehistory.

The Dravidian language-culture group, represented by Numbers 14–19, is another ancient group that predates the Indo-European invasions by a thousand years or more. Today, Dravidian languages are found mostly in southern India, but a small remnant of the extensive Dravidian past can still be found in the Indus Valley in south-central Pakistan.

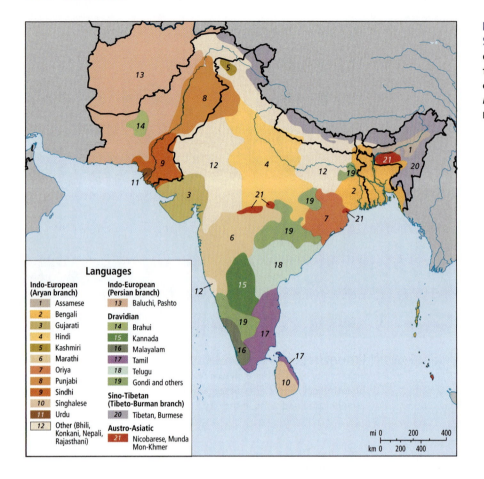

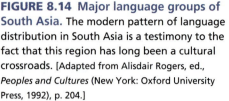

FIGURE 8.14 Major language groups of South Asia. The modern pattern of language distribution in South Asia is a testimony to the fact that this region has long been a cultural crossroads. [Adapted from Alisdair Rogers, ed., *Peoples and Cultures* (New York: Oxford University Press, 1992), p. 204.]

Languages

Indo-European (Aryan branch)
- 1 Assamese
- 2 Bengali
- 3 Gujarati
- 4 Hindi
- 5 Kashmiri
- 6 Marathi
- 7 Oriya
- 8 Punjabi
- 9 Sindhi
- 10 Singhalese
- 11 Urdu
- 12 Other (Bhili, Konkani, Nepali, Rajasthani)

Indo-European (Persian branch)
- 13 Baluchi, Pashto

Dravidian
- 14 Brahui
- 15 Kannada
- 16 Malayalam
- 17 Tamil
- 18 Telugu
- 19 Gondi and others

Sino-Tibetan (Tibeto-Burman branch)
- 20 Tibetan, Burmese

Austro-Asiatic
- 21 Nicobarese, Munda Mon-Khmer

By the time of British colonization, Hindi—an amalgam of Persian-based and Sanskrit-based northern Indian languages—was the *lingua franca* of northern India and what is now Pakistan. Today, variants of Hindi serve as national languages for both India and Pakistan (there, called Urdu), though it is the first, or native, language of only a minority. English is a common second language throughout the region. For years, it was the language of the colonial bureaucracy, and it remains a language used at work by professionals. Between 10 and 15 percent of South Asians speak, read, and write English.

Religion, Caste, and Conflict

The main religious traditions of South Asia are Hinduism, Buddhism, Sikhism, Jainism, Islam, and Christianity (see Figure 8.16 on page 362). (For a discussion of Islam, Christianity, and Judaism, refer to Chapter 6, pages 261–265.)

Hinduism **Hinduism** is a major world religion practiced by approximately 900 million people, 800 million of whom live in India. It is a complex belief system, with roots in both ancient literary texts (known as the Great Tradition) and in highly localized folk traditions (known as the Little Tradition).

A major tenet of classical Hindu philosophy, as described in the 4000-year-old scriptures called the *Vedas*, is that all gods are merely illusory manifestations of the ultimate divinity, which is formless and infinite. Some devout Hindus worship no gods at all, but may engage in meditation, yoga, and other spiritual practices. Such practices are designed to liberate worshipers from self-centered illusions, such as sadness or happiness, and bring them closer to the ultimate reality, described as infinite consciousness. The average person, however, is thought to need the help of personified divinities in the form of gods and goddesses. While some deities are recognized by all Hindus, many are found only in one region, one village, or even one family (the Little Tradition).

Some beliefs are held in common by nearly all Hindus. One is the belief in reincarnation, the idea that any living thing that desires the illusory pleasures (and pains) of life will be reborn after it dies. A reverence for cows, which are seen as only slightly less spiritually advanced than humans, also binds all Hindus together. This attitude, along with the Hindu prohibition on eating beef, may stem from the fact that cattle have been tremendously valuable in rural economies as the primary source of transport, field labor, dairy products, fertilizer, and fuel (animal dung is often burned). All of these beliefs, plus the caste system discussed below, have countless implications for daily life, affecting personal relationships, consumption patterns, diet and medicinal practices, resource use, and local and national politics.

Caste Hinduism includes the **caste system**, a complex and ancient system of dividing society into hereditary hierarchical categories (**Figure 8.15**). One is born into a given subcaste, or community

---

Hinduism a major world religion practiced by approximately 900 million people, 800 million of whom live in India; a complex belief system, with roots both in localized folk traditions (known as the Little Tradition) as well as in a broader system based on literary texts (known as the Great Tradition)

caste system an ancient Hindu system for dividing society into hereditary hierarchical classes

FIGURE 8.15 Caste and marriage. Caste remains a particularly powerful force with respect to marriage. **(A)** A Brahmin priest officiates at a Hindu wedding ceremony. The specifics of the ceremony vary considerably depending on the caste of the bride and groom. **(B)** A barber, often a member of a "barber caste" or *jati*, practices his trade in Rajasthan, India. In many areas, members of the barber caste act as go-betweens when families are arranging marriages for their children. **(C)** A *hijra* performs at a wedding ceremony in Delhi, India. *Hijras* are men who leave their families, often at a young age, to join a "hijra family." Here they undergo a gradual transition toward femininity, sometimes culminating in castration. They also must learn the *hijra* trades, which include singing and dancing for weddings and birth ceremonies, sometimes fortune telling, and often prostitution. Because of their unique identity, *hijras* are allowed to move between different castes with greater ease than most.

(called a ***jati***). Traditionally, that chance event largely defined one's experience for a lifetime—where one would live, where and what one could eat and drink, with whom one would associate, one's marriage partner, and often one's livelihood. The classical caste system has four main divisions or tiers, called ***varna***, within which are many hundreds of *jatis* and sub-*jatis*, which vary from place to place.

Brahmins, members of the priestly caste, are the most privileged in ritual status. Thus, they must conform to those behaviors that are considered most ritually pure (for example, strict vegetarianism and abstention from alcohol). Brahmins may be large landowners, but not necessarily so, nor are they always wealthy. Then, in descending rank, are *Kshatriyas*, who are warriors and rulers; *Vaishyas*, who are landowning (small-plot) farmers and merchants; and *Sudras*, who are low-status laborers and artisans. A fifth group, the *Dalits*—"the oppressed," or untouchables—is actually considered to be so lowly as to have no caste. Dalits perform those tasks that caste Hindus consider the most despicable and ritually polluting: killing animals, tanning hides, cleaning, and disposing of refuse. A sixth group, also outside the caste system, is the *Adivasis*, who are thought to be descendants of the region's ancient aboriginal inhabitants.

Although *jatis* are associated with specific subcategories of occupations, in modern economies this aspect of caste is more symbolic than real. Members of a particular *jati* do, however, follow the same social and cultural customs, dress in a similar manner, speak the same dialect, and tend to live in particular neighborhoods or villages. This spatial separation arises from the higher-caste communities' fears of ritual pollution through physical contact or sharing of water or food with lower castes. When one stays in the familiar space of one's own *jati*, one is enclosed in a comfortable circle of families and friends that becomes a mutual aid society in times of trouble. The social and spatial cohesion of *jatis* helps explain the persistence and respect paid to a system that, to outsiders, seems to put a burden of shame and poverty on the lower ranks.

It is important to note that caste and class are not the same thing. Class refers to economic status, and there are class differences within caste groups because of differences in wealth. Historically, upper-caste groups (Brahmins and Kshatriyas) owned or controlled most of the land, and lower-caste groups (Sudras) were the laborers, so caste and class tended to coincide. There were many exceptions, however. Today, as a result of legally mandated expanding educational and economic opportunities, caste and class status are less connected. Some Vaishyas and Sudras have become large landowners and extraordinarily wealthy businesspeople, while some Brahmin families struggle to achieve a middle-class standard of living. By and large, however, Dalits remain very poor.

Caste, Politics, and Culture In the twentieth century, Mohandas Gandhi began an organized effort to eliminate discrimination against "untouchables." As a result, India's constitution bans caste

jati in Hindu India, the subcaste into which a person is born, which largely defines the individual's experience for a lifetime

varna the four hierarchically ordered divisions of society in Hindu India underlying the caste system: *Brahmins* (priests), *Kshatriyas* (warriors/kings), *Vaishyas* (merchants/landowners), and *Sudras* (laborers/artisans)

Buddhism a religion of Asia that originated in India in the sixth century B.C.E. as a reinterpretation of Hinduism; it emphasizes modest living and peaceful self-reflection leading to enlightenment

Jainism originally a reformist movement within Hinduism, Jainism is a faith tradition that is more than 2000 years old; found mainly in western India and in large urban centers throughout the region, Jains are known for their educational achievements, nonviolence, and strict vegetarianism

discrimination. However, in recognition that caste is still hugely influential in society, upon independence from Britain, India began an affirmative action program. The program reserves a portion of government jobs, places in higher education, and parliamentary seats for Dalits and Adivasis. Together, the two groups now constitute approximately 23 percent of the Indian population and are guaranteed 22.5 percent of government jobs. In 1990, this program was extended to include other socially and educationally "backward castes" (the term used in India), such as disadvantaged *jatis* of the Sudras caste. An additional 27 percent of government jobs were reserved for these groups. However, reserving nearly half of government jobs in this way has resulted in considerable controversy. In 2006, medical students successfully protested against quotas in elite higher education institutions for lower-caste applicants. The Indian Supreme Court ruled in favor of the students.

At the local level, most political parties design their vote-getting strategies to appeal to subcaste loyalties. They often secure the votes of entire *jati* communities with such political favors as new roads, schools, or development projects. These arrangements fly in the face of the official ideologies of the major political parties and of Indian government policies, which actively work

to eliminate discrimination on the basis of caste. Nonetheless, currently, the role of caste in politics seems to be increasing, as several new political parties that explicitly support the interests of low castes have emerged. This assertion of political rights by low castes has met with a backlash from upper-caste groups, resulting in a number of violent clashes in recent years. It seems, then, that caste has been woven into the political system in ways that create and maintain tension and conflict.

Nevertheless, among educated people in urban areas, the campaign to eradicate discrimination on the basis of caste has been remarkably successful. Throughout the country there are now at least some Dalits in powerful positions. Members of high and low castes ride city buses side by side, eat together in restaurants, and attend the same schools and universities. For some urban Indians—especially educated professionals who meet in the workplace—caste is deemphasized as the crucial factor in finding a marriage partner. Nonetheless, it would be incorrect to conclude that caste is now irrelevant in India. Less than 5 percent of registered marriages cross even *jati* lines. Nearly everyone notices the tiny social clues that reveal an individual's caste, and in rural areas, where the majority of Indians still reside, the divisions of caste remain prevalent and provide a certain comfort, as well as constraints.

Geographic Patterns of Religious Beliefs The geographic pattern of religions in South Asia is uneven in distribution and overlapping, as **Figure 8.16** shows. Hindus, as we have observed, are the most numerous, and the majority live in India. Other religions are important in various parts of the region.

The 420 million *Muslims* in South Asia form the majority in Afghanistan, Pakistan, Bangladesh, and the Maldives; and Muslims are a large and important minority in India, where at 120 million, they form about 12 percent of the population. They live mostly in the northwestern and central Ganga River plain.

Buddhism began about 2600 years ago as an effort to reform and reinterpret Hinduism. Its origins are in northern India, where it flourished early in its history before spreading eastward to East and Southeast Asia. Only 1 percent of South Asia's population—about 10 million people—are Buddhists. They are a majority in Bhutan and Sri Lanka.

Jainism, like Buddhism, originated as a reformist movement within Hinduism more than 2000 years ago. Jains (about 6 million people, or 0.6 percent of the region's population) are found mainly in cities and in western India. They are known for their educational achievements, promotion of nonviolence, and strict vegetarianism.

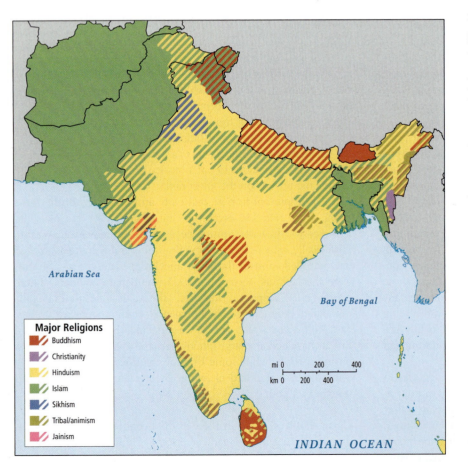

Major Religions
- Buddhism
- Christianity
- Hinduism
- Islam
- Sikhism
- Tribal/animism
- Jainism

Arabian Sea

Bay of Bengal

mi 0   200   400
km 0   200   400

INDIAN OCEAN

FIGURE 8.16 Major religions in South Asia. Notice the overlapping patterns in many parts of the region. [Adapted from Gordon Johnson, *Cultural Atlas of India* (New York: Facts on File, 1996), p. 56.]

Sikhism was founded in the fifteenth century as a challenge to both Hindu and Islamic systems. Sikhs believe in one God, hold high ethical standards, and practice meditation. Philosophically, Sikhism accepts the Hindu idea of reincarnation but rejects the idea of caste. (In everyday life, however, caste plays a role in Sikh identity.) The 18 million Sikhs in the region live mainly in Punjab, in northwestern India. Their influence in India is greater than their numbers because throughout India many Sikhs hold positions in the government, military and security forces, and the police.

The first Christians in the region are thought to have arrived in the far southern Indian state of Kerala with St. Thomas, the Apostle of Jesus, in the first century C.E. Today, Christians and Jews are influential but tiny minorities along the west coast of India. A few Christians live on the Deccan Plateau and in northeastern India.

Animism, the most ancient religious tradition, is practiced throughout South Asia, especially in central and northeastern India, where there are indigenous people whose occupation of the area is so ancient that they are considered aboriginal inhabitants. (For a discussion of animism, see Chapter 7, pages 333–334.)

The Hindu–Muslim Relationship The great Indian independence leaders Mohandas Gandhi and Jawaharlal Nehru (first Prime Minister of India, 1947–1964) both emphasized the common cause that once united Muslim and Hindu Indians: throwing off British colonial rule. Since independence, members of the Muslim upper class have been prominent in Indian national government and the military. Muslim generals have served India willingly, even in its wars with Pakistan after Partition. Hindus and Muslims often interact amicably, and they occasionally marry each other.

But there is a darker side to the Hindu–Muslim relationship. At the community level in South Asia, relations between the region's two largest religious groups are often quite tense. Especially in Indian villages, some Hindus regard Muslims as members of low castes. Religious rules about food are often the source of discord because dietary habits are a primary means of distinguishing caste. Because Hindus forbid killing cows, consumption of beef and the processing of cow hides into leather is permitted for only the lowest castes. Muslims, on the other hand, run slaughterhouses and tanneries (though discreetly), eat beef, and use cowhide to make shoes and other items. To some Hindus, therefore, some Muslims appear to have offensive customs. Also fueling this perception is the occasional conversion to Islam (or sometimes Christianity) of entire low-caste or tribal Hindu villages seeking to escape the hardships of being members of a disadvantaged social category.

The Hindu–Muslim relationship is no less complex in Bangladesh. After Partition in 1947, some Hindu landowners remained. In some Bangladeshi villages today, while Muslims may be a majority, Hindus are often somewhat wealthier. Although the two groups may coexist amicably for many years, they view each other as different, and conflict resulting from religious or economic disputes—euphemistically called **communal conflict**—can erupt over seemingly trivial events, as described in the following vignette.

Personal Vignette The sociologist Beth Roy, who studies communal conflict in South Asia, recounts an incident that she refers to as "some trouble with cows" in the village of Panipur (a pseudonym), Bangladesh (**Figure 8.17**). The incident started when a Muslim farmer either carelessly or provocatively allowed one of his cows to graze in the lentil field of a Hindu. The Hindu complained, and when the Muslim reacted complacently, the Hindu seized the offending cow. By nightfall, Hindus had allied themselves with the lentil farmer and Muslims with the owner of the cow. More Muslims and Hindus converged from the surrounding area, and soon there were thousands of potential combatants lined up facing each other. Fights broke out. The police were called. In the end, a few people

Sikhism a religion of South Asia that combines beliefs of Islam and Hinduism

communal conflict a euphemism for religiously based violence in South Asia

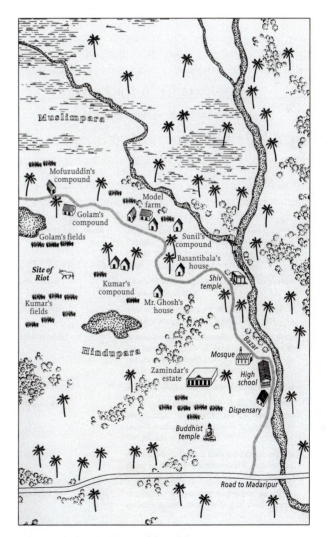

FIGURE 8.17 Some trouble with cows. This map of the village of Panipur (a pseudonym) illustrates how intimately the separate Muslim and Hindu communities were connected. The map shows Muslim and Hindu areas, the area where the riot took place, and numerous other features of village life. [Courtesy of the University of California Press.]

died when the police fired into the crowd of rioters. [*Source: Beth Roy, Some Trouble with Cows—Making Sense of Social Conflict (Berkeley: University of California Press, 1994), pp. 18–19.*] ■

THINGS TO REMEMBER

1. About 70 percent of South Asians live in rural villages, yet the region has four cities (metropolitan areas) with more than 10 million people. Mumbai has more than 22 million residents.

2. There are many distinct ethnic groups in South Asia, each with its own language or dialect. Today, variants of Hindi are the principal languages of India and Pakistan, while Bengali is the official language of Bangladesh. English is the common second language throughout South Asia.

3. Caste and class are not the same thing: "class" refers to economic status, and "caste" to hierarchical categories. There are class differences within caste groups because of differences in wealth. Even though India's constitution bans caste discrimination, caste is still hugely influential in Indian society.

4. There is a geographic pattern to religion in South Asia, but it is not absolute. People of different religions often live in close proximity, and both communal conflict and communal peace are common.

Geographic and Social Patterns in the Status of Women

Across South Asia a number of women hold and have held very high positions of power, and compared to a generation ago, young women today have greater educational and employment opportunities. However, the overall status of women in the region is notably lower than the status of men. Women's literacy rates, social status, earning power, and welfare are lowest in the belt that stretches from the northwest in Afghanistan across Pakistan, western India, Nepal and Bhutan, and the Ganga Plain into Bangladesh. Women fare better in eastern and central India and considerably better in southern India and in Sri Lanka. In these latter regions, where literacy rates are higher (**Figure 8.18**), different marriage, inheritance, and religious practices give women greater access to education and resources.

187. Sufi Rock Singer Falu Blends Old with New

Urban women in South Asia generally enjoy greater individual freedom than rural women, with many now pursuing professional careers and some becoming involved in politics. Generally speaking, rural women are less free than urban women. In rural India, middle- and upper-caste Hindu women are more restricted in their movements than are lower-caste women because they have a status to maintain. Meanwhile, lower-caste women who go into public spaces may have to contend with sexual harassment and exploitation from upper-caste men.

The socioeconomic status of Muslim women in India is notably lower than that of their Hindu and Christian counterparts. A recent national survey in India reported that Muslims on the whole have an average standard of living well below that of most Hindus. This disparity translates into educational levels for Muslim women that are significantly below the national average. In India, Muslim women's formal workforce participation rates also tend to be the lowest. Low rates of education and workforce participation for women also prevail in Muslim-dominated countries such as Pakistan and Afghanistan, although Muslim Bangladesh has a notably better record.

191. Remembering Pakistan's Former Prime Minister Benazir Bhutto

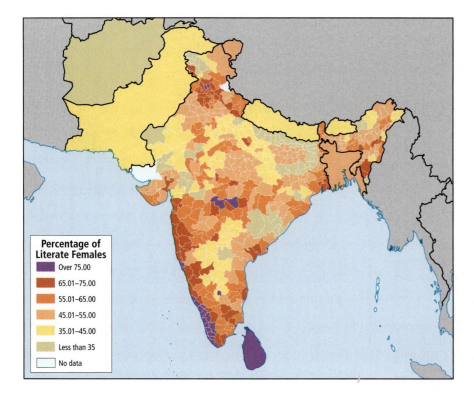

Percentage of Literate Females

- Over 75.00
- 65.01–75.00
- 55.01–65.00
- 45.01–55.00
- 35.01–45.00
- Less than 35
- No data

FIGURE 8.18 Female literacy in India, 2005, and in adjacent countries. Female literacy lags behind male literacy in nearly all parts of South Asia. Female literacy is crucial to improving the lives of children: women who can read often seize opportunities to earn some income and nearly always use this income to help their children. The white lines indicate district boundaries in India. Female literacy in India as a whole is 54.5 percent. [Adapted from "District Wise Female Literacy Rate of India," Maps of India, at http://www.mapsofindia.com/census2001/femaleliteracydistrictwise.htm; and *United Nations Human Development Report 2009* (New York: United Nations Development Programme), Table J, "Gender-Related Development Index and Its Components," at http://hdr.undp.org/en/reports/global/hdr2009/.]

Purdah The practice of concealing women from the eyes of non-family men, especially during their reproductive years, is known as *purdah*. It is observed in various ways across the region. The practice is strongest in Afghanistan and across the Indo-Gangetic Plain, where it takes the form of seclusion of women and of veiling or head covering in both Muslim and Hindu communities (see the story of Mishri on page 358). Purdah is weaker in central and southern India, but even there, separation between unrelated men and women is maintained in public spaces. The custom is generally not observed by low-caste Hindus, but that is changing. In recent decades, as low-status households increase their economic standing, their ability to seclude women signals surplus wealth and increased ritual purity. Unfortunately for women, the effect of this status-conscious trend is to limit their economic independence and autonomy.

Purdah practices have influenced the architecture of South Asia. Homes are often in walled compounds that seclude kitchens and laundries as women's spaces. In grander homes, windows to the street are usually covered with lattice screens known as *jalee* that allow in air and light but shield women from the view of outsiders (Figure 8.19).

Dowry, Marriage, Motherhood, and Widowhood Throughout South Asia, most women are partners in marriages arranged for them. Especially in wealthier, better-educated families, the wishes of the bride and groom are considered, but in some cases they are not. This is especially true of child marriage, when a young girl (often as young as 12 and in some places yet younger) is married off to a much older man. In most cases, the bride's family pays the groom's family a sum of money called a **dowry** at the time of the marriage. Dowries originated as an exchange of wealth between land-owning, high-caste families. With her ability to work reduced by purdah, an upper-caste female was considered a liability for the family that took her in.

Usually a bride goes to live in her husband's family compound, where she becomes a source of domestic labor for her mother-in-law. Most brides work at domestic tasks for many years until they have produced enough children to have their own crew of small helpers, at which point they gain some prestige and a measure of autonomy.

Motherhood in South Asia determines much about a woman's life. A woman's power and mobility increase when she has grown children and becomes a mother-in-law herself. On the other hand, in some communities, the death of a husband, regardless of cause, is a disgrace to a woman and can completely deprive her of all support and even of her home, children, and reputation. Widows may be ritually scorned and blamed for their husbands' deaths. Widows of higher caste rarely remarry, and in some areas, they become bound to their in-laws as household labor or may be asked to leave the family home. Most simply become low-status members of extended families and help with household duties of all sorts.

Dowry, Bride Price, and Violence Against Females Changing dowry customs appear to be a cause of the growing incidence of various kinds of domestic violence against females in Pakistan, India, and Bangladesh. Until the last several decades, the custom

FIGURE 8.19 Material culture of purdah. Lattice screens in Lahore, Pakistan. Known as *jalee*, lattice screens are often found in parts of South Asia where women are secluded. Like the louvers and latticed windows of Southwest Asia (see Figure 6.13 on page 329), *jalee* allow ventilation and let in light but shield women from the view of strangers.

when a young woman married was for the groom to pay a *bride price* to the family of the bride. The bride price was a relatively small sum that symbolized the loss of their daughter's work to her family's economy and the gain of her labor by his. It was only wealthy or high-caste girls whose families gave the groom a *dowry*—a substantial sum that symbolized the family wealth and was meant to give a daughter a measure of security in her new family.

dowry a price paid by the family of the bride to the groom (the opposite of *bride price*); formerly a custom practiced only by the rich

Increasing education for males and increasing family affluence have reinforced the custom of dowry and made it much more common for all families. Young men came to feel that their diplomas increased their worth as husbands and gave them the power to demand larger and larger dowries. Soon, the practice spread through lower-caste families wanting to upgrade their status.

Taliban an archconservative Islamist movement that gained control of the government of Afghanistan in the mid-1990s

Now the dowries they must pay to get their daughters married can cripple poor families. A village proverb captures this dilemma: "When you raise a daughter, you are watering another man's plant." Some poor families view the birth of a daughter as such a calamity that they are led to the desperate act of female infanticide—killing second and third daughters soon after birth.

The spread of the dowry into the poorer echelons of society also appears to be a cause of the practice known as *dowry killing*. In this phenomenon, a husband and his relatives stage an "accident," such as a kitchen fire, that kills his wife. The wife's death enables the widower to marry again and collect another dowry. Bride burning has been reported for many years, especially in India; the National Crime Records Bureau of India reported that 7026 such murders occurred in 2005. In Pakistan and Bangladesh, each with about one-tenth the population of India, in 2005 there were 300 and 165, respectively.

Earning Power, Population, and the Status of Women Efforts to increase the status of women are gaining momentum in South Asia, in part because it is believed this will slow population growth and increase economic well-being. Research across the globe shows that if women are afforded the opportunity to achieve education and a paying job, they not only tend to have fewer children but their families will benefit from better health and increased educational opportunities.

Personal Vignette In Mazār-e-Sharif, Afghanistan, in 2005 the state governor supported the establishment of a women's bazaar, where all the shops were to be owned and operated by women, but the customers would be of both sexes. Raqiba Barmaki was the first to open her shop, then based on handmade clothing for women and children. For a while, four women were brave enough to join her but all had to withstand the harassment of men, who resented their presence in the bazaar. Fariba Majid, the Women's Ministry representative who founded the bazaar, hoped that at least 20 women shopkeepers would eventually participate. She remarks that, for the bazaar to survive, the women had to quickly turn a profit; otherwise their husbands would put a stop to it. By 2009, possibly because of the global recession, only Ms. Barmaki's business survived, largely because her husband gave her moral support. She runs her handmade clothing shop out of a cargo container, and has added cutlery and dried fruit to her wares. With her husband's consent, she employs their daughters as seamstresses. [Source: Woman's Shop Breaks Afghan Mold, BBC News, July 1, 2009, at http://news.bbc.co.uk/2/hi/south_asia/8126116.stm.] ■

A minority of upper- and middle-class urban professional women in India, Pakistan, Sri Lanka, and Bangladesh may have more in common with their counterparts in Europe and America than they do with village women in their own countries. In the region's major cities, growing numbers of women are highly successful in business, as bankers and financial managers, directors of companies, highly qualified medical doctors and technicians, high-ranking academics, and mid-level managers in government.

TABLE 8.2	Women in Parliament in South Asia, February 2009	
Country	Percent of parliament	HDI rank
Sri Lanka	6	102
Bhutan	14	132
Pakistan	21	141
Nepal	33	144
Bangladesh	6	146
India	9	134
Afghanistan	26	181

Sources: United Nations Human Development Report 2009 (New York: United Nations Development Programme), Table K.

The daughters of these women are usually free to pursue careers, and also may choose their own spouses.

Gender and Democratization The status of women has been advanced throughout South Asia by moves toward greater democracy. India, Pakistan, Bangladesh, and Sri Lanka have all had female heads of state (prime ministers). However, it is important to note that all of these women were either wives or daughters of previous heads of state. Women have been notably less successful in elections, and at the parliamentary level, Indian and Sri Lankan women remain very poorly represented (Table 8.2).

As of 2009, only 9 percent of the Indian parliament was female. A confederation of Muslim and Hindu women's groups has lobbied for legislation that would reserve one-third of the seats in the lower house of Parliament and in state assemblies for women for a 15-year trial period. Such quotas are already in place in Pakistan (21 percent) and Nepal (33 percent). Bangladesh, where women currently make up only 6 percent of the parliament (Table 8.2), allocated one-third of the seats for women in March 2009, which will take effect next election.

Women and the Taliban in Afghanistan Women in Afghanistan have suffered sometimes brutal repression since a conservative Islamist movement, the **Taliban**, gained control of the government there in the mid-1990s. Prior to that time, rights for women in Afghanistan were slowly but steadily improving, and upper-class women enjoyed many freedoms. The Taliban support strict and radical interpretations of Islamic law, forcing females, even urban professional women, to live in seclusion. In regions where the Taliban retain control, girls and women are not allowed to work outside the home or attend school, and in virtually all parts of the country, despite the decline of Taliban control, women must wear a heavy, completely concealing garment, called a *burqa* (or burka), whenever they are out of the house. (Men also must follow a dress code, though a less restrictive one.) Although the Taliban were driven from official power in November of 2001, they maintain control of the southern provinces and mountainous zones near Pakistan, where cultural and religious conservatism continues to adversely affect Afghan women. Efforts to improve formal education for girls and women are numerous, but all operate in an environment that is precarious, with violence a constant threat. Even efforts to provide often-illiterate women a basic public education, such as that described in the vignette below about Radio Sahar, run into hostility. Reformers of gender roles in Afghanistan risk severe punishment and even death.

📺 **189. Report: Domestic Violence Widespread in Afghanistan**

📺 **190. "Frontrunner" Documentary Tells Story of Afghan Politician Who Inspires Women**

Personal Vignette From behind her microphone at Radio Sahar ("Dawn"), Nurbegum Sa'idi speaks to a female audience on a wide range of topics. Located in the city of Herat, Radio Sahar is one in a network of independent women's community radio stations that have sprung up in Afghanistan since early 2003. Radio Sahar provides 13 hours of daily programming consisting of educational items that address cultural, social, and humanitarian matters as well as music and entertainment. For example, one recent broadcast, aimed at informing women of their legal rights, followed the life of a young woman who was physically abused by her husband and his entire family. The woman took the brave step of asking for a divorce and was as a result forced into hiding, where she was counseled on what steps she might take next. Another program explored the various concepts of what democracy is and how it might work in Afghanistan. A reported 600,000 Afghan women and youth listen to Radio Sahar while they do their chores.

The broadcasts allow women to connect with one another in this conservative, male-dominated society. As Sa'idi attests, "It's great when you feel you can bring about change. The feedback we have been getting from listeners tells us that Sahar is providing new hope for the women in Herat." *[Source: Internews Afghanistan, March 2008, at http://www.internews.org/bulletin/afghanistan/Afghan_200803.html.]* ∎

THINGS TO REMEMBER

1. Of the more than 900 million Hindus living in South Asia, most (800 million) live in India; of the 420 million Muslims living in the region, 120 million live in India. Relations between these religious groups, the region's two largest, are often quite tense, and the rise of religious nationalism has led to violent confrontation. Other religious traditions also play prominent roles in the life of South Asia.

2. South Asia has had a number of women in very high positions of power, and young women today have greater educational and employment opportunities than those of a generation ago. However, the overall status of women in the region is notably lower than the status of men, and women's influence on policy remains low even when the number of women in national parliaments rises above a tiny minority.

3. The socioeconomic status of Muslim women is significantly lower than that of their Hindu and Christian counterparts; this is especially so in those parts of Afghanistan and Pakistan where tribal customs include deeply ingrained patterns of violence against women and Taliban influence remains high.

Population Patterns

South Asia is the most densely populated region in the world (see Photo Essay 8.4 Map on page 368). The region already has more people (1.59 billion) than China (1.33 billion), which has almost twice the land area of South Asia. U.S. Census Bureau projections indicate that by 2050, India alone, with 1.7 billion people, will overtake China's 1.4 billion. By 2050, China's population (if the one-child policy persists) will be shrinking at the rate of about 20 million people per decade, while India's will still be growing. And at current rates of growth, Pakistan and Bangladesh will have nearly doubled their populations.

Population densities in this region are greatest in the cities that lie just south of the Hindu Kush and Himalayas, and the very greatest density is in the Ganga-Brahamaputra delta (see the Photo Essay 8.4 Map), much of which is rural and, yet, extremely densely occupied.

Population Growth Factors in South Asia

Many factors have encouraged South Asia's high population growth over the past century. In rural areas, incentives for large families (Photo Essay 8.4A) continue because children from an early age contribute their labor to farming, and hence contribute to family income (Photo Essay 8.4B). Large families are also a response to poor access to health care, which means that many babies die in infancy. As a result, couples often choose to have many children to ensure that at least some will survive to adulthood. A further incentive for large families is that children are the only retirement plan that most South Asians will ever have. This makes it imperative that at least one child, and preferably more, reach maturity and be able to care for their elderly parents. Another factor that fuels population growth is the youthfulness of South Asia's population. Region-wide, about 35 percent of the population is under the age of 15; as this large group reaches reproductive age, population growth is assured even if radical fertility-reduction efforts are undertaken. The population pyramids for Sri Lanka and Pakistan (see Figure 8.21 on page 369) show the relative size of the population under age 15; notice how the pyramid for Sri Lanka is tapering toward the bottom. Pakistan's pyramid is also beginning to taper, meaning that year-to-year, fewer children are being born. The cause of the slight rise in childbirths in Sri Lanka in the last year or two is not yet understood (see Figure 8.21B).

📺 **185. Child Labor Persists in India Despite New Laws**

The long-term picture for this region shows population growth slowing. While South Asian nations have been trying to reduce population growth since 1952, it was not until the 1970s that growth rates actually declined. One factor in this decline was improved health care (see Photo Essay 8.4C); another was investment in education for women (see Photo Essay 8.4D). Today, growth rates continue to decline in nearly every country. India spends more than a billion dollars a year on population-control programs. Everywhere in South Asia, with the exception of Sri Lanka already mentioned (where a recent increase in birth rates is probably a temporary phenomenon), the

South Asia is the most densely populated world region. While its population growth is slowing as most countries begin the demographic transition, all countries will keep adding large numbers of people for at least another 40 years.

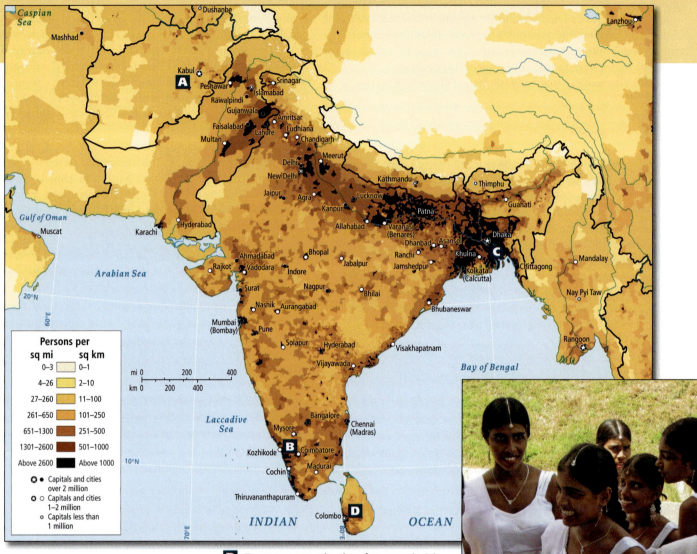

Persons per

sq mi	sq km
0–3	0–1
4–26	2–10
27–260	11–100
261–650	101–250
651–1300	251–500
1301–2600	501–1000
Above 2600	Above 1000

✪ ● Capitals and cities over 2 million
✪ ○ Capitals and cities 1–2 million
○ Capitals less than 1 million

D Teenagers at a school performance in Sri Lanka, where women's development has helped slow population growth.

A A family in Kabul, Afghanistan, where the demographic transition has only begun.

B A child plows a field in rural India, where families' need for labor provides an incentive to have many children.

C A nurse at a diarrhea research center in Bangladesh. Better health care encourages smaller families.

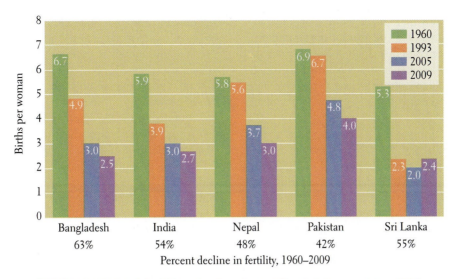

FIGURE 8.20 **Total fertility rates in selected South Asia countries, 1960–2009.** Most South Asian countries have had substantial declines in fertility since 1960. Many factors have contributed to this decline. Observe, however, that Sri Lanka's fertility rate has increased slightly since 2005, possibly in response to civil disorder or the death toll of the tsunami in 2004.

average number of children per woman has declined significantly (Figure 8.20).

Many factors are contributing to lower population growth rates. A comparison of Sri Lanka (Figure 8.21B), which is far along but not completely through the *demographic transition* (see Figure 1.12 on page 20), with Pakistan, which is still growing

rapidly (Figure 8.21A), highlights a number of differences. Sri Lanka has a much higher GDP per capita (U.S.$4480) than Pakistan (U.S.$2700). Health care is generally better in Sri Lanka, as indicated by its much lower infant mortality rate (in 2009, Sri Lanka had 15 deaths per 1000 live births versus Pakistan's 67 per 1000). Sri Lanka has also made greater efforts to educate and

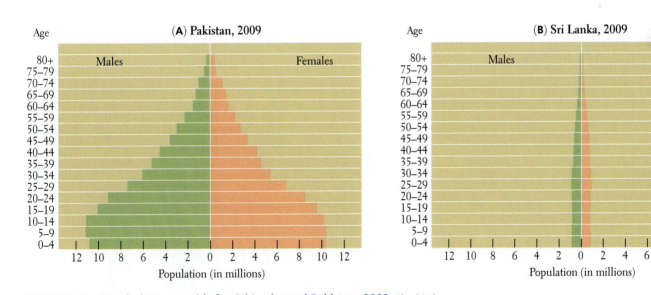

FIGURE 8.21 **Population pyramids for Sri Lanka and Pakistan, 2009.** The birth rates in both Pakistan and Sri Lanka have declined since 1960. However, the great size and youth of Pakistan's population ensure that it will continue to grow for years, while Sri Lanka's population is more or less stable. Observe that the scales on the two pyramids are the same; Pakistan's population is about 9 times larger than Sri Lanka's. [Adapted from U.S. Census Bureau, Population Division, "Population Pyramids for Pakistan" and "Population Pyramids for Sri Lanka" (Washington, D.C.: U.S. Census Bureau, International Data Base), at http://www.census.gov/ipc/www/idb/country.php.]

economically empower its women. Three key indicators of women's education and empowerment that are associated with reduced fertility rates are much higher in Sri Lanka than in Pakistan: female literacy (90 percent versus 30 percent), the percentage of young women who attend high school (89 percent versus 19 percent), and the percentage of women who work outside the home (36 percent versus 16 percent).

Gender and Population Change A significant gender imbalance exists in South Asian populations because of cultural customs that make sons more likely than daughters to contribute to a family's wealth (see **Thematic Concepts Part D** on page 342). A popular toast to a new bride is "May you be the mother of a hundred sons." Many middle-class couples who wish to have sons hire high-tech laboratories that specialize in identifying the sex of an unborn fetus. The intention is to abort a female fetus. This practice is now illegal (see **Thematic Concepts Part E** on page 342), but it persists. Poorer South Asians may neglect the health of girl children, with some even committing female infanticide (the dowry implications are discussed above on page 365).

🎬 **186. Girls Pay Price for India's Preference for Boys**

Such practices have resulted in men significantly outnumbering women throughout this region (see **Thematic Concepts Part F** on page 342). Elsewhere on earth, adult women normally outnumber adult men, because of their longer natural life spans; but the 2001 Indian census showed only 933 females for every 1000 males. After the 2001 census, India took a stronger stand against selecting for male children. A follow-up sex-ratio survey in five Indian states from 2004 to 2006 showed some improvement in all five states, with Tamil Nadu coming closest to a normal sex ratio, probably due to an especially aggressive campaign for women's health and against sex-selective abortions.

Gender imbalance of the magnitude India is facing could create serious problems, such as a surge in crime and drug abuse, due to the presence of so many young men with no prospect of having a family. The efforts of the state of Kerala to overcome gender imbalance through greater attention to women's development bear watching. Kerala's government has made women's development a priority, funding education for women well beyond basic levels. The results are reflected in its female literacy rate, which is the highest in India at 87.8 percent (the average for India as a whole is 54.5 percent), and in the numbers of women who work outside the home. In Kerala, daughters, far from being viewed as an economic liability to the family, are seen as an asset. Women can travel the public streets alone or in groups and commonly work in public places, many in high positions in government, education, health care, and other professions.

Urbanization Some of the recent decline in fertility rates throughout South Asia may be related to an ongoing shift in population from rural to urban areas (**Photo Essay 8.5**). Fertility rates are generally lower in urban areas for a number of reasons. Improved access to health care means that children are more likely to survive into adulthood. There is less economic incentive for large families, as urban children are more likely to go to school and thus represent a cost to the family rather than the contribution

to family incomes that rural children represent. Food and clothing are more expensive in cities. And as women see educational and employment options in cities, childbearing becomes less central to their lives.

🎬 **188. Global Population Boom Puts 'Mega' Pressure on Cities in Developing World**

Although only 30 percent of the region's population is urban, South Asia has several of the world's largest cities (metropolitan areas). In 2010, Mumbai has 21.9 million people; Kolkata, 15.6 million; Delhi, 20.9 million; Dhaka, 14.3 million; and Karachi, 13.2 million. All of these cities are growing quickly, and South Asia's current urban population of around 460 million could expand to as much as 712 million by 2025.

Many middle-class South Asians move to cities for education, training, or business opportunities (**Photo Essay 8.5C**). These reasons for urbanization are reflected in the higher literacy rates in large cities. Mumbai and Delhi both have literacy rates above 80 percent, approximately 25 percent above the average for the country; Dhaka in Bangladesh and Karachi in Pakistan both have 63 percent literacy, again over 15 percent above the rate for each country as a whole.

These urban literacy rates would be higher but for the fact that many poor rural people have been pushed into urban migration by agricultural modernization. Other urban migrants are refugees who have left drought-stricken or flooded landscapes or have been displaced by development projects, such as building of the dam on the Narmada discussed in the vignette that opens this chapter.

🌸 **Personal Vignette** One consequence of South Asia's rapid urban growth is a lack of affordable housing. Many people simply live on the streets. In a National Public Radio interview some years ago, an Indian journalist reported that she asked a bicycle rickshaw driver (**Figure 8.22**) about himself as

FIGURE 8.22 A bicycle rickshaw and driver in Delhi.

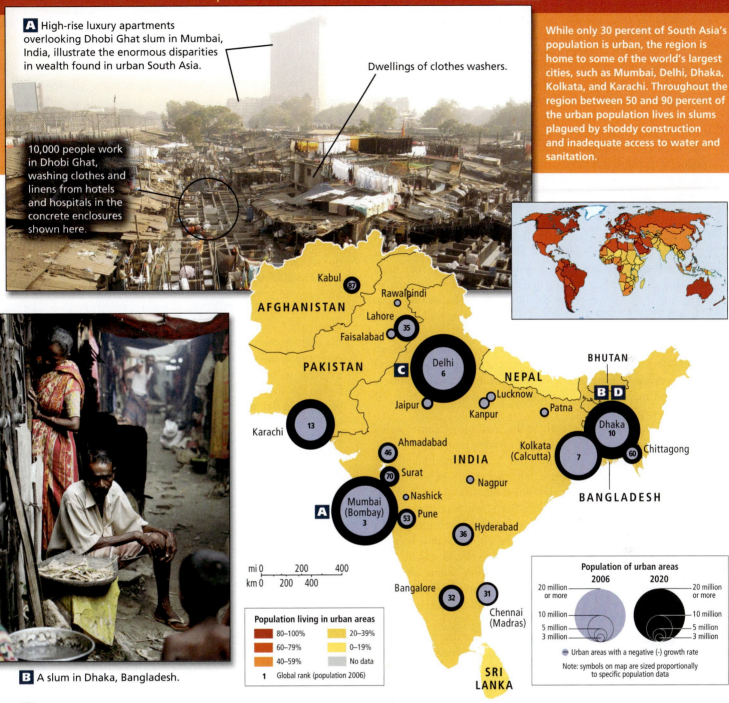

A High-rise luxury apartments overlooking Dhobi Ghat slum in Mumbai, India, illustrate the enormous disparities in wealth found in urban South Asia.

Dwellings of clothes washers.

10,000 people work in Dhobi Ghat, washing clothes and linens from hotels and hospitals in the concrete enclosures shown here.

While only 30 percent of South Asia's population is urban, the region is home to some of the world's largest cities, such as Mumbai, Delhi, Dhaka, Kolkata, and Karachi. Throughout the region between 50 and 90 percent of the urban population lives in slums plagued by shoddy construction and inadequate access to water and sanitation.

Kabul 87
AFGHANISTAN
Rawalpindi
Lahore 35
Faisalabad
PAKISTAN
Delhi 6 **C**
NEPAL
BHUTAN
Jaipur
Lucknow
Kanpur
Patna
B D
Dhaka 10
Karachi 13
Ahmadabad 46
INDIA
Kolkata (Calcutta) 7
Chittagong 60
Surat 70
Nagpur
BANGLADESH
Mumbai (Bombay) 3 **A**
Nashick
Pune 53
Hyderabad 36
mi 0   200   400
km 0 200 400
Bangalore
32   31
Chennai (Madras)
SRI LANKA

Population living in urban areas
- 80–100%
- 60–79%
- 40–59%
- 20–39%
- 0–19%
- No data

1 Global rank (population 2006)

Population of urban areas
2006 | 2020
20 million or more
10 million
5 million
3 million
⊖ Urban areas with a negative (-) growth rate
Note: symbols on map are sized proportionally to specific population data

B A slum in Dhaka, Bangladesh.

C A computer workshop at the University of Delhi. Many middle-class South Asians come to cities, and stay there, to take advantage of educational opportunities.

D Workers unload sand from a barge in Dhaka, Bangladesh. Most poor rural South Asians who come to cities work in low-paying, low-skill jobs.

he peddled her through Delhi. He replied that he had no family and that his belongings—a second set of clothes, a bowl, and a sleeping mat—were under the seat where she was sitting. He had come to Delhi from the countryside 14 years before and had never found a home. He knew virtually no one, had few friends, and no one had ever inquired about him before. He worked virtually around the clock and slept here and there for 2 hours at a time. He made so little he was ashamed to go back to his village. In the years since this report, the numbers of bicycle rickshaw drivers has ballooned, but their income has not, nor have their life conditions changed. [Source: Gagan Gill, "Weekend Edition," National Public Radio, August 16, 1997.] ∎

Risks to Children in Urban Settings As already discussed, in rural areas children are considered an important part of the labor supply. By age 5, a child can perform useful tasks and contribute economically to family well-being. But what about children in urban areas? A 2008 report by the International Labor Organization states that 23 million children between the ages of 5 and 14 are working in South Asia, many of them in urban areas. They work primarily in the informal sector as domestic servants, in export-oriented factories, as dump scavengers, sex workers, and carpet weavers. Clearly many of these occupations are entirely unsafe and inappropriate for children, who should be protected from such exploitation.

But if we look more closely at the example of carpet weaving, some serious issues arise. Carpet weaving is an ancient artistic and economic enterprise in South Asia; and traditionally, it has been a family-run enterprise, with women and children the weavers and men the merchants. For thousands of years, young children have learned weaving skills from their parents and have become proud members of the family's home-based production unit. Yet, in today's modernizing society, children must balance their role as family workers with the need to attend school. One of the chief benefits of urban life is education, where children learn the skills that will enable them to survive and prosper in a modern economy. All citizens need to read and write and do math.

As global trade has increased the demand for fine hand-woven carpets, most of the profit has gone, not to the weaving families, but to South Asian middlemen and foreign traders from Europe and America. Some unscrupulous carpet merchants have even set up factories where kidnapped children are forced to labor to produce carpets. Clearly, kidnapping and enslavement must be stopped, but the question remains: Is there room for cottage industry employment of children within the family circle?

The United Nations, South Asian governments, and NGOs like RugMark are now addressing this issue. They have instituted an active program to curb child labor abuses, while remaining open to the positive experience that learning a skill and being part of a family production unit can be for a child. India has established a national system to certify that exported carpets are made in shops where the children go to school, have an adequate midday meal, and receive basic health care. Such carpets will bear the label "Kaleen" or "RugMark."

Human Well-Being The maps of human well-being in South Asia (Figure 8.23) show that the overall state of human well-being in South Asia is marginal, with nearly the entire region ranking in the medium-low levels. In average annual GDP per capita PPP (see Figure 8.23A), no part of the region has an average above U.S.$5000. Of course the huge country of India has many cities and a few provinces (Kerala, for example) that rank above this level, but these variances are lost in the countrywide average. A bright spot in this data is that disparity of wealth (the gap between rich and poor) is quite narrow in South Asia. Some will say this is related to the broad-based democracies found in this region. Certainly the provision of services is better and more consistent here than in Africa, for example. In per capita GDP, however, Nepal ranks in the lowest category (less than $1000) and Bangladesh does only slightly better, while Afghanistan is not accounted for at all because data about it is so sketchy. Estimates in 2009 show Afghanistan with a per capita GDP of about $800 a year.

On the Human Development Index (HDI) there is even less variation (Figure 8.23B). All are in the medium-low rank, with the slight exception of Sri Lanka (medium), which means that all countries in this region have a lower-than-average record of providing basic income, education, and health for their citizens. In fact, India, Pakistan, and Bangladesh have all lost ground in recent years, a fact that may be related to the global economic recession.

Figure 8.23C shows that in gender equality as measured by female earned income as a percent of male earned income, South Asia ranks in the very lowest categories. Based on the global thumbnail map, it is clear that in equity of pay for females, South Asia is in a league with North Africa and Southwest Asia and virtually nowhere else except Mexico.

While South Asia's record in human development is not impressive and may even be backsliding a bit during the global recession, there has been economic progress across the region and in fact, some remarkable technological and theoretical innovations in the region have already contributed to development elsewhere, as the next section shows.

THINGS TO REMEMBER

1. A significant, and often extreme, gender imbalance exists in South Asia, probably originating in cultural customs that make sons more likely to contribute to a family's wealth than daughters. Too often this imbalance results from outright discrimination and oppression that affects women in every stage of their lives.

2. Population growth is definitely slowing across the region but the momentum provided by a young population means growth will continue for the foreseeable future.

3. South Asia's current urban population of around 460 million could expand to as much as 712 million by 2025.

4. The overall average state of human well-being in South Asia is marginal, with nearly the entire region ranking at the medium-low levels.

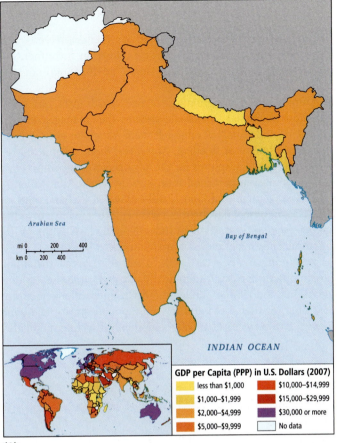

(A)

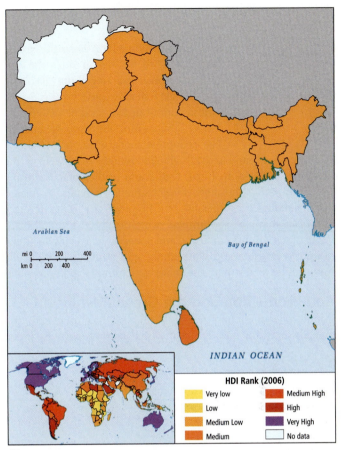

(B)

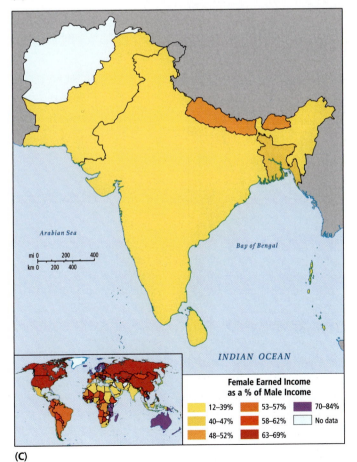

(C)

GDP per Capita (PPP) in U.S. Dollars (2007)

less than $1,000	$10,000–$14,999
$1,000–$1,999	$15,000–$29,999
$2,000–$4,999	$30,000 or more
$5,000–$9,999	No data

HDI Rank (2006)

Very low	Medium High
Low	High
Medium Low	Very High
Medium	No data

Female Earned Income
as a % of Male Income

12–39%	53–57%	70–84%
40–47%	58–62%	No data
48–52%	63–69%	

FIGURE 8.23 Maps of human well-being. (A) Gross domestic product (GDP). **(B)** Human Development Index (HDI). **(C)** Female earned income as a percent of male earned income (F/MEI). [Maps adapted from data: "Human development indices," (Maps A and B): Table 2, pages 28–32; (Map C): Table 5, pages 41–44 at http://hdr.undp.org/en/media/HDI_2008_EN_Tables.pdf.]

Economic Issues

South Asia is a region of startling economic incongruities. India is a good example: It is home to hundreds of millions of desperately poor people, but it is also a global leader in the computer software industry. It is celebrated for being the largest democracy in the world and yet its poor are often left out of the political process and bypassed or hurt by economic reforms.

Economic Trends

Agriculture still employs more than 50 percent of the workers in South Asia, but the contribution of agriculture to most national economies is much lower, averaging below 20 percent of the GDP for the region. The industrial sector employs far fewer people, yet produces between one-quarter and one-third of GDP in all countries except Nepal, Bhutan, and Afghanistan. The service sector has expanded across the region more rapidly than either agriculture or industry and now produces more than half of the GDP in every country except Afghanistan and Bhutan. Rapid development

green revolution increases in food production brought about through the use of new seeds, fertilizers, mechanized equipment, irrigation, pesticides, and herbicides

and self-sufficiency have been the dream for all parts of the region since the end of the colonial era (for India, Pakistan, and Bangladesh this came via the end of British control in 1947 and in 1948 for Sri Lanka). Successful measures to encourage economic modernization include an emphasis on IT and other high-tech industries as well as the automotive industry in India, textile and leather manufacturing in Pakistan, and innovative microfinancing strategies pioneered in Bangladesh. Less successful, but still promising, is the development of modernized agribusiness, which really involves all three sectors of agriculture, industry, and services. Especially in India, but also in Pakistan and Bangladesh, agribusiness has drastically improved the food supply and made agricultural exports possible.

Food Production and the Green Revolution

Agricultural production per unit of land has increased dramatically over the past 50 years, especially in parts of India; nonetheless, agriculture remains the least efficient economic sector, meaning that it gets the lowest return on investments of land, labor, and cash. Figure 8.24 shows the distribution of agricultural zones in South Asia.

Until the 1960s, agriculture across South Asia was based largely on traditional small-scale systems that managed to feed families in good years, but often left them hungry or even starving in years of drought or flooding. Moreover, these systems did not produce sufficient surpluses for the region's growing cities. Even now, cities must import food from outside South Asia. Much of South Asia's agricultural land is still cultivated by hand, and for decades, agricultural development was neglected in favor of industrial development, especially in India. Nonetheless, by the 1970s, important gains in agricultural production had begun.

Beginning in the late 1960s, a so-called **green revolution** boosted grain harvests dramatically through the use of new agricultural tools and techniques. Such innovations included seeds bred for high yield and for resistance to disease and wind damage; fertilizers; mechanized equipment; irrigation; pesticides;

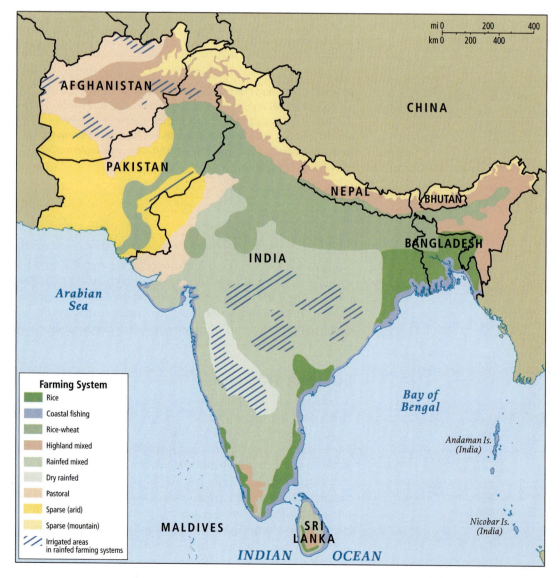

Farming System
- Rice
- Coastal fishing
- Rice-wheat
- Highland mixed
- Rainfed mixed
- Dry rainfed
- Pastoral
- Sparse (arid)
- Sparse (mountain)
- Irrigated areas in rainfed farming systems

FIGURE 8.24 Major farming systems of South Asia. [Adapted from John Dixon and Aidan Gulliver with David Gibbon, *Farming Systems and Poverty: Improving Farmers' Livelihoods in a Changing World* (Rome and Washington, DC: FAO and World Bank, 2001), at http://www.fao.org/farmingsystems/FarmingMaps/SAS/01/FS/index.html.]

herbicides; and double-cropping (two crops produced consecutively per year). To a lesser extent, an increase in the amount of land under cultivation also contributed to a greater yield. Where the new techniques were used, yield per unit of farmland improved by more than 30 percent between 1947 and 1979, and both India and Pakistan became food exporters.

The benefits of the green revolution have been uneven, however. Some Indian states, such as Punjab and Haryana, which have extensive irrigation networks, have gained tremendously. Others have lagged behind. Many poor farmers who were unable to afford the special seeds, fertilizers, pesticides, and new equipment could no longer compete and had to give up farming. Most migrated to the cities, where their skills matched only the lowest-paying jobs.

Green revolution technologies have also inadvertently reduced the utility of many crops for the rural poor, especially women. For example, the new varieties of rice and wheat yield more grain. However, they produce fewer of the other plant components previously used by women, such as the wheat straw used to thatch roofs, make brooms and mats, and feed livestock. Moreover, women's already low status in agricultural communities is often eroded further as their labor is replaced by new technologies and they are excluded from learning new skills by purdah customs.

There is a need for alternatives to standard green revolution strategies because of concerns about South Asia's long-term ability to maintain current levels of food production. The green revolution's chemical fertilizers, pesticides, and high levels of irrigation all contribute to pollution of waterways, increased erosion, and the loss of soil fertility over time through the buildup of salt in soils. *Soil salinization* (see page 256 in Chapter 6) is already reducing yields in many areas, such as the Pakistani Punjab, the country's most productive, but highly irrigated, agricultural zone.

▶ 192. Technology Key to Producing More Food

A potential remedy for some of the failings of the green revolution style of agriculture is **agroecology**. Traditional methods are being revived, such as fertilizing crops with animal manure, intercropping (planting several species together) with legumes to add nitrogen and organic matter, and using natural predators to control pests. Unlike green revolution techniques, the methods of agroecology are not disadvantageous to poor farmers because the necessary resources are readily available in most rural areas.

Malnutrition remains a problem despite the green revolution. Increased food supplies have reduced, but not eliminated, hunger and malnutrition in the region, because food goes to those with money to spend and, as noted previously, agricultural modernization can push the unskilled off the land and into precarious underemployment. Between 1970 and 2001, the amount of food produced per capita in South Asia increased 18 percent, and the proportion of undernourished people dropped from 33 percent of the population to 22 percent. Nonetheless, 22 percent of nearly 1.6 billion is 352 million people—roughly half of the world's total undernourished population. Because of corruption and social discrimination, government programs that provide food to the poor have generally failed to reach those most in need. For example, in India, despite massive resources devoted to improve child nutrition, about half of the children show signs of malnutrition.

Microcredit: A South Asian Innovation for the Poor

Over the past four decades, a highly effective strategy for helping lift people out of extreme poverty has been pioneered in South Asia. **Microcredit** makes very small loans (generally under U.S.$100) available to poor would-be business owners. Throughout South Asia, as in most of the world, banks are generally not interested in administering the small loans that poor people, especially poor women, need. Instead, the poor must rely on small-scale moneylenders, who often charge interest rates as high as 30 percent or more *per month*. In the late 1970s, Muhammad Yunus, an economics professor in Bangladesh, started the Grameen Bank, or "Village Bank," which makes small loans, mostly to people in rural villages who wish to start businesses.

▶ 195. Nobel Peace Prize Goes to Bangladesh's "Banker to the Poor"

The microcredit loans often pay for the start-up costs of small enterprises such as cell phone–based services, chicken raising, small-scale egg production, or the construction of pit toilets. Potential borrowers (more than 90 percent are women) are organized into small groups that are collectively responsible for repaying any loans to group members (**Figure 8.25**). If one member fails to repay a loan, then everyone in the group will be denied loans until the loan is repaid. This system, reinforced with weekly meetings, creates incentives to repay. The repayment rate on the loans is extremely high, averaging around 98 percent, much higher than most banks achieve.

So far, the Grameen Bank has been an enormous success in Bangladesh, where it has loaned over U.S.$8 billion to more than

> **agroecology** the practice of traditional, nonchemical methods of crop fertilization and the use of natural predators to control pests
>
> **microcredit** a program based on peer support that makes very small loans available to very low-income entrepreneurs

FIGURE 8.25 Microcredit for poor entrepreneurs. A microcredit group meets in a rural area near Kanchipuram, Tamil Nadu, India. Following the model established by the Grameen Bank, such groups gather weekly to discuss business and pay their loan installments. For many, this is a treasured social outing.

8 million borrowers. Similar microcredit projects have been established in India and Pakistan and throughout Africa, Middle and South America, North America, and Europe. In 2006, Dr. Yunus was awarded the Nobel Peace Prize for his work in microcredit. In 2009, he received the Presidential Medal of Freedom from President Obama.

Personal Vignette In a small hamlet in Bangladesh, not too far from the Indian border, is the house of Mosamad Shonabhan, a 32-year-old married woman whose life has been changed by her 11-year participation in the Grameen Bank. Everyone agreed that she had been the smartest of her family's children. However, because her father earned only 50 cents a day as a farm laborer, she could not go to school, and was instead married at the age of 14 to a young barber. For a year, she lived in her father-in-law's house, but then financial problems forced her to move back into her father's home. There she faced increasingly dire circumstances as her father's health deteriorated.

After a few years, a local political leader suggested that Mosamad join the Grameen Bank's lending program. Eventually, she took out a loan for $40.00 that would allow her to set up a small rice-husking operation in her father's backyard.

Eleven years and 11 loans later, Mosamad earns about $1.50 every day—three times what her father once made—and is a pillar of the local community. Her main source of income is a small shop inside her father's old house. She also leases an acre of land, which produces enough rice to feed her family and the numerous guests and friends who now come by to see her. [*Source: Adapted from Alex Pulsipher's field notes, 2000.*] ∎

Industry over Agriculture: A Vision of Self-Sufficiency

After independence from Britain in 1947, the new leaders in India, Pakistan, Bangladesh, and Sri Lanka tended to favor industrial development over agriculture. Influenced by socialist ideas, they believed that government involvement in industrialization was necessary to ensure the levels of job creation that would cure poverty. Another of their goals was to reduce the need to import manufactured goods from the industrialized world (see Chapter 3, page 136 for a discussion of *import substitution industrialization*). To accomplish this goal, governments took over the industries they believed to be the linchpins of a strong economy: steel, coal, transportation, communications, and a wide range of manufacturing and processing industries.

South Asian industrial policies in the decades after independence generally failed to meet their goals. The emphasis on industrial self-sufficiency was ill suited to countries that had such large agricultural populations. In India, for example, governments invested huge amounts of money in a relatively small industrial sector—even today industry employs only 14 percent of the population compared with the 52 percent employed in agriculture. Since such a small portion of the population directly benefited from this investment, industrialization failed to significantly increase South Asia's overall prosperity.

Another problem was that the measures intended to boost employment often contributed to inefficiency. One policy en-couraged industries to employ as many people as possible, even if they were not needed. So, for example, until recently, it took more than 30 Indian workers to produce the same amount of steel as 1 Japanese worker. Consequently, for years, Indian steel was not competitive in the world market. In addition, as in the former Soviet Union, decisions about which products should be produced were made by ill-informed government bureaucrats and were not driven by consumer demand. Until the 1980s, items that would improve daily life for the poor majority, such as cheap cooking pots or simple tools, were produced in only small quantities and were inferior in quality. At the same time, there was a relative abundance of large kitchen appliances and cars that only a tiny minority could afford.

Economic Reform: Globalization and Competitiveness

During the 1990s, much of South Asia began to undergo a wave of economic reforms. In many world regions structural adjustment programs (SAPs; see Chapter 3, pages 136–138) were mandated by the International Monetary Fund and World Bank; but in contrast, in India, economic reforms were initiated by the Indian government itself in response to an earlier financial crisis in the 1980s. Although privatization of India's public sector industries and banks has proceeded slowly, marketization of the economy has been significant. International investment has poured into the country in recent years, much of it from successful emigrants, and now a wide range of foreign consumer goods is available. The result has been rapid economic growth over the past decade, with recent years seeing India register high rates of yearly economic growth—9.6 percent in 2007—just prior to the global recession. Growth was down to 7.4 percent in 2008 and to an estimated 6.5 percent in early 2009, but then South Asians living abroad seeking a safe place to invest their savings sent a rush of cash.

Freed from a maze of regulations by the economic reforms, by 2009 both foreign and Indian companies were investing heavily in manufacturing and other industries (Figure 8.26). Drawn by India's large and cheap workforce, its excellent educational infrastructure (for the middle and upper classes), and especially by large pent-up domestic demand for manufactured goods of all sorts, many companies are setting up global manufacturing headquarters in major Indian cities. For example, in addition to Indian auto companies (e.g., Tata Motors), nearly every major global automobile company is currently establishing significant manufacturing facilities somewhere in India—to produce economical first cars for Indian families and for export; but also to produce luxury brands such as Mercedes-Benz for Indian elites. So many global manufacturers have flocked to India in recent years that the country is challenging China as an exporter of manufactured goods. As the current recession subsides, many of India's urban poor will see their incomes rise as manufacturing jobs increase and then they too will increase their consumption of Indian-made products. Since 2006, GDP per capita (PPP) has risen in all Indian states (Figure 8.26) and the average (PPP) income in India in 2009 was U.S.$2960, nearly ten times the income in 1990.

India's current manufacturing boom is benefiting from a previous boom in offshore outsourcing that started in the 1990s.

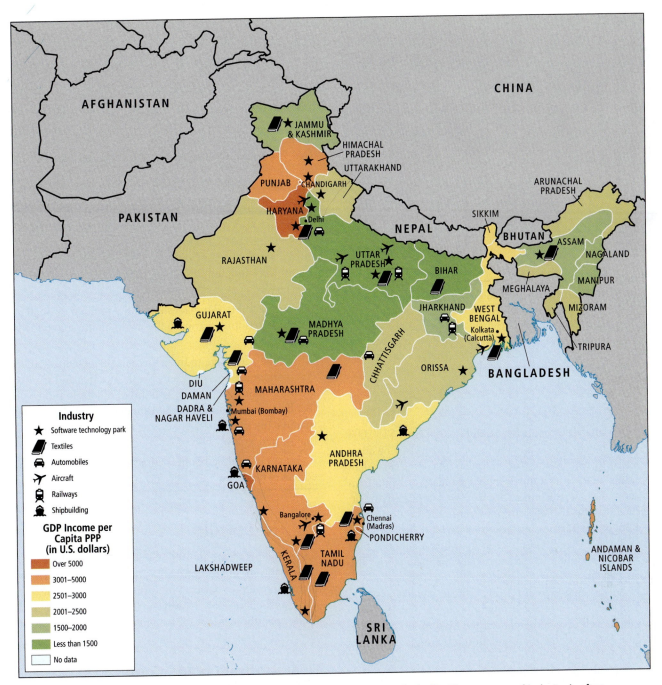

Industry

★ Software technology park

◣ Textiles

🚗 Automobiles

✈ Aircraft

🚆 Railways

⛴ Shipbuilding

GDP Income per Capita PPP (in U.S. dollars)

- Over 5000
- 3001–5000
- 2501–3000
- 2001–2500
- 1500–2000
- Less than 1500
- No data

FIGURE 8.26 GDP income per capita PPP and industrial and IT centers in India. The presence of industry is often associated with higher incomes, and because of this, planners may try to bring industry to low-income places. In some areas, poverty may be so great that even fairly intensive industry is able to raise average incomes above base levels only slowly. (The per capita income PPP information is primarily from 2007–2009 data, but for five states the data is from 2005–2007.) [Data from the Directorate of Economics & Statistics of each of the respective state governments, 2006; GDP data from the Reserve Bank of India, Table 8, "Per Capita Net State Domestic Product at Factor Cost—State-Wise," at http://rbidocs.rbi.org.in/rdocs/Publications/PDFs/8T_HB150909.pdf.]

Offshore outsourcing occurs when a company contracts to have some of its business functions performed in a country other than the one where its products or services are actually developed, manufactured, and sold. To take advantage of India's large, college-educated, and relatively low-cost workforce, companies in North America and Europe are outsourcing an increasing number of jobs to cities such as Bangalore, Mumbai, and Ahmadabad. Many types of jobs are being outsourced, including IT, engineering, telephone support, pharmaceutical research, and "back office" work. Many of the workers in these jobs are women.

Major global finance firms are increasingly seeking out the highly skilled workers of Mumbai's "Wall Street"—Dalal Street—to provide finance and accounting services. Stiff global competition makes cost cutting imperative for finance firms, and India's relatively low salaries provide a solution.

offshore outsourcing the contracting of certain business functions or production functions to providers where labor and other costs are lower

FIGURE 8.27 Protestors in Bangalore, workers in the high tech sector of India's growing service economy. Computer programmers in Bangalore protest government efforts to make policy changes that will reduce their ability to participate in software innovation. Bangalore is known as the "Silicon Valley" of India because of the large number of IT companies based there. Thirty-three percent of India's software exports (which were worth about $50 billion in 2009) come from Bangalore.

While a junior analyst from an Ivy League school costs $150,000 a year in the United States, a graduate of a top Indian business school costs only $35,000 a year in India. Yet, for that Indian employee, this salary buys a much higher standard of living than the U.S. employee would enjoy. In fact, well-educated Indian migrants in the United States are now returning home to take these seemingly low salaries because they can still live well and join this exciting development phase in their home country. This trend of return-migration could eventually happen across the South Asian region.

India's outsourcing growth reflects a broad pattern of steady expansion in the service sector throughout South Asia. Services now contribute well over 50 percent of the GDP in all countries of the region except Afghanistan and Bhutan. Bangladesh has a service sector that is proportionately nearly as large as India's, and it is increasingly the site of international investment.

Within the service sector, facilities that engage in trade, transportation, storage, and communication (including all facets of IT) show the most growth. Finance, insurance, real estate, business services, and tourism have also grown quickly. All these activities are connected in some way with international commerce and benefit from India's success at developing information technology. Yet because so much of this sector is linked to the global economy, all who are connected to it are vulnerable to downturns in demand such as those experienced during the global recession beginning in 2007 (**Figure 8.27**).

regional conflict especially in South Asia, a conflict created by the resistance of a regional ethnic or religious minority to the authority of a national or state government

The impact on wealth distribution of India's self-implemented economic reforms is the subject of much debate. The highly skilled and educated urban middle and upper classes have gained the most by far. India's middle class now stands at

50 million and is likely to grow dramatically in the near future. However, like SAPs in Middle and South America and Africa, the new economic policies will very likely produce wider disparities in income. And while many of India's more than 300 million urban poor people are gaining from recent growth in manufacturing and other industries, such growth will only indirectly benefit the 72 percent of India's population still living in rural areas, many of whom remain exceedingly poor.

194. Bhutan Strives to Develop "Gross National Happiness"

THINGS TO REMEMBER

1. Many poor farmers in India, unable to afford special seeds, fertilizers, pesticides, and new equipment, can no longer compete and have had to give up farming; some of them moved to urban areas hoping to find jobs. Other farmers have successfully adopted agroecology methods.

2. The Grameen Bank and its strategy of microcredit has been very successful in Bangladesh, and the method of small-loan financing has spread around the world, including to the United States.

3. To take advantage of India's large, college-educated, low-cost workforces, companies in North America and Europe are outsourcing an increasing number of jobs to Indian cities like Bangalore, Mumbai, and Ahmadabad. India's current manufacturing boom is benefiting from the offshore outsourcing that started in the 1990s.

4. India's economic boom is attracting the return of highly educated migrants.

Political Issues

Since independence in 1947, South Asian countries have peacefully resolved many conflicts, smoothed numerous potentially bloody transfers of power, and nurtured vibrant public debate over the issues of the day. However, South Asians have also missed many opportunities to resolve conflicts through democratic means, favoring the use of force on occasion.

Democratization and Conflict

The most intense armed conflicts in South Asia today are **regional conflicts**, in which nations dispute territorial boundaries or a minority actively resists the authority of a national or state government (**Photo Essay 8.6**). Most represent failed opportunities to resolve problems through the democratic processes.

Conflict in Punjab Punjab, a large area in northwestern India and northeastern Pakistan is the ancestral home of the Sikh community (see Figure 8.8 on page 354). When India and Pakistan were partitioned in 1947, the state of Punjab was divided between the two countries (see Figure 8.9 on page 355). The Indian Punjab has a more moist climate and is relatively prosperous. The part of Punjab in Pakistan, the country's largest and most populous province, is less prosperous; its southern part is arid and very poor, and increasingly this part of Pakistan is a recruitment area for Al Qaeda and the launching pad for terrorist assaults.

Most armed conflicts in South Asia were sparked by breakdowns in the democratic process and authoritarian behavior by governments. Supporters of political opposition groups have faced disenfranchisement, imprisonment, and sometimes execution. However, in some cases, democratization has paved the way for peaceful reconciliation between former combatants.

A Afghan troops, trained and supplied by the US, fire on Taliban positions near Kabul. The U.S.-led war in Afghanistan is largely a response to the September 11th attacks of 2001, though supporting democratization in Afghanistan is also a major stated goal.

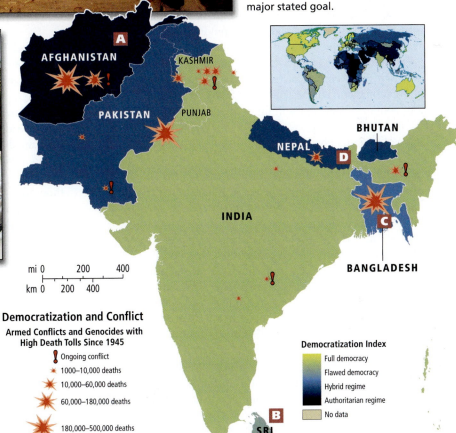

Democratization and Conflict
Armed Conflicts and Genocides with High Death Tolls Since 1945

Ongoing conflict
1000–10,000 deaths
10,000–60,000 deaths
60,000–180,000 deaths
180,000–500,000 deaths
500,000–1,000,000 deaths

Democratization Index
Full democracy
Flawed democracy
Hybrid regime
Authoritarian regime
No data

B A Tamil refugee of Sri Lanka's civil war. The war was sparked in part by antidemocratic policies that discriminated against Tamils. Tamil insurgents went on to disrupt numerous elections, though some of their supporters have now become peaceful elected politicians.

C In Dhaka, Bihari refugees of Bangladesh's 1971 war of independence from Pakistan look down from a building they occupy. The war was sparked by the refusal of West Pakistan (now Pakistan) to recognize elections that gave East Pakistan (now Bangladesh) control of the government. The Bihari community (some 600,000 people) sided with West Pakistan during the war and have since been denied citizenship in Bangladesh.

D A Nepali rupee depicting the former monarch, King Gyanendra, whose removal from power paved the way for democratization and the end of Nepal's civil war. Since then warring parties, such as the Maoists, have become peaceful political parties.

379

But there is an older source of conflict in the Indian Punjab. Many Sikhs chose to live in the Indian part of Punjab, where they thought their unique identity would be better protected than in Pakistan. Since Partition, however, Sikhs have often felt alienated from the rest of India, despite their wealth and influence in many cities outside the Punjab. In the 1970s, moderate Sikh elected officials issued demands for more water rights and greater respect for local democratic processes. The demands were generally ignored by India's national government.

In the early 1980s, militant Sikh extremists barricaded themselves in the holiest Sikh shrine, the Golden Temple in Amritsar. They then used the shrine as a base of operations to agitate for an independent Sikh nation. In 1984, Indian government forces attacked the shrine, damaging the temple and killing numerous innocent pilgrims as well as the militants. Shortly thereafter, Indian Prime Minister Indira Gandhi, who had called for the attack, was assassinated by two of her Sikh bodyguards. Riots and organized mob violence spread throughout India over the next few days, resulting in the deaths of more than 2700 Sikhs.

Since then, agreements acceding to many Sikh demands for water rights and control of religious sites have been signed but not implemented. This has alienated Sikhs yet further and led to more political violence, resulting in the deaths of 25,000 people. The situation in both parts of the Punjab remains sensitive and a source of friction between India and Pakistan.

Conflict in Kashmir Since 1947, between 60,000 and 100,000 people have been killed in violence in Kashmir in a dispute that probably could have been resolved democratically (see the map in Photo Essay 8.6).

Kashmir has long been a Muslim-dominated area, and in 1947, some Kashmiris and all of Pakistan's leaders believed that it should be turned over to Pakistan. However, the maharaja (king) of Kashmir at the time, a Hindu, wanted Kashmir to remain independent. The most popular Kashmiri political leader and significant portions of the populace, on the other hand, favored joining India. They preferred India's stated ideals of secular, nonreligiously based government to Pakistan's less robust safeguards for secularism. When Pakistan-sponsored raiders invaded western Kashmir in 1947, the maharaja quickly agreed to join India. A brief war between Pakistan and India resulted in a cease-fire line that became a tenuous boundary.

A popular vote for or against joining India was never held. Pakistan attempted another invasion of Kashmir in 1965, but was defeated. The two countries are technically still waiting for a UN decision about the final location of the border. In the meantime, Pakistan effectively controls the thinly populated mountain areas north and west of the densely populated valley known as the Vale of Kashmir. India holds nearly all the rest of the territory, where it maintains more than 500,000 troops. The Ladakh region of Kashmir (see Figure 8.1 on page 338) is the object of a more limited border dispute between India and China.

After years of military occupation, many Kashmiris now support independence from both India and Pakistan. However, neither country is willing to hold a vote on the matter. As in Punjab, much of the conflict in Kashmir has centered around the right of people to run their own affairs through local democratic processes.

India's national leaders have often appointed their own favorites in an attempt to maintain strong central control. Anti-Indian Kashmiri guerrilla groups, equipped with weapons and training from Pakistan, have carried out many bombings and assassinations of these appointees. Blunt counterattacks launched by the Indian government have killed large numbers of civilians and alienated many Kashmiris. Meanwhile, sporadic fighting between India and Pakistan continues along the boundary line, at an altitude of 20,000 feet (6000 meters)—the world's highest battle zone.

Another complication in the Kashmir dispute is the fact that both India and Pakistan—which came close to war against each other in 1999 and again in 2002—have nuclear weapons. Because of the nationalistic fervor of the protagonists, many see the conflict in Kashmir as more likely to result in the use of nuclear weapons than any other conflict in the world. Analysts agree that any use at all of nuclear weapons would have severe repercussions for all on earth.

War and Reconstruction in Afghanistan In the 1970s, political debate in Afghanistan became polarized. On one side were several factions of urban elites, who favored modernization and varying types of democratic reforms. Opposing them were rural conservative religious leaders, whose positions as landholders and ethnic leaders were threatened by the proposed reforms. Divisions intensified as successive governments, all of which came to power through military coups, became more and more authoritarian. Political opponents were imprisoned, tortured, and killed by the thousands, resulting in a growing insurgency outside the major cities. Some of the more authoritarian urban elites, inclined toward violence, allied themselves with the Soviet Union, which supported some of their goals and promised aid.

In 1979, fearing that a civil war in Afghanistan would destabilize neighboring Soviet republics in Central Asia, the Soviet Union invaded Afghanistan. Rural conservative leaders (often erroneously called "warlords") and their followers formed an anti-Soviet resistance group, the *mujahedeen*. As resistance to Soviet domination increased, the mujahedeen became ever more strongly influenced by militant Islamist thought and by Gulf-Arab activists who provided funding and arms. The United States and its regional ally Pakistan, choosing to ignore this development because it was still the Cold War era and they considered the mujahedeen an anti-Soviet movement, supported the mujahedeen with billions of dollars for equipment and weapons. Moderate, educated Afghans who favored democratic reforms fled the country during this turbulent time, hoping to go back eventually when peace returned.

The mujahedeen proved to be tenacious fighters, and in 1989, the Soviets, after heavy losses—14,000 Soviet soldiers killed and billions of dollars wasted—gave up and left the country in humiliating defeat. Back in Afghanistan where deaths of civilians and combatants numbered 1.5 million, anarchy prevailed for a time as the Afghan factions fought one another, but the rural conservatives eventually defeated the remaining reformist urban elites.

In the early 1990s, the radical religious-political-military movement called the *Taliban* emerged from among the mujahedeen. For the most part, the Taliban were (and remain) illiterate young men from remote villages, led by students from the *talibs*

(Islamist schools of philosophy and law). The Taliban wanted to control corruption and crime and minimize Western ways introduced in earlier decades by the urban elites and reinforced or made more extreme by the Russian occupation. Many of the most disliked Western ways related to the role, status, and dress of women. The behavior of Russian women accompanying the Russian army was seen as especially licentious by rural Afghans. The Taliban also wanted to strictly enforce shari'a, the Islamic social and penal code (see Chapter 6, page 266). Efforts by the Taliban to purge Afghan society of non-Muslim influences included greatly restricting women (see pages 366–367), promoting only fundamentalist Islamic education, and publicly banning the production of opium, to which many Afghan men had become addicted, while privately promoting its sale to raise funds for their side. By 2001, the Taliban controlled 95 percent of the country, including the capital, Kabul.

198. Taliban Insurgency Fueled by Poppy Cultivation

The events of September 11, 2001, focused the United States and its allies on removing the Taliban, who were apparently giving shelter to Osama bin Laden and his international Al Qaeda network. By late 2001, the Taliban were overpowered by an alliance of Afghans, supported heavily by the United States and the United Kingdom and eventually NATO (see Photo Essay 8.6A and Map). The United Nations stepped in to help establish an interim coalition government. A national assembly was convened to designate a new national government and appoint a head of state in 2002, and to ratify a new constitution in 2004.

In 2003, the United States launched the war in Iraq that diverted national attention, troops, and financial resources away from Afghanistan. Conditions in Afghanistan deteriorated to desperate levels during a time when humane strategies might have won the support of the Afghan people. Seven years after the Taliban were ousted, they were back again, effectively thwarting the ability of Afghanistan's new government to ensure security and to meet the needs of people outside Kabul. Based in rural areas in both Pakistan and Afghanistan, the Taliban are now aided by the fact that the Afghan public is widely distrustful of both the government in Kabul, which is seen as corrupt, and the international military forces and private security personnel stationed in Afghanistan since 2001, seen as hostile intruders. Civilian casualties incurred during military operations fuel growing resentment against foreign troops. Meanwhile, the vast majority of the people in the country remain desperately poor and uneducated despite the greatly increased attention from international donors since 2001. It appears that most people favor democratic government based on Muslim principles, but an election held in 2009 was so flawed that confidence in the Afghan government (especially President Karzai) was undermined at home and abroad.

The election of President Obama in the United States refocused attention on Afghanistan and the need for what is called *capacity building* in conjunction with military aide to that country. A number of private NGOs, the UN, and hundreds of returning Afghan emigrants are working to improve the Afghani people's capacity to participate in a modernizing economy by providing literacy classes and training centers for agriculture, artisan, and business skills. Education and jobs for females have been emphasized (Figure 8.28). Men are in no less need of improved opportunities.

FIGURE 8.28 Economic recovery in Afghanistan. Afghan women sort raisins, produced from Afghan grapes, destined for Afghan and foreign markets. This is one of many projects funded by foreign governments to help Afghanistan's economy recover from decades of war.

Increasingly, capacity-building activities appear to be crucial to winning the war and are a focus of allied efforts in Afghanistan.

Sri Lanka's Civil War In Sri Lanka, violence between the majority Singhalese and minority Tamil communities has left 68,000 people dead, produced more than a million refugees, and severely impeded Sri Lanka's economic development. Though both groups coexisted peacefully for centuries, beginning in the 1970s, antidemocratic government policies that favored the Singhalese over Tamils resulted in violence.

The Singhalese have dominated Sri Lanka since their migration from Northern India several thousand years ago. Today they make up about 74 percent of Sri Lanka's population of 20.5 million. Most Singhalese are Buddhist. Tamils, a Hindu ethnic group from South India, make up about 18 percent of the total population of Sri Lanka. About half of these Tamils have been in Sri Lanka since the thirteenth century, when a Tamil Hindu kingdom was established in the northeastern part of the island. The other half were brought over by the British in the nineteenth century to work on tea, coffee, and rubber plantations. Some Tamils have done well, especially in urban areas, where they dominate the commercial sectors of the economy. However, many others have remained poor laborers isolated on rural plantations.

Upon its independence in 1948, Sri Lanka had a thriving economy, led by a vibrant agricultural sector and a government that made significant investments in health care and education. It was poised to become one of Asia's most developed economies. But Singhalese nationalism was already alienating many Tamils. Singhalese was made the only official language, and Tamil plantation workers were denied the right to vote. Efforts were also made to deport hundreds of thousands of Tamils to India. Protests against these moves were brutally repressed by the government. Conditions in rural areas took a drastic turn for the worse in the 1960s when global prices for the country's chief agricultural exports declined. In response, the government shifted investment away from agricultural

religious nationalism the association of a particular religion with a political unit

development and toward urban manufacturing and textile industries, which were dominated by Singhalese.

In 1983, the Tamil minority, lacking the political power and influence to demand attention to their grievances, chose guerilla warfare against the Singhalese. A variety of Tamil groups united behind a guerilla army known as the Tamil Tigers. India intervened in 1987 at the request of the Sri Lankan government, but this failed to end the violence.

While the main battleground was in the north where the Tamils were hoping to establish an independent state, for over 30 years the entire island was subjected to repeated terrorist bombings and kidnappings (see Photo Essay 8.6B on page 379). Peace agreements were attempted several times, but in the end it was an overwhelming show of force by the government—which resulted in the death of the Tamil insurgency leader and included the destruction of large arms-smuggling operations—and a crackdown on international funding for the Tamils that forced their surrender in May of 2009.

Despite long years of violence that severely curtailed the tourist industry, economic growth in other sectors has been surprisingly robust in Sri Lanka. Driven by strong growth in food processing, textiles, and garment making, Sri Lanka is today one of the wealthiest nations in South Asia on a per capita GDP basis and is among those that provide best for the human well-being of their citizens.

Nepal's Rebels In Nepal, a dramatic expansion of democracy that appeared to be paving the way toward a tentative peaceful reconciliation between warring parties began to disintegrate in 2009. An elected legislature and multiparty democracy were introduced into Nepal in 1990; but until 2008, Nepal was governed by a royal family that paid only superficial respect to these changes.

In 1996, Maoist revolutionaries inspired by the ideals of the late Chinese leader Mao Zedong (but with no apparent support from China) waged a "people's war" against the Nepalese monarchy. After a decade of civil war, during which 13,000 Nepalese died, the Maoists had both military control of much of the countryside and strong political support from most Nepalese, who objected to the dictatorial rule of the latest monarch, King Gyanendra. Massive protests forced Gyanendra to step down in 2006, and soon thereafter the Maoists declared a ceasefire with the government (see Photo Essay 8.6C on page 379). In 2008, the Maoists won sweeping electoral victories that gave them a majority in parliament and made their former rebel leader prime minister. Then, in May of 2009, when his many conditions for reforming Nepalese society remained unmet, the Maoist prime minister resigned and took his party into opposition against a new prime minister and his weak 22-party coalition. Amnesty International has established that while the Maoists committed many human rights violations, including the recruitment of child soldiers, the governments of King Gyanendra and the present prime minister also used killings, torture, and rape as a matter of course. The Maoist opposition recently agreed to participate in writing a new constitution. Observers agree that if the various factions can see that democratic processes will give them a voice, a return to war will be unlikely.

▶ **200. Future of Nepal's King Gyanendra in Question**

Religious Nationalism

Increasingly, people frustrated by government inefficiency, corruption, caste politics, and the failure of governments to deliver on their promises of broad-based prosperity are joining religious nationalist movements. **Religious nationalism** is the association of a particular religion with a political unit—be it a neighborhood, a city, or an entire country. Political control over a given territory is often the ultimate goal of such movements.

Although both India and Pakistan were formally created as secular states, religious nationalism has long been a reality, shaping relations between people and their governments in those countries. Rather than embracing the idea of multiculturalism, India is increasingly thought of as a Hindu state, while Pakistan calls itself an Islamic Republic and Bangladesh a People's Republic. In each country, many people in the dominant religious group strongly associate their religion with their national identity.

Hindu nationalism in India (sometimes called Hindutva) is supported predominantly by urban men from middle- and upper-caste groups. Its proponents fear the erosion of their castes' political influence and particularly resent the extension of the quota system for government jobs and seats in universities to lower-caste groups (see pages 361–362). Meanwhile, politically mobilized lower castes are no longer willing to follow the dictates of the dominant castes.

Political parties based on religious nationalism have gained popularity throughout South Asia. Although their members think of these parties, such as the BJP in India, as forces that will purge their country of corruption and violence, they are usually no less corrupt or violent than other parties.

Terrorist Attacks in Mumbai in 2008 A 3-day terrorist attack in Mumbai in November 2008 left 160 dead in luxury hotels, a Jewish center, and a railroad station. To link this attack directly with religious nationalism as it has played out since 1947 would be incorrect, however. The attack appears, on the one hand, to be related to Islamic militancy centered in Afghanistan and Pakistan. But on the other hand, investigations have revealed that the terrorists were impoverished young men from Punjab recruited by Lashkar-e-Taiba, a Pakistan-based militant group, with origins based in the Kashmir dispute, that is suspected of links with Al Qaeda. There is little doubt, however, that unchecked religious nationalism in the region helps create the antagonistic context for such violence.

〰️ THINGS TO REMEMBER 〰️

1. Many conflicts have been worsened by an unwillingness on the part of governments and warring parties to recognize the results of elections, or even to allow people to vote. Some conflicts have been diffused, at least for the short term, by incorporating former combatants into the democratic process.

2. Most armed conflicts in South Asia were sparked by breakdowns in the democratic process and authoritarian behavior by governments. Supporters of political opposition groups have faced disenfranchisement, imprisonment, and sometimes execution. However, in some cases, democratization has paved the way for peaceful reconciliation between former combatants.

Reflections on South Asia

Despite the number, scale, and complexity of problems facing South Asia, optimism regarding the future is not unwarranted. While this region's enormous population is straining resources, population growth is also slowing as women gain education and economic empowerment. Changes in agriculture are addressing food shortages and low productivity, but also creating environmental problems and forcing more rural people into the cities. At the same time, urbanization increases access to education, jobs, and health care, and these changes help slow population growth overall.

South Asian innovations such as microcredit have brought millions out of extreme poverty, and a recent boom in manufacturing promises to transform the incomes of India's vast population of urban poor. Globalization has already boosted the ranks of India's middle class with jobs outsourced to India from abroad. Although the global recession has shrunk these opportunities, it has also brought a flood of investment from South Asian emigrants that promises to go to job creation, and some migrants are themselves returning to participate in development as experts and entrepreneurs.

South Asia's long history of democratic politics is threatened by rampant corruption, religious nationalism, terrorism, and communal conflict, but in general democracy seems to be expanding, if haltingly. In India a more competitive multiparty system has emerged alongside efforts to reduce corruption, and there is now a certain maturity to public debate that acknowledges that the best antidote for ethnic and religious strife is for all to feel they have a voice in the national conversation.

Pakistan remains highly precarious, with an alarming increase in Islamic extremist activity and a collapse of civil society in several parts of the country. Meanwhile, Bangladesh, once referred to by an American official as an economic "basket case," has given the world microcredit, and after years of military dictatorship, democratic elections have been held with some regularity. Government corruption is a recurring cause for public protest, however, and a paramilitary mutiny in February of 2009 threatened to halt progress toward creating a civil society.

Following years of civil war and authoritarian rule, both Afghanistan and Nepal are now nominal democracies, although political stability in both countries is a distant dream.

The tiny Himalayan kingdom of Bhutan granted its people the right to elect local representatives in 2002 and national leaders in 2008. In Sri Lanka, however, the recent end to a long-standing civil war was not the result of democratic processes, but rather of state force.

Of any world region, South Asia has the largest populations vulnerable to the effects of global climate change and the risks are only likely to increase. While the region is pursuing many important solutions to the climate crisis, such as electric vehicles, alternative energy, and water conservation, it is also on the verge of an explosion of motor vehicle ownership that will very likely overwhelm any energy savings and pollution control achievements. India's surging middle class now has the disposable income to afford Indian-produced economy cars.

In this time of tumultuous change and expanding global connections, South Asia's problems are daunting but no more so than those faced by Mohandas Gandhi, the leader of India's independence movement, when he said, "We must be the change we wish to see." The emerging solutions to pressing global problems today suggest that many in the region have taken these words to heart.

Critical Thinking Questions

1. Dams and the reservoirs they create are sources of irrigation water and generators of electricity. Name some of the factors that have rendered so many places in South Asia, as well as elsewhere around the world, in need of markedly more water and electricity.

2. Explain why some would say that India has been part of globalization for thousands of years. Tie this history to what is happening in the present.

3. What are some of the lingering features of the British colonial era in South Asia? To what extent is globalization reinforcing or erasing these features?

4. How are changes in agriculture resulting in urbanization?

5. Describe microcredit and its impact on extreme poverty in South Asia.

6. Describe how South Asia is vulnerable to climate change. Identify some responses to the climate crisis that are emerging from this region.

7. Describe the main challenges to democracy in South Asia. In what ways is South Asian democracy strengthening?

8. Identify the factors that have led to a recent boom in manufacturing industries in South Asia.

9. What factors have encouraged high population growth in South Asia in the past? What factors may encourage slower population growth in the future?

10. Describe the main factors that create South Asia's gender imbalance.

Chapter Key Terms

agroecology, 375
Buddhism, 362
caste system, 360
civil disobedience, 356
communal conflict, 363
dowry, 365
green revolution, 374
Harappa culture, 353

Hinduism, 360
Indus Valley civilization, 353
Jainism, 362
jati, 361
microcredit, 375
monsoon, 345
Mughals, 353
offshore outsourcing, 377

Partition, 356
purdah, 358
regional conflict, 378
religious nationalism, 382
Sikhism, 363
subcontinent, 344
Taliban, 366
varna, 361

R U S S I A

KAZAKHSTAN

W. Sayan Mts.

Lake
Baikal

Ulan-Ude

Yablonovyy Range

Saryesik Atyrau
Desert

Tannu Ola

Uvs
Nuur

Hövsgöl
Nuur

Hentiyn
Nuruu

Shilka

Argun

Hulun
Nur

Lake
Balkhash

Zhangar

Hovd

Har Us
Nuur

Selenge

Hangayn Mts.

Ulan Bator

Choybalsan

Buir
Nur

Orhon

Kerulan

Greater Khingan Range

Bishkek

Almaty

Yining

Junggar
Basin

Altai Mts.

MONGOLIA

B

Dzamin Uud

NEI MONGOL
(INNER MONGOLIA)

Liao

KYRGYZSTAN

Lake
Issyk

Victory Peak
elev. 24,700

Urumqi

Turfan
Depression

Gobi Desert

Narym

Tian Shan

Kuqa

Bosten
Hu

Old Beds

Hohhot

Zhangjiakou

Beijing

Tangshan

TAJ.

Muztagata
elev. 24,757

Kashi

Tarim

Tarim Basin

Lop
Nor

Huang He (Yellow)

Ordos
Desert

Baotou

Datong

Yinchuan

SHANXI

Taiyuan

Luliang Shan

Shijiazhuang

Baoding

North China Plain

TIANJIN
Tianjin

HEBEI

Bo

PAKISTAN

K2
(Godwin Austen)
elev. 28,250

Taklimakan
Desert

XINJIANG UYGUR

Yutian

Yumen

NINGXIA
HUIZU

Loess
Plateau

Taihang Shan

Handan

Zibo

Jinan

SHANDONG

Qaidam
Basin

Qinghai
Hu

QINGHAI

GANSU

Lanzhou

CHINA

Yellow

Xianyang

Wei

Xian

Luoyang

Zhengzhou

Xuzhou

HENAN

Hongze
Hu

Bengbu

Golmud

Huang He

Kunlun

TIBET
(XIZANG)

Plateau of Tibet

A

Nu Salween

SHAANXI

Qin Ling

Daba Shan

Haianh

Xinyang

Hefei

Nanjing

ANHUI

Lucknow

Himalayas

Xigaze

Lhasa

Gyangze

SICHUAN

Chengdu

HUBEI

Yichang

Three Gorges
Dam

Wuhan

C

Dabie Shan

NEPAL

Kathmandu

Mt. Everest
elev. 29,028

Thimphu

BHUTAN

Gongga Shan
elev. 24,700

Qiaotou

Chongqing

Sichuan
Basin

Yangtze

Changsha

Pingxiang

Pingxiang

Dongting
Hu

Nanchang

Poyang
Hu

JIANGXI

Kanpur

Ganga Plain

Brahmaputra

Dalou Shan

HUNAN

Patna

Ganges

Guiyang

Wujiang Shan

Xiang

Wuyi Shan

Kolkata
(Calcutta)

Dhaka

BANGLADESH

Chittagong

INDIA

Yunnan-Guizhou
Plateau

GUIZHOU

Dian
Hu

Kunming

YUNNAN

GUANGXI
ZHUANGZU

Nanning

Nan Ling

Chu Jiang (Pearl)

GUANGDONG

Guangzhou (Canton)

Rongcheng

Shantou

FUJIAN

D

Hong Kong
(China S.A.R.)

Macao
(China S.A.R.)

Mouths of the Ganga

LAOS

Hanoi

Gulf of
Tonkin

Zhanjiang

Haikou

HAINAN

South China
Sea

Vientiane

THAILAND

Bangkok

VIETNAM

CAMBODIA

Phnom Penh

10°N

110°E

A Plateau of Tibet, China

THEMATIC CONCEPTS

**Population • Gender • Development
Food • Urbanization • Globalization
Democratization • Climate Change • Water**

East Asia

Map labels:

Sea of Okhotsk

Tatar Strait

Kuril Islands

La Perouse Strait

Sikhote Alin Range

Lesser Khingan Range

HEILONGJIANG

Qiqihar

Ranghulu

Harbin

Sungari

Khanka

Hokkaido

Sapporo

Tsugaru Strait

40°N

Changchun

Jilin

JILIN

Northeast China Plain

Changbai Shandi

Sungari Hu

Sea of Japan

Shenyang

Fushun

Benxi

Anshan

Pai T'ou Shan

Yaebuek Sanmack

Sendai

LIAONING

Supung Res.

Dandong Sinuiju

NORTH KOREA

P'yongyang

Korea Bay

Dalian

Hai

Seoul

SOUTH KOREA

Taejon

Taegu

Changwon

Ulsan

Pusan

Kwangju

Hiroshima

Kitakyushu

Okayam

Kobe

Osaka

Nagoya

Shizuoka

Mt. Fuji

Kiso Sammyaku

Mikuni Sammyaku

Honshu

JAPAN

Tokyo

Yokohama

Hamamatsu

Qingdao

Yellow Sea

Shikoku

Fukuoka

Korea Strait

Korean Archipelago

Cheju

Kyushu

JIANGSU

Grand Canal

Zhenjiang

Wuxi

Suzhou

Shanghai

SHANGHAI

Hangzhou

Tai Hu

Ningbo

ZHEJIANG

East China Sea

Osumi Islands

Amimi Islands

Ryukyu Islands (Japan)

Wenzhou

30°N

140°E

Fuzhou

Okinawa Islands

PACIFIC OCEAN

Taipei

Taichung

Sakishima Islands

TAIWAN

Taiwan Strait

Philippine Sea

Kaohsiung

20°N

Bashi Channel

10°N

B Mongolia

C Three Gorges Dam, Yangtze River, China

D Hong Kong Island and South China Sea

E Mount Fuji, Japan

FIGURE 9.1 Regional map of East Asia.

Global Patterns, Local Lives In 2000, at age 18, Li Xia (Li is her family name) left her farming village in China's Sichuan Province for the city of Dongguan in the Guangdong Province of Southern China. She was accompanied by two friends. A few months earlier, the government had taken their families' farmland for an urban real estate project, paying compensation of only U.S.$2000 per family. The three young women accepted an offer to work in a Dongguan toy factory so they could send money back to their families, who now have to pay cash for food and housing.

When the young women arrived in the city, they joined 5 million other recent internal migrants. Sixty percent, like Li Xia, were illegal migrants without residency rights, and thus were dependent on their employers for housing. As did many others, Li Xia and her friends soon found that the labor recruiters had lied about their wages. Per month they would be paid not U.S.$45, as promised, but U.S.$30. In addition, they would work 12-hour shifts in 100°F heat, receive no overtime pay, and have only one day off a month. But there was no point in protesting. Their families in Sichuan needed the money they would send home, and the recruiters purposely brought in thousands of extra workers to replace any complainers.

Xia felt better when she saw that the toy factory was a clean, modern building known locally as the "Palace of Girls" (nearly all 3500 employees were female). Within a day, she had completed her training, signed a 3-year contract, and mastered her task of putting eyes on stuffed animals destined for toddlers in the United States and Europe. Her enthusiasm faded, however, when she learned that she and her friends would be spending much of their money on the expensive but low-quality food provided by the company, and would be sharing one small room and a tiny bath with eight other women and a rat or two.

In less than 6 months, Xia broke her contract and returned home to her village in Sichuan. There, using ideas and assertiveness she had gained from her time in Dongguan, she opened a snack stand. In just one month, she made ten times her investment of U.S.$12. But Xia yearned to return to the southern coast to try again. A few months later, she returned to Dongguan with her sister, leaving the stand to her sister's husband (who also cared for the couple's baby).

Since her first arrival, the city had grown by 20 percent and now had 1400 foreign companies trying to hire thousands of workers. Through connections, Xia and her sister found jobs requiring midlevel skills in a Taiwanese-owned fiber optics factory at twice the wages Xia had earned at the toy factory. One year after her first trip to Dongguan, Xia was making a bit more than U.S.$100 a month, enough to live relatively comfortably with only three roommates. She had prospects for a raise, and she was sending money home and once again saving, this time to open a bar in her home village.

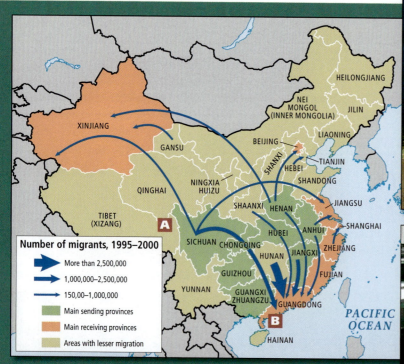

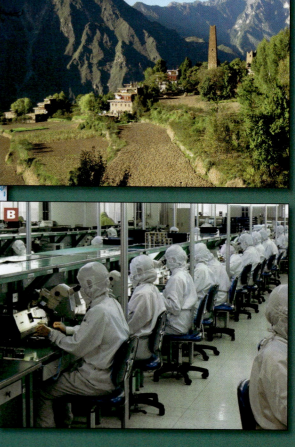

FIGURE 9.2 Workers, development, and urbanization in the new Chinese economy. By 2003, 114 million new workers—one-third of whom were women—had moved from rural areas to urban jobs in China. By 2025 another 243 rural migrants will join them, giving China an overall urban population of over one billion. The map identifies the interior provinces from which most the rural workers are coming and the primarily coastal areas where they find work. **(A)** A village in rural Sichuan province. **(B)** Workers in Dongguan, Guangdong province, assemble and test fiber-optic systems.

TABLE 9.1	Migrant workers and urban employment, 2000–2007		
Year	Rural migrant workers (millions)	Urban employment (millions)	Percent of rural to urban employed
2000	78.49	212.74	36.9
2001	83.99	239.4	35.1
2002	104.7	247.8	42.3
2003	113.9	256.39	44.4
2004	118.23	264.76	44.7
2005	125.78	273.31	46.0
2006	132.12	283.1	46.7
2007	136.49	293.5	46.5

Between 2000 and 2007, the number of people migrating from rural to urban areas in China in search of work has increased 57.5 percent. This number is approaching half of the working people in China's cities.

[Adapted from Cai Fang, Du Yang, and Wang Meiyan, "Human Development Research Paper 2009/09: Migration and Labor Mobility in China" (United Nations Development Programme, April 2009), p. 4, at http://hdr.undp.org/en/reports/global/hdr2009/papers/HDRP_2009_09.pdf.]

[Adapted from Peter S. Goodman, "In China's Cities, a Turn from Factories," Washington Post, September 25, 2004, at http://www.washingtonpost.com/wp-dyn/articles/A48818-2004Sep24.html; Louisa Lim, "The End of Agriculture in China," Reporter's Notebook, National Public Radio, May 19, 2006; with background information from Kathy Chen, "Boom-Town Bound," Wall Street Journal, October 29, 1996, p. A6; and "Life Lessons," Wall Street Journal (July 9, 1997)]. ■

The experiences of Li Xia and her sister illustrate first how the needs of rural areas are being subverted to the needs of China's burgeoning cities. Rural land grabs by developers have increased 15-fold since the mid-1990s, and farmers are rarely given a fair deal. Meanwhile, urbanization, globalization, and changes in gender roles are transforming East Asia as millions of rural young adults flock to work in factories in East Asia's coastal cities, producing goods for sale on global markets, learning new skills, and gaining new confidence (Figure 9.2).

Despite its popularity, in China most rural-to-urban migration is illegal because the **hukou** (household registration) **system,** an ancient practice reinforced in the Maoist era and only now being liberalized, effectively ties rural people to the place of their birth. In a desperate search for work, over 136 million people (Table 9.1) like Li Xia (nearly half of the urban workforce) are ignoring the *hukou* system, but by doing so, they become part of what is called the **floating population,** a term used in China to indicate those who have no rights to housing, schools, or health care. These migrants generally work in menial, low-wage jobs and make agonizing sacrifices to send money home to children and spouses living in rural areas.

China is part of the region of East Asia, home to nearly one-fourth of humanity. This vast territory stretches from the Taklimakan Desert in far western China to Japan's rainy Pacific coastline, and from the frigid mountains of Mongolia in the north to the subtropical forests of China's southeastern coastal provinces (Figure 9.1). East Asia (Figure 9.3) includes the countries of China, Mongolia, North Korea, South Korea, Japan, and Taiwan (the last has operated as an independent country since World War II but is claimed by China as a province). These countries are grouped together because

of their cultural and historical roots, many of which are in China. Because of China's great size, historical influence, enormous population, and huge economy, this chapter gives it particular emphasis. Japan, whose large and prosperous economy makes it a major player on the world stage, is also emphasized.

hukou system the system in China by which citizens' permanent residence is registered

floating population the Chinese term for jobless or underemployed people who have left economically depressed rural areas for the cities and move from place to place looking for work

THINGS TO REMEMBER

1. The shift of workers from rural China to cities includes at least 136 million workers who have moved without legal papers and who now constitute nearly half (46.5 percent) of China's urban labor force.

2. Most rural workers who have moved to China's cities without official permission are in low-paying jobs with little or no access to social services or education.

FIGURE 9.3 Political map of East Asia.

Urbanization, Pollution, Development, and Globalization:
Rapid urban economic development combined with weak environmental protections has resulted in some of the fastest-growing and most polluted cities on the planet. In China, urbanization has been driven in large part by the opening of China's economy to the global economy.

(A) Steelworkers arrive at work in Guangzhou, China. Urban populations in China have grown by 450 million since 1980.

(B) The busy port of Shanghai, where shipping containers are loaded onto vessels destined mainly for the United States, Japan, Korea, and Europe.

(C) Smog hangs over the city of Xian, China, where air pollution comes from a mix of poorly regulated industries and vehicles.

Climate Change and Water:
East Asia has long suffered from droughts and devastating floods, and scientists fear that these hazards will intensify with climate change. Meanwhile, water flows in major Chinese rivers could diminish over the next few decades as glaciers on the Tibetan plateau melt.

(D) A flash flood in Beijing after a sudden downpour. Shifting rainfall patterns could cause flooding throughout East Asia.

(E) The Li River in southern China at an extremely low level. Flows in many rivers have diminished in recent years.

(F) A glacier in Tibet and the lake beneath it that is growing as the glacier melts. Tibet is experiencing increased flooding and landslides.

Development, Globalization, and Democratization:
As levels of economic development and globalization have increased throughout East Asia, pressures for democratization have also risen. Japan, South Korea, and Taiwan are now among the world's more democratized places, and pressures are increasing in China and North Korea for greater democratization.

(G) Japanese students protest against government foreign policy in Kyoto.

(H) A civil servant in North Korea guides voting. All "candidates" run unopposed in North Korea's non-competitive elections.

(I) Chinese police patrol a Tibetan town. Pressure for democracy is growing in Tibet and among Tibetans living outside China.

Food, Globalization, and Development:
As East Asian countries have developed economically, they have become more able to buy food on the global market and less able to produce enough food domestically. Much agricultural land has been lost to urban and industrial expansion or to mismanagement. Meanwhile, people are demanding more foods that must be imported.

(J) Urban encroachment on farmland outside Beijing. Since 1996, 12 percent of farmland has been lost to urbanization, and another 8 to 21 percent may be lost by 2030.

(K) Kebabs for sale in Xian, China. Meat consumption has quadrupled since 1980, and China is now the world's largest importer of animal feed.

(L) Tuna from across the globe are inspected at Tokyo's Tsukiji fish market. Japan's 2 percent of the world's population consumes 15 percent of the global fish catch.

Population and Aging:
Across East Asia, birth rates are slowing and life expectancy is increasing. Economies may be strained because of this, as fewer working-age people must support more retirees. China is struggling to train enough nurses and doctors, and build more nursing homes and other infrastructure needed by the elderly. Meanwhile, Japan is turning to technology for solutions.

(M) Elderly retirees keep fit by practicing martial arts in a Beijing park.

(N) An expanding hospital in Hong Kong, China.

(O) A display showcasing a robotic interface for elderly patients in Japan. The robots "watch" patients, monitoring how long it takes them to respond to questions or perform tasks, and also provide an interface for patients to speak with caregivers.

Population and Gender in China:
China's "one-child policy" has created a shortage of women. Cultural preferences for male children mean that many families choose an abortion if their "one child" is female, or else put her up for adoption. The shortage of women willing to marry is even greater, as so many women want to postpone family life and pursue careers instead.

(P) An orphanage in Xian, Shaanxi Province, full of female babies available for adoption. Foreigners will adopt many of them.

(Q) Graduates of a medical school in Nanning, Guangxi Province. Many women are postponing or forgoing marriage to pursue careers.

(R) Bachelors playing pool in Beijing. There are approximately 32 million more boys than girls under age 20 in China.

I THE GEOGRAPHIC SETTING

Terms in This Chapter

East Asian place-names can be very confusing to English-speaking readers. We give place-names in English transliterations of the appropriate Asian language, taking care to avoid redundancies. For example, *he* and *jiang* are both Chinese words for *river*. Hence the Yellow River is the Huang He, and the Long River, also called the Yangtze in English, is the Chang Jiang in China; it is redundant to add the term *river* to either name. The word *shan* appears in many place-names and usually means "mountain."

Pinyin (a spelling system based on Chinese sounds) versions of Chinese place-names are now commonplace. For example, the city once called Peking in English is now Beijing, and Canton is Guangzhou.

The region popularly known as Manchuria is here referred to as China's Far Northeast to emphasize its geographical location. Although China refers to Tibet as Xizang, people around the world who support the idea of Tibetan self-government avoid using that name. This text uses Tibet for the region (with Xizang in parentheses), and Tibetans for the people who live there.

Physical Patterns

A quick look at the regional map of East Asia (see Figure 9.1 on pages 384–385) reveals that its topography is perhaps the most rugged in the world. East Asia's varied climates result from a dynamic interaction between huge warm and cool air masses and the land and oceans. The region's rapidly expanding human populations have affected the variety of ecosystems that have evolved there over the millennia and that still contain many important and unique habitats.

tsunami a large sea wave caused by an earthquake

Landforms

The complex topography of East Asia is partially the result of the slow-motion collision of the Indian subcontinent with the southern edge of Eurasia over the past 60 million years. This tremendous force created the Himalayas and lifted up the Plateau of Tibet (depicted in gray and gold in Figure 9.1), which can be considered the highest of four descending steps that define the landforms of mainland East Asia, moving roughly west to east.

The second step down from the Himalayas is a broad arc of basins, plateaus, and low mountain ranges (depicted in yellowish tan in Figure 9.1). These landforms include the broad, rolling highland grasslands and deep, dry basins and deserts of Western China (Taklimakan Desert, Junggar Basin) and Mongolia and also include the Sichuan Basin and the rugged Yunnan–Guizhou Plateau to the south, which is dominated by a system of deeply folded mountains and valleys that bend south through the Southeast Asian peninsula.

The third step, directly east of this upland zone, consists mainly of broad coastal plains and the deltas of China's great rivers (shown in shades of green in Figure 9.1). Starting from the south, this step is defined by three large lowland river basins: the Zhu Jiang (Pearl River) basin, the massive Chang Jiang basin, and the lowland basin of the Huang He on the North China Plain. Each of these rivers has a large delta. Despite deltas being subject to periodic flooding, they have long been used for agriculture; but now coastal cities have spread into these deltas, filling in wetlands. Each is now a zone of dense population and industrialization (see Photo Essay 9.6 Map on page 422). Low mountains and hills (shown in light brown) separate these river basins. China's far northeast and the Korean Peninsula are also part of this third step.

The fourth step consists of the continental shelf, covered by the waters of the Yellow Sea, the East China Sea, and the South China Sea. Numerous islands—including Hong Kong, Hainan, and Taiwan—are anchored on this continental shelf; all are part of the Asian landmass.

The islands of Japan have a different geological origin: they are volcanic, not part of the continental shelf. They rise out of the waters of the northwestern Pacific in the highly unstable zone where the Pacific, Philippine, and Eurasian plates grind against one another. Lying along a portion of the Pacific Ring of Fire (see Figure 1.27 on page 52), the entire Japanese island chain is particularly vulnerable to disastrous eruptions, earthquakes, and **tsunamis** (seismic sea waves). The volcanic Mount Fuji, perhaps Japan's most recognizable symbol (Figure 9.4), last erupted in 1707. However, the mountain is still classed as active and deep internal rumblings have been detected since 2001.

The East Asian landmass has few flat portions, and most flat land is either very dry or very cold. Consequently, the large

FIGURE 9.4 Mount Fuji. Japan's highest peak, at 12,388 feet (3776 meters), provides a stately distant backdrop to a woman working in tea fields. At one time considered a sacred mountain, Mount Fuji today attracts some 200,000 climbers annually, 30 percent of them foreigners.

numbers of people who occupy the region have had to be particularly inventive in creating spaces for agriculture. They have cleared and terraced entire mountain ranges, until recently using only simple hand tools (see Photo Essay 9.2A on page 394). They have irrigated drylands with water from melted snow, drained wetlands using elaborate levees and dams, and applied their complex knowledge of horticulture and animal husbandry to help plants and animals flourish in difficult conditions.

Climate

East Asia has two principal contrasting climate zones (see Photo Essay 9.1 on page 392): the dry interior west and the wet (monsoon) east. Recall from Chapter 8 that the term *monsoon* refers to the seasonal reversal of surface winds that flow from the Eurasian continent to the surrounding oceans during winter and from the oceans inland during summer.

The Dry Interior Because land heats up and cools off more rapidly than water, the interiors of large landmasses in the midlatitudes tend to experience intense cold in winter and powerful heat in summer. Western East Asia, roughly corresponding to the first two topographic steps described above, is an extreme example of such a midlatitude continental climate because it is very dry. With little vegetation or cloud cover to retain the warmth of the sun after nightfall, summer daytime and nighttime temperatures may vary by as much as 100°F (55°C).

Grasslands and deserts of several types cover most of the land in this dry region. Only scattered forests grow on the few relatively well-watered mountain slopes and in protected valleys supplied with water by snowmelt. In all of East Asia, humans and their impacts are least conspicuous in the large, uninhabited portions of the deserts of Tibet (Xizang), the Tarim Basin in Xinjiang, and the Mongolian Plateau.

The Monsoon East The monsoon climates of the east are influenced by the extremely cold conditions of the huge Eurasian landmass in the winter and the warm temperatures of the surrounding seas and oceans in the summer. During the dry winter monsoon, descending frigid air sweeps south and east through East Asia, producing long, bitter winters on the Mongolian Plateau, on the North China Plain, and in China's Far Northeast subregion. While occasional freezes may reach as far as southern China, winters there are shorter and less severe. The cold air of the dry winter monsoon is partially deflected by the east-west mountain ranges of the Qin Ling, and the warm waters of the South China Sea moderate temperatures on land.

During the summer monsoon, as the continent warms, the air above it rises, pulling in wet tropical air from the adjacent seas (see Photo Essay 9.1). The warm, wet air from the ocean deposits moisture on the land as seasonal rains. As the summer monsoon moves northwest, it must cross numerous mountain ranges and displace cooler air. Consequently, its effect is weakened toward the northwest. Thus the Zhu Jiang basin in the far southeast is drenched with rain and enjoys warm weather for most of the year, whereas the Chang Jiang basin, which lies in central China to the north of the Nan Ling range, receives only about 5 months of summer monsoon weather. The North China Plain, north of the Qin Ling and Dabie Shan ranges, receives only about 3 months of monsoon rain. Very little monsoon rain reaches the dry interior.

China's Far Northeast subregion is wet in summer, and neighboring Korea and Japan have wet climates year-round because of their proximity to the sea. All of these areas still have hot summers and cold winters because of their northerly location and exposure to the continental effects of the huge Eurasian landmass. Japan and Taiwan actually receive monsoon rains twice: once in spring, when the main monsoon moves toward the land, and again in autumn, as the winter monsoon forces warm air off the continent. This retreating warm air picks up moisture over the coastal seas, which is then deposited on the islands. Much of Japan's autumn precipitation falls as snow.

Natural Hazards The entire coastal zone of East Asia is intermittently subjected to **typhoons** (tropical storms, hurricanes). Japan's location along the northwestern edge of the Pacific Ring of Fire means that it has many volcanoes, and it also experiences earthquakes and tsunamis. These natural hazards are a constant threat in Japan; the heavily populated zone from Tokyo southwest through the Inland Sea (between Shikoku Island and southern Honshu) is particularly endangered. Earthquakes are also a serious natural hazard in Taiwan and in China's mountainous interior.

typhoon a tropical storm or hurricane

THINGS TO REMEMBER

1. There are four main topographical zones, or "steps," that form the East Asian continent.

2. Japan was created by volcanic activity along the Pacific Ring of Fire.

3. East Asia has two principal contrasting climates: the dry continental interior (west) and the monsoon east.

Environmental Issues

Today, East Asia's worst environmental problems result from high population density combined with rapid urbanization and environmentally insensitive economic development (see Photo Essay 9.2 on page 394).

Climate Change: Emissions and Vulnerability in East Asia

Climate change, especially global warming, is a growing concern in East Asia. China now leads the United States as the world's largest overall producer of greenhouse gases (but not on a per capita basis; see Chapter 1, page 47), and Japan is also a major emitter. China has recently given attention to the need to limit its emissions, perhaps out of recognition of its responsibility as an increasingly influential global economic power; but leading scientists say that even with large increases in efficiency through new technologies, China's greenhouse gas emissions will double by 2030 as its urban households adopt Western lifeways: cars, air conditioning, appliances, computers.

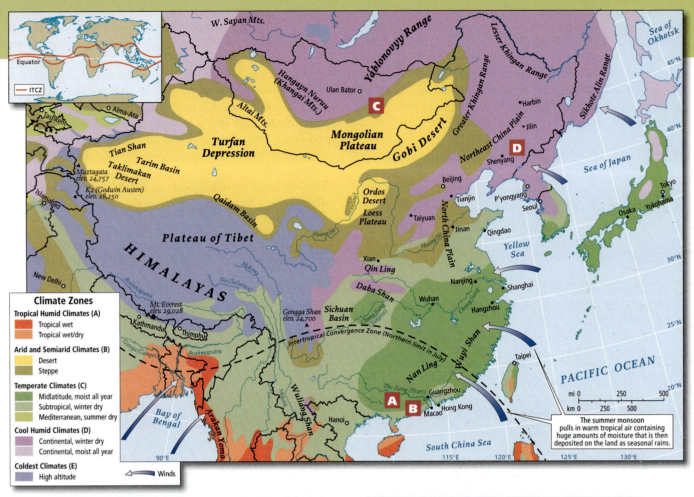

Climate Zones

Tropical Humid Climates (A)
- Tropical wet
- Tropical wet/dry

Arid and Semiarid Climates (B)
- Desert
- Steppe

Temperate Climates (C)
- Midlatitude, moist all year
- Subtropical, winter dry
- Mediterranean, summer dry

Cool Humid Climates (D)
- Continental, winter dry
- Continental, moist all year

Coldest Climates (E)
- High altitude

→ Winds

The summer monsoon pulls in warm tropical air containing huge amounts of moisture that is then deposited on the land as seasonal rains.

A Midlatitude, moist all year, Guilin, China

B Subtropical, winter dry, Macao, China

C Steppe, Mongolia

D Continental, winter dry, Liaoning, China

China's Vulnerability to Glacial Melting Glaciers on the Tibetan Plateau capture monsoon moisture and store it frozen for slow release. Two of China's largest rivers, the Huang He and the Chang Jiang, are partially fed by these glaciers, which are now melting so rapidly that scientists predict they will eventually disappear (see Photo Essay 9.2A). One result could be significantly lower flows in these two great rivers during the winter when little rain falls. Both have already begun to run low during winter, and trade is suffering as river boats and barges become stranded on sandbars. Irrigation for dry-season farming could also suffer.

Water Shortages Nearly every year, abnormally low rainfall or abnormally high temperatures result in a drought somewhere in China. These droughts often cause more suffering and damage than any other natural hazard.

Droughts can be worsened by human activity on a local or regional scale. When people begin to live or farm in dry environments, as many millions have done in China and Mongolia during the twentieth century, *desertification* (see Chapter 6, page 258) can result. Natural vegetation is cleared to grow crops, which have higher water requirements than the natural vegetation, so the crops must be irrigated with water pumped to the surface from underground aquifers. Many dry areas in China are subject to strong winds that can blow away topsoil once natural vegetation is removed. Further, irrigated crops are often less able to hold the soil than the natural vegetation. As a result, huge dust storms have plagued China in recent years. High dunes of dirt and sand have appeared almost overnight in some areas that border the desert, threatening crops, roads, and homes (see Photo Essay 9.2D). Particulate matter from these dust storms circles the globe in the upper atmosphere.

Personal Vignette In Ningxia Huizu Autonomous Region on the Loess Plateau in northwest China, Wang Youde squints out at what are now sand-colored low hills barren of vegetation. He explains how these vast tracts of former farmland have been transformed into deserts by a combination of human error and climate change.

The area was already prone to drought, and then agricultural expansion into this fragile, wind-deposited soil led to the removal of thick, natural, deep-rooted grasses. One can still see the agricultural terraces on the arid slopes where now not even grass grows. Wang Youde's family and 30,000 others fled the area when Wang was 10 years old because one day a sand dune covered their village. From his early childhood, he remembers flowers, bird song, and occasional snowfalls; all have vanished. Now Wang is back, heading up a project to revegetate thousands of hectares with drought-resistant plants that will hold the soil. Squares of braided straw are laid down by hand to protect the seedlings. The seedling survival rate is only 30 to 50 percent. Yet, Wang says, "Every time we see an oasis that we have created we are very satisfied... because we have poured sweat and blood into our work." His adult children are helping him, hoping to remain with the family and escape the hardships of migrating to find urban factory work. [Adapted from Maria Siow, "Desertification: one of the challenges faced by China." Asia Pacific News, October 29, 2009. http://www.channelnewsasia.com/stories/eastasia/view/1014326/1/.html.] ■

Water shortages also have been particularly intense in the North China Plain, which produces half of China's wheat and a third of its corn. Here the water table is falling more than 10 feet a year due to increased use of groundwater for irrigation and urban needs. Meanwhile, withdrawals of water from the Huang He often make its lower sections run completely dry during the winter and spring.

Japan, the Koreas, and Taiwan are less vulnerable to drought than China and Mongolia due to monsoon rainfall patterns, which gives them a generally wetter climate.

Water Issues, Climate Change, and Urbanization in Mongolia Mongolia's location on a dry, high, landlocked plateau between China and Russia means that access to water has always been an issue for the inhabitants. The climate of Mongolia consists of mostly desert and steppe, with a somewhat moister zone in the upland central north (see Photo Essay 9.1C and Map), where the capital, Ulan Bator, is located. Global climate change appears to be decreasing the amount and regularity of what is already scarce rainfall in Mongolia.

Mongolia's traditional economy of nomadic herding was essentially a strategy to deal with scarce water. Living with their families in collapsible portable homes called *gers*, the herders study seasonal variations in the availability of water across wide spans of grazing lands and move their herds and their families on a cyclical basis to take advantage of available water, fresh grasslands, and other natural resources.

Herding is still the means of livelihood for about 50 percent of the population, but social changes brought about first by central planning during Mongolia's experiment with communism and now by marketization in the post-communist era mean that a difficult water situation has gotten worse. Herding has become more precarious, partly because of drought but also because it does not produce the amount of cash that families now need. So herding families are bringing their *ger* homes and settling in ever greater numbers around the fringes of Ulan Bator, hoping to find work in the city. These *ger* suburbs are said to house 40 percent of the entire country's population. Only a few neighborhoods have piped water; the vast majority (65 percent) must collect their household and animal water from sparsely located standpipes and must pay for what they collect even though it is not guaranteed to be safe or consistently available.

Flooding in Central China The same shifting patterns of rainfall that may worsen droughts can also worsen flooding. Under usual conditions, the huge amounts of rain deposited on eastern China during the summer monsoon periodically can cause catastrophic floods along the major rivers. If global warming leads to even slight changes in rainfall patterns, flooding could be much more severe (see Photo Essay 9.2C). Engineers have constructed elaborate systems of dikes, dams, reservoirs, and artificial lakes to help control flooding. However, these systems failed in 2004, when heavy rains in the Chang Jiang basin caused some of the worst flooding in 2 centuries. Two hundred and forty million people were affected, with 3656 drowned and 14 million left homeless.

China is particularly vulnerable to drought, desertification, flooding, and other hazards that climate change may intensify.

A The terraced hillsides of China's eastern Gansu Province are in a dry upland zone where agriculture depends on rainfall or irrigation water taken from rivers or underground aquifers. Rainfall is already unpredictable and could become more so with climate change. Melting glaciers on the Tibetan Plateau threaten to reduce river flows, and overuse of groundwater for irrigation is depleting aquifers. Large poor rural populations are increasingly being forced to relocate elsewhere in China.

B A resident of Beijing takes water from a canal. Beijing faces severe water shortages, and projected massive increases in population are prompting the government to consider massive canal projects to divert water from distant rivers in southern and western China.

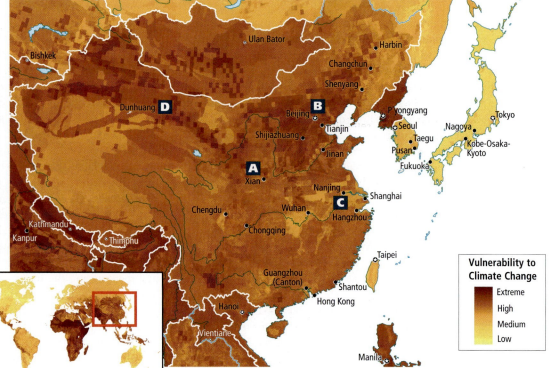

Vulnerability to Climate Change

Extreme
High
Medium
Low

C Flooding hits the Chang Jiang at Nanjing. Climate change could result in changes in rainfall patterns that would increase flooding. This is especially likely in eastern China, where large amounts of rain fall during the summer monsoon.

D Sand dunes loom over Dunhuang, China, where ongoing desertification could intensify if climate change brings more droughts. Past attempts to farm dry areas in western China have led to desertification, especially when droughts are accompanied by strong winds that blow sand in from nearby deserts.

Responses to the Climate Crisis East Asia is increasingly focused on responding to the warming aspect of global climate change. Japan has led such efforts for decades. In 1997, the Japanese city of Kyoto hosted the meeting in which countries first committed to reduce their greenhouse gas emissions. Japanese automakers such as Toyota were among the first to develop and sell hybrid gas-electric vehicles, and Japan is now second only to Germany in its installed solar power–generating capacity. However, Japan has seen little actual reduction of its greenhouse gas emissions.

China's options for handling its water shortages are more limited than Japan's, which receives plentiful rain. With few untapped sources of water, China is focusing on conservation. Already, 30 percent of China's urban water is recycled, and many cities are trying to raise this percentage. Official support for these conservationist policies is indicative of a major new national effort to remove pollutants from wastewater discharged by industry and farming.

Food Security and Sustainability in East Asia

East Asia's **food security**—the capacity of the people in a geographic area to consistently provide themselves with adequate food—is increasingly linked to the global economy. The wealthiest countries in the region are already heavily dependent on food imports from around the globe. About three-quarters of the food consumed in Japan, South Korea, and Taiwan is imported. In China, on the other hand, self-sufficiency in grain production is important to national identity because hideous famines were a recurring problem before the Communist Revolution. China is now nearly self-sufficient with regard to basic necessities, but it relies on imports of important commodities to supply growing demands for luxury food items including grain for animal feed and soybeans for high-quality cooking oil (see below).

From one perspective, high levels of imported food are a sign of increasing food security in East Asia because countries with competitive globalized economies can afford to bring in food from elsewhere. However, recent dramatic increases in global food prices illustrate the perils of dependence on food imports, especially for the poor. In 2007, for example, the world market price for corn shot up because of proposals to use corn to produce ethanol fuel as a replacement for gasoline in the Americas and elsewhere. 🎞 **217. Food Prices Skyrocketing in China**

Food Production Only a relatively small portion of East Asia's vast territory can support agriculture (see Figure 9.13 on page 409). In much of this area, food production has been pushed well beyond what can be sustained over the long term. As a result, East Asia's fertile zones are shrinking. In China, roughly one-fifth of the agricultural land has been lost since 1949 due to urban and industrial expansion (see Thematic Concepts Part J on page 389) and to agricultural mismanagement that created soil erosion and desertification (see the vignette on page 393 and Photo Essay 9.2D). Also, as urban populations grow more affluent, their consumption of meat and other animal products requires more land and resources than the plant-based national diet of the past. Soybeans, which used to be important as an export for China, are illustrative of this change in diet. More than 70 percent of the soybeans used in China are now imported, primarily from the United

States, Brazil, and Argentina. Soybeans are no longer primarily used for human food but for animal feed, for farmed fish food, and especially for high-grade cooking oil. China prefers imported soybeans for oil because these beans are genetically modified (GMO) to produce more oil than locally grown varieties.

Rice Cultivation The region's most important grain is rice, and over the millennia its cultivation has dramatically transformed landscapes throughout central and southern China, Japan, Korea, and Taiwan. In these areas, rainfall is sufficient to sustain **wet rice cultivation**, which is capable of being highly productive.

Wet rice cultivation requires elaborate systems of water management, as the roots of the plants must be submerged in water early in the growing season. Centuries of painstaking human effort have channeled rivers into intricate irrigation systems, and whole mountainsides have been transformed into descending terraces that evenly distribute the water. Writing about wet rice cultivation in Sichuan Province, geographer Chiao-Min Hsieh describes how "[e]verywhere one can hear water gurgling like music as it brings life and growth to the farms." However, these same cultivation techniques, combined with extensive forestry and mining, have led to the loss of most natural habitat in all but the most mountainous, dry, or remote areas. For these reasons, further expansion of the area under wet rice cultivation is unlikely.

Fisheries and Globalization Many East Asians depend heavily on ocean-caught fish for protein. The Japanese in particular have had a huge impact on the seas of not only this region but of the entire world. With only 2 percent of the world's population, Japan consumes 15 percent of the global wild fish catch. There are some 4000 coastal fishing villages in Japan, sending out tens of thousands of small crafts to work nearby waters each day. Japan also sends out large fishing ships, complete with onboard canneries and freezers, to harvest oceans around the world. Japan has long been criticized by environmentalists for overfishing tuna around the globe and for its consumption of endangered marine species, especially whales. Japan is also criticized by those concerned with sustainable local fisheries for dominating the waters off Africa to the extent that African fish catches are substantially reduced and fishers have been forced to migrate to Europe for work. With more than 75 percent of the world's fisheries either fully exploited or in decline, there is little room for Japan to expand its fish intake.

Three Gorges Dam: The Power of Water

The Chinese environmental activist, Dai Qing, notes that China has 22,000 large dams, all of which have displaced people—perhaps as many as 60 million—without any attention to their rights as stakeholders in the projects. The Three Gorges Dam (see Photo Essay 9.3A on page 398) is at this time the largest dam in the world at 600 feet (183 meters) high and 1.4 miles (2.3 kilometers) wide and the second largest engineering project in China's history. It is designed to improve navigation on the Chang Jiang (Yangtze) and control flooding, but it is most lauded for its role in generating hydroelectricity for this energy-hungry country that longs to improve its greenhouse gas emissions record.

Many experts involved with the Three Gorges project see serious design flaws. Of greatest concern is the dam's position above a seismic fault. The dam was built at the east end of the Three Gorges because the deep canyons provided a prodigious reservoir for water to power turbines, providing electricity for all of Central China from the sea to the Tibetan plateau. Unfortunately, the enormous weight of the water and its percolation through geological fissures in the 370-mile-long (600-kilometer-long) reservoir behind the dam could trigger earthquakes. Already at issue are huge landslides along the gorges lubricated by the rising reservoir water. As many as 80 visible cracks in the dam raise doubts about its structural integrity. Similar defects led to the failure of China's much smaller Banqiao Dam in 1975, which was responsible for 171,000 deaths due to flooding and an ensuing famine. Even if the dam holds, its power generation potential will probably be reduced by the buildup of eroded silt behind the dam.

Any failure of the dam would be a financial as well as human disaster. Official construction costs are $25 billion, but the real costs may be three times this figure, due in part to unforeseen negative environmental and social impacts and to theft from the project by corrupt officials.

Also of concern are the incalculable costs associated with relocating the 1.23 million people who once lived where the dam now forms a reservoir. Thirteen major cities have been submerged, along with 140 large towns, hundreds of small villages, 1600 factories, and 62,000 acres (25,000 hectares) of farmland. The reservoir has destroyed important archaeological sites, as well as some of China's most spectacular natural scenery. There are significant environmental costs as well. The giant sturgeon, for example, a fish that can weigh as much as three-quarters of a ton and is as rare as China's giant panda, may become extinct. Sturgeon used to swim more than 1000 miles (1600 kilometers) up the Chang Jiang past the location of the dam to spawn. Now the sturgeon's reproductive process has been irretrievably interrupted.

▶ **209. Three Gorges Dam Leaves Some Chinese Swamped**

The plan to build Three Gorges came just as UN development specialists were deciding that the benefits dams could bring did not sufficiently outweigh the many problems they caused. Decades ago, international funding sources such as the World Bank withdrew their support of the dam because of concerns over the social and environmental costs and other shortcomings of the project. However, Chinese industrialists who need the energy, construction companies that have prospered from building the dam and its many ancillary projects, and government officials eager to impress the world and leave their mark on China continue not only to support the *Da Ba* (the Big Dam), but to seek other global locations where China can gain influence and profit by building dams.

Air Pollution: Choking on Success

Air pollution is often severe throughout East Asia, but the air quality in China's cities is the worst in the world. Coal burning is the primary cause (see **Photo Essay 9.3C**). China is the world's largest consumer and producer of coal, accounting for 40 percent of all the coal burned on the planet each year. Between 1975 and 2005, China's coal consumption more than quadrupled. By 2006, China was bringing a new coal-fired power plant online every week.

Air Pollution in China The combustion of coal releases high levels of two pollutants—suspended particulates and sulfur dioxide, both of which can cause respiratory ailments (**Figure 9.5**). In Chinese cities, these emissions can be ten times higher than World Health Organization guidelines. Air pollution is worst in urban areas, where industries are concentrated and where there are large numbers of buildings to be heated. The worst pollution is in northeastern China, where homes are in close proximity to industries that also depend heavily on coal for fuel.

Personal Vignette Every winter the elderly couple got through the biting Beijing winters by feeding 1200 one-kilogram coal bricks into a small iron stove. The ashes and coal dust blackened their belongings, and they worried about carbon monoxide poisoning. Then, in 2009 they were suddenly given a new electric space heater by the city government. In a move to clean up Beijing's air and reduce the city's contribution to China's greenhouse gas emissions, the city replaced nearly 100,000 old coal stoves with electric heaters and cut electric nighttime rates to just 3 cents per kilowatt-hour. The couple was astonished with the change the new cheap heater made in their lives. But of course, 100,000 more homes were now dependent on electrical power, most of it generated by coal-fired power plants. [*Source: Michael Wines, Beijing's Air Is Cleaner, but Far from Clean, New York Times/International Herald Tribune, Global Edition, October 16, 2009. http://www.nytimes.com/2009/10/17/world/asia/17beijing.html?partner=rss&emc=rss.?*] ∎

Sulfur dioxide from coal burning also contributes to acid rain, which is displaced to the northeast by prevailing winds, reaching as far as Korea, Japan, Taiwan, and beyond. **Photo Essay 9.3D** and the map above it show the zones of heavy pollution, depicting Japan and Taiwan as particularly afflicted (see also **Photo Essay 9.3B**). Particulates from China's coal burning are transported globally by high-altitude west-to-east flowing jet streams, thus affecting air quality in North America and Europe.

Air pollution from vehicles is also severe, even though the use of cars for personal transportation in China is just beginning. For years, China's vehicles have had very high rates of lead and carbon dioxide emissions. For example, Beijing has much worse air quality than Los Angeles, even though it has only one-tenth the number of vehicles. Very recently, the government has taken action. As of 2008, new Chinese cars had to meet EU standards for mileage and emissions—much stricter than those for U.S. cars. Even so, to provide decent air quality during the 2008 summer Olympics, China had to remove half of Beijing's cars from the road.

▶ **203. Worldwatch Institute:**
16 of 20 of World's Most-Polluted Cities in China

Air Pollution Elsewhere in East Asia Public health risks related to air pollution are serious in the largest cities of Japan (Tokyo and Osaka), Taiwan (Taipei), Mongolia (Ulan Bator), and South Korea (Seoul), and in adjacent industrial zones. Even with antipollution

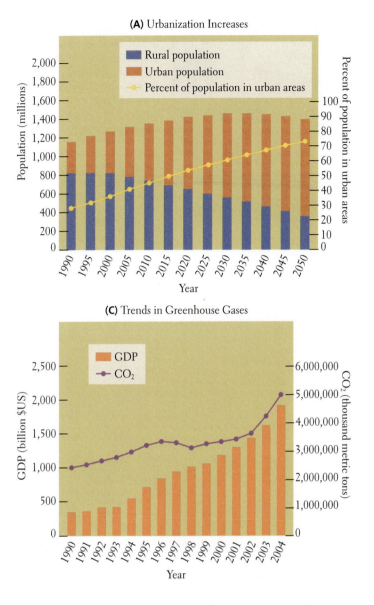

(A) Urbanization Increases

(B) Energy Consumption

(C) Trends in Greenhouse Gases

FIGURE 9.5 Urbanization, economic growth, and CO₂ emissions in China. As urban areas grow larger, GDP is increasing and the cities put increased pressure on available resources and on clean air. As China's GDP increases especially in urban areas, demands for energy for households, industry, and transportation also go up. **(A)** By 2050, projections indicate that 73 percent of China's population will live in cities. **(B)** As China's GDP continues to grow, so does its energy consumption. Between 1990 and 2005, GDP rose about 385 percent and energy consumption rose 148.5 percent, and is projected to rise another 131.2 percent by 2030. **(C)** As GDP increases, CO₂ emissions also rise. Between 1990 and 2004, CO₂ emissions rose 108.7 percent. Even if China improves its environmental regulation and enforcement, CO₂ emissions are projected to increase another 10 percent by 2020. [Adapted from "Clean Air in the People's Republic of China: Summary of Progress on Improving Air Quality," Country Network China, CAI-Asia Center China Program, November 2008, at http://www.cleanairnet.org/caiasia/1412/articles-70822_PRC.pdf.]

legislation and increased enforcement, high population densities and rising expectations for better living standards make it difficult to improve environmental quality. Taiwan is a case in point.

Taiwan has some of the dirtiest air on earth (see **Photo Essay 9.3 Map**). Some of the main causes are the island's extreme population density of 1600 people per square mile (615 per square kilometer), its high rate of industrialization, and its close proximity to industrialized south China. In Taiwan there are now 4 motor vehicles (cars or motorcycles) for every 5 residents—more than 16.5 million exhaust-producing vehicles on this small island. In addition, there are nearly 8 registered factories for every square mile (3 per square kilometer), all emitting waste gases. The government of Taiwan acknowledges that the air is six times dirtier than that of the United States or Europe.

Both North and South Korea depend on hydro- or nuclear-generated electricity to run factories and heat buildings, unlike elsewhere in East Asia. North Korea has relatively few industries and uses very few cars, so its air pollution levels are thought to

be low for the region. South Korea uses fossil fuels in its many industries and has many gas-powered cars. These are the sources of most of its internally generated air pollution. Mongolia generally has the cleanest air, but in Ulan Bator, NGOs have funded some imaginative projects—for example, centrally located boilers for hot-water-heating systems for whole sections of the city, and general insulation projects—to address the problems that pollution from coal heaters and car exhaust have produced.

THINGS TO REMEMBER

1. East Asia has long suffered from lengthy searing droughts and devastating floods. These disruptions are likely to increase with global climate change.

2. China's main river systems will be affected by the melting of its glaciers.

3. About three-quarters of the food consumed in Japan, South Korea, and Taiwan is imported. China is currently self-sufficient

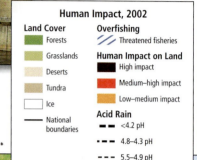

Economic development in East Asia has often proceeded without adequate environmental protection or safeguards for human health. Here we focus on the Three Gorges Dam, air pollution, and energy issues.

A China's Three Gorges Dam is the largest dam in the world. It is capable of supplying about 3 percent of China's electricity needs, resulting in significantly less air pollution than the several large coal-fired power plants that it replaces. However, the dam has also resulted in significant environmental damage due to its 370 mile (600 km)-long reservoir, which has submerged habitat for endangered species and the dwellings of 1.2 million people. Structural and design flaws in the dam threaten further environmental damage.

B A commuter in Tokyo wears a mask for protection against the city's air pollution. Recent restrictions on the use of diesel vehicles in Tokyo have resulted in improved local air quality.

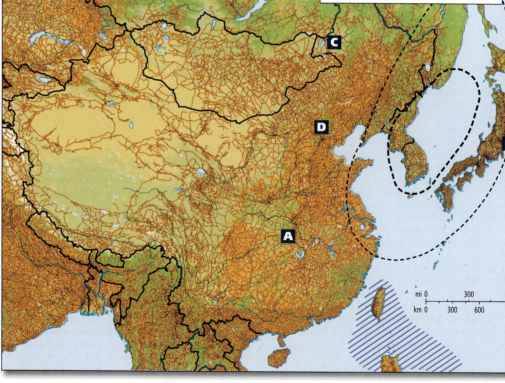

Human Impact, 2002

Land Cover
- Forests
- Grasslands
- Deserts
- Tundra
- Ice
- —— National boundaries

Overfishing
- Threatened fisheries

Human Impact on Land
- High impact
- Medium–high impact
- Low–medium impact

Acid Rain
- ▬ ▬ <4.2 pH
- ▪ ▪ ▪ 4.8–4.3 pH
- – – – 5.5–4.9 pH

mi 0   300   600
km 0   300   600

C An open pit coal mine in China's province of Inner Mongolia. China is already the world's largest producer and consumer of coal, and its demand for coal is projected to double by 2030, making it then by far the largest contributor of greenhouse gases.

D A toxic haze of air pollution engulfs Beijing and the North China Plain, often blowing eastward to reach Korea, Japan, and even the United States and Europe. China's coal-fired power plants, industries, and vehicles are emerging as major global environmental concerns.

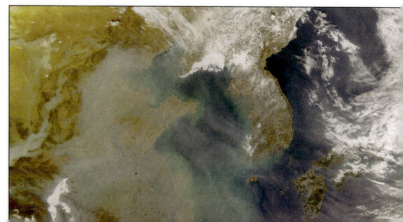

with regard to basic necessities such as rice, but it is highly dependent on imports for other foods.

4. In an effort to address droughts, floods, and the burgeoning need for electricity, China has built many dams, the largest of which is the Three Gorges Dam, which itself now poses many potential problems.

5. Rapid urban economic development in East Asia, combined with weak environmental protections, has resulted in some of the fastest-growing and most polluted cities on the planet.

Human Patterns over Time

East Asia is home to some of the most ancient civilizations on earth. Settled agricultural societies have flourished in China for more than 7000 years.

Chinese civilization evolved from several hearths, including the North China Plain, the Sichuan Basin, and the lands of interior Asia that were inhabited by Mongolian nomadic pastoralists. On East Asia's eastern fringe, the Korean Peninsula and the islands of Japan and Taiwan were profoundly influenced by the culture of China, but they were isolated enough that each developed a distinctive culture and maintained political independence most of the time. In the early twentieth century, Japan industrialized rapidly by integrating European influences that China disdained.

Bureaucracy and Imperial China

Although humans have lived in East Asia for hundreds of thousands of years, the region's earliest complex civilizations appeared in various parts of what is now China about 4000 years ago. Written records exist only from the civilization that was located in north-central China. There, a small, militarized, feudal aristocracy controlled vast estates on which the majority of the population lived and worked as semi-enslaved farmers and laborers. The landowners usually owed allegiance to one of the petty kingdoms that dotted northern China. These kingdoms were relatively self-sufficient and well defended with private armies.

An important move away from feudalism came with the Qin empire (beginning in 221 B.C.E.), which instituted a trained and salaried bureaucracy in combination with a strong military to extend the monarch's authority into the countryside (**Figure 9.6**).

The Qin system proved more efficient than the old feudal allegiance system it replaced. The estates of the aristocracy were divided into small units and sold to the previously semi-enslaved farmers. The empire's agricultural output increased because the people worked harder to farm land they now owned. In addition, the salaried bureaucrats were more responsible than the aristocrats they replaced, especially about building and maintaining levees, reservoirs, and other tax-supported public works that reduced the threat of flood, drought, and other natural disasters. Although the Qin empire was short-lived, subsequent empires maintained Qin bureaucratic ruling methods, which have proved essential in governing a united China.

Confucianism Molds East Asia's Cultural Attitudes

> **Confucianism** a Chinese philosophy that teaches that the best organizational model for the state and society is a hierarchy based on the patriarchal family

Closely related to China's bureaucratic ruling tradition is the philosophy of **Confucianism**. Confucius, who lived from 551 to 479 B.C.E., was an idealist who was interested in reforming government and eliminating violence from society. He thought human relationships should involve a set of defined roles and mutual obligations. Confucian values include courtesy, knowledge, integrity, and respect for and loyalty to parents and government officials. These values diffused across the region and are still widely shared throughout East Asia.

(A)

(B)

FIGURE 9.6 An emperor's army. After unifying China in 221 B.C.E., Qin Shi Huang became its emperor. One of his many public works was the creation of a life-sized terra-cotta army of thousands of lifelike soldiers, each with a distinctive face and body. They were arranged in a vast underground chamber that became his mausoleum. Since 1974, archaeologists have been excavating, restoring, and preserving this site, which is located near modern-day Xian. (A) An archer. (B) An overview of the site, which draws more than 2 million visitors each year.

The Confucian Bias Toward Males The model for Confucian philosophy was the patriarchal extended family. The oldest male held the seat of authority and was responsible for the well-being of everyone in the family. All other family members were aligned under the patriarch according to age and gender. Beyond the family, the Confucian patriarchal order held that the emperor was the grand patriarch of all China, charged with ensuring the welfare of society. Imperial bureaucrats were to do his bidding and commoners were to obey the bureaucrats.

Over the centuries, Confucian philosophy penetrated all aspects of East Asian society. Concerning the ideal woman, for example, a student of Confucius wrote: "A woman's duties are to cook the five grains, heat the wine, look after her parents-in-law, make clothes, and that is all! When she is young, she must submit to her parents. After her marriage, she must submit to her husband. When she is widowed, she must submit to her son." These concepts about limited roles for women affected society at large, where the idea developed that sons were the more valuable offspring, with public roles; while daughters were primarily servants within the home.

The Bias Against Merchants For thousands of years, Confucian ideals were used to maintain the power and position of emperors and their bureaucratic administrators at the expense of merchants. In parable and folklore, merchants were characterized as a necessary evil, greedy and disruptive to the social order. At the same time, the services of merchants were sorely needed. Such conflicting ideas meant the status of merchants waxed and waned. At times, high taxes left merchants unable to invest in new industries or trade networks. At other times, however, the anti-merchant aspect of Confucianism was less influential, and trade and entrepreneurship flourished. Under communism, merchants again acquired a negative image; then when marketization was encouraged, the social status of merchants once again rose.

Cycles of Expansion, Decline, and Recovery Although the Confucian bureaucracy at times facilitated the expansion of Imperial China (**Figure 9.7**), its resistance to change also led to periods of decline. Heavy taxes were periodically levied on farmers, bringing about farmer revolts that weakened imperial control. Threats from outside, particularly invasions by nomadic people from what are today Mongolia and western China, inspired the creation of massive defenses, such as the Great Wall, built along China's northern border. Nevertheless, the Confucian bureaucracy always recovered, and China's cultural and economic sophistication usually

FIGURE 9.7 The extent of Chinese empires, 221 B.C.E.–1850 C.E. The Chinese state has expanded and contracted throughout its history. [Adapted from *Hammond Times Concise Atlas of World History* (Maplewood, NJ: Hammond, 1994).]

overwhelmed the invaders. After a few generations, the nomads were indistinguishable from the Chinese.

One nomadic invasion did result in important links between China and the rest of the world. In the 1200s, the Mongolian military leader Genghis Khan and his descendants were able to conquer all of China. They then pushed west across Asia as far as Hungary and Poland (see Chapter 5, page 219). It was during the time of this Mongol empire (also known as the Yuan empire) that traders such as the Venetian Marco Polo made the first direct contacts between China and Europe. These connections proved much more significant for Europe, which was dazzled by China's wealth and technologies, than for China, which saw Europe as backward and barbaric.

Indeed, from 1100 to 1600, China remained the world's most developed region, despite enduring several cycles of imperial expansion, decline, and recovery. It had the largest economy, the highest living standards, and the most magnificent cities. Improved strains of rice allowed dense farming populations to expand throughout southern China and supported large urban industrial populations. Nor was innovation lacking: Chinese inventions included papermaking, printing, paper currency, gunpowder, and improved shipbuilding techniques.

Why Did China Not Colonize an Overseas Empire?

During the well-organized Ming dynasty, 1368–1644, Zheng He, a Chinese Muslim admiral in the emperor's navy, ventured far into the outside world. From 1405 to 1433, he sailed a fleet of 250 ships that were bigger and more technologically advanced than those of Columbus. Zheng He took his fleet throughout Southeast Asia, across the Indian Ocean, and along the coasts of India, continuing into the Persian Gulf and then down the east coast of Africa.

The lavish voyages of Zheng He were funded for almost 30 years, but they never resulted in an overseas empire like those established by European countries a century or two later. These newly explored regions simply lacked trade goods that China wanted. Moreover, back home the empire was continually threatened by the armies of nomads from Mongolia, so any surplus resources were needed for upgrading the Great Wall (Figure 9.8). Eventually the emperor decided that Zheng He's explorations were not worth the effort, and China as a whole turned inward. In the years following Zheng He's voyages, contacts with the rest of the world were minimized as the emperor focused on repelling invaders from Mongolia. As a result, the pace of technological change slowed, leaving China ill prepared to respond to growing challenges from Europe after 1600.

European and Japanese Imperialism

By the mid-1500s, during Europe's Age of Exploration, Spanish and Portuguese traders interested in acquiring China's silks, spices, and ceramics found their way to East Asian ports. To exchange, they brought a number of new food crops from the Americas, such as corn, peppers, peanuts, and potatoes. These

FIGURE 9.8 The Great Wall. Built and rebuilt between the fifth century B.C.E. and the sixteenth century C.E., the purpose of the Great Wall was to counter the threat of invasion by the nomads of Mongolia and Manchuria. The wall, which was breached by invaders repeatedly throughout its history, stretches for 3,889.5 miles (6,259.6 km).

new sources of nourishment contributed to a spurt of economic expansion and population growth during the Qing, or Manchurian, dynasty (1644–1912) and by the mid-1800s, China's population was more than 400 million.

By the nineteenth century, European merchants gained access to Chinese markets and European influence increased markedly. In exchange for Chinese silks and ceramics, British merchants supplied opium from India, which was one of the few things that Chinese merchants would trade for. The emperor attempted to crack down on this drug trade because of its debilitating effects on Chinese society. The result was the Opium Wars (1839–1860), in which Britain badly defeated China. Hong Kong became a British possession, and British trade, including opium, expanded throughout China.

The final blow to China's long preeminence in East Asia came in 1895, when a rapidly modernizing Japan won a spectacular naval victory over China in the Sino-Japanese War. After this first defeat by the Japanese, the Qing dynasty made only halfhearted attempts at modernization, and in 1912, it was overthrown

FIGURE 9.9 "Twelfth-Night Cake of Kings and Emperors." This French cartoon lithograph, made in 1898, shows Britain, Germany, Russia, and Japan imperialistically carving up China as France watches. A stereotypical Chinese official throws up his hands to show he is powerless to stop them. [The Granger Collection, New York.]

by an internal revolt and collapsed. Beginning with the decline of the Qing empire (1895) until China's Communist Party took control in 1949, much of the country was governed by provincial rulers in rural areas and by a mixture of Chinese, Japanese, and European administrative agencies in the major cities (Figure 9.9).

China's Turbulent Twentieth Century

Two rival reformist groups arose in China in the early twentieth century. The Nationalist Party, known as the Kuomintang (KMT), was an urban-based movement that appealed to workers as well as the middle and upper classes. The Chinese Communist Party (CCP), on the other hand, found its base in rural areas among peasants. At first the KMT gained the upper hand, uniting the country in 1924. However, Japan's invasion of China in 1931 changed the dynamic.

By 1937, Japan had control of most major Chinese cities. The KMT did not

Great Leap Forward an economic reform program under Mao Zedong intended to quickly raise China to the industrial level of Britain and the United States

resist the Japanese effectively and were confined to the few deep interior cities not in Japanese control. The CCP, however, waged a constant guerilla war against the Japanese throughout rural China. This resistance gained the CCP heroic status. Japan's brutal occupation caused 10 million Chinese deaths. When Japan finally withdrew in 1945, defeated at the end of World War II by the United States, Russia, and other Allied forces, the vastly more popular CCP pushed the KMT out of the country and into exile in Taiwan. In 1949, the CCP, led by Mao Zedong, proclaimed the country the "People's Republic of China," with Mao as president.

Mao's Communist Revolution Mao Zedong's revolutionary government became the most powerful China ever had. It dominated all the outlying areas of China—the Far Northeast subregion, Inner Mongolia, and western China (Xinjiang Uygur)—and launched a brutal occupation of Tibet (Xizang). The People's Republic of China was in many ways similar to past Chinese empires. The Chinese Communist Party replaced the Confucian bureaucracy and Mao Zedong became a sort of emperor with unquestioned authority.

The early chief beneficiaries of the revolution were the masses of Chinese farmers and landless laborers. On the eve of the revolution, huge numbers lived in abject poverty. Famines were frequent, infant mortality was high, and life expectancy was low. The vast majority of women and girls held low social status and spent their lives in unrelenting servitude.

The revolution drastically changed this picture. All aspects of economic and social life became subject to central planning by the Communist Party. Land and wealth were reallocated, often resulting in an improved standard of living for those who needed it most. Heroic efforts were made to improve agricultural production and to reduce the severity of floods and droughts. The masses, regardless of age, class, or gender, were mobilized to construct almost entirely by hand huge public works projects—roads, dams, canals, whole mountains terraced into fields for rice and other crops. "Barefoot doctors" with rudimentary medical training dispensed basic medical care, midwife services, and nutritional advice to people in the remotest locations. Schools were built in the smallest of villages. Opportunities for women became available, and some of the worst abuses against them were stopped—such as the crippling binding of women's feet to make their feet small and childlike. Most Chinese people who are old enough to have witnessed these changes say that the revolution did a great deal to improve overall living standards for the majority.

Mao's Missteps Nonetheless, progress came at enormous human and environmental costs. During the **Great Leap Forward** (a government-sponsored program of massive economic reform initiated in the 1950s), 30 million people died from famine brought on by poorly planned development objectives. Meanwhile, deforestation, soil degradation, and agricultural mismanagement became widespread. In the aftermath of the Great Leap Forward, some Communist Party leaders tried to correct the inefficiencies of the centrally planned economy only to be demoted or jailed as Mao Zedong remained in power.

FIGURE 9.10 Communist China. A young female member of the People's Liberation Army holds a book of quotations from Mao Zedong, chairman of the Communist Party. Behind her is a young female member of the "Red Guard," a mass movement of students and young people mobilized by Mao Zedong during the Cultural Revolution.

In 1966, the **Cultural Revolution**, a series of highly politicized and destructive mass campaigns, enforced support for Mao and punished dissenters. Everyone was required to study the "Little Red Book" of Mao's sayings (**Figure 9.10**). Educated people and intellectuals were a main target of the Cultural Revolution because they were thought to instigate dangerously critical evaluations of Mao and Communist Party central planning. Tens of millions of Chinese scientists, scholars, and students were sent out of the cities to labor in mines and industries or to jail, where as many as 1 million died. Children were encouraged to turn in their parents. Petty traders were punished for being capitalists, as were those who adhered to any type of organized religion. The Cultural Revolution so disrupted Chinese society that by Mao's death in 1976, the communists had been seriously discredited.

Changes After Mao Two years after Mao's death, a new leadership formed around Deng Xiaoping. Limited market reforms were instituted, but the Communist Party retained tight political control. In 2009, after more than 30 years of reform and remarkable levels of economic growth, China's economy became the third largest in the world behind the European Union and the United States. It passed Japan when China managed to weather the world recession beginning in 2008 better than expected. However, the disparity of wealth in China has been increasing for some years, human rights are still often abused, and political activity remains tightly controlled even as discontent boils over into open protests against government foibles.

> **Cultural Revolution** a series of highly politicized and destructive mass campaigns launched in 1966 to force the entire population of China to support the continuing revolution

Japan Becomes a World Leader

Although China's influence was dominant in East Asia for thousands of years, for much of the twentieth century, Japan, with only one-tenth the population and 5 percent of the land area of China, controlled East Asia economically and politically. Japan's rise as a modern global power results largely from its response to challenges from Europe and North America.

Beginning in the mid-sixteenth century, active trade with Portugal colonists brought new ideas and technology that strengthened Japan's wealthier feudal lords (*shoguns*), allowing them to unify the country under a military bureaucracy. However, the shoguns monopolized contact with the European outsiders, allowing no Japanese people to leave the islands, on penalty of death.

A second period of radical change came with the arrival in Tokyo Bay in 1853 of a small fleet of U.S. naval vessels. The foreigners, carrying military technology far more advanced than Japan's, forced the Japanese government to open the economy to international trade and political relations. In response, a group of reformers (the Meiji) seized control of the Japanese government, setting the country on a crash course of modernization and industrial development. They sent Japanese students abroad and recruited experts from around the world, especially from Western nations, to teach everything from foreign languages to modern military technology.

Between 1895 and 1945, Japan fueled its economy with resources from a vast colonial empire. Equipped with imported European and North American military technology, its armies occupied first Korea, then Taiwan, then coastal and eastern China, and eventually Indonesia and much of Southeast Asia, as shown in **Figure 9.11**. Many of the people of these areas still harbor resentment about the brutality they suffered at Japanese hands. Japan's imperial ambitions ended with its defeat in World War II and its subsequent occupation by U.S. military forces until 1952.

📹 **204. Japanese Still Resolving Feelings About the War**

In the immediate post-war era, the U.S. government imposed many social and economic reforms. Japan was required to create a democratic constitution and reduce the emperor to symbolic status. Its military was reduced dramatically, forcing it to rely on U.S. forces to protect it from attack. With U.S. support, Japan rebuilt rapidly after World War II, and it eventually became a giant in industry and global business, exporting automobiles, electronic goods, and many other products. In September 2010, Japan's economy was still among the world's largest and most technologically advanced,

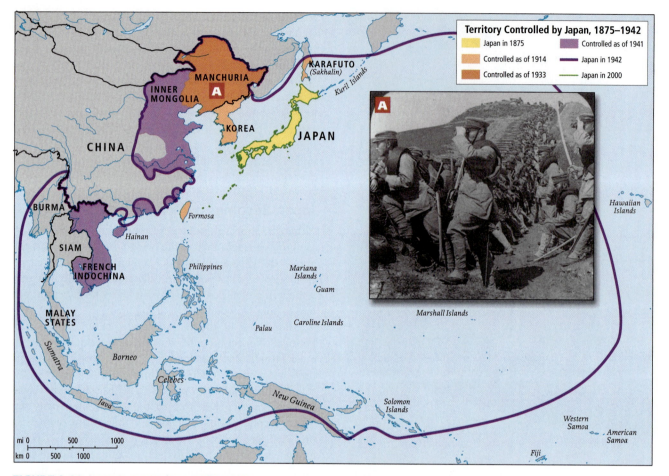

FIGURE 9.11 Japan's expansions, 1875–1942. Japan colonized Korea, Taiwan (then known as Formosa), Manchuria, China, parts of Southeast Asia, and several Pacific islands to further its program of economic modernization and to fend off European imperialism in the early twentieth century. **(A)** Japanese soldiers await an attack by the Russian army in Manchuria. Russia and Japan competed for dominance in Manchuria in the early twentieth century. [Adapted from *Hammond Times Concise Atlas of World History* (Maplewood, NJ: Hammond, 1994).]

although China has replaced it as the next-largest economy after those of the European Union and the United States.

Chinese and Japanese Influences on Korea, Taiwan, and Mongolia

The history of the entire East Asia region is largely grounded in what transpired in China and Japan.

The Korean War and Its Aftermath Korea was a unified but poverty-stricken country until 1945. At the end of World War II, in an effort to extend its global influence, the Soviet Union declared war against Japan and invaded Manchuria (northeast China) and the northern part of Korea; it was prevented from taking the whole peninsula only by U.S. military intervention. The United States took control of the southern half of the Korean peninsula, where it instituted reforms similar to those in Japan. The Soviet Union took the northern half, where it implemented communist development strategies. After the United States withdrew its troops in the late 1940s, North Korea attacked South Korea. The United States returned to defend the south, leading a 3-year war against North Korea and its allies, the Soviet Union and Communist China.

After great loss of life on both sides and devastation of the peninsula's infrastructure, the Korean War ended in 1953 in a truce and the establishment of a demilitarized zone (DMZ) at the 38th parallel. North Korea closed itself off from the rest of the world, and to this day it remains isolated, impoverished, and defensive, occasionally gaining international attention by hinting at its nuclear potential. Meanwhile, South Korea has developed into a prosperous and technologically advanced market economy.

📹 **201. Inter-Korean Cooperation Grows**

📹 **210. High-Tech Korea Eyes "Ubiquitous" Future**

Taiwan's Uncertain Status For thousands of years, Taiwan was a poor agricultural island on the periphery of China; then between 1895 and 1945 it became part of Japan's regional empire. In 1949, when the Chinese nationalists (the Kuomintang) were pushed out of mainland China by the Chinese Communist Party, they set up an anticommunist government in Taiwan, naming it the Republic of China (ROC). For the next 50 years, with U.S. aid and encouragement, the ROC became a modern industrialized economy which quickly dwarfed that of China and remained dominant until the 1990s, serving as a properous icon of capitalism right next door to massive communist China.

Today, Taiwan remains an economic powerhouse. Taiwanese investors have been especially active in Shanghai and the cities of China's southeast coast. Yet throughout this time, mainland China has never relinquished claim to Taiwan and considers it such a rebellious province that it warrants China's maintaining a substantial military force and occasionally using threats of invasion and military occupation against it. As China's economic power has increased, the United Nations, the World Bank, and most other countries, agencies, and institutions have judiciously tiptoed around the issue of whether Taiwan should continue as an independent country or be downgraded to simply a province of China. Taiwan itself remains divided over just how strongly it should hold on to its sovereignty.

Mongolia Seeks Its Own Way For millennia, Mongolia's nomadic horsemen periodically posed a threat to China, so much so that the Great Wall was built and reinforced and extended to combat them. China has long been obsessed with both deflecting and controlling its northern neighbor; and China did control Mongolia from 1691 until the 1920s. Revolutionary communism spread to Mongolia soon thereafter, and Mongolia continued as an independent communist country under Soviet, not Chinese, guidance until the breakup of the Soviet Union in 1989. Communism brought education and basic services. Literacy for both men and women rose above 95 percent. Deeply suspicious of both Russia and China, since 1989, Mongolia has been on a difficult road to a market economy. In need of cash to participate in the modern world,

families have elected to abandon nomadic herding and permanently locate their portable ger homes near Ulan Bator and search for paid employment. This rapid and drastic change in lifestyle has resulted in broken homes, increasing personal debt, and poverty, in a society that formerly took pride in an egalitarian if not prosperous standard of living.

THINGS TO REMEMBER

1. The histories of East Asian countries are deeply intertwined, and Confucian thought still permeates the entire region.

2. For six centuries, from 1100 to 1600, China was the world's most developed region despite cycles of imperial expansion, decline, and recovery.

3. In recent times, the countries of East Asia have followed very different philosophical paths toward development.

4. Japan has long-term cultural ties to the mainland of East Asia and has also exercised influence over the region, primarily through conquest. Since World War II, it has been the dominant economy in the region, only recently losing that status to China.

5. Taiwan and South Korea emerged after World War II as rapidly industrializing countries, while Mongolia until recently retained communist connections to the USSR. North Korea remains a communist country, and is extremely defensive and cut off from the wider world.

II CURRENT GEOGRAPHIC ISSUES

Although the countries of East Asia adopted new economic systems only after World War II, most of them are making progress toward creating a better life for their citizens. In fact, Japan, South Korea, and China's Hong Kong Special Administrative Unit have among the highest standards of living in the world.

Economic and Political Issues

After World War II, the countries of East Asia established two basic types of economic systems. The communist regimes of China, Mongolia, and North Korea relied on central planning by the government to set production goals and to distribute goods among their citizens. In contrast, first Japan, and then Taiwan and South Korea, established **state-aided market economies** with the assistance and support of the United States and Europe. In this type of economic system, market forces, such as supply and demand and competition for customers, determine many economic decisions. However, the government intervenes strategically, especially in the financial sector, to make sure that certain economic sectors develop in a healthy fashion. Investment in the country by foreigners is also limited so that the government can retain greater control over the direction of the economy. In the cases of Japan, South Korea, and Taiwan, government intervention was designed to enable **export-led growth.** This economic development strategy relies heavily on the production of manufactured goods

destined for sale abroad, primarily to the large economies of North America and Europe, while limiting imports for local consumers.

More recently, the differences among East Asian countries have diminished as China and Mongolia have set aside strict central planning and adopted reforms that rely more on market forces. China now also relies heavily on exports of its manufactured goods to North America and Europe. Politically, however, contrasts remain stark. While Japan, South Korea, Taiwan, and Mongolia have become democracies, unelected governments in China and North Korea maintain a tight grip on politics and the media. In China, there is some experimentation with democracy at the local level, but the central government is primarily a force against widespread democratic participation.

state-aided market economy an economic system based on market principles such as private enterprise, profit incentives, and supply and demand, but with strong government guidance; in contrast to the free market (limited government) economic system of the United States and Europe

export-led growth an economic development strategy that relies heavily on the production of manufactured goods destined for sale abroad

The Japanese Miracle

Throughout the nineteenth century, the economies of Japan, Korea, and Taiwan were minuscule compared with China's. Then, during the twentieth century, all three grew tremendously.

Japan Rises from Ashes Japan's recovery after its crippling defeat at the end of World War II is one of the most remarkable tales in modern history. Except for Kyoto, which was spared because of its historical and architectural significance, all of Japan's major cities were destroyed by the United States. Most notably, the United States bombed Hiroshima and Nagasaki with nuclear weapons, and leveled Tokyo with incendiary bombs.

📹 **205. Survivors Recall the Nuclear Bombing of Hiroshima**

Key to Japan's rapid recovery was government guidance of private investors in creating new export-based industries. The Japanese government negotiated trade deals with the United States and Europe, providing Japan large and wealthy foreign markets in which to sell its products. Internally, Japan's government made arrangements with labor unions to guarantee lifetime employment by a single company for most workers in return for relatively modest pay.

The government-engineered trade and labor deals produced explosive economic growth of 10 percent or more annually between 1950 and the 1970s. The leading sectors were export-oriented automobile and electronics manufacturing. Japanese brand names such as Sony, Panasonic, Nikon, and Toyota became household words in North America and Europe and these products sold at much higher volumes than would have been possible in the then relatively small Japanese and nearby Asian economies. Although growth slowed considerably both during the 1990s and since 2007 (discussed below), Japan's postwar "economic miracle" continues to have an immense worldwide impact as a model, and Japan remains a significant actor in the world economy. Japan purchases resources from all parts of the world for its industries and domestic use. These purchases and investments in various local economies continue to create jobs for millions of people around the globe.

Productivity Innovations in Japan Over the years, Japan has made two major innovations in manufacturing that have boosted its productivity and diffused to other industrial economies, changing the spatial arrangements of industries. The **kanban**, or "just in time," **system** clusters together companies that are part of the same production process so that they can deliver parts to each other when they are needed (**Figure 9.12**). For example, factories that make automobile parts are clustered around the final assembly plant, delivering parts literally minutes before they will be used. This saves money by making production more efficient and reducing the need for warehouses.

A related innovation is the **kaizen system**, or "continuous improvement system." This system ensures that fewer defective parts are produced because production lines are constantly surveyed for errors. Production lines are also constantly adjusted and improved to save time and energy. Both kanban and kaizen systems have been imitated by companies around the world and have been taken overseas by Japanese companies that invest abroad. For example, Toyota uses kaizen and kanban in all of its U.S. plants (see Chapter 2, page 95).

FIGURE 9.12 Japan's kanban system. In this example of the kanban system, related industries are clustered together on a human-made island in Yokohama, Japan. They supply each other with various chemical inputs needed for their production processes.

Future Economic Prospects In the 1980s, Japan's economy began to slow considerably, as did other state-aided Asian market economies; but for some time Japan continued as an important foreign investor in the Americas, Southeast Asia, and elsewhere. The economic slowdown worsened in the 1990s when, in retrospect, it appears the economy may have grown by as little as 1 percent or less per year. Analysts generally fault the long-standing close relationship between government and industry for nurturing favoritism, corruption, and inefficiency, all of which made Japan increasingly less competitive. Factories closed, banks failed, standards of living fell.

Very slow growth persisted to 2002, when a brief recovery began, based largely on the post-9/11 consumer binge and housing bubble in the United States. Factories producing for export expanded, expecting that U.S. consumers would keep spending and would be joined by other new consumers in China and India. In 2008, though, foreign demand collapsed. By the middle of 2009, unemployed executives were sleeping in Tokyo city parks, ashamed to go home. Soup kitchens were opened, and poverty among children and the elderly rose sharply. Perhaps most important symbolically was the fact that sometime during 2009, the size of Japan's economy began to be exceeded by that of China.

Significantly, the government that played such a central role during the post–World War II rise of the Japanese economy was controlled from 1955 to 2009 by one political party, the Liberal Democratic Party (LDP). In late August 2009, for the first time in 54 years, the Japanese people elected a government led by the opposition party, the Democratic Party of Japan (DPJ). With government debt at 200 percent of GDP, the new government faced the prospect of cutting many social services that had supported Japan's high standard of living. The DPJ-headed government

kanban system the "just-in-time" system pioneered in Japanese manufacturing that clusters companies that are part of the same production system close together so that they can deliver parts to each other precisely when they are needed

kaizen system the "continuous improvement system" pioneered in Japanese manufacturing; it ensures that fewer defective parts are produced because production lines are constantly surveyed for errors

appears to favor closer relations with China, recognizing that it will likely be the market of the future for Japanese exports, but it is also trying to reduce Japan's dependence on exports for income and to spur consumption of domestic products by lowering household taxes.

Mainland Economies: Communists in Command

After World War II, economic development on East Asia's mainland proceeded on a dramatically different course than Japan's. Communist economic systems transformed poverty-ridden China, Mongolia, and North Korea. Private property was abolished, and the state took full control of the economy, loosely following the example of the Soviet *command*, or *centrally planned* economy (see the discussion on pages 402–403 and in Chapter 5 on page 222). These sweeping changes transformed life for the poor majority, but ultimately proved less resilient and successful than was hoped.

By design, most people in the communist economies were not allowed to consume more than the bare necessities. On the other hand, the "iron rice bowl" policy guaranteed nearly everyone a job for life, sufficient food, basic health care and housing that was better than what they had before. One drawback was that overall productivity remained low.

The Commune System When the Communist Party first came to power in China in 1949, its top priority was to make monumental improvements in both agricultural and industrial production. Similar early goals were held by the communist regimes in North Korea and Mongolia, though they had much smaller populations and resource bases to work with. In China, the first strategy was **land reform**. The initial effort was to take large, unproductive tracts out of the hands of landlords and put them into the hands of the millions of landless farmers. By the early 1950s, much of China's agricultural land was divided into tiny plots.

In the years following World War II, it soon became clear that these small plots were not going to produce enough to feed the people who were leaving agriculture to work in the industrializing cities. In response, Communist leaders joined small landholders together into cooperatives so that they could pool their labor and resources to increase agricultural production. In time, the cooperatives became full-scale communes, with an average of 1600 households each. The communes, at least in theory, took care of all aspects of life for the inhabitants. They provided health care and education, and built rural industries to supply such items as simple clothing, fertilizers, small machinery, and eventually even tractors. The rural communes also had to fulfill the ambitious expectations that the leaders in Beijing had of better flood control, expanded irrigation systems, and especially, increased food production.

The Chinese commune system met with several difficulties. Rural food shortages developed because farmers had too little time to farm. They were required to spend much of their time building roads, levees, terraces, and drainage ditches, or working in the new rural industries. Local Communist Party administrators often compounded the problem by overstating harvests in their communes to impress their superiors in Beijing. The leaders in Beijing

responded by requiring larger food shipments to the cities, which created yet greater food shortages in the countryside.

Although the Chinese agricultural communes were inefficient and resulted in spectacular food scarcities during the Great Leap Forward, they did eventually result in a stable food supply that kept Chinese people well fed and even able to export food, which remains the case today (see page 409).

In North Korea, the Communists pushed quirky and secretive forms of industrial development, often futilely aimed at impressing the outside world with episodic flamboyant shows of accomplishment, such as rocket launchings and low-level nuclear tests. They also focused on mining, using unsafe methods and outdated technology. Agriculture has been so neglected that rural development, such as electrification and farm mechanization, remains primitive. Farm production is chronically precarious, with famine a common recurrence.

In Mongolia, Communist policy followed the Soviet model of collectivization, but was tailored to Mongolia's then largely herding agricultural economy. Changes in the nomadic pastoralist way of life were minimal, but emphasis was placed on mining and industry, and the contribution to GDP by herding and forestry declined to less than 20 percent.

land reform a policy that breaks up large landholdings for redistribution among landless farmers

Focus on Heavy Industry The Communist leadership in China, North Korea, and Mongolia believed that investment in heavy industry would dramatically raise living standards. Massive investments were made in the mining of coal and other minerals and in the production of iron and steel. In China, heavy machinery produced equipment to build roads, railways, dams, and other infrastructural improvements that would increase overall economic productivity. However, funds for the development of heavy industry came from the already strained agricultural sector. Farmers were required to sell their products to the state at artificially low prices; the state then resold these farm products at a profit and directed those funds toward industrialization. Other funds for industry came from profits in mining and forestry.

Unfortunately, the focus on heavy industry in all three countries failed to achieve the desired effect. Much as in India (see Chapter 8, pages 376–377), the vast majority of the population remained poor agricultural laborers who received little benefit from industries that created jobs mainly in urban areas. Not enough attention was paid to producing consumer goods (such as cheap pots, pans, hand tools, and other household items) that would have driven modest internal economic growth and improved living standards for the rural poor. Even in the urban areas, growth remained sluggish because, as also was the case in the Soviet command economy, small miscalculations by bureaucrats resulted in massive shortages and production bottlenecks that constrained economic growth.

Communist Struggles with Regional Disparity

For centuries, China's interior west has been poorer and more rural than its coastal east. The interior west has been locked into agricultural and herding economies, while even in the most restricted times, the economies of the east have benefited from trade and industry. This spatially uneven development continues to plague the Chinese even today and can also be seen on the provincial level where rural–urban disparities are sometimes extreme (see Figure 9.15 on page 411). The first effort to address regional disparities, right after the revolution, was an economic policy focused on **regional self-sufficiency.** Each region was encouraged to develop as an independent entity with both agricultural and industrial sectors that would create jobs and produce food and basic necessities.

Market Reforms and Globalization in China

In the 1980s, China's leaders enacted market reforms that changed the country's economy in four ways. First, economic decision making was decentralized and given the name **responsibility system,** which meant that agricultural decision making was returned to the farm household level, subject to the approval of the commune. Second, market reforms allowed farmers and small businesses to become **petty capitalists,** meaning that they were permitted to sell their produce and goods in competitive markets. Third, **regional specialization,** rather than regional self-sufficiency, was encouraged in order to take advantage of regional variations in climate, natural resources, and location, and thereby encourage national economic integration. Finally, the government allowed **foreign investment** in Chinese export-oriented enterprises and the sale of foreign products in China. This four-part shift to a more market-based economy dramatically improved the efficiency with which food and goods were produced and distributed, and to some extent addressed problems related to regional disparities.

China's market reforms have transformed not only China's economy but also those of wider East Asia and indeed the whole world. Today China is the world's largest producer of manufactured goods, supplying stores and street vendors across the globe. Meanwhile, of the other Communist-led countries, Mongolia has participated in this revolution only modestly, and North Korea has not participated at all.

Regional Specialization The decentralization of decision-making has encouraged *regional specialization.* Because managers of many state-owned enterprises were allowed to set production levels and prices for goods and services according to the demands of consumers in the open market, some managers and entrepreneurs (sometimes with assistance from the state) began taking advantage of the different resources and opportunities offered by their particular area of the country. The old colonial city of Shanghai, for example, has once again become a center of trade and finance. And the Zhu Jiang (Pearl River) delta, partly due to its proximity to the trading center of Hong Kong (see page 413), has evolved into a massive

regional self-sufficiency an economic policy in Communist China that encouraged each region to develop independently in the hope of evening out the wide disparities in the national distribution of production and income

responsibility system in the 1980s, a decentralization of economic decision making in China that returned agricultural decision making to the farm household level, subject to the approval of the commune

petty capitalists in the 1980s, farmers and small businesses that were allowed to sell their produce and goods in competitive markets as a part of China's market reforms

regional specialization the encouragement of specialization rather than self-sufficiency in order to take advantage of regional variations in climate, natural resources, and location

foreign investment investment in Chinese business and manufacturing enterprises by foreign firms

industrial center for the production of export goods. Meanwhile, with decreasing support from the government, some old state-run industries and enterprises have collapsed, especially in China's far northeast and in remote interior areas. Although inefficient, they once offered stable employment to millions of workers in small cities and villages, who now suffer from their deterioration.

Market Reforms in Rural Economies One of the most remarkable developments resulting from China's decentralization of economic control is the rapid growth of new *rural enterprises*, which generally operate outside the state centralized planning process; although most are registered and pay taxes, some are in the informal economy. The hope has been that such enterprises would not only stem migration to cities but would also even out regional disparities by producing wealth in the western interior as well as in the rural areas of every province. Rural enterprises now constitute one-quarter of the Chinese economy, produce 40 percent of its exports, and employ more than 128 million people, more than the Chinese government itself! Rural enterprises provide a wide variety of goods and services—from processing food, operating mines, making cooking pots, and assembling electronic equipment, to providing snack food, as described in the opening vignette. Such enterprises are increasingly owned privately, although they may still be village collectives or communes. All of them leave major decisions to managers and price their products according to market demand. International reports reviewing rural policy in China estimate that since 2000, the value and profits of rural enterprises have grown at an astonishing annual rate of 20 percent.

208. Chinese Village a Living Replica of Collective Past

Personal Vignette Wang Chunqiao is a button millionaire. His factories are in Qiaotou, a small Chinese town southeast of Shanghai that has about 200 factories and 20,000 migrant workers. Buttons used to be made in small factories all over the world from seashells, wood, leather, brass, and copper. Italy, Turkey, and France were important producers. But like the garment industry, button production has largely relocated to Asia. And Qiaotou has cornered 60 percent of the world production of buttons. Now Qiaotou's button manufacturers have to compete with themselves to gain more customers, cannibalizing one another through buyouts and drastic price cuts. Price cutting is painful because operating costs are rising. So many foreign firms have rushed to China recently that good workers can demand higher wages—skilled workers in the button factories now earn at least U.S.$120 per month. To attract and keep them, bosses must also provide food, housing, and cultural and sports activities. Furthermore, energy costs have increased, as have costs of supplies; copper prices have more than doubled in a year. Still savoring his rags-to-riches rise with a big smile, Wang Chunqiao tells a reporter that he had not anticipated that his profit margin could get so narrow so quickly. [Adapted from Louisa Lim, "Chinese 'Button Town' Struggles with Success," *Morning Edition, National Public Radio, August 22, 2006, http://www.npr. org/templates/story/story.php?storyId=5686805.*] ∎

Despite the importance of rural enterprises in the economy, their growth has been accompanied by some problems. They are significant contributors to environmental stress. In Qiaotou, the button town, the river, which formerly was used for irrigating crops, is now heavily polluted with colorful plastic waste. Corruption and an increase in organized crime are also problems. And increasingly, rural enterprises are clustering in the coastal provinces rather than in the western interior where they are so needed; and within provinces, they tend to huddle close to cities rather than lie dispersed in the hinterland where jobs are needed.

Agricultural Change in the Global Context Patterns of food production and consumption in China are changing, and global effects are likely. Until the Communist Revolution in 1949, famines were legend in China, and fear of food shortages was an important impetus for the revolution. As recently as 1990, 80 percent of the Chinese population lived in rural areas; most were agricultural workers producing food for the ever-expanding population (**Figure 9.13**).

Food security remains a focal point of central planning, but since the start of market reforms in the 1980s, farmers have been less locked into producing only for official quotas. Instead, they have been able to profit from the vegetables they grow and the

animals they raise on the margins of large communal grain (rice, wheat, corn) farms. They market this privately grown produce to urbanites, some of it informally to their own kinfolk and neighbors who have migrated to cities. Over time, formal systems for efficiently getting high-quality farm products to urban centers have improved, and now market gardening is common, especially close to urban areas. The spaces for this gardening are under constant threat from urban expansion.

East Asia's Role as Mega-Financier When Americans were consuming so rampantly during the 2000 decade, they were borrowing both from themselves and others. By 2009, foreign banks owned 25 percent of U.S. public debt, and of that 25 percent, Chinese and Japanese banks owned 44 percent, or about U.S.$1.5 trillion (see **Figure 9.14** on page 410). China and Japan both had a great deal of cash because of their successful export economies, and they lent the money partly to ensure that consumers in the United States would continue to buy their goods.

Elsewhere in the world, East Asians are able to use their reserves to finance development and thereby secure privileged trade deals. China's activities in Africa are a case in point. In the fall of

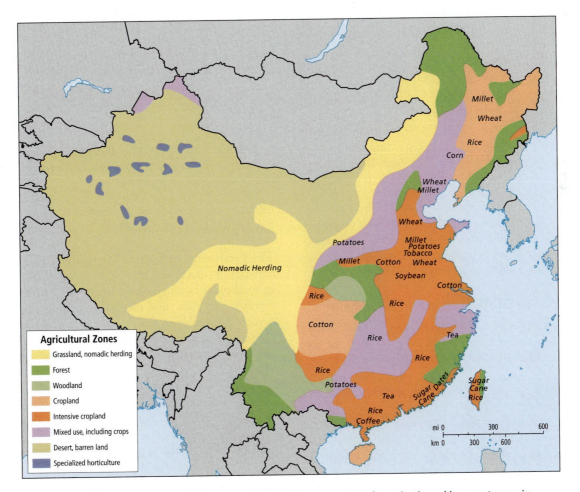

Agricultural Zones

- Grassland, nomadic herding
- Forest
- Woodland
- Cropland
- Intensive cropland
- Mixed use, including crops
- Desert, barren land
- Specialized horticulture

FIGURE 9.13 China's agricultural zones. As part of the economic reforms instituted in recent years in China, greater regional specialization in agricultural products is taking place. [Adapted from "World Agriculture," *National Geographic Atlas of the World,* 8th ed. (Washington, D.C.: National Geographic Society, 2005), p. 19; and "China: Economic, Minerals" map, *Goode's World Atlas,* 21st ed. (New York: Rand McNally, 2005), pp. 39 and 207.]

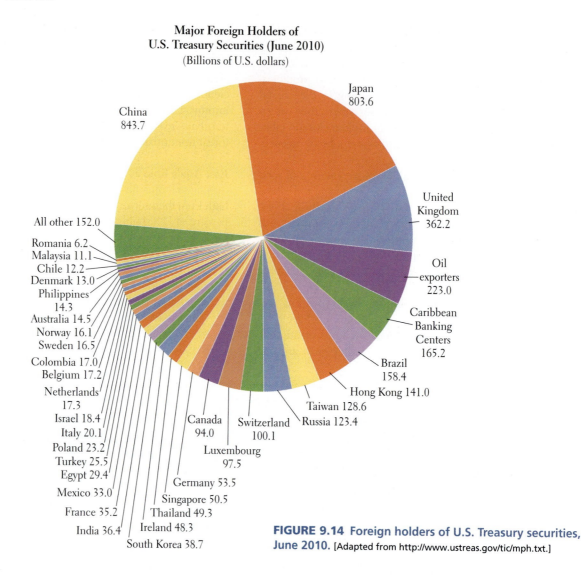

Major Foreign Holders of
U.S. Treasury Securities (June 2010)
(Billions of U.S. dollars)

Japan 803.6

China 843.7

United Kingdom 362.2

Oil exporters 223.0

Caribbean Banking Centers 165.2

Brazil 158.4

Hong Kong 141.0

Taiwan 128.6

Russia 123.4

All other 152.0
Romania 6.2
Malaysia 11.1
Chile 12.2
Denmark 13.0
Philippines 14.3
Australia 14.5
Norway 16.1
Sweden 16.5
Colombia 17.0
Belgium 17.2
Netherlands 17.3
Israel 18.4
Italy 20.1
Poland 23.2
Turkey 25.5
Egypt 29.4
Mexico 33.0
France 35.2
India 36.4
South Korea 38.7
Ireland 48.3
Thailand 49.3
Singapore 50.5
Germany 53.5
Luxembourg 97.5
Canada 94.0
Switzerland 100.1

FIGURE 9.14 Foreign holders of U.S. Treasury securities, June 2010. [Adapted from http://www.ustreas.gov/tic/mph.txt.]

2009 at a China–Africa trade summit in Cairo, China's premier pledged U.S.$10 billion in low-interest loans to African countries. As part of this package, China offered to forgive some outstanding loans to the poorest countries, to build "green" energy projects, and to open up China to zero-tariff trade with some African countries.

Urbanization and Globalization in East Asia

Across East Asia, cities have grown rapidly over the last several decades as urban industries have focused on production for the global market (see Photo Essay 9.4 Map on page 414). Urban areas in China have undergone the most massive urbanization in the world's history, a change that is difficult to accurately assess because of the floating population (see page 387). Since 1980, when market reforms were initiated, the urban population has more than quadrupled, and is now estimated at nearly 600 million. This explosive growth is fueled largely by the highly globalized economies of China's big coastal cities, whose urban factories now supply consumers throughout the world. Despite the rapid growth, in China urbanites still represent only 46 percent of China's total population, so more growth is likely; whereas in every other country in the region—even North Korea—the majority has been urban for at least two decades. Most East Asian urban landscapes exhibit

extreme pressure on space and environmental quality (see Photo Essay 9.4A,B,C,D on page 414).

The Persistence of Spatial Disparities The patterns of public and private investment that accompany urbanization have led to rural areas lagging behind urban areas in access to jobs and income, education, and medical care. The map in Figure 9.15 depicts the problem. GDP per capita is significantly lower in China's interior provinces than it is in coastal provinces, and within each province there is a disparity between rural and urban places, here shown in pie diagrams of rural versus urban incomes. Notice that rural/urban GDP disparities, while still significant, are less extreme in northeastern and coastal provinces than in interior and western provinces. Similar rural-to-urban disparity patterns are found in all other countries in East Asia, including North Korea.

📹 **206. Old Beijing Making Way for Modern Development**

📹 **219. Some Chinese Fear Private Property Law Will Cost Them Their Homes**

International Trade and Special Economic Zones One way China has tried to address regional disparities is to increase economic growth in underdeveloped areas by opening them up to foreign investment and trade. Wary of the disruption that could

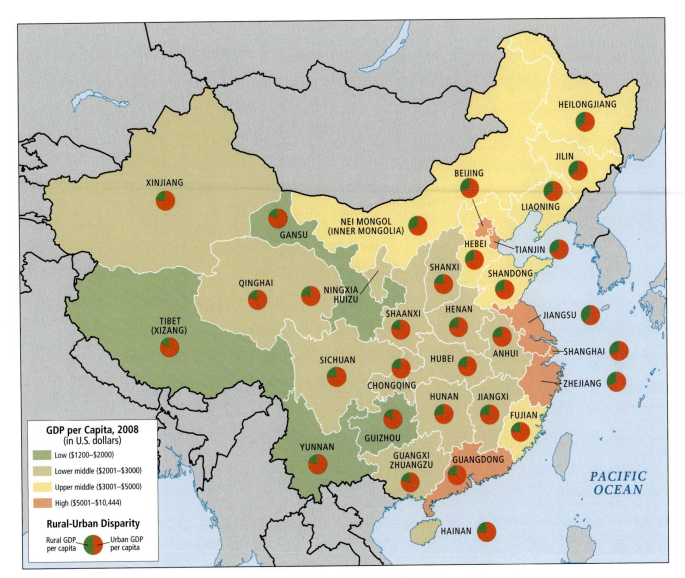

FIGURE 9.15 China's rural–urban GDP per capita disparities, 2008–2009. The province colors represent the range of GDP per capita for the year 2008. The rural–urban pie diagrams are from data for the third quarter of 2009 (China does not provide concurrent data by province for 2008 or 2009). Notice the disparity in GDP per capita across China as indicated by the colors of the provinces as well as the rural–urban income disparity in each province as represented by the pie diagrams. [Adapted from data at http://www.fdi.gov.cn/pub/FDI_EN/Economy/Investment%20Environment/Macro-economic%20Indices/Population%20&%20GDP/t20091103_113868.htm; and from http://www.fdi.gov.cn/pub/FDI_EN/Economy/Investment%20Environment/Macro-economic%20Indices/Population%20&%20GDP/t20091120_114779.htm.]

result from abruptly opening the economy to international trade, China first selected five coastal cities to function as free trade zones, or **special economic zones (SEZs)** (see **Figure 9.16** on page 412). Industries in these cities were allowed to recruit foreign investors and use capitalist management methods that had not yet been permitted in the rest of the country. In the late 1990s, the program was expanded to 32 other cities, many of them in the interior. These new locations were designated *economic and technology development zones* (ETDZs in Figure 9.16). Like SEZs, the ETDZs provide footholds for international investors and multinational companies eager to establish operations in the country (see **Figure 9.17** on page 413).

This program was successful. Today the SEZs and ETDZs are China's greatest **growth poles**, meaning that their development, like a magnet, is drawing yet more investment and migration. The first coastal SEZs were spectacularly successful. In only 25 years, many coastal cities, including Dongguan (the city Li Xia migrated to in the vignette that opens this chapter) grew from medium-sized towns or even villages into some of the largest urban areas in the world. SEZs and ETDZs in the interior have had higher rates of growth lately than those on the coast; during the global recession their economies also shrank less, but despite the encouraging

special economic zones (SEZs) free trade zones within China

growth poles zones of development whose success draws more investment and migration to a region

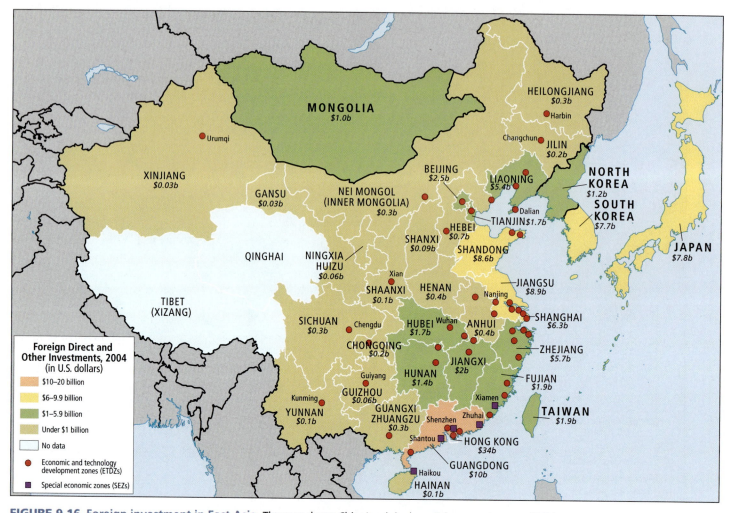

FIGURE 9.16 Foreign investment in East Asia. The map shows China's original special economic zones (SEZs) and more recently designated economic and technology development zones (ETDZs). The colors on the map reflect levels of foreign investment (direct and otherwise) in each of the countries of the region and in each of China's provinces. [FDI data from FDI Invest in China, at http://www.fdi.gov.cn/common/info.jsp?id5ABC00000000000022787. Specific Web site no longer available without registering at http://www.fdi.gov.cn/pub/FDI_EN/Statistics/default.htm.]

growth trends, foreign investment remains concentrated on the eastern seaboard, which accounts for 94 percent of China's exports. As shown in Figure 9.15, GDP per capita PPP rates remain noticeably lower in China's interior than on the eastern coast.

Urban and Suburban Japan Nearly all of Japan's major cities are located along the coastal perimeter. The coastal cities are where Japan's rapid industrialization has taken place, and their ports facilitate the import of raw materials and the export of finished products. Ideas from the outside world can also penetrate easily, aiding technological advancement and cultural synthesis. Tokyo, for example, has the world's second-largest stock exchange, numerous centers for research and development, and some of the world's most beautiful modern architecture, and it is also a major international cultural center.

Despite these assets, Japanese cities also suffer from overcrowding and pollution. In Tokyo, it is not uncommon for a middle-class family of four, plus a grandparent or two, to live in a one-bedroom

apartment. Japanese cities cannot easily expand to relieve crowding because they are limited by the surrounding mountains and ocean, by building codes minimizing potential damage from earthquakes, and by regulations protecting Japan's scarce agricultural lands. Every bit of land for suburban development is hard won. Also contributing to overcrowding are corruption and a lack of competition in the construction industry, which keep housing in short supply. Although there is growing pressure for change, most Japanese stoically endure minuscule, expensive apartments and long commutes to work. In the Tokyo–Yokohama metropolitan area, it is common to travel 2 or 3 hours one way in trains that are so overcrowded that stations employ "shovers" who physically push as many passengers as possible into a single car.

Japanese homes are known for their simple, elegant aesthetics and their highly functional use of space. The three or four rooms of a typical middle-class home are used for many purposes during the course of the day, transformed as needed by sliding doors of paper and wood or by folding decorative room dividers. Furniture

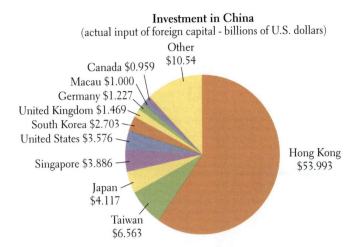

Investment in China
(actual input of foreign capital - billions of U.S. dollars)

- Other $10.54
- Canada $0.959
- Macau $1.000
- Germany $1.227
- United Kingdom $1.469
- South Korea $2.703
- United States $3.576
- Singapore $3.886
- Japan $4.117
- Taiwan $6.563
- Hong Kong $53.993

FIGURE 9.17 Foreign investment capital in China, 2009. Foreign investment in China reached more than $90 billion. The graph depicts the actual amount for the top 10 investors, which composed 88.3 percent of the total FDI, and several other countries whose investments make up the remaining amount. [Adapted from "News Release of National Assimilation of FDI from January to December 2009," Sinolize.com, January 18, 2010, at http://sinolize.com/index.php/government-release/42-general/1278-news-release-of-national-assimilation-of-fdi-from-january-to-december-2009.]

is simple, with much of family life centered on a low table and floor cushions. For sleeping, a firm futon is often rolled out on whatever floor space is available.

It has long been the practice in Japan that most older people who cannot fully care for themselves are cared for by their adult children (Japan's aging society is discussed on page 420). This pattern of home care is being eroded, however, as younger generations find that they lack space in their dwellings for their parents or grandparents or prefer privacy to traditional family obligations. Thus a relatively new industry of nursing homes and other elder-care facilities is growing in Japan.

Hong Kong's Unique Role Hong Kong is one of the most densely populated cities on earth, and its residents have China's highest per capita income. In 2007, its annual GDP per capita (adjusted for PPP) was more than U.S.$42,000. Hong Kong was a British crown colony until July 1997, when Britain's 99-year lease ran out and Hong Kong became a special administrative region (SAR) of China. Many wealthy citizens fled, worried that China would absorb Hong Kong and no longer allow it economic and political freedom.

While Hong Kong's democracy has been curtailed, political freedom is somewhat more evident there than elsewhere in China, and its role as China's unofficial link to the global economy has continued. Before 1997, some 60 percent of foreign investment in China was funneled through Hong Kong, and since then Hong Kong has remained the financial hub for China's booming southeastern coast (see Figure 9.17). Because so much of this investment comes from Japan, Korea, and Taiwan, Hong Kong is an important regional financial hub as well.

📹 **220. As Hong Kong Enters Second Decade Under China, City Ponders Place in Economic Giant**

Shanghai's Latest Transformation Shanghai has a long history as a trendsetter. Its opening to Western trade in the early nineteenth century spawned a period of phenomenal economic growth and cultural development that led it to be called "the Paris of the East." As a result of China's recent reentry into the global economy, Shanghai is undergoing another boom, which has enriched some people and dislocated others.

In less than a decade, the city's urban landscape has been remade by the construction of more than a thousand business and residential skyscrapers; subway lines and stations; highway overpasses; bridges; and tunnels. For hundreds of miles into the countryside, suburban development linked to Shanghai's economic boom is gobbling up farmland, and displaced farmers have rioted.

Pudong, the new financial center for the city, sits across the Huangpu River from the Bund—Shanghai's famous, elegant row of big brownstone buildings that served as the financial capital of China until half a century ago. Previously, Pudong was a maze of dirt paths and sprawling neighborhoods of simple tile-roofed houses, but its former residents were pushed out to make way for soaring high-rises (see Photo Essay 9.4A).

Shanghai's long-standing role as a window on the outside world has meant that it often is host to quirky behavior that is less common elsewhere in the country. For example, people in Shanghai have long relaxed in public in their pajamas—light, loose cotton tops and bottoms stamped with images of puppies or butterflies. The city government has tried to squelch the custom, but the citizens have proved recalcitrant. The police seem to understand that images of them arresting pudgy grandmothers for wearing pajamas would be ludicrous, so for now the issue is unresolved.

As unwilling as the people of Shanghai are to obey dress codes, they calmly put up with astounding decrees on other matters. Many in the city, often single, well-heeled young bureaucrats or older retirees, have pet dogs—often quite big dogs—even though they live in high-rise apartment buildings. In August 2009, the city government announced new dog rules. No longer would dogs be allowed in public places, or in public transportation, including elevators in high-rises. The rules are not yet enforced, but thousands of people are puzzling out how they will manage to descend 25 or 50 floors on foot to walk their dogs in some place that is not defined as a public space.

Shanghai's reputation as a cosmopolitan trendsetter has disappointed China's tens of millions of gays and lesbians who hoped that the rather lustrous gay nightlife there and the support groups and gay Web sites indicated a growing acceptance of gay lifestyles across the country. This is not the case. The government only tolerates, but does little to protect gay rights, and thus gays and lesbians are left vulnerable.

Urban Labor Surpluses and Shortages Spectacular urban growth in East Asia was based on hundreds of millions of new urban migrants who were willing to put up with almost any abuse to earn a little cash. However, so many factories, businesses, and shopping malls were built or were under construction that by 2005, experienced and skilled workers of all types were increasingly in short supply. Those with management skills were especially scarce.

Some of the fastest growing, largest, and wealthiest urban areas in the world are in East Asia. China is experiencing particularly rapid growth.

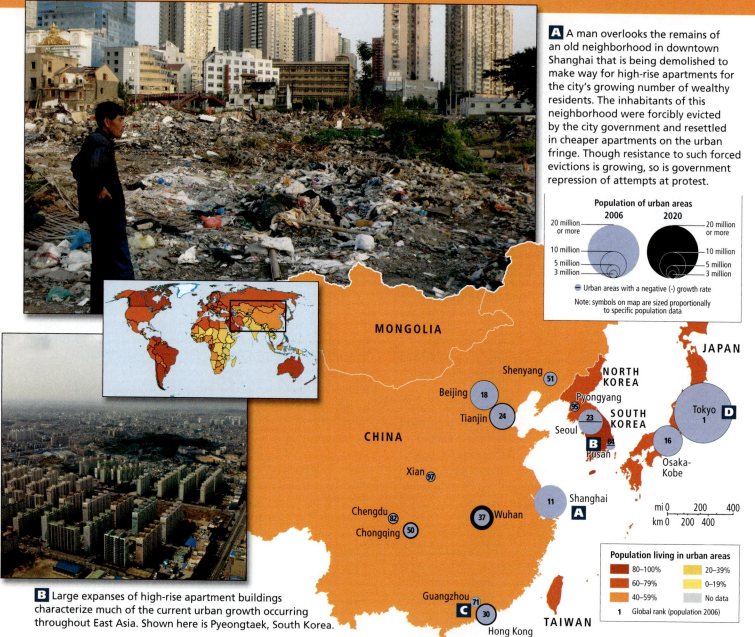

A A man overlooks the remains of an old neighborhood in downtown Shanghai that is being demolished to make way for high-rise apartments for the city's growing number of wealthy residents. The inhabitants of this neighborhood were forcibly evicted by the city government and resettled in cheaper apartments on the urban fringe. Though resistance to such forced evictions is growing, so is government repression of attempts at protest.

Population of urban areas

2006	2020
20 million or more	20 million or more
10 million	10 million
5 million	5 million
3 million	3 million

— Urban areas with a negative (-) growth rate

Note: symbols on map are sized proportionally to specific population data

MONGOLIA

Shenyang 51

Beijing 18

Tianjin 24

CHINA

NORTH KOREA

Pyongyang 95

SOUTH KOREA

Seoul 23

Pusan 84 **B**

JAPAN

Tokyo 1 **D**

Osaka-Kobe 16

Xian 97

Chengdu 82

Chongqing 50

Wuhan 37

Shanghai 11 **A**

mi 0 200 400
km 0 200 400

Population living in urban areas

80–100%	20–39%
60–79%	0–19%
40–59%	No data

1 Global rank (population 2006)

Guangzhou 71

C 30

Hong Kong

TAIWAN

B Large expanses of high-rise apartment buildings characterize much of the current urban growth occurring throughout East Asia. Shown here is Pyeongtaek, South Korea.

C Migrants in Guangzhou, Chinese head home for the spring holidays. China's urban growth is based on migration from rural areas.

D Ginza is the most lavish shopping district in Tokyo, the world's wealthiest and most populous city.

To attract employees, some factory owners in China offered higher pay, better working conditions, and shorter workdays or increased time off. The extra costs these changes imposed meant that China was no longer the cheapest place to manufacture products. Some factories moved to Vietnam, the Philippines, and countries in Africa with even cheaper labor. The more technologically sophisticated items, such as cell phones and computers, which sell for more money, could absorb the higher wages for workers, at least for a while (Figure 9.18). Then the global recession, which was felt in Asia by 2008, changed the dynamic yet again. Demand for China's products dropped, and factories quickly laid off workers or shut down entirely. By early 2009, throngs of urban workers returned to their farms and villages, dejected about not being able to help their families and confused about what the future held for them. These ups and downs in China's labor market added a new twist to the story of Li Xia, whom we met in the chapter opening vignette (page 386).

Personal Vignette Li Xia and her sister returned home to rural Sichuan a second time. With the money she had saved from her second job in Dongguan, Xia tried to open a bar in the front room of her parents' house. She hoped to introduce the popular custom of karaoke singing she had enjoyed in the city, but people in her village could not stand the noise, and family tensions rose. Xia's sister, delighted to be home with her husband and baby, quickly found a job in a small new fruit-processing factory—a rural enterprise. Her husband began farming again to fill the new demand for organic vegetables among the middle class in the Sichuan city of Chongqing.

In early 2007, amid all this success in her family and the failure of her own bar, news that training was now available in Dongguan for skilled electronics assemblers convinced Xia to try again. Past experience with the bureaucracy and her knowledge of Dongguan helped Xia to sign up for the electronics training. The cost of tuition was to come out of her wages, which were only U.S.$350 a month rather than the U.S.$400 she had thought she would be paid. Also, the waiting list for an apartment was long. She needed several years to repay her tuition, so she went back to shared quarters with her friends.

In November of 2008, Xia heard rumors that a global recession was causing orders for electronics to be cut and that she might soon be laid off. But her factory limped along into 2009 with a reduced staff. Then miraculously in June of 2009, orders picked up. Consumers in America had continued to buy electronics even as they downsized their homes and cars. Electronic inventories in America were down so orders for products from Xia's factory were up significantly. And with so many laid-off workers returning home, the apartment waiting list had shortened drastically. Now the apartment developers were only too happy to give her a firm move-in date.
[Adapted from the following sources: Paul Wiseman, "Chinese Factories Struggle to Hire," USA Today, April 11, 2005, at http://www.usatoday.com/money/world/2005-04-11-china-labor_x.htm; Louisa Lim, "The End of Agriculture in China" on Reporter's Notebook, National Public Radio, May 19, 2006, at http://www.npr.org/templates/story/story.php?storyId=5411325; Mei Fong, "A Chinese Puzzle: Surprising Shortage of Workers Forces Factories to Add Perks; "Pressures on Pay—and Prices," Wall Street Journal, August 16, 2004, p. B1; David Barboza, "Labor Shortage in China May Lead to a

FIGURE 9.18 Manufacturing near Shanghai, China. Workers test computer hard drives made for Seagate, a California technology company. Seagate now makes most of its products in China, where it also exports the products worldwide.

Trade Shift," New York Times, April 3, 2006, Business Section, p. 1, at http://www.nytimes.com/2006/04/03/business/03labor.html?pagewanted=1&_r=1; Qiu Quanlin, "Labor Shortage Hinders Guangdong Factories," China Daily, August 25, 2009, at http://www.chinadaily.com.cn/china/2009-08/25/content_8612599.htm; Nina Ying Sun, "Labor Shortage Returns to China," Plastics News.Com, August 10, 2009, at http://plasticsnews.com/china/english/chinablog/2009/08/labor_shortage_returns_to_chin.html.] ∎

THINGS TO REMEMBER

1. Key to Japan's rapid economic recovery after World War II was close cooperation between government and private investors to create new export-based industries, primarily automotive and electronic.

2. Japanese management innovations known as the kanban and kaizen systems also were important to economic recovery and were so successful that they diffused globally.

3. In the 1980s, China's leaders enacted market reforms that changed the country's economy in significant ways, and

subsequently those market reforms transformed the region and indeed the whole world.

4. One of the most remarkable developments resulting from China's decentralization of economic control is the rapid growth of new rural enterprises. These now constitute one-quarter of the Chinese economy, produce 40 percent of its exports, and employ more than 128 million people.

5. China's cities are among the fastest growing on earth. The Shanghai metropolitan agglomeration now accounts for more than 25 percent of China's GDP. Urbanization has resulted in a large and growing middle class.

Democratization in East Asia

Pressures for greater democracy are growing throughout East Asia. Japan's democracy was established after World War II, South Korea's in the late 1980s, and Taiwan's in the mid-1990s. Mongolia has had a dramatic expansion of democracy since abandoning socialism in 1992. North Korea and China, however, remain under the tight control of undemocratic regimes. On the map in Photo Essay 9.5 (page 418), the colors of the countries roughly indicate the level of democracy enjoyed by the populace. The red starbursts indicate where civil unrest has broken out.

With China now a globalized economy, many wonder how much longer the Communist Party can remain in control without instituting countrywide democracy. The Communist Party officially claims that China is a democracy, and indeed some elections have long been held at the village level and within the Communist Party. However, representatives to the *National People's Congress*, the country's highest legislative body, are appointed by the Communist Party elite, who maintain tight control throughout all levels of government.

While radical change is unlikely in the near future, most experts on China agree that a steady shift toward greater democracy is underway and might be inevitable as the population becomes more educated and travels more widely, and thus is exposed to places that have greater freedoms and better government. Demands for political change built to a crescendo less than a decade after market reforms began, culminating in a series of pro-democracy protests that drew hundreds of thousands to Beijing's *Tiananmen Square* in 1989 (see Photo Essay 9.5A). These protests were brutally repressed, with thousands (the precise number is uncertain) of students and labor leaders massacred by the military. Since then, pressure for change has continued to mount both within China and internationally, but it seems the link between economic development and democratization is only tenuous.

International Pressures China's admission to the World Trade Organization in 2001 brought pressures for political change. Informed consumers and environmentalists in developed countries have long criticized China for its "no holds barred" pursuit of economic growth. Much of China's growth has been built on environmentally destructive activities and abuses of workers, both of which effectively lower production costs so that goods can sell at a lower price.

The 2008 Olympics in Beijing highlighted a number of issues that have also created pressures for greater democracy. Before the games, the international media shined its spotlight on human rights abuses of workers, protesters, prisoners, ethnic minorities, and spiritual/mystical groups such as Falun Gong. Particular attention was given to protests in Tibet and elsewhere around the world against China's ongoing repression of Tibetan Buddhist monasteries and the mass importation of non-Tibetans into the province. More recently, China's repressive treatment of the Uygur people in Xinjiang Uygur Autonomous Region (discussed on page 425) has received international attention.

With the approach of the 2008 Olympic games, the world's attention was focused on the lavish sports facilities and spectacular opening and closing ceremonies of the games (Figure 9.19). China spent U.S.$44 billion for the games as a whole, more than had been spent on the previous ten summer Olympic games

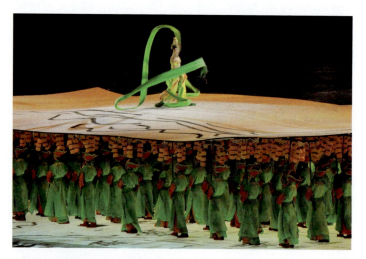

FIGURE 9.19 The 2008 Olympic Games in Beijing. China spent more than U.S.$44 billion on the 2008 Olympic Games, mostly on infrastructure, energy, transportation, and water supply projects in the city of Beijing. The opening ceremonies alone cost U.S.$300 million.

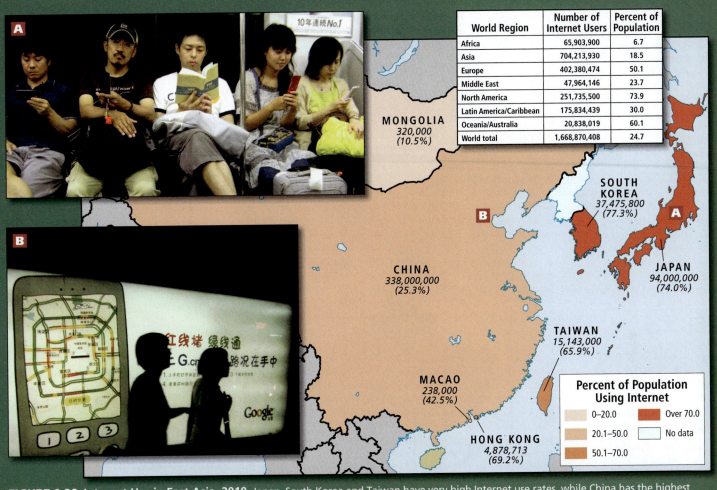

World Region	Number of Internet Users	Percent of Population
Africa	65,903,900	6.7
Asia	704,213,930	18.5
Europe	402,380,474	50.1
Middle East	47,964,146	23.7
North America	251,735,500	73.9
Latin America/Caribbean	175,834,439	30.0
Oceania/Australia	20,838,019	60.1
World total	1,668,870,408	24.7

MONGOLIA
320,000
(10.5%)

SOUTH KOREA
37,475,800
(77.3%)

JAPAN
94,000,000
(74.0%)

CHINA
338,000,000
(25.3%)

TAIWAN
15,143,000
(65.9%)

MACAO
238,000
(42.5%)

HONG KONG
4,878,713
(69.2%)

Percent of Population Using Internet
0–20.0
20.1–50.0
50.1–70.0
Over 70.0
No data

FIGURE 9.20 Internet Use in East Asia, 2010. Japan, South Korea and Taiwan have very high Internet use rates, while China has the highest number of Internet users of any country in the world (338,000,000). Mongolia has the lowest Internet use at 10.5 percent. **(A)** Japanese subway riders use their cell phones, known as "ketai" to access the Internet, read and send email, buy train tickets, and play games while commuting. Japan often leads the world in deployment of the latest cell phone and Internet technology and cell phones have now become so ubiquitous, especially among young people, that many Japanese speak of a "ketai culture" or "cell phone culture" that is transforming Japan. **(B)** An advertisement by Google in Beijing for an Internet-based service offering navigation assistance via cell phones. Internet use is growing rapidly in China.

combined. This extravagance became a subject of criticism by the global media. Many journalists pointed out that no democracy would ever be able to devote so many tax dollars to such an event, especially not in a country with as many pressing human and environmental problems as China has.

🎦 **213. Beijing Olympics: Political Battleground?**

🎦 **211. Olympic Relay Cut Short by Paris Protests**

🎦 **221. Reports of Sale of Executed Falun Gong Prisoners' Organs in China Called "Shocking"**

Information Technology and Democracy The spread of information via the Internet has increased pressures for democratization from within China. Since the revolution in 1949, China's central government has controlled the news media in the country. By the late 1990s, however, the expanding use of electronic communication devices was loosening central control over information. By 2001, twenty-three million people were connected to the Internet

in China, and by 2009, more than 338 million (or 25 percent of the population) were connected (Figure 9.20). Just a few years ago, telephones were very rare and required permission, but now anyone can buy temporary cell phone access without showing identification. As a result, millions of Chinese have anonymous access to an international network of information. It is now much more difficult for the government to give inaccurate explanations for problems caused by inefficiency and corruption. Reporters can easily check the accuracy of government explanations by phoning witnesses or the principal actors directly and then posting the explanations on the Internet. Analysts both inside and outside China see the availability of the Internet to ordinary Chinese citizens as a watershed event that supports democracy.

Nevertheless, the Internet in China is not the open forum that it is in most Western countries. For example, if people writing blogs in China use the words "democracy," "freedom," or "human rights," they may get the following reminder: "The title must not

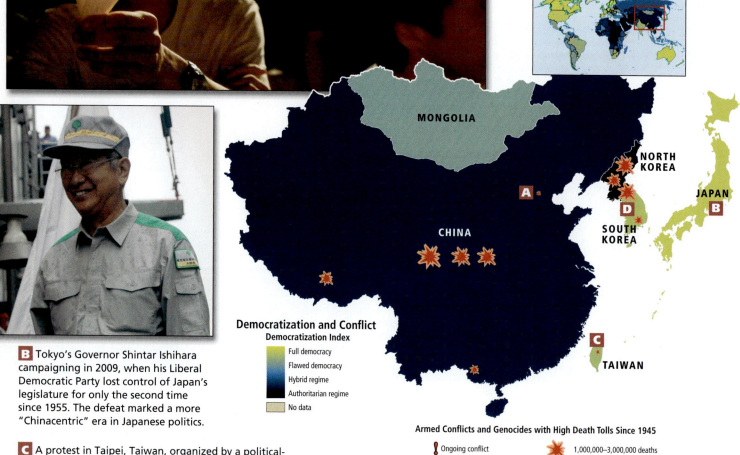

Democratization is far along in Japan, Taiwan, and South Korea, but China and North Korea have fewer outlets for citizens to directly influence government. Nevertheless, pressure for democratization is growing.

A The annual vigil commemorating those who died in the 1989 Tiananmen Square massacre in Beijing has become a focus for China's pro-democracy movement. Up to 150,000 people have shown up for the event in recent years. Hong Kong is the only part of China where the right to political protest is protected by law.

B Tokyo's Governor Shintar Ishihara campaigning in 2009, when his Liberal Democratic Party lost control of Japan's legislature for only the second time since 1955. The defeat marked a more "Chinacentric" era in Japanese politics.

C A protest in Taipei, Taiwan, organized by a political-opposition party against the government's moves to strengthen ties with China. The issue of Taiwan's continued independence from China, and hence its maintenance of democratic political freedoms, plays a central role in the island's electoral politics.

Democratization and Conflict
Democratization Index
- Full democracy
- Flawed democracy
- Hybrid regime
- Authoritarian regime
- No data

Armed Conflicts and Genocides with High Death Tolls Since 1945
- Ongoing conflict
- 1000–10,000 deaths
- 10,000–100,000 deaths
- 100,000–1,000,000 deaths
- 1,000,000–3,000,000 deaths
- 30,000,000 deaths

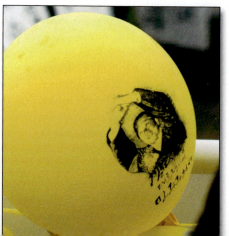

D A memorial balloon bearing the image of the late Kim Dae Jung, former president of South Korea (1998–2003). Jung received the Nobel Peace Prize in 2000 for his work in support of democracy and human rights during several periods of military dictatorship in South Korea during the 1970s and 1980s. He was also recognized for his efforts to reconcile with North Korea's authoritarian government.

contain prohibited language, such as profanity. Please type a different title." Such censorship has been aided by U.S. technology firms such as Yahoo and Microsoft that allow the Chinese government to use software that blocks access to certain Web sites for users in China. Google, now in dispute with the Chinese government over state-sponsored censorship, invasion of human rights activists' email accounts, and cyber attacks, has moved operations to Hong Kong, where somewhat more freedom is allowed (see page 413). The final chapter of Google's Internet ideology "war" with the Chinese government over freedom of Internet access and privacy has yet to be written.

Urbanization, Protests, and Technology May Enhance Democracy Protests by workers for better pay and living conditions, or riots by farmers and urban dwellers displaced by new real estate developments, have exposed average citizens to the depth of dissatisfaction felt by their compatriots. In 2003, China's government reported that there were 58,000 public protests in the country. This number rose to 74,000 in 2004, and 87,000 in 2005. Since then, the government has stopped issuing complete statistics on protests.

Once those experiencing this collective discontent gained access to communications technology, it became possible to form *interest groups*, which are associations of individuals with particular complaints that require government action and perhaps compensation. These groups communicate face-to-face or via cell phones (or, to a lesser extent, on the Internet), formulate a position, and then coalesce into larger organizations with increasing power and momentum to demand governmental action. One of the most publicized of such protests was by parents whose children died during the earthquake (which registered 7.8 on the Richter scale) in Chengdu in May 2008. They demanded redress for poorly constructed schools that collapsed, killing thousands.

Groups that pressure for compensation are actually less threatening to the central government than those that work for political reform. Those demanding compensation have met with modest success; those pressing for political reform, less so. In December 2008, people seeking more political freedoms signed a manifesto called the *08 Charter*, calling for a decentralized federal system of government (that is, more power to the provinces), democratic elections, and the end of the political monopoly by the Communist Party. Liu Xiaobo, coauthor of the 08 Charter, was arrested, and in December 2009 was sentenced to 11 years in prison, even as prominent international human rights activists lobbied on his behalf. Three hundred other signatories were arrested, interrogated, and released. Political protests by minorities such as the Tibetans and the Uygurs are seen as a very serious threat to the Chinese state. The government has used extreme repression against Tibetans and Uygurs. The effect that the protests by minorities will ultimately have on government policies is not yet clear.

In an effort to maintain control over China's increasingly articulate protestors, the government has allowed elections to be held for "urban residents' committees." The idea seems to be to provide a peaceful outlet for voicing frustrations and achieving limited change at the local level. Many of these elections are hardly democratic, with candidates selected by the Communist Party. However, in cities where unemployment is high and where protests have been particularly intense, elections tend to be more free, open, and truly democratic. It could be that interest group protests are an important first step toward actual participatory democracy, even when some, such as the 08 Charter, are harshly stifled.

THINGS TO REMEMBER

1. While radical change is unlikely in the near future, most China experts agree that a steady shift toward greater democracy is underway and might be inevitable.

2. The Chinese state holds that human rights and democratization are purely internal affairs.

3. The international community, often led by the United States, argues that China should reduce its use of repressive political control and shift toward decentralization and elections on the federal and provincial level.

4. By 2009, China had 338 million people using the Internet, but they represent only 25 percent of the country's population. China has strict controls on Internet access and use.

5. Chinese citizens who share a grievance against the government are increasingly cooperating in their protests. In December 2008, a manifesto for change, the 08 Charter, was signed and submitted to the government by 300 protestors.

Sociocultural Issues

East Asia's economic progress has led to social and cultural change throughout the region. Population growth, long a dominant issue, has slowed, while other demographic issues are receiving greater attention. Modernized economies, bearing some features of capitalism, are changing work patterns and family structures.

Population Patterns

Although East Asia remains the most populous world region, families there are having many fewer children than in the past. Of all world regions, only Europe has a lower rate of natural increase (<0.0 percent increase per year, compared to East Asia's 0.5 percent). In China this is partially due to government policies that harshly penalize families for having more than one child. But in China as well as elsewhere, urbanization and changing gender roles are also encouraging smaller families, regardless of official policy. Women in Mongolia (with 2.7 million people) and North Korea (with 22.7 million) are still averaging two or more children each, but even in these two countries family size is shrinking.

Responding to an Aging Population

Low birth rates mean that fewer young people are being added to the population, and improved living conditions mean that people are living longer across East Asia. The overall effect is that the average age of the populations is rising. Put another way, populations are aging. For Mongolia, South Korea, North Korea, and Taiwan, it will be several decades before the financial and social costs of supporting numerous elderly people will have to be addressed. China faces especially serious future problems with elder care because the one-child policy and urbanization have so drastically reduced family kin-groups. Japan, on the other hand, has already confronted the problem of having a large elderly population that requires support and a reduced number of young to do the job.

Japan's Options Japan's population is growing slowly and aging rapidly, raising concerns about economic productivity and humane ways to care for dependent people. The demographic transition (see Chapter 1, page 20) is well underway in this highly developed country where 86 percent of the population live in cities. Japan's rate of natural increase is in the negative range (<0.0), the lowest in East Asia, and on par with Europe's. If this trend continues, Japan's population will plummet from the current 127.6 million to 95.2 million by 2050.

At the same time, the Japanese have the world's longest life expectancy at 83 years. As a result, Japan also has the world's oldest population, 23 percent of which is over the age of 65. By 2055, this age group will account for 40 percent of the population. By 2050, Japan's labor pool could be reduced by more than a third, but it would still need to produce enough to take care of more than twice as many retirees as it does now. Clearly these demographic changes will have a momentous effect on Japan's economy, and the search for solutions is underway. One possibility is increasing immigration to bring in younger workers who will fill jobs and contribute to the tax rolls, as the United States and Europe have done. Another is to keep the elderly fit so they can work if necessary and take care of themselves.

In Japan, recruiting immigrant workers from other countries is a very unpopular solution to the aging crisis. Many Japanese consider foreigners a source of "pollution," and the few small minority populations with cultural connections to China or Korea have long faced discrimination. The children of foreigners born in Japan are not granted citizenship, and some communities that have been in the country for generations are still thought of as foreign. Today, immigrants are fingerprinted and photographed upon entering Japan and must carry an "alien registration card" at all times.

Nonetheless, foreign workers are dribbling into Japan in a multitude of legal and illegal guises. Many are "guest workers" from South Asia brought in to fill the most dangerous and low-paying jobs with the understanding that they will eventually leave. Others are the descendants of Japanese people who once migrated to South America (Brazil and Peru). Regardless, Japan's foreign population remains tiny, making up only 1.2 percent of the total population (approximately 1.5 million people). A recent United Nations report estimates that Japan would have to import over 640,000 immigrants per year just to maintain its present workforce and avoid a 6.7 percent annual drop in its GDP.

In a novel approach to Japan's demographic changes, the government has invested enormous sums of money in robotics over the past decade (see Thematic Concepts Part O on page 389). Robots are already widespread in Japanese industries such as auto manufacturing, and their industrial use is growing. Now they are also being developed to care for the elderly, to guide patients through hospitals, to look after children, and even to make sushi. By 2025, the government plans to replace up to 15 percent of Japan's workforce with robots.

China's Options The proportion of China's population over 65 is now only 8 percent, but this will change rapidly as conditions improve and life expectancies increase by 10 or more years, as they have done elsewhere. However, a crisis in elder care is already upon China for two other reasons: the high rate of rural-to-urban migration and the shrinkage of family support systems because of the one-child policy (discussed below).

When hundreds of millions of young Chinese were lured into cities to work, most thought rural areas would benefit from remittances, and this has happened. However, few anticipated that the one-child family would mean that for every migrant two aging parents would be left to fend for themselves, often in rural, underdeveloped areas. As migrations stretch into 10 and 20 years, elderly parents must deal with loneliness and increasing infirmities. The Chinese are being proactive: when throngs of elderly Chinese are seen exercising in public parks, few outsiders realize that this movement to keep the elderly fit is linked to far-sighted policies aimed at lowering the costs of supporting the aged (see Thematic Concepts Part M on page 389).

China's One-Child Policy

In response to fears of overpopulation and environmental stress, China has had a one-child-per-family policy since 1979 (Figure 9.21). The policy is enforced with rewards for complying and with monetary penalties and other sanctions for having more than one child. As a result, China's rate of natural increase (0.5 percent) is slightly lower than that of the United States (0.6 percent), and

FIGURE 9.21 China's one-child-per-family policy. A man walks past a sculpture in Beijing that promotes China's one-child-per-family policy. Enforcement of the policy, introduced in 1979 to control China's population growth, has relaxed a bit in recent years.

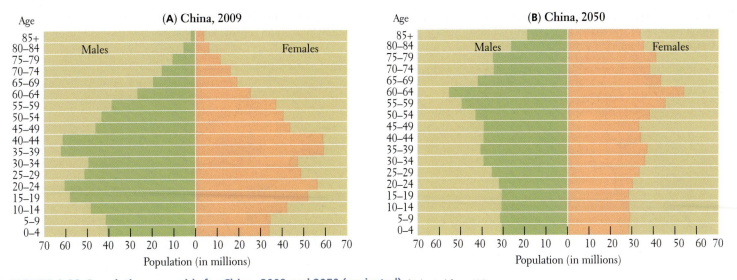

FIGURE 9.22 Population pyramids for China, 2009 and 2050 (projected). [Adapted from U.S. Census Bureau, Population Division, "Population Pyramids of China," International Data Base, 2009, at http://www. census.gov/ipc/www/idb/country.php.]

roughly half the world average (1.2 percent). If the one-child family pattern continues, China's population will start to shrink sometime between 2025 and 2050 (Figure 9.22B), creating many of the same economic and social problems that Japan is now facing.

The one-child policy transformed Chinese families and Chinese society at large. For example, an only child has no siblings, so within two generations, the kinship categories of brother, sister, cousin, aunt, and uncle have disappeared from most families, meaning that any individual has very few if any related age peers with whom to share family responsibilities. The effect on society of the one-child family is that most children are doted on by several adults, and children are not taught by siblings to share. Conscious efforts must be made to instill self-sufficiency in only children. The one-child policy has sometimes been enforced brutally. In various times and places, the government has waged a campaign of forced sterilizations and forced abortions for mothers who already have one child.

Cultural Preference for Sons The prospect of a couple's only child being a daughter, without the possibility of having a son in the future, is the aspect of the one-child policy that has caused families the most despair. For years, the makers of Chinese social policy have sought to eliminate the old Confucian-based preference for males by empowering women economically and socially. In many ways the policymakers are succeeding in empowering young women, but the preference for sons persists.

East Asia's Missing Females and Lonely Males The population pyramid in Figure 9.22A illustrates the extent to which the preference for sons has created a gender imbalance in China's population. The normal sex ratio at birth is 105 boys to 100 girls, which evens out by the age of 5. In China, however, for nearly every category until age 70, males outnumber females. The census data show that for those born in the 20-year period from 1980 to 2000, there are

22 million more men than women. For those born since 2000, the gender imbalance appears to be, if anything, more extreme. In 2009, there were roughly 40 million girls in the 0 to 5 category but roughly 46 million boys, indicating a deficit of about 6 million girls. Viewed another way, this is a ratio of 114 boys to every 100 girls rather than the usual 105 to 100. What happened to the missing girls?

There are several possible answers. Given the preference for male children, the births of these girls may simply have gone unreported as families hoped to conceal their daughter and try again for a son. There are many anecdotes of girls being raised secretly or even disguised as boys. Also, adoption records indicate that girls are given up for adoption much more often than boys (see Thematic Concepts Part P on page 389). Or the girls may have died in early infancy, either through neglect or infanticide. Finally, some parents have access to medical tests that can identify the sex of a fetus. There is evidence that in China, as elsewhere around the world, some of these parents choose to abort a female fetus.

The cultural preference for sons persists elsewhere in East Asia as well. A deficit of girls appears on the 2000 population pyramids for Japan, the Koreas, Mongolia, and Taiwan. Nonetheless, evidence shows that attitudes may be changing. In Japan, South Korea, and Mongolia, the percentage of women receiving secondary education equals or exceeds that of men.

A Shortage of Brides A side effect of the preference for sons is that there is now a growing shortage of women of marriageable age throughout East Asia. China alone had an estimated deficit of 10 million women aged 20 to 35 in 2000. Females are also effectively "missing" from the marriage rolls because many educated young women are too busy with career success to meet eligible young men (see Photo Essay 9.6C on page 422).

A recent study, "The Consequences of the 'Missing Girls' of China" by A. Y. Ebenstein and E. J. Sharygin (*World Bank Econ. Rev.* Vol. 23, No. 3, 2009), noted that at least 10 percent of young

More people live in East Asia than in any other world region, but population growth is now slowing as family size decreases. Most of the region now faces the challenge of caring for large elderly populations. Meanwhile, China is grappling with an unexpected consequence of its "one-child policy": a shortage of women.

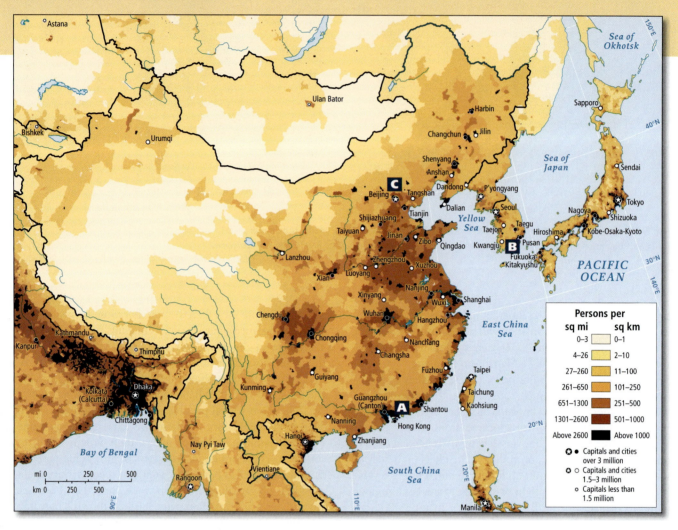

Persons per

sq mi	sq km
0–3	0–1
4–26	2–10
27–260	11–100
261–650	101–250
651–1300	251–500
1301–2600	501–1000
Above 2600	Above 1000

✪ ● Capitals and cities over 3 million
✪ ○ Capitals and cities 1.5–3 million
○ Capitals less than 1.5 million

A A child takes a break at a Hong Kong museum exhibit depicting life in Hong Kong in the 1950s, when families were much larger and children had more playmates.

B An elderly woman in South Korea, where the population is projected to shrink by 13 percent by 2050. By then, the population will have aged as people live longer but have fewer children.

C A bride in Beijing. With so many families preferring male children, there is now a shortage of women. Many women are pursuing careers and delaying or forgoing family life, resulting in a lack of brides.

Chinese men will fail to find a mate; and poor, rural, uneducated men will have the greatest difficulty. The shortage of women will lower the birth rate yet further, causing a more rapid shrinkage of the population over the next century. When single men age, without spouses, children, or even siblings, there will be no one to care for them. Already observable are even darker sides of the shortage of women; in China, cases of kidnapping and forced prostitution of young girls and women are increasing. Furthermore, China's growing millions of single young men are emerging as a threat to civil order. The Ebenstein and Sharygin article makes the observation that societies with large numbers of unattached males often experience a surge in drug abuse, violent crime, HIV infection, and even political radicalism.

Population Distribution

In East Asia, people are not evenly distributed on the land (see Photo Essay 9.6 Map on page 422). China, with 1.34 billion people, has more than one-fifth of the world's population. However, 90 percent of these people are clustered on the roughly one-sixth of the total land area that is suitable for agriculture and 46 percent live in urban areas. People are concentrated especially densely in the eastern third of China in the North China Plain, the coastal zone from Tianjin to Hong Kong that includes the delta of the Zhu Jiang (Pearl River) in the southeast, the Sichuan Basin, and the middle and lower Chang Jiang (Yangtze) basin.

The west and south of the Korean Peninsula are also densely settled, as are northern and western Taiwan. In Japan, settlement is concentrated in a band that stretches from the cities of Tokyo and Yokohama on Honshu island south through the coastal zones of the Inland Sea to the islands of Shikoku and Kyushu. This urbanized region is one of the most extensive and heavily populated metropolitan zones in the world, accommodating 86 percent of Japan's total population. The rest of Japan is mountainous and more lightly settled. Mongolia is only lightly settled, with one modest urban area.

HIV-AIDS in East Asia

Information on HIV-AIDS in East Asia is sketchy, with many governments not reporting data to international agencies. According to the Joint United Nations Program on HIV-AIDS, East Asia is undergoing a concentrated and potentially explosive HIV-AIDS epidemic. Nonetheless, at the present time, reported rates of infection are relatively low by general Asian and global standards, with 0.1 percent of the population aged 15 to 49 infected. This figure, however, may mask a large and growing number of infections. The stigma against HIV-AIDS is very strong in this region, preventing people from being tested and treated and thus increasing the likelihood that the syndrome will spread.

·The majority of HIV infections are in China (estimated at 700,000 in 2007, but perhaps as high as 1,000,000), where localized epidemics have been under way for many years in certain provinces—Yunnan, Xinjiang, Guangxi, Sichuan, Henan, and Guangdong—and are poised to spread rapidly in several others. The number of reported HIV-AIDS cases has increased significantly in recent years, particularly among migrants, intravenous drug users, and sex workers.

There is a steady increase in the rate of new HIV infections in Japan; the number of new HIV-AIDS cases reported annually has doubled since the 1990s to more than 900 per year in 2005. This rise has been accompanied by an increase in other sexually transmitted infections over the same period and may be related to more widespread sexual activity among Japanese youth. There are perhaps 10,000 total cases of HIV infection in the country (less than 0.01 percent of the population aged 15 to 49).

The incidence of HIV-AIDS in South Korea is somewhat higher than in Japan. Surprisingly, in Mongolia, where 50 percent of the population is under 23 and drug use and sex work are increasing with greater global interaction, fewer than 1000 adult cases were reported as of 2007.

215. AIDS Budgets in Asian Countries Called Insufficient

Human Well-Being

"The objective of development is to create an enabling environment for people to enjoy long, healthy and creative lives."

MAHBUB UL HAQ, FOUNDER OF THE UNITED NATIONS HUMAN
DEVELOPMENT REPORT

The best ways to measure how well specific countries enable their citizens to enjoy healthy and rewarding lives is disputed, but it is generally agreed that GDP per capita (PPP) is inadequate as a measure, although income is relevant to well-being. In addition to GDP, we include here maps on human development and on gender differences in pay (Figure 9.23). On all maps, Japan and South Korea (plus Taiwan on Map A) stand out as quite different from China and Mongolia. (Taiwan is not covered on Maps B and C because in UN data, Taiwan is included with China and because Taiwan is so wealthy that it raises China's ranking to some extent). North Korea is not included on any of the maps because it does not belong to the United Nations and does not submit data.

On Map A, Japan, South Korea, and Taiwan all have high GDP per capita adjusted for purchasing power parity (PPP). China and Mongolia rank quite low on Map A, each with an average per capita GDP (PPP) of less than U.S.$5000. What the GDP figures mask is the extent to which there is disparity of wealth in a country. Surprisingly, wealth disparity in China is now nearly three times what it is in Japan. In fact, the Chinese government openly says in its own development report (2007–2008) that its "move away from egalitarianism" has prompted a debate about how central equity and social justice should be in the new socialist economy. The government now aims for equality of opportunity, not equality of outcomes: "…each [person] should have a fair chance of succeeding in the new economy—people should not face large obstacles or enjoy excessive advantages from the start due to the conditions of their upbringing, place of birth, ethnicity, political connections, and so on" (*Human Development Report*, China, 2007–2008 (Beijing: United Nations Development Programme, 2008), pp. iii–iv).

Map B represents a formula that ranks countries according to how well they provide for the basic needs (income, health, and education) of their citizens. Again, there are stark differences between the very high global ranking for Japan (10) and that for South Korea (26), and the barely medium global rankings for China (92) and Mongolia (115). And yet, China and Mongolia

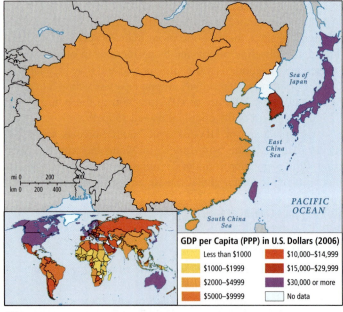

(A)

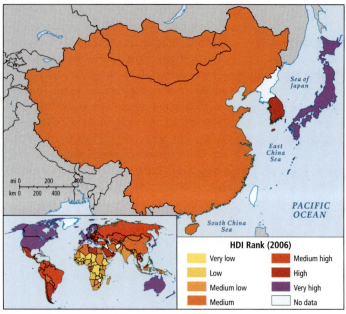

(B)

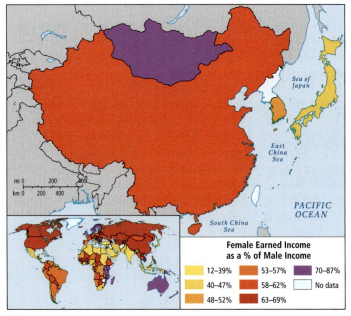

(C)

FIGURE 9.23 Maps of human well-being. (A) Gross domestic product (GDP). **(B)** Human Development Index (HDI). **(C)** Female earned income as a percent of male earned income (F/MEI). [Maps adapted from data: "Human development indices," (Maps A and B): Table 2, pages 28–32; (Map C): Table 5, pages 41–44 at http://hdr.undp.org/en/media/HDI_2008_EN_Tables.pdf.]

keep women's pay low compared to that of men. Given Confucian-based cultural patterns and present trajectories, it seems likely that pay inequity for females in all of East Asia will get worse before it gets better.

THINGS TO REMEMBER

1. More people live in East Asia than in any other world region, but population growth is now slowing as family size decreases. Most of the region now faces the challenge of caring for large elderly populations.

2. The Japanese have the world's longest life expectancy at 83 years and the world's oldest population—23 percent of Japanese are over the age of 65. By 2055, this age group will account for 40 percent of the population.

3. China is grappling with an unexpected consequence of its "one-child policy": a shortage of women and the demise of sibling relationships and the extended family.

4. Female workers earn about two-thirds of what male workers earn in China, and about half of what males earn in South Korea and Japan.

have made significant progress over the last 20 years. Both countries say that their specific goals now are to provide equitable access to basic services for all of their people, so these rankings should improve in the years to come.

Map C, which shows female income as a percentage of male income, is one way to depict the extent to which there are pay disparities based on gender. Although the pay for women is lower than for men in all countries, Japan and South Korea, which rank high on Maps A and B, rank lower on this particular scale than do China and Mongolia. (China has the same rating as the United States.) Undoubtedly, China and especially Mongolia's better records on pay equity (remember that actual amounts of pay are very low; see **Figure 9.23A**) are the result of the emphasis on gender equity under communism, while in Japan and Korea, the market economy and relic Confucian attitudes toward women

Cultural Diversity in East Asia

Most countries in East Asia have one dominant ethnic group, but all countries have considerable cultural diversity. In China, for example, 93 percent of Chinese citizens call themselves "people of the Han." The name harks back about 2000 years to the Han empire, but it gained currency only in the early twentieth century, when nationalist leaders were trying to create a mass Chinese identity. The term *Han* simply connotes people who share a general way of life, pride in Chinese culture, and a sense of superiority to ethnic minorities and outsiders. The main language spoken by the Han is Mandarin, although it is only one of many Chinese dialects.

China's non-Han minorities number about 117 million people in more than 55 different ethnic or culture groups scattered across the country. Most live outside the Han heartland of eastern China (Figure 9.24). Some of these areas have been designated autonomous regions, where minorities theoretically manage their own affairs. In practice, however, the Han-dominated Communist Party in Beijing controls the fate of the minorities, especially those considered to be security risks or who have resources of economic value. We profile a few of China's ethnic groups here.

Western China's Muslims Muslims of various ethnic origins have long been prominent minorities in China. All are originally of Central Asian origins; most are Turkic people who historically have been seminomadic herders. Others specialized in trading. They tend to be concentrated in China's northwest and to think of themselves as quite separate from mainstream China.

In the autonomous region of Xinjiang Uygur in the far northwest (see Figure 9.1 on pages 384–385), live the Uygurs and Kazakhs who are Turkic-speaking Muslims. Historically, these peoples were nomadic herders, and some still are. Contact with culturally related peoples in Central Asia has been revived since

FIGURE 9.25 The Silk Road market in Kashi. Uygur men inspect currency before selling their goats at the Sunday market in Kashi, Xinjiang Uygur Autonomous Region, China. Kashi has been an important trading center along the Silk Road for at least 2000 years.

China's market reforms began (Figure 9.25). The Beijing government wishes to claim this area's oil, other mineral resources, and irrigable agricultural land for national development; it also wants to intercept any Central Asian moves to infiltrate this western zone of China. As a result, it has sent troops and by now many millions of Han settlers to Xinjiang. The Han settlers fill most managerial jobs in the bureaucracy, in mineral extraction, the military, and power generation. An important secondary role of the Han is to dilute the power of Uygurs and Kazakhs within their own lands. In Xinjiang Uygur, there are now as many Han as Uygurs (9 million each), plus small numbers of other minorities.

The Beijing government has rushed to develop Xinjiang and its capital Urumqi with special development zones (ETDZs), but the Uygurs have been left out of most policy-making roles and indeed have been excluded from participating in the economic boom. For example, one young Uygur man in Urumqi writes of his discontent: "I am a strong man, and well-educated. But [Han] Chinese firms won't give me a job. Yet go down to the railroad station and you can see all the [Han] Chinese who've just arrived. They'll get jobs. It's a policy to swamp us."

In addition to being left out economically, the Uygurs are concerned about Han prejudices against Uygur ethnicity and their Muslim religion, which since 9/11 have often been used to label the Uygurs as separatists and terrorists. Insulting treatment of Uygur sacred sites by Han developers—such as building a gas station on top of a shrine, suppressing the Uygur (Turkic) language, or administrative rules that ban Uygur children from entering mosques—are taken by the Uygur to mean the intention is to stamp out their culture. Urumqi is fast changing as the Han modernize the city to fit their tastes and in so doing they are destroying centuries-old quarters of the city beloved by the Uygurs as essential components of their heritage. The Uygurs view this as politically motivated *cultural cleansing*, in this case the destruction of all physical vestiges of Uygur ways of life.

Until recently, the Uygur people of Xinjiang expressed their resistance to Han dominance merely by reinvigorating their Islamic

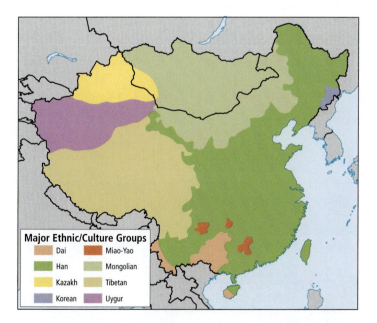

FIGURE 9.24 Major ethnic groups of China. This map shows the areas traditionally occupied by the Han and the ethnic minorities. It does not show the recent resettling of Han in Xinjiang and Tibet, nor does it show the Hui, people from many ethnic groups whose ancestors converted to Islam and who are found in disparate locations along the old Silk Road and in coastal southeastern China. [Adapted from Chiao-min Hsieh and Jean Kan Hsieh, *China: A Provincial Atlas* (New York: Macmillan, 1995), p. 12. The Web site http://www.index-china.com/minority/minority-english.htm supplies a comprehensive survey of minorities in China.]

Major Ethnic/Culture Groups

Dai		Miao-Yao	
Han		Mongolian	
Kazakh		Tibetan	
Korean		Uygur	

culture. Islamic prayers were increasingly heard publicly, more Muslim women were wearing Islamic dress, Uygur was spoken rather than Chinese, and Islamic architectural traditions were being revived. Then more active resistance groups were formed. The Beijing government responded by harshly punishing the resisters and broadcasting the accusation that they are Islamic fundamentalists bent on terrorism. In July 2009, violence erupted in the streets between Uygurs and the Han and about 200 were killed.

Distinct from the Uygurs and Kazakhs are the *Hui*, who altogether number about 8 million. The original Hui people were descended from ancient Turkic Muslim traders who traveled the Silk Road from Europe across Central Asia to Kashgar (now Kashi) and on to Xian in the east (see the map of the Silk Road in Figure 5.8 on page 219). Although today most Hui are descended from non-Turkic peoples (mostly Han) who converted to Islam, they are still culturally distinct. Subgroups of Hui live in the Ningxia Huizu Autonomous District and throughout northern, western, and southwestern China. There they continue as traders, farmers, and artisans. There are numerous famous Hui civil servants, philosophers, and historians, and the great navigator, Zheng He, who took a flotilla of ships as far as East Africa in the early 1400s, was Hui. The long tradition of commercial activity among the Hui has facilitated their success in China. Many are active in the new free market economies of southeastern China as businesspeople, technicians, and financial managers, using their money not only to buy luxury goods, but also to revive religious instruction and to fund their mosques, which are now more obvious in the landscape.

The Tibetans In contrast to the prosperous assimilated Hui are the Tibetans, an impoverished ethnic minority of nearly 5 million individuals scattered thinly over a huge, high mountainous region in western China. The history of Tibet's political status vis-à-vis China is long and complex, characterized by both cordial relations and conflict. During China's imperial era (prior to the twentieth century), Tibet maintained its own government and sent representatives to the Chinese imperial court, but faced the constant threat of invasion and of Chinese meddling in its affairs. In the early 1900s, Tibet declared itself separate and free from China and conducted its affairs as an independent country. It was able to maintain this status until 1949–1950, when the Chinese Communist army "liberated" Tibet, promising a "one-country, two-systems" structure. A Tibetan uprising in 1959 led China to abolish the Tibetan government and violently reorder Tibetan society. Since the 1950s, the Chinese government has referred to Tibet as the Xizang Autonomous Region. The Chinese government suppressed the Tibetan Buddhist religion, long the mainstay of most Tibetans' daily lives, by destroying thousands of temples and monasteries and massacring many thousands of monks and nuns. In 1959, the spiritual and political leader of Tibet, the Dalai Lama, was forced into exile in India along with thousands of his followers.

216. Dalai Lama Calls for Autonomy but Not Independence

By the 1990s, the Beijing government's strategy was to overwhelm the Tibetans with secular social and economic modernization and with Han Chinese settlers rather than outright military force (though China maintains a military presence in Tibet). To attract trade and quell foreign criticism of their treatment of Tibetans, China is spending hundreds of millions of dollars on housing and on roads, railroads, and a tourism infrastructure that capitalizes on European and American interest in Tibetan culture. China presents its actions in Tibet as part of its overall strategy to integrate the entire country economically and socially. Schools are being built and jobs opened up to young Tibetans. A

FIGURE 9.26 The Qinghai–Tibet Railway.
Opened in July 2006, the world's highest railway travels nearly 596 miles (1000 km) at an altitude above 13,123 feet (4000 meters). It now connects Beijing to Lhasa in less than 48 hours, transporting goods and people at a much lower cost than the existing highway. However, many Tibetans worry that it will be come a conduit for more Han migration to Tibet. **(A)** The Qinghai–Tibet train as it passes through Tibet's Chang Tang Nature Reserve.

new railway link connecting Tibet more conveniently to the rest of China was completed in 2006. The Han in Tibet see the railway as a public service that will promote Tibetan development, but Tibetan activists see it as a conveyor belt for more Han dominance over the Tibetan economy and culture (Figure 9.26).

▶ **214. Tibetan Buddhist Nuns' Protest Songs Bring Punishment from China**

Within Tibetan culture, women have held a relatively high position. Among the nomadic herders, they were free to have more than one husband, just as men were free to have more than one wife. In addition, the cultural custom that in marriage a husband joins the wife's family allowed Tibetan women to attain a higher status than Han women. Although Buddhism introduced patriarchal attitudes from outside Tibet, it continued to encourage female independence, because at any given time, up to one-third of the male population was living a short-term monastic life, so Tibetan Buddhist women had to be particularly self-sufficient. Han Chinese culture has typically regarded the women of the western minorities as barbarian precisely because their roles were not circumscribed: they were not secluded, they rode horses, they worked alongside the men in herding and agriculture, and they were assertive.

Indigenous Diversity in Southern China In Yunnan Province in southern China, more than 20 groups of ancient native peoples live in remote areas of the deeply folded mountains that stretch into Southeast Asia. These groups speak many different languages, and many have cultural and language connections to the indigenous people of Tibet, Burma, Thailand, or Cambodia. Women and men are treated more equally in this area than among the Han. A crucial difference may be that among several groups, most notably the Dai, the husband moves in with the wife's family at marriage and provides her family with labor or income. A husband inherits from his wife's family rather than from his birth family, and female children are valued just as highly as males.

Taiwan's Many Minorities In Taiwan and the adjacent islands, the Han account for 95 percent of the population, but Taiwan is also home to 60 indigenous minorities. Some have cultural characteristics—languages, crafts, and agricultural and hunting customs—that indicate a strong connection to ancient cultures in far Southeast Asia and the Pacific. The mountain dwellers among these groups have resisted assimilation more than the plains peoples. Both groups may live on reservations set aside for indigenous minorities if they choose, but most are now being absorbed into mainstream urbanized Han-influenced Taiwanese life.

The Ainu in Japan There are several indigenous minorities in Japan and most have suffered considerable discrimination. A small and distinctive minority group is the **Ainu**, characterized by their light skin, heavy beards, and thick, wavy hair. Now numbering only about 16,000, the Ainu are a racially and culturally distinct group thought to have migrated many thousands of years ago from the northern Asian steppes. They once occupied Hokkaido and northern Honshu, living by hunting, fishing, and some cultivation, but they are now being displaced by forestry and other development activities. Few full-blooded Ainu remain because, despite prejudice, they have been steadily assimilated into the mainstream Japanese population (Figure 9.27).

FIGURE 9.27 The Ainu of Japan. A woman demonstrates a traditional instrument in the Ainu living history village in the Akan National Park in eastern Hokkaido, Japan.

East Asia's Largest Cultural Export: The Overseas Chinese

Ainu an indigenous cultural minority group in Japan characterized by their light skin, heavy beards, and thick, wavy hair who are thought to have migrated thousands of years ago from the northern Asian steppes

China has had an impact on the rest of the world not only through its global trade, but also through the migration of its people to nearly all corners of the world. The first recorded emigration by the Chinese took place over 2200 years ago. Since then, China's contacts spread eastward to Korea and Japan, westward into Central and Southwest Asia via the Silk Road, and by the fifteenth century, to Southeast Asia, coastal India, Arabia, and even Africa.

Trade was probably the first impetus for Chinese emigration. The early merchants, artisans, sailors, and laborers came mainly from China's southeastern coastal provinces. Taking their families with them, some settled permanently on the peninsulas and islands of what are now Indonesia, Thailand, Malaysia, and the Philippines. Today they form a prosperous urban commercial class known as the "Overseas Chinese."

In the nineteenth century, economic hardship in China and a growing international demand for labor spawned the migration of as many as 10 million Chinese people to countries all over the world. By the middle of the twentieth century, they were joined by many others fleeing the repression of China's Communist Revolution. As a result, "Chinatowns" are present in places as widely scattered as Singapore, Trieste, London, São Paulo, San Francisco, Havana, and Melbourne. The term *Overseas Chinese* has been extended to apply to Chinese emigrants and their descendants in all such locations.

1. There are several distinct minority groups in East Asia: the Uygurs, Kazakhs, and Tibetans in western China; the Hui, scattered throughout central China; smaller indigenous groups in southern China and Taiwan; and the Ainu in Japan.

2. There is resentment and sometimes open resistance to Han Chinese domination in the various minority homelands.

3. Throughout East Asia, minorities have experienced discrimination. Some are now protected as heritage symbols.

4. Millions of "Overseas Chinese" live in cities and towns around the world. Most have settled permanently in their foreign locations, and many (perhaps most) maintain relationships with China.

Reflections on East Asia

Just as the astounding economic rise of Japan, South Korea, and Taiwan became the model for developing countries over the past 50 years, the opening of China to the global economy will continue to transform East Asia and the world for the next 50. The changes underway in this region today are immense. Incomes have risen and cities have boomed, but at the cost of some of the worst air and water pollution in the world. Greenhouse gas emissions have increased, but so has the ability to use new technologies to create cleaner energy economies. The ability of countries to feed their own people with domestic agricultural resources is declining, but the ability to pay for food imported from abroad is increasing.

Wealth disparities in China have increased dramatically, with the income of millions hurt by the reformulation of the communist command economy and the loss of the safety nets it provided, while the income of others has been helped by new economic opportunities created by the forces of free market globalization. Shortages of skilled labor are bringing better wages and working conditions, and may also hasten a shift toward the production of more technologically sophisticated and higher-quality goods.

A major question remains regarding China's growing and future prosperity. Where will China find the natural resources to supply increasing living standards for so many people? China's massive importation of food, minerals, and oil is already affecting many regions of the world (see Chapter 3, page 126, Chapter 7, pages 314–315, and Chapter 10, pages 451–452). These areas may not be willing or able to sustain such massive resource exports for long.

Throughout East Asia, as birth rates decline and populations age, some economies may suffer if they fail to either import labor or increase mechanization. This shift is already occurring in Japan, South Korea, and Taiwan, whose existing broad prosperity has given them more breathing room. China's economy may take a bigger hit because the one-child-per-family policy has reduced population growth drastically before broader prosperity has been achieved. This, combined with the country's gender imbalance, may produce considerable social instability. And as more women throughout East Asia pursue careers instead of family life, birth rates will continue to fall, bringing further social and economic changes.

The combined impact of all of these transformations may eventually force China to adopt the democratic systems that already prevail among East Asia's wealthier countries. Such a shift could improve China's dismal human rights record, and would no doubt be celebrated by the millions who have suffered from the tragic failures of judgment of the Communist Party political elite. Still, many of China's problems will remain even if and when strong democratic institutions are in place. Democracy has not solved the problems of an aging population in Japan, Taiwan, and South Korea, or of severe air pollution in Taiwan. In this ever-more-globalized world, the ways in which China and East Asia as a whole solve these problems will influence all of our lives.

Critical Thinking Questions

1. Why is the interior west of continental East Asia (western China and Mongolia) so dry and subject to extremes in temperature?

2. In what ways is China more vulnerable to global warming than South Korea, Japan, and Taiwan?

3. Historically, China has been afflicted with recurring famines. What are the present concerns about its food security?

4. How might a more democratic political system have resulted in changes in the overall conception and design of the Three Gorges Dam?

5. Why might a Confucian praise or criticize China's current political system and its approach to economic and political change?

6. In what ways might globalization have taken a different course if the voyages of Admiral Zheng He had inspired China to conquer a vast colonial empire, like that of Britain or the Netherlands?

7. Contrast Japan's pre–World War II policies in East Asia with its present role in the region. Describe the principal similarities or differences.

8. In what ways has the one-child-per-family policy helped maintain stability in China? How might it destabilize China in the future?

9. Which problems related to East Asian urbanization and globalization might lead to greater political changes than have occurred already?

10. Which forces are likely to change gender roles in East Asian countries in the future?

Chapter Key Terms

Ainu, 427
Confucianism, 399
Cultural Revolution, 403
export-led growth, 405
floating population, 387
food security, 395
foreign investment, 408
Great Leap Forward, 402

growth poles, 411
hukou system, 387
kaizen system, 406
kanban system, 406
land reform, 407
petty capitalists, 408
regional self-sufficiency, 408
regional specialization, 408

responsibility system, 408
special economic zones (SEZs), 411
state-aided market economy, 405
tsunami, 390
typhoon, 391
wet rice cultivation, 395

BHUTAN

INDIA

CHINA

Hkakabo Razi
elev 19,295

Congga Shan
elev 24,700

Yunnan–Guizhou Plateau

Putao

Guiyang

Chittagong

Kunming

Mandalay

20°N

BURMA
(MYANMAR)

Phong Saly

LaoCai

Ha Giang

Fan Si Pan
elev 10,312

Dien Bien Phu

Hanoi

Nyakan Yoma

Bago Mts.

Nay Pyi Taw

Loikaw

Chiang Rai

Muong Sai

Samneua

Ban Ban

Haiphong

Luang
Prabang

LAOS

Thanh Hoa

*Gulf of
Tonkin*

**Bay of
Bengal**

Chiang Mai

Vientiane

B

Vinh

VIETNAM

Dong Hoi

Rangoon

*Gulf of
Martaban*

Tak

Udon Thani

Khon Kaen

Ban Nape

Mekong

Quang Tri

Hue

*Preparis
North Channel*

THAILAND

Savannakhet

Da Nang
Hoi An

*Andaman
Islands*

Lop Buri

Surin

Dangrek Range

Pakxe

Kon Tum

Play Ku

**Andaman
Sea**

Saraburi

Chao Phraya

Siem Reap

Khone Falls

Lomphat

Qui Nhon

Bangkok

Batdambang

CAMBODIA

*South China
Sea*

10°N

Ten Degree Channel

Kratie

Da Lat

Nha Trang
Cam Ranh

Phnom Penh

Kompong Som

Ho Chi Minh City
(Saigon)

Phan Rang

*Nicobar
Islands*

Surat Thani

*Gulf of
Thailand*

Can
Tho

*Mekong
Delta*

Great Channel

Phuket

Malay Peninsula

Songkhla

Strait of Malacca

Pattani

Langkawi

Alor Setar

George Town
(Penang)

Banda Aceh

Ipoh

E

Medan

Kuantan

MALAYSIA

Kuala Lumpur

Melaka
(Malacca)

Barisan Mountains

Johor Baharu

0° Equator

Padang

*Kerinci
elev 12,484*

SINGAPORE

Riau Archipelago

Kuching

Sumatra

Palembang

Pontianak

Borneo

Kalimantan

**INDIAN
OCEAN**

Bandar
Lampung

Java Sea

Jakarta

Krakatoa
elev 2,667

Cirebon

Banjarmasin

Tegal

Bandung

Semarang

Surakarta

Mt. Merapi
elev 9,724

Yogyakarta

Malang

D

Java

Denpasar

Lombok

10°S

Lesser Sunda Islands

East
China
Sea

Ryukyu Islands

*Sakishima
Islands*

Luzon Strait

Luzon

Aparri

**Philippine
Sea**

Baguio

Sierra Madre

Mt. Pinatubo
elev 6,683

Quezon City
Pasig

Olongapo

Manila

C

Mt. Taal
elev 987

Lucena

Mt Mayon
elev 9,810

Mindoro

Samar

PHILIPPINES

Palawan

Panay

Cebu

Leyte

**Sulu
Sea**

Negros

Zamboanga

Davao

Mindanao

Mt. Kinabalu
elev 13,451

Kota Kinabalu

Bandar Seri Begawan

Sabah

BRUNEI

**Celebes
Sea**

Niah

MALAYSIA

Tarakan

Sarawak

Manado

Samarinda

*Molucca
Sea*

Balikpapan

Makassar Strait

*Sulawesi
(Celebes)*

Makassar
(Ujungpandang)

Banda

Madura

Surabaya

I N D O N E

Bali

Mt. Tambora
elev 9,354

Mataram

Flores

Dili

**TIMOR-
LESTE**

B Mekong River Plains, Laos

C Philippine Archipelago

PACIFIC
OCEAN

MICRONESIA

PALAU
Ngerulmud

Land Elevations

meters	feet
4877	16,000
3353	11,000
2134	7000
914	3000
305	1000
152	500
0	0

mi 0   100   200   300
km 0 100 200 300 400 500

1:17,000,000
Azimuthal Equidistant Projection

Halmahera

Sorong

Moluccas

Ceram

Buru

Sea

S I A

West Papua

New Guinea

Puntiak Jaya
elev. 16,499

Maoke Mountains

Jayapura

PAPUA
NEW
GUINEA

Aru
Islands

*Arafura
Sea*

Merauke

Timor

Timor Sea

AUSTRALIA

130°E   140°E

THEMATIC CONCEPTS
Population · Gender · Development
Food · Urbanization · Globalization
Democratization · Climate Change · Water

Southeast Asia

D Volcanoes, Java, Indonesia

E Earthquake/Tsunami, Sumatra, Indonesia

FIGURE 10.1 Regional map of Southeast Asia.

431

Global Patterns, Local Lives In December 2005, a group of indigenous people in the Malaysian state of Sarawak, on the island of Borneo (**Figure 10.1**), attended a public meeting wearing orangutan masks. They carried signs informing onlookers that, although the government protects orangutan territories, it ignores the basic right of indigenous people to live on their own ancestral lands.

Over the years, Sarawak forest dwellers have tried many tactics to save their lands from deforestation by logging companies and expanding oil palm plantations. In the mid-twentieth century, the state government began licensing logging companies to cut down forests occupied by indigenous peoples. By the 1980s, 90 percent of Sarawak's lowland forests had been degraded and 30 percent had been clear-cut, with much of the cleared area replaced by oil palm plantations. As a consequence, indigenous people found that even on uncleared land their hunts declined. Meanwhile, streams and rivers became polluted with eroded sediment and by fertilizers and the pesticides applied to the palm trees.

In the 1990s, a group of citizens in Berkeley, California, who were concerned about news reports of the deforestation in Sarawak, organized the Borneo Project. They offered to become a "sister city" to one indigenous group in Sarawak, the Uma Bawang, giving help wherever it was needed. With the help of the Borneo Project, the Uma Bawang began a community-based mapping project in 1995. Using rudimentary compass and tape techniques, they began mapping both the extent and content of their forest home (**Figure 10.2**). Since then, indigenous people from across Sarawak have learned how to use global positioning systems (GPS), geographic information systems (GIS), and satellite imagery to make more sophisticated maps. The power of these maps was demonstrated in 2001, when they helped win a precedent-setting court case that protected indigenous lands from an encroaching oil palm plantation.

This favorable court ruling was challenged in 2005, when the government appealed, but in 2009, the Malaysian federal court ruled in favor of the Uma Bawang, stating that native customary land rights had been protected since 1939, when British colonial officials had directed the district lands and survey departments to map the boundaries of native lands. Now indigenous people have the right to sue the government for past illegal leases to logging companies, and some 203 such cases are in litigation. *[Adapted from The Borneo Wire (newsletter of the Borneo Project), Jessica Lawrence, ed., Spring 2006, at http://borneoproject.org/article.php?id=623 "Community Stops Illegal Logging and Bulldozing Towards Protected Rainforests" The Borneo Wire (newsletter of the Borneo Project), Jessica Lawrence, ed., Fall 2006 http://borneoproject.org/article.php?list=type&type=39; Mark Bujang, "A Community Initiative: Mapping Dayak's Customary Lands in Sarawak," presented at the Regional Community Mapping Network Workshop, Nov. 8–10, 2004, Diliman, Quezon City, Philippines; "Malaysia Highest Court Affirms Tribes' Land Rights," Julia Zappei, Associated Press, May 10, 2009, available at http://borneoproject.org/article.php?id=762; and "Actualization of the UN Declaration on the Rights of Indigenous Peoples (UNDRIP)," International Forum on Globalization: Indigenous Rights Programs, at http://www.ifg.org/programs/indigenousrights.htm.]* ∎

This account of the global tactics that indigenous groups are using to secure their rights to ancestral lands highlights the extent to which local or national issues are becoming global issues. As

(A)

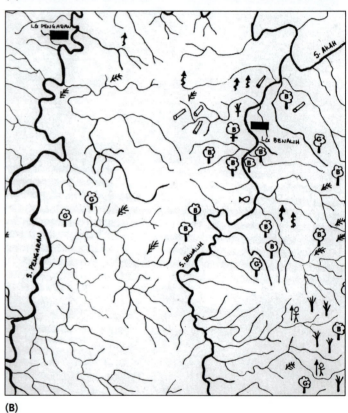

(B)

FIGURE 10.2 Villagers mapping their home territory. (A) These villagers first gathered to discuss and draw the boundaries of their lands in Penan, Sarawak. **(B)** Then they added important features such as long-houses (communal dwellings for extended kin-groups), some with spiritual significance (some longhouses are now abandoned), sago palm trees, hunting areas, and graveyards, as well as physical features such as rivers.

problems like climate change become better understood, ever more issues are being addressed at the global level. This presents both challenges and opportunities for the world's indigenous people (Figure 10.3). It is important to notice that it was only with international support that the Uma Bawang were able to see their land claims validated. The 2009 United Nations report on the state of the world's indigenous peoples is available online at http://www.un.org/esa/socdev/unpfii/documents/SOWIP_web.pdf.

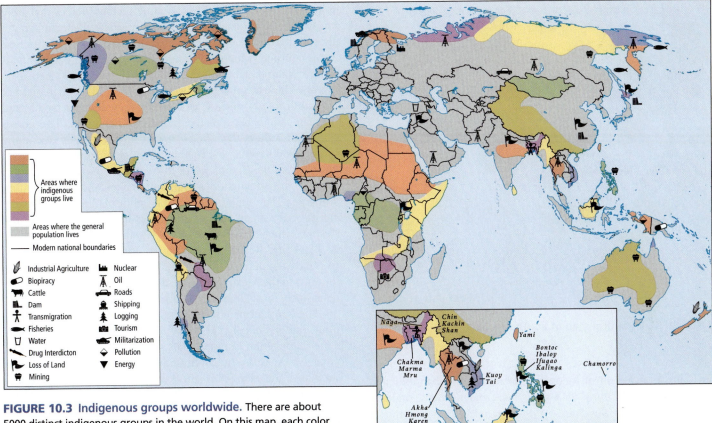

FIGURE 10.3 Indigenous groups worldwide. There are about 5000 distinct indigenous groups in the world. On this map, each color represents one or more of these groups that are related by language, culture, or an affinity to a geographic location. Many of these peoples have participated in community mapping projects, similar to those of the Sarawak forest dwellers, in order to identify and protect their rights to their traditional lands. The symbols reflect global issues that affect a particular group. [Adapted from "Struggling Cultures," *National Geographic Atlas of the World,* 8th ed. (Washington, DC: National Geographic Society, 2005), p. 15; and from the "Globalization: Effects on Indigenous Peoples" map, International Forum on Globalization, at http://www.ifg.org/programs/indig.htm.]

The nine **Thematic Concepts** covered in this chapter on pages 436–437 are illustrated here with photo essays that show examples of how each concept is experienced in this region. In all cases, interactions between two or more of these concepts are illustrated. The captions explain the pictures and how some of these interactions work in specific places.

THINGS TO REMEMBER

1. Indigenous land claims are no longer just local issues. Indigenous groups have shifted their efforts to secure their rights to ancestral lands to the global scale, highlighting a trend in which many once local or national issues are now becoming global issues.

I THE GEOGRAPHIC SETTING

Terms in This Chapter

Many governments in Southeast Asia (see **Figure 10.4** on page 434) choose to dispense with place-names that originated in their colonial past. However, when the governments that make these changes earn broad disrespect in the international community by violating the human rights of their citizens, their chosen name may not be acknowledged. Such is the case with Burma, where a military government seized control in a coup d'état in 1990, changing the country's name to Myanmar. In this text, we use the country's traditional name of Burma instead of Myanmar to acknowledge the repression of the people of that country by a government they had no role in choosing.

Another potential point of confusion is Borneo, a large island that is shared by three countries. The part of the island known

FIGURE 10.4 Political map of Southeast Asia.

as Kalimantan is part of Indonesia; Sarawak and Sabah are part of Malaysia; and Brunei is a very small, independent, oil-rich country.

Physical Patterns

The physical patterns of Southeast Asia have a continuity that is not immediately obvious on a map. A map of the region shows a unified mainland region that is part of the Eurasian continent and a vast and complex series of islands arranged in chains and groups. These landforms are actually related in origin. Climate is another source of continuity, with most of the region tropical or subtropical.

Landforms

Southeast Asia is a region of peninsulas and islands (see Figure 10.1 on pages 430–431). Although the region stretches over an area larger than the continental United States, most of that space is ocean; the area of all the region's land amounts to less than half that of the contiguous United States. The large mainland peninsula that extends to the south of China is occupied by Burma, Thailand, Laos, Cambodia, and Vietnam. This peninsula itself sprouts the long, thin Malay peninsular that is shared by outlying parts of Burma and Thailand, a part of Malaysia, and the city-state of Singapore, which is built on a series of islands at the southern tip. The **archipelago** (a series of large and small islands) that fans out to the south and east of the mainland is grouped into the countries of Indonesia, Malaysia, Brunei, Timor-Leste (East Timor), and the Philippines. Indonesia alone has some 17,000 islands, and the Philippines has 7000.

The irregular shapes and landforms of the Southeast Asian mainland and

archipelago a group, often a chain, of islands

archipelago are the result of the same tectonic forces that were unleashed when India split off from the African Plate and gradually collided with Eurasia (see Figure 1.26 page 51). As a result of this collision, which is still under way, the mountainous folds of the Plateau of Tibet reach heights of almost 20,000 feet (6100 meters). These tectonic folds, which bend out of the high plateau, turn south into Southeast Asia. There, they descend rapidly and then fan out to become the Indochina peninsula. The gorges widen into valleys that stretch toward the sea, each containing hills of 2000 to 3000 feet (600 to 900 meters) and a river or two flowing from the mountains of the Yunnan–Guizhou Plateau of China to the Andaman Sea, the Gulf of Thailand and the South China Sea. The major rivers of the peninsula are the Irrawaddy and the Salween in Burma; the Chao Phraya in Thailand; the Mekong, which flows through Laos, Cambodia, and Vietnam; and the Black and Red rivers of northern Vietnam. Several of these rivers have major delta formations, especially the Irrawaddy, Chao Phraya, and the Mekong, that are densely used for agriculture and increasingly for settlement (see discussion on pages 442 and 462).

The curve formed by Sumatra, Java, the Lesser Sunda Islands (from Bali to Timor), and New Guinea conforms approximately to the shape of the Eurasian Plate's leading edge (see Figure 1.26 on page 51). As the Indian-Australian Plate plunges beneath the Eurasian Plate along this curve, hundreds of earthquakes and volcanoes occur, especially on the islands of Sumatra and Java. Volcanoes and earthquakes also occur in the Philippines, where the Philippine Plate is pushing against the eastern edge of the Eurasian Plate. The volcanoes of the Philippines are part of the Pacific Ring of Fire (see Figure 1.27 on page 52).

Volcanic eruptions, and the mudflows and landslides that occur in their aftermath, endanger and complicate the lives of many Southeast Asians. Over the long run, though, the volcanic material creates new land and provides minerals that enrich the soil for farmers. Earthquakes are especially problematic because of the tsunamis they can set off. The tsunami of December 2004, triggered by a giant earthquake (9.3 in magnitude) just north of Sumatra, swept east and west across the Indian Ocean, taking the lives of 230,000 people and injuring many more (Figure 10.5). It is thought to be one of the deadliest natural disasters in recorded history. A series of strong earthquakes have occurred since, with several in August and September of 2009 and April and May 2010 along coastal Sumatra, but they did not generate notable tsunamis.

The now-submerged shelf of the Eurasian continent that extends under the Southeast Asian peninsulas and islands was above sea level during the recurring ice ages of the Pleistocene epoch, during which much of the world's water was frozen in glaciers. The exposed shelf, known as Sundaland (Figure 10.6), allowed ancient people and Asian land animals (such as elephants, tigers, rhinoceroses, and orangutans) to travel south to what, when sea levels rose, became the islands of Southeast Asia.

Climate and Vegetation

The largely tropical climate of Southeast Asia is distinguished by continuous warm temperatures in the lowlands—consistently

(A)

(B)

FIGURE 10.5 **Banda Aceh, Indonesia, before and after the December 2004 tsunami.** When the tsunami hit, this community at the western end of Sumatra was totally destroyed. **(A)** Banda Aceh shoreline before June 2004. **(B)** After the tsunami, the original shoreline is gone.

above 65°F (18°C)—and heavy rain (see Photo Essay 10.1 on page 438). The rainfall is the result of two major processes: the monsoons (seasonally shifting winds) and the intertropical convergence zone (ITCZ), the band of rising warm air that circles the earth roughly around the equator (see page 298). The wet summer season extends from May to October, when the warming of the Eurasian landmass sucks in moist air from the surrounding seas. Between November and April, there is a long dry season on the mainland, when the seasonal cooling of Eurasia causes dry air from the interior of the continent to flow out toward the sea. On the many islands, however, the winter can also be wet because the air that flows from the continent picks up moisture as it passes south and east over the seas. The air releases its moisture as rain after ascending high enough over landforms to cool. With rains coming from both the monsoon and the ITCZ, the island part of Southeast Asia is one of the wettest areas of the world.

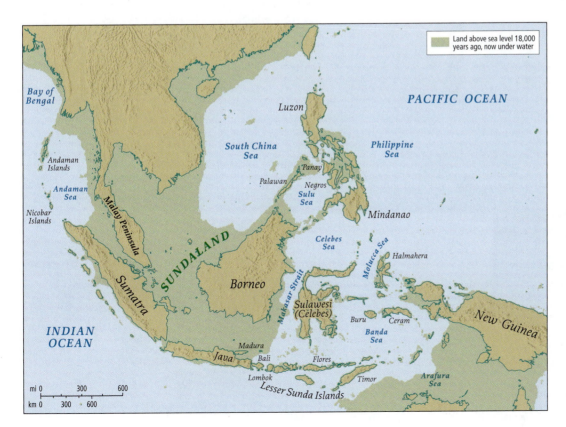

FIGURE 10.6 **Sundaland 18,000 years ago, at the height of the last ice age.** The now-submerged shelf of the Eurasian continent that extends under Southeast Asia's peninsulas and islands was exposed during the last ice age and remained above sea level until about 16,000 years ago, when that ice age was ending. [Adapted from Victor T. King, *The Peoples of Borneo* (Oxford: Blackwell, 1993), p. 63.]

THEMATIC CONCEPTS: Population • Gender • Development • Food • Urbanization
Globalization • Democratization • Climate Change • Water

Climate Change, Development, Food, and Water:
This region is a major contributor to greenhouse gas emissions via widespread deforestation. The conversion of forests to subsistence farms, oil palm plantations, and other uses is a major aspect of economic development. Meanwhile, many here are highly vulnerable to hazards such as flooding that climate change may intensify.

(A) The last patch of rainforest, already surrounded by oil palm plantations, is cleared for more plantations in Sarawak, Malaysia.

(B) Subsistence farmers in Laos plant beans on a recently deforested hillside. Small-scale farming drives much deforestation in the region.

(C) Flooding along the Mekong River in Vietnam. Flooding could worsen over the short term as the glaciers that feed the Mekong melt.

Urbanization, Food, and Development:
Changes in agriculture and industry are driving rapid urbanization throughout this region. Large mechanized corporate farms are replacing small family farms, forcing many farmers into cities. Jobs are available in burgeoning industries in the cities, but inadequate urban infrastructure means that rural migrants often live in slums.

(D) Tractors are used to cultivate rice paddies outside of a growing beach resort on Langkawi, Malaysia.

(E) Dock workers at a container shipping facility in Bangkok, Thailand, where economic growth has been rapid.

(F) A slum area in Chau Doc, Vietnam, a small but growing town in the Mekong Delta.

Globalization and Development:
Globalization has produced long periods of strong economic growth punctuated by brief but dramatic periods of decline. Urban incomes have increased significantly during times of growth, but many poor people remain highly vulnerable to periods of global economic decline that threaten their access to jobs, adequate living quarters, and food.

(G) Billboards in Kuala Lumpur, Malaysia's increasingly wealthy capital city, advertise U.S. and Italian brands.

(H) Dozens of ships anchor outside of the island nation of Singapore, home to the world's largest shipping container port.

(I) Children in a slum in Phnom Penh, Cambodia. Many Cambodians have lost jobs in the recent global economic crisis.

436

Democratization:
While significant barriers remain throughout the region, some countries have recently had dramatic expansions of democratization. Others, however, have also reversed the trend toward democratization. Moreover, some pro-democracy movements have been intertwined with violence directed at specific ethnic groups.

(J) Burmese citizens and Buddhist monks march in Rangoon in 2007 in a series of protests, some of them explicitly pro-democracy. They were brutally repressed.

(K) Thai soldiers outside an office building in Bangkok during the military coup of 2007 that ousted elected Prime Minister Thaksin Shinawatra.

(L) Chinese-owned shops are looted during riots in May 1998 that led to the resignation of President Suharto. An expansion of democracy followed his resignation.

Population:
Slower rates of population growth have been driven by economic development, urbanization, and efforts by governments and NGOs to encourage the use of contraceptives. Some wealthy and urbanized countries are now well into the "demographic transition" and are coping with the challenges of very low fertility, while poor and rural countries still have high fertility.

(M) Children in Timor-Leste, where poverty is widespread and fertility rates are among the highest in the world.

(N) A billboard in Bangkok advertising Thailand's highly effective campaign to encourage condom use.

(O) Bangladeshi immigrants supply much-needed labor in wealthy Singapore, where fertility rates are among the world's lowest.

Gender:
Gender roles are shifting in this region as more women work outside the home. In some areas, women's economic and political empowerment builds on cultural traditions that give more power to women. However, some countries have also become centers for the global "sex trade," which puts many women at risk of kidnapping, slavery, and disease.

(P) A dance in West Sumatra, Indonesia, where numerous aspects of society give women more power.

(Q) Construction workers in Hanoi, Vietnam, where women are increasingly working outside the home.

(R) A club in the "red light" district of Bangkok, Thailand. This region's sex industry is linked to global criminal networks.

The inset shows the intertropical convergence zone (ITCZ), which is partly responsible for making Southeast Asia one of the wettest regions in the world.

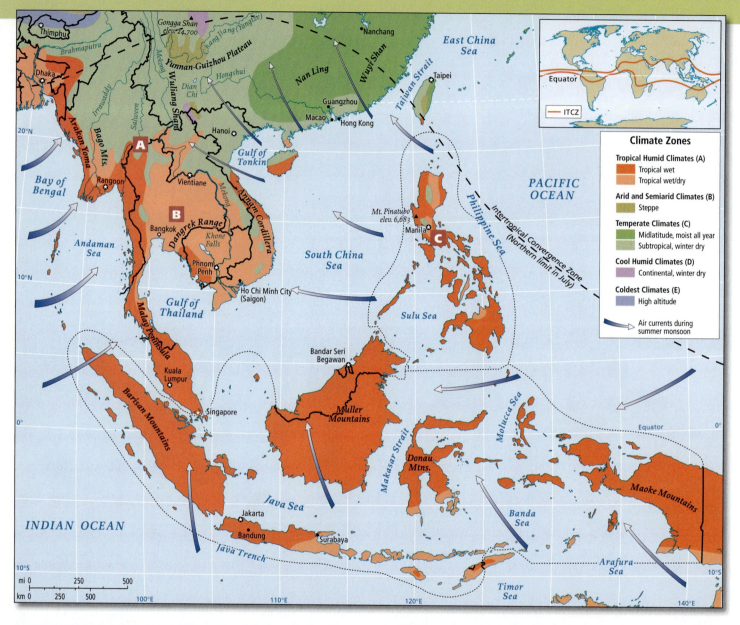

Climate Zones

Tropical Humid Climates (A)
- Tropical wet
- Tropical wet/dry

Arid and Semiarid Climates (B)
- Steppe

Temperate Climates (C)
- Midlatitude, moist all year
- Subtropical, winter dry

Cool Humid Climates (D)
- Continental, winter dry

Coldest Climates (E)
- High altitude

→ Air currents during summer monsoon

A Subtropical, winter dry, Pai, Thailand

B Tropical, wet/dry, Khao Yai, Thailand

C Tropical wet, Calauan, Philippines

Irregularly every 2 to 7 years, the normal patterns of rainfall are interrupted, especially in the islands, by the **El Niño** phenomenon (see Figure 11.6 on page 481). In an El Niño event, the usual patterns of air and water circulation in the Pacific are reversed. Ocean temperatures are cooler than usual in the western Pacific near Southeast Asia. Instead of warm, wet air rising and condensing as rainfall, cool, dry air sits on the ocean surface. The result is severe drought, with often catastrophic results for farmers and tinder-dry forests often catch fire. The cool El Niño air can trap smoke or other pollutants at the earth's surface, creating unusually toxic smog and low visibility.

The soils in Southeast Asia are typical of the tropics. Although not particularly fertile, they will support dense and prolific vegetation when left undisturbed for long periods. The warm temperatures and damp conditions promote the rapid decay of **detritus** (dead organic material) and the quick release of useful minerals. These minerals are taken up directly by the living forest rather than enriching the soil. Because rainfall is usually abundant during the summer wet season (when drought is only episodic), this region has some of the world's most impressive forests in both tropical and subtropical zones. On the mainland, human interference, coupled with the long winter dry season can negatively affect forest cover. Human–environment relations that affect the forests are discussed further below.

THINGS TO REMEMBER

1. The irregular shapes and landforms of the Southeast Asian mainland and archipelago are the result of the same tectonic forces that were unleashed when India split off from the African Plate and crashed into Eurasia.

2. The tropical climate of Southeast Asia is distinguished by continuous warm temperatures in the lowlands and heavy rain, except during those irregular years of the El Niño phenomenon.

Environmental Issues

Many of the environmental issues in Southeast Asia are in some way related to climate change. However, as we saw in the chapter introduction, deforestation has multiple negative effects (see Photo Essay 10.2 A–D and Figure 10.7 on pages 440–441). One is the loss of living space and resources for indigenous people of the forest. Another is the loss of habitat for orangutans, the Sumatran tiger, the Sumatran rhinoceros, and tens of thousands of other less spectacular species that are lost when rain forests are converted to agricultural uses. Finally, the emissions of greenhouse gases that accompany deforestation processes foster climate change. 📺 **222. Indonesia's Polluted Environment Threatens Hawksbill Turtle**

Climate Change and Deforestation

Deforestation is a major contributor to global climate change. Enormous amounts of CO_2 are released when forests are logged and the underbrush is burned. Fewer trees mean that less carbon dioxide

is absorbed from the atmosphere. According to United Nations Food and Agriculture Organization (UNFAO) reports (2010), although Southeast Asia contains only 5 percent of the world's forests, in the last 10 years it accounted for nearly 25 percent of deforestation globally. The region has the world's second-highest rate of deforestation, after sub-Saharan Africa. Every day, 13 to 19 square miles (34 to 50 square kilometers) of Southeast Asia's rain forests are destroyed, much of it done illegally. A great deal of this takes place in Malaysia and Indonesia, where the logging of tropical hardwoods for sale on the global market is a major activity. Another contributor to CO_2 emissions is the conversion of forests into oil palm plantations, especially the burning associated with clearing the land. In 2004, these activities made Indonesia the world's third-largest contributor to global climate change, after China and the United States. Figure 10.8 on page 442 shows the activities that result in legal deforestation across the region. Illegal logging may be even more widespread but is difficult to measure.

Logging, Legal and Illegal Susilo Bambang Yudhoyono (known in Indonesia as "SBY") was elected president of Indonesia in 2004 and reelected in 2009. He is widely credited with leading the fight to reduce deforestation, starting with Operation Sustainable Forestry, which was launched in 2005. Recognizing that controlled logging could be a long-term source of income for the country, President Yudhoyono cracked down on illegal logging by holding provincial forestry officials responsible, fining and jailing them when evidence of corrupt dealing in timber was found. Previous enforcement had focused on low-level workers such as logging laborers and truck drivers. Additionally, because the military had long been suspected of running illegal timber concessions, the president rotated military personnel in and out of forested zones on short terms of duty in order to interrupt cozy relationships with logging companies.

Observers report a recent decrease in rampant deforestation in Indonesia, although this is difficult to measure scientifically. Since 1950, tropical forest cover on the Indonesian island of Borneo alone has decreased by about 60 percent. Oil palm plantations now consitute the predominant use of the land.

Oil Palm Plantations Palm oil, which is used across the globe as a cooking oil, in food products, in soaps and cosmetics, and as a fine machine oil, is at the center of debates over global climate change in Southeast Asia. Indonesia and Malaysia are the world's largest palm oil producers, and the expansion of their oil palm plantations has resulted in extensive deforestation. Much of this has occurred in lowland peat swamps, which in a natural state are capable of storing huge amounts of carbon in their soils. To create space for oil palm plantations, the swamps are drained and most of their vegetation burned. Often, the fires spread underground to the peat beneath the forests, smoldering for years after the surface fires have been extinguished. These subsurface fires release

El Niño a recurring phenomenon during which the usual patterns of air and water circulation in the Pacific are reversed. Ocean temperatures are cooler than usual in the western Pacific near Southeast Asia and warmer than usual along the southern Pacific coast of South America

detritus dead organic matter that falls to forest floors

Deforestation and the conversion of land to agricultural uses in Southeast Asia are having a profound impact on both regional ecosystems and the entire biosphere.

A Land recently cleared of forest to make way for an oil palm plantation in Malaysia. Deforestation in much of Malaysia and Indonesia is now driven by the expansion of oil palm plantations.

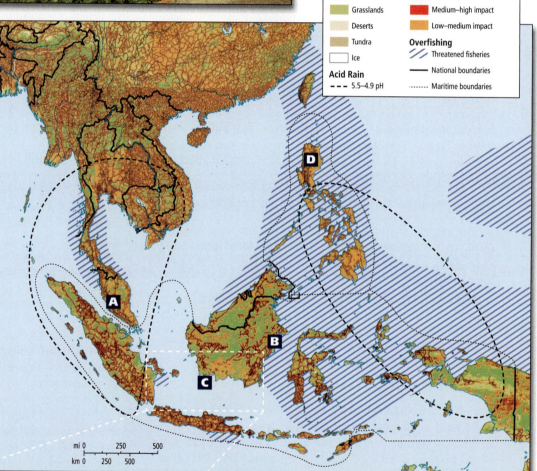

Human Impact, 2002

Land Cover		Human Impact on Land	
Forests		High impact	
Grasslands		Medium–high impact	
Deserts		Low–medium impact	
Tundra		**Overfishing**	
Ice		Threatened fisheries	
Acid Rain		National boundaries	
5.5–4.9 pH		Maritime boundaries	

B A wild orangutan eats leaves in Kutai National Park, Kalimantan, Indonesia. Deforestation and fragmentation of habitat by roads have made orangutans an endangered species.

C A satellite image of fires used to clear forest for agriculture in Indonesia and Malaysia. The CO_2 produced contributes to climate change and degrades local air quality.

D Rice terraces in the Philippines. Agricultural expansion is a major driver of deforestation in the region, and wet rice agriculture is a significant contributor of greenhouse gas emissions through the release of methane.

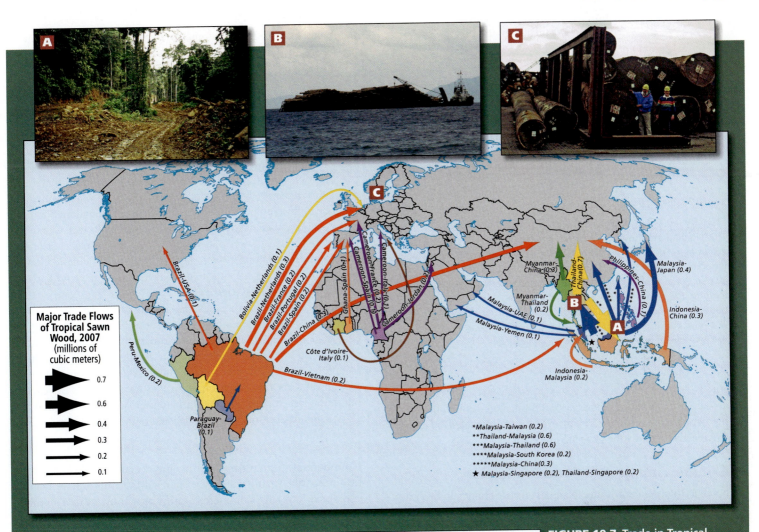

Major Trade Flows of Tropical Sawn Wood, 2007 (millions of cubic meters)

0.7
0.6
0.4
0.3
0.2
0.1

Brazil-USA (0.1)
Peru-Mexico (0.2)
Bolivia-Netherlands (0.1)
Brazil-Netherlands (0.3)
Brazil-France (0.2)
Brazil-Portugal (0.2)
Brazil-Spain (0.2)
Brazil-China (0.3)
Brazil-Vietnam (0.2)
Ghana-Spain (0.1)
Côte d'Ivoire-Italy (0.1)
Cameroon-France (0.2)
Cameroon-Spain (0.3)
Cameroon-Italy (0.2)
Cameroon-Jordan (0.2)
Paraguay-Brazil (0.1)
Myanmar-China (0.3)
Myanmar-Thailand (0.2)
Malaysia-UAE (0.1)
Malaysia-Yemen (0.1)
Thailand-China (0.7)
Philippines-China (0.1)
Malaysia-Japan (0.4)
Indonesia-China (0.3)
Indonesia-Malaysia (0.2)

*Malaysia-Taiwan (0.2)
**Thailand-Malaysia (0.6)
***Malaysia-Thailand (0.6)
****Malaysia-South Korea (0.2)
*****Malaysia-China (0.3)
★ Malaysia-Singapore (0.2), Thailand-Singapore (0.2)

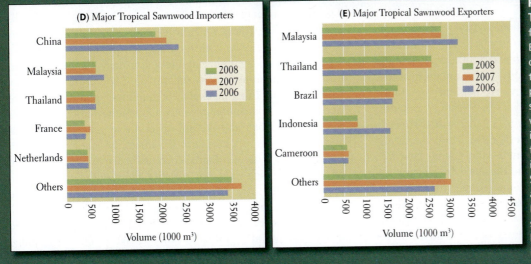

(D) Major Tropical Sawnwood Importers

China
Malaysia
Thailand
France
Netherlands
Others

2008
2007
2006

Volume (1000 m³)

(E) Major Tropical Sawnwood Exporters

Malaysia
Thailand
Brazil
Indonesia
Cameroon
Others

2008
2007
2006

Volume (1000 m³)

FIGURE 10.7 Trade in Tropical Timber Most of the consumer demand for tropical timber is in North America, Europe, and Japan. Increasingly large quantities of tropical timber sold to China are made into furniture, plywood, and flooring, which are then sold to consumers in the developed world. The trade shown on the map is mostly legal, but much of the wood fueling China's wood-processing industries is harvested illegally in Southeast Asia, North Korea, Russia, and Africa. **(A)** A logging road in Kalimantan, Indonesia. **(B)** A logging ship in the gulf of Thailand. **(C)** Tropical hardwoods being inspected at the port in Hamburg, Germany.

enormous amounts of carbon into the atmosphere. In recent years, smoke from burning forests has periodically covered much of the region, dramatically reducing air quality. During El Niño events, the smoke is so bad that airplanes have difficulty landing in the cities, and even in rural areas, people are urged to wear masks.

There is some support for palm oil as a potential solution to global climate change because it can be converted into fuel for

automobiles. The claim is that because the palm trees supposedly absorb carbon dioxide from the atmosphere when they are growing, palm oil could be considered a "carbon neutral" fuel. However, critics point out that palm trees do not absorb carbon at a rate equivalent to natural rain forest and if forests are burned to clear land for plantations, it can take decades to counteract the initial carbon emissions of deforestation. If a peat swamp is cleared for

Sources of Legal Deforestation

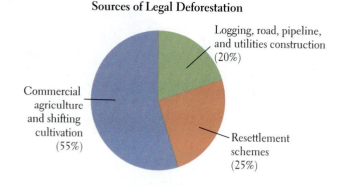

FIGURE 10.8 Legal deforestation in Southeast Asia, 2004. This pie diagram shows an estimate of various legal deforestation activities in Southeast Asia. Much deforestation is the result of illegal logging. The 2004 World Wildlife Fund for Nature estimates that 83 percent of timber production in Indonesia stems from illegal logging. [Data from "Forestry Issues—Deforestation: Tropical Forests in Decline," at http://www.canadian-forests.com/Deforestation_Tropical_Forests_in_Decline.pdf; http://www.panda.org/about_our_earth/about_forests/deforestation/forestdegradation/forest_illegal_logging/]

an oil palm plantation, more carbon is released through the burning of subsurface peat than can ever be counteracted by carbon dioxide absorbed by the growing oil palms.

Climate Change and Food Production

Food production contributes to global climate change in two ways. First, it is a major contributor to deforestation, and second, some types of cultivation actually produce significant greenhouse gases.

Shifting Cultivation Shifting cultivation, also known as *slash and burn* or *swidden* cultivation, has been practiced for thousands of years in the hills and uplands of mainland Southeast Asia and in many parts of the islands (Figure 10.9C). To maintain soil fertility in these warm, wet environments where nutrients are quickly lost to decay, farmers move their fields every 3 years or so, letting old plots lie *fallow* for 15 years or more. The regrowth of forest on once-cleared fields not only regenerates the soil, it also absorbs significant amounts of carbon dioxide from the atmosphere. However, if fallow periods are shortened, or are disrupted by logging, soil fertility can collapse, making cultivation impossible. In some cases, fertility can be restored through the use of chemical or organic fertilizers, but these can be too expensive for most farmers and chemical fertilizers may be ineffective after a year or be too unhealthy for food cultivation. Tropical soils left bare of forest for too long will eventually turn into hard, infertile, sun-baked clay.

Where population densities are relatively low, subsistence farmers can practice sustainable shifting cultivation indefinitely, but to allow for long fallow periods and still support human populations, this system requires larger areas than other types of agriculture. As population density increases, farmers may be forced to shorten fallow periods, thus inhibiting forest regrowth. Even though plots are small, because shifting cultivation requires clearing forest at each move, it accounts for a significant portion of

the region's deforestation. Moreover, because the forests are usually cleared by burning, shifting cultivation can result in wildfires. This is especially true during an El Niño period, when rainfall is low. These wildfires have increased in recent years, particularly in Indonesia, further contributing to deforestation there.

Wet Rice Cultivation Another major contributor to global climate change is Southeast Asia's most productive form of agriculture. *Wet rice cultivation* (sometimes called *paddy rice*) entails planting rice seedlings by hand in flooded terraced fields that are first cultivated with hand-guided plows pulled by water buffalo. Wet rice cultivation has transformed landscapes throughout Southeast Asia. It is practiced throughout this generally well-watered region, but especially on rich volcanic soils and in places where rivers and streams bring a yearly supply of silt (see Thematic Concepts Part D on page 436, Photo Essay 10.2D on page 440, and Photo Essay 10.3D on page 444).

The flooding of rice fields also results in the production of methane, a powerful greenhouse gas responsible for about 20 percent of global climate change. It is estimated that up to one-third of the world's methane is released from flooded rice fields where organic matter in soil undergoes fermentation as oxygen supplies are cut off. Wet rice has been cultivated for thousands of years, but in the last 25 years, growing human populations have driven a 17 percent expansion of the area devoted to this crop. Figure 10.9 shows patterns of field and forest crops across the region.

Commercial Agriculture During the recent decades of relative prosperity, small farms once operated by families have been combined into large commercial farms owned by local or multinational corporations. These farms produce cash crops for export, such as rubber, palm oil, bananas, pineapples, tea, and rice (see Thematic Concepts Parts A, D on page 436; Figure 10.9A, B, and Map). This style of agriculture entails combining huge tracts of land into one system; clear-cutting remaining trees that once separated farms; deep plowing the soil; usually planting one species of crop plant over many square miles; bolstering soil fertility with chemicals; using mechanized equipment; and in the case of rice, using large quantities of water. Commercial farming reduces the need for labor, and the objectives are to generate big yields and quick profits, not long-term sustainability. Many commercial farmers have achieved dramatic boosts in harvests (especially of rice) by using high-yield crop varieties, the result of green revolution research that has been applied in many parts of the world (see Chapter 8 on pages 374–375 and Photo Essay 10.3D).

The dramatically increased yields, especially of rice, a staple in the region and a profitable export, make it difficult to judge the ultimate value and sustainability of commercial agriculture. As we have noted in relation to other world regions, such large-scale commercial farming has significant negative environmental effects, including the loss of wildlife habitat and hence biodiversity, increased soil erosion, flooding, chemical pollution, and depletion of groundwater resources. In addition, poor farmers usually cannot afford to become green revolution farmers because the new technologies are too expensive for them. Often they cannot compete with large commercial farms and are forced to migrate to cities to look for work.

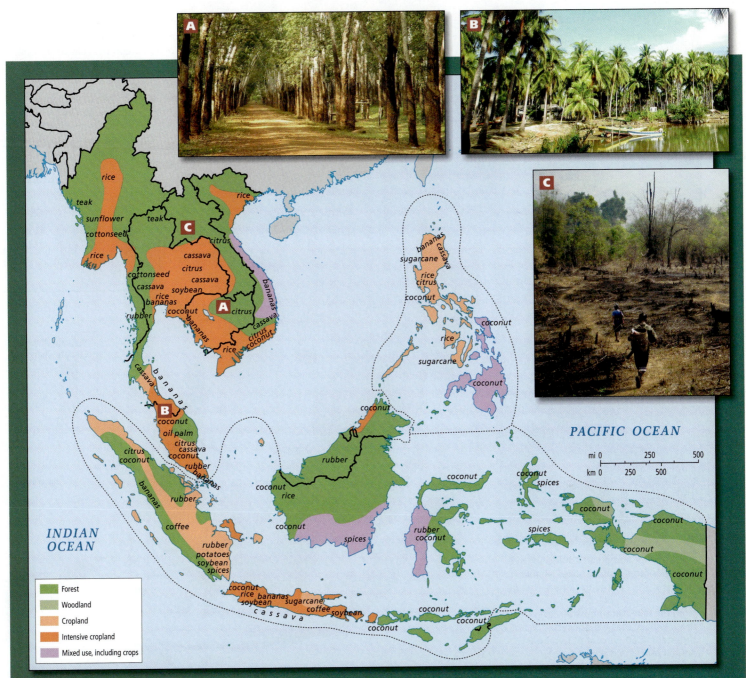

FIGURE 10.9 Agricultural patterns in Southeast Asia. Tropical forests and crops, rice production, and shifting cultivation dominate the agricultural patterns of Southeast Asia. **(A)** A rubber plantation in Cambodia. **(B)** A coconut plantation in Malaysia. **(C)** Slash and burn agriculture in Thailand.

Climate Change and Water

Many of Southeast Asia's potential vulnerabilities to global climate change are related to water resources. Four areas of vulnerability may affect the region's economy and food supply: glacial melting, increased evaporation, coral reef bleaching, and storm surge flooding.

Melting Glaciers and Increased Evaporation Like much of Asia, mainland Southeast Asia's largest rivers (the Irrawaddy, Salween,

Mekong, and Red rivers) are fed during the dry season (November–March) by glaciers high in the Himalayas. These glaciers are now melting so rapidly that they may eventually disappear.

As glacial melting accelerates, the immediate risk is flooding, which is already dramatic in some areas (see Thematic Concepts Part C on page 436). The longer-term concern is reduced dry-season flows in the rivers. As much as 15 percent of this region's rice harvest depends on the dry-season flows of the major rivers. The loss of these harvests would strain many farmers' incomes. In

Much of Southeast Asia's vulnerability to climate change relates to water. Here we explore vulerabilities related to tropical storms, flooding, and coral reefs.

A Residents of Manila, Philippines caught in floods created by typhoon Ketsana. Typhoons (known in the Atlantic as hurricanes) are projected to increase in frequency as a result of the warmer temperatures created by climate change.

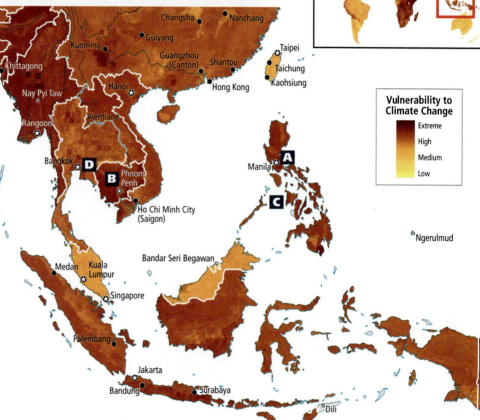

Vulnerability to Climate Change

Extreme
High
Medium
Low

Ngerulmud

B A girl in Cambodia's Tonle Sap Lake helps her family catch snakes. Higher temperatures related to climate change may be increasing rates of evaporation from the lake, causing water levels to drop. Fish catches are down significantly, threatening 60 percent of Cambodia's annual protein intake.

C Partially bleached coral on Tubbataha reef, Philippines. Climate change is raising ocean temperatures, resulting in more coral bleaching events that severely degrade the ability of reefs to provide fish with habitat.

D A researcher in Thailand inspects a variety of rice that can withstand severe flooding. This region is a center for research into how tropical agricultural systems can adapt to climate change. This research is due in part to the work of the International Rice Research Institute in the Philippines.

addition, it could possibly lead to food shortages in cities. Coming on top of a global rise in food prices in recent years, this would place further strain on the incomes of poor people throughout the region.

The higher temperatures associated with present trends in global climate change mean evaporation rates will rise, resulting in drier conditions in fields, lower lake levels, and lower fish catches because of changing habitats for aquatic animals. Evaporation resulting in reduced river and groundwater flows can also cause saltwater intrusions into estuaries and fresh water aquifers.

Coral Reef Bleaching Global climate change is expected to increase sea temperatures in Southeast Asia, threatening the coral reefs that sustain much of the region's fishing and tourism. A coral reef is an intricate structure composed of the calcium-rich skeletons of millions of tiny living creatures called coral polyps. The polyps are subject to **coral bleaching**, or color loss, which results when photosynthetic algae that live in the corals are expelled by a variety of human-instigated changes (rising water temperature, sedimentation). Bleaching may also occur when corals become polluted from a nearby city or industrial activity, or when they are overfished. Under normal conditions, the coral will recover within weeks or months. However, severe or repeated bleaching can cause corals to die. Unprecedented global coral bleaching events occurred in 1998 and again in 2002. Roughly half of the world's coral reefs were affected, causing significant coral die-offs in some areas. Scientists generally agree that these events are the result of global climate change (see **Photo Essay 10.3C**).

📹 **224. New Species of Undersea Life Found Near Indonesia**

Many of the fish caught in Southeast Asia's seas are dependent on healthy coral reefs for their survival. Hence, the thousands of rural communities throughout coastal Southeast Asia that depend on these fish for food are also threatened by coral bleaching. So far, however, the greatest observable impacts on humans are in tourism. In the Philippines, the coral bleaching event of 1998 brought a dramatic decline in tourists who come to dive the country's usually spectacular reefs, resulting in a loss of about U.S.$30 million to the economy.

Storm Surges and Flooding Although the relationship is not yet entirely understood, violent tropical storms seem to be increasing as the climate warms. Normally the Philippines can expect to encounter six typhoons (known in the Atlantic as hurricanes) in a year. Yet in the month of October 2009 alone, four typhoons struck these islands. Climatologists studying data on tropical storms in the region predict that the entire Southeast Asian region will witness a somewhat higher rate of typhoons over the next few years, and that, in particular, the duration of peak winds along coastal zones will increase. This is significant because many poor urban migrants have crowded into precarious dwellings in low-lying coastal cities such as Manila, Bangkok, Rangoon, and Jakarta, where coping with high winds, flooding, and the aftermath of storms will be a common experience (see **Photo Essay 10.3A**).

Responses to Climate Change

The reduction of fossil fuel consumption is a goal of all governments in this region, and some are delivering on those goals

despite start-up costs. Both the Philippines and Indonesia have significant potential for generating electricity from *geothermal energy* (heat stored in the earth's crust). This energy is particularly accessible near active volcanoes, which both countries have in abundance. Already, the Philippines generates 27 percent of its electricity from geothermal energy, and is second only to the United States in the amount of geothermal power it generates. By some estimates, geothermal energy could eventually provide a majority of Indonesia's energy needs. Solar energy is another attractive option, given that the entire region lies near the equator, the part of the earth that receives the most solar energy. For most countries, however, wind is the most cost-effective option, especially in Laos and Vietnam, where many population centers are in high-wind areas.

> **coral bleaching** color loss that results when photosynthetic algae that live in the corals are expelled
>
> **Australo-Melanesians** a group of hunters and gatherers who moved from the present northern Indian and Burman parts of southern Eurasia into the exposed landmass of Sundaland about 60,000 to 40,000 years ago

THINGS TO REMEMBER

1. Southeast Asia has the world's second-highest rate of deforestation, after sub-Saharan Africa, and is a major contributor to greenhouse gas emissions via widespread deforestation.

2. Much of this region's rice harvest depends on the dry-season flows of the major rivers of mainland Southeast Asia, which are threatened by glacial melting.

3. Climate change poses special challenges for people in low-lying cities or in typhoon zones.

4. Alternative energy options in this region are numerous but expensive to develop.

Human Patterns over Time

First settled in prehistory by migrants from the Eurasian continent, Southeast Asia was later influenced by Chinese, Indian, and Arab traders. Later still, it was colonized by Europe (1500s to early 1900s), the United States (1898 to 1946 in the Philippines), and was occupied by Japan (during World War II). By the late twentieth century, domination by outsiders had ended and the region was profiting from the sale of manufactured goods to its former colonizers.

The Peopling of Southeast Asia

The modern indigenous populations of Southeast Asia arose from two migrations widely separated in time. In the first migration, about 40,000 to 60,000 years ago, **Australo-Melanesians**, a group of hunters and gatherers from the present northern Indian and Burman parts of southern Eurasia, moved into the exposed landmass of Sundaland and into Australia. Their descendants still live in Indonesia's easternmost islands and in small, usually remote pockets on other islands and the Malay Peninsula, and in Australia and parts of Oceania.

In the second migration (about 10,000 years ago, at the end of the last ice age), people from southern China began moving into Southeast Asia (**Figure 10.10**). Their migration gained momentum about 5000 years ago, when a culture of skilled farmers and

FIGURE 10.10 Austronesian migrations. Members of the Eng tribe in Burma's Shan state. The Eng language is of the Austronesian family, suggesting that the Eng tribe's ancestors migrated out of southern China several thousand years ago.

Both Hinduism and Buddhism arrived thousands of years ago via Indian monks and traders traveling by sea and along overland trade routes that connected India and China through Burma. Many early Southeast Asian kingdoms and empires switched back and forth between Hinduism and Buddhism as their principal religion. Spectacular ruins of these Hindu-Buddhist empires are scattered across the region; the most famous is the city of Angkor in present-day Cambodia. At its zenith in the 1100s, Angkor was among the largest cities in the world, and its ruins are now a World Heritage Site (see page 453 Figure 10.16A on page 454). Today, Buddhism dominates mainland Southeast Asia, while Hinduism is dominant only on the Indonesian islands of Bali and Lombok.

In Vietnam, people practice a mix of Buddhist, Confucian, and Taoist beliefs that reflect the thousand years (ending in 938 c.e.) when it was part of various Chinese empires. China's traders and laborers also brought cultural influences to scattered coastal zones throughout Southeast Asia.

Islam is now dominant in the islands of Southeast Asia. Islam came mainly through South Asia after India fell to Muslim (Mughal) conquerors in the fifteenth century. Muslim spiritual leaders and traders converted many formerly Hindu-Buddhist kingdoms in Indonesia, Malaysia, and parts of the southern Philippines, where Islam is still dominant. Roman Catholicism is the predominant religion in Timor-Leste, which was colonized by Portugal, and in most of the Philippines, which was colonized by Spain.

seafarers from southern China, the **Austronesians**, migrated first to Taiwan, then to the Philippines, and then into the islands of Southeast Asia and the Malay Peninsula. Some of these sea travelers eventually moved westward to southern India and to Madagascar (off the east coast of Africa), and eastward to the far reaches of the Pacific islands (see Chapter 11).

Austronesians a Mongoloid group of skilled farmers and seafarers from southern China who migrated south to various parts of Southeast Asia between 10,000 and 5000 years ago

Diverse Cultural Influences

Southeast Asia has been and continues to be shaped by a steady circulation of cultural influences, both internal and external. Overland trade routes and the surrounding seas brought traders, religious teachers, and sometimes even invading armies from China and India, as well as Arab armies from southwest Asia. These newcomers brought religions, trade goods (such as cotton textiles), and food plants (such as mangoes and tamarinds) deep into the Indonesian and Philippine archipelagos and throughout the mainland. The monsoon winds, which blow from the west in the spring and summer, facilitated access by merchant ships from South Asia and the Persian Gulf. The ships sailed home on winds blowing from the east in the autumn and winter. These winds carried people, spices, bananas, sugarcane, silks, and other Southeast Asian items to the wider world.

Religious Legacies Spatial patterns of religion in Southeast Asia reveal an island–mainland division that reflects the history of influences from India, China, Southwest Asia, and Europe.

European Colonization

Over the last five centuries, several European countries established colonies or quasi-colonies in Southeast Asia (Figure 10.11). Drawn by the region's fabled spice trade, the Portuguese established the first permanent European settlement in Southeast Asia at the port of Malacca, Malaysia, in 1511. Although better ships and weapons gave the Portuguese an advantage, their anti-Islamic and pro-Catholic policies provoked strong resistance in Southeast Asia. Only in Timor-Leste did the Portuguese establish Catholicism as the dominant religion.

By 1540, the Spanish had established trade links across the Pacific between the Philippines and their colonies in the Americas. Like the Portuguese, they practiced a style of colonial domination grounded in Catholicism, but they met less resistance because of their greater tolerance of non-Christians. The Spanish ruled the Philippines for more than 350 years, and as a result, the Philippines is the most deeply Westernized and certainly the most Catholic part of Southeast Asia.

The Dutch were the most economically successful of the European colonial powers in Southeast Asia. From the sixteenth to the nineteenth centuries, they extended their control of trade over most of what is today called Indonesia, known previously as the Dutch East Indies (see Figure 10.11A). The Dutch became

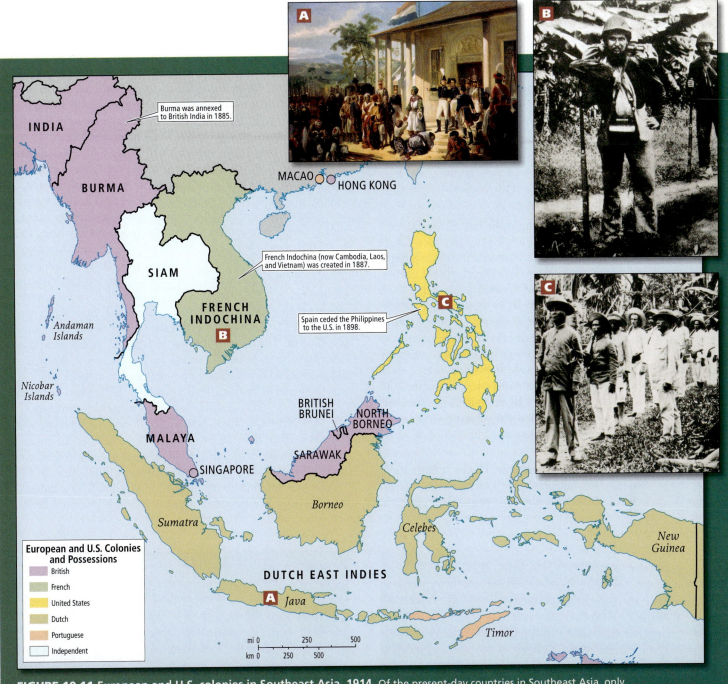

FIGURE 10.11 European and U.S. colonies in Southeast Asia, 1914. Of the present-day countries in Southeast Asia, only Thailand (formerly called Siam) was never colonized. **(A)** A painting showing the Dutch defeat of the Javanese in 1830. **(B)** A French soldier in Vietnam in 1888. **(C)** Members of the Philippine army fighting the U.S. takeover of the Philippines in 1898.

Map labels:
- Burma was annexed to British India in 1885.
- French Indochina (now Cambodia, Laos, and Vietnam) was created in 1887.
- Spain ceded the Philippines to the U.S. in 1898.

INDIA
BURMA
MACAO · HONG KONG
SIAM
Andaman Islands
FRENCH INDOCHINA
Nicobar Islands
BRITISH BRUNEI
NORTH BORNEO
MALAYA
SARAWAK
SINGAPORE
Borneo
Sumatra
Celebes
New Guinea
DUTCH EAST INDIES
Java
Timor

European and U.S. Colonies and Possessions
- British
- French
- United States
- Dutch
- Portuguese
- Independent

mi 0   250   500
km 0   250   500

interested in growing cash crops for export. Between 1830 and 1870, they forced indigenous farmers to leave their own fields and work part time without pay in Dutch coffee, sugar, and indigo plantations. The resulting disruption of local food production systems caused severe famines and provoked resistance that often took the form of Islamic religious movements. Such movements hastened the spread of Islam throughout Indonesia, where the Dutch had made little effort to spread their Protestant version of Christianity.

Beginning in the late eighteenth century, the British established colonies at key ports on the Malay Peninsula. They held these ports both for their trade value and to protect the Strait of Malacca, the passage for sea trade between China and Britain's empire in India. In the nineteenth century, Britain extended its rule over the rest of modern Malaysia to benefit from Malaysia's tin mines and plantations. Britain also added Burma to its empire, which provided access to forest resources and overland trade routes into southwest China.

The French first entered Southeast Asia as Catholic missionaries in the early seventeenth century. They worked mostly in the eastern mainland area in the modern states of Vietnam, Cambodia, and Laos. In the late nineteenth century, spurred by rivalry with Britain and other European powers for greater access to the markets of nearby China, the French formally colonized the area, which became known as French Indochina (see Figure 10.11B).

In all of Southeast Asia, the only country not to be colonized was Thailand (then known as Siam). Like Japan, it protected its sovereignty through both diplomacy and a vigorous drive toward European-style modernization.

Struggles for Independence

Agitation against colonial rule began in the late nineteenth century when Filipinos fought first against Spain. They then fought against the United States, which took control of the Philippines in 1898 after the Spanish-American War (see Figure 10.11C). However, the Philippines and the rest of Southeast Asia did not win independence until after World War II. By then, Europe's ability to administer its colonies had been weakened by internal strife, and its attention and resources were diverted by the devastation of the war. Furthermore, Japan had conquered most of European-controlled Southeast Asia in the early 1940s and held it until defeated by the United States several years later (see Figure 9.11 on page 404). By the mid-1950s, the colonial powers had granted self-government to most of the region.

domino theory a foreign policy theory that used the idea of the domino effect to suggest that if one country "fell" to communism, others in the neighboring region would also fall

The Vietnam War The most bitter battle for independence took place in French Indochina (the territories of Vietnam, Laos, and Cambodia). Although all three became nominally independent in 1949, France retained political and economic power over them. Various nationalist leaders, most notably Vietnam's Ho Chi Minh, headed resistance movements against continued French domination. The resistance leaders accepted military assistance from Communist China and the Soviet Union, even though they were not doctrinaire communists and despite ancient antipathies toward China for its previous millennia of domination. In this way, the Cold War was brought to mainland Southeast Asia.

In 1954, the French were defeated by Ho Chi Minh at Dien Bien Phu, in northern Vietnam. Although the United States was against aiding the French continuation of quasi-colonial powers, it stepped in because of increasing worry about the spread of international communism should the resisters, now supported by communists, succeed. The **domino theory**—the geographically based idea that if one country fell to communism other nearby countries would follow—was a major influence in this decision, because both North Korea and China had recently become communist. The Vietnamese resistance, which controlled the northern half of the country, attempted to wrest control of the southern half from the United States and a U.S.-supported and quite corrupt South Vietnamese government. The pace of the war accelerated in the mid-1960s. After many years of brutal conflict, public opinion in the United States forced U.S. withdrawal from the conflict in 1973. The civil war continued in Vietnam, finally ending in 1975, when the North defeated the South and established a new national government.

More than 4.5 million people died during the Vietnam War, including more than 58,000 U.S. soldiers. Another 4.5 million on both sides were wounded, and bombs, napalm, and defoliants ruined much of the Vietnamese environment (Figure 10.12). Land mines continue to be a hazard to this day. The withdrawal from Vietnam in 1973 ranks as one of the most profound defeats

(A)

(B)

FIGURE 10.12 Legacies of the Vietnam War: Agent Orange. Approximately 4.8 million Vietnamese people and at least 650,000 U.S. soldiers were exposed to Agent Orange, a highly toxic defoliant used by the U.S. military to clear vegetation during the Vietnam War. In Vietnam, 400,000 deaths and disabilities, as well as birth defects in more than 500,000 children, have been linked to Agent Orange. **(A)** A U.S. Army helicopter sprays Agent Orange over the Mekong Delta in 1969. **(B)** A man in Ho Chi Minh City, Vietnam, born with severe deformities as a result of his mother's exposure to Agent Orange.

in U.S. history. After the war, the United States crippled Vietnam's recovery by imposing severe economic sanctions that lasted until 1993. Since then, the United States and Vietnam have become significant trading partners.

The "Killing Fields" in Cambodia In Cambodia, where the Vietnam War had spilled over the border, a particularly violent revolutionary faction called the Khmer Rouge seized control of the government in the mid-1970s. Inspired by the vision of a rural communist society, they attempted to destroy virtually all traces of European influence. They targeted Western-educated urbanites in particular, forcing them into labor camps, where more than 2 million Cambodians—one-quarter of the population—starved or were executed in what became known as the "killing fields."

In 1978, Vietnam deposed the Khmer Rouge and ruled Cambodia through a puppet government until 1989. A 2-year civil war then ensued. Despite a major UN effort to establish a multiparty democracy in Cambodia throughout the 1990s, the country remains plagued by political tensions between rival factions and by government corruption. In March 2009, the first of the Khmer Rouge leaders was put on trial before a UN-backed tribunal and forced to listen to and watch lengthy accounts of the torture of men, women, and children that he is accused of supervising. Others will be tried, but most operatives in the "killing fields" will never be prosecuted.

🎥 **223. Cambodia Hip Hop Artist Tells Story through Rap**

Personal Vignette One night in 1977, soldiers came to the home of Samrith Phum in Cambodia and took away her husband. She thought he was just going to a meeting, but he never came home. Samrith was then only 20 years old and had three young children, one a newborn. With her infant in her arms, she went to talk to Choch, the Khmer Rouge village chief, and asked him, "Brother, do you know where my husband is?" The village chief told her not to worry about other people's business. A short while later, Choch appeared at Samrith's door and said he would take her to see her husband. Instead, he drove to the nearby prison and locked her up with her baby. She was released a year later, after the Vietnamese drove the Khmer Rouge from power. Today, Samrith still lives just down the street from Choch. For his part, Choch denies any involvement in the killings or even ever being at the prison. [Adapted from the Frontline/World video, "Cambodia: Pol Pot's Shadow" and Amanda Pike's "Reporter's Diary: In Search of Justice," at http://www.pbs.org/frontlineworld/stories/cambodia/diary03.html.] ∎

THINGS TO REMEMBER

1. Over the last five centuries, several European countries, and later the United States and Japan, established colonies or quasi-colonies that covered almost all of Southeast Asia.

2. Colonial rule in Southeast Asia came to violent ends in Vietnam and Cambodia, where war took the lives of millions, including more than 58,000 U. S. soldiers.

II CURRENT GEOGRAPHIC ISSUES

Like Middle and South America, Africa, and South Asia, Southeast Asia is expanding its links to the global economy. However, it has had greater success than other areas in achieving widespread prosperity, largely by following the example of some of East Asia's most successful countries.

Economic and Political Issues

Initially, trade among countries within the region was inhibited by the fact that they all exported similar goods—primarily food and raw materials—and traditionally imposed tariffs against one another. They imported consumer products, industrial materials, machinery, and fossil fuels mostly from the developed world. Several decades ago, the economic and political situation in the region changed dramatically. Southeast Asian countries had some of the highest economic growth rates in the world based on an economic strategy that emphasized the export of manufactured goods—initially clothing and then more sophisticated technical products. Then, in the late 1990s, economic growth stagnated and political instability increased as people's expectations were dashed. The high level of corruption revealed by the crisis strengthened calls for democracy, which has expanded only unevenly across the region.

Strategic Globalization: State Aid and Export-Led Growth

From the 1960s to the 1990s, some national governments in Southeast Asia achieved strong and sustained economic expansion by emulating two strategies for economic growth pioneered earlier by Japan, Taiwan, and South Korea (see Chapter 9, pages 405–406). One was the formation of *state-aided market economies*. National governments in Indonesia, Malaysia, Thailand, and, to some extent, the Philippines intervened strategically in the financial sector to make sure that certain economic sectors developed. Investment by foreigners was limited so that the governments could have more control over the direction of the economy. The other strategy was *export-led growth*, which focused investment on industries that manufactured products for export, primarily to developed countries. These strategies amounted to a limited and selective embrace of globalization in that global markets for the region's products were sought but foreign sources of capital were not.

These strategies were a dramatic departure from those used in other developing areas. In Middle and South America and parts of Africa, post-colonial governments relied on import substitution industries that produced manufactured goods mainly for local use. By contrast, export-led growth allowed Southeast Asia's industries

feminization of labor the increasing representation of women in both the formal and the informal labor force

to earn much more money in the vastly larger markets of the developed world. Standards of living increased markedly, especially in Malaysia, Singapore, Indonesia, and Thailand. Other important results of Southeast Asia's success were a decrease in disparities of wealth (Table 10.1) and improvements in vital statistics—lower population growth rate, infant mortality, and maternal mortality, and longer life expectancy.

Export Processing Zones In the 1970s, some governments in the region began adopting an additional strategy for encouraging economic development. This time foreign sources of capital were sought, but the places they could invest were limited to specially designated free trade areas. Such Export Processing Zones (EPZs) are places in which foreign companies can set up industries using inexpensive, locally available labor to produce items only for export (maquiladoras in Mexico, for example). Taxes are eliminated or greatly reduced. Since the 1970s, EPZs have expanded economic development in Malaysia, Indonesia, Vietnam, and the Philippines, and are now used in China and Middle and South America.

The Feminization of Labor Between 80 and 90 percent of the workers in the EPZs are women, not only in Southeast Asia but in other world regions as well (Figure 10.13). The **feminization of labor** has been a distinct characteristic of globalization over the past three decades (see Chapter 9, pages 386–387). Employers prefer to hire young, single women because they are perceived as the cheapest, least troublesome employees. Statistics do show that, generally, women will work for lower wages than men, will not complain about poor and unsafe working conditions, will accept being restricted to certain jobs on the basis of sex, and are not as likely as men to agitate for promotions.

Working Conditions In general, the benefits of Southeast Asia's "economic miracle" have been unequally apportioned. In the region's new factories and other enterprises, it is not unusual for assembly-line employees to work 10 to 12 hours per day, 7 days per week, for less than the legal minimum wage, and without benefits. Labor unions that typically would address working conditions and wage grievances are frequently repressed by governments, and international consumer pressure to improve working conditions at U.S. companies like Nike has been only partially effective. By 2010, for example, Nike had sidestepped the entire issue of customer complaints by outsourcing the manufacturing to non-U.S. contractors in the region. U.S. protestors now have a more difficult time tracing abuse of workers back to Nike.

A much more powerful force that is driving up pay and improving working conditions is the service sector, which is growing throughout the region. In Singapore, the Philippines, and Timor Leste, the service sector already dominates. In Malaysia and Cambodia, the service sector contributes more to the country's GDP than the other sectors. In all other countries in the region except for Brunei, which is dominated by the industrial (oil) sector, the service sector is approaching parity with the industrial or agricultural sectors. Many service sector jobs require at least a high school education, and competition for the smaller number of educated workers means that wages and working conditions are already better than in manufacturing and are likely to improve faster.

Economic Crisis and Recovery: The Perils of Globalization

The economic crisis that swept through Southeast Asia in the late 1990s forced millions of people into poverty and changed the political order in some countries. A major cause of the crisis was the lifting of controls on Southeast Asia's once highly regulated financial sector. There were some geographic elements to the crisis: the banks involved were located primarily in Singapore and the main cities of Thailand, Malaysia, and Indonesia. These were also the countries to feel the immediate effects of the crisis. Eventually the effects filtered into the hinterland, and workers in remote areas lost jobs and access to essentials.

Deregulating Investment As part of a general push by the IMF to open national economies to the free market, Southeast Asian governments relaxed controls on the financial sector in the 1990s. Soon, Southeast Asian banks were indeed flooded with

TABLE 10.1	**Wealth disparities in selected countries in Southeast Asia and the Americas**		
Country	Income-spread ratio, 2003	Income-spread ratio, 2005	Income-spread ratio, 2009
Cambodia	11.6	11.6	11.5
Indonesia	7.8	7.8	10.8
Laos	9.7	9.7	7.3
Malaysia	22.1	22.1	11.0
Philippines	16.5	16.5	14.1
Singapore	17.7	17.7	17.7
Thailand	13.4	13.4	13.1
Vietnam	8.4	9.4	9.7
Brazil	65.8	68.0	40.6
Chile	43.2	40.6	26.2
Costa Rica	20.7	25.1	23.4
Mexico	34.6	45.0	21.0
Peru	22.3	49.9	26.1
United States	16.6	15.9	15.9

Disparities are shown through income-spread ratios that compare the wealthiest 10 percent to the poorest 10 percent of the population. The higher the number, the greater the discrepancy between the wealthy and the poor.

Sources: United Nations Human Development Report [UNHDR] 2003, Table 13; *UNHDR 2005*, Table 15; *UNHDR 2009*, Table M (New York, United Nations Development Programme).

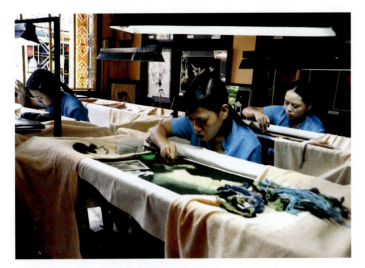

FIGURE 10.13 The feminization of labor. Women in Vietnam work at a small garment factory in Hoi An, Vietnam.

money from investors in the rich countries of the world who hoped to profit from the region's growing economies. Flush with cash and newfound freedoms, however, the banks often made reckless decisions. For example, bankers made risky loans to real estate developers, often for high-rise office building construction. As a result, many Southeast Asian cities soon had a glut of office space. Building projects were halted unfinished.

Lifting the controls on investments was also made problematic by a kind of corruption known as **crony capitalism.** In most Southeast Asian countries, as elsewhere, corruption is encouraged by the close personal and family relationships between high-level politicians, bankers, and wealthy business owners. In Indonesia, for example, the most lucrative government contracts and business opportunities were reserved for the children of former president Suharto, who ruled the country from 1967 to 1997. His children became some of the wealthiest people in Southeast Asia. This kind of corruption expanded considerably with the new foreign investment money, much of which was diverted to bribery or unnecessary projects that brought prestige to political leaders.

The cumulative effect of crony capitalism and the lifting of controls on banks was that many ventures failed to produce any profits at all. In response, foreign investors panicked, withdrawing their money on a massive scale. In 1996, before the crisis, there was a net inflow of U.S.$94 billion to Southeast Asia's leading economies. In 1997, inflows had ceased and there was a net outflow of U.S.$12 billion.

The IMF Bailout and Its Aftermath The International Monetary Fund (IMF) made a major effort to keep the region from sliding deeper into recession by instituting reforms designed to make banks more responsible in their lending practices. The IMF also required *structural adjustment policies* (see Chapter 3, pages 136–138), which required countries to cut government spending (especially on social services) and abandon policies intended to protect domestic industries.

After several years of economic chaos, and much debate over whether the IMF bailout helped or hurt a majority of Southeast

Asians, economies began to recover. In the largest of the region's economies, growth resumed by 1999, and by 2006 the crisis, while still serving as an ominous reminder of the risks of globalization, had been more or less overcome. Then by late 2008, the region was hit again by the effects of the global recession; growth slowed markedly because high oil and food prices restricted disposable cash worldwide and consumers in wealthy countries, faced with crippling debt, seriously curtailed their spending. However, while the recession dragged on in North America and Europe, by late 2009, East and Southeast Asia appeared to be recovering. In part the recovery was based on the pent-up demand for goods and services within the domestic economies of China and Southeast Asia. Southeast Asia's ability to respond to this demand was facilitated by policies that opened up intraregional trade and access to China. The recovery was carefully modulated by tight trade and monetary regulations aimed at controlling inflation. Older strategies, such as the establishment of EPZs, were expanded to attract additional multinational corporations to Southeast Asia.

▱ 230. Bankers, Analysts See Resurgent Asia 10 Years After Economic Crisis

The Impact of China's Growth During the Southeast Asian financial crisis of the 1990s, Singapore, Malaysia, and Thailand lost out to China in attracting new industries and foreign investors. By the early 2000s, China attracted more than twice as much foreign direct investment (FDI) as Southeast Asia. Skilled labor and start-up costs then rose in China, and Singapore, Malaysia, Thailand, and the Philippines began positioning themselves as locations offering more highly skilled labor and more high-tech infrastructure than China.

However, China's growth also became an opportunity for Southeast Asia. Singapore, Malaysia, Indonesia, Thailand, and the Philippines began to "piggy-back" on China's growth by winning large contracts to upgrade China's infrastructure in areas such as wastewater treatment, gas distribution, and shopping mall development. Meanwhile, in Southeast Asia's poorer countries, such as Vietnam, wages have remained considerably lower than in China. Vietnam has thus been able to attract investment in low-skill manufacturing that had been going to China.

Regional Trade and ASEAN

During the 1980s and 1990s, Southeast Asian countries traded more with China and the rich countries of the world than they did with each other. This issue of insufficient reciprocal trade within Southeast Asia was the reason behind the creation of the **Association of Southeast Asian Nations (ASEAN),** an increasingly strong organization of all ten Southeast Asian nations (Brunei, Burma, Cambodia, Indonesia, Laos, Malaysia, the Philippines, Singapore, Thailand, and Vietnam).

Regional Integration Although started in 1967 as an anti-communist, anti-China association, ASEAN now focuses on agreements that strengthen regional cooperation, including agreements

> **crony capitalism** a type of corruption in which politicians, bankers, and entrepreneurs, sometimes members of the same family, have close personal as well as business relationships
>
> **Association of Southeast Asian Nations (ASEAN)** an organization of Southeast Asian governments established to further economic growth and political cooperation

with China. One example is the Southeast Asian Nuclear Weapons–Free Zone Treaty signed in December 1995. Another is the ASEAN Economic Community, or AEC, a trade bloc patterned after the North American Free Trade Agreement and the European Union. Tariffs between countries are being reduced in order to lower production costs and make ASEAN's manufacturing industries more efficient. And increasingly, ASEAN is focused on increasing trade between all 10 members of the association. By 2008, the total intraregional trade between ASEAN countries was more than trade with any one outside country or region. Exports between ASEAN countries represented 27.6 percent of total exports and 25.9 percent of total imports (Figure 10.14).

The ASEAN charter urges work toward cooperative programs similar to those of the European Union, such that politically, ASEAN promotes uniform adherence to democratic principles, the rule of law, and human rights. Economically, regional integration is to be accelerated, making possible the free movement of goods, services, investment capital, and skilled labor (but not unskilled

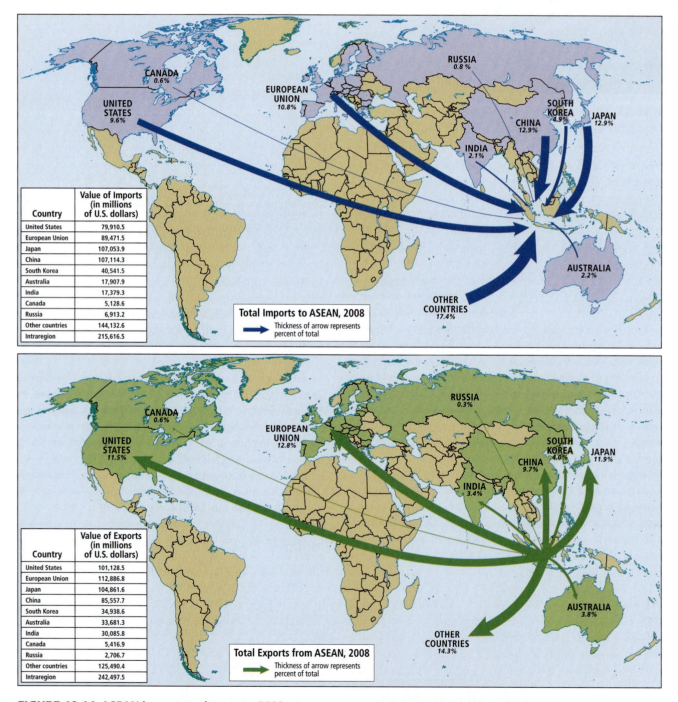

Total Imports to ASEAN, 2008

Country	Value of Imports (in millions of U.S. dollars)
United States	79,910.5
European Union	89,471.5
Japan	107,053.9
China	107,114.3
South Korea	40,541.5
Australia	17,907.9
India	17,379.3
Canada	5,128.6
Russia	6,913.2
Other countries	144,132.6
Intraregion	215,616.5

Total Exports from ASEAN, 2008

Country	Value of Exports (in millions of U.S. dollars)
United States	101,128.5
European Union	112,886.8
Japan	104,861.6
China	85,557.7
South Korea	34,938.6
Australia	33,681.3
India	30,085.8
Canada	5,416.9
Russia	2,706.7
Other countries	125,490.4
Intraregion	242,497.5

FIGURE 10.14 ASEAN imports and exports, 2008. The ASEAN countries are very active in world trade as importers and exporters. However, the largest share of trade is among the ASEAN countries themselves, with imports at 25.9 percent and exports 27.6 percent. [Adapted from http://www.aseansec.org/22122.htm and ASEAN trade statistics, 2008, Table 19, at http://www.aseansec.org/Stat/Table19.pdf.]

labor). And socially, ASEAN would encourage solidarity and unity among the peoples of ASEAN by forging a common identity and building an inclusive and harmonious society where the goal would be the well-being, livelihood, and welfare of all. Two interesting geographic problems are addressed and resolutions attempted in the ASEAN charter: in recognition of the diversity in the region, the use of the English language is to be promoted despite what this will mean in terms of cultural change and homogenization; and because the region is sprawled across a wide expanse of land and ocean, Internet technology is to be broadly implemented to facilitate communication and general integration.

At the ASEAN meeting in 2009, attended by China, Japan, and the United States, China and Japan jockeyed for the top leadership position among Asian countries. Seeking to extend their influence with *soft power*, they offered to finance infrastructure projects in Southeast Asia—roads in Burma and Cambodia, a bridge over the Mekong between Laos and Thailand. Meanwhile, leaders in the largest ASEAN economies were urging their fellow members to abandon the old growth model based on selling Southeast Asian goods and services to Western consumers and to look instead for other global markets and other products.

Tourism International tourism is an important and rapidly growing economic activity in most Southeast Asian countries. Between 1991 and 2001, the number of international visitors to the region doubled to more than 40 million, and by 2008 international visitors numbered more than 65 million. As in other trade matters, Southeast Asians are themselves increasingly touring neighboring countries. By 2008, close to 50 percent of tourists in ASEAN countries were from within the region (**Figure 10.15**). This is a positive trend because familiarity between neighbors lays the groundwork for various forms of regional cooperation, such as infrastructure improvements.

In response to its popularity with global and regional tourists, ASEAN members have been working to improve the region's transportation infrastructure. One such project is the Asian Highway, a web of standardized roads that loop through the mainland and connect it with Malaysia, Singapore, and Indonesia (the latter via ferry) (see **Figure 10.16** on page 454). Eventually, the Asian Highway will facilitate ground travel through 32 Eurasian countries from Moscow to Indonesia and from Turkey to Japan.

The surge of tourism in Southeast Asia has also raised concerns about too great a dependency on an industry that leaves economies vulnerable to events that precipitously stop the flow of visitors or damage the tourism infrastructure, or damage the citizenry (see discussion of sex tourism on pages 469–470). Some examples of these are natural disasters such as the tsunami of December 2004 that killed several thousand international tourists in Thailand and Indonesia, and human-made disasters such as the terrorist bombings in Bali (2002, 2005) and in southern Thailand (June and September 2006). In addition, mass tourism threatens the long-term survival of cultural heritage sites and local cultures, as the following vignette illustrates.

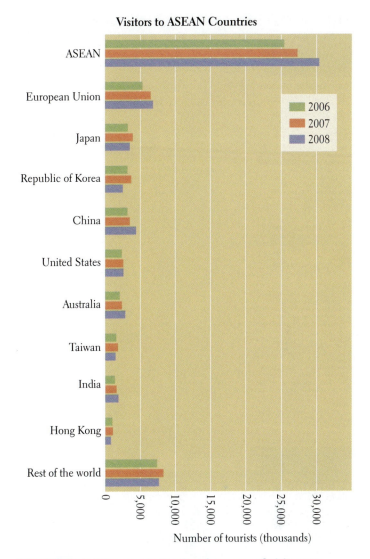

FIGURE 10.15 Top country/regional sources of visitors to Southeast Asian countries, 2006–2008. Like trade, the largest share of visitors to Southeast Asia come from countries within the region—they account for 46.5 percent of all visitors in 2008. [Adapted from ASEAN tourism statistics, Table 30, at http://www.aseansec.org/stat/Table30.pdf.]

humankind. The Greater Mekong Basin is the new "hot spot" in Southeast Asian tourism. Airline routes, boat tours from Thailand, and now the Asian Highway (see Figure 10.16) bring more than 500 tourists a day to the town, making it a key contributor to the tourism dollars that Laos earns. But critics ask "Development for whom?" and argue that the people themselves have little choice in deciding the pace or direction of this development. How can a small geographic area, isolated for centuries, deal with a sudden influx of outsiders who, by their sheer numbers, will strain the infrastructure and services? The presence of these tourists could also damage cultural assets and make local people feel unwelcome in their own home spaces.

In Luang Prabang, the effects are already visible. The buildings in the center of Luang Prabang's old town are now restaurants, shops, bakeries, guest houses, or Internet cafés. Plastic bottles and garbage litter the area, and shreds of plastic cling to the tall weeds

Personal Vignette Luang Prabang is a UNESCO World Heritage Site—meaning that it is of outstanding cultural or natural importance to the common heritage of

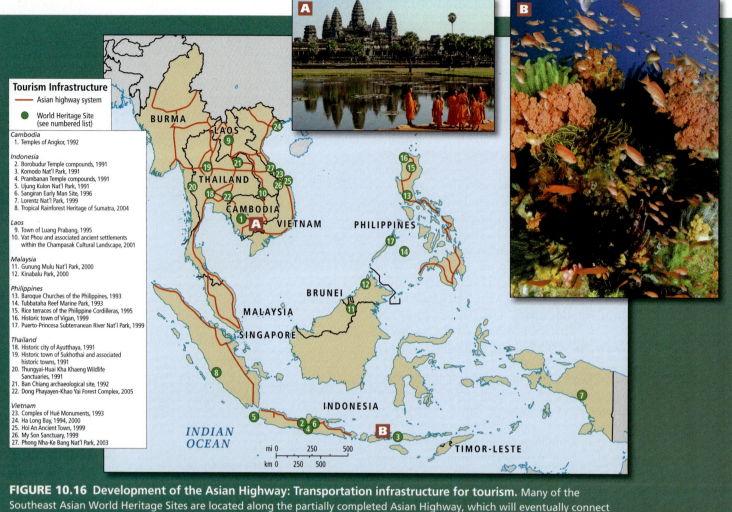

FIGURE 10.16 Development of the Asian Highway: Transportation infrastructure for tourism. Many of the Southeast Asian World Heritage Sites are located along the partially completed Asian Highway, which will eventually connect Europe to Indonesia. See "Development of the Asian Highway" and a map of the entire system at http://www.unescap.org/ttdw/common/TIS/AH/files/AH_2008.pdf. **(A)** Buddhist monks visiting an Angkor Wat temple complex in Cambodia. **(B)** A coral reef in Komodo National Park, Indonesia. [Adapted from "World Heritage List," UN World Heritage Convention, at http://whc.unesco.org/en/list/; "Tourism Attractions Along the Asian Highway," UN Economic and Social Commission for Asia and the Pacific, 2004, at http://www.unescap.org/ttdw/common/tis/ah/tourism%20attractions.asp; and Asia Times Online, at http://www.atimes.com/atimes/Asian_Economy/images/highways.html.]

lining the bank of the Mekong River. Tensions are rising between those who profit from tourism and those, such as farmers, who do not. Before, young people in the town aspired to be teachers or doctors; now they want to be tour guides. [Sources: Teena Amrit Gill, "Locals Lose Out as Tourism Booms," Asia Times Online, March 12, 2002, at http://www.atimes.com/se-asia/DC12Ae02.html; Bui Nguyen Cam Ly, "As Hordes of Tourists Come, Heritage Goes," Inter Press Service News, 2003, at http://www.ipsnews.net/mekong/stories/heritage.html; Gulfer Cezayirli, "Fast-Growing Asian Tourism Should Enlist Help of the Urban Poor," Asian Development Bank, April 14, 2003, at http://www.adb.org/Media/Articles/2003/2009_Regional_Asian_Tourism_Should_Enlist_Help_of_the_Urban_Poor/default.asp; and Raja M., "Asian Highway Network Gathers Speed," Asia Times Online, June 14, 2006, at http://www.atimes.com/atimes/Asian_Economy/HF14Dk01.html.] ∎

Pressures For and Against Democracy

Progress toward a greater public voice in the political process in Southeast Asia has been uneven. Significant barriers to democratic participation still exist across the region (Photo Essay 10.4 Map), but the types of barrier vary. Several countries are plagued with violent conflict, and democracy, once established, is not always durable. For example, Thailand had seemed to have constructed a democratic system that allowed for protest and provided mechanisms for constitutional adjustments and smooth transitions after regular elections. Recently, however, Thailand has come close to losing its standing as a democratic country due to a 2006 coup d'état against an apparently corrupt but popular prime minister, Thaksin Shinawatra.

📹 **322. Thailand's Protesters Highlight Rifts, Political Participation**

Can Democracy Work in Indonesia? The greatest recent shift toward democracy occurred in Indonesia in the wake of the economic crisis of the late 1990s. After three decades of semidictatorial rule by President Suharto, the economic crisis spurred massive demonstrations that forced Suharto to resign. Since then, democratic parliamentary and presidential elections have initiated a new political era in the country.

Significant barriers to democratization exist in this region, such as high levels of political violence and authoritarian political cultures. However, demands for greater public participation in governance are growing.

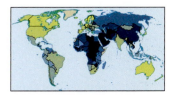

A Burmese refugees in Malaysia protest the 2007 government repression of demonstrators in Rangoon, Burma, and the ongoing imprisonment of Burmese pro-democracy activist and Nobel Peace Prize winner, Aung San Suu Ki.

B A protest in Jakarta, Indonesia against the government's agricultural policies. Such activity was repressed in the past but has been permitted since democratization took a major step forward with the resignation in 1998 of Suharto, Indonesia's dictator for 31 years.

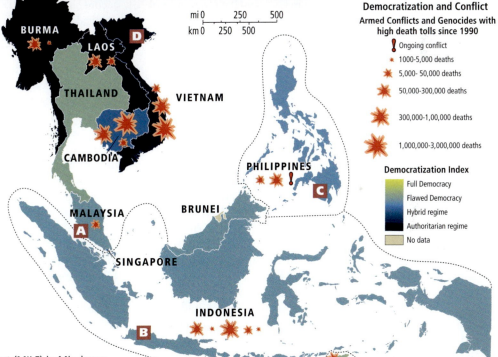

Democratization and Conflict

Armed Conflicts and Genocides with high death tolls since 1990

- ❗ Ongoing conflict
- ✳ 1000-5,000 deaths
- ✳ 5,000- 50,000 deaths
- ✳ 50,000-300,000 deaths
- ✳ 300,000-1,00,000 deaths
- ✳ 1,000,000-3,000,000 deaths

Democratization Index

- Full Democracy
- Flawed Democracy
- Hybrid regime
- Authoritarian regime
- No data

C A patrol of the Moro Islamic Liberation Front (MILF) in Mindanao, Philippines. In response to what they see as discrimination against Muslims by the Catholic-dominated government, the MILF demands a separate Islamic state in Mindanao. Violence subsided recently after the Muslim-dominated areas of Mindanao were given greater political autonomy by the central government.

D A sign in Hanoi, Vietnam features a portrait of Ho Chi Minh along with Communist symbols and icons. Vietnam's government remains Communist and authoritarian, allowing few political freedoms.

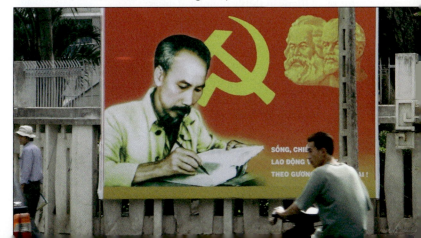

Indonesia is the largest country in Southeast Asia and the most fragmented—physically, culturally, and politically. It comprises more than 17,000 islands (3000 of which are inhabited), stretching over 3000 miles (8000 kilometers) of ocean. It is also the most culturally diverse, with dozens of ethnic groups and multiple religions. But periodic instability in Indonesia has many people wondering whether this multi-island country of 243 million might be headed for disintegration.

Until the end of World War II, Indonesia was not a nation at all but rather a loose assemblage of distinct island cultures, which Dutch colonists managed to hold together as the "Dutch East Indies." When Indonesia became an independent country in 1945, its first president, Sukarno, hoped to forge a new nation out of these many parts. To that end, he articulated a national philosophy known as *Pancasila*, which was aimed at holding the disparate nation together primarily through religious tolerance.

Encouraging Cohesiveness with Government Policies *Pancasila* embraces five precepts: belief in God, and the observance of *conformity*, *corporatism* (often defined as organic social solidarity with the state), *consensus*, and *harmony*. These last four precepts could be interpreted as discouraging dissent or even loyal opposition, and they seem to require a perpetual stance of boosterism. For some people, the strength of *Pancasila* is the emphasis on harmony and consensus that seems to ensure there will never be either an Islamic or a communist state. Others note that conformity and corporatism counteract the extreme ethnic diversity and geographic dispersion of the country. But conformity and corporatism have also had a chilling effect on participatory democracy and on criticism of the government, president, and the army. The first orderly democratic change of government did not take place until national elections in 2004. Since then there have been several peaceful elections, and the government is stable enough to allow citizens to publicly protest policies (see Photo Essay 10.4B).

In recent years, separatist movements have sprouted in four distinct areas. The only one to succeed was in Timor-Leste, which became an independent country in 2002. However, its case is unique in that this area was under Portuguese control until 1975, when it was forcibly integrated into Indonesia. Two other separatist movements have grown largely in response to Indonesia's **resettlement schemes**. Also known as *transmigration schemes*, between 1965 and 1995 these programs relocated approximately 8 million people from crowded islands such as Java to less densely settled islands. The policies were originally initiated under the Dutch in 1905 to relieve crowding and provide agricultural labor for plantations in thinly populated areas.

After independence, Indonesia used resettlement schemes for the same purposes and also to bring outlying areas under closer control of the central government in Jakarta. However, the government ultimately lost rather than gained control as the newcomers inspired separatist movements in two main areas of resettlement: in the Indonesian half of the island of New Guinea (the Indonesian provinces of Papua and West Papua) and in the Malucca islands. In 2001, three years after the expansion of Indonesian

resettlement schemes government plans to move large numbers of people from one part of a country to another to relieve urban congestion, disperse political dissidents, or accomplish other social purposes; also called *transmigration*

democracy following Suharto's resignation in 1998, the resettlement schemes were cancelled. This was in response to both violence in the resettlement areas and lack of funds to continue the programs. The separatist movements were further diffused by the extension of greater local autonomy to the Maluccas, Papua, and West Papua. Violence has declined as local and provincial governments have become more responsive to their own people and less subject to the whims of national leaders in Jakarta.

The most dramatic turnaround was in the far western province of Aceh in Sumatra, where in 2005, separatists who had battled Indonesian security forces for decades laid down their arms. Conflicts had originally developed because most of the wealth yielded by Aceh's resources, especially oil, was going to the central government in Jakarta. The conflict began to decline after the democratization that followed Suharto's resignation. A final boost came from the tsunami of 2004. Recovery efforts following the disaster created a powerful incentive for separatists and the government to cooperate in order to receive outside aid. A peace accord signed in 2005 brought many former combatants into the political process as democratically elected local leaders. Violence has decreased dramatically since then.

Southeast Asia's Authoritarian Tendencies Despite the democratization that has occurred in Indonesia, sporadically in Thailand, and to a lesser extent in Malaysia, authoritarianism (see Chapter 7, pages 320–322) is still a powerful force in Southeast Asia. Undemocratic socialist regimes still control Laos and Vietnam (see Photo Essay 10.4D), and a military dictatorship runs Burma. Cambodia's democracy is precarious and violence frequent. Brunei is an authoritarian sultanate, and from 2006 to 2008, Thailand's government was taken over by the military, and subsequent elections have not quieted protesters. Powerful and corrupt leaders have subverted the democratic process even in the region's oldest democracy, the Philippines, as well as occasionally in the wealthier countries of Malaysia and Singapore.

📺 **230. Bankers, Analysts See Resurgent Asia 10 Years after Economic Crisis**

Some Southeast Asian leaders, such as Singapore's former prime minister, Lee Kuan Yew, have argued that Asian values are not compatible with Western ideas of democracy. Yew and other leaders assert that Asian values are grounded in the Confucian view that individuals should be submissive to authority. Hence, Asian countries should avoid the highly contentious public debate of electoral politics. Nevertheless, when confronted with governments that abuse their power, people throughout Southeast Asia have repeatedly rebelled, often in the form of pro-democracy movements (see Photo Essay 10.4A, B).

📺 **319. Aung San Suu Kyi Trial Under Way in Burma**

Some pro-democracy movements have resulted in real change, as in Thailand (before 2007) and Indonesia, but others have not. For more than two decades, people in Burma have been futilely protesting the rule of a corrupt and undemocratic military regime. The regime refused to step aside when the people elected Aung San Suu Kyi to lead a civilian reformist government in 1990. Suu Kyi has been under house arrest ever since, despite having won the Nobel Peace Prize in 1991. Widespread pro-democracy protests throughout the country in 2007 were brutally repressed (see Photo Essay 10.4A).

📹 **234. New Technology Beams Burma Protests Across Globe**
📹 **235. Jimmy Carter Calls for More International Pressure on Burma**

Terrorism and Democracy Like authoritarianism, terrorism has long loomed as a counterforce to democracy in this region. During the late 1990s, a series of bombs exploded in the Philippines and across Indonesia, and small terrorist cells were discovered in Malaysia, southern Thailand, and Singapore. Until the bombing of the Sari Hotel in Bali (Indonesia) in October 2002, which killed nearly 200 foreign tourists, terrorist activity in the region was local in nature. It was carried out by local militant groups pursuing domestic political agendas and grievances—most often revenge for government campaigns against Muslim separatists. The Bali bombing, the bombing of the Marriott Hotel in Jakarta in August 2003, and ongoing violence in Muslim southern Thailand have drawn attention to apparent connections between local groups and international Islamist terrorist networks.

📹 **225. Terror and Islamic Struggle in Indonesia**
📹 **232. Violence in Thailand's Muslim South Intensifies**

Terrorist violence short-circuits the public debate that is at the heart of the democratic processes. The 2004 election of secular political parties in Malaysia, Indonesia, and Thailand showed that most Southeast Asians, who are accustomed to a tolerant version of Islam, do not support Islamist militants who engage in terrorism. However, rather than trying to address local conditions that fuel terrorism, some states are resorting to political repression. Malaysia and Singapore have both been criticized for using counterterrorism as a cover to crack down on peaceful opposition groups that merely challenge the government's policies through democratic processes. Another route is suggested by Indonesia's recent success at bringing former combatants into the democratic process, a strategy that has also been successful in North Africa and Southwest Asia (see Chapter 6, pages 279–280).

Terrorism in the Philippines has a different quality. There, warring family clans, some connected to high government officials, attack each other, but are also involved in controlling supposed Muslim fundamentalists by depriving them of land and resources for living.

THINGS TO REMEMBER

1. Globalization has produced long periods of strong economic growth in Southeast Asia punctuated by brief but dramatic periods of decline.

2. Urban incomes have increased significantly during times of growth, but many poor people remain highly vulnerable to periods of global economic decline that threaten their access to jobs, food, and adequate living quarters.

3. While significant barriers remain, some countries throughout the region have recently seen dramatic expansion of democratization. However, some of these countries also have had a reversal of the trend toward democratization.

4. The chief foes of democracy are extreme diversity, powerful family dynasties, the lack of cohesive identities, authoritarian tendencies, and terrorism.

Sociocultural Issues

Southeast Asia is home to 586 million people who occupy a land area that is about one-half the size of the United States. Due to a long and complex history, these people exhibit a great diversity of cultures and religious traditions.

Population Patterns

Southeast Asia's population is large and growing, but urbanization and population control efforts are slowing the rate of growth. At present rates of growth, Southeast Asia's population is projected to reach 826 million by 2050, by which time much of this population will live in cities (less than half do now). However, population projections could be inaccurate, both because rates of natural increase are slowing markedly and because many Southeast Asians are migrating to find employment outside the region.

Population Distribution The population map in Photo Essay 10.5 (page 459) reveals that relatively few people live in the rugged upland reaches of Burma, Thailand, and northern Laos, or in much of Cambodia. Light settlement is also found in much of Malaysia, Indonesia, and the Philippines, in wetlands, dense forests, mountains, and geographically remote areas. For thousands of years, small groups of indigenous people have lived in forested uplands, supported by shifting cultivation and by hunting, gathering, and small-plot permanent agriculture. Despite resettlement schemes, much of Sumatra, Kalimantan (on Borneo), Sulawesi (formerly known as Celebes), the Moluccas, and West Papua remain lightly settled.

About 60 percent of the people of Southeast Asia live in patches of particularly dense rural settlement along coastlines, on the floodplains of major rivers, and in the river deltas of the mainland (see Thematic Concepts Part F on page 436). On the islands, settlement is most concentrated on Luzon (in the northern Philippines) and on Java. These places are attractive because the rich and well-watered volcanic soils allow intensive agriculture.

Population Dynamics There is considerable variety in the population dynamics of this region. All countries are nearing the last stage of the *demographic transition*, where births and deaths are low and growth is minuscule or slightly negative; but some are much closer than others, and the disparities of wealth between countries as well as religious and cultural practices greatly influence reproduction behavior. In the last several decades, overall fertility rates in Southeast Asia have dropped by more than half (see Figure 10.17 on page 458). Whereas women formerly had 5 to 7 children, they now have 2 or 3. The only major exception is Timor-Leste, where fertility rates are still over 7 (see Thematic Concepts Part M on page 437), and possibly Laos (3.5) and the Philippines (3.3) (see Photo Essay 10.5C). Nonetheless, in most countries populations are young, with between one-quarter and one-third of the people aged 15 years or younger. Thus, for these countries steady population growth is ensured for several decades because so many are just coming into their fertile years.

On the other hand, Brunei, Singapore, and Thailand have reduced their fertility rates so steeply—below replacement levels—that they must already begin to think about how they will

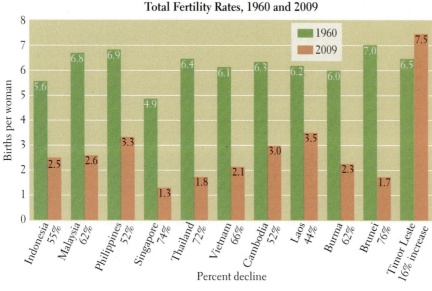

Total Fertility Rates, 1960 and 2009

FIGURE 10.17 Total fertility rates, 1960 and 2009. Total fertility rates declined during this period for all Southeast Asian countries except Timor-Leste. [Data from Sasha Loffredo, *A Demographic Portrait of South and Southeast Asia* (Washington, D.C.: Population Reference Bureau, 1994), p. 9; Globalis, at http://globalis.gvu.unu.edu/; and *2009 World Population Data Sheet* (Washington, DC: Population Reference Bureau, 2009).]

cope with aging and shrinking populations even though those trends are just visible in the statistics (see Figure 10.17). The Singapore government is now so concerned about the low fertility rate that it offers young couples various incentives for marrying and procreating (see Photo Essay 10.5A). A greater source of population growth for Singapore is the steady stream of highly skilled immigrants that its vibrant economy attracts. Thailand's low fertility rate of 1.8 children per adult woman was achieved via a government-sponsored condom campaign attuned to rapid economic change and urbanization, all of which worked to make most couples feel that smaller families would be best. As women moved to cities, they had more chances to work and study outside the home, opportunities that tend to reduce fertility rates in most societies. High literacy rates for both men and women and Buddhist attitudes that accept the use of contraception have also been credited for the decline in Thailand's fertility rate.

Some of these same conditions prevail in Vietnam, but there the government, deciding not to leave things to chance, took more assertive action. Worried that a rapidly growing population would jeopardize its upswing in development, the government of Vietnam recently introduced a two-child family policy (see Photo Essay 10.5B).

The poorest and most rural countries in the region show the usual correlation between poverty, high fertility, and infant mortality. In Cambodia and Laos, fertility rates average between 3.0 and 3.5 children respectively. Infant mortality rates are 62 per 1000 births for Cambodia and 64 for Laos. In Vietnam, where people are only slightly more prosperous and urbanized, the fertility rate (2.1 children per adult woman) and infant mortality rate (15 per 1000 births) are much lower. Vietnam's lower rates are explained by the fact that this socialist state provides basic education and health care to all of its people, regardless of income. In Vietnam, literacy rates are more than 94 percent for men and 87 percent for women, whereas only 68 percent of women in Cambodia and only 63 percent in Laos can read. In addition, Vietnam's rapidly growing and urbanizing economy is attracting foreign investment, which provides more employment

for women. Hence, careers are replacing child rearing as the central focus of many women's lives.

Personal Vignette In Roman Catholic Philippines, Gina Judilla, who works outside the home, has had six children with her unemployed husband. They wanted only two, but because of the strong role of the Catholic Church and the political pressure it exerts, birth control was not available to the poor and abortion is illegal, so with every succeeding pregnancy she tried folk methods of inducing an abortion. None worked. Now she can afford to send only two of her six children to school.

A move by family planners to provide national reproductive health services and sex education is underway. A recent survey showed that 54 percent of all pregnancies in the Philippines in 2008 were unintended, and only one-third of Philippine women have access to modern birth control methods. [Source: Carlos H. Conde, "Bill to Increase Access to Contraception Is Dividing Filipinos," New York Times, October 25, 2009, at http://www.nytimes.com/2009/10/26/world/asia/26iht-phils.html?hpw.] ∎

Population Pyramids The youth and gender features of Southeast Asian populations is best appreciated by looking at the population pyramids for Indonesia, 2009 and projected to 2050 (see Figure 10.18 on page 460). The wide bottom of the 2009 pyramid shows that most are under age 30, but the projections to 2050 show that eventually, with declining birth rates, Indonesia will accumulate ever larger numbers in the upper age groups and the pyramid will eventually be more box shaped as those for Europe are now. The gender disparities (more males than females) can be seen by carefully examining the sizes of each age group on the male and female sides of the pyramid. The difference is slight but significant, because there is a rather consistent deficit of females, apparently selected out before birth.

Southeast Asia's Encounter with HIV-AIDS As in sub-Saharan Africa (see Chapter 7), HIV-AIDS constitutes a significant public health issue in Southeast Asia. Infection rates are growing across

Population growth in Southeast Asia is slowing, due largely to economic development, urbanization, changing gender roles, and government policies. Fertility rates have declined sharply in all countries since the 1960s, but are still high in the poorest areas.

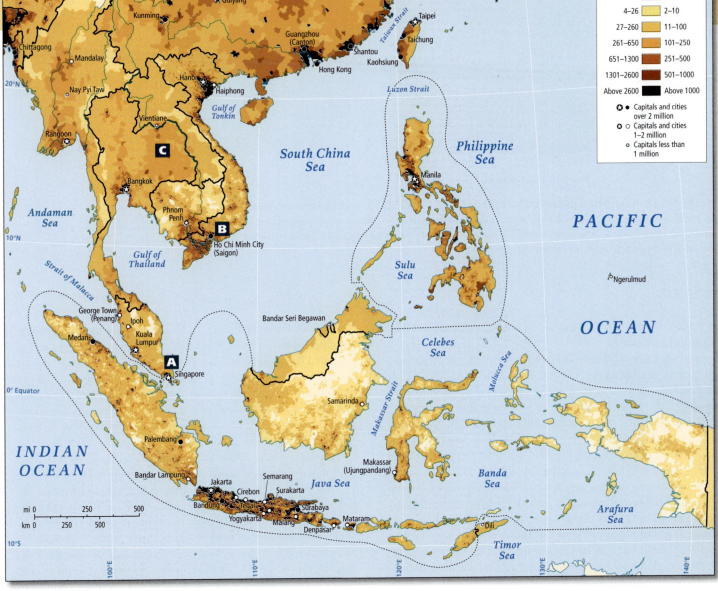

Persons per	
sq mi	sq km
0–3	0–1
4–26	2–10
27–260	11–100
261–650	101–250
651–1300	251–500
1301–2600	501–1000
Above 2600	Above 1000

✪ ● Capitals and cities over 2 million
✪ ○ Capitals and cities 1–2 million
○ Capitals less than 1 million

A A young couple in Singapore, where the government is trying to counteract the "demographic transition" by encouraging earlier marriage in hopes of a higher birth rate.

B A father and his two children in Ho Chi Minh City, Vietnam, where the government has implemented a "two-child policy" in response to rising birth rates.

C A mother and her four children in Vientiane, Laos. Birth rates remain high in Laos, where economic and women's development is relatively low for the region and the demographic transition is in the early stages.

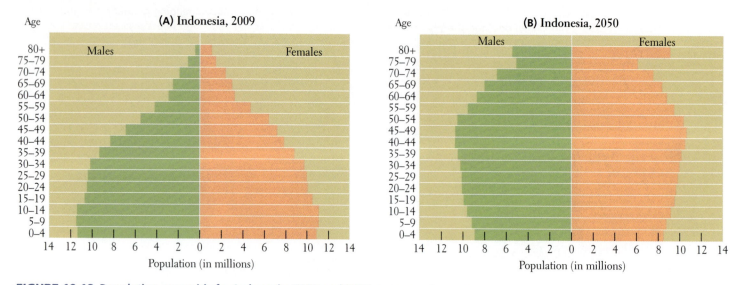

FIGURE 10.18 Population pyramids for Indonesia, 2009 and 2050. In 2009, Indonesia had 240.3 million people, while in 2050 Indonesia is projected to have 313 million people. [Adapted from U.S. Census Bureau, Population Division, "Population Pyramids of Malaysia" and "Population Pyramids of Indonesia," International Data Base, 2009, at http://www.census.gov/ipc/www/idb/pyramids.html.]

the region; at present, Cambodia, Thailand, Burma, and West Papua in Indonesia have the highest rates. In Thailand, AIDS is the leading cause of death, overtaking accidents, heart disease, and cancer, and estimates are that more than a million people are infected. Men between the ages of 20 and 40 are the most common victims, but the number of women victims is rising.

HIV-AIDS infections may soon increase rapidly in rural areas and in secondary cities that are just coming into closer economic and social contact with the region's big cities, where infection is most prevalent. Several reasons explain this gloomy forecast. Conservative religious leaders and faith-based international agencies restrict public sex education and AIDS-prevention programs, such as the promotion of condom use, because these activities are viewed as promoting promiscuity. At the same time, popular customs that support sexual experimentation (at least among men), the high mobility of young adults, the reluctance of women to insist that their husbands and boyfriends use condoms, and intravenous drug use (primarily by men) make aggressive prevention programs all the more essential.

📹 **233. Activists: AIDS Linked to Women's Rights Abuses**

The development of sex tourism and associated human trafficking has also contributed to the spread of HIV among young women (discussed further on page 469). Just as in South Asia and Africa, truck drivers in Southeast Asia spread HIV because they may have sexual partners in several different locales. In Thailand, a promising government education campaign to promote condom use lost momentum due to cuts in international funding, and now HIV is spreading to more diverse populations. The greatest increase is among young women. Buddhist nuns have led the movement to take care of people with AIDS, and when antiretroviral drugs were developed, Buddhists were among the first to make them available to destitute patients.

Geographic Patterns of Human Well-Being If human well-being is defined as the ability to enjoy long, healthy, and creative lives, as suggested by Mahbub ul Haq, the founder of the United Nations development reports, then personal income statistics will not be the best way to measure success. Nonetheless, income is relevant to well-being, as evidenced by a series of three maps of Southeast Asia. Map A shows GDP per capita (PPP) by country, Map B shows levels of human development (HDI). Because gender is recognized as having a significant impact on well-being, we include Map C, which depicts gender differences in pay (Figure 10.19A, B, C).

On Map A, two small countries, Singapore and Brunei, are the only places where annual per capita GDP (PPP) is in the highest category. For both countries, the figure is about U.S.$50,000, which is a bit higher than in the United States and Australia and comparable to the richest countries in Europe (see the inset map of the world). Elsewhere in the region, per capita GDP (PPP) is considerably lower, with Malaysia the only country in the medium-high range at U.S.$13,518. Thailand is at the top of the low range; and just below it are Vietnam, Indonesia, and the Philippines. The Philippines has lost ground in this category in recent years because of the global recession, which has cut the size of remittances and because of political troubles in the southern islands. At the bottom are Laos, Cambodia, Timor-Leste, and, finally, Burma.

Map B (HDI) again shows that Singapore and Brunei do well at providing for the overall well-being of their citizens as they rank in the highest category. Malaysia and Thailand are in the high and medium-high categories, although both have fallen off a little since 2005, probably due to the global recession. Indonesia is in the medium category and has held the same level as it did in 2005, meaning it may not have suffered as much from the global recession as feared, but it has also not progressed. The rest of the

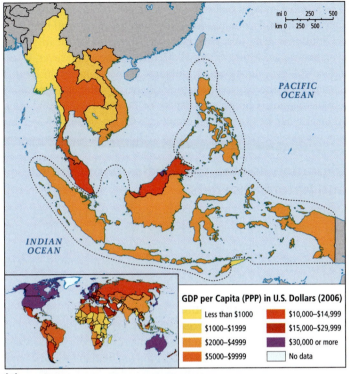

(A)

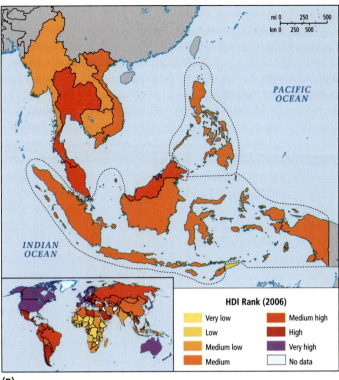

(B)

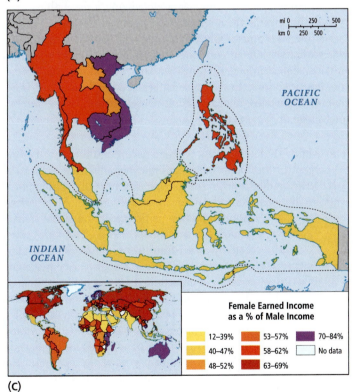

(C)

FIGURE 10.19 Maps of human well-being. (A) Gross domestic product (GDP). **(B)** Human Development Index (HDI). **(C)** Female earned income as a percent of male earned income (F/MEI). [Maps adapted from data: "Human development indices," (Maps A and B): Table 2, pages 28–32; (Map C): Table 5, pages 41–44 at http://hdr.undp.org/en/media/HDI_2008_EN_Tables.pdf.]

GDP and HDI—show the least gender pay disparities. Meanwhile, the richest countries, with high HDI rankings—Singapore and Brunei—and also several countries that are in the medium ranks of per capita GDP and HDI, show the most gender disparity in income. These particular patterns are the result of several phenomena: glass ceilings in Singapore and Brunei keep qualified women from reaching upper-echelon positions in companies and the government. Also, those countries with the narrowest gender pay disparities—Vietnam and Cambodia—are agricultural countries, where pay scales are very low for everyone. Finally, the gender disparities are rather extreme across the board, with women earning at best only 84 percent of men, and in Malaysia and Indonesia, which have over half of the region's population, women earn less than 50 percent of what men earn.

Urbanization

Southeast Asia as a whole is only 43 percent urban, but the rural–urban balance is shifting steadily in response to declining agricultural employment and booming urban industries. The forces driving farmers into the cities are called the *push factors* in rural-to-urban migration. They include the rising cost of farming caused by the use of new technologies. *Pull factors*, in contrast, are those that attract people to the city, such as abundant manufacturing jobs and

countries, including Burma, have remained in the medium-low category for some years, and Timor-Leste continues to rank the lowest in the region.

Map C shows one measure of gender equality: female income as a percentage of male income. Here the patterns are startlingly different from the other two maps, because two countries—Vietnam and Cambodia—with relatively low rankings in per capita

foreign exchange foreign currency that countries need to purchase imports

education opportunities. In Southeast Asia, as in all other regions, these factors have come together to create steadily increasing urbanization. Malaysia is already 68 percent urban; the Philippines, 63 percent; Brunei, 72 percent; and Singapore, 100 percent.

Throughout Southeast Asia, employment in agriculture has been declining since the introduction of new production methods that increase the need for labor-saving equipment and reduce the need for human labor. Meanwhile, the use of chemical pesticides and fertilizers has also spread. While such additives can increase harvests dramatically, they also drive the cost of production beyond what most farmers can afford. Many family farmers have sold their land to more prosperous farmers or to corporations and moved to the cities. These people skilled at traditional farming but with little formal education end up in the most menial of urban jobs that offer none of the rewards of making things grow (Photo Essay 10.6C).

Labor-intensive manufacturing industries, such as garment and shoe making, are expanding in cities and towns of the poorer countries, such as Cambodia, Vietnam, and parts of Indonesia and Timor-Leste. In the urban and suburban areas of the wealthier countries—Singapore, Malaysia, Thailand, and parts of Indonesia and the northern Philippines—technologically sophisticated manufacturing industries are also growing. These include automobile assembly, chemical and petroleum refining, and assembly of computers and other electronic equipment (see Photo Essay 10.6D).

Push and pull factors have sent rural migrants streaming into cities like Jakarta, Manila, and Bangkok, which are among the most rapidly growing metropolitan areas on earth (Table 10.2). Such cities are *primate cities*—cities that are at least two times the size of the second-largest city in a given country. Bangkok is over 20 times larger than Thailand's next-largest metropolitan area, Udon Thani, and Manila is over 9 times larger than Davao, the second-largest city in the Philippines. Thanks to their strong industrial base, political power, and the massive immigration they attract, primate cities can dominate whole countries. Table 10.2 indicates all primate cities in the region.

Rarely can such cities provide sufficient housing, water, sanitation, or even decent jobs for all the new arrivals. Of all the cities in Southeast Asia, only Singapore provides well for nearly all of its citizens (see Photo Essay 10.6B). Even there, however, a significant illegal, noncitizen population lives in poverty on islands surrounding the city. More typical is the experience of rural-to-urban migrants who go to Bangkok or Jakarta. Living conditions in Jakarta can be especially difficult (see Photo Essay 10.6A). The city is expected to grow by 50 percent in the next 10 years.

Migration Related to Globalization, Conflict, and Natural Disasters

Millions of Southeast Asians are moving to cities beyond the borders of their home countries. Flows of workers within this world region and to others are an important aspect of globalization. Other Southeast Asians are refugees forced to migrate by violent conflict or natural disasters.

Emigration and Globalization The same push and pull factors driving urbanization are also driving some people to migrate out of Southeast Asia. These migrants are a major force of globalization as they supply much of the world's growing demand for low- and mid-wage workers who are willing to travel or live temporarily in foreign countries. They are also a globalizing force within their home countries as their remittances (monies sent home) boost family incomes and supply governments with badly needed **foreign exchange** (foreign currency) that countries need to purchase imports. For example, Filipinos working abroad are that country's largest source of foreign exchange, sending home over U.S. $6 billion annually, and increasing household annual income by an average of 40 percent.

The Maid Trade Recently, women have constituted well over 50 percent of the more than 8 million migrants from Southeast Asia. Many skilled nurses and technicians from the Philippines work in European, North American, and Southwest Asian cities. About 3 million participate in the global "maid trade" (Figure 10.20 Map). Most are educated women from the Philippines and Indonesia who work under 2- to 4-year contracts in wealthy homes throughout Asia. An estimated 1 to 3 million Indonesian maids now work outside of the country, mostly in the Persian Gulf.

The maid trade has become notorious for abusive working conditions and employers who often do not pay what they promise. In Saudi Arabia, the NGO Human Rights Watch is monitoring the cases of Muslim Indonesian women who were brutally abused—two of whom were killed—by members of a privileged Saudi family. The Philippines went so far as to ban the maid trade in 1988, but reestablished it in 1995 after better pay and working conditions were negotiated with countries receiving the workers.

TABLE 10.2	**Recent growth in Southeast Asia's metropolitan areas with over 5 million inhabitants**	
Largest city	**Population (Year)***	**Population in 2010 (est.)**
Manila, Philippines	9,906,048 (2005)	20,654,307
Jakarta, Indonesia	15,961,014 (2000)	19,231,919
Bangkok, Thailand	8,450,000 (2000)	10,132,974
Kuala Lumpur, Malaysia	4,428,836 (2000)	8,063,230
Bandung, Indonesia	3,416,707 (1995)	6,724,301
Singapore, Singapore	4,500,000 (2001)	5,998,943
Ho Chi Minh City, Vietnam	3,924,435 (1989)	5,381,158
Rangoon, Burma (Myanmar)	3,361,741 (1993)	5,032,190
Phnom Penh, Cambodia	999,800 (1998)	1,485,661

* Year of most recent data. Note that metropolitan areas may have expanded.
Source: World Gazetteer, at http://world-gazetteer.com/.

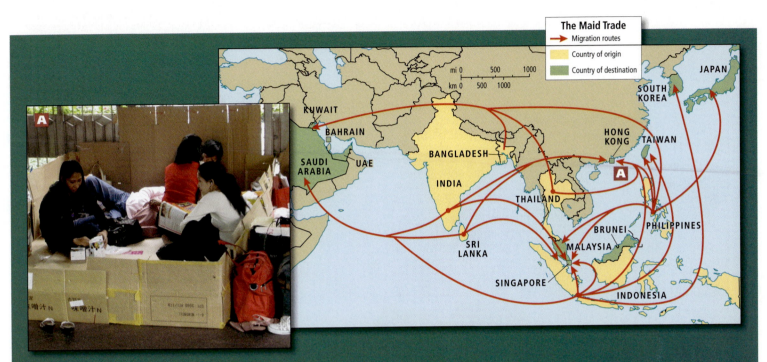

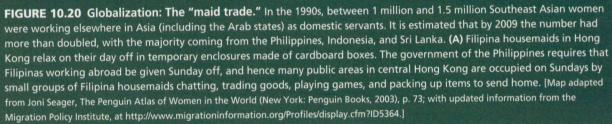

FIGURE 10.20 Globalization: The "maid trade." In the 1990s, between 1 million and 1.5 million Southeast Asian women were working elsewhere in Asia (including the Arab states) as domestic servants. It is estimated that by 2009 the number had more than doubled, with the majority coming from the Philippines, Indonesia, and Sri Lanka. **(A)** Filipina housemaids in Hong Kong relax on their day off in temporary enclosures made of cardboard boxes. The government of the Philippines requires that Filipinas working abroad be given Sunday off, and hence many public areas in central Hong Kong are occupied on Sundays by small groups of Filipina housemaids chatting, trading goods, playing games, and packing up items to send home. [Map adapted from Joni Seager, The Penguin Atlas of Women in the World (New York: Penguin Books, 2003), p. 73; with updated information from the Migration Policy Institute, at http://www.migrationinformation.org/Profiles/display.cfm?ID5364.]

Personal Vignette Every Sunday is Amah (nanny) day in Hong Kong. Gloria Cebu and her fellow Filipina maids and nannies stake out temporary geographic territory on the sidewalks and public spaces of the central business district. Informally arranging themselves according to the different dialects of Tagalog (the official language of the Philippines) they speak, they create room-like enclosures of cardboard boxes and straw mats where they share food, play cards, give massages, and do each other's hair and nails (see **Figure 10.20A**). Gloria says it is the happiest time of her week, because for the other 6 days she works alone caring for the children of two bankers.

Gloria, who is a trained law clerk, has a husband and two children back home in Manila. Because the economy of the Philippines has stagnated, she can earn more in Hong Kong as an Amah than in Manila in the legal profession. Every Sunday she sends most of her income (U.S.$125 a week) to her family. [Sources: Adapted from notes by Kirsty Vincin, an Australian teacher in Hong Kong, November 17, 2009, and "The Filipina Sisterhood," Economist, December 20, 2001, at http://www.economist.com/world/asia/displaystory.cfm?story_id=E1_RRPJDJ.] ■

Skilled Male Seamen Certainly males have experienced hardships on the international labor market, but for the many skilled male workers from Southeast Asia who work in the international merchant marine, conditions are much better. They typically work onboard international freighters or on luxury cruise ships as seamen, cooks, and engine mechanics; a few are officers. They generally work for 6 months or more at a stretch with only a few hours in the day for breaks. At the end of their tour of duty they return home to their families for 4 to 6 months. Workers from different countries doing the same job can earn quite different amounts because in the merchant marine one is paid according to the pay scales of one's home country, but generally a sailor's earnings are sufficient to provide a middle-class lifestyle and education for his family back home.

Refugees from Conflict and Natural Disaster Those who must move out of necessity often confront serious difficulties. Most of the refugees in this region tend to be the victims of either governmental violence or natural disasters. During the last half of the twentieth century, millions of mainland Southeast Asians fled into neighboring states to escape protracted conflict. Thailand, Laos, and Cambodia, in particular, received many refugees during and after the Vietnam War. Currently, repression by Burma's military government has resulted in at least 680,000 refugees, with 500,000 displaced within Burma and 180,000 having fled to Thailand

Southeast Asia is rapidly urbanizing as the growth of manufacturing and service-sector industries pull in people from rural areas and as changes in agriculture push farmers to the cities. Many cities are struggling to cope with rapid growth.

A A slum area in Jakarta, Indonesia, where 62 percent of the population lives in slums. Jakarta will grow by almost fifty percent by 2020.

Population of urban areas

2006	2020
20 million or more	20 million or more
10 million	10 million
5 million	5 million
3 million	3 million

Note: symbols on map are sized proportionally to specific population data

Population living in urban areas

80–100%	20–39%
60–79%	0–19%
40–59%	No data

1 Global rank (population 2006)

mi 0 250 500
km 0 250 500

BURMA
LAOS
63 Hanoi
C
64 Rangoon
VIETNAM
THAILAND
Bangkok 33
D
CAMBODIA
49
Ho Chi Minh City (Saigon)
Manila 20
PHILIPPINES
MALAYSIA
BRUNEI
Kuala Lumpur
55
SINGAPORE
B
A
INDONESIA
65 Bandung
Jakarta 8
TIMOR-LESTE

B Public housing in Singapore, where 85 percent of the population lives in apartment buildings managed by the government.

C A man in Hanoi, Vietnam, pushes a bicycle converted to carry heavy loads. Many migrants start out in low-wage, urban manual-labor jobs like this.

D Women working construction in Bangkok, Thailand. Gender roles are changing as more women move to the cities.

(Figure 10.21A), Bangladesh, India, and Malaysia, where many have lived for several decades in refugee camps.

 Personal Vignette "Basically I think [America] will be better than a refugee camp," said Hsar Say, a little doubtfully. "In a refugee camp, you have no rights. You are put in a cage. It's illegal to travel outside the camp, so it's very different from being a human."

Hsar Say was leaving on a bus for resettlement in the United States, after 20 years as a Burmese refugee in the Mae La camp in Thailand (see Figure 10.21). He had grown up there and become a teacher in the camp school, married, and had two children. He had

misgivings about going to America because, after being penned up all his life, he had little experience operating on his own. And maybe the Americans wouldn't realize how much he wanted to help others, perhaps as a social worker. He heard that an educated cousin had ended up in Kentucky packing boxes in a clothing store. *[Source: Adapted from Jesse Wright, "Burmese Refugees Fearful of New Life in USA," USA Today, January 22, 2009, at http://usatoday.com/news/world/2009-01-22-burmarefugees_N.htm.]* ■ 📺 **238. Army Offensive in Eastern Burma Creates Growing Humanitarian Crisis**

The tsunami of December 2004 complicated the refugee picture when it displaced well over 130,000 people in Sumatra alone

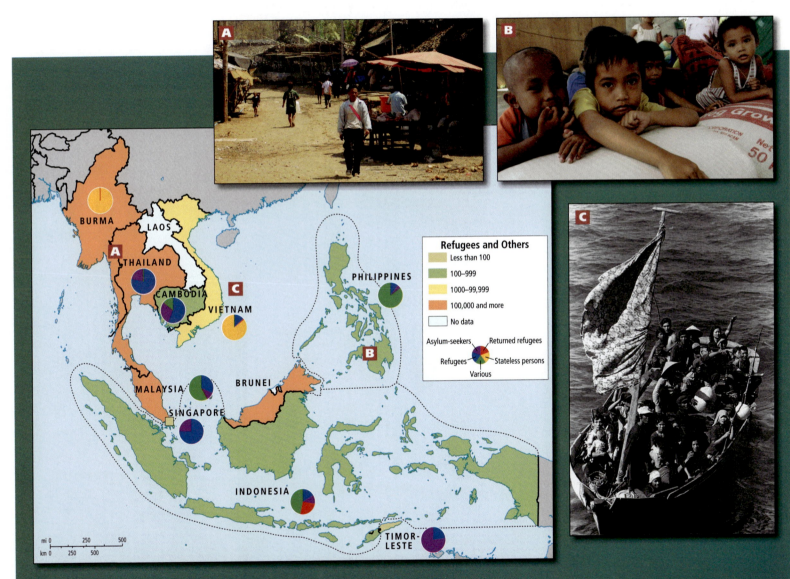

FIGURE 10.21 Refugees and other persons of concern. Nearly every country in this region has refugees. Statistics on refugees are provided by governments, based on their own definitions and data collection methods and hence are not always completely accurate. **(A)** The Mae La refugee camp in Thailand is home to 50,000 refugees from Burma. **(B)** Children in Mindanao, Philippines, take refuge from fighting between Muslim separatist militias and the Philippine government that has left 100,000 homeless. **(C)** Thirty-five Vietnamese refugees about to be taken aboard a U.S. Navy vessel after 8 days at sea. Over 1.2 million Vietnamese refugees were granted asylum in foreign countries due to persecution by the Vietnamese government during the 1980s and 1990s. Most went to the United States, but many also went to Australia, Canada, China, and Western Europe. [Source: United Nations High Commission for Refugees, at http://www.unhcr.org/cgi-bin/texis/vtx/page?page=49e489296.]

(in addition to killing an equal number of people on that island plus several thousand in Thailand and Burma). The tsunami refugees went temporarily to Malaysia or to a safe haven in their own home country, and have now returned home. The map in Figure 10.21 shows the numbers of refugees (from all causes) reported by each country, and their present status.

THINGS TO REMEMBER

1. About 43 percent of Southeast Asia's population is urban, and this proportion is increasing because of changes in agriculture and industry.

2. Often the focus of migration is the capital of a country, which may become a primate city—one at least two times the size of the second-largest city.

3. Fertility rates have declined sharply in all countries except Timor-Leste since the 1960s due largely to economic growth, urbanization, and literacy and education for women.

4. Thousands of people emigrate from the region each year to seek employment, temporary or long-term, in the Middle East, Europe, North America, or Japan. Millions more have become refugees, been displaced due to internal political and/or ethnic conflicts.

Religious Pluralism

The major religious traditions of Southeast Asia include *Hinduism, Buddhism, Confucianism and Taoism, Islam, Christianity,* and *animism* (**Figure 10.22**). The patterns of religious practice

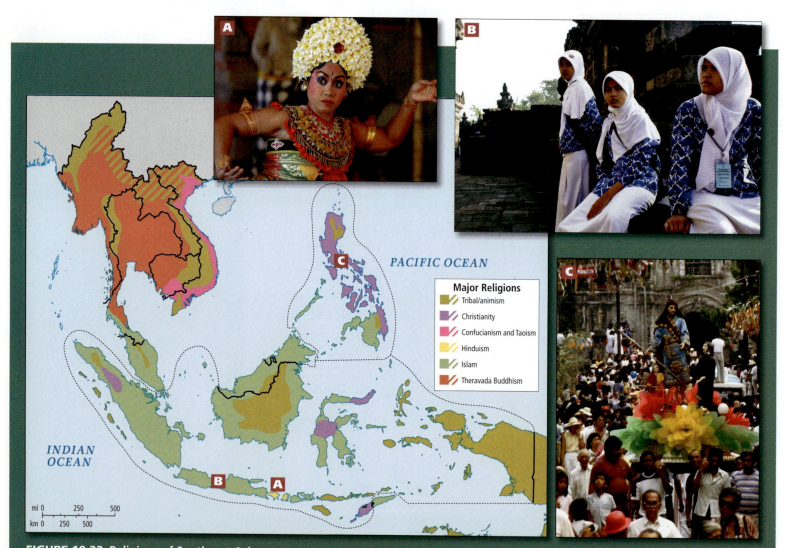

FIGURE 10.22 Religions of Southeast Asia. Southeast Asia is religiously very diverse: five of the world's six major religions are practiced there. Animism, the oldest belief system, is found in both island and mainland locations and has many subtle influences. **(A)** A dancer in Bali, Indonesia, where traditional Hindu performances incorporate Buddhist and animist symbols and stories. **(B)** Muslim students visit Borobudur, an ancient Buddhist monument on the island of Java, Indonesia. Many Islamic traditions on Java incorporate Buddhist ideas and practices. **(C)** San Isidro Labrador, patron saint of farmers, is carried by Catholics during the Pahiyas festival in Lucban, Quezon, Philippines. The Pahiyas festival is a Catholic festival that was originally an animist harvest celebration. [Adapted from *Oxford Atlas of the World* (New York: Oxford University Press, 1996), p. 27.]

are complex. All originated outside the region, with the exception of the animist belief systems of the indigenous peoples. Animism takes many different forms in this region. In general, in animism such natural features as trees, rivers, crop plants, and the rains all carry spiritual meaning. These natural phenomena are the focus of festivals and rituals to give thanks for bounty and to mark the passing of the seasons, and these ideas touch in some way all religious traditions in this region.

Personal Vignette Although arranged marriages were long the tradition across Southeast Asia, now in most urban and rural areas marriages are love matches. Such is the case for Harum and Adinda, who live in Tegal, on the island of Java in Indonesia. They met in high school and some 10 years later, after saving a considerable sum of money, they decided to formally ask both sets of parents if they could marry. Harum, an accountant, would normally be expected to pay the wedding costs, which could run to many thousands of dollars; however, Adinda has contributed from her salary as a teacher. Like nearly all Javanese, both are Muslims but because Islam does not have elaborate marriage ceremonies, colorful rituals from Christianity and Buddhism will enhance the elaborate and festive occasion.

In preparation, both bride and groom undergo rituals that remain from the days when marriages were arranged and the bride and groom did not know each other. A *pemaes,* a woman who prepares a bride for her wedding and whose role is to inject mystery and romance into the marriage relationship, bathes and perfumes the bride. She also puts on the bride's makeup and dresses her, all the while making offerings to the spirits of the bride's ancestors and counseling her about how to behave as a wife and how to avoid being dominated by her husband (**Figure 10.23**). The groom also undergoes ceremonies meant to prepare him for marriage. Both are counseled that their relationship is

FIGURE 10.23 The Dahar Klimah (Dahar Kembul). In this phase of a traditional Javanese wedding, the *pemaes* gives the bride a bundle of food (yellow rice, fried eggs, soybeans, and meat). The bridegroom makes three small balls of the food and feeds them to the bride; she then does the same for him. The ritual reminds them that they should share joyfully whatever they have.

bound to change over the course of the decades as they mature and as their family grows older.

Despite the elaborate preparations for marriage, divorce in Indonesia (and also in Malaysia) is fairly common among Muslims, who often go through one or two marriages early in life before they settle into a stable relationship. Although the prevalence of divorce is lamented by society, it is not considered outrageous. Apparently, ancient indigenous customs predating Islam allowed for mating flexibility early in life, and this attitude is still tacitly accepted. *[Sources: Adapted from personal communications with the anthropologist Jennifer W. Nourse, University of Richmond, a specialist in Southeast Asia, 2010; and Walter Williams, Javanese Lives (Piscataway, NJ: Rutgers University Press, 1991), pp. 128–134.]* ▪

Cultural Pluralism

Southeast Asia is a place of **cultural pluralism** in that it is inhabited by groups of people from many different backgrounds. Over the past 40,000 years, migrants have come to the region from India, the Tibetan plateau, the Himalayas, China, Southwest Asia, Japan, Korea, and the Pacific. Until very recently, many of these groups have remained distinct, partly because they lived in isolated pockets separated by rugged topography or seas. However, urbanization is now bringing many of these groups together, resulting in much peaceful cultural exchange but also some violent conflict.

> **cultural pluralism** is a term describing a situation where groups of people from many different backgrounds have lived together for a long time but have remained distinct

The Globalizing Urban and the Traditional Rural Southeast Asia's cities are some of the most diverse parts of the region, but in some ways this diversity is decreasing as many groups are exposed to each other and to global culture. Free market capitalism, communism, nationalism, consumerism, and environmentalism have all modified life and landscapes. This is especially true in urban areas, where their appeal often cuts across societal divisions. For example, Malaysian teens of many different ethnicities (Chinese, Tamil, Malay, Bangladeshi) spend much of their spare time following the same European soccer teams, playing the same video games, visiting the same shopping centers, eating the same fast food, and talking to each other in English.

Meanwhile, rural areas retain more traditional influences—in language, for example. While one main language usually dominates trade and politics in a city, dozens of different languages may be spoken in rural areas. Indeed, of the world's 6000 or so still actively spoken languages, 1000 can be found in this region, mainly in rural areas and small towns. Today, many linguistic and other cultural barriers that divide groups are falling as more people move into the cities.

The Overseas Chinese One group that is prominent beyond its numbers in Southeast Asia is the Overseas (or ethnic) Chinese (see Chapter 9, pages 427–428). Small groups of traders from southern and coastal China have been active in Southeast Asia for thousands of years, and over the centuries, there has been a constant trickle of immigrants from China. The ancestors of most of today's Overseas Chinese, however, began to arrive in large numbers during the nineteenth century, when the European colonizers

FIGURE 10.24 Overseas Chinese in Malaysia. Women of Chinese descent light incense sticks as a form of worship at a Chinese temple in Kuala Lumpur, Malaysia.

needed labor for their plantations and mines. Later, many of those who fled China's Communist Revolution after 1949 sought permanent homes in Southeast Asian trading centers. Today, more than 26 million Overseas Chinese live and work in Southeast Asia (Figure 10.24); they tend to be shopkeepers or run other types of small businesses, though some are wealthy financiers, and a significant number are still engaged in agricultural labor.

Chinese commercial and business activity throughout the region has reinforced the perception that the Chinese are diligent, clever, and extremely frugal, often working very long hours. At the same time, despite the fact that most have modest incomes and lead quiet lives, the Chinese in Southeast Asia have the reputation of being rich and influential in government and commerce. The Overseas Chinese are indeed industrious, and with their region-wide connections and access to start-up money, they have been well positioned to take advantage of the new growth sectors in the modernizing and globalizing economies of the region. Sometimes externally funded, new Chinese-owned enterprises have put out of business older, more traditional establishments depended upon by local people of modest incomes (both ethnic Malay and ethnic Chinese).

Many low- and middle-income Southeast Asians who were hurt by the financial crisis of the late 1990s blamed their problems on the Overseas Chinese. A wave of violence resulted, with Chinese people assaulted, their temples desecrated, and their homes and businesses destroyed. Conflicts involving the Overseas Chinese have occurred in Vietnam, Malaysia, and many places in Indonesia (Sumatra, Java, Kalimantan, and Sulawesi) as well. Some Overseas Chinese (occasionally aided by projects funded by the Chinese government) have attempted to diffuse tensions through public education about Chinese culture and the historical contributions made by the Chinese to the Southeast Asian countries in which they live. Others, to show their civic awareness, are financing economic and social aid projects to help their poorer neighbors, usually of local ethnic origins (Malay, Thai, Indonesian). Still others, following the strategies of Overseas Chinese through the ages, are creating global business and financial networks that are designed in part to provide a refuge should anti-Chinese sentiment again result in violence.

Gender Patterns in Southeast Asia

Gender roles are being transformed throughout Southeast Asia by urbanization and the changes it brings to family organization and employment. Moving to the city shifts people away from extended families and toward the nuclear family. Women have made significant gains in political empowerment and in education, but they still lag well behind men.

Family Organization, Traditional and Modern Throughout the region, it has been common for a newly married couple to reside with, or close to, the wife's parents. Along with this custom is a range of behavioral rules that empower the woman in a marriage, despite some basic patriarchal attitudes. For example, a family is headed by the oldest living male, usually the wife's father. When he dies, he passes on his wealth and power to the husband of his oldest daughter, not to his own son. (A son goes to live with his wife's parents and inherits from them.) Hence, a husband may live for many years as a subordinate in his father-in-law's home. Instead of the wife being the outsider, subject to the demands of her mother-in-law—as is the case, for example, in South Asia—it is the husband who must show deference. The inevitable tension between the wife's father and the son-in-law is resolved by the custom of ritual avoidance—in daily life they simply arrange to not encounter each other much. The wife manages communication between the two men by passing messages and even money back and forth. Consequently, she has access to a wealth of information crucial to the family and has the opportunity to influence each of the two men.

Urbanization has brought a shift to the nuclear family that has transformed these traditional relationships. Young couples now frequently live apart from the extended family, an arrangement that takes the pressure off the husband in daily life. Because this nuclear family unit is often dependent entirely on itself for support, wives usually work for wages outside the home. Although married women lose the power they would have if they lived among their close kin, they are empowered by the opportunity to have a career and an income. The main drawback of this compact family structure, as many young families have discovered in Europe and the United States, is that there is no pool of relatives available to help working parents with child care and housework. Further, no one is left to help elderly parents maintain the rural family home.

 Personal Vignette Buaphet Khuenkaew, age 35, lives in Ban Muang Wa, a village near the northern Thai city of Chiang Mai. She married at 18 and has two

FIGURE 10.25 Sharing the morning meal. Buaphet Khuenkaew with her family before setting off on her motor scooter to her job in Chiang Mai, Thailand.

children: a son, 10, and a daughter, 17 (**Figure 10.25**). A Buddhist with a sixth-grade education, Buaphet is both a homemaker and a seamstress. Six days a week she drives the family motor scooter 30 minutes to her job in Chiang Mai, where she sews buttonholes in men's shirts for 2800 baht (U.S.$118) per month. The children perform weekday household chores when they return from school.

Buaphet's husband, Boontham, is a farmer who is about 5 years older than she. Although Boontham believes that men are rightly regarded as superior in Thai society, Buaphet reports that she and her husband have an egalitarian marriage in which all decisions are made jointly. As is common for married couples in the region, they do not spend much of their leisure time together. She regularly spends time with her female friends and relatives, and he with his male friends and family. Buaphet says she is happy with her life, but also regularly complains about not having the appliances and up-to-date furnishings her friends have. [*Source: Adapted from Faith D'Aluisio and Peter Menzel,* Women in the Material World *(San Francisco: Sierra Club Books, 1996), pp. 228–239.]* ∎

Political and Economic Empowerment of Women Women have made some impressive gains in politics in Southeast Asia. Economically, they still earn less money than men and work less outside the home, but this will likely change if their level of education in relation to that of men continues to increase (see Table 10.3 on page 470).

Southeast Asia has had several prominent female leaders over the years, most of whom have risen to power in times of crisis as the leaders of movements opposing corrupt or undemocratic regimes. In the Philippines, Corazon Aquino, a member of a large and powerful family, became president in 1986 after leading the opposition to Ferdinand Marcos, whose 21-year reign was infamous for its corruption and authoritarianism. Gloria Macapagal-Arroyo, from another powerful family, became president in 2001 after opposing a similarly corrupt president, and then she was accused of corruption herself.

In Indonesia, Megawati Sukarnoputri became president in 2001 after decades of leading the opposition to Suharto's notoriously corrupt 31-year reign. And in Burma, for more than two decades, opposition to the military dictatorship has been led by a woman, Aung San Suu Kyi. All of these women were wives or daughters of powerful political leaders, which raises some questions of nepotism. However, family favoritism cannot account for the several countries where the percentage of female national legislators is well above the world average of 18 percent: Vietnam (26 percent), Laos (25 percent), Singapore (24 percent), and the Philippines (20 percent).

Despite their successes in politics and their acknowledged role in managing family money, women still lag well behind men in economic empowerment in the wider society. Throughout the region, men have a higher rate of employment outside the home than women. But changes may be on the way. In Brunei, Malaysia, the Philippines, and Thailand, significantly more women than men are completing training beyond secondary school (Table 10.3, columns IV and V). Hence, if training qualifications were the sole consideration for employment, women would appear to have an advantage over men. This advantage may be significant if service sector economies, which generally require more education, become dominant in more countries. The service economy already dominates in the Philippines and Singapore.

📹 **318. Thailand's "Third Sex" Wants Acceptance, Legal Support**

Globalization and Gender: The Sex Industry

Southeast Asia has become one of several global centers for the sex industry, supported in large part by international visitors willing to pay for sex. **Sex tourism** in Southeast Asia grew out of the sexual entertainment industry that served foreign military troops stationed in Asia during World War II, the Korean War, and the Vietnam War. Now, primarily civilian men arrive from around the globe to live out their fantasies during a few weeks of vacation. The industry is found throughout the region but is most prominent in Thailand. In 2008, fourteen million tourists visited Thailand alone, up from 250,000 in 1965, and some observers estimate that as many as 70 percent were looking for sex. Even though the industry is officially illegal, some Thai government officials have even publicly praised sex tourism for its role in helping the country weather the economic crisis of 1997, because it created jobs. Some corrupt officials also support sex tourism because it provides them with a source of untaxed income from bribes.

> **sex tourism** the sexual entertainment industry that services primarily men who travel for the purpose of living out their fantasies during a few weeks of vacation

One result of the "success" of sex tourism is a high demand for sex workers, and this demand has attracted organized crime (see **Thematic Concepts Part R** on page 437). Estimates of the numbers of sex workers vary from 30,000 to more than a million in Thailand alone. Once they have been first tricked into the trade, girls and women are coerced by gangs into remaining in sex work. Demographers estimate that 20,000 to 30,000 Burmese girls taken against their will—some as young as 12—are working in Thai brothels. Their wages are too low to enable them to buy their own freedom. In the course of their work, they must service more than 10 clients per day, and they are routinely exposed to physical abuse and sexually transmitted diseases, especially HIV-AIDS.

TABLE 10.3 Gender comparisons for Southeast Asia and selected countries: Income and education level differentials

HDI rank	Country Southeast Asia	Estimated earned income, female, 2007 (PPP U.S.$)	Estimated earned income, male, 2007 (PPP U.S.$)	Female earned income as a percent of male earned income, 2007	Combined gross enrollment ratio in education (all levels), 2007 Percent female	Combined gross enrollment ratio in education (all levels), 2007 Percent male
30	Brunei	36,838	62,631	59	79.1	76.5
138	Burma (Myanmar)	640	1043	61	No data	No data
137	Cambodia	1465	2158	68	54.8	62.1
111	Indonesia	2263	5163	44	66.8	69.5
133	Laos	1877	2455	77	54.3	64.8
66	Malaysia	7972	18,886	42	73.1	69.8
105	Philippines	2506	4293	58	81.6	77.8
23	Singapore	34,554	64,656	53	No data	No data
87	Thailand	6341	10,018	63	79.6	76.6
162	Timor-Leste	493	934	53	62.1	64.2
116	Vietnam	2131	3069	69	60.7	63.9
	Selected countries for comparison					
13	United States	34,996	56,536	62	96.9	88.1
53	Mexico	8375	20,107	42	79.0	81.5
92	China	4323	6375	68	68.5	68.9

Source: United Nations Human Development Report 2009 (New York: United Nations Human Development Programme), Table J.

Personal Vignette

Twenty-five-year-old Watsanah K. (not her real name) awakens at 11:00 every morning, attends afternoon classes in English and secretarial skills, and then goes to work at 4:00 P.M. in a bar in Patpong, Bangkok's red light district. There she will meet men from Europe, North America, Japan, Taiwan, Australia, Saudi Arabia, and elsewhere, who will pay to have sex with her. She leaves work at about 2:00 A.M., studies for a while, and then goes to sleep.

Watsanah was born in northern Thailand to an ethnic minority group who are poor subsistence farmers. She married at 15 and had two children shortly thereafter. Several years later, her husband developed an opium addiction. She divorced him and left for Bangkok with her children. There, she found work at a factory that produced seat belts for a nearby automobile plant. In 1997, Watsanah lost her job as the result of the economic crisis that ripped through Southeast Asia. To feed her children, she became a sex worker.

Although the pay, between U.S.$400 and U.S.$800 a month, is much better than the U.S.$100 a month she earned in the factory, the work is dangerous and demeaning. Sex work, though widely practiced and generally accepted in Thailand, is illegal, and the women who do it are looked down on. As a result, Watsanah

must live in constant fear of going to jail and losing her children. Moreover, she cannot always make her clients use condoms, which puts her at high risk of contracting AIDS or other sexually transmitted diseases. "I don't want my children to grow up and learn that their mother is a prostitute," says Watsanah. "That's why I am studying. Maybe by the time they are old enough to know, I will have a respectable job." [Sources: Adapted from the field notes of Alex Pulsipher and Debbi Hempel, 2000; coverage of the HIV-AIDS conference in Thailand, July 11–16, 2004, by the Kaiser Family Foundation, at http://www.kaisernetwork.org/aids2004/kffsyndication.asp?show=guide. html; Life as a Thai Sex Worker, BBC News, February 22, 2007, at http://news.bbc.co.uk/2/hi/asia-pacific/6360603.stm.] ∎

THINGS TO REMEMBER

1. The major religious traditions of Southeast Asia include Hinduism, Buddhism, Confucianism and Taoism, Islam, Christianity, and animism. All originated outside the region, with the exception of the animist belief systems.

2. The religious pluralism of the region indicates that it has been settled by people with different cultural roots.

3. The Overseas Chinese are now prominent in commerce and technology or are small business owners; many first came to the region as agricultural workers.

4. Gender roles are shifting in this region as more women work outside the home. In some areas, women's economic and political empowerment builds on cultural traditions that give special responsibilities to women. However, some countries have also become centers for the global "sex trade," which puts many women at risk of kidnapping and disease.

Reflections on Southeast Asia

Southeast Asia is a geographically complex region that is often held up as a model for other developing regions for its "miraculous" economic development. While the economic success of this region is indeed remarkable, so are the environmental and social crises that have accompanied this success. The enormous amounts of carbon released into the atmosphere by deforestation have put this region at the center of efforts to deal with global climate change. Indigenous peoples, threatened by rapacious deforestation, are using new technologies to chart effective strategies for challenging corporations seeking to appropriate their resources. Many cities are also facing crises as changes in rural food production systems, combined with the growth of employment in urban-based, export-oriented manufacturing, have created a massive tide of rural-to-urban migrants. The region's largest cities have found their infrastructures swamped by ever-increasing demands for water, housing, and sanitation.

Yet, in the face of these challenges, there have been many positive changes. Population growth is slowing due to urbanization, economic growth, and the success of some government policies to reduce birth rates. Despite the continuation of military dictatorships or undemocratic rule in a number of countries, democracy is strengthening throughout the region. The bright side of the economic crisis of the 1990s was that it exposed corruption and brought reform to both Thailand and Indonesia. The global recession that began in 2008 hit the region later than elsewhere and seems to be waning more quickly. Even on issues of gender, there are signs of change. Women's political empowerment and access to education are increasing, and greater attention is being paid to the abuse of women in the region's huge sex industry.

Looking toward the future, China looms ever larger in trading relationships of this region. As the China–Southeast Asian free trade agreement comes into effect in 2010, this region is under greater pressure to strengthen the areas where it can have an edge over China—high-value, technologically sophisticated industries. Otherwise, Southeast Asia will be stuck playing second fiddle to its much larger neighbor, able to attract investment only when wages and other production costs in China rise beyond those in Southeast Asia. And yet, trade with China also offers many opportunities, and Southeast Asian companies are already profiting from some of them. Whether the majority of Southeast Asians can similarly benefit from China's growth is a key question for this region's future.

Critical Thinking Questions

1. What do you think are the most serious threats to Southeast Asia posed by global climate change? What are the most promising responses to global climate change emerging from this region?

2. Which countries in the region have had the greatest success in controlling population growth? How do population issues differ among Southeast Asian countries? How do the population issues of the region as a whole compare with those of Europe or Africa?

3. How is urbanization influencing gender roles in families?

4. Describe the changes in agriculture that have produced migration toward cities.

5. In what ways did state aid to market economies and export-led growth amount to a strategic and limited embrace of globalization?

6. How did the economic crisis of the 1990s result in expanded democracy in some countries? Which countries in this region are the least democratic? Why do you think that is the case?

7. Describe the spatial distribution of major religious traditions of this region. How have these religious traditions influenced each other over time?

8. What factors have produced major flows of refugees in Southeast Asia?

9. Gender roles and mating relationships in traditional Southeast Asian families are not necessarily what an outsider might expect. Describe the characteristics of these relationships that interested you most. Why? Discuss some of the ways in which gender roles vary from those in South Asia or East Asia.

10. How has the role of the Overseas Chinese evolved over time? Why have some people in this region resented them? To what extent is this resentment understandable, if unfair? Where have the Overseas Chinese been most and least sensitive to public opinion?

Chapter Key Terms

archipelago, 434

Association of Southeast Asian Nations (ASEAN), 451

Australo-Melanesians, 445

Austronesians, 446

coral bleaching, 445

crony capitalism, 451

cultural pluralism, 467

detritus, 439

domino theory, 448

El Niño, 439

feminization of labor, 450

foreign exchange, 462

resettlement schemes, 456

sex tourism, 469

CHINA
Nanjing
Shanghai
Hangzhou
Changsha
Nanchang
JAPAN

Guangzhou (Canton)
Taipei
Taichung
TAIWAN
Hong Kong
Kaohsiung

South China Sea

PHILIPPINES
Manila

BRUNEI DARUSSALAM
Bandar Seri Begawan
MALAYSIA

Borneo

Sulu Sea

Celebes Sea

Sulawesi (Celebes)

Java

Bali

INDONESIA

Surabaya

Dili
TIMOR-LESTE

Banda Sea

Arafura Sea

Timor Sea

Northern Mariana Islands (U.S.)

Saipan
Tinian
Agana
Guam (U.S.)

Philippine Sea

Yap

PALAU
Ngerulmud

Caroline Islands

A Australia's Eastern Highlands, Blue Mountains

FEDERATED STATES OF MICRONESIA

Pohnpei
Truk
Palikir
Kosrae

MARSHALL ISLANDS
Majuro

M i c r o n e s i a

M e l a n e s i a

NAURU
Yaren

Tarawa
KIRIBATI (GILBERT ISLANDS)

Phoenix Islands

West Papua
Jayapura
New Guinea
Admiralty Islands
Wewak
Bismarck Archipelago
Kavieng
New Ireland
Rabaul

PAPUA NEW GUINEA
Salamaua
Daru
Port Moresby

Buka
Bougainville
New Britain

SOLOMON ISLANDS
Malaita
Honiara

TUVALU (ELLICE ISLANDS)
Funafuti

Tokelau (New Zealand)

Wallis & Futuna (France)
Mata Utu

SAMOA (WESTERN SAMOA)
Apia
Pago Pago
American Samoa (U.S.)
Alofi
Niue (New Zealand)

INDIAN OCEAN

Darwin
Jabiru
Katherine
Daly Waters

King Leopold Ranges
Derby

Hamersley Range

Carnarvon

WESTERN AUSTRALIA

Geraldton

Perth

Great Sandy Desert

Gibson Desert

Great Victoria Desert

Nullarbor Plain

Albany
Esperance

NORTHERN TERRITORY
Tennant Creek
Macdonnell Ranges
Alice Springs
Uluru (Ayers Rock)
B Musgrave Ranges

Thursday Island
Torres Strait
Cape York Peninsula
Cooktown
Cairns

Gulf of Carpentaria

Selwyn Range

QUEENSLAND

Great Artesian Basin

AUSTRALIA

SOUTH AUSTRALIA

Flinders Range
Broken Hill
Elliston
Adelaide

Great Australian Bight

INDIAN OCEAN

NEW SOUTH WALES

Great Dividing Range

Coral Sea Islands (Australia)

E

Coral Sea

VANUATU (NEW HEBRIDES)
Port-Vila

FIJI
Vanua Levu
Viti Levu
Suva

Mackay

New Caledonia (France)
Noumea

TONGA
Nuku'alofa

Brisbane

Darling

Murray

Great Dividing Range

Canberra
Australian Alps
Sydney
A

VICTORIA
Melbourne

Launceston
TASMANIA
Hobart

B Australian Desert

Norfolk (Australia)
Kingston

Kermadec Islands (New Zealand)

Tasman Sea

NEW ZEALAND

Auckland
North Island
Gisborne

Westport
South Island
Wellington
Christchurch
Southern Alps
D
Dunedin
Invercargill

Chatham Islands (New Zealand)

Bounty Islands (New Zealand)

Auckland Islands (New Zealand)

Antipode Islands (New Zealand)

Campbell Island (New Zealand)

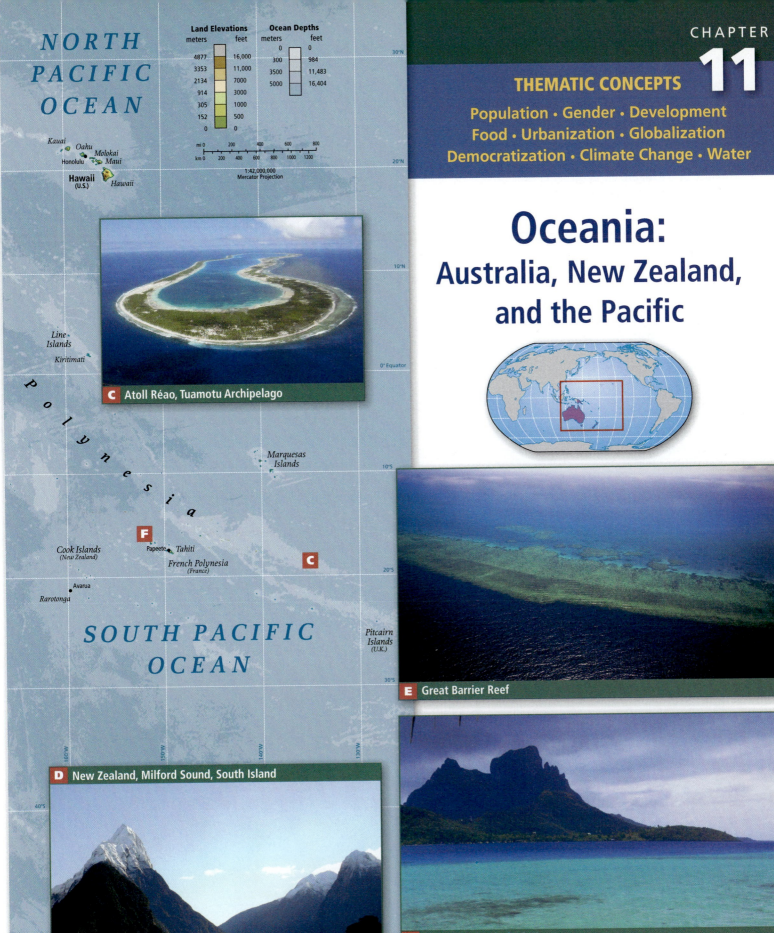

NORTH PACIFIC OCEAN

Land Elevations

meters	feet
4877	16,000
3353	11,000
2134	7000
914	3000
305	1000
152	500
0	0

Ocean Depths

meters	feet
0	0
300	984
3500	11,483
5000	16,404

mi 0 200 400 600 800
km 0 200 400 600 800 1000 1200

1:42,000,000
Mercator Projection

Kauai
Oahu
Molokai
Honolulu Maui
Hawaii (U.S.) Hawaii

Line Islands
Kiritimati

C Atoll Réao, Tuamotu Archipelago

Polynesia

Marquesas Islands

F

Cook Islands (New Zealand)
Papeete Tahiti
French Polynesia (France)

C

Avarua
Rarotonga

SOUTH PACIFIC OCEAN

Pitcairn Islands (U.K.)

D New Zealand, Milford Sound, South Island

THEMATIC CONCEPTS

Population · Gender · Development
Food · Urbanization · Globalization
Democratization · Climate Change · Water

Oceania:
Australia, New Zealand, and the Pacific

E Great Barrier Reef

F The High Island of Tahiti

FIGURE 11.1 Regional map of Oceania.

473

Global Patterns, Local Lives In October 2003 in Brisbane, Australia, a crowd of 47,000 waited eagerly for the Haka as New Zealand's national rugby team, the All Blacks, took the field opposite the team from Tonga, a Polynesian archipelago in the South Pacific (**Figure 11.1**).

The Haka is a highly emotional and physical dance traditionally performed by the **Maori,** the indigenous (Polynesian) people of New Zealand, to motivate fellow warriors and intimidate opponents before entering battle. Dances like this have long been a part of many cultures in the islands of Oceania (**Figure 11.2A**), but the Haka has now become an integral part of rugby, the region's most popular sport. Before almost every international match for the past century, the All Blacks have performed the Haka in unison: chanting, screaming, jumping, stomping their feet, poking out their tongues, widening their eyes to show the whites, and beating their thighs, arms, and chests.

Maori Polynesian people indigenous to New Zealand

Until 2003, only the New Zealand team performed the Haka, but that night it was a different story. Halfway through New Zealand's Haka, the Tongan team responded with its own Haka. The crowd roared its approval of this scene, a revival of the traditional prelude to battle throughout Oceania's long history.

The Haka has now spread in the world of rugby and beyond. Outside of Oceania, those who perform the Haka include the rugby teams at Jefferson High in Portland, Oregon, and Middlebury College in Vermont (see **Figure 11.2B**), and the football teams at Brigham Young University and the University of Hawaii. All of these teams have players who are of Polynesian heritage. Most practitioners speak of the Haka as filling them with the necessary exuberance, aggression, and spirituality to play a vigorous and successful game.

To see videos of a Haka, go to http://www.YouTube.com and type in "haka." [Adapted from articles by Phil Wilkins, "Tonga Can Only Match the Kiwis in the Haka," Brisbane, Australia, October 25, 2003; and Mark Falcous, "The Decolonizing National Imaginary: Promotional Media Constructions During the 2005 Lions Tour of Aotearoa," New Zealand, Journal of Sport and Social Issues, November 2007, Vol. 31, No. 4, pp. 374–393.] ■

The Haka is an example of how, in the postcolonial modern era, indigenous culture in Oceania is being revived, celebrated, and appropriated, in this case by those who wish to project a multi-cultural national image for New Zealand. And yet the Haka itself seems to be a globalizing phenomenon. The fact that a Maori war dance is being performed before a game of rugby, which was brought to Oceania by the British (**Figure 11.3**), shows how deeply globalization has penetrated this region. Oceania, which comprises Australia, New Zealand, Papua New Guinea, and the myriad Pacific islands, was dominated politically and economically by people of European descent for more than 200 years. Now, Oceania finds itself subject to a new wave of globalization emanating from Asia—a situation that has far-reaching economic implications.

The nine **Thematic Concepts** covered in this chapter are illustrated on pages 478–479 with photo essays that show examples of how each concept is experienced in this region. Interactions between two or more of these concepts are shown with explanatory captions.

THINGS TO REMEMBER

1. For more than two centuries, European peoples dominated Oceania economically, politically, and to some extent, culturally.

2. Today, European influence is challenged by reinvigorated native traditions and by economic globalization.

3. The Haka today is an example of globalization—the merging of a Maori war dance with the European game rugby—that is spreading beyond Oceania.

(A)

(B)

FIGURE 11.2 The Haka, a Maori tradition. **(A)** A Haka performed by Maori in Christchurch, New Zealand. **(B)** The Middlebury College men's rugby team performing their Haka before a match in Middlebury, Vermont.

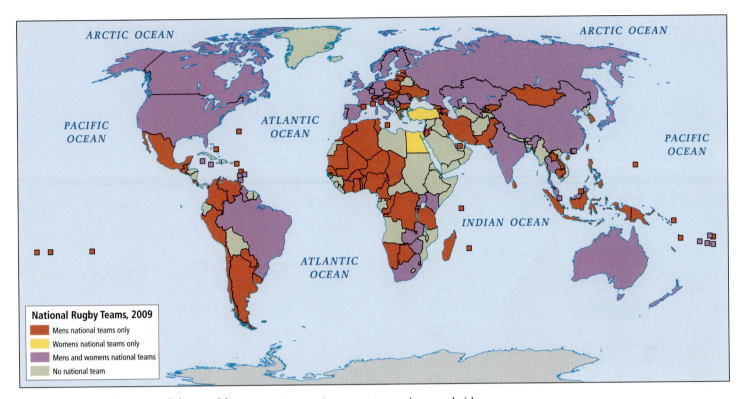

FIGURE 11.3 Rugby around the world. In over 136 countries, women, men, boys, and girls play rugby: 136 countries have men's national rugby teams, and 58 countries have women's national teams. The men's World Cup rugby competition began in 1987 and the women's World Cup competition began in 1991. In April 2010, New Zealand had the top team for both men and women, but the ranking of the men's teams can change weekly. [For additional information, see "http://www.irb.com/EN/IRB1Organisation/"http://www.irb.com/EN/IRB1Organisation/.]

I THE GEOGRAPHIC SETTING

Oceania is made up of Australia, New Zealand, Papua New Guinea, and the many small islands scattered across the Pacific Ocean (see Figure 11.4 on page 476). It is a unique world region in that it is composed primarily of ocean and covers the largest portion of the earth's surface of any region, yet has only 36 million people, the vast majority of whom live in Australia (21.9 million), Papua New Guinea (6.6 million), and New Zealand (4.3 million).

Terms in This Chapter

Some maps in this chapter show different place names in similar locations. This reflects the political evolution of this region, where some islands previously were grouped under one name that remains in use but are now regrouped into country units with another name. For example, the Caroline Islands, located north of New Guinea, still go by that name on maps and charts, but they have been divided into two countries: Palau (a small group of islands at the western end of the Caroline Islands) and the Federated States of Micronesia, which extends over 2000 miles west to

east, from Yap to Kosrae. In addition, there are three commonly used names for island groupings—Micronesia, Melanesia, and Polynesia—that are not political units, but are based on ancient ethnic and cultural links (see Figure 11.13 on page 490).

Physical Patterns

The huge expanses of the Pacific Ocean in Oceania serve as both a link and a barrier in this region. For some living things, the Pacific links widely separated lands. Plants and animals have found their way from island to island by floating on the water, swimming, or flying, and humans have long used the ocean as a way to make contact with other peoples for visiting, trading, and raiding. Note too that in modern times some animals—brown snakes in Guam, for example—have managed to sneak on board aircraft and arrive at new places where they have wreaked havoc on local bird populations. Sea life is a source of food, and today it is also a source of income, as catches are sold in the global market. Also, the movement of the water in the Pacific, with its varying temperatures,

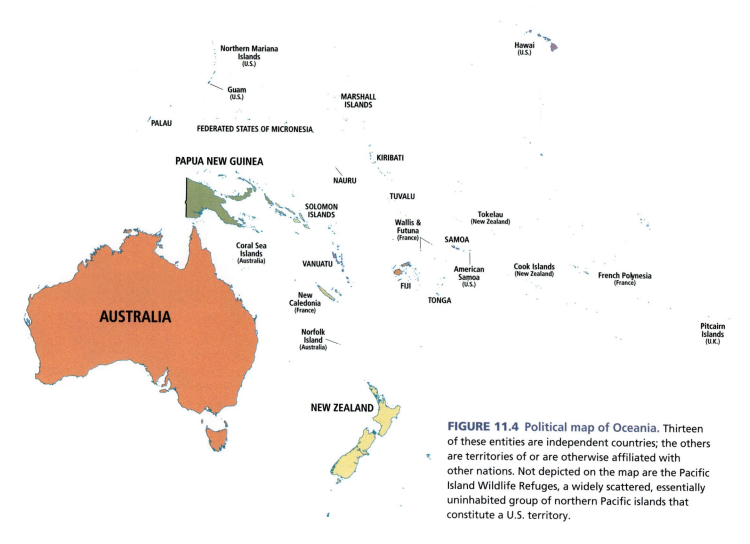

FIGURE 11.4 Political map of Oceania. Thirteen of these entities are independent countries; the others are territories of or are otherwise affiliated with other nations. Not depicted on the map are the Pacific Island Wildlife Refuges, a widely scattered, essentially uninhabited group of northern Pacific islands that constitute a U.S. territory.

influences climates across all the landmasses of the Pacific Rim. But in its role as a barrier, the wide expanses of water have also profoundly limited the natural diffusion of plant and animal species and kept Pacific Islanders relatively isolated from one another. The vast ocean has imposed solitude and fostered self-sufficiency and subsistence economies well into the modern era.

Continent Formation

The largest landmass in Oceania is the ancient continent of Australia at the southwestern perimeter of the region (see **Figure 11.1 Map** on page 472). The Australian continent is partially composed of some of the oldest rock on earth and has been relatively stable for more than 200 million years, with very little volcanic activity and only an occasional mild earthquake. Australia was once a part of the great landmass, called **Gondwana,** which formed the southern part of the ancient supercontinent Pangaea (see Figure 1.26 on page 51). What became present-day Australia broke free from Gondwana and drifted until it eventually collided with the part of the Eurasian Plate on which Southeast Asia sits.

That impact created the mountainous island of New Guinea to the north of Australia.

Australia is shaped roughly like a dinner plate with a lumpy, irregular rim and two bites taken out of it: one in the north (the Gulf of Carpentaria) and one in the south (the Great Australian Bight). The center of the plate is the great lowland Australian desert, with only two hilly zones and rocky outcroppings (see **Figure 11.1B** on page 472). The lumpy rim of Australia is composed of uplands; the highest and most complex of these are the long, curving Eastern Highlands (labeled "Great Dividing Range" on the map in **Figure 11.1** and shown in **Photo A**). Over millennia, the forces of erosion—both wind and water—have worn most of Australia's landforms into low, rounded formations, some, like Uluru, quite spectacular (**Figure 11.5**).

Off the northeastern coast of the continent lies the **Great Barrier Reef,** the largest coral reef in the world and a World Heritage Site since 1981 (see **Figure 11.1E**). It stretches in an irregular arc for more than 1250 miles (2000 kilometers) along the coast of Queensland, covering 135,000 square miles (350,000 square kilometers). The Great Barrier Reef is so large that it influences Australia's climate. It interrupts the westward-flowing ocean currents in the mid–South Pacific circulation pattern, shunting warm water to the south, where it warms the southeastern coast

Gondwana the great landmass that formed the southern part of the ancient supercontinent Pangaea

Great Barrier Reef the longest coral reef in the world, located off the northeastern coast of Australia

FIGURE 11.5 Uluru (also known as Ayers Rock). This land formation is a smooth remnant of ancient mountains. The site is held sacred by central Australian Aborigines. It is also among Australia's most popular tourist destinations.

of Australia. Threats to the health of the Great Barrier Reef are discussed on page 483.

Island Formation

The islands of the Pacific were created (and are being created still) by a variety of processes related to the movement of tectonic plates. The islands found in the western reaches of Oceania—including New Guinea, New Caledonia, and the main islands of Fiji—are remnants of the Gondwana landmass; they are large, mountainous, and geologically complex. Other islands in the region are volcanic in origin and form part of the Ring of Fire (see Figure 1.27 on page 52). Many of this latter group are situated in boundary zones where tectonic plates are either colliding or pulling apart. For example, the Mariana Islands east of the Philippines are volcanoes that were formed when the Pacific Plate plunged beneath the Philippine Plate. The two much larger islands of New Zealand were created when the eastern edge of the Indian-Australian Plate was thrust upward by its convergence with the Pacific Plate.

The Hawaiian Islands were produced through another form of volcanic activity associated with **hot spots,** places where particularly hot magma moving upward from Earth's core breaches the crust in tall plumes. Over the past 80 million years, the Pacific Plate has moved across one of these hot spots, creating a string of volcanic formations 3600 miles (5800 kilometers) long. The youngest volcanoes, only a few of which are active, are on or near the islands known as Hawaii.

Volcanic islands exist in three forms: volcanic high islands, low coral atolls, and coral platforms raised or uplifted by volcanism known as *makatea*. High islands are usually volcanoes that rise above the sea into mountainous, rocky formations that contain a rich variety of environments. New Zealand, the Hawaiian Islands, Tahiti, and Easter Island are examples of high islands (see Figure 11.1D, F on page 473). An **atoll** is a low-lying island, or chain of islets, formed of coral reefs that have built up on the circular or oval rim of a submerged volcano (see Figure 11.1C on page 473). These reefs are arranged around a central lagoon that was once the

volcano's crater. As a consequence of their low elevation, atoll islands tend to have only a small range of environments and very limited supplies of fresh water.

Climate

Although the Pacific Ocean stretches nearly from pole to pole, most of Oceania is situated within the tropical and subtropical latitudes of that ocean. The tepid water temperatures of the central Pacific bring mild climates year-round to nearly all the inhabited parts of the region (Photo Essay 11.1 on page 480). The seasonal variation in temperature is greatest in the southernmost reaches of Australia and New Zealand.

Moisture and Rainfall With the exception of the vast arid interior of Australia, much of Oceania is warm and humid nearly all the time. New Zealand and the high islands of the Pacific receive copious rainfall and once supported dense forest vegetation, although much of that forest is gone after 1000 years of human impact (see Photo Essay 11.2D on page 484). Travelers approaching New Zealand, either by air or by sea, sometimes notice a distinctive long, white cloud that stretches above the north island. A thousand years ago, the Maori settlers also noticed this phenomenon, and they named that place *Aotearoa,* "land of the long white cloud," a name that is now applied to the whole country.

The distinctive mass of moisture that is represented by Aotearoa is brought in by the legendary **roaring forties** (named for the 40th parallel south), powerful air and ocean currents that speed around the far Southern Hemisphere virtually unimpeded by landmasses. These westerly winds (blowing west to east) deposit a drenching 130 inches (330 centimeters) of rain per year in the New Zealand highlands and more than 30 inches (76 centimeters) per year on the coastal lowlands. At the southern tip of New Zealand's

hot spots individual sites of upwelling material (magma) that originate deep in the mantle of the earth and surface in a tall plume; hot spots tend to remain fixed relative to migrating tectonic plates

atoll a low-lying island, formed of coral reefs that have built up on the circular or oval rims of a submerged volcano

roaring forties powerful air and ocean currents at about 40° S latitude that speed around the far Southern Hemisphere virtually unimpeded by landmasses

Globalization and Development: A long-term shift in patterns of globalization and development is well under way in Oceania. While Europe and the United States have exerted a profound influence over this region for over two centuries, Oceania is now reorienting toward Asia. The region's two largest economies, Australia and New Zealand, now trade primarily with Japan and China, two countries that are also increasingly important to Papua New Guinea and Pacific island economies.

(A) The world's largest coal port at Newcastle, Australia, where coal is loaded onto ships bound for Japan and China.

(B) A boat in Lyttleton, New Zealand, is loaded with logs destined for Japan. Most New Zealand logs go to Japan or South Korea.

(C) Japanese tourists ride a "banana boat" in Guam. Japan dominates the tourist trade throughout much of the Pacific.

Climate Change: Sea level rise threatens many Pacific islands with submersion, and water supplies throughout the region may be strained as rainfall patterns shift and temperatures rise. Higher ocean temperature and acidification (due to increasing carbon dioxide in the atmosphere) present a potentially devastating threat to fishing throughout the region. Numerous responses to the climate crisis are emerging, with New Zealand establishing itself as a world leader in the generation of renewable energy.

(D) Small-scale fishers in Micronesia depend on reefs threatened by rising ocean temperatures and acidification.

(E) A bushfire near Melbourne, Australia, that killed 173 people in 2009. Fire risk increases as temperatures rise.

(F) The Te Apiti wind farm in New Zealand, a country that plans to use renewable energy to supply 95 percent of its electricity by 2025.

Population, Development, and Urbanization: Two patterns are emerging in Oceania. While Australia, New Zealand, and Hawaii are prosperous, highly urbanized, and have slow-growing and increasingly elderly societies, the rest of the region is generally more rural, poorer, and has younger and often more rapidly growing populations.

(G) A mother and grandmother in a park in Sydney, Australia. Australia and New Zealand have the lowest fertility rates in the region.

(H) Molokai, Hawaii's, senior gala night. Hawaii has the oldest population in Oceania with 13.3 percent of people over 65.

(I) Young people in the Solomon Islands, which is 83 percent rural and has Oceania's highest fertility rate at 4.5 children per woman.

Food Production and the Environment: This region's unique and fragile ecosystems have been transformed by food production systems brought in from outside the region, the products of which are generally exported to areas outside the region. Australia and New Zealand are among the world's largest exporters of food, and fisheries off the coast of many Pacific islands are now being intensely fished by large ships from Japan, Europe, and elsewhere.

(J) Cattle in southern Australia. Beef is Australia's most valuable agricultural product, and over 65 percent of it is exported.

(K) Sheep in Whitecliffs, New Zealand. Ranching has brought massive deforestation and pollution of streams and rivers.

(L) A Japanese-owned tuna-processing facility in Fiji, where locally caught tuna are frozen and packaged for Japanese sushi restaurants.

Gender and Democratization: Women's participation in the political process varies dramatically across the region. Australia and New Zealand were among the first countries in the world to grant women the right to vote and run for office. Women's participation in politics is much lower in Papua New Guinea and the Pacific islands, where women gained the right to vote more recently.

(M) Edith Cowan, a founder of Australia's women's suffrage movement and the country's first female parliamentarian.

(N) Helen Clark, New Zealand's second female prime minister and now head of the United Nations Development Programme.

(O) A "matai," or village chief, in Samoa. Matai are the only people allowed to run for parliament in Samoa, and very few are female.

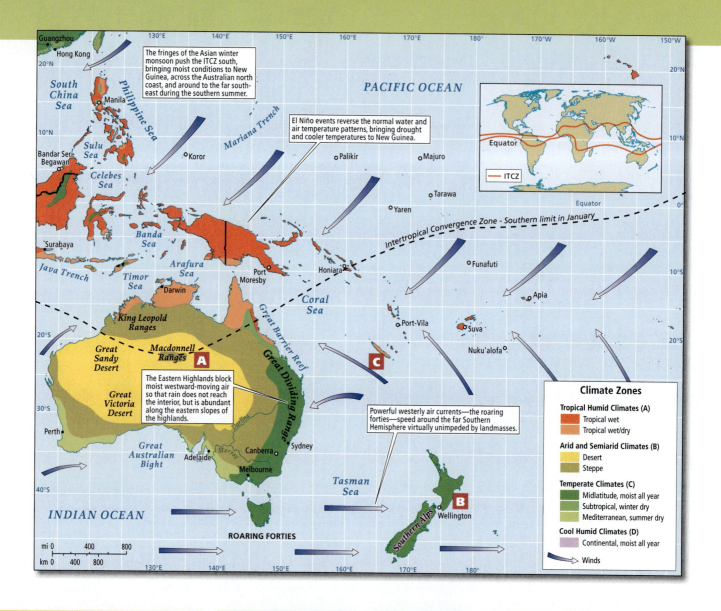

The fringes of the Asian winter monsoon push the ITCZ south, bringing moist conditions to New Guinea, across the Australian north coast, and around to the far southeast during the southern summer.

El Niño events reverse the normal water and air temperature patterns, bringing drought and cooler temperatures to New Guinea.

The Eastern Highlands block moist westward-moving air so that rain does not reach the interior, but is abundant along the eastern slopes of the highlands.

Powerful westerly air currents—the roaring forties—speed around the far Southern Hemisphere virtually unimpeded by landmasses.

PACIFIC OCEAN

Intertropical Convergence Zone - Southern limit in January

Equator

ITCZ

Climate Zones

Tropical Humid Climates (A)
- Tropical wet
- Tropical wet/dry

Arid and Semiarid Climates (B)
- Desert
- Steppe

Temperate Climates (C)
- Midlatitude, moist all year
- Subtropical, winter dry
- Mediterranean, summer dry

Cool Humid Climates (D)
- Continental, moist all year

→ Winds

mi 0   400   800
km 0   400   800

ROARING FORTIES

A Desert, Alice Springs, Australia

B Midlatitude, moist all year, New Zealand

C Tropical wet/dry, New Caledonia

North Island, the wind averages more than 40 miles per hour (64 kilometers per hour) about 118 days a year. Farmers in the area stake their cabbages to the ground so they will not blow away.

By contrast, two-thirds of the continent of Australia is overwhelmingly dry. The dominant winds affecting Australia (see the map in Photo Essay 11.1) are the north and south easterlies (blowing east to west) that converge east of the continent. The Great Dividing Range blocks the movement of moist, westward-moving air, so that rain does not reach the interior (an orographic pattern, see Figure 1.28, page 53). As a result, a large portion of Australia receives less than 20 inches (50 centimeters) of rain per year, and humans have found rather limited uses for this interior territory. But the eastern (windward) slopes of the highlands receive more abundant moisture. This relatively moist eastern rim of Australia was favored as a habitat by both the indigenous people and the Europeans who displaced them after 1800. During the southern summer, the fringes of the monsoon that passes over Southeast Asia and Eurasia bring moisture across Australia's northern coast. There, annual rainfall varies from 20 to 80 inches (50 to 200 centimeters).

Overall, Australia is so arid that it has only one major river system, which is in the temperate southeast where most Australians live. There, the Darling and Murray rivers drain one-seventh of the continent, flowing west and south into the Indian Ocean near Adelaide. One measure of the overall dryness of Australia is that the entire average *annual* flow of the Murray-Darling river system is equal to just *one day*'s average flow of the Amazon in Brazil.

In the island Pacific, mountainous high islands also exhibit orographic rainfall patterns, with a wet windward side and a dry leeward side. Low-lying islands across the region vary considerably in the amount of rainfall they receive. Some of these islands lie directly in the path of trade winds, which deliver between 60 and 120 inches of rain per year on average. These islands support a remarkable variety of plants and animals. Other low-lying islands, particularly those on the equator, receive considerably less rainfall and are dominated by grasslands that support little animal life.

El Niño Recall from Chapter 3 the *El Niño* phenomenon, a pattern of shifts in the circulation of air and water in the Pacific that occurs irregularly every 2 to 7 years. Although these cyclical shifts, or oscillations, are not yet well understood, scientists have worked out a model of how the oscillations may occur (Figure 11.6).

The El Niño event of 1997–1998 illustrates the effects of this phenomenon. By December of 1997, the island of New Guinea (north of Australia; see Figure 11.1 on page 472) had received very little rainfall for almost a year. Crops failed, springs and streams dried up, and fires broke out in tinder-dry forests. The cloudless sky allowed heat to radiate up and away from elevations above 7200 feet (2200 meters), so temperatures at high elevations dipped below freezing at night for stretches of a week or more. Tropical plants died, and people unaccustomed to chilly weather sickened. Meanwhile, along the Pacific coasts of North, Central, and South America, the warmer-than-usual weather brought unusually strong storms, high ocean surges, and damaging wind and rainfall (see Figure 11.6C). Recently, an opposite pattern, in which normal conditions become unusually strong, has been identified and named *La Niña*, though scientists have barely begun to study it.

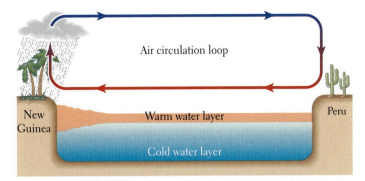

(A) Normal equatorial conditions. Water in the equatorial western Pacific (the New Guinea/Australia side) is warmer than water in the eastern Pacific (the Peru side). Due to prevailing wind patterns, the warm water piles up in the west. Warm air rises above this warm-water bulge in the western Pacific and forms rain clouds. The rising air cools and, once in the higher atmosphere, moves in an easterly direction. In the east, the dry cool air descends, bringing little rainfall to Peru.

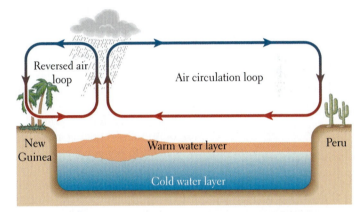

(B) Developing El Niño conditions. As an El Niño event develops, the ocean surface's warm-water bulge (orange) begins to move east. The air rising above it splits into two formations, one circulating east to west in the upper atmosphere and one west to east.

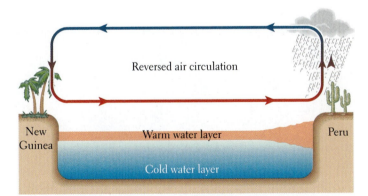

(C) Fully developed El Niño. Slowly, as the bulge of warm water at the surface of the ocean moves east, it forces the whole system into the fully developed El Niño with air at the water surface and in the upper atmosphere flowing in reverse of normal (A). Instead of warm, wet air rising over the mountains of New Guinea and condensing as rainfall, cool, dry, cloudless air descends to sit at the earth's surface. Meanwhile, in the east, the normally dry, clear coast of Peru experiences clouds and rainfall.

FIGURE 11.6 A model of the El Niño phenomenon. [Adapted from Environmental Dynamics Research, Inc., 1998; Ivan Cheung, George Washington University, Geography 137, Lecture 16, October 29, 2001.]

Fauna and Flora

The fact that Oceania comprises an isolated continent and numerous islands has affected its animal life (*fauna*) and plant life (*flora*). Many of its species are **endemic**, meaning they exist in a particular place and nowhere else on earth. This is especially true of Australia, but many Pacific islands also have endemic species.

Animal and Plant Life in Australia The uniqueness of Australia's animal and plant life is the result of the continent's long physical isolation, large size, relatively homogeneous landforms, and arid climate. Since Australia broke away from Gondwana more than 65 million years ago, its animal and plant species have evolved in isolation. One spectacular result of this long isolation is the presence of more than 144 living species of endemic marsupial animals. **Marsupials** are mammals that give birth to their young at a very immature stage and then nurture them in a pouch equipped with nipples. The best-known marsupials are the kangaroos; other species include wombats, koalas, and bandicoots. The various marsupials fill ecological niches that in other regions of the world are occupied by rats, badgers, moles, cats, wolves, ungulates (grazers), and bears. The **monotremes**, egg-laying mammals that include the duck-billed platypus and the spiny anteater, are endemic to Australia and New Guinea (**Figure 11.7**). Some of the 750 species of birds known in Australia migrate in and out, but more than 325 species are endemic.

Most of Australia's endemic plant species are adapted to dry conditions. Many of the plants have deep taproots to draw moisture from groundwater and small, hard, pale green or shiny leaves to reflect heat and to hold moisture. Much of the continent is grassland and scrubland with bits of open woodland; there are only a few true forests, found in pockets along

endemic belonging or restricted to a particular place

marsupials mammals that give birth to their young at a very immature stage and nurture them in a pouch equipped with nipples

monotremes egg-laying mammals, such as the duck-billed platypus and the spiny anteater

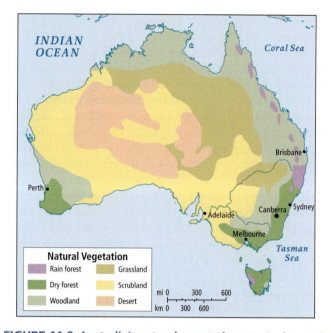

FIGURE 11.8 Australia's natural vegetation. Much of Australia is grassland and scrubland; a few forests can be found in the Eastern Highlands, in the far southwest, and in Tasmania. [Adapted from Tom L. McKnight, *Oceania* (Englewood Cliffs, N.J.: Prentice Hall, 1995), p. 28.]

the Eastern Highlands, the southwestern tip, and in Tasmania (**Figure 11.8**). Two plant genera account for nearly all the forest and woodland plants: *Eucalyptus* (450 species, often called "gum trees") and *Acacia* (900 species, often called "wattles").

Plant and Animal Life in New Zealand and the Pacific Islands Naturalists and evolutionary biologists have long been interested in the species that inhabited the Pacific islands before humans arrived. Charles Darwin formulated many of his ideas about evolution after visiting the Galápagos Islands of the eastern Pacific (see Figure 3.1 on pages 114–115) and the islands of Oceania.

Islands gain plant and animal populations from the sea and air around them as organisms are carried from larger islands and continents by birds, storms, or ocean currents. Once these organisms "colonize" their new home, they may evolve over time into new species that are unique to one island. High, wet islands generally contain more varied species because their more complex environments provide niches for a wider range of wayfarers and thus greater opportunities for evolutionary change.

The flora and fauna of islands are also modified by human inhabitants once they arrive. In prehistoric times, Asian explorers in oceangoing sailing canoes brought plants such as bananas and breadfruit, and animals such as pigs, chickens, and dogs. Today, human activities from tourism to military exercises to urbanization continue to change the flora and fauna of Oceania.

240. Breadfruit Advocates Say It Could Solve Hunger in Tropical Regions

Generally, the diversity of land animals and plants is richest in the western Pacific, near the larger landmasses. It thins out to the east, where the islands are smaller and farther apart. The natural rain forest flora is rich and abundant on New Zealand, New

FIGURE 11.7 A duck-billed platypus. The platypus is a truly unusual animal that is found in Australia and New Guinea. It is an egg-laying mammal. It has the tail of a beaver and is covered in fur, but it has a bill and webbed feet like a duck. It has powerful venom, like a reptile, which is released from a spur, not from the mouth.

Guinea, and also on the high islands of the Pacific. However, the natural fauna is much more limited on these islands. While New Guinea has fauna comparable to Australia, to which it was once connected via Sundaland (see Figure 10.6 on page 435), New Zealand and the Pacific islands have no indigenous land mammals, almost no indigenous reptiles, and only a few indigenous species of frogs. New Zealand and the islands were never connected to Australia and New Guinea by a land bridge that land animals could cross. On the other hand, indigenous birds have been numerous and varied. Two examples are New Zealand's kiwi and the huge moa (a bird that grew up to 12 feet [3.7 meters] tall) and was a major source of food for the Maori people until they hunted it to extinction before Europeans arrived. Today, New Zealand may well be the country with the most introduced species of mammals, fish, and fowl, nearly all brought in by European settlers.

THINGS TO REMEMBER

1. This largest region of the world is primarily water and has a very small population of just 37 million.

2. The largest land areas are the continent of Australia and the two island nations of New Zealand and the eastern half of New Guinea, known as Papua New Guinea.

3. The thousands of islands of the Pacific are made up of two main types of volcanic formations.

4. The climate of the region is mostly tropical and subtropical.

5. The fauna and flora of the region are unique in many distinctive ways that have informed knowledge about evolution. Partly because they have been isolated for lengthy periods, many species are endemic to a specific locale.

Environmental Issues

Oceania faces a host of environmental problems, and despite its relatively small population, public awareness of environmental issues is keen. Global climate change, primarily warming, has brought a broad array of threats to the region's ecology, as has human introduction of many nonnative species. The expansion of herding, agriculture, fishing fleets, and the human population itself has also impacted much of the region (see Photo Essay 11.2 on page 484).

Global Climate Change

Oceania is a minor contributor of greenhouse gases, except for Australia, which has some of the world's highest greenhouse gas emissions on a per capita basis (see Figure 1.25 on page 47). Much like the United States, its emissions result from the use of automobile-based transportation systems to connect a widely dispersed network of cities and towns. Also, as in the United States, heavy dependence on coal for electricity generation leads to high emissions. However, because Australia has a relatively small population (22 million people in 2010), it accounts for only slightly more than 1 percent of global emissions. Despite negligible contributions of greenhouse gases by the islands of the Pacific, they, as well as Australia, are highly vulnerable to the effects of global climate change (Photo Essay 11.3 on page 486).

Sea Level Rise As we have seen, global warming may raise sea levels by melting glaciers and ice caps. Obviously, this issue is of great concern to residents of islands that already barely rise above the waves (see Photo Essay 11.3A). If sea levels rise the 4 inches (10 centimeters) per decade predicted by the International Panel on Climate Change, many of the lowest-lying Pacific atolls, such as Tuvalu, will disappear under water within 50 years. Other islands, some with already very crowded coastal zones, will be severely reduced in area and will become more vulnerable to storm surges and cyclones.

Personal Vignette Nicholas Hakata is a community leader and youth worker on the Carteret Islands that are administered by Papua New Guinea. The islands rise less than 2 meters above the sea. Since salty seawater began washing over their cultivation plots, he and his fellow citizens have often gone hungry, subsisting on fish and coconuts. As the sea rises, standing water encourages mosquito infestations and now malaria is spreading among the children already weakened by hunger. The children used to do well in school, but now, sick and hungry, they have lost interest. Hakata organizes community meetings to study the situation and the people are reluctantly reaching a consensus to try to migrate to the New Guinea mainland; but New Guinea, already dealing with others wishing to migrate, is not willing to take them. [Source: Neil MacFarquhar, "Refugees Join List of Climate-Change Issues," International Herald Tribune, May 28, 2009; UNU Channel, "Local Solutions on a Sinking Paradise, Carterets Islands, Papua New Guinea," at http://www.vimeo.com/4177527.] ■

Other Water-Related Vulnerabilities Much of Oceania is vulnerable to changes in the region's climate that could result from global warming. Parts of Australia and some low, dry Pacific islands are already undergoing prolonged droughts and freshwater shortages that are requiring major changes in daily life and livelihoods. The fear is that the severe droughts are not merely periodic dry spells but may represent permanent alterations in rainfall patterns that could also worsen wildfires. Such fires emerged as a major issue in Australia in February 2009 when 173 people in the state of Victoria died in a rural firestorm near Melbourne (see Thematic Concepts Part E on page 478).

Global warming is also causing the oceans to warm, which can bring stronger tropical storms (see Photo Essay 11.3B on page 486). Warmer ocean temperatures threaten coral reefs and the fisheries that depend on them by causing *coral bleaching* (see Chapter 10, page 445), a phenomenon affecting all reefs in this region in recent years, but especially the Great Barrier Reef. Because so many fish depend on reefs, coral bleaching also threatens many fishing communities (see Thematic Concepts Part D on page 478). This is especially the case in some of the Pacific islands that have few other local food resources.

As fresh water becomes more scarce across the region, especially in Australia and on dry Pacific islands, *virtual water* (see Chapter 1, page 42) becomes an issue. All export-related activities, including tourism, that permanently consume or degrade fresh water in their extraction or production processes (such as iron,

Despite its vast size and relatively small population, Oceania has been severely impacted by human activity. Much damage has been wreaked by people from distant countries, as well as by countries in Oceania that export resources outside the region.

A One of the first underwater tests of a nuclear weapon and its effects on naval vessels took place on Bikini Atoll in the Marshall Islands (then a U.S. territory) in 1946. Over 300 nuclear tests have been conducted in Oceania since then.

B Once hunted to near extinction to protect sheep herds, the Tasmanian devil is now threatened by low genetic diversity, which leaves it vulnerable to disease.

C Part of a Ukrainian fishing fleet at port for repairs in Lyttleton, New Zealand. Fleets from around the world come to Oceania to take advantage of its fisheries, many of which are now over-exploited. Nevertheless, many Pacific islands are still selling fishing rights to foreign fleets because they need the money.

Human Impact, 2002

Land Cover
- Forests
- Grasslands
- Deserts
- Tundra
- Ice

Overfishing
- Threatened fisheries

Human Impact on Land
- High impact
- Medium–High impact
- Low–Medium impact

Acid Rain
- 5.5–4.9 pH

mi 0 400 800
km 0 400 800

D Logging in Marlborough, New Zealand. Despite increasing moves toward conservation of remaining forests and wildlife habitat, logs and wood products are still major exports for New Zealand.

coal, gold and bauxite mining; oil and gas extraction; meat, wool and wheat production; and hotel and golf course construction) are essentially exporting virtual water from places that have few water reserves and are already under water stress. Were the true costs of this freshwater depletion counted and added to the price of the products, these exports might no longer be competitive on the world market—not until all producers saw such virtual water accountability to be in their best interests and also raised their prices accordingly.

Responses to Crises Oceania is pursuing a number of alternative energy and water technologies that are reducing its contributions to climate change and making better use of the region's water resources. New Zealand has set a goal of obtaining 95 percent of its energy from renewable sources by 2025 (see **Thematic Concepts Part F** on page 478). Much of this will come from wind power, for which this region has excellent potential, especially in areas near the roaring forties. Geothermal energy (see Chapter 10, page 445) is either already being used or planned for future use throughout Oceania, with the exception of some of the low, nonvolcanic Pacific islands. Solar energy is widely used in some remote Pacific islands where the cost of importing fuel is prohibitive.

Oceania is a world leader in the implementation of water technologies. Some are age-old methods that are simple but effective, such as harvesting rainwater from roofs and the ground itself for household use (see **Photo Essay 11.3C**). Most buildings in rural Australia, New Zealand, and many Pacific islands get at least part of their water this way, relieving surface and groundwater resources. Australia and New Zealand are now stretching water resources further with extensive use of highly efficient drip irrigation technologies in agriculture (see Chapter 6, page 256, and **Photo Essay 11.3D**) and new low-cost water-filtration techniques.

Invasive Species and Food Production

The many unique endemic plants and animals of Oceania have been displaced by **invasive species**, organisms that spread into regions outside of their native range, adversely affecting economies or environments. Many exotic plants and animals were brought to this region by Europeans to support their food production systems. Ironically, many of these same species are now major threats to food production.

Australia When Europeans first settled the continent, they brought many new animals and plants with them, sometimes intentionally, sometimes unintentionally. European rabbits are among the most destructive of the introduced species. Rabbits were brought to Australia by early British settlers who enjoyed eating them. Many were released for hunting, and with no natural predators, they multiplied quickly, consuming so much of the native vegetation that many indigenous animal species starved. Moreover, rabbits became a major source of agricultural crop loss and reduced the capacity of many grasslands to support herds of introduced sheep and cattle. Attempts to control the rabbit population by introducing European foxes and cats backfired as these animals became major invasive species themselves. Foxes and cats have driven

several native Australian predator species to extinction without having much effect on the rabbit population. Intentionally introduced diseases have proven more effective at controlling the rabbit population, though rabbits have repeatedly developed resistances to them.

> **invasive species** organisms that spread into regions outside of their native range, adversely affecting economies or environments

Herding has also had a huge impact on Australian ecosystems. Because the climate is arid and soils in many areas are relatively infertile, the dominant land use in Australia is the grazing of introduced domesticated animals—primarily sheep, but also cattle. More than 15 percent of the land has been given over to grazing, and Australia leads the world in exports of sheep and cattle products.

Dingoes, the indigenous wild dogs of Australia, prey on the introduced sheep and young cattle. To separate the wild dogs from the herds, the Dingo Fence—the world's longest fence—was built, extending 3200 miles (5800 kilometers) (**Figure 11.9**). The fence is also a major ecological barrier to other wild species, although kangaroos, the natural prey of dingoes, have learned to live on the sheep side of the fence, where their population has boomed beyond sustainable levels.

FIGURE 11.9 The Dingo Fence. The world's longest fence snakes 3200 miles (5800 kilometers) between Yalata in South Australia and Jandowae, near the coast of Queensland. It is an effort to separate the dingo, Australia's indigenous dog, from sheep herds. A single dingo can kill as many as 50 sheep in one night of marauding. The fence requires constant patrolling and mending, but it has dramatically reduced killing of sheep by dingoes.

Oceania is vulnerable to a wide variety of hazards related to climate change, including sea level rise, increased tropical storm intensity, and less certain water availability. Fortunately, several countries in this region are already implementing solutions that are increasing resilience to climate hazards.

A A low-lying atoll in Tuvalu. With its highest point only 14.7 feet (4.5 meters) above sea level, Tuvalu is highly vulnerable to sea level rise. Combined with increased flooding during tropical storms, higher sea levels could make many low islands in this region uninhabitable. Tuvalu's government is already negotiating the future resettlement of parts of its population in nearby nations, such as New Zealand.

B A child in Guam after Typhoon Pongsona destroyed his home. Higher temperatures are bringing stronger tropical storms to this region.

C A rainwater harvesting system provides drinking water to a house in Ceres, a sustainable residential community near Melbourne, Australia. Such systems provide resilience in the face of drought and reduce potential for flooding.

Map labels: Taipei, Hong Kong, Guangzhou, Manila, **B**, Ngerulmud, Palikir, Majuro, Bandar Seri Begawan, Tarawa, Yaren, Surabaya, Dili, Port Moresby, Honiara, **A** Funafuti, Apia, Port-Vila, Suva, Nuku'alofa, Canberra, Sydney, Melbourne, **C**, **D** Wellington

Vulnerability to Climate Change

- Extreme
- High
- Medium
- Low

D A vineyard fitted with a drip irrigation system in Marlborough, New Zealand. These systems provide resilience in the face of drought and use substantially less water than other irrigation methods.

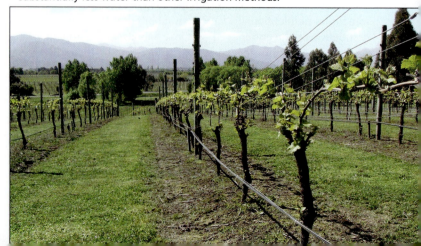

New Zealand The environment of New Zealand has been transformed by introduced species and food production systems even more extensively than has Australia's environment. No humans lived in New Zealand until about 700 years ago, when the Polynesian Maori people settled there. When they arrived, dense midlatitude rain forest covered 85 percent of the land. The Maori were cultivators who brought in yams and taro as well as other nonnative plants, pacific rats, and birds. By the time of European contact (1642), forest clearing and overhunting by the Maori had already degraded many environments and driven several species of birds to extinction.

European settlement in New Zealand dramatically intensified environmental degradation associated with food production (Figure 11.10) and invasive species. Attempts to recreate European farming and herding systems in New Zealand resulted in environments that are actually hostile to many native species, a growing number of which are now extinct. Today, only 23 percent of the country remains forested, with ranches, farms, roads, and urban areas claiming more than 90 percent of the lowland area.

Most of the cleared land is used for export-oriented farming and ranching. Grazing has become so widespread that today there are 15 times as many sheep as people, and 3 times as many cattle. Both of these activities have severely degraded environments. Soils exposed by the clearing of forests proved infertile, forcing farmers and ranchers to augment them with agricultural chemicals. The chemicals, along with feces from sheep and cattle, have severely

polluted many waterways, causing the extinction of some aquatic species.

Pacific Islands In the Pacific islands, many unique species of plants and animals have been driven to extinction as islands were deforested and converted to agriculture. In Hawaii, for example, extensive conversion of forests to land used for agriculture caused the extinction of numerous plant, bird, and snail species. Hawaii is home to more threatened or endangered species than any other U.S. state, despite having less than 1 percent of the U.S. landmass.

🎬 **239. Hawaii Considered America's Endangered Species Capital**

Globalization and the Environment in the Pacific Islands

As the Pacific islands have become more connected to the global economy over the years, flows of both resources and pollutants have increased dramatically. Mining, nuclear pollution, and tourism are all examples of how globalization has transformed environments in the Pacific islands.

Mining in Papua New Guinea and Nauru Two mines have had particularly devastating effects on the environment. Both were operated by foreign-owned mining companies that took advantage of poorly enforced or nonexistent environmental laws. In the case of the Ok Tedi Mine on Papua New Guinea, huge amounts of mine waste devastated river systems (see Figure 11.11 on page 488). Tens of thousands of indigenous subsistence cultivators were forced into new mining market towns, where their skills were of little use and where they needed cash to buy food and pay rent.

Eventually, 30,000 of these people sued the Australian parent mining company, which was then BHP Billiton, for U.S.$4 billion. Two villagers, Rex Dagi and Alex Maun, traveled to Europe and the United States to explain their cause and meet with international environmental groups. They and their supporters convinced U.S. and German partners in the Ok Tedi mine to divest their shares.

The most extreme case of environmental disaster due to mining took place on the once densely forested Melanesian island of Nauru (one-third the size of Manhattan, located northeast of the Solomon Islands. See Figure 11.1 on pages 472–473). During most of the last half of the twentieth century, the people of Nauru lived in prosperity. In fact, for a few years, the country had the highest per capita income in the world, thus attracting many immigrants. Nauru's wealth was based on the proceeds from the strip-mining of high-grade phosphates used in the manufacture of fertilizer, derived from eons of bird droppings (guano). The mining companies were owned first by Germany, then by Japan, and finally by Australia.

The phosphate reserves are now depleted, the proceeds ill-spent, and the environment destroyed. Junked mining equipment sits on miles of bleached white sand where forest once stood. To avoid financial disaster, the government has been exploring shady ways of making money, including money laundering for the Russian mafia, and clandestine offshore banking services for illegal rain forest loggers in Africa, South America, and Southeast Asia.

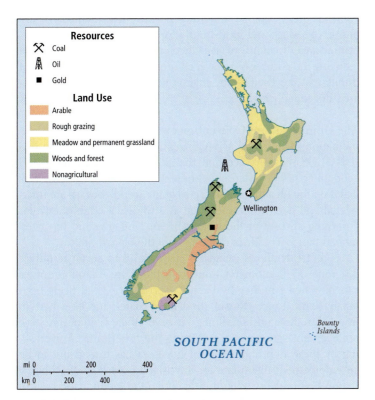

FIGURE 11.10 Land uses and natural resources of New Zealand. As a result of European settlement and the clearing of land for farming, only 23 percent of New Zealand remains forested. [Adapted from Richard Nile and Christian Clerk, *Cultural Atlas of Australia, New Zealand, and the South Pacific* (New York: Facts on File, 1996), p. 194.]

(A)   **(B)**   **(C)**   **(D)**

Figure 11.11 **The Ok Tedi Mine.** **(A)** The Ok Tedi open-pit copper and gold mine. A large hole now exists where Mount Fubila once stood. **(B)** An aerial view of the Ok Tedi River, which flows downstream of the mine. Each year since its opening in 1984, the mine has discharged 80 million tons of contaminated mine tailings and mine-induced erosion, resulting in a once deep and slow-moving river becoming a shallow, wide, and sometimes fast-moving river. **(C)** A close-up of the Ok Tedi riverbed, which is now contaminated with copper and other chemicals that kill fish and produce illness in the local human population. Sediment from the polluted river has contaminated 500 square miles (1300 square kilometers) of farmland. **(D)** A meeting about the mine and potential court settlements in the village of Serki. The Ok Tedi Mine has affected 50,000 people in 120 villages. Litigation against the mine owners is ongoing.

Nuclear Pollution The geopolitical aspects of globalization have hit this region especially hard. From the 1940s to the 1960s, the United States exploded 106 nuclear bombs in tests, primarily over the Marshall Islands. Similarly, Mururoa in French Polynesia has been the site of 180 nuclear weapons tests and the recipient of numerous shipments of nuclear waste from France. The weapons tests and imported waste have become major environmental issues for the Pacific islands.

In July 1985, the ship, *Rainbow Warrior*, owned by the environmental group Greenpeace, was blown up in New Zealand by the French secret service. In response, the 1986 Treaty of Rarotonga established the South Pacific Nuclear Free Zone. Most independent countries in Oceania have signed this treaty, which bans nuclear weapons testing and nuclear waste dumping on their lands. Because of political pressure from France and the United States, however, French Polynesia and U.S. territories such as the Marshall Islands have not signed the treaty.

Tourism Even tourism, which until recently was considered a "clean" industry, has been shown to create environmental problems. Foreign-owned tourism enterprises have often accelerated the loss of wetlands and worsened beach erosion by clearing coastal vegetation for construction, golf courses, and waterfront-related entertainment.

Tourism has also strained island water resources. Tourists increase the population dependent on this scarce resource, and they inevitably consume more water per capita than they do at home, because of increased showering, laundering, and other services that consume fresh water. Furthermore, inadequate methods of disposing of sewage and trash from resorts has polluted many once-pristine areas. Ecotourism (see Chapter 3, pages 126–128) aimed at reducing these impacts is now a common element of development throughout the Pacific, but environmental impacts from tourism are still generally high.

The United Nations Convention on the Law of the Sea Implementation of the 1994 UN Convention on the Law of the Sea (UNCLOS) has revealed how the globalization of Pacific island economies has thwarted environmental protection. UNCLOS is based on the idea that all problems of the world's oceans are interrelated and need to be addressed as a whole. It establishes rules governing all uses of the world's oceans and seas and has been ratified by 157 countries (although not the United States).

The treaty allows islands to claim rights to ocean resources 200 miles (320 kilometers) out from the shore. Island countries can now make money by licensing privately owned fleets from Japan, South Korea, Russia, the United States, and elsewhere to fish within these offshore limits. However, there is no overarching enforcement agency, and protecting the fisheries from overfishing by these rich and powerful licensees has turned out to be an enforcement nightmare for tiny island governments with few resources. Similarly, it has proved difficult to monitor and control the exploitation of seafloor mineral deposits by foreign companies.

241. Endangered Hawaiian Monk Seal Population Continues to Decline

242. New Species of Undersea Life Found Near Indonesia

243. Scientists Warn of Depletion of Ocean Fish in 40 Years

1. Global climate change will affect Oceania perhaps more than other regions.

2. Sea level rise threatens some Pacific islands with submersion, and water supplies throughout the region may be strained as rainfall patterns change. If sea levels rise the predicted 4 inches per decade, many of the lowest-lying atolls will disappear under water, leaving a multitude of refugees.

3. As part of its efforts to combat global warming, New Zealand has set a goal of obtaining 95 percent of its energy from renewable sources by 2025.

4. Throughout Oceania, the introduction of food production systems from elsewhere has resulted in the spread of ecologically and economically damaging invasive species.

5. Globalization and the patterns of consumption by people who live far from the Pacific are seriously affecting life in this region.

Human Patterns over Time

Personal Vignette *"With courage, you can travel anywhere in the world and never be lost. Because I have faith in the words of my ancestors, I'm a navigator."*

—*Mau Piailug*

In 1976, Mau Piailug made history by sailing a traditional Pacific island voyaging canoe across the 2400 miles (3860 kilometers) of deep ocean between Hawaii and Tahiti. He did so without a compass, charts, or other modern instruments, using only methods passed down through his family. He relied mainly on observations of the stars, the sun, and the moon to find his way. When clouds covered the sky, he used the patterns of ocean waves and swells, as well as the presence of seabirds, to tell him of distant islands over the horizon.

Piailug reached Tahiti 33 days after leaving Hawaii and made the return trip in 22 days. His voyage resolved a major scholarly debate over how people settled the many remote islands of the Pacific without navigational instruments, thousands of years before the arrival of Europeans. Some thought that navigation without instruments was impossible and argued that would-be settlers simply drifted about on their canoes at the mercy of the winds, most of them starving to death on the seas, with a few happening upon new islands by chance. It was hard to refute this argument because local navigational methods had died out almost everywhere. However, in isolated Micronesia, where Piailug lives, indigenous navigational traditions survive.

Since the successful 1976 voyage, Piailug has trained several students in traditional navigational techniques, which have become a symbol of cultural rebirth and a source of pride throughout the Pacific. In 2007, these protégés of Mau Piailug sailed from Hawaii through the Marshall Islands to Yokohama, Japan (**Figure 11.12**), as a gesture to celebrate peace and the human need to stay connected with nature. *[Source: Richard Nile and Christian Clerk,* Cultural Atlas of Australia, New Zealand, and the South Pacific *(New York: Facts on File, 1996), pp. 63–65; Voyage Weblog, Polynesian Voyaging Society, 2007, at http:// pvs.kcc.hawaii.edu/2007voyage/index.html.]*

The Peopling of Oceania

The longest-surviving inhabitants of Oceania are Australia's **Aborigines**, who migrated from Southeast Asia 50,000 to

Aborigines the longest-surviving inhabitants of Oceania, whose ancestors, the Australoids, migrated from Southeast Asia possibly as early as 50,000 years ago over the Sundaland landmass that was exposed during the ice ages

FIGURE 11.12 Hōkōle'a, a traditional Pacific Island voyaging canoe, visits Yokohama, Japan. In 2007, thirty-one years after its inaugural voyage between Tahiti and Hawaii, the Hōkōle'a sailed from Hawaii to the Marshall Islands in Micronesia, and on to Yokohama, Japan (shown above). Along the way, it stopped by Mau Piailug's home on Satawal Atoll in Micronesia where he was presented with the Alingano Maisu, a voyaging canoe similar to the Hōkōle'a, in recognition of his revival of traditional Pacific navigation techniques.

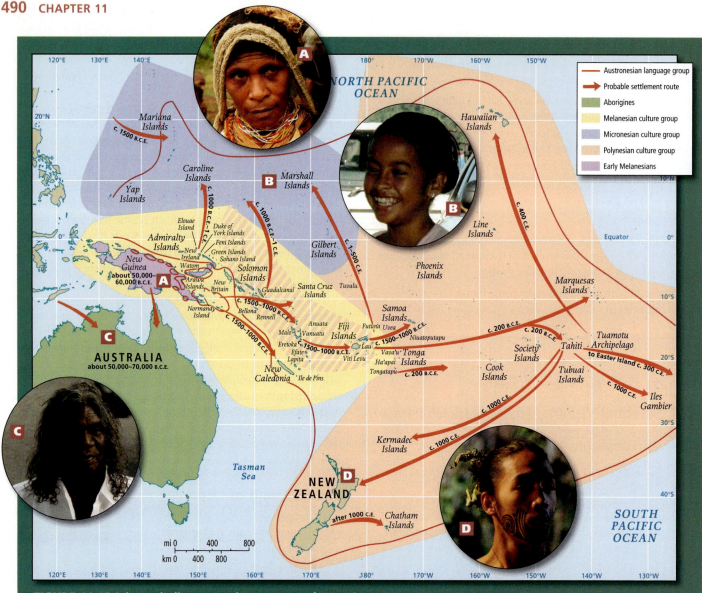

FIGURE 11.13 Primary indigenous culture groups of Oceania. By 50 to 70,000 years ago, humans had come to New Guinea and Australia. 25,000 years ago people had begun moving across the ocean to nearby Pacific islands. Movement into the more distant islands began with the arrival of Austranesians, who went on to inhabit the farthest reaches of Oceania. [Adapted from Richard Nile and Christian Clerk, *Cultural Atlas of Australia, New Zealand, and the South Pacific* (New York, Facts on File, 1996, pp 58–59.] **(A)** A Melanesian woman in New Guinea. **(B)** A Micronesian girl on Lelu island, Micronesia. **(C)** An Australian Aboriginal from Northern Australia. **(D)** A young Maori (Polynesian culture group) man in Rotorua, New Zealand.

Melanesians a group of Australoids named for their relatively dark skin tones, a result of high levels of the protective pigment melanin; they settled throughout New Guinea and other nearby islands

Melanesia New Guinea and the islands south of the equator and west of Tonga (the Solomon Islands, New Caledonia, Fiji, and Vanuatu)

70,000 years ago (Figure 11.13C). Amazingly, some memory of this ancient journey may be preserved in Aboriginal oral traditions, which recall mountains and other geographic features that are now submerged under water. At about the same time that the Aborigines were settling Australia, related groups were settling nearby areas.

Melanesians, so named for their relatively dark skin tones, a result of high levels of the protective pigment *melanin,*

migrated throughout New Guinea and other nearby islands, giving this area its name, **Melanesia** (see Figure 11.13A). Archeological evidence indicates that they first arrived more than 50,000 to 60,000 years ago from Sundaland (see Figure 10.6 on page 435), a shelf exposed during the Pleistocene epoch. They lived in isolated pockets, which resulted in the evolution of hundreds of distinct yet related languages. Like the Aborigines, the Melanesians survived mostly by hunting, gathering, and fishing, although some groups—especially those inhabiting the New Guinea highlands—also practiced agriculture.

Much later, between 5000 to 6000 years ago and as recently as 1000 years ago, linguistically related *Austronesians* settled

Micronesia and **Polynesia**, sometimes mixing with the Melanesian peoples they encountered. Micronesia consists of small islands that lie east of the Philippines and north of the equator. Polynesia is made up of numerous islands situated inside a large irregular triangle formed by New Zealand, Hawaii, and Easter Island (see **Figure 11.13B, D**). (Easter Island is a tiny speck of land in the far eastern Pacific, at 109° W, 27° S, not shown in the figures in this chapter.) Although Europeans long questioned the skills of Pacific navigators, experiments run by Mau Piailug (see the vignette on page 489) proved that ancient sailors could navigate over vast distances, using seasonal winds, astronomic calculations, bird and aquatic life, and wave patterns to reach the most far-flung islands of the Pacific. They were fishers, hunter-gatherers, and cultivators who developed complex cultures and maintained trading relationships among their widely spaced islands.

In the millennia that have passed since first settlement, humans have continued to circulate throughout Oceania. Some apparently set out because their own space was too full of people and conflict, food reserves were declining, or they wanted a life of greater freedom. It is also likely that Pacific peoples were enticed to new locales by the same lures that later attracted some of the more romantic explorers from Europe and elsewhere: sparkling beaches, magnificent blue skies, beautiful people, scented breezes, and lovely landscapes.

Arrival of the Europeans

The earliest recorded contact between Pacific peoples and Europeans took place in 1521, when the Portuguese navigator Ferdinand Magellan (exploring for Spain), landed on the island of Guam in Micronesia. The encounter ended badly. The islanders, intrigued by European vessels, tried to take a small skiff. For this crime, Magellan had his men kill the offenders and burn their village to the ground. A few months later, Magellan was himself killed by islanders in what became the Philippines, which he had claimed for Spain. Nevertheless, by the 1560s, the Spanish had set up a lucrative Pacific trade route between Manila in the Philippines and Acapulco on the west coast of Mexico. Explorers from other European states followed, first taking an interest mainly in the region's valuable spices. The British and French explored extensively in the eighteenth century.

The Pacific was not formally divided among the colonial powers until the nineteenth century, and by that time, the United States, Germany, and Japan had joined France and Britain in taking control of various island groups. European colonization of Oceania proceeded according to the models developed in Latin America, Africa, South Asia, and Southeast Asia, with major emphasis on extractive agriculture and mining. Because native people were often displaced from their lands or exposed to exotic diseases to which they had no immunity, their populations declined sharply.

Many enduring notions about the Pacific arose from the European explorations of the eighteenth and nineteenth centuries. During this time, European thinkers were debating whether or not civilization actually improves the quality of life for human beings. Some argued that civilization corrupts and debases people, and they glorified the "**noble savages**" who lived primitive lives in distant places supposedly untouched by corrupting influences. Explorers of the Pacific who were influenced by such ideas were caught off guard when from time to time the islanders armed themselves and rebelled, attacking those who were taking their lands and resources. The surprised Europeans quickly revised their opinions and recast the "noble savages" as brutish and debased.

The realities of life in Oceania were much more balanced and complex than the positive and negative extremes that Europeans perceived. In Australia and on New Guinea and the other larger Melanesian islands, relatively plentiful resources made it possible for people to live in small, simple societies, less subject to the stratification and class tensions seen in so much of the world. On the smaller islands of Micronesia and Polynesia, land and resources were scarcer, and many people coexisted in a state of moderate antagonism. Although warfare occurred, hostilities were often settled ritualistically and by means of annual tribute-paying ceremonies, rather than by resorting to combat. Individual rulers rarely amassed large territories or controlled them for long. On these islands, many societies were and still are hierarchical, with layers of ruling elites at the top and undifferentiated commoners at the bottom.

The Colonization of Australia and New Zealand

Although all of Oceania has been under European or American rule at some point, the most Westernized parts of the region are Australia and New Zealand. The colonization of these two countries by the British has resulted in many parallels with North America. In fact, the American Revolution was a major impetus for "settling" Australia because once the North American colonies were independent, the British needed somewhere else to send their convicts. In early nineteenth-century Britain, a relatively minor theft—for example, of a piglet—might be punished with 7 years of hard labor in Australia (see **Figure 11.14** on page 492).

A steady flow of English and Irish convicts arrived in Australia until 1868. Most of the convicts chose to stay in the colony after their sentences were served. They are given credit for Australia's rustic self-image and egalitarian spirit. They were joined by a much larger group of voluntary immigrants from the British Isles who were attracted by the availability of inexpensive farmland. Waves of these immigrants arrived until World War II. New Zealand was settled somewhat later than Australia, in the mid-1800s. Although its population also derives primarily from British immigrants, New Zealand was never a penal colony.

Another similarity among Australia, New Zealand, and North America was the treatment of indigenous peoples by European settlers. In both Australia and New Zealand, native peoples were killed outright, annihilated by infectious diseases, or shifted to the margins of society. The few who lived on territory the Europeans

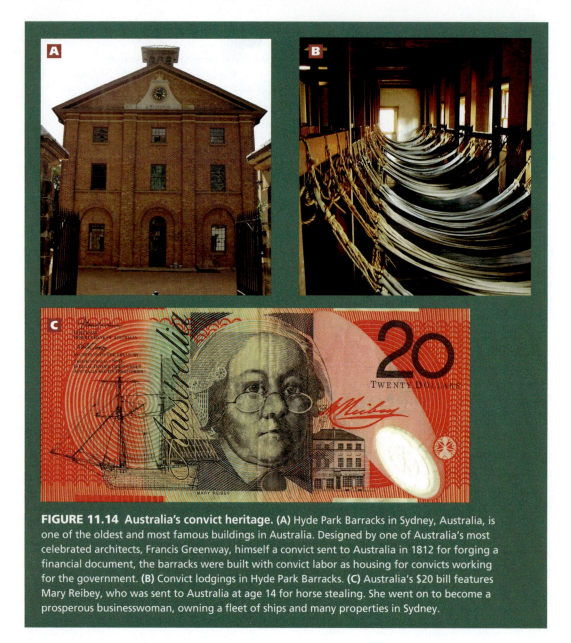

FIGURE 11.14 Australia's convict heritage. (A) Hyde Park Barracks in Sydney, Australia, is one of the oldest and most famous buildings in Australia. Designed by one of Australia's most celebrated architects, Francis Greenway, himself a convict sent to Australia in 1812 for forging a financial document, the barracks were built with convict labor as housing for convicts working for the government. **(B)** Convict lodgings in Hyde Park Barracks. **(C)** Australia's $20 bill features Mary Reibey, who was sent to Australia at age 14 for horse stealing. She went on to become a prosperous businesswoman, owning a fleet of ships and many properties in Sydney.

thought undesirable were able to maintain their traditional way of life. However, the vast majority of the survivors lived and worked in grinding poverty, either in urban slums or on cattle and sheep ranches. Today, native peoples still suffer from pervasive discrimination and maladies such as alcoholism and malnutrition. Even so, some progress is being made toward improving their lives (as described on pages 503–505). In 2008, the then newly elected prime minister of Australia, Kevin Rudd, officially apologized to Aboriginal people for the treatment they had received since the land had first been colonized.

Closely related to attitudes toward indigenous people were attitudes toward any people of color as immigrants. Although Pacific island people (indigenous) were brought in as laborers during the nineteenth century, by 1901 a whites-only policy governed Australian immigration. The favored migrants were those from the British Isles; then after World War II, southern Europeans were encouraged. This discrimination persisted until the mid-1970s, when the White Australia policy on immigration was ended. In New Zealand, where similar racist attitudes prevailed, there was not an official whites-only policy, and by the 1970s, students and immigrants were arriving from Asia and the Pacific islands.

Oceania's Shifting Global Relationships

During the twentieth century, Oceania's relationship with the rest of the world went through three phases: from a predominantly European focus to identification with the United States and

Canada to the currently emerging linkage with Asia. Until roughly World War II, the colonial system gave the region a European orientation. In most places, the economy depended largely on the export of raw materials to Europe. Thus, even when a colony gained independence from Britain, as Australia did in 1901 and New Zealand did in 1907, people remained strongly tied to their mother countries. Even today, the Queen of England remains the titular head of state in both countries. During World War II, however, the European powers provided only token resistance to Japan's invasion of much of the Pacific and its bombing of northern Australia. Such European impotence began a change in political and economic orientation.

📺 **244. Veterans Remember Tragedy of War in Pacific**

After the war, the United States, with an already strong foothold in the Philippines, became the dominant power in the Pacific, and U.S. investment became increasingly important to the economies of Oceania. Australia and New Zealand joined the United States in a Cold War military alliance, and both fought alongside the United States in Korea and Vietnam, suffering considerable casualties and experiencing significant antiwar activity at home. U.S. cultural influences were strong, too, as North American products, technologies, movies, and pop music penetrated much of Oceania.

By the 1970s, another shift was taking place as many of the island groups were granted self-rule by their European colonizers, and Oceania became steadily drawn into the growing economies of Asia. Since the 1960s, Australia's thriving mineral export sector has become increasingly geared toward supplying Asian manufacturing industries (first Japan in the 1960s, and increasingly China since the 1990s). Similarly, since the 1970s, New Zealand's wool and dairy exports have gone mostly to Asian markets. Despite occasional backlashes against "Asianization," Australia, New Zealand, and the rest of Oceania are becoming increasingly transformed by Asian influences. Fisheries have been particularly impacted by Asian markets (Thematic Concepts Part L on page 479). Many Pacific islands have significant Chinese, Japanese, Filipino, and Indian minorities, and the small Asian minorities of Australia and New Zealand are increasing. On some Pacific islands, such as Hawaii, Asians now constitute the largest portion (42 percent) of the population.

🎨 **THINGS TO REMEMBER** 🎨

1. Oceania can be divided into four distinct cultural regions: Australia and Tasmania, settled originally by Aborigines; Melanesia, settled by Melanesians, so identified because of their skin tone derived from melanin; Micronesia, whose original inhabitants came from coastal Asia and from the more northern Melanesian islands; and Polynesia, settled by a variety of Austronesian people.

2. Through colonization, Europeans dominated Oceania from the early sixteenth century until the end of World War II. During the 50 years after the war, the United States was the principal power in the region. Beginning in the 1970s, Asian countries—China and Japan in particular—have had increasing influence throughout Oceania.

Population Patterns

Although Oceania occupies a huge portion of the planet, its population is only 36 million people, close to that of the state of California (36.7 million) (see Photo Essay 11.4 on page 494). They live on a total land area slightly larger than the contiguous United States but spread out in bits and pieces across an ocean larger than the Eurasian landmass. The Pacific islands, including Hawaii, have nearly 4.5 million people; Australia has 21.9 million; Papua New Guinea, 6.6 million; and New Zealand, 4.3 million.

Disparate Population Patterns in Oceania

Different population patterns exist in Oceania. Australia, New Zealand, and Hawaii, like many wealthy countries, have highly urbanized, relatively older, and more slowly growing populations, with life expectancies of about 80 years. The Pacific islands and Papua New Guinea, like many developing countries, have much more rural, younger, and rapidly growing populations, with life expectancies in the 60s or low 70s. The overall trend throughout the region, however, is toward smaller families, aging populations, and urbanization.

Population densities remain low in Australia, at 7.8 people per square mile (3 per square kilometer), and New Zealand, at 41.4 per square mile (16 per square kilometer). Densities vary widely in the Pacific islands; some are sparsely settled or uninhabited, while others—including some of the smallest and poorest, such as the Marshall Islands and Funafuti, the capital of Tuvalu, which have 772 and 4847 people respectively per square mile (298 and 1871 respectively per square kilometer)—are densely populated.

Urbanization in Oceania

The global trend of migration from the countryside to cities is highly visible in Oceania, where 73 percent of the population live in urban areas (see Table 11.1 on page 495). The shift from agricultural and resource-based economies toward service economies is a major driver of urbanization, especially in Australia, New Zealand, and some of the wealthier Pacific islands such as Hawaii and Guam. These trends are weakest in Papua New Guinea and many smaller Pacific Islands. Nauru is a special case of extreme over-settlement (see page 487).

Australia and New Zealand Australia and New Zealand have among the highest percentages of city dwellers outside Europe. More than 83 percent of Australians live in a string of cities along the country's well-watered and relatively fertile eastern and southeastern coasts. Similarly, 86 percent of New Zealanders live in urban areas. The vast majority of the people in these two countries live in modern comfort, work in a range of occupations typical of highly industrialized societies, and have access to tax-supported leisure facilities (see Photo Essay 11.5 Map, A, B on page 496). Vibrant, urban-based service economies have expanded to employ about three-quarters of the population in both countries. Declining employment in mining and agriculture, where mechanization has dramatically reduced the number of workers needed, has also contributed to urbanization.

Population growth is slowing throughout Oceania as more people move to the cities, where health care is better and women are more likely to pursue careers—factors that make large families less likely. Australia, New Zealand, and Hawaii are furthest along in this process and now are highly urbanized with smaller families and increasingly old populations.

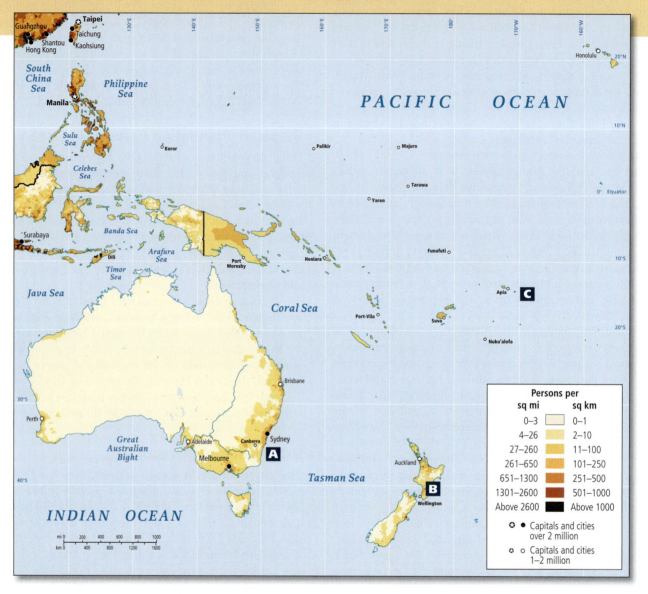

Persons per

sq mi	sq km
0–3	0–1
4–26	2–10
27–260	11–100
261–650	101–250
651–1300	251–500
1301–2600	501–1000
Above 2600	Above 1000

✪ ● Capitals and cities over 2 million

✪ ○ Capitals and cities 1–2 million

A An elderly woman in Newtown, Australia, waves the national flag to a passing train. Australia has a relatively old population, with 13 percent over the age of 65.

B A father and son go for a walk by the docks in Wellington, New Zealand. One- or two-child families are the norm in New Zealand, Australia, and Hawaii.

C A family in Samoa sits down to an evening meal. Large families are common throughout many Pacific islands and Papua New Guinea, although birth rates are declining as more people move to the cities where the incentives for small families are stronger.

TABLE 11.1	Population data for cities and towns in selected places in Oceania		
Country	Total Population	Urban Population	Percent Urban
Australia	21,900,000	18,177,000	83
Federated States of Micronesia	100,000	22,000	22
Fiji	800,000	408,000	51
French Polynesia	300,000	159,000	53
Guam	200,000	186,000	93
Hawaii	1,300,000	910,000	70
Kiribati	100,000	44,000	44
Marshall Islands	100,000	67,000	67
Nauru	10,000	10,000	100
New Caledonia	300,000	174,000	58
New Zealand	4,300,000	3,698,000	86
Palau	20,000	15,400	77
Papua New Guinea	6,600,000	858,000	13
Samoa	200,000	44,000	22
Solomon Islands	500,000	85,000	17
Tonga	100,000	24,000	24
Tuvalu	10,000	4,700	47
Vanuatu	200,000	42,000	21
Totals	34,840,000	24,928,100	71.5

Source: 2009 World Population Data Sheet (Washington, DC: Population Reference Bureau, 2009).

Pacific Islands Throughout the Pacific, urban centers have transformed natural landscapes, and in some small countries, such as Guam, Palau, and the Marshall Islands, they have become the dominant landscape. Although cities are places of opportunity, they can also be sites of cultural change, conflict, and environmental hazards (Photo Essay 11.5C, D).

The great majority of Pacific island towns, and all the capital cities, are located in ecologically fragile coastal settings. Many of these waterfront towns were established during the colonial era as ports or docking facilities and were situated in places suitable for only limited numbers of people. Consequently, little land is available for development, and access to housing is limited. Squatter settlements have been a visible feature of the region's urban areas for several decades.

Multiculturalism has resulted from urbanization across the region. Although many urban residents were born outside the towns and maintain close connections to their rural home communities, more and more urban islanders are letting go of rural lifeways, ethnic identity, and cultural commitments. Increasingly, people are marrying in town and across language divisions, creating new patterns of social alliances and networks. Along with the adoption of urban lifestyles, such cultural blending results in new social tensions, changing the very nature of social life in the island Pacific. Urban unemployment and crime are on the rise, and low economic growth restricts the revenue available to governments to manage urban development.

Human Well-Being in Oceania

Human well-being varies dramatically across this region as measured by the usual indicators used in this book—gross domestic product per capita (GDP per capita PPP), rank on the United Nations Human Development Index (HDI), and female income as a percentage of male income. Over the last decade, Australia typically has ranked among the top 20 countries in GDP per capita PPP, and New Zealand has been among the top 25. Australia usually occupies one of the top 5 slots on the HDI, and New Zealand falls at the lower end of highly developed countries. Hawaii, part of the United States, typically has a GDP per capita PPP that is $10,000 higher than New Zealand's, and because it is a state with an unusually well-developed social welfare system, it would rank higher than the overall U.S. level of thirteenth on the HDI.

Two patterns of urbanization exist in Oceania. While Australia, New Zealand, and Hawaii are already highly urbanized places with high standards of living, Papua New Guinea and many Pacific islands are much more rural and have lower standards of living.

Population of urban areas

2006   2020

20 million or more
10 million
5 million
3 million

Note: symbols on map are sized proportionally to specific population data

Population living in urban areas

■ 80–100%	■ 20–39%
■ 60–79%	■ 0–19%
■ 40–59%	■ No data

1 Global rank (population 2006)

A Sydney, Australia, the largest city in Oceania and one that is consistently ranked among the most liveable cities in the world, along with Melbourne and Perth (two other large Australian cities), and Auckland, New Zealand.

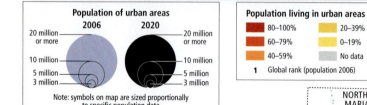

B A public park in Auckland, New Zealand's largest city.

C Part of Funafuti atoll, capital of Tuvalu. Funafuti is one of the most densely populated islands in the world, with 4,847 people per square mile (1,871 per km²). Tuvalu is among the poorest nations in Oceania, with GDP (PPP) per capita of around U.S.$1600.

D Children in Ravi Ravi, a village in rural Fiji where incomes are relatively low as is access to health care and education. The home behind them was constructed by a non-profit funded by foreign aid.

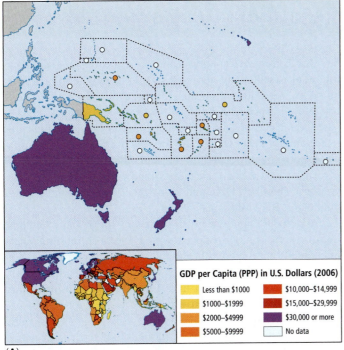

(A)

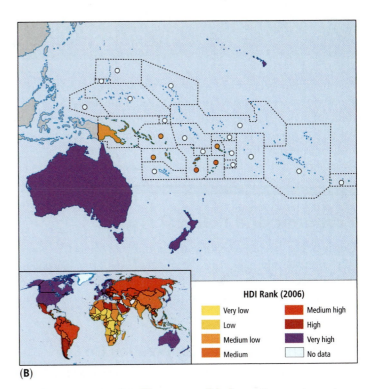

(B)

FIGURE 11.15 Maps of human well-being. (A) Gross domestic product (GDP). **(B)** Human Development Index (HDI). **(C)** Female earned income as a percent of male earned income (F/MEI). [Maps adapted from data: "Human development indices," (Maps A and B): Table 2, pages 28–32; (Map C): Table 5, pages 41–44 at http://hdr.undp.org/en/media/HDI_2008_EN_Tables.pdf.]

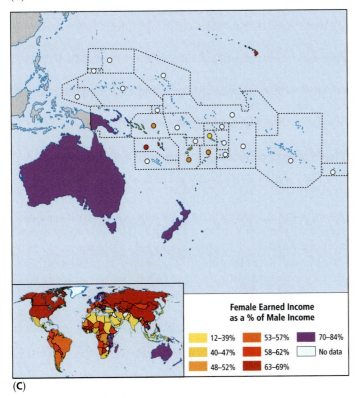

(C)

The maps of human well-being (Figure 11.15) illustrate these rankings. Map A shows that except for Australia, New Zealand, and Hawaii, Oceania has low levels of GDP per capita. Map B shows medium to low HDI rankings. Although the Pacific islands have very low levels of income and well-being as measured in official statistics, it should be remembered that *subsistence affluence* (discussed on page 500) and strong communitarian values (see the discussion on page 505 about the Pacific Way) can result in higher-than-expected actual well-being.

Map C shows similar patterns for female-to-male income ratios, with Australia and New Zealand having the best record. Hawaii, however, ranks lower on this scale, closer to the level of the United States as a whole. All of the islands rank yet lower, with the exceptions of Papua New Guinea and Vanuatu. However, the apparently more equal pay for females and males in these two places should be assessed in light of the overall very low incomes on those islands. And overall, across the region few countries have reported sufficient statistics to determine the gap in pay between men and women.

THINGS TO REMEMBER

1. Urbanization is the dominant settlement pattern in Oceania.

2. The shift to life in cities has brought broad-based affluence to Australia, New Zealand, and islands such as Hawaii; many Pacific islands remain much more rural and have lower standards of living.

3. The great majority of Pacific island towns, and all the capital cities, are located in ecologically fragile coastal settings.

4. Human well-being varies dramatically across the region.

II CURRENT GEOGRAPHIC ISSUES

Many current geographic issues in Oceania are related to the transition under way from European to Asian and inter-Pacific cultural influences. Oceania's old relationships were built on historical factors, such as the settlement of Australia and New Zealand by Europeans and the depth of ancient Pacific cultural affiliations. By contrast, its new relationships are influenced by economic and geographic considerations, particularly physical proximity to Asia.

Economic and Political Issues

Oceania has been powerfully transformed by global relationships over the past 200 years. Now, new forces of globalization, driven largely by Asia's growing affluence and enormous demand for resources, are shifting trade, migration, and tourism within Oceania.

Asia-Pacific Economic Cooperation (APEC)

Established in 1989 to enhance economic growth and prosperity and to strengthen the Asia–Pacific community, APEC is the only intergovernmental group in the world that operates on the basis of nonbinding commitments and open dialogue among all participants. Unlike the WTO or NAFTA or even the European Union (after which it is partially patterned), APEC does not oblige its members to participate in any treaties. Decisions made within APEC are reached by consensus and commitments are undertaken on a voluntary basis.

APEC is composed of 21 members: Australia, Brunei, Canada, Chile, China, Hong Kong, Indonesia, Japan, South Korea, Malaysia, Mexico, New Zealand, Papua New Guinea, Peru, the Philippines, Russia, Singapore, Taipei, Thailand, the United States, and Vietnam. These "member economies" account for approximately 40 percent of the world's population, just over 50 percent of global production, and more than 40 percent of global trade.

The potential of APEC to make a difference to life in Oceania is considerable, if only because it is a forum for discussing issues that range well beyond economic concerns. All of APEC's promised services are needed in Oceania: securing the Asia–Pacific food supply, reducing vulnerability to global climate change, finding energy-efficient means of transportation, developing financial institutions for small businesses, training teams to handle pandemics, forming working groups to interpret free trade agreements for small businesses, helping communities fend off the worst effects of periodic recessions. If identifying needs is the first step toward addressing these issues, APEC is off to a good start. For additional information, see http://www.apec.org/.

Globalization and Oceania's New Asian Orientation

APEC is only the most recent example of a globalizing entity in Oceania. One could say that globalization actually began when the first European explorers came into the region, beginning the trend of settlement and cultural and economic influence by outsiders, especially Europeans. More recently, the United States has exerted a powerful influence on trade and politics. For the past several decades, however, globalization has reoriented this region toward Asia, which buys more than 76 percent of Australia's exports (mainly coal, iron ore, and other minerals), nearly 35 percent of New Zealand's exports (mainly meat, wool, and dairy products, (**Figure 11.16**) and many other products and services from across island Oceania.

Asia is also increasingly a source of the region's imports. Because there is little manufacturing in Oceania, most manufactured goods are imported from China, Japan, South Korea, Singapore, and Thailand—Oceania's leading trading partners. Both Australia and New Zealand have free trade relationships either completed or in negotiation with Asia's two largest economies, China and Japan.

The Pacific islands are even further along in their reorientation toward Asia than either Australia or New Zealand. Not only are coconut, forest, and fish products from the Pacific islands sold to Asian markets, but Asian companies increasingly own these industries (Thematic Concepts Part L on page 479). Fishing fleets from Asia regularly ply the offshore waters of Pacific island nations. Asians also dominate the Pacific island tourist trade, both as tourists and as investors in tourism infrastructure. And increasing numbers of Asians are taking up residence in the Pacific islands, exerting widespread economic influence.

The Stresses of Asia's Economic Rise for Australia and New Zealand

For Australia and New Zealand, Asia's global economic rise has meant not only increased trade but also increased competition with Asian economies in foreign markets. Throughout the region, local industries used to enjoy protected or "preferential" trade with Europe. They have lost that advantage because new European Union regulations stemming from the EU's membership in the World Trade Organization prohibit such arrangements. In their trade with Europe, these industries now face stiff competition from larger companies in Asia that benefit from much cheaper labor.

Since the 1970s, competition from Asian companies has meant that increasing numbers of workers in Australia and New Zealand have lost jobs and seen their hard-won benefits scaled back or eliminated. This has been especially traumatic for the labor movements of these countries, which historically have been among the world's strongest. Australian coal miners' unions successfully agitated for the world's first 35-hour workweek. Other labor unions won a minimum wage, pensions, and aid to families with children long before such programs were enacted in many other industrialized countries. For decades, these arrangements were highly successful. Both Australia and New Zealand enjoyed

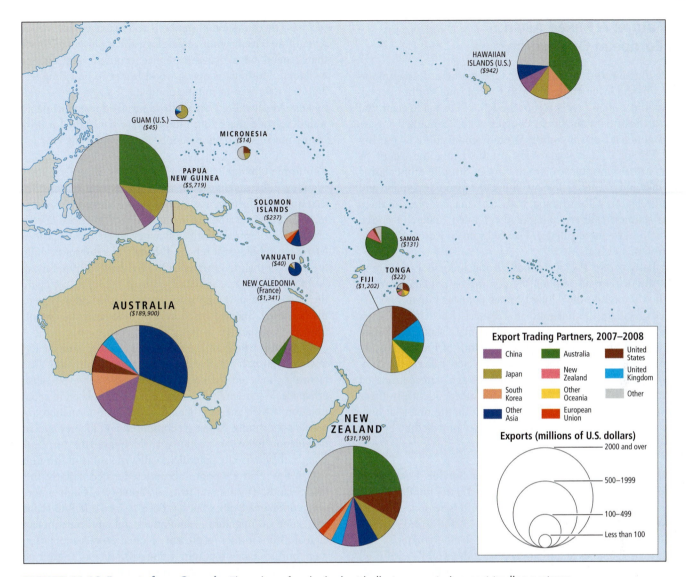

FIGURE 11.16 Exports from Oceania. The colors of each pie chart indicate a country's export trading partners. The "other" sections can include trade with Canada, Mexico, the Caribbean, non-EU Europe, sub-Saharan Africa, and other locales, some of them new trading partners. (Figures for Hawaii do not include exports to other parts of the United States.) [Sources: Central Intelligence Agency, *The World Factbook*, at "https://www.cia.gov/library/publications/the-world-factbook/"https://www.cia.gov/library/publications/the-world-factbook/; U.S. Department of State, "Background Note: Kiribati," at "http://www.state.gov/r/pa/ei/bgn/1836.htm"http://www.state.gov/r/pa/ei/bgn/1836.htm; "http://www.dfat.gov.au/geo/guam/guam_brief.html"http://www.dfat.gov.au/geo/guam/guam_brief.html; Samoa Bureau of Statistics, "Economic Statistics," at "http://www.sbs.gov.ws/Portals/138/PDF/Trade/Export%20trading%20partners.pdf"http://www.sbs.gov.ws/Portals/138/PDF/Trade/Export%20trading%20partners.pdf; http://extsearch.worldbank.org/servlets/GoogleCachedLinkProvider?CID=iitfrMW4K0gJ&full_url=http://info.worldbank.org/etools/wti/docs/wti2008/mainpaper.pdf&domain=info.worldbank.org; Encyclopedia of the Nations, "Tuvalu—International Trade," at "http://www.nationsencyclopedia.com/economies/Asia-and-the-Pacific/Tuvalu-INTERNATIONAL-TRADE.html#ixzz0XctzTLD8" http://www.nationsencyclopedia.com/economies/Asia-and-the-Pacific/Tuvalu-INTERNATIONAL-TRADE.html#ixzz0XctzTLD8; U.S. Census Bureau, "Total U.S. Exports (Origin of Movement) via Hawaii," at "http://www.census.gov/foreign-trade/statistics/state/data/hi.html#ctry" http://www.census.gov/foreign-trade/statistics/state/data/hi.html#ctry.]

living standards comparable to those in North America but with a more egalitarian distribution of income.

Competition from Asian companies also led to lower corporate profits. As corporate profits fell, so did government tax revenues, which necessitated cuts to previously high rates of social spending on welfare, health care, and education. The loss of social support, especially for those who have lost jobs, has contributed to rising poverty in recent years. Australia now has the second-highest poverty rate in the industrialized world, after that of the United States.

The Future: A Mixed Asian and European Orientation?

Despite the powerful forces pushing Oceania toward Asia, important factors still favor strong ties with Europe and North America. The Asian economic recession of the late 1990s made clear that Oceania must maintain broad contacts with economies outside Asia. Another factor supporting Western connections is the lingering fear of Chinese aggression, justified to some extent by China's expansionist moves toward Taiwan. In spite of increasing trade links and a recent effort by China to expand diplomatic and cultural relations with Australia, both Australia and New Zealand remain staunch military allies of the United States. Over the years, both have participated in U.S.-led wars in Korea, Vietnam, Afghanistan, and Iraq.

In parts of the Pacific islands, strong links to Europe and North America are also upheld by continuing political domination. In Micronesia, the United States controls Guam and the Northern Mariana Islands, and in Polynesia, the United States controls American Samoa. Just as the Hawaiian Islands are a state of the United State, similarly, the 120 islands of French Polynesia—including Tahiti and the rest of the Society Islands, the Marquesas Islands, and the Tuamotu Archipelago—are considered Overseas Lands of France. Any desire for independence in these possessions has not been sufficient to override the financial benefits of aid, subsidies, and investment money provided by France and the United States.

Maintaining Raw Materials Exports as Service Economies Develop For decades, natural resources and agricultural products supported the two dominant economies of this region, Australia and New Zealand. Australia, for example, is the world's largest exporter of coal, bauxite, and a number of other minerals and metals. It also supplies about 50 percent of the world's wool used for clothing. New Zealand specializes in dairy products, meat, fish, wool, and timber products. Neither country has been a major supplier of manufactured goods—which are usually more profitable exports—to the world market.

Although the capacity for manufacturing and materials processing is growing, the shift toward trading more with Asia than with Europe has had limited effect on the pattern of exporting unprocessed raw materials, because most Asian economies have a much greater need to import raw materials than manufactured goods. Nevertheless, although their dollar contribution to national economies remains high, raw materials export industries are of decreasing prominence in the economies of Australia and New Zealand, in that they now employ fewer people because of mechanization. This shift to lower labor requirements has been essential for these industries to stay globally competitive with other countries that have much cheaper labor.

Today, the economies of both countries are dominated by diverse and growing service sectors. However, their service economies have links to their export sectors. The extraction of minerals and management of herds and agriculture have become technologically sophisticated enterprises that depend on a dynamic service economy and an educated workforce. Australia is now a world leader in providing technical and other services to mining companies, sheep farms, and winemakers. Meanwhile, New Zealand's well-educated workforce and well-developed marketing infrastructure has helped it break into luxury markets for dairy products, meats, and fruits.

Economic Change in the Pacific Islands In general, the Pacific islands are also making a shift away from extractive industries, such as mining and fishing, and toward service sector industries such as tourism and government. On many islands, the stress of economic change is cushioned by self-sufficiency and resources from abroad. Many households still rely on fishing and subsistence cultivation for much of their food supply. On the islands of Fiji, for example, part-time subsistence agriculture engages more than 60 percent of the population, although it accounts for just under 17 percent of the economy. Statistics also rarely include remittances sent home from Pacific Islanders working abroad.

Islanders who can be self-sufficient in food and shelter while saving extra cash for travel and occasional purchases of manufactured goods are sometimes said to have achieved **subsistence affluence.** Where there is poverty, it is often related to geographic isolation, which means a lack of access to information and economic opportunity. Although computers and the new global communication networks are not yet widely available in the Pacific islands, they have the potential to alleviate some of this isolation.

In the relatively poor and undereducated nations (the Solomon Islands, Tuvalu, and parts of Papua New Guinea, for example), conditions typify what has been termed a **MIRAB economy**—one based on migration, remittances, aid, and bureaucracy. Foreign aid from former or present colonial powers supports government bureaucracies that supply employment for the educated and semiskilled. This type of economy has little potential for growth.

The Advantages and Stresses of Tourism

Tourism is a growing part of the economy throughout Oceania, with tourists coming largely from Japan, Korea, Taiwan, Southeast Asia, the Americas, and Europe (Figure 11.17). In 2008 (the latest year for which complete figures are available), 17.7 million tourists arrived in Oceania: 23 percent of the tourists came from Asia—16 percent from Japan, alone; just 11 percent came from Europe, down from 17 percent in recent years; 33 percent came from North America; 19 percent were from within Oceania; and 12 percent came from other global locations.

In some Pacific island groups, the number of tourists far exceeds the island population. Guam, for example, annually receives tourists equivalent to five times its population. Palau and the Northern Marianas Islands annually receive more than four times their populations. Such large numbers of visitors, expecting to be entertained and graciously accommodated, can place a special stress on local inhabitants. And although they bring money to the islands' economies, these visitors create problems for island ecology, place extra burdens on water and sewer systems, and

subsistence affluence a lifestyle whereby self-sufficiency is achieved for most necessities, while some opportunities to earn cash allow for travel and occasional purchases of manufactured goods

MIRAB economy an economy based on migration, remittance, aid, and bureaucracy

(A)

(B)

(C)

FIGURE 11.17 Tourism in Oceania. Tourism plays a major role in the economies of all countries in Oceania. Between 2004 and 2008, over 17.7 million tourists visited the region, just under one-third going to Australia. In 2007 alone, over 10.7 million visitors spent nearly $32 billion in local economies, with some $22 billion going to Australia. The origins of the tourists reflect changing trade patterns in the region, with more and more coming from Asia. **(A)** Samoans taking visitors to a favorite picnic spot. **(B)** "Samoan" culture as represented to tourists in Hawaii. **(C)** Hotels on Honolulu's Waikiki Beach provide lodging for a majority of Hawaii's 6 million yearly visitors.

require a standard of living that may be far out of reach for local people. Perhaps nowhere in the region are the issues raised by tourism clearer than in Hawaii.

A Case Study of Civil Conflict: Hawaii

Since the 1950s, travel and tourism has been the largest industry in Hawaii, producing nearly 18 percent of the gross state product in 2008. Tourism is related in one way or another to nearly 75 percent of all jobs in the state. (By comparison, travel and tourism accounts for 9 percent of GDP worldwide.) In 2008, tourism employed one out of every six Hawaiians and accounted for 25 percent of state tax revenues.

During the early 1990s, the point of origin of Hawaii's visitors had been shifting from North America to Asia, and by 1995, over 40 percent of all visitors to Hawaii came from Asia. Thus the dramatic slump in Asian economies in the late 1990s, which led fewer people to travel for pleasure, had a major effect on the economy of Hawaii. The decline in tourism hurt not only the tourism industry itself, but also the construction industry, which had been thriving by building condominiums, hotels, and resort and retirement facilities. The terrorist attacks of September 11, 2001, also hurt Hawaii's economy by discouraging air travel. Although the industry had recovered by June 2002, these slumps illustrate the vulnerability of tourism to economic downturns and political events. The economic recession of 2008–2009 also created a slump in Hawaii's tourism, exacerbated by a gradual shift over the last few years of Asian tourists away from Hawaii to other places. In 2008, Asian visitors accounted for only 19 percent of total visitors to the island (66 percent came from the U.S. mainland).

Sometimes the mass tourism can seem like an invading force to ordinary citizens. For example, an important segment of the Honolulu tourist infrastructure—hotels, golf courses, specialty shopping centers, import shops, and nightclubs—is geared to visitors from Japan, and many such facilities are owned by Japanese investors. Hawaiian citizens and other non-Japanese shoppers and vacationers can be made to feel out of place.

Another example of the impositions of mass tourism is the demand by tourists for golf courses on many Hawaiian islands, which resulted in what Native (indigenous) Hawaiians view as desecration of sacred sites. Land that in precolonial times was communally owned, cultivated, and used for sacred rituals was then confiscated by the colonial government and more recently sold by the state to Asian golf course developers. Now the only people with access to the sacred sites are fee-paying tourist golfers. By 2007, there were 83 golf courses in Hawaii, and the golf industry alone contributed $1.6 billion to the state's economy—more than twice the amount from agriculture. Golf's total impact is $2.5 billion, which represents about 12.5 percent of the tourism sector income. Relocation to Hawaii by retired Americans looking for a sunny spot—often called "residential tourism"—has also had an impact on property values and the use of sacred lands for local citizens. ■

Sustainable Tourism Some islands have attempted to deal with the pressures of tourism by adopting the principle of sustainable tourism, which aims to decrease the imprint of tourism and minimize disparities between hosts and visitors. Samoa, for example, has created the Samoan Tourism Authority in conjunction with the South Pacific Regional Environment Programme. (The name *Samoa* refers to the independent country that was formerly known as Western Samoa; that country is politically distinct from American Samoa, a U.S. territory.) With financial aid from New Zealand, the Authority develops and monitors sustainable tourism components: beaches, wetlands, and forested island environments, and provides knowledge-based tourism experiences for visitors (information-rich explanations of island political, social, and environmental issues).

THINGS TO REMEMBER

1. Globalization is reorienting Oceania (especially Australia and New Zealand) toward Asia as a major destination for exports and an increasing source of Oceania's imports.

2. Service industries are becoming the dominant income source for most of the region's economies, although extractive industries remain important.

3. Tourism is a significant and growing part of the economies in Oceania.

Sociocultural Issues

The cultural sea change in Oceania away from Europe and toward Asia and the Pacific has been accompanied by new respect for indigenous peoples. In addition, a growing sense of common economic ground with Asia has heightened awareness of the attractions of Asian culture.

Ethnic Roots Reexamined

Until very recently, most people of European descent in Australia and New Zealand thought of themselves as Europeans in exile.

Many considered their lives incomplete until they had made a pilgrimage to the British Isles or the European continent. In her book *An Australian Girl in London* (1902), Louise Mack wrote: "[We] Australians [are] packed away there at the other end of the world, shut off from all that is great in art and music, but born with a passionate craving to see, and hear and come close to these [European] great things and their home[land]s."

These longings for Europe were accompanied by racist attitudes toward both indigenous peoples and Asians. Most histories of Australia written in the early twentieth century failed to even mention the Aborigines, and later writings described them as amoral. From the 1920s to the 1960s, whites-only immigration policies barred Asians, Africans, and Pacific Islanders from migrating to Australia and discouraged them from entering New Zealand. And, as we have seen, trading patterns further reinforced connections to Europe.

Weakening of the European Connection When migration from the British Isles slowed after World War II, both Australia and New Zealand began to lure immigrants from southern and eastern Europe, many of whom had been displaced by the war. Hundreds of thousands came from Greece, what was then Yugoslavia, and Italy. The arrival of these non-English-speaking people began a shift toward a more multicultural society. With the demise of the whites-only immigration policy, there was an influx of Vietnamese refugees in the early 1970s following the United States' withdrawal from Vietnam. More recently, growing demand for information technology specialists throughout the service sector has been met by recruiting skilled workers from India.

People of Asian birth or ancestry remain a minor percentage of the total population in both Australia and New Zealand, but new immigration policies are increasing the numbers of Asian immigrants, especially from China, Vietnam, and India. In 2006, the latest year for which complete statistics are available, 42 percent of Australia's foreign-born residents were from Europe; 15 percent were from Asia (**Figure 11.18**). Although most immigrants to New Zealand continue to come from the United Kingdom (76.8 percent of New Zealand's population is of European descent; 15 percent are Native Maori), by 2006 the second-largest influx shifted from Pacific peoples to Asians, who comprise over 8 percent. People of European descent are declining as a proportion of the population, although they are expected to remain the most numerous segment throughout the twenty-first century. Auckland now has the largest Polynesian population (including Native Maori) of any city in the world.

The Social Repositioning of Indigenous Peoples in Australia and New Zealand The makeup of the populations of Australia and New Zealand is also changing in another respect. In Australia, for the first time in 200 years, the number of people who claim indigenous origins is increasing. Between 1991 and 1996, the number of Australians claiming Aboriginal origins rose by 33 percent. By 2009, the indigenous population was estimated at 528,600 (2.3 percent of the total), 39 percent of whom were 15 years of age or younger. In New Zealand, the number claiming Maori background rose by 20 percent between 1991 and 2009. In

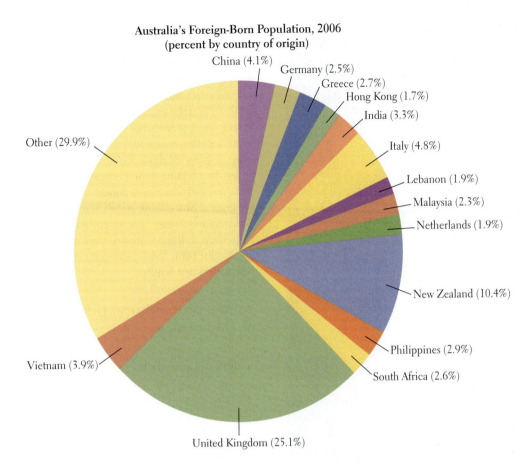

Australia's Foreign-Born Population, 2006
(percent by country of origin)

- China (4.1%)
- Germany (2.5%)
- Greece (2.7%)
- Hong Kong (1.7%)
- India (3.3%)
- Italy (4.8%)
- Lebanon (1.9%)
- Malaysia (2.3%)
- Netherlands (1.9%)
- New Zealand (10.4%)
- Philippines (2.9%)
- South Africa (2.6%)
- United Kingdom (25.1%)
- Vietnam (3.9%)
- Other (29.9%)

FIGURE 11.18 Australia's cultural diversity in 2006. Over 24 percent (4.95 million) of Australia's people were born in other places, making Australia one of the world's most ethnically diverse nations. [Data from Australian Bureau of Statistics, "Main Countries of Birth," *Year Book Australia*, 2008, Table 7.39, at "http://www.abs.gov.au/ausstats/abs@.nsf/bb8db737e2af84b8ca2571780015701e/F1C38FAE9E5F2B82CA2573D200110333?opendocument" http://www.abs.gov.au/ausstats/abs@.nsf/bb8db737e2af84b8ca2571780015701e/F1C38FAE9E5F2B82CA2573D200110333?opendocument.]

2009, there were 652,900 Maori; they comprise about 15 percent of New Zealand's population.

These increases are not the result of a population boom, but rather more positive attitudes toward indigenous peoples, which encourage more people to acknowledge their Aborigine or Maori ancestry. It is broadly recognized now that discrimination has been the main reason for the low social standing and impoverished state of indigenous peoples. Marriages between European and indigenous peoples are also more common and more open, and so the number of people with mixed heritage is increasing.

Recent decades have also witnessed increased respect for Aboriginal and Maori culture. Aborigines base their way of life on the idea that the spiritual and physical worlds are intricately related (see Figure 11.19 on page 504). The dead are everywhere present in spirit, and they guide the living in how to relate to the physical environment. Much Aboriginal spirituality refers to the *Dreamtime*, the time of creation when the human spiritual connections to rocks, rivers, deserts, plants, and animals were made clear. Unfortunately, the social repositioning of Aboriginal and Maori people remains incomplete. Very few Aboriginal people practice their own cultural traditions or live close to ancient homelands. Instead, many live in impoverished urban conditions.

Aboriginal Land Claims In 1988, during a bicentennial celebration of the founding of white Australia, a contingent of some 15,000 Aborigines protested that they had little reason to celebrate. During the same 200 years, they had been excluded from their ancestral lands, had lost basic civil rights, and had effectively been erased from Australian national consciousness. Into the 1960s, Aborigines had only limited rights of citizenship, and it was even illegal for them to drink alcohol.

Until 1993, Aborigines were assumed to have no prior claim to any land in Australia. British documents indicate that during colonial settlement, all Australian lands were deemed to be available for British use. The Aborigines were thought to be too primitive to have concepts of land ownership since their nomadic cultures had "no fixed abodes, fields or flocks, nor any internal hierarchical differentiation." The Australian High Court declared this position void in 1993. After that, Aboriginal groups began to win some land claims, mostly for land in the arid interior

FIGURE 11.19 Aboriginal rock art. A "Mimi spirit" painted on a rock at Kakadu National Park, Australia. To the Aboriginals, Mimi spirits are teachers who pass between this world and another dimension via crevices in rocks. They are responsible for many teachings on hunting, food preparation, use of fire, dance, and sexuality.

British, on the other hand, assumed that the treaty had transferred Maori lands to them, giving them *exclusive* rights to settle the land with British migrants and to extract wealth through farming, mining, and forestry.

By 1950, the Maori had lost all but 6.6 percent of their former lands to European settlers and the government. Maori numbers shrank from a probable 120,000 in the early 1800s to 42,000 in 1900, and they came to occupy the lowest and most impoverished rung of New Zealand society. In the 1990s, however, the Maori began to reclaim their culture, and they established a tribunal that forcefully advances Maori interests and land claims through the courts. Since then, nearly half a million acres of land and several major fisheries have been transferred back to Maori control. Nonetheless, the Maori still have notably higher unemployment, lower education levels, and poorer health than the New Zealand population as a whole.

Politics and Culture: Different Definitions of Democracy

In recent decades, stark divisions have emerged in Oceania over definitions of democracy—the system of government that dominates in New Zealand and Australia and the political and cultural

previously controlled by the Australian government. Court cases to restore Aboriginal rights and lands continue. Figure 11.20 shows the Aboriginal Embassy, a permanent installation in Canberra that advocates for Aborigines.

Maori Land Claims In New Zealand, relations between the majority European-derived population and the indigenous Maori have proceeded only somewhat more amicably. In 1840, the Maori signed the Waitangi Treaty with the British, assuming they were granting only rights of land usage, not ownership. The Maori did not regard land as a tradable commodity, but rather as an asset of the people as a whole, used by families and larger kin groups to fulfill their needs. The geographer Eric Pawson, in *Inventing Places-Studies in Cultural Geography*, writes: "To the Maori the land was sacred . . . [and] the features of land and water bodies were woven through with spiritual meaning and the Maori creation myth." The

FIGURE 11.20 The Aboriginal tent embassy in Canberra, Australia. Intermittently since 1972, and continuously since 1992, Aboriginal activists have camped out on the grounds of Australia's parliament in Canberra, Australia. Considered by many to be the most effective political action ever taken by Australian Aborigines, the first "tent embassy" was a response to the Australian government's denial of land ownership and other "land rights" by Aboriginals regarding territories they had continuously occupied for thousands of years. With increasing recognition of Aboriginal land rights, other causes have been championed by the tent embassy, including opposition to mining activity that threatens Aboriginal communities and cultural sites, and the plight of Aboriginal slums in major cities, such as the community of Redfern in Sydney, Australia. The tent embassy remains controversial; in 2003 anti-Aboriginal arsonists burned down parts of it. The Australian government plans a more permanent structure and a ban on camping at the embassy.

philosophy that influences governments throughout the rest of the region known as the **Pacific Way** (see Photo Essay 11.6 on page 506).

Democracy as it is practiced in Australia and New Zealand is a parliamentary system based on universal suffrage, debate, and majority rule. The Pacific Way is based on traditional notions of power and problem solving and refers to a way of settling issues based in the traditional culture of many Pacific Islanders. It favors consensus and mutual understanding over open confrontation, and respect for traditional leadership (especially the usually patriarchal leadership of families and villages) over free speech, personal freedom, and democracy.

The Pacific Way is a political and cultural philosophy developed in Fiji around the time of its independence from the United Kingdom in 1970. It subsequently gained popularity in many Pacific islands, most of which gained independence in the 1970s and 1980s. The Pacific Way carries a flavor of resistance to Europeanization and has often been invoked to uphold the notion of a regional identity shared by Pacific islands that grows out of their own unique history and social experience. It was particularly important to academics given the task of writing new textbooks to replace those used by their former colonial masters. The new texts focused students' attention on their own cultures and places before they studied Britain, France, or the United States. Appeals to the Pacific Way have also been used to uphold attempts by Pacific island governments to control their own economic development and solve their own political and social problems, even when their methods are criticized by outsiders, foreign governments, and their own citizens.

In politics, the Pacific Way has occasionally been invoked as a philosophical basis for overriding democratic elections that challenge the power of indigenous Pacific islanders. In 1987, 2000, and 2006, indigenous Fijians used the Pacific Way to justify coups d'état against legally elected governments (see Photo Essay 11.6A). All three of the overthrown governments were dominated by Indian Fijians, the descendants of people from India who were brought to Fiji more than a century ago by the British to work on sugar plantations.

Fiji's population is now about evenly divided between indigenous Fijians and Indian Fijians. Indigenous Fijians are generally less prosperous and tend to live in rural areas where community affairs are still governed by traditional chiefs. By contrast, Indian Fijians hold significant economic and political power, especially in the urban centers and in areas of tourism and sugar cultivation. In response to the coups, many Indian Fijians left the islands, resulting in a loss of badly needed skilled labor, which has slowed economic development.

Political responses around Oceania to the coups have been divided. Australia, New Zealand, and the United States (via the state government of Hawaii) have demanded that the election results stand and the Indian Fijians be returned to office. The rest of Oceania has appealed to the Pacific Way in arguments supporting the coup leaders. Like Fiji, most islands are governed by leaders of indigenous descent who have not always had the strongest respect for democracy, especially when it could threaten their hold on power.

On the global stage, Fiji has been made to suffer officially for subverting majority rule democracy by being suspended from the Commonwealth of Nations (a union of former British colonies). As a result, it will be ineligible for Commonwealth aid and will not be allowed to participate in Commonwealth sports events for the foreseeable future. Because sports play a particularly central role in Pacific identity (see page 507) this latter sanction carries significant weight.

Regardless of its global political status, the Pacific Way is likely to endure, especially as a concept that upholds regional identity and traditional culture. Further, some organizations now use the Pacific Way as the basis of an integrated approach to economic development and environmental issues. For example, the South Pacific Regional Environmental Programme builds on traditional Pacific island economic activities—such as fishing and local traditions of environmental knowledge and awareness—to promote grassroots economic development and environmental sustainability.

Forging Unity in Oceania

A sense of unity is growing throughout Oceania as people develop more appreciation for the region's cultural complexity. Although the great diversity of languages in the region may sometimes make communication difficult, travel and sports are two forces that help to bring people closer together.

Interisland Travel One way in which unity is manifested in Oceania is through interisland travel. Today, people travel in small planes from the outlying islands to hubs such as Fiji, where jumbo jets can be boarded for Auckland, Melbourne, or Honolulu. Cook Islanders call these little planes "the canoes of the modern age." New Zealanders can migrate to Australia to teach or train; a businessperson from Kiribati in Micronesia can fly to Fiji to take a short course at the University of the South Pacific; or a Cook Islands teacher can take graduate training in Hawaii.

Languages in Oceania The Pacific islands—most notably Melanesia—have a rich variety of languages. In some cases, the islands in a single chain have several different languages. A case in point is Vanuatu, a chain of 80 mostly high volcanic islands to the east of northern Australia. At least 108 languages are spoken by a population of just 180,000—an average of 1 language for every 1600 people! Another example is New Guinea, the largest, most populous, and most ethnically diverse island in the Pacific. No fewer than 800 languages are spoken on New Guinea by a populace of 5.5 million.

Languages are both an important part of a community's cultural identity and a hindrance to cross-cultural understanding. In Melanesia and elsewhere in the Pacific, the need for communication with the wider world is served by a number of **pidgin** languages that are sufficiently similar to be mutually intelligible

Pacific Way the idea that Pacific islanders have a regional identity and a way of handling conflicts peacefully that grows out of their particular social experience

pidgin a language used for trading; made up of words borrowed from the several languages of people involved in trading relationships

Levels of democratization vary significantly between Australia, New Zealand, and Hawaii (which are among the world's more democratized places) and New Guinea and many Pacific islands (which are much less democratized).

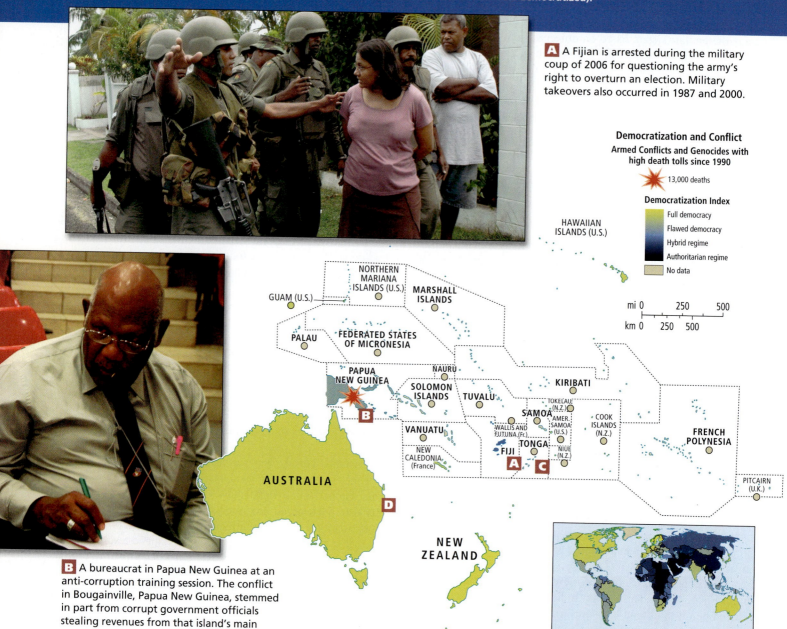

A A Fijian is arrested during the military coup of 2006 for questioning the army's right to overturn an election. Military takeovers also occurred in 1987 and 2000.

Democratization and Conflict
Armed Conflicts and Genocides with high death tolls since 1990

✴ 13,000 deaths

Democratization Index
- Full democracy
- Flawed democracy
- Hybrid regime
- Authoritarian regime
- No data

mi 0   250   500
km 0 250 500

HAWAIIAN ISLANDS (U.S.)

NORTHERN MARIANA ISLANDS (U.S.)
GUAM (U.S.)
MARSHALL ISLANDS
PALAU
FEDERATED STATES OF MICRONESIA
PAPUA NEW GUINEA
NAURU
SOLOMON ISLANDS
KIRIBATI
TUVALU
TOKELAU (N.Z.)
SAMOA
AMER. SAMOA (U.S.)
COOK ISLANDS (N.Z.)
VANUATU
WALLIS AND FUTUNA (Fr.)
FRENCH POLYNESIA
NEW CALEDONIA (France)
FIJI
TONGA
NIUE (N.Z.)
AUSTRALIA
PITCAIRN (U.K.)
NEW ZEALAND

B A bureaucrat in Papua New Guinea at an anti-corruption training session. The conflict in Bougainville, Papua New Guinea, stemmed in part from corrupt government officials stealing revenues from that island's main copper mine.

C The streets of Nuku'alofa, capital of the Kingdom of Tonga, after riots that broke out when the legislature refused to enact democratic reforms that would reduce the role of the royal family in politics.

D A march in Brisbane, Australia, protesting present and past discrimination against Aborigines, who were not allowed to vote throughout Australia until 1967.

FIGURE 11.21 A poster written in pidgin English. A poster written in pidgin by the Solomon Islands Ministry of Natural Resources. A literal translation of the sign's pidgin is "Suppose you cut 'em one fellow [a counting unit, in this case a tree] down, plant 'em [a] stack more."

(Figure 11.21). Pidgins are made up of words borrowed from several languages of people involved in trading relationships. Over time, they can grow into fairly complete languages, capable of fine nuances of expression. When a particular pidgin is in such common use that mothers talk to their children in it, then it can literally be called a "mother tongue." In Papua New Guinea, a version of pidgin English is the official language.

Sports as a Unifying Force Sports and games are a major feature of daily life throughout Oceania. The region has both shared sports traditions with, and borrowed them from, cultures around the world. Long-distance sailing, now a world-class sport, was an early skill in this region. Outrigger sailing vessels and canoes were once the main means of water transportation, and canoeing has been revived as a competitive sport (Figure 11.22). Surfing evolved in Hawaii and, like outrigger sailing and canoeing, derives from ancient navigational customs that matched human wits against the power of the ocean. On hundreds of Pacific islands and in Australia and New Zealand, sports originally borrowed from Europe—rugby, volleyball, soccer, and cricket—are important community-building activities. Baseball is a favorite in the parts of Micronesia that were U.S. trust territories. Women compete in the popular sport of netball, which is similar to basketball but without a backboard. They also compete in rugby and basketball; in the 2008 Olympics, Australia's womens' basketball team finished with the silver medal and New Zealand's team was tenth.

Sports competitions (including native dances such as the Haka described in this chapter's opening vignette) are the single most common and resilient link among the countries of Oceania. Such competitions encourage regional identity, and they provide opportunities for ordinary citizens to travel extensively around the region and to sports venues in other parts of the world. The South Pacific Games—featuring soccer, boxing, tennis, golf, and netball, among other sports—are held every 4 years. The Micronesians hold periodic games that incorporate tests of many traditional skills, such as spearfishing, climbing coconut trees, and racing outrigger canoes.

Gender Roles in Oceania

Perceptions of Oceania are colored by many myths about how men and women are and should be. As always, the realities are more complex.

Gender Myths and Realities Because of Oceania's cultural diversity, many different roles for men and women exist. In the Pacific islands, men traditionally were cultivators, deepwater

FIGURE 11.22 Sports as a unifying force. An outrigger canoe race in Bora Bora, French Polynesia. Once a main method of transport connecting Pacific islands, outrigger canoe rowing has now become a popular sport throughout Oceania.

fishers, and masters of seafaring. In Polynesia, they also were responsible for many aspects of food preparation, including cooking. In the modern world men fill many positions, but idealized male images continue to be associated with vigorous activities.

In Australia and New Zealand, the super-masculine, white, working-class settler has long had prominence in the national mythologies (Figure 11.23A). In New Zealand, he was a farmer and herdsman. In Australia, he was more often a many-skilled laborer—a stockman, sheep shearer, cane cutter, or digger (miner)—who possessed a laconic, laid-back sense of humor. Labeled a "swagma" for the pack he carried, he went from station (large farm) to station or mine to mine, working hard but sporadically, gambling, and then working again until he had enough money or experience to make it in the city. There, he often felt ill at ease and chafed to return to the wilds. Now immortalized in songs, novels, and films, these men are portrayed as a rough and nomadic tribe whose social life was dominated by male camaraderie and frequent brawls. No small part of this characterization of males derived from the fact that many of Australia's first immigrants were convicts.

Today, as part of larger efforts to recognize the diversity of Australian society, new ways of life for men are emerging and are breaking down the national image of the tough male loner. For example, the Australian city of Sydney annually hosts one of the world's largest gay pride parades (see Figure 11.23B). Nonetheless, the old model persists and remains prominent in the public images of Australian businessmen, politicians, and movie stars.

Perhaps the most enduring myth Europeans created regarding Oceania was their characterization of the women of the Pacific islands as gentle, simple, compliant love objects. (Tourist brochures still promote this notion.) There is ample evidence to suggest that Pacific Islanders did have more sexual partners in a lifetime than Europeans did. However, the reports of unrestrained sexuality related by European sailors were no doubt influenced by the exaggerated fantasies one might expect from all-male crews living at sea for months at a time. The notes of Captain James Cook are typical: "No women I ever met were less reserved. Indeed, it appeared to me, that they visited us with no other view, than to make a surrender of their persons." Over the years, such notions about Pacific island women have been encouraged by the paintings and prints of Paul Gauguin (Figure 11.24), the writings of novelist Herman Melville (*Typee*), and the studies of anthropologist Margaret Mead (*Coming of Age in Samoa*), as well as by movies and musicals such as *Mutiny on the Bounty* and *South Pacific*.

In reality, women's roles in the Pacific islands varied considerably from those in Europe, but not in the ways European explorers imagined. Women often exercised a good bit of power in family and clan, and their power increased with motherhood and advancing age. In Polynesia, a woman could achieve the rank of ruling chief in her own right, not just as the consort of a male chief. Women were primarily craftspeople, but they also contributed to subsistence by gathering fruits and nuts and by fishing. And in some places—Micronesia, for example—lineage was established through women, not men.

Gender, Democracy, and Economic Empowerment Today, there is a trend toward equality across gender lines throughout Oceania, but there is persistent inequality as well. A striking disparity is emerging between Australia and New Zealand (where women are gaining political and economic empowerment) and Papua New Guinea and the Pacific islands (where change is much slower).

In Australia and New Zealand, women's access to jobs and policy-making positions in government has improved, particularly over the last few decades. New Zealand and the Australian province of South Australia were among the first places in the world to grant European women full voting rights (1893 and 1895, respectively). New Zealand has elected two female prime ministers, and Australia, one deputy prime minister (similar to the vice president in the United States) (see Thematic Concepts Parts M, N on page 479). In 2010 Australia elected a woman Prime Minister, Julia Gillard. Moreover, in both countries, according to the 2009 UNHDR, the proportion of females in national legislatures (34 percent in New Zealand, 30 percent in Australia) is well above the global average of 18 percent. In Papua New Guinea and the Pacific islands, women are generally less empowered politically and economically. No woman has yet been elected

(A)   **(B)**

FIGURE 11.23 Australia's growing diversity in gender identity. (A) A "swagman," or itinerant worker, still a powerful icon of Australian maleness, in 1901. **(B)** Sydney's annual Mardi Gras Parade in 2008, one of the world's largest lesbian, gay, bisexual, and trangender (LGBT) events.

FIGURE 11.24 "Arearea" ("Amusement") by Paul Gauguin. In this 1892 painting, Tahitian women are rendered in a European Romantic pastoral style that emphasizes their gentle, compliant demeanor.

to a top-level national office, and women are a tiny minority in national legislatures, if present at all.

In both Australia and New Zealand, young women are pursuing higher education and professional careers and postponing marriage and childbearing until their thirties. (This is also a trend in Hawaii, Guam, and those islands with French affiliation.) Nonetheless, both societies continue to reinforce the housewife role for women in a variety of ways. For example, the expectation is that women, not men, will interrupt their careers to stay home to care for young or elderly family members. As noted above in the discussion of human well-being in the region, women in Australia receive on average only about 70 percent of the pay (69 percent in New Zealand) that men receive for equivalent work. This is, however, a smaller gender pay gap than in many developed countries.

Throughout Papua New Guinea and the Pacific islands, gender roles and relationships vary greatly over the course of a lifetime. Today, many young women fulfill traditional roles as mates and mothers and practice a wide range of domestic crafts, such as weaving and basketry. Then in middle age, they may return to school and take up careers. Some Pacific Island women, with the aid of government scholarships, pursue higher education or job training that takes them far from the villages where they raised their children. Accumulating age and experience may boost women into positions of considerable power in their communities.

Throughout their lives, women in Papua New Guinea and the Pacific islands contribute significantly to family assets through the formal and informal economies. Most traders in marketplaces are women, and the items they sell are usually made and transported by women. Yet like women everywhere, they have trouble obtaining credit to expand their businesses. The World Bank in 2007-2008 ranks Papua New Guinea 131 out of 181 for entrepreneurs to get credit and only women with a salary would qualify for a formal bank loan; otherwise they need a salaried husband as a guarantor. Recently, however, the Asian Development Bank notes that micro-credit loans to women are increasing significantly and women are using these loans to benefit their families. The repayment rate is very high.

THINGS TO REMEMBER

1. The number of Asian immigrants into Australia and New Zealand has been increasing over the past two decades, while the number of Europeans has been declining. Asians, however, still compose less than 20 percent of Australia's population and less than 10 percent of New Zealand's.

2. The indigenous populations of both Australia (Aborigines) and New Zealand (Maori) are relatively small, but represent an increasing percentage of each country: 2.4 percent in Australia and 14.9 percent in New Zealand. People in Oceania are now more willing to claim native roots than they once were.

3. Because sports are a unifying force in Oceania, they also can be used to enforce conformity to officially accepted rules on democracy, as in Fiji's case.

4. Gender roles and relationships vary considerably in Oceania. In Australia and New Zealand, compared with elsewhere in the region, women generally are better off in terms of political power and pay equity with men. In some traditional societies, however, women inherit or accrue considerable power in their own communities over the course of a lifetime.

Reflections on Oceania

In this region that has been so powerfully shaped by the globalizing influence of outsiders, several ongoing waves of transformation will define much of what happens in the near future. The economic reorientation away from Europe and toward Asia is likely to continue. The shift in employment away from mining and agriculture and toward service economies is equally profound, and in the case of Australia and New Zealand, may ameliorate the transition toward greater involvement with Asia because service jobs will be created. The boom in Asian tourism in the Pacific islands will strengthen the shift toward Asia. Thriving service economies are likely to sustain high levels of urbanization, and hence average living standards will probably continue to rise but disparities in income and well-being may increase. Population growth will likely continue to slow, and eventually all populations will age.

Positive ongoing transformations set Oceania on a course toward weathering new challenges. Democracy is well established in Australia, New Zealand, and Hawaii, as is the political and economic empowerment of women. However, both of these trends are less well established in the Pacific islands, especially Papua New Guinea. Nevertheless, these areas also have traditions of cooperation and consensus that may help societies navigate future stresses.

It is difficult to make detailed predictions about the future of Oceania because of the uncertainties brought by climate change. While some low-lying islands are clearly threatened with submersion, the precise nature of the impacts facing Australia and New Zealand is less certain. Water stress could constrain agricultural expansion, and fire could become a more pressing concern. Changes in Oceania's regional climate could further threaten the many unique ecosystems of this region, much as the importing of European food systems did. Advances in water technologies, though, could reduce the vulnerability of many areas to water shortage, drought, and food shortage. Further, New Zealand's ambitious goals for renewable energy are among the most advanced in the world. However, the future of Oceania, much like its past, will likely remain powerfully linked to events in distant places. Perhaps more than any other world region, Oceania is not the master of its own destiny.

Critical Thinking Questions

1. As Australia and New Zealand move away from intense cultural and economic involvement with Europe, new policies and attitudes have had to evolve to facilitate increased involvement with Asia. If you were a college student in Australia or New Zealand, how might you experience these changes? Think about fellow students, career choices, language learning, and travel choices.

2. Discuss the emerging cultural identity of the Pacific islands, taking note of the extent to which Australia and New Zealand share or do not share in this identity. What factors are helping to forge a sense of unity across Polynesia and beyond? (First, review the spatial extent of Polynesia.)

3. Discuss the many ways in which Asia has historic, and now increasingly economic, ties to Oceania. Include in your discussion patterns of population distribution, mineral exports and imports, technological interactions, and tourism.

4. Describe the main concerns in Oceania related to global warming. Which parts of the region are likely to be most affected? To what extent can the countries of Oceania exercise control over their likely future as the climate warms?

5. Australia and New Zealand differ from each other physically. Compare and contrast the two countries in relation to water, vegetation, and prehistoric and modern animal populations.

6. Indigenous peoples worldwide are beginning to speak out on their own behalf. Discuss how the indigenous peoples of Australia, New Zealand, and the Pacific islands are serving as leaders in this movement and what measures they are taking to reconstitute a sense of cultural heritage.

7. Oceania is one of the most urbanized regions on earth. Discuss how and why this fact varies from popular impressions of the region.

8. How is tourism both boosting economies and straining environments and societies throughout the Pacific islands? Describe the solutions that are being proposed to reduce the negative impacts of tourism.

9. Compare how women have or have not been empowered, politically and economically, in Australia, New Zealand, Papua New Guinea, and the Pacific islands.

10. Compared with other regions, Australia and New Zealand are somewhat unusual in having achieved broad prosperity on the basis of raw materials exports. How would you account for this achievement?

Chapter Key Terms

Aborigines, 489
atoll, 477
endemic, 482
Gondwana, 476
Great Barrier Reef, 476
hot spots, 477
invasive species, 485

Maori, 474
marsupials, 482
Melanesia, 490
Melanesians, 490
Micronesia, 491
MIRAB economy, 500
monotremes, 482

noble savage, 491
Pacific Way, 505
pidgin, 505
Polynesia, 491
roaring forties, 477
subsistence affluence, 500

Despite its remote location and few human inhabitants, environmental change is accelerating in Antarctica as temperatures rise and surrounding fisheries are exploited.

A Tourists on a Southern Ocean cruise observe the break up of one of Antarctica's many coastal ice shelves. While most sea level–rise predicted for the near future is due to thermal expansion of oceans, sea levels would rise 200 feet (60 meters) if Antarctica's massive glaciers were all to melt.

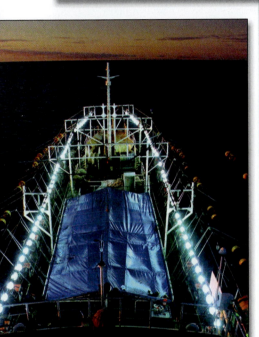

B A fishing vessel in the Southern Ocean, one of thousands placing increasing pressure on Antarctica's fisheries. Species like the Patagonian toothfish (marketed as Chilean sea bass) are severely overfished, and tiny shrimp known as "krill" that form the basis of many oceanic ecosystems, are also increasingly sought after.

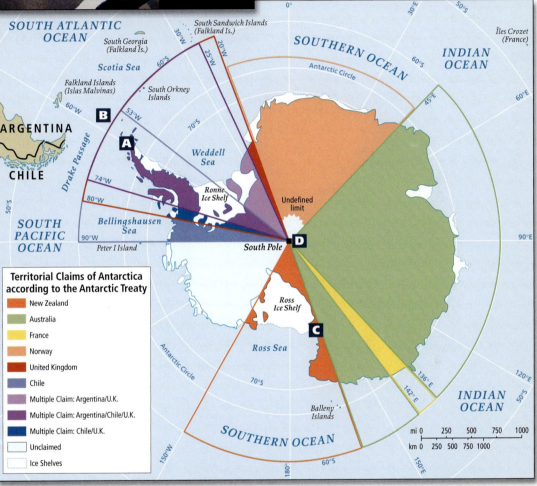

Territorial Claims of Antarctica according to the Antarctic Treaty

- New Zealand
- Australia
- France
- Norway
- United Kingdom
- Chile
- Multiple Claim: Argentina/U.K.
- Multiple Claim: Argentina/Chile/U.K.
- Multiple Claim: Chile/U.K.
- Unclaimed
- Ice Shelves

C An emperor penguin dives beneath a hole in some sea ice on a fishing expedition. Emperor penguins are found only in Antarctica and are proving extremely sensitive to climate change. In warmer years declines in sea ice, their ideal hunting habitat, have resulted in widespread starvation. Meanwhile, in colder years fewer penguin chicks hatch.

D Amundsen-Scott South Pole Station, the dome-shaped structure, is a U.S. research station. Outside fly the flags of the first countries to sign the Antarctic Treaty in 1959, which bans all military or resource extraction-related activity, making the continent a scientific and nature reserve.

PHOTO CREDITS

GLOSSARY

Aborigines (p. 489) the longest-surviving inhabitants of Oceania, whose ancestors, the Australoids, migrated from Southeast Asia possibly as early as 50,000 years ago over the Sundaland landmass that was exposed during the ice ages

acculturation (p. 151) adaptation of a minority culture to the host culture enough to function effectively and be self-supporting; cultural borrowing

acid rain (p. 75) falling precipitation that has formed through the interaction of rainwater or moisture in the air with sulfur dioxide and nitrogen oxides emitted during the burning of fossil fuels, making it acidic

agribusiness (p. 91) the business of farming conducted by large-scale operations that produce, package, and distribute agricultural products

agriculture (p. 56) the practice of producing food through animal husbandry, or the raising of animals, and the cultivation of plants

agroecology (p. 375) the practice of traditional, nonchemical methods of crop fertilization and the use of natural predators to control pests

agroforestry (p. 300) the raising of economically useful trees

Ainu (p. 427) an indigenous cultural minority group in Japan characterized by their light skin, heavy beards, and thick, wavy hair who are thought to have migrated thousands of years ago from the northern Asian steppes

animist (p. 333) a belief system in which natural features carry spiritual meaning

apartheid (p. 308) a system of laws mandating racial segregation in South Africa, in effect from 1948 until 1994

aquifers (p. 67) natural underground reservoirs

archipelago (p. 434) a group, often a chain, of islands

assimilation (pp. 151, 195) the loss of old ways of life and the adoption of the lifestyle of another culture

Association of Southeast Asian Nations (ASEAN) (p. 451) an organization of Southeast Asian governments established to further economic growth and political cooperation

atoll (p. 477) a low-lying island, formed of coral reefs that have built up on the circular or oval rims of a submerged volcano

Australo-Melanesians (p. 445) a group of hunters and gatherers who moved from the present northern Indian and Burman parts of southern Eurasia into the exposed landmass of Sundaland about 60,000 to 40,000 years ago

Austronesians (p. 446) a Mongoloid group of skilled farmers and seafarers from southern China who migrated south to various parts of Southeast Asia between 10,000 and 5000 years ago

authoritarianism (p. 38) a political system based on the power of the state or of elitist regional and local leaders

Aztecs (p. 129) indigenous people of high-central Mexico noted for their advanced civilization before the Spanish conquest

biodiversity (p. 121) the variety of life forms to be found in a given area

biosphere (pp. 27, 168) the global ecological system that integrates all living things and their relationships

birth rate (p. 18) the number of births per 1000 people in a given population, per unit of time, usually per year

Bolsheviks (p. 220) a faction of communists who came to power during the Russian Revolution

brain drain (p. 154) the migration of educated and ambitious young adults to cities or foreign countries, depriving the communities from which the young people come of talented youth in whom they have invested years of nurturing and education

brownfields (p. 100) old industrial sites whose degraded conditions pose obstacles to redevelopment

Buddhism (p. 362) a religion of Asia that originated in India in the sixth century B.C.E. as a reinterpretation of Hinduism; it emphasizes modest living and peaceful self-reflection leading to enlightenment

capitalism (pp. 40, 180) an economic system based on the private ownership of the means of production and distribution of goods, driven by the profit motive and characterized by a competitive marketplace

capitalists (p. 220) usually a wealthy minority that owns the majority of factories, farms, businesses, and other means of production

carbon sequestration (p. 300) the removal and storage of carbon taken from the atmosphere

carrying capacity (p. 28) the maximum number of people that a given territory can support sustainably with food, water, and other essential resources

cartel (p. 278) a group of producers strong enough to control production and set prices for its products

cash economy (p. 20) an economic system in which the necessities of life are purchased with monetary currency

caste system (p. 360) an ancient Hindu system for dividing society into hereditary hierarchical classes

Caucasia (p. 233) the mountainous region between the Black Sea and the Caspian Sea

central planning (p. 180) a communist economic model in which a central bureaucracy dictates prices and output with the stated aim of allocating goods equitably across society according to need

centrally planned, or socialist, economy (p. 221) an economic system in which the state owns all real estate and means of production, while government bureaucrats direct all economic activity, including the locating of factories, residences, and transportation infrastructure

Christianity (p. 262) a monotheistic religion based on the belief in the teachings of Jesus of Nazareth, a Jew, who described God's relationship to humans as primarily one of love and support, as exemplified by the Ten Commandments

civil disobedience (p. 356) the breaking of discriminatory laws by peaceful means

civil society (pp. 40, 310) the totality of voluntary, civic, and social organizations and institutions that form the basis of a functioning society and encourage a sense of unity and informed common purpose among the general population

clear-cutting (p. 73) the cutting down of all trees on a given plot of land, regardless of age, health, or species

climate (p. 51) the long-term balance of temperature and precipitation that characteristically prevails in a particular region

climate change (p. 45) a slow shifting of climate patterns due to the general cooling or warming of the atmosphere

cold war (pp. 180, 222) the contest that pitted the United States and Western Europe, espousing free market capitalism and democracy, against the USSR and its allies, promoting a centrally planned economy and socialist state

commodification (p. 128) turning something not previously thought of as having economic value into a good, or commodity, which can be bought and sold

commodities (p. 311) raw materials that are traded, often to other countries, for processing or manufacturing into more valuable goods

Common Agricultural Program (CAP) (p. 187) a European Union program, meant to guarantee secure and safe food supplies at affordable prices, that places tariffs on imported agricultural goods and gives subsidies to EU farmers

communal conflict (p. 363) a euphemism for religiously based violence in South Asia

communism (pp. 180, 220) an ideology, based largely on the writings of the German revolutionary Karl Marx, that calls on workers to unite to overthrow capitalism and establish an egalitarian society in which workers share what they produce

Communist Party (p. 220) the political organization that ruled the USSR from 1917 to 1991; other communist countries, such as China, Mongolia, North Korea, and Cuba, also have communist parties

Confucianism (p. 399) a Chinese philosophy that teaches that the best organizational model for the state and society is a hierarchy based on the patriarchal family

contested space (p. 141) any area that several groups claim or want to use in different and often conflicting ways, such as the Amazon or Palestine

cool humid continental climate (p. 168) a midlatitude climate pattern in which summers are fairly hot and moist, and winters become longer and colder the deeper into the interior of the continent one goes

coral bleaching (p. 445) color loss that results when photosynthetic algae that live in the corals are expelled

coup d'état (p. 143) a military- or civilian-led forceful takeover of a government

Creoles (p. 133) people mostly of European descent born in the Caribbean

crony capitalism (p. 451) a type of corruption in which politicians, bankers, and entrepreneurs, sometimes members of the same family, have close personal as well as business relationships

cultural homogenization (p. 162) the tendency toward uniformity of ideas, values, technologies, and institutions among associated culture groups

cultural pluralism (p. 467) the cultural identity characteristic of a region where groups of people from many different backgrounds have lived together for a long time but have remained distinct

Cultural Revolution (p. 403) a series of highly politicized and destructive mass campaigns launched in 1966 to force the entire population of China to support the continuing revolution

culture (p. 57) all the ideas, materials, and institutions that people have invented to use to live on Earth that are not directly part of our biological inheritance

currency devaluation (p. 314) the lowering of a currency's value relative to the U.S. dollar, the Japanese yen, the European euro, or another currency of global trade

czar (p. 220) title of the ruler of the Russian empire; derived from the word "caesar," the title of the Roman emperors

death rate (p. 18) the ratio of total deaths to total population in a specified community, usually expressed in numbers per 1000 or in percentages

delta (p. 51) the triangular-shaped plain of sediment that forms where a river meets the sea

democratization (p. 38) the transition toward political systems guided by competitive elections

demographic transition (p. 20) the change from high birth and death rates to low birth and death rates that usually accompanies a cluster of other changes, such as change from a subsistence to a cash economy, increased education rates, and urbanization

desertification (pp. 258, 304) a set of ecological changes that converts nondesert lands into deserts

detritus (p. 439) dead organic material (such as plants and insects) that collects on the ground

development (p. 24) usually used to describe economic changes like greater productivity of agriculture and industry that lead to better standards of living or simply to increased mass consumption

diaspora (p. 262) the dispersion of Jews around the globe after they were expelled from the eastern Mediterranean by the Roman Empire beginning in 73 C.E.; can now refer to other dispersed culture groups

dictator (p. 143) a ruler who claims absolute authority, governing with little respect for the law or the rights of citizens

digital divide (p. 93) the discrepancy in access to information technology between small, rural, and poor areas and large, wealthy cities that contain major government research laboratories and universities

divide and rule (p. 319) the deliberate intensification of divisions and conflicts by European colonial powers

domestication (p. 56) the process of developing plants and animals through selective breeding to live with and be of use to humans

domino theory (p. 448) a foreign policy theory that used the idea of the domino effect to suggest that if one country "fell" to communism, others in the neighboring region would also fall

double day (p. 196) the longer workday of women with jobs outside the home who also work as caretakers, housekeepers, and/or cooks for their families

dowry (p. 365) a price paid by the family of the bride to the groom (the opposite of *bride price*); formerly a custom practiced only by the rich

dual economy (p. 312) an economy in which the population is divided by economic disparities into two groups, one prosperous and the other near or below the poverty level

early extractive phase (p. 133) a phase in Central and South American history, beginning with the Spanish conquest and lasting until the early twentieth century, characterized by a dependence on the export of raw materials

ecological footprint (p. 45) the biological productive area needed to sustain a human population at its current standard of living, usually calculated per person

economic core (p. 81) the dominant economic region within a larger region

economic diversification (p. 279) the expansion of an economy to include a wider array of activities

economies of scale (p. 184) reductions in the unit cost of production that occur when goods or services are efficiently mass produced, resulting in increased profits per unit

ecotourism (p. 126) nature-oriented vacations, often taken in endangered and remote landscapes, usually by travelers from industrialized nations

El Niño (pp. 124, 439) periodic climate-altering changes, especially in the circulation of the Pacific Ocean, now understood to operate on a global scale

endemic (p. 482) belonging or restricted to a particular place

erosion (p. 30) the process by which fragmented rock and soil are moved over a distance, primarily by wind and water

ethnic cleansing (p. 40) the deliberate removal of an ethnic group from a particular area by forced migration

ethnic group (p. 57) a group of people who share a set of beliefs, a way of life, a technology, and usually a geographic location

ethnicity (p. 103) the quality of belonging to a particular culture group

euro (p. 184) the official (but not required) currency of the European Union as of January 1999

European Union (EU) (p. 162) a supranational organization that unites most of the countries of West, South, North, and Central Europe

evangelical Protestantism (p. 157) a Christian movement that focuses on personal salvation and empowerment of the individual through miraculous healing and transformation; some practitioners preach the "gospel of success"—that a life dedicated to Christ will result in prosperity for the believer— to the poor

export-led growth (p. 405) an economic development strategy that relies heavily on the production of manufactured goods destined for sale abroad

Export Processing Zones (EPZs) (p. 138) specially created legal spaces or industrial parks within a country where, to attract foreign-owned factories, duties and taxes are not charged

extended family (p. 152) a family that consists of related individuals beyond the nuclear family of parents and children

external debts (p. 136) debts a country owes to foreign banks or governments that are repayable only in foreign currency

fair trade (p. 37) trade that values equity throughout the international trade system; now proposed as an alternative to free trade

fair trade movement (p. 140) a global movement to distribute profits more fairly to producers by upgrading their knowledge of markets and hence to increase their competitiveness

favelas (p. 154) Brazilian urban slums and shantytowns built by the poor; called colonias, barrios, or barriados in other countries

female earned income as a percent of male earned income (F/MEI) (p. 25) a measure of pay equity that shows average female earned income as a percent of average male earned income

female genital mutilation (FGM) (p. 332) removing the labia and the clitoris and sometimes stitching the vulva nearly shut

female seclusion (p. 267) the requirement that women stay out of public view

feminization of labor (p. 450) the increasing representation of women in both the formal and the informal labor force

Fertile Crescent (p. 260) an arc of lush, fertile land formed by the uplands of the Tigris and Euphrates river systems and the Zagros Mountains, where nomadic peoples began the earliest known agricultural communities

floating population (p. 387) the Chinese term for jobless or underemployed people who have left economically depressed rural areas for the cities and move from place to place looking for work

floodplain (p. 51) the flat land around a river where sediment is deposited during flooding

food security (pp. 28, 395) the ability of a state to consistently supply a sufficient amount of basic food to the entire population

foreign direct investment (FDI) (p. 139) the amount of money invested in a country's business by citizens, corporations, or governments of other countries

foreign exchange (p. 462) foreign currency that countries need to purchase imports

foreign investment (p. 408) investment in Chinese (or another country's) business and manufacturing enterprises by foreign firms

formal economy (p. 25) all aspects of the economy that take place in official channels

fossil fuel (p. 248) a source of energy formed from the remains of dead plants and animals

free trade (p. 37) the movement of goods and capital without government restrictions

Gazprom (p. 225) in Russia, the state-owned energy company; it is the tenth-largest oil and gas entity in the world

gender (p. 21) the sexual category of a person

gender roles (p. 22) the socially assigned roles of males and females

genetic modification (GM) (p. 30) in agriculture, the practice of splicing together the genes from widely divergent species to achieve particular desirable characteristics

genetically modified organisms (GMOs) (p. 91) animals and crop plants in which the DNA is modified

genocide (pp. 40, 321) the deliberate destruction of an ethnic, racial, or political group

gentrification (p. 101) the renovation of old urban districts by middle-class investment, a process that often displaces poorer residents

Geographic Information Science (p. 7) the body of science that underwrites multiple spatial analysis technologies and keeps them at the cutting edge

geopolitics (p. 40) the use of strategies by countries to ensure that their best interests are served

glasnost (p. 231) literally, "openness"; the opening up of public discussion of social and economic problems that occurred in the Soviet Union under Mikhail Gorbachev in the late 1980s

global economy (p. 34) the worldwide system in which goods, services, and labor are exchanged

global scale (p. 9) the level of geography that encompasses the entire world as a single unified area

global warming (p. 45) the predicted warming of the earth's climate as atmospheric levels of greenhouse gases increase

globalization (p. 32) the growth of interregional and worldwide linkages and the changes these linkages are bringing about

Gondwana (p. 476) the great landmass that formed the southern part of the ancient supercontinent Pangaea

grassroots economic development (p. 318) economic development projects designed to help individuals and their families achieve sustainable livelihoods

Great Barrier Reef (p. 476) the longest coral reef in the world, located off the northeastern coast of Australia

Great Leap Forward (p. 402) an economic reform program under Mao Zedong intended to quickly raise China to the industrial level of Britain and the United States

Green (p. 172) environmentally conscious

green revolution (pp. 28, 374) increases in food production brought about through the use of new seeds, fertilizers, mechanized equipment, irrigation, pesticides, and herbicides

greenhouse gases (p. 45) harmful gases, such as carbon dioxide and methane, released into the atmosphere by human activities

gross domestic product (GDP) per capita (p. 19) the market value of all goods and services produced by workers and capital within a particular country's borders and within a given year, divided by the number of people in the country

gross domestic product (GDP) per capita PPP (p. 25) the market value of all goods and services produced by modern workers and capital within a particular country's borders and within a given year, divided by the number of people in the country and adjusted for purchasing power parity

Group of Eight (G8) (p. 227) an organization of highly industrialized countries: France, the United States, Britain, Germany, Japan, Italy, Canada, and Russia

growth poles (p. 411) zones of development whose success draws more investment and migration to a region

guest workers (p. 192) legal workers from outside a country who help fulfill the need for temporary workers but who are expected to return home when they are no longer needed

Gulf States (p. 268) Saudi Arabia, Kuwait, Bahrain, Oman, Qatar, And The United Arab Emirates

haciendas (p. 134) a large agricultural estate in Middle or South America, more common in the past; usually not specialized by crop and not focused on market production

hajj (p. 266) the pilgrimage to the city of Makkah (Mecca) that all Muslims are encouraged to undertake at least once in a lifetime

Harappa culture (p. 353) see Indus Valley civilization

Hinduism (p. 360) a major world religion practiced by approximately 900 million people, 800 million of whom live in India; a complex belief system, with roots both in localized folk traditions (known as the Little Tradition) as well as in a broader system based on literary texts (known as the Great Tradition)

Hispanic (p. 67) term used to refer to all Spanish-speaking people from Middle and South America, although their ancestors may have been black, white, Asian, or Native American

Holocaust (p. 179) during World War II, a massive execution by the Nazis of 6 million Jews and 5 million Roma (Gypsies); disabled and mentally ill people; gays, lesbians, and transgendered people; political dissidents, and ethnic Poles and other Slavs

Horn of Africa (p. 298) the triangular peninsula that juts out from northeastern Africa below the Red Sea and wraps around the Arabian Peninsula

hot spots (p. 477) individual sites of upwelling material (magma) that originate deep in the mantle of the earth and surface in a tall plume; hot spots tend to remain fixed relative to migrating tectonic plates

hukou system (p. 387) the system in China by which citizens' permanent residence is registered

human geography (p. 50) the study of various aspects of human life that create the distinctive landscapes and regions of the world

human well-being (p. 25) various measures of the extent to which people are able to obtain a healthy life in a community of their choosing

humanism (p. 176) a philosophy and value system that emphasizes the dignity and worth of the individual

import substitution industrialization (ISI) (p. 136) policies that encouraged local production of machinery and other items that previously had been imported at great expense from abroad

Incas (p. 129) indigenous people who ruled the largest pre-Columbian state in the Americas, with a domain stretching from southern Colombia to northern Chile and Argentina

income disparity (p. 133) the gap in wealth between the richest 10 percent and the poorest 10 percent of a country's population

indigenous (p. 117) native to a particular place or region

Indus Valley civilization (p. 353) the first substantial settled agricultural communities, which appeared about 4500 years ago along the Indus River in modern-day Pakistan and northwest India

informal economy (pp. 25, 315) all aspects of the economy that take place outside official channels

information technology (IT) (p. 93) the part of the service sector that relies on the use of computers and the Internet to process and transport information; includes banks, software companies, medical technology companies, and publishing houses

infrastructure (p. 80) road, rail, and communication networks and other facilities necessary for economic activity and human well-being

interregional linkages (p. 32) economic, political, or social connections between regions, whether contiguous or widely separated

intertropical convergence zone (ITCZ) (p. 298) a band of atmospheric currents that circle the globe roughly around the equator; warm winds from both north and south converge at the ITCZ, pushing air upward and causing copious rainfall

intifada (p. 288) a prolonged Palestinian uprising against Israel

invasive species (p. 485) organisms that spread into regions outside of their native range, adversely affecting economies or environments

iron curtain (p. 179) a long, fortified border zone that separated western Europe from (then) eastern Europe during the Cold War

Islam (p. 248) a monotheistic religion that emerged in the seventh century C.E. when, according to tradition, the archangel Gabriel revealed the tenets of the religion to the Prophet Muhammad

Islamism (p. 248) a grassroots religious revival in Islam that seeks political power to curb what are seen as dangerous non-Muslim influences; also seeks to replace secular governments and civil laws with governments and laws guided by Islamic principles

Jainism (p. 362) originally a reformist movement within Hinduism, Jainism is a faith tradition that is more than 2000 years old; found mainly in western India and in large urban centers throughout the region, Jains are known for their educational achievements, nonviolence, and strict vegetarianism

jati (p. 361) in Hindu India, the subcaste into which a person is born, which largely defines the individual's experience for a lifetime

Judaism (p. 262) a monotheistic religion characterized by the belief in one god, Yahweh, a strong ethical code summarized in the Ten Commandments, and an enduring ethnic identity

kaizen system (p. 406) the "continuous improvement system" pioneered in Japanese manufacturing; it ensures that fewer defective parts are produced because production lines are constantly surveyed for errors

kanban system (p. 406) the "just-in-time" system pioneered in Japanese manufacturing that clusters companies that are part of the same production system close together so that they can deliver parts to each other precisely when they are needed

Kyoto Protocol (p. 46) an amendment to a United Nations treaty on global warming, the Protocol is an international agreement, adopted in 1997 and in force in 2005, that sets binding targets for industrialized countries for the reduction of emissions of greenhouse gases

land reform (pp. 136, 407) a policy that breaks up large landholdings for redistribution among landless farmers

landforms (p. 50) physical features of the earth's surface, such as mountain ranges, river valleys, basins, and cliffs

liberation theology (p. 157) a movement within the Roman Catholic Church that uses the teachings of Jesus to encourage the poor to organize to change their own lives and the rich to promote social and economic equity

lines of latitude (p. 6) the distance in degrees north or south of the equator; lines of latitude run parallel to the equator, and are also called parallels

lines of longitude (p. 3) the distance in degrees east and west of Greenwich, England; lines of longitude, also called meridians, run from pole to pole (the line of longitude at Greenwich is 0° and is known as the prime meridian)

lingua franca (p. 336) a common language used to communicate by people who do not speak one another's native languages; often a language of trade

living wages (p. 37) minimum wages high enough to support a healthy life

local scale (p. 10) the level of geography that describes the space where an individual lives or works; a city, town, or rural area

machismo (p. 152) a set of values that defines manliness in Middle and South America

Maori (p. 474) Polynesian people indigenous to New Zealand

map projections (p. 6) the various ways of showing the spherical earth on a flat surface

maquiladoras (p. 138) foreign-owned, tax-exempt factories, often located in Mexican towns just across the U.S. border from U.S. towns, that hire workers at low wages to assemble manufactured goods which are then exported for sale

marianismo (p. 152) a set of values based on the life of the Virgin Mary, the mother of Jesus, that defines the proper social roles for women in Middle and South America

marketization (p. 136) the development of a free market economy in support of *free trade*

marsupials (p. 482) mammals that give birth to their young at a very immature stage and nurture them in a pouch equipped with nipples

material culture (p. 61) all the things, living or not, that humans use

Mediterranean climate (p. 168) a climate pattern of warm, dry summers and mild, rainy winters

megalopolis (p. 98) an area formed when several cities expand so that their edges meet and coalesce

Melanesia (p. 490) New Guinea and the islands south of the equator and west of Tonga (the Solomon Islands, New Caledonia, Fiji, and Vanuatu)

Melanesians (p. 490) a group of Australoids named for their relatively dark skin tones, a result of high levels of the protective pigment melanin; they settled throughout New Guinea and other nearby islands

mercantilism (pp. 133, 176) the policy by which European rulers sought to increase the power and wealth of their realms by managing all aspects of production, transport, and commerce in their colonies

Mercosur (p. 140) a free trade zone created in 1991 that links the economies of Brazil, Argentina, Uruguay, and Paraguay to create a common market

mestizos (p. 133) people of mixed European, African, and indigenous descent

metropolitan areas (p. 98) cities of 50,000 or more and their surrounding suburbs and towns

microcredit (p. 375) a program based on peer support that makes very small loans available to very low-income entrepreneurs

Micronesia (p. 491) the small islands that lie east of the Philippines and north of the equator

Middle America (p. 120) in this book, a region that includes Mexico, Central America, and the islands of the Caribbean

migration (p. 18) movement of people from a place or country to another, often for safety or economic reasons

MIRAB economy (p. 500) an economy based on migration, remittance, aid, and bureaucracy

mixed agriculture (p. 300) the raising of a variety of crops and animals on a single farm, often to take advantage of several environmental riches

Mongols (p. 219) a loose confederation of nomadic pastoral people centered in East and Central Asia, who by the thirteenth century had established by conquest an empire stretching from Europe to the Pacific

monotheistic (p. 261) pertaining to the belief that there is only one god

monotremes (p. 482) egg-laying mammals, such as the duck-billed platypus and the spiny anteater

monsoon (pp. 52, 345) a wind pattern in which in summer months, warm, wet air coming from the ocean brings copious rainfall, and in winter, cool, dry air moves from the continental interior toward the ocean

Mughals (p. 353) a dynasty of Central Asian origin that ruled India from the sixteenth to the nineteenth century

multiculturalism (p. 62) the state of relating to, reflecting, or being adapted to diverse cultures

multinational corporation (p. 34) a business organization that operates extraction, production, and/or distribution facilities in multiple countries

Muslims (p. 262) followers of Islam

nationalism (p. 179) devotion to the interests or culture of a particular country, nation, or cultural group; the idea that a group of people living in a specific territory and sharing cultural traits should be united in a single country to which they are loyal and obedient

(to) nationalize (p. 139) to seize private property and place under government ownership, with some compensation

Neolithic Revolution (p. 56) a period 20,000 to 8000 years ago characterized by the transition from hunting and gathering to agriculture, accompanied by the making of polished stone tools, often called the first agricultural revolution

noble savage (p. 491) a term coined by European Romanticists to describe what they termed the "primitive" peoples of the Pacific who lived in distant places supposedly untouched by corrupting influences

nomadic pastoralists (p. 219) people whose way of life and economy are centered on the tending of grazing animals who are moved seasonally to gain access to the best grasses

nongovernmental organization (NGO) (p. 40) an association outside the formal institutions of government in which individuals, often from widely differing backgrounds and locations, share views and activism on political, social, economic, or environmental issues

nonpoint sources of pollution (p. 209) diffuse sources of environmental contamination, such as untreated automobile exhaust, raw sewage, or agricultural chemicals that drain from fields into water supplies

North American Free Trade Agreement (NAFTA) (pp. 92, 140) a free trade agreement made in 1994 that added Mexico to the 1989 economic arrangement between the United States and Canada

North Atlantic Drift (p. 168) the easternmost end of the Gulf Stream, a broad warm-water current that brings large amounts of warm water to Europe

North Atlantic Treaty Organization (NATO) (p. 187) a military alliance between European and North American countries that was developed during the Cold War to counter the influence of the Soviet Union; since the breakup of the Soviet Union, NATO has expanded membership to include much of Eastern Europe and Turkey, and is now focused mainly on providing the international security and cooperation needed to expand the European Union

nuclear family (p. 106) a family consisting of a married father and mother and their children

Occupied Palestinian Territories (OPT) (p. 252) Palestinian lands occupied by Israel in 1967

offshore outsourcing (p. 377) the shifting of jobs from a relatively wealthy country to one where labor or other production costs are lower

oligarchs (p. 223) in Russia, those who acquired great wealth during the privatization of Russia's resources and who use that wealth to exercise power

OPEC (Organization of Petroleum Exporting Countries) (p. 278) a cartel of oil-producing countries—including Algeria, Angola, Indonesia, Iran, Iraq, Kuwait, Libya, Nigeria, Qatar, Saudi Arabia, the United Arab Emirates, and Venezuela—that was established to regulate the production, and hence the price, of oil and natural gas

orographic rainfall (p. 52) rainfall produced when a moving moist air mass encounters a mountain range, rises, cools, and releases condensed moisture that falls as rain

Pacific Rim (p. 109) a term referring to all the countries that border the Pacific Ocean

Pacific Way (p. 505) the idea that Pacific islanders have a regional identity and a way of handling conflicts peacefully that grows out of their particular social experience

Partition (p. 356) the breakup following Indian independence that resulted in the establishment of Hindu India and Muslim Pakistan

pastoralism (p. 304) a way of life based on herding; practiced primarily on savannas, on desert margins, or in the mixture of grass and shrubs called open bush

patriarchal (p. 266) relating to a social organization in which the father is supreme in the clan or family

perestroika (p. 231) literally, "restructuring"; the restructuring of the Soviet economic system in the late 1980s in an attempt to revitalize the economy

permafrost (p. 208) permanently frozen soil just a few feet beneath the surface

petty capitalists (p. 408) in the 1980s, farmers and small businesses that were allowed to sell their produce and goods in competitive markets as a part of China's market reforms

physical geography (p. 50) the study of the earth's physical processes: how they work, how they affect humans, and how they are affected by humans

pidgin (p. 505) a language used for trading; made up of words borrowed from the several languages of people involved in trading relationships

plantation (p. 135) a large factory farm that grows and partially processes a single cash crop

plate tectonics (p. 50) the scientific theory that the earth's surface is composed of large plates that float on top of an underlying layer of molten rock; the movement and interaction of the plates create many of the large features of the earth's surface, particularly mountains

political ecologist (p. 27) a geographer who studies power allocations in the interactions among development, human well-being, and the environment

polygyny (pp. 268, 332) the taking by a man of more than one wife at a time

Polynesia (p. 491) the numerous islands situated inside an irregular triangle formed by New Zealand, Hawaii, and Easter Island

population pyramid (p. 18) a graph that depicts the age and gender structures of a country

populist movements (p. 156) popularly based efforts, often seeking relief for the poor

primate city (p. 152) a city, plus its suburbs, that is vastly larger than all others in a country and in which economic and political activity is centered

privatization (pp. 136, 229) the sale of industries that were formerly owned and operated by the government to private companies or individuals

proletariat (p. 221) the working class; the lowest social or economic class

purchasing power parity (PPP) (p. 25) the amount that the local currency equivalent of U.S.$1 will purchase in a given country

purdah (p. 358) the practice in South Asia of concealing women, especially during their reproductive years, from the eyes of nonfamily men

push/pull phenomenon of urbanization (p. 31) conditions, such as political instability, that encourage (push) people to leave rural areas, and urban factors, such as job opportunities, that encourage (pull) people to move to the urban area

Québecois (p. 67) French Canadians living in Québec; an ethnic group distinct from the rest of Canada, they all are citizens of Canada

Qur'an (or Koran) (p. 254) the holy book of Islam, believed by Muslims to contain the words Allah revealed to Muhammad through the archangel Gabriel

race (p. 61) a social or political construct that is based on apparent characteristics such as skin color, hair texture, and face and body shape, but that is of no biological significance

rate of natural increase (RNI) (p. 16) the rate of population growth measured as the excess of births over deaths per 1000 individuals per year without regard for the effects of migration

recession (p. 136) a slowing of economic activity

region (p. 7) a unit of the earth's surface that contains distinct patterns of physical features and/or of human development

regional conflict (p. 378) especially in South Asia, a conflict created by the resistance of a regional ethnic or religious minority to the authority of a national or state government

regional self-sufficiency (p. 408) an economic policy in Communist China that encouraged each region to develop independently in the hope of evening out the wide disparities in the national distribution of production and income

regional specialization (p. 408) the encouragement of specialization rather than self-sufficiency in order to take advantage of regional variations in climate, natural resources, and location

religious nationalism (p. 382) the association of a particular religion with a political unit; political control of a territory is often the ultimate goal of such a movement

resettlement schemes (p. 456) government plans to move large numbers of people from one part of a country to another to relieve urban congestion, disperse political dissidents, or accomplish other social purposes; also called *transmigration*

responsibility system (p. 408) in the 1980s, a decentralization of economic decision making in China that returned agricultural decision making to the farm household level, subject to the approval of the commune

Ring of Fire (p. 50) the tectonic plate junctures around the edges of the Pacific Ocean; characterized by volcanoes and earthquakes

roaring forties (p. 477) powerful air and ocean currents at about 40° S latitude that speed around the far Southern Hemisphere virtually unimpeded by landmasses

Roma (p. 179) the now-preferred term in Europe for Gypsies

Russian Federation (p. 208) Russia and its political subunits, which include 30 internal republics and more than 10 so-called autonomous regions

Russification (p. 224) the assimilation of all minorities to Russian (Slavic) ways

Sahel (p. 298) a band of arid grassland, where steppe and savanna grasses grow, that runs east-west along the southern edge of the Sahara

salinization (p. 256) a process that occurs when large quantities of water are used to irrigate areas where evaporation rates are high, leaving behind dissolved salts and other minerals

scale (of a map) (p. 3) the proportion that relates the dimensions of the map to the dimensions of the area it represents; also, variable-sized units of geographical analysis from the local scale to the regional scale to the global scale

Schengen Accord (p. 191) an agreement signed in the 1990s by the European Union and many of its neighbors that called for free movement across common borders

seawater desalination (p. 256) the removal of salt from seawater, usually accomplished through the use of expensive and energy-intensive technologies, making the water suitable for drinking or irrigating

secular states (p. 281) countries that have no state religion and in which religion has no direct influence on affairs of state or civil law

self-reliant development (p. 318) small-scale development schemes in rural areas that focus on developing local skills, creating local jobs, producing products or services for local consumption, and maintaining local control so that participants retain a sense of ownership

service sector (p. 92) economic activity that involves the sale of services

sex (p. 21) the biological category of male or female

sex tourism (p. 469) the sexual entertainment industry that services primarily men who travel for the purpose of living out their fantasies during a few weeks of vacation

shari'a (p. 266) literally, "the correct path"; Islamic religious law that guides daily life according to the interpretations of the Qur'an

shifting cultivation (pp. 129, 303) a productive system of agriculture in which small plots are cleared in forestlands, the dried brush is burned to release nutrients, and the clearings are planted with multiple species; each plot is used for only 2 or 3 years and then abandoned for many years of regrowth

Shi'ite (or Shi'a) (p. 266) the smaller of two major groups of Muslims, with different interpretations of shari'a; Shi'ites are found primarily in Iran and southern Iraq

Sikhism (p. 363) a religion of South Asia that combines beliefs of Islam and Hinduism

silt (p. 121) fine soil particles

Slavs (p. 219) a group of farmers who originated between the Dnieper and Vistula Rivers in modern-day Poland, Ukraine, and Belarus

slums (p. 31) densely populated areas characterized by crowding, run-down housing, and poverty

social safety net (p. 96) the services provided by the government—such as welfare, unemployment benefits, and health care—that prevent people from falling into extreme poverty

social welfare (in the European Union, social protection) (p. 199) in Europe, tax-supported systems that provide citizens with benefits such as health care, pensions, and child care

South America (p. 120) the continent south of Central America

sovereign-wealth funds (p. 280) in the Gulf States, state-owned savings from oil and gas income, which are then invested globally in a range of income-producing ventures

Soviet Union (p. 207) see Union of Soviet Socialist Republics

special economic zones (SEZs) (p. 411) free trade zones within China

state-aided market economy (p. 405) an economic system based on market principles such as private enterprise, profit incentives, and supply and demand, but with strong government guidance; in contrast to the free market (limited government) economic system of the United States and Europe

steppes (p. 208) semiarid, grass-covered plains

structural adjustment policies (SAPs) (p. 136) policies that require economic reorganization toward less government involvement in industry, agriculture, and social services; sometimes imposed by the World Bank and the International Monetary Fund as conditions for receiving loans

subcontinent (p. 344) a term often used to refer to the entire Indian peninsula, including Nepal, Bhutan, India, Pakistan, and Bangladesh

subduction zone (p. 121) a zone where one tectonic plate slides under another

subregions (p. 10) smaller divisions of the world regions delineated to facilitate the study of patterns particular to the areas

subsidies (p. 187) monetary assistance granted by a government to an individual or group in support of an activity, such as farming, that is viewed as being in the public interest

subsistence affluence (p. 500) a lifestyle whereby self-sufficiency is achieved for most necessities, while some opportunities to earn cash allow for travel and occasional purchases of manufactured goods

subsistence agriculture (p. 300) farming that provides food for only the farmer's family and is usually done on small farms

subsistence economy (p. 20) circumstances in which a family produces most of its own food, clothing, and shelter

suburbs (p. 98) populated areas along the peripheries of cities

Sunni (p. 266) the larger of two major groups of Muslims, with different interpretations of shari'a

sustainable agriculture (p. 28) farming that meets human needs without poisoning the environment or using up water and soil resources

sustainable development (p. 27) improvement of standards of living in ways that will not jeopardize those of future generations

taiga (p. 209) subarctic forests

Taliban (p. 366) an archconservative Islamist movement that gained control of the government of Afghanistan in the mid-1990s

temperate midlatitude climate (p. 168) as in south-central North America, China, and Europe, a climate that is moist all year with relatively mild winters and long, hot summers

temperature-altitude zones (p. 137) regions of the same latitude that vary in climate according to altitude

theocratic states (p. 281) countries that require all government leaders to subscribe to a state religion and all citizens to follow rules decreed by that religion

total fertility rate (TFR) (p. 18) the average number of children that women in a country are likely to have at the present rate of natural increase

trade deficit (p. 94) the extent to which the money earned by exports is exceeded by the money spent on imports

trade winds (p. 124) winds that blow from the northeast and the southeast toward the equator

tsunami (p. 390) a large sea wave caused by an earthquake

tundra (p. 268) a treeless area, between the ice cap and the tree line of arctic regions, where the subsoil is permanently frozen

typhoon (p. 391) a tropical cyclone or hurricane

underemployment (p. 229) the condition in which people are working too few hours to make a decent living or are working at menial jobs even though highly trained

Union of Soviet Socialist Republics (USSR) (p. 207) the nation formed from the Russian empire in 1922 and dissolved in 1991

United Nations (UN) (p. 40) an assembly of 192 member states that sponsors programs and agencies that focus on scientific research, humanitarian aid, planning for development, fostering general health, and peacekeeping assistance

United Nations Human Development Index (HDI) (p. 25) the ranking of countries based on three indicators of well-being: life expectancy at birth, educational attainment, and income adjusted to purchasing power parity

urban sprawl (p. 73) the encroachment of suburbs on agricultural land

urbanization (p. 31) the movement of people from rural areas to cities

varna (p. 361) the four hierarchically ordered divisions of society in Hindu India underlying the caste system: Brahmins (priests),

Kshatriyas (warriors/kings), Vaishyas (merchants/landowners), and Sudras (laborers/artisans)

veil (p. 268) the custom of covering the body with a loose dress and/or of covering the head—and in some places the face—with a scarf

virtual water (p. 42) the volume of water used to produce all that a person consumes in a year

water footprint (p. 42) the water used to meet a person's basic needs for a year, added to the person's annual virtual water

weathering (p. 50) the physical or chemical decomposition of rocks by sun, rain, snow, ice, and the effects of life-forms

welfare state (p. 179) a government that accepts responsibility for the well-being of its people, guaranteeing basic necessities such as education, employment, and health care for all citizens

West Bank barrier (p. 289) a 25-foot-high concrete wall in some places and a fence in others that now surrounds much of the West Bank and encompasses many of the remaining Jewish settlements there

wet rice cultivation (p. 395) a prolific type of rice production that requires the submersion of the plant roots in water for part of the growing season

world region (p. 9) a part of the globe delineated according to criteria selected to facilitate the study of patterns particular to the area

World Trade Organization (WTO) (pp. 37, 140) a global institution made up of member countries whose stated mission is to lower of trade barriers and to establish ground rules for international trade

Zionists (p. 287) those who have worked, and continue to work, to create a Jewish homeland (Zion) in Palestine

INDEX

boldface indicates a definition; *italic* indicates a figure.